AF335121

Quantitative Cardiac Electrophysiology

Quantitative Cardiac Electrophysiology

edited by

Candido Cabo

*Columbia University and
City University of New York
New York, New York, U.S.A.*

David S. Rosenbaum

*MetroHealth Campus
Case Western Reserve University
Cleveland, Ohio, U.S.A.*

MARCEL DEKKER, INC.　　　　NEW YORK · BASEL

ISBN: 0-8247-0774-5

This book is printed on acid-free paper.

Headquarters
Marcel Dekker, Inc.
270 Madison Avenue, New York, NY 10016
tel: 212-696-9000; fax: 212-685-4540

Eastern Hemisphere Distribution
Marcel Dekker AG
Hutgasse 4, Postfach 812, CH-4001 Basel, Switzerland
tel: 41-61-260-6300; fax: 41-61-260-6333

World Wide Web
http://www.dekker.com

The publisher offers discounts on this book when ordered in bulk quantities. For more information, write to Special Sales/Professional Marketing at the headquarters address above.

PART TWO   ELECTRICAL MEASUREMENTS AND MAPPING

PART THREE   SIGNAL PROCESSING

PART FOUR   OPTICAL MAPPING

Contents

processing techniques that can be applied to extract meaningful information from cardiac signals recorded during complex arrhythmias, and Chapter 13 examines how to automate the discrimination between different cardiac rhythms. Chapters 14 to 17 cover emerging technologies for imaging electrical activity in the heart using voltage-sensitive dyes.

The book will be useful to a broad audience interested in cardiovascular medicine and physiology, including clinicians, students, and researchers in the fields of biomedical engineering, applied physics and mathematics, and computational biology. Clinical cardiac electrophysiologists and arrhythmologists will benefit from the groundwork provided on the technological basis for measurements made in clinical practice.

Our primary thanks go to our contributors, key leaders in their fields, without whom this book would not have been possible. Their time and enthusiasm is greatly appreciated. We would also like to thank our families, especially our wives, Teresa Hervada and Anita B. Rosenbaum, for their support and understanding in this and other projects.

Candido Cabo
David S. Rosenbaum

Preface

In the last twenty years, advances in computer technologies, nonlinear dynamics, signal acquisition and processing, and voltage-sensitive dyes have permeated the field of cardiac electrophysiology and redefined the methods and technologies that are used in basic and clinical cardiac electrophysiology. As is often the case, the introduction of new technologies has led to a better understanding of basic mechanisms, resulting in improved diagnostic methods and more effective therapeutics in the clinical setting.

This book provides in-depth coverage of the theoretical and engineering principles behind those methods and technologies. It also illustrates practical applications to problems in basic and clinical cardiac electrophysiology such as the dynamics of cardiac arrhythmias, electrical stimulation of cardiac tissue, and defibrillation mechanisms. The book is organized into four sections: computer modeling (Chapters 1–6), electrical measurements and mapping (Chapters 7–11), signal processing (Chapters 12 and 13), and optical mapping (Chapters 14–17). The first six chapters cover methods for the development of computer models of ion channels, action potential, and structure of cardiac tissue and show how to apply these models to increase our understanding of electrical stimulation and defibrillation mechanisms. Chapter 7 reviews the technologies for the construction of electrodes and catheters for stimulation and recording of cardiac electrical signals. Chapters 8 discusses how to measure electrical properties of cardiac tissue. Application of electrical mapping techniques to the understanding of cardiac arrhythmias is discussed in Chapter 9. Chapters 10 and 11 illustrate two different applications of electrical mapping technology to clinical electrophysiology. Chapter 12 discusses signal

Contributors

Fadi G. Akar, Ph.D. The Heart and Vascular Research Center and the Departments of Medicine and Biomedical Engineering, Case Western Reserve University, Cleveland, Ohio, U.S.A.

Cory Anderson, M.S. Department of Biomedical Engineering, Tulane University, New Orleans, Louisiana, U.S.A.

Felipe Aguel, Ph.D. Department of Biomedical Engineering, Tulane University, New Orleans, Louisiana, U.S.A.

Isabelle Banville, Ph.D. Department of Biomedical Engineering, University of Alabama at Birmingham, Birmingham, Alabama, U.S.A.

Philip V. Bayly, Ph.D. Department of Mechanical Engineering, Washington University, St. Louis, Missouri, U.S.A.

Tamara C. Baynham, Ph.D. Department of Biological and Agricultural Engineering, The University of Georgia, Athens, Georgia, U.S.A.

Glenna C. L. Bett, Ph.D. Department of Physiology and Biophysics, State University of New York at Buffalo, Buffalo, New York, U.S.A.

Candido Cabo, Ph.D. Department of Pharmacology, College of Physicians and Surgeons, Columbia University, and Department of Computer Systems, New York City Technical College, City University of New York, New York, New York, U.S.A.

Wayne E. Cascio, M.D.   Division of Cardiology, Department of Medicine, University of North Carolina, Chapel Hill, North Carolina, U.S.A.

Peng-Sheng Chen, M.D.   Division of Cardiology, Department of Medicine, Burns and Allen Research Institute, Cedars-Sinai Medical Center, and the University of California, Los Angeles, Los Angeles, California, U.S.A.

Yuanna Cheng, M.D., Ph.D.   Department of Cardiovascular Medicine, The Cleveland Clinic Foundation, Cleveland, Ohio, U.S.A.

Anthony W. C. Chow, M.R.C.P.   Department of Cardiology, Imperial College School of Medicine and St. Mary's Hospital, London, United Kingdom

David W. Davies, M.D.   Department of Cardiology, Imperial College School of Medicine and St. Mary's Hospital, London, United Kingdom

James Eason, Ph.D.   Department of Biomedical Engineering, Tulane University, New Orleans, Louisiana, U.S.A.

Igor R. Efimov, Ph.D., F.A.H.A.   Department of Biomedical Engineering, Case Western Reserve University, Cleveland, Ohio, U.S.A.

Matthew G. Fishler, Ph.D.   Cardiac Rhythm Management Division, St. Jude Medical, Inc., Sunnyvale, California, U.S.A.

Lior Gepstein, M.D., Ph.D.   Cardiovascular Research Laboratory, The Bruce Rappaport Faculty of Medicine, Technion–Israel Institute of Technology, and the Cardiology Department, Rambam Medical Center, Haifa, Israel

Richard A. Gray, Ph.D.   Department of Biomedical Engineering, University of Alabama at Birmingham, Birmingham, Alabama, U.S.A.

Craig S. Henriquez, Ph.D.   Department of Biomedical Engineering, Duke University, Durham, North Carolina, U.S.A.

Hrayr S. Karagueuzian, Ph.D.   Division of Cardiology, Department of Medicine, Burns and Allen Research Institute, Cedars-Sinai Medical Center, and the University of California, Los Angeles, Los Angeles, California, U.S.A.

Stephen B. Knisley, Ph.D.   Division of Cardiology, Department of Medicine, University of North Carolina, Chapel Hill, North Carolina, U.S.A.

Wanda Krassowska, Ph.D.   Department of Biomedical Engineering, Duke University, Durham, North Carolina, U.S.A.

Robert A. Malkin, Ph.D. Department of Biomedical Engineering, The University of Memphis, Memphis, Tennessee, U.S.A.

John C. Neu, Ph.D. Department of Mathematics, University of California at Berkeley, Berkeley, California, U.S.A.

Bradford D. Pendley, Ph.D. Department of Chemistry, Rhodes College, Memphis, Tennessee, U.S.A.

Nicholas S. Peters, M.D. Department of Cardiology, Imperial College School of Medicine and St. Mary's Hospital, London, United Kingdom.

Randall L. Rasmusson, Ph.D. Department of Physiology and Biophysics, State University of New York at Buffalo, Buffalo, New York, U.S.A.

Jack M. Rogers, Ph.D. Department of Biomedical Engineering, The University of Alabama at Birmingham, Birmingham, Alabama, U.S.A.

Stephan Rohr, M.D. Department of Physiology, University of Bern, Bern, Switzerland

Kristina M. Ropella, Ph.D. Department of Biomedical Engineering, Marquette University, Milwaukee, Wisconsin, U.S.A.

David S. Rosenbaum, M.D. The Heart and Vascular Research Center and the Departments of Medicine and Biomedical Engineering, MetroHealth Campus, Case Western Reserve University, Cleveland, Ohio, U.S.A.

Ziad S. Saad, Ph.D. Scientific and Statistical Computing, National Institute of Mental Health, National Institutes of Health, Bethesda, Maryland, U.S.A.

Richard J. Schilling, M.D. Department of Cardiology, Imperial College of Medicine and St. Mary's Hospital, London, United Kingdom

Bradley A. Stone, M.S. Medtronic U.S.A., Inc., Columbus, Ohio, U.S.A.

Joseph V. Tranquillo, M.S. Department of Biomedical Engineering, Duke University, Durham, North Carolina, U.S.A.

Natalia Trayanova, Ph.D. Department of Biomedical Engineering, Tulane University, New Orleans, Louisiana, U.S.A.

Quantitative Cardiac Electrophysiology

1
Computer Models of Ion Channels

Glenna C. L. Bett and Randall L. Rasmusson
State University of New York at Buffalo, Buffalo, New York, U.S.A.

I. INTRODUCTION

Electrical activity in the heart is generated at the molecular level by specialized membrane-spanning proteins that control the movement of ions either by passive electrodiffusion through transmembrane pores (i.e., channels) or translocation across the membrane by carrier proteins (pumps, exchangers, and transporters). As a first approximation, ion channels can be thought of as mediating the dynamic portions of the action potential, such as the upstroke and repolarization, and also providing the entry of trigger calcium to initiate excitation–contraction coupling. In contrast, pumps, exchangers, and transporters can be thought of as steadily working in the background to establish and maintain ionic gradients. Obviously, this is only an approximation: pumps, exchangers, and transporters can and do contribute to the overall behavior of the action potential, particularly in pathophysiological conditions, but they have slower effects than the rapidly opening and closing channels. Nonetheless, channels dominate depolarization and repolarization, and the process of repolarization is largely understood as the dynamic interaction of membrane ion channels. Consequently, in many situations the action potential can be approximated well using a model containing only channels. This chapter describes some of our current understanding of voltage-gated ion channel biophysics and the mathematical modeling of these processes. We hope it will serve as an introduction to engineers and scientists from other disciplines who are relatively new to the field, and who need a brief and simple explanation of the rationale behind many of the mathematical formulations routinely used in current cellular models.

II. IONIC SELECTIVITY AND THE ION TRANSFER FUNCTION

The need for a cell to maintain relatively high concentrations of intracellular solutes such as proteins, nucleotides, and ATP presents a problem for cells, for they must do this while maintaining osmotic balance. Freely permeable solutes will redistribute between the intracellular and extracellular spaces until their activity (i.e., concentration) is equal, or, more precisely, until the electrochemical potential difference is zero. Intracellular solutes, by definition, cannot move out of the cell to equalize their concentration. Consequently, water will tend move into the cell and cause it to swell until the intracellular concentration has been diluted to the concentration of extracellular impermeable solutes, or until sufficient hydrostatic pressure develops to oppose the flow of water. It is this movement of water and cell swelling that causes turgidity in plants.

Animal cells lack the tough cellulose wall of plants, but still manage to preserve a dynamic equilibrium with their environment by means of a constant expenditure of metabolic energy to pump ions across the membrane against their concentration gradient. The "pump-leak" model of volume maintenance was proposed over 40 years ago to explain how lysis is avoided in animal cells [1–3]. Briefly, the sodium potassium ATPase pump (Na/K ATPase pump) uses energy to pump 3 sodium ions out of the cell and 2 potassium ions into the cell for each molecule of ATP hydrolyzed [4]. The Na/K ATPase pump establishes the concentration gradients of the permeable ions which drive ions through the channels. Potassium can "leak" out across the membrane relatively easily, whereas resting sodium permeability is low. Effectively, sodium is excluded from the cell and acts as though it were a counterbalancing impermeable solute. The pump-leak model, expanded and elaborated as the molecular basis of the various "leak" components have been discovered, remains the fundamental basis for understanding how ionic gradients are established and maintained.

The pump-leak mechanism establishes ionic concentration gradients and therefore provides a store of energy in the form of chemical potential. If a permeable membrane separates two solutions, A and B, and if ion X is present on either side of the membrane, the relative probability of finding a particle in either solution A or solution B is given by the Boltzmann equation:

$$\frac{P_B}{P_A} = \exp\left(-\frac{u_B - u_A}{kT}\right) \tag{1}$$

where

$$u_A = \text{the energy of a particle in solution A (state A)}$$
$$u_B = \text{the energy of a particle in solution B (state B)}$$
$$P_A = \text{the probability of a particle being in state A}$$
$$P_B = \text{the probability of a particle being in state B}$$
$$k = \text{Boltzmann's constant}$$
$$T = \text{absolute temperature}$$

Equation (1) can be framed in terms of molar energies and concentrations, to take account of the properties of the bulk solutions rather than individual components:

$$\frac{[X]_B}{[X]_A} = \exp\left(-\frac{U_B - U_A}{RT}\right) \tag{2}$$

where

$$[X]_A = \text{concentration of ion X in state (solution) A}$$
$$[X]_B = \text{concentration of ion X in state (solution) B}$$
$$U_A = \text{molar energy of state (solution) A}$$
$$U_B = \text{molar energy of state (solution) B}$$
$$R = \text{gas constant}$$

Rearranging Eq. (2) and taking logs gives the following equation:

$$U_2 - U_1 = RT \ln\left(\frac{[X]_B}{[X]_A}\right) \tag{3}$$

This shows the molar energy difference due to the concentration gradient. If the ion is charged, there will be not only a chemical force, but also an electromotive force. The electrical potential acting on anion of valence z in a potential field of Ψ is $zF\Psi$. In the steady state, there will be no net flux of ions across the membrane, i.e., the sum of the electrical and chemical forces is zero. The potential across the membrane at which there is no net movement of ions is termed the equilibrium potential of that ion, and is calculated as follows:

$$0 = RT \ln\left(\frac{[X]_B}{[X]_A}\right) + zFE_{\text{Eqm}}$$

$$E_{\text{Eqm}} = -\left(\frac{RT}{zF}\right) \ln\left(\frac{[X]_B}{[X]_A}\right) \tag{4}$$

where

$$E_{\text{Eqm}} = \text{the equilibrium potential}$$

This is the Nernst relationship [5], and indicates the net direction that an ion will electrodiffuse (i.e., into or out of the cell) when both chemical and electrical gradients are present. Ions move through open channels via electrodiffusion. The simplest model available for describing the uncoupled movement of a charged species through an open channel is given by the equation

$$I_x = \gamma_x \left(E_m - E_{\mathrm{Eqm},x} \right) \tag{5}$$

where

$$\gamma_x \quad = \text{conductance of the channel to ion X}$$
$$I_x \quad = \text{net current due to movement of ion X through the channel}$$
$$E_m \quad = \text{transmembrane potential}$$
$$E_{\mathrm{Eqm},x} = \text{equilibrium or Nernst potential for ion X}$$

This equation describes an ohmic conductor, as there is a linear relationship between current and voltage. Even though the Nernst equation can be used to calculate the correct reversal potential for an ion and the net driving force for an ion, the net flux is not always linearly related to the voltage difference, as implied by this equation.

The electrodiffusion of ions across the membrane occurs through channels that are membrane-spanning proteins containing a water-filled pore providing a continuous aqueous environment from the intracellular to the extracellular spaces. Ions diffuse freely through much of the length of the channel but are subjected to interactions with sites inside the channel. These sites of interaction determine the selectivity of the open channel. The rate at which ions can pass through a channel is high ($\sim 10^6$ per second) [6–8], so the sites must operate through physical mechanisms which provide some selectivity, but still allow rapid movement of the ions through the pore. Ion channels vary in their degree of selectivity and the degree to which they also pass other ions.

The lack of perfect selectivity has an important implication for the channel reversal potential: it is *not* the same as the Nernst potential for its dominant ion. When ionic conditions are held steady, this can be dealt with in a model by the simple expedient of altering the Nernst potential to agree with experimental data. When dealing with changing ionic concentrations, there are two ways that this problem is usually handled. The simplest is the parallel conductance approach, where the channel is modeled as having two or more conductances, each with a separate Nernst battery driving each ion species permeating the channel. Another approach is to use the Goldman-Hodgkin-Katz equation [9,10] for the reversal potential:

$$E_{\text{rev}} = \left(\frac{RT}{zF}\right) \ln \left(\frac{\sum^{+} P_x[X]_o + \sum^{-} P_y[Y]_i}{\sum^{-} P_y[Y]_o + \sum^{+} P_x[X]_i}\right) \tag{6}$$

where

P_x = relative permeability of positively charged ions
P_y = relative permeability of negatively charged ions
$[X]$ = concentration of positively charged ions
$[Y]$ = concentration of negatively charged ions

The relationship between the open channel current and the voltage is called the ion transfer function. Open channel currents are seldom ohmic in nature as described in Eq. (6), and many channels can pass current more easily in one direction than the other. The ability to pass current more readily from the inside of the cell to the outside is called outward rectification, and, conversely, the ability to pass ions more rapidly from the extracellular space to the inside of the cell is called inward rectification. By convention in experimental electrophysiology, inward current is defined as being negative and outward current as positive. Notable exceptions to this rule are the Nobel prize-winning papers of Hodgkin and Huxley, which used the opposite convention [11–14].

Before moving on to a discussion of why the ion transfer function displays different forms of rectification, it is important to discuss another factor which plays into the ability of Eq. (6) to predict current as a function of ionic concentration. Since conductance through ion channels is essentially the result of aqueous diffusion, conductance is strongly influenced by the permeant ion concentration. In aqueous solution alone, the conductance–activity relationship might be expected to be roughly linear. However, channels act in a manner similar to enzymes to catalyze the reaction of moving ions across the normally nonconducting cell membrane. Experimentally, the conductance–activity relationship of the ion transfer function of open channels has been shown to saturate with increasing concentration [15–18]. In essence, a non-concentration-dependent step in ion permeation becomes rate limiting for conductance, i.e., the transit time for crossing the membrane. Electrophysiologists have borrowed from the world of enzymology and use various forms of the Hill equation to modify the concentration dependence of conduction with respect to the maximum current, I_{max}:

$$I = I_{\text{max}} \left(\frac{[X]^n}{[X]^n + K_{1/2}^n}\right) \tag{7}$$

where $K_{1/2}$ is the concentration of ion X resulting in half-maximal current, and n is the Hill coefficient. When the Hill coefficient is 1, this is equivalent to a simple model in which ions interact with the channel one at a time, independently.

The assumption that ions interact with the channel independently and that conductance is dependent on ion concentration can provide an explanation for some forms of rectification. If the concentration of an ion is low on the outside and high on the inside, current will pass more easily from the inside to the outside of the cell, thus leading to outward rectification.

Goldman, Hodgkin, and Katz [4,10] developed an ion channel permeation model (the GHK model) based on the assumption that ions pass independently through channels, and that channels are long, water-filled pores over which the transmembrane potential drops uniformly across its length (i.e., the net electrical field is constant over the length of the channel and the energy barrier has a square profile).

In the GHK model, each type of ion that passes through the channel is subjected to different chemical and electrical forces, and the net effect is summed to obtain the characteristic flux through the whole channel. The separate components of the current are calculated with respect to the concentration gradient and the electric field they experience:

$$I_x = z_x F D_x \left(\frac{dc_x}{dt} + \frac{z_x F c_x}{RT} \frac{d\Psi}{ds} \right) \tag{8}$$

where

$$
\begin{aligned}
I_x &= \text{current due to ion X} \\
z_x &= \text{valence of ion X} \\
F &= \text{Farady constant} \\
D_x &= \text{diffusion coefficient of ion X} \\
c_x &= \text{local concentration of ion X} \\
R &= \text{gas constant} \\
T &= \text{absolute temperature} \\
\Psi &= \text{local potential in the membrane} \\
s &= \text{membrane position}
\end{aligned}
$$

This can be integrated to give the following equation:

$$I_x = -z_x F \beta_x \left\{ \frac{[X]_i \exp(z_x EF/RT) - [X]_o}{\int_{s=0}^{l} ([\exp(z_x F\Psi/RT)]/D_x)\, ds} \right\} \tag{9}$$

where

$$\beta_x [X]_i = \text{concentration of ion X, just inside the membrane}$$
$$\text{at the intracellular surface, i.e., at } s = 0$$

$\beta_x[X]_o$ = concentration of ion X, just inside the membrane
at the extracellular surface, i.e., at $s = 1$
l = thickness of the membrane

Final integration of Eq. (10) and substitution of parameters gives the GHK constant field equation for conductance:

$$I_x = P_x \frac{z_x^2 EF^2}{RT} \left\{ \frac{[X]_i - [X]_o \exp(z_x EF/RT)}{1 - \exp(z_x EF/RT)} \right\} \tag{10}$$

where

P_X = permeability of ion X
E = transmembrane voltage

This equation takes into account the asymmetric distribution of the permeant ion species. Also, because of the independence assumptions, multiple ionic species can be included and a concentration dependence of conductance (albeit nonsaturating) is built in to the formula. It is used in many action potential models today.

The conductance–activity relationship for many channels shows evidence of saturation and the Hill coefficient frequently takes nonunity values, e.g., 0.5, 1.5, or 2. In enzymology the value of the Hill coefficient is interpreted as a measure of the degree of cooperativity. When applied to ion channels, a nonunity Hill coefficient is interpreted as being evidence that more than one ion occupies the channel at the same time and that the ions are interacting (e.g., electrostatically). When the assumption of independence does not hold, our concept of permeability becomes somewhat tenuous. Up until now, all models of conduction have defined permeability as the ability of ions to move rapidly through the pore, and the more permeable the ion in a particular channel, the stronger its contribution to the reversal potential will be. Thus, in the literature one frequently encounters two measures of relative permeability: one is the conductance measured in a solution containing only one permeant species; the other is determined in the presence of two ions simultaneously, often with one on each side of the membrane. This second method is referred to as the bi-ionic reversal potential.

When the current is measured with only one permeant ion species present, the current–voltage relationship is determined in the presence of a single concentration of each ionic species. This measurement is often done under approximately physiological conditions, so other ions are present as well. The permeability ratio is calculated from the relative conductance of each ion. Using this method, the L-type calcium channel has a much higher conductance for monovalent ions when the calcium concentration is severely reduced (less than 1 μM) than it does for calcium [19–22]. To measure the bi-ionic reversal potential, the solution on one side of the membrane contains ions of one species, while on the opposite side only the other species is present. If the independence principle holds, these two methods should give similar results for linear conductances, with the relationship:

$$\frac{I'}{I} = \frac{\sum P_x[X_x]'_i - \sum P_x[X_x]'_o \exp(z_x EF/RT)}{\sum P_x[X_x]_i - \sum P_x[X_x]_o \exp(z_x EF/RT)} \tag{11}$$

Unfortunately, the two measures can be *qualitatively* very different, as in the case of sodium and calcium in the L-type calcium channel [19–22], protons and sodium in the sodium channel [23], and protons and potassium in the potassium channel [24]. In these cases the permeability ratios can be inverted using different measuring techniques. Obviously, a detailed discussion of the biophysics of ionic diffusion in a channel can become rather complicated. When modeling from published data, it becomes extremely important to understand the ionic conditions under which measurements were made.

Although ionic gradients and the intrinsic properties of a channel permeation pathway can cause significant rectification, the strong rectification of the inwardly rectifying potassium channel, I_{K1}, deserves special attention. The ion transfer relationship for this channel shows a strong negative slope, with less current being produced at more positive potentials.

Empirically, Hagiwara and Takahashi [25,26] determined the potassium conductance of the inward rectifier current in echinoderm oocytes to be dependent on both the transmembrane voltage and the extracellular potassium concentration:

$$G_K = B[K_o]^{1/2}\left\{\frac{1}{(1 + \exp[(\Delta V - \Delta V_h)/\mu]}\right\} \tag{12}$$

where

$$G_K = \text{potassium conductance}$$
$$\Delta V = \text{driving force}$$
$$B, \Delta V_h, \text{ and } \mu = \text{constants}$$

Rectification of I_{K1} is steeply dependent on the extracellular potassium concentration, with rectification disappearing altogether when there are high levels of extracellular potassium. There is a strong similarity between the effects of tetraethylammonium (TEA) block of the potassium channel and inward rectification [27,28], which suggests that inward rectification may result, in part, from block of the pore by cations. Magnesium ions and polyamines (excised patches, which lack polyamines, lose rectification) have both been shown to be responsible for rectification of potassium channels [29–36]. Inward rectification with a negative-slope region also occurs in potassium-permeable HERG (human ether-a-go-go) channels. In these channels intrinsic rectification occurs via a gating process, which is also dependent on the extracellular potassium concentration [37–39].

A high concentration of extracellular protons reduces sodium current in a concentration- and voltage-dependent manner similar to the rectification of potassium channels [24,40–43]. Protons do not permeate all the way across the membrane, but remain bound to a site within the channel for a relatively long time, physically preventing sodium flow through the pore [44–46]. As the proton-binding site is located within the pore, the binding, and therefore block, are affected by membrane potential [24,40,42]. The number of open channels, as determined by the tail current magnitudes, is unaffected, so this is not a gating process, but rather just block of a channel that remains open, though not permeable to ions [42].

Woodhull [42] developed a model to describe this phenomenon, based on Eyring rate theory [47]. Sodium ions were considered to be traveling through the pore that had an energy profile with three energy wells and two barriers, as shown in Fig. 1. This leads to the following equation:

$$\text{OUT} \overset{k_1}{\underset{k_{-1}}{\rightleftharpoons}} \text{BOUND}_{\text{intrachannel}} \overset{k_2}{\underset{k_{-2}}{\rightleftharpoons}} \text{IN} \tag{13}$$

In the steady state, the probability that there is ion flow in the channel (i.e., that the binding site is unoccupied, is

$$P = k_{-1} + \frac{k_2}{(k_1 + k_2 + k_{-1} + k_{-2})} \tag{14}$$

The reaction rates depend on the heights of the energy barriers separating the energy minima, and are given by

$$k_{-1} = A \exp\left(\frac{U}{RT}\right) \tag{15}$$

It is important to note that the energy barriers in this formalism are superimposed on a gradient of membrane electrical potential. The heights of the energy barriers and wells are functions of both their physical positions in the transmembrane field and the net charge on the permeating ionic species. This becomes particularly important for blocking particles that cannot permeate all the way through the channel, such as antiarrhythmic drugs, and gives rise to the concept of fractional electrical distance as discussed later in this chapter.

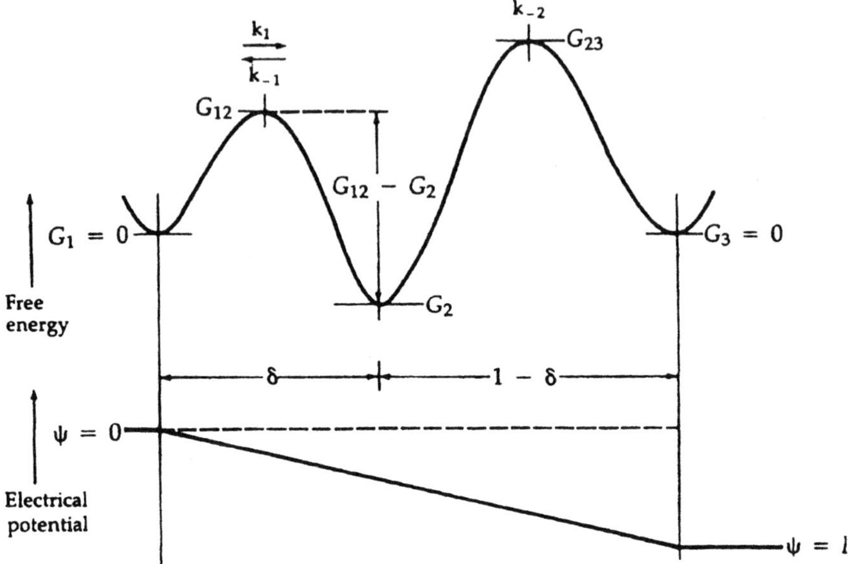

Figure 1   The energy profile of a pore in the membrane modeled with two barriers and three wells. The free energy of the barriers determines the magnitude of the rate constants, k_x. The electrical potential is shown as a linear drop across the membrane. The effective electrical distance, (δ) is the fraction of the electric field crossed by the ion before reaching the binding site. This is not necessarily the same as the physical fraction of the membrane crossed, as the potential drop across the membrane may not be linear. (From Ref. 6.)

III.   HODGKIN AND HUXLEY MODELS OF VOLTAGE-DEPENDENT GATING

Although the changes in membrane potential which form the action potential are the result of the dynamic changes in voltage-gated channels, the best way to determine the biophysical properties of ion channels is to treat the cell as an electrical circuit, and observe the behavior of channels in response to a controlled change in an imposed voltage applied across the cell membrane. The potential across the cell membrane can be measured and/or controlled by impaling an isolated cell with a microelectrode and comparing the intracellular potential to the that of the bath solution, which is defined as 0 mV. The resting potential of a cell is usually around -70 to -90 mV (close to the potassium reversal potential, as the membrane is permeable to potassium at rest), and the action potential peaks at around $+50\,\text{mV}$, so the behavior of channels over this range is of particular interest.

Hodgkin and Huxley [11,14] took advantage of the then newly invented voltage clamp technique to investigate the voltage dependency of the "active patches" in the squid giant axon membrane that they had found to be permeable to sodium and potassium and developed a series of equations to describe their behavior. The Hodgkin-Huxley equations are derived solely from an empirical description of the behavior of ion channels, and are not based on molecular mechanisms: at the time the equations were published the existence of ions channels had yet to be confirmed, and their structure and physiology was completely unknown. Nonetheless, the equations still provide a good representation of channel behavior, even as our understanding of the structure and biophysics of ion channels increased since Hodgkin and Huxley's time.

Ion channels respond to, and are responsible for, dynamic changes in the membrane potential, hence the kinetics of opening and closing are of great interest. Activation is the rapid increase in the probability of a channel opening, as a result of energy put in to the system. Voltage-gated channels by their very definition use the energy supplied by a change in potential across the membrane, and therefore the channel, to change the conformation of the ion channel from a nonconducting to conducting state. The probability of the individual channels being open on the microscopic scale can be expressed macroscopically as changes in the magnitude of the whole cell current.

Some ion channels, e.g., sodium, enter a state that no longer conducts ions even in the presence of a continued stimulus. This is termed inactivation. The channels will remain in the inactivated state until the membrane is repolarized for a sufficiently long time, allowing the channels to recover from the inactivation in a time-dependent manner. The rate of recovery from inactivation can be measured by applying two identical stimuli at varying interstimulus intervals and determining the relative magnitudes of the peak currents elicited.

Once the stimulus is removed (e.g., the membrane is no longer depolarized), the remaining activated channels close, i.e., they are deactivated. This closing of channels can be seen as "tail currents" following a step change in membrane potential. The rate at which channels activate, inactivate, and deactivate define the biophysical properties of the channels, and it is these properties that are modeled in the Hodgkin-Huxley equations.

Activation of an ionic current clearly is the result of a conformational change in the ion channel from a nonconducting to conducting state. If one considers the channel to be a large conduit through which ions flow, then activation could be represented as a gate opening in the channel and allowing the passage of ions. This interpretation is not to be taken literally, and although the conformational changes within the channel during activation are likely to be quite subtle, the term "gating" is used to describe the process associated with steep change in the probability of the channel being open following a stimulus. Gating is independent of the permeability of a channel.

In the Hodgkin-Huxley formulation, gating is treated as a stochastic process. Channels can be only in either the closed or open state, with a rapid transition between the two and a negligible number of channels in transition (single-channel analysis has shown this assumption to be valid). Therefore, if the fraction of channels in the open state is n, the fraction of closed channels must be $(1-n)$:

$$1 - n \underset{\beta}{\overset{\alpha}{\rightleftharpoons}} n$$
$$\text{closed} \qquad \text{open} \tag{16}$$

The dynamic responses of the currents are controlled by the rate constants α and β, which are time- and voltage-dependent. If the maximum whole-cell conductance for ion X is $\bar{i}_x$, the current through an ohmic channel permeable to X, I_x, would be given by

$$I_x = n\bar{i}_x \tag{17}$$

where n is the fraction of open channels. This can be expanded for a simple linear ohmic current [see Eq. (5)] to be

$$I_x = n\gamma_x(E_m - E_{\text{Eqm},x}) \tag{18}$$

Clearly, the dynamic qualities of the current are due solely to changes in the number of open channels, n. If the channels are modeled as having only two states, as in Eq. (16), the rate of change of open channels is given by

$$\frac{dn}{dt} = \alpha_n(1 - n) - \beta_n n \tag{19}$$

In the steady state, the rate of change of open channels will be equal to zero, therefore:

$$0 = \alpha_n(1 - n_\infty) - \beta_n n_\infty$$
$$0 = \alpha_n - n_\infty(\alpha_n + \beta_n) \tag{20}$$
$$n_\infty = \frac{\alpha_n}{\alpha_n + \beta_n}$$

where n_∞ is the number of channels open in the steady state.

Rearranging Eq. (19) and using n_∞ from Eq. (20),

$$\frac{dn}{dt} = \frac{n_\infty - n}{\tau_n} \tag{21}$$

where

$$\tau_n = \frac{1}{\alpha_n + \beta_n} \tag{22}$$

This can be integrated to give n as a function of time:

$$n(t) = n_\infty - (n_\infty - n_0)\exp\left(\frac{t}{\tau_n}\right) \tag{23}$$

where n_0 is the value of $n(t)$ at time $t = 0$. The time constant, τ_n, is a function of the rate constants α and β so, like them, it is time- and voltage-dependent.

Hodgkin and Huxley determined that the major components of the squid giant axon action potential were the sodium and potassium currents, so it was necessary to determine the kinetic parameters of these two channels. Both potassium and sodium channels are voltage activated and open in a time-dependent manner following depolarization of the membrane. In response to a step depolarization, the potassium channel has a sigmoidal increase in conductance, which rises to a steady-state value. The sodium channel has more complex kinetics, with a rapid increase in conductance being followed by a decrease as the majority of channels enter the inactivated state and can no longer conduct ions.

The fraction of channels in the open state (represented by $n(t)$ in Eq. (23)) will depend on the gating properties of the channel. Hodgkin and Huxley used the concept of "gating particles" to determine the probability that a channel was in the open state. A channel could have several independent gating particles, each of which would have to be in the correct state to allow current flow. A single gating particle would give a current a with simple exponential activation and deactivation rates.

The experimental data from potassium and sodium channels are not so simple; the probability of opening both rises in a sigmoid manner rather than following a simple exponential. The potassium channel is best fitted with four gating particles, so n in Eq. (18) is replaced with g^4, giving the current as

$$I_K = g^4 \gamma_K (E - E_K) \tag{24}$$

where γ_K is the conductance of the potassium channel and E_K is its reversal potential.

The sodium current kinetics are more complex than those of the potassium channel: not only does the channel activate in a sigmoid manner, it also inactivates rapidly. The probability of the channel being open depends on the probability of the channel being activated, but also that it is *not* inactivated, so the gating particle requires two components. The sodium channel is best fitted with activation being represented by three gating particles (m^3), and inactivation by one (h). Equation (18) therefore becomes

$$I_{Na} = m^3 h \gamma_{Na}(E - E_{Na}) \tag{25}$$

where γ_{Na} is the conductance of the sodium channel, and E_{Na} is the reversal potential.

Hodgkin and Huxley added a third, ungated channel to their model to represent a generic leak current, I_L, with conductance γ_L and reversal potential E_L:

$$I_L = \gamma_L(E - E_L) \tag{26}$$

The complete Hodgkin-Huxley representation of the ionic component of the membrane current is therefore

$$I_{\text{Ionic}} = \gamma_{Na} m^3 h(E - E_{Na}) + \gamma_K g^4(E - E_K) + \gamma_L(E - E_L) \tag{27}$$

This simple representation of the ionic currents can reproduce the activity of the potassium and sodium currents, and the central features of the action potential, given appropriate rate constants for m, h, and n. The rate constants α_x and β_x are derived from Eyring rate theory [47, 48]. Gating can be considered as the channel moving from one conformationally stable state to another, by crossing over a single energy barrier. The energy difference between the two states depends on an intrinsic (voltage-independent) energy difference between the states and the (voltage-dependent) work done by moving charges in the transmembrane field. The energy difference, ΔU, between the two states with energy U_{open} and U_{closed}, is given by

$$\begin{aligned} \Delta U &= U_{\text{open}} - U_{\text{closed}} \\ &= \Delta U_o + QV \end{aligned} \tag{28}$$

where ΔU_o is the intrinsic voltage-independent energy difference between the open and closed states, and Q is the charge that has to be moved through a potential field V. According to Eyring rate theory [47,48], the rate of reaction is proportional to the exponential of the height of the energy barrier. The rate constants are therefore given by

$$\begin{aligned} k &= A \exp\left(-\frac{\Delta U}{RT}\right) \\ &= A \exp\left[-\frac{Q(V - V_o)}{RT}\right] \end{aligned} \tag{29}$$

where A is a constant and $QV_o = \Delta U_o$.

This type of exponential rate constant is suitable for describing gating reactions in which transmembrane charge movement is the rate-limiting step. However, when modeling the gating process of ion channels, this simple exponential relationship is unsatisfactory for two reasons. The first is practical, as exponentials become very large with large driving force, presenting difficulties for the numerical methods used to solve ordinary differential equations. The second problem with the exponential rate constant is that for strong voltage-dependent driving forces, the translocation of charge is no longer the rate-limiting step and voltage-insensitive processes become rate limiting. This deviation from the exponential is clearly demonstrated experimentally, for the rate constants become saturated at large voltages.

Hodgkin and Huxley plotted α_x and β_x against voltage to determine the best fit to the experimental data for the all the rate constants controlling the gating reactions in the potassium and sodium currents, and arrived at the following modified exponential relationships:

$$\alpha_n = 0.01 \left\{ \frac{V + 10}{\exp[(V + 10)/10] - 1} \right\} \tag{30}$$

$$\beta_n = 0.125 \exp\left(\frac{V}{80}\right) \tag{31}$$

$$\alpha_m = 0.1 \left\{ \frac{V + 25}{\exp[(V + 25)/10] - 1} \right\} \tag{32}$$

$$\beta_m = 4 \exp\left(\frac{V}{18}\right) \tag{33}$$

$$\alpha_h = 0.07 \exp\left(\frac{V}{20}\right) \tag{34}$$

$$\beta_h = \left\{ \frac{1}{\exp[(V + 30)/10] + 1} \right\} \tag{35}$$

These equations provide an approximation of the exponential dependence on voltage with relatively small voltages, but limit exponential growth in the presence of high driving forces. This enables the notion of gating being controlled by a charge particle moving within an electric field to be preserved, but unlimited exponential growth at large voltages is prevented.

Even with these amendments, the original form of these equations produces problems at voltages where the denominator of the rate constant is equal to zero (e.g., $-10 \, \text{mV}$ for n). In programming such equations, an "if then else" statement must be inserted to maintain computational integrity. An alternative numerical approximation [49] that can be used to circumvent these problems is to use the following style of rate-constant relationship, which will never have a zero denominator:

$$k = A \left\{ \frac{\exp(V/K_{\text{slope}})}{1 + \exp[(V + \text{offset})/K_{\text{slope}}]} \right\} \tag{36}$$

IV. MOLECULARLY BASED MODELS OF ACTIVATION AND INACTIVATION

The Hodgkin-Huxley (HH) equations are purely empirical, for they were developed long before channel structure was known. Advances in electrophysiological techniques and molecular biology have allowed us to build new models of the biophysical properties of ion channels based on their molecular structure.

A. Activation Models

Hodgkin and Huxley proposed that ion channels are controlled by gating particles that act independently, and are charged, thus conferring voltage sensitivity on the gating process. The potassium channel was governed by the activation of four gating particles, whereas the sodium channel had three activation particles and one inactivation particle. The number and type of gating particles assigned to each channel was determined from analysis of the activation properties, particularly the sigmoid onset of activation. When the structure of potassium channels were first shown to be a tetramer of four α-subunits by both cloning [50–53] and electron miscroscopy [54], it was immediately assumed that the four subunits corresponded directly to the gating particles of Hodgkin and Huxley. It was assumed that each subunit acted independently, and only when all four subunits were in the activated state would the channel conduct ions. The fact that the sodium channel was composed of a single α-subunit with four homologous but not identical domains, and therefore potentially fewer activation gating particles, seemed to strengthen this argument further [55,56].

Analysis of channel kinetics such as sigmoid activation is a poor measure of properties such as independence and stoichiometry. To demonstrate this point, the Hodgkin-Huxley two-state activation/inactivation model of the sodium channel can be expanded to an eight-state Markov model [57] with seven closed states ($C_{i,j}$) and one open state (O):

$$C_{0,0} \underset{\beta_m}{\overset{3\alpha_m}{\rightleftharpoons}} C_{1,0} \underset{2\beta_m}{\overset{2\alpha_m}{\rightleftharpoons}} C_{2,0} \underset{3\beta_m}{\overset{\alpha_m}{\rightleftharpoons}} C_{3,0}$$

$$\alpha_h \updownarrow \beta_h \qquad \alpha_h \updownarrow \beta_h \qquad \alpha_h \updownarrow \beta_h \qquad \alpha_h \updownarrow \beta_h$$

$$C_{0,1} \underset{\beta_m}{\overset{3\alpha_m}{\rightleftharpoons}} C_{1,1} \underset{2\beta_m}{\overset{2\alpha_m}{\rightleftharpoons}} C_{2,1} \underset{3\beta_m}{\overset{\alpha_m}{\rightleftharpoons}} O$$

The $C_{0,0}$ state represents a channel in which all three gating particles are in the closed conformation. $C_{1,1}$ corresponds to a channel with one particle open (i.e., two gating particles remain in the closed state) and the inactivation particle removed. O corresponds to the open channel, in which all three gating particles are activated, and the inactivation particle indicates the channel is *not* inactivated. The rate constants correspond to the HH rate constants. The rate constant for the transition from $C_{0,0}$ to $C_{1,0}$ is equal to $3\alpha_m$, since there are three independent particles which can effect this change. Conversely, the rate from $C_{1,0}$ to $C_{0,0}$ is just β_m, since there is only one open particle in the $C_{1,0}$ state that can make a closing transition. The rest of the rate constants in the Markov model can be extracted similarly. The rates for removal and addition of the inactivation particle are always just α_h and β_h, as there is just one particle. Considering only activation, the state diagram becomes

$$C_1 \overset{3\alpha}{\rightleftharpoons} C_2 \overset{2\alpha}{\rightleftharpoons} C_3 \overset{\alpha}{\rightleftharpoons} O$$

which gives a mathematically identical time course of activation as

$$C_1 \overset{\alpha}{\rightleftharpoons} C_2 \overset{2\alpha}{\rightleftharpoons} C_3 \overset{3\alpha}{\rightleftharpoons} O$$

Only slight differences are seen when the rate constants are all equal (reviewed in [57]):

$$C_1 \overset{1.67\alpha}{\rightleftharpoons} C_2 \overset{1.67\alpha}{\rightleftharpoons} C_3 \overset{1.67\alpha}{\rightleftharpoons} O$$

Although sigmoidicity indicates that activation is a multistep process, it reveals only limited information about the organization of these steps. The HH model of gating postulates that opening of an ion channel is preceded by the activation process in which movement of a charged gating particle changes the state of the channel from nonconducting to conducting. As this charged particle moves through an electric field, it is by very definition a current, and is known as the gating current. Gating currents are not the same as ionic currents through the membrane: they have a very much smaller amplitude, as they are due to the small outward movement of voltage sensors in a depolarizing field [58,59]. The sensors are thought to be located in the S4 segment of the voltage-gated channels, i.e., the fourth of six transmembrane segments [60–70]. On a molecular level, gating currents arise from the conformational shift of positively charged groups (lysine and arginine) in the transmembrane potential [71–74]. These gating currents were measured prior to the cloning of voltage-dependent potassium channels [58,59].

Analysis of gating currents has revealed that for many of the early steps in activation, independence is relatively consistent, but the final steps may involve nonindependent conformational changes [75,76]. Charge movement may be unevenly distributed between these steps. Furthermore, in some ion channels, some of the activation steps may be voltage insensitive [37,77–80]. Thus, in the literature, we find several non-HH models of activation that use the Markov process formalism [81–86]. Markov processes have proved a useful general tool in describing ion channel kinetics which do not lend themselves to the HH formalism.

B. Inactivation and the Ball and Chain Model

Inactivation was modeled by Hodgkin and Huxley as a process that was just in reciprocal of activation. In their formalism, inactivation was identical to activation except that it was mediated by a slower gating particle with an oppositely directed charge movement. In reality, the molecular mechanisms of inactivation are not only quite distinct from activation, there is a diverse range of inactivation mechanisms. Gating currents can be immobilized by inactivation, which shows that not only are the mechanisms different, but inactivation is a voltage-independent process [87–89].

Of all the various inactivation mechanisms, the "ball and chain" model is perhaps the best understood and most thoroughly described. The ball and chain mechanism was first suggested in a series of studies by Armstrong [28] using proteases in squid axon sodium channels, and the molecular basis for this model was later demonstrated in *Shaker* potassium channels [90]. This inactivation was called "N-type" because the ball and chain were located on the N terminal of the channel. N-type inactivation occurs relatively quickly (on the order of milliseconds to tens of milliseconds) and is mediated by a segment of about 20 amino acids at the N-terminus of the channel protein (the "tethered ball") which binds at the intracellular mouth of the channel pore [90–94]. A cartoon depiction of this process is shown in Fig. 2.

Deletion of the "ball" domain from the N-terminal of the channel by enzymatic cleavage leaves only the "chain" segment, which is unable to block the open channel [91,95]. Inactivation can be restored to channels with N-terminal deletions by applying short peptides derived from the N-terminal to an inside-out patch [95,96]. If N-type inactivation is truly a simple ball and chain mechanism, deletions that shorten the tether region should increase the rate of inactivation, for this will reduce the effective diffusional distance between the ball and the blocking site in the pore. However, some deletion mutations of the chain in *Shaker* and Kv1.4 channels actually slows inactivation, which suggests that chain may have a secondary or tertiary structure that constrains the movement and orientation of the inactivation ball, which results in an increased likelihood of block with a longer chain [97,98].

Events that occur at the extracellular mouth of the pore, such as drug binding or changes in extracellular potassium concentration, do not affect N-type inactivation, whereas those that occur at the intracellular mouth of the pore do [99–101]. N-type inactivation is also insensitive to point mutations at the outer mouth of the channel pore and the outer region of the sixth transmembrane segment, S6 [91,100,101].

One test for the development of N-type inactivation is, therefore, a sensitivity to intracellular tetraethylammonium ions (TEA$^+$) contrasted with an insensitivity to extracellular TEA$^+$ [99,102]. Although the ball is constructed from amino acids in the N-terminal region, which contains basic residues (positively charged lysines and arginines), the inactivation binding site is presumably near the channel surface and so is not far from the edge of the transmembrane electrical field. Consequently, N-type inactivation is voltage insensitive at positive potentials [90].

C. Partially Coupled Models

Hodgkin and Huxley modeled activation and inactivation as completely independent processes, even though they were aware that activation and inactivation might not be independent. However, they decided to use the m^3h formalism in part because of its simplicity, and in part because they did not have any data with which to build a more physiologically correct model. Subsequently, many experimenters have shown that inactivation is linked activation, and so far, no voltage-gated channels have yet been found in which inactivation is a completely independent process from activation [103–108].

The ball and chain mechanism of inactivation has several important properties to be considered when generating a mathematical model of gating. Perhaps the most important is that the channel must open or at least become partially activated before it can enter the inactivated state, i.e., activation and inactivation are not independent processes. This immediately designates it as part of a broader class of models, namely, "coupled" or "partially coupled" inactivation models.

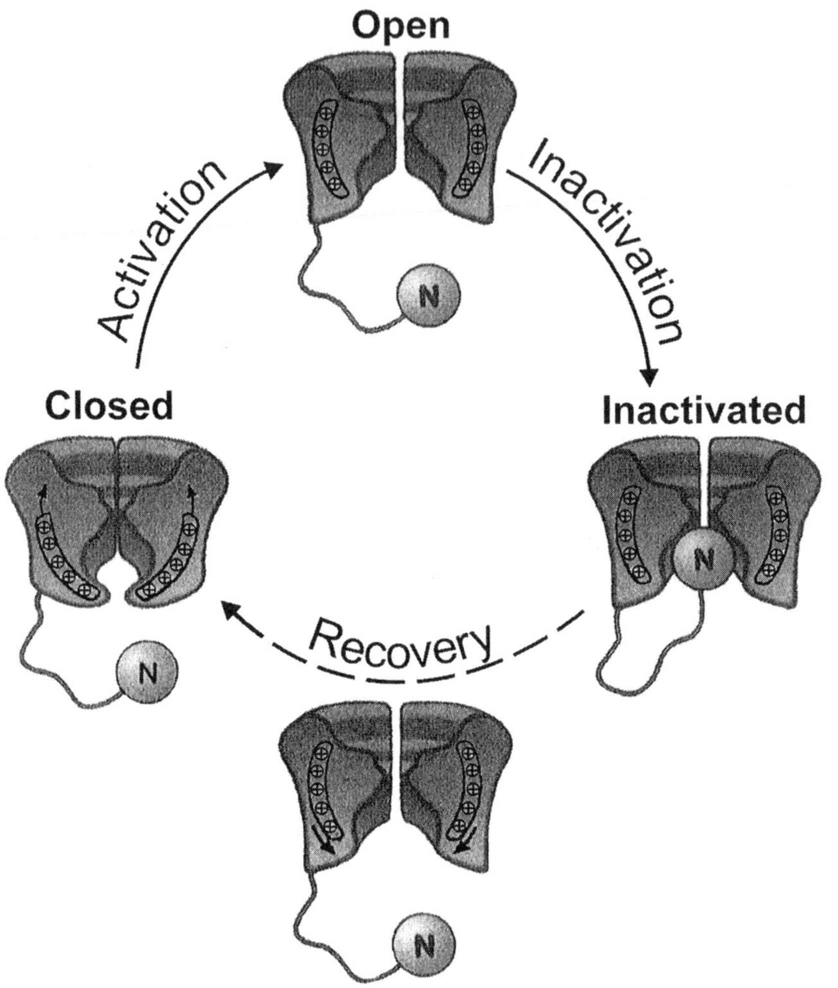

N-Type Inactivation

Even channels that inactivate via an intrinsic voltage-dependent mechanism rather than a ball and chain exhibit some degree of coupling between activation and inactivation [37]. This is not surprising, for activation involves a conformational change [109,110], which can affect the conformation of the inactivation-producing regions of the channel too. For most channels inactivation is a voltage-independent process once activation is complete [87,90,11].

The simplest model of coupled inactivation is has just three states: closed (C), open (O), and inactivated (I):

$$C_1 \underset{\beta_1}{\overset{\alpha_1}{\rightleftharpoons}} O \underset{\beta_2}{\overset{\alpha_2}{\rightleftharpoons}} I$$

The rate of channel opening, α_1, is voltage-dependent. Even though the rate of inactivation, α_2, is voltage-independent, the number of channels that will be inactivated in the steady state will have an apparent voltage dependency because of the voltage dependency of α_1. For example, at voltages near the threshold for activation, the probability of the channel being in the open state is low and thus the probability of it reaching the inactivated state is also low. At large driving voltages the number of channels in the open state will become saturated, so the number of channels entering the inactivated state will saturate too, which is what is observed experimentally. In both cases there is an apparent voltage dependence of inactivation due solely to coupling of inactivation with activation.

In a fully coupled model, such as the one above, the channel must open before it can inactivate, and it must reopen upon recovery from inactivation. Single-channel records, however, show that many ion channels can inactivate and recover *without* the channel opening. This can be described by a partially coupled model of inactivation, in which some of the rate constants of activation, α_x and β_x, are voltage-dependent (minimally,

◀───

Figure 2 Schematic representation of voltage-gated channel activation, N-type inactivation, and recovery. The N-terminal of the channel forms a "chain" with a "ball" attached. In the closed state there is no pore for ion flux, and the ball and chain are in the cytoplasm. When the membrane is depolarized, charged residues on the S4 segment move, leading to a conformational change that opens the pore and reveals the ball binding site. When the ball binds, the channel is blocked and the channel is inactivated. Following repolarization, the channel recovers from inactivation in a time-dependent manner. Many channels have a voltage-dependent recovery rate: the "push-off" theory suggests that the movement of the S4 charges back to the resting state helps to dislodge the ball from the binding site. (From Ref. 116.)

the transition from C_3 to O and one of the C_x to C_y transitions must be voltage sensitive), but the rate of inactivation, k_f, and the recovery from inactivation, k_b, are not:

$$C_1 \underset{\beta_1}{\overset{\alpha_1}{\rightleftharpoons}} C_2 \underset{\beta_2}{\overset{\alpha_2}{\rightleftharpoons}} C_3 \underset{\beta_3}{\overset{\alpha_3}{\rightleftharpoons}} O$$
$$k_b \updownarrow k_f \qquad\qquad k_b \updownarrow k_f$$
$$I_1 \underset{\beta_3}{\overset{\alpha_3}{\rightleftharpoons}} I_o$$

The two inactivated states represented an inactivated closed channel (I_1), and an inactivated open channel (I_o). Transition between the two inactivated states mirrors the voltage-sensitive transition between the partially activated closed state (C_3) and the open state (O). This model is only partially coupled, since the channel can become inactivated directly from the partially activated closed state, C_3, without the channel actually opening. Similarly, recovery from inactivation does not require the channel to open.

The model predicts that the rate of recovery from inactivation will be much slower than the rate of development of inactivation, since k_b must be much slower than k_f if inactivation is to be relatively complete. The rate of exit from the inactivated state to the open or partially activated closed states is voltage insensitive, so the model predicts that there will be a saturation in the voltage sensitivity of recovery from inactivation as k_b becomes the rate-limiting step.

Many inactivating channels recover from inactivation at negative potentials in a relatively fast voltage-dependent manner, so there must be a source of voltage-dependent energy for recovery from inactivation. This can be understood in the context of the ball and chain model by the ball "push-off" mechanism (see Fig. 2). The hypothesis is that when the membrane potential is repolarized, the voltage sensor moves "backwards" to the resting position, thus destabilizing the inactivated channel, and "pushing" the "ball" out of its binding site. Exit from the inactivated state is, therefore, driven by the energy derived from deactivation. To represent this in a model, an extra voltage-dependent pathway is needed connecting the inactivated state and the closed state. This can be modeled by adding a single voltage-dependent transition between I_1 and C_2:

$$C_1 \underset{\beta_1}{\overset{\alpha_1}{\rightleftharpoons}} C_2 \underset{\beta_2}{\overset{\alpha_2}{\rightleftharpoons}} C_3 \underset{\beta_3}{\overset{\alpha_3}{\rightleftharpoons}} O$$
$$\beta_4 \nwarrow \alpha_4 \quad k_b \updownarrow k_f \qquad\qquad k_b \updownarrow k_f$$
$$I_1 \underset{\beta_3}{\overset{\alpha_3}{\rightleftharpoons}} I_o$$

or by inserting an entirely new inactivated state, I_2, which is a low-affinity, partially coupled state:

$$C_1 \underset{\beta_1}{\overset{\alpha_1}{\rightleftharpoons}} C_2 \underset{\beta_2}{\overset{\alpha_2}{\rightleftharpoons}} C_3 \underset{\beta_3}{\overset{\alpha_3}{\rightleftharpoons}} O$$

$$k_b \updownarrow k_f \qquad k_b \updownarrow k_f \qquad k_b \updownarrow k_f$$

$$I_2 \underset{\beta_2}{\overset{\alpha_2}{\rightleftharpoons}} I_1 \underset{\beta_3}{\overset{\alpha_3}{\rightleftharpoons}} I_o$$

Both of these formulations yield similar results, but recovery from inactivation reaches a saturating maximum rate only in the second model. This is because every pathway out of the inactivated state in the second model has a step controlled by a voltage-insensitive rate constant, whereas in the first model there is a voltage-sensitive pathway that has no such limit.

Channel inactivation is complex, and more than one type of inactivation may be present in a single channel type. In a Hodgkin-Huxley type of model, this is dealt with by adding several gating particles with differing time constants. In reality, modes of inactivation can be coupled, in a similar way to the coupling of inactivation with activation.

A relatively slow type of inactivation resulting from changes in conformation at the extracellular mouth of the channel [112] is called C-type inactivation, after the C-terminal splice variant *Shaker* potassium channel in which it was first seen [113,114]. The molecular basis of C-type inactivation is not as well understood as N-type is. It is not removed by N-terminal deletion, but is sensitive to changes in the extracellular concentration of the permeant ion, or external application of drugs such as TEA$^+$ [99,100].

A relationship between N-type and C-type inactivation was suggested from the very beginning, as it was observed that mutations that enhanced the rate of C-type inactivation slowed the rate of recovery from N-type inactivation [91]. Conversely, C-type inactivation is significantly slowed in the in the absence of the N-terminus, and, at all potentials, is incomplete.

Two distinct mechanisms have been proposed to explain the interaction between N-type and C-type inactivation. One group working with *Shaker* channels proposed that the predominant effect of N-type inactivation was inhibition of the potassium flow in the channel, thus making C-type inactivation more likely [115]. Another group working with Kv1.4 channels proposed that N-type inactivation immobilizes channel gating, leaving it in a conformation which makes C-type inactivation more favorable due to steric factors [101,116].

These two proposed mechanisms may coexist in a channel, with their relative dominance being dependent on the channel type, physiological or experimental conditions, etc. A channel may be observed to enter inactivation in a characteristically N-type manner (e.g., insensitivity to potassium). Subsequent to N-type inactivation, C-type inactivation may develop, and then the channel will be observed to recover from inactivation via a C-type mechanism, with markedly different properties to the N-type mechanism by which it entered inactivation.

Although the Hodgkin-Huxley model cannot be justified in terms of molecular mechanisms, the equations still provide a fair representation of the currents. In many cases, activation and inactivation of channels has been demonstrated to involve more steps than the simple one closed and one open state in the Hodgkin-Huxley formulation. However, many of these more complex relationships can be reduced to a two-state model in which the experimentally determined rate constants represent an averaged contribution from the multiple rate constants. When the model cannot be reduced to an equivalent one open/one closed state model (e.g., cyclical or nonlinear models), or when the situation is more complex (e.g., presence of blockers or other modulators, as discussed below), the Hodgkin-Huxley model fails.

V. DIFFUSION THEORY AND MODELING ION CHANNELS

In an effort to produce models that are more readily linked to physical processes operating within the channel at an atomic level, some researchers have proposed ion channel models based on Brownian motion and diffusion theory [117]. Instead of representing the gating process and ion channel kinetics as Markovian processes, i.e., progression in a series of discrete steps, these models use Brownian dynamics to solve the chaotic and diffusional motions of the individual ions. An atomic model includes mathematical descriptions of the channel protein, the water within the pore, the lipid surrounding the membrane protein, and a representation of the bulk water at either end of the channel. Ions are then introduced to the model, and their motion is calculated in discrete and very small time intervals based on the local electric field, ion–ion interactions, ion–channel interactions, the concentration difference between bulk solutions, and the temperature.

The advantage of this type of modeling is that it can offer a more precise result, in terms of the passage of an ion through the channel, than could ever be achieved with a model based on the Hodgkin–Huxley equations. However, with current computational approaches a single transmembrane journey of just one ion is about the limit of what can be calculated within a reasonable time frame, as it takes several hours to calculate the exact molecular dynamic solution to the movement of an ion subjected to these forces for only a picosecond [118].

Assumptions and simplifications bring the calculations to within more reasonable calculation times. First, if the protein is assumed to be immobile and the water molecules can be represented by a continuum, then the movement of an ion i calculated by integrating the following equation in discrete time steps:

$$m_i \frac{dv_i}{dt} = m_i f_i v_i + F_R(t) + q_i E_i \qquad (37)$$

where, for particle i, m_i is the mass, v_i is the velocity, f_i is the frictional coefficient, q_i is the charge, E_i is the total electric field experienced by the particle (including contributions from the protein, other ions, and the results of the changing dielectric constant at the protein–water–lipid boundaries), and F_R is a random thermal force representing collisions of the ion with water molecules and the protein.

Further assumptions and simplifications that speed up the computational process include using a mean field approximation for E_i, considering a 1-D concentration function rather than a radially varying 3-D one, or sectioning off the channel into various compartments, then determining the kinetics of an ion moving from section to section.

The Brownian motion and molecular dynamics approach has been most successful with theoretical channels [119–122] or ion permeation of gramicidin channels [123–126,126a,127], as this small simple channel was the first to have its complete structure revealed on the atomic scale.

With the recent increase in information concerning the atomic structure of more complex channels [128], more elaborate channel models (though still idealized) are now being developed: e.g., the acetylcholine receptor [129], voltage-gated sodium channel [130], KcsA potassium channel [131], and calcium channel [132–134].

With the brief interval over which molecular dynamics can be modeled, long-term channel kinetics (especially inactivation) cannot be reproduced with this type of modeling. Nonetheless, models of ion channel activation have been proposed for the gating of a few channels over a nanosecond time scale. A surface tension variable was included in a model of the mechanosensitive channel MscL, which induced conformational change in the model opening a large pore in the protein [135]. A gating model was also proposed for the bacterial potassium channel KcsA, which showed transient increases in the diameter of the intracellular mouth [136].

Overinterpretation of results in such a situation is not appropriate. Although these models may soon provide insights into the exact atomic interactions in activation and inactivation, this approach is not suitable for building up a model of current through a single channel, let alone all the currents present in a particular cell type.

VI. DRUG BINDING, LIGAND BINDING, AND OTHER MODIFIERS OF CHANNEL GATING

Any pharmacological modulation of channel behavior involves binding of a drug to a receptor site and its subsequent effect on the channel. Therefore, the simplest and most basic property to be reproduced when considering pharmacological modulation of ion channel properties is the dose–response relationship. Drugs and other ligands can modulate ion channel gating by binding covalently to sites on or in the channel. If a concentration of a ligand [X] results in a current I, the relationship between [X] and I is given by the Hill equation:

$$\frac{I}{I_{max}} = \left(\frac{[X]^n}{[X]^n + K^n_{1/2}} \right) \tag{38}$$

where I_{max} is the current flowing in the absence of the drug, $K_{1/2}$ is the concentration at which the current is reduced to half-maximal, and n (the Hill coefficient) is the number of totally cooperative bindings sites on the channel. Rearranging Eq. (38) and taking logs gives the following expression:

$$\log \left[\frac{I}{I_{max}} \left(\frac{1}{1 - I/I_{max}} \right) \right] = n \log[X] - n \log(K_{1/2}) \tag{39}$$

Therefore, a plot of $\log\{I/[I_{max}(1 - I/I_{max})]\}$ against $\log[X]$ will give a line with gradient n and intercept $K_{1/2}$. For a given experimental system, the minimum number of binding sites required for effective block and the degree of cooperativity can be estimated by constructing a Hill plot and determining the Hill coefficient.

Frequently, models incorporate the Hill equation as though it were just another gating variable, i.e., drug binding is calculated as being independent of conformation. This is adequate for many situations, but this approximation becomes problematic when considering time- and voltage-dependent drug effects.

For many compounds of interest, drug binding is both time- and voltage-dependent. In general, the voltage-dependent properties of drug binding arise from two different mechanisms. One is the indirect result of voltage-dependent gating, which is discussed below. The other mechanism results from the fact that most blockers have a net charge and must enter the transmembrane electric field to reach the binding site. The fraction of the field crossed by the drug before reaching the binding site is called the effective electrical distance (δ). It does not necessarily correspond to the physical fraction of the binding-site location in the membrane, for the drop of potential across the membrane-bound protein is not uniform. The relative depths of various binding sites can be compared by calculating the apparent electrical distance as a fraction of the total transmembrane voltage.

If a charged compound is too large to permeate through the channel, the compound must exit the channel on the same side it entered. In this case, an applied membrane potential can either enhance entry and diminish exit or vice versa, depending on the charge of the compound, the side of the membrane from which entry occurs, and the polarity of the membrane potential [42]. At equilibrium, the ratio of blocked to open states will remain steady, though voltage dependent. If the equilibrium binding constant for a compound of valence z which blocks a channel at a single site location is $K_D(V)$, then the effective electrical distance is given by

$$\delta = \left(\frac{RT}{zFV}\right)\left[-\ln\left(\frac{K_D(V)}{K_{D,0\,\text{mV}}}\right)\right] \tag{40}$$

where F, R, and T have their usual meanings, and $K_{D,0\,\text{mv}}$ is the equilibrium binding affinity at 0 mV. Rearranging Eq. (40) gives an expression for $K_D(V)$:

$$K_D(V) = K_{D,0\,\text{mV}} \exp\left[-\left(\delta\frac{zFV}{RT}\right)\right] \tag{41}$$

which can then be used to calculate the fraction of open channels that are blocked in the presence of a blocking compound at concentration [B]:

$$\frac{P_{\text{blocked}}}{P_{\text{open}}} = \frac{[B]}{[B] + K_{D,0\,\text{mv}}\exp[-\delta(zFV/RT)]} \tag{42}$$

where P_{blocked} and P_{open} are the numbers of blocked and open channels, respectively. These equations are based on equilibrium assumptions. Several reviews deal with the subject of fractional electrical distance and kinetic models [6,137,138].

A. Conformation-Dependent Block

In addition to direct effects of membrane potential on drug binding to the channel, changes in voltage can also result in alterations in the physical conformation of the channel. Voltage-gated channels by their very definition assume different stable conformations in response to changes in transmembrane potential, as the channel switches between conducting and non-conducting modes. The two conformations have different relative free energies, reflecting the relative stabilities of the states. The change in membrane potential is translated into a change in the free energy in the protein by the movement of charges (either displacement of static charge or reorientation of dipole moments) in the electric field. The probability of a channel being in the open or closed state varies with the energy put into the system, i.e., the membrane potential. The correlation between gating (i.e., the probability of a channel being open, P_{open}, or closed, P_{closed}) and voltage is described by a Boltzmann relationship:

$$\frac{P_{open}(V)}{P_{closed}(V)} = K_{0\,mV}\exp\left(-\frac{qFV}{RT}\right) \tag{43}$$

where F, R, and T have their usual meanings, $K_{0\,mV}$ is the equilibrium value at $0\,mV$, and q is the elementary charge which represents the minimum effective charge movement necessary to effect a change in the equilibrium between the open and closed states.

A total of approximately 12 unitary charges cross the membrane during the activation process [73,79,139]. If q is 12, the voltage changes during an action potential result in an exceptionally large change in the total free energy, so there is ample free energy in the gating process for the channel to modulate the binding affinity by several orders of magnitude.

Activation initiates a host of large-scale conformational changes in a channel protein. The most important of these changes is the formation of an ion-conducting pore forming a continuous aqueous domain through which ions can travel from one side to the other. In the *Shaker* potassium channel, a critical step in activation is the formation of an open, stable, intracellular vestibule. The opening of this vestibule not only permits transmembrane ion flux, it also provides access to previously hidden binding sites. For example, the N-terminal domain binding site that leads to N-type inactivation is revealed, as are binding sites for blockers that bind to the open conformation of the channel. The longer the channel is in the open conformation with the binding sites exposed, the greater the likelihood that the channel will become blocked. This is called use dependence. Depending on the drug and channel involved, the drug may have to unbind before deactivation can take place (conventional open channel block) or the channel may close around the blocker that remains bound (trapping block).

Voltage-gated potassium channels exposed to 4-aminopyridine (4-AP) not only show open channel block and use dependence [140], but also closed-state block, or *reverse* use dependence, in which activation of the current removes the ability of 4-AP to block the channel [141–144]. Activation and deactivation may have direct effects on drug binding and may also provide the energy needed to "push off" drugs that might occlude the intracellular mouth of the channel, as evidenced by bi-stable block of the Kv1.4 channel [145].

1. Conventional Open Channel Block

Open channel block is perhaps the most common form of block encountered experimentally. This mechanism arises from a need for the channel to be activated, or opened, prior to drug binding, presumably reflecting the requirement for the opening of the intracellular mouth to expose the binding site. The simplest Markov model representing open state block is

$$C \Leftrightarrow O \Leftrightarrow B$$

which is analogous to the model of the simple inactivated state. Depending on the kinetics of drug binding, open channel blockade can resemble inactivation. Such behavior has been described for internally applied quaternary ammonium ions on some potassium channel types (e.g., squid axon delayed rectifier [28,146]).

Inactivation-like behavior following application of a drug is only one of four major criteria used to establish open channel blockade. The degree of block of the tail currents is dependent on the length of the preceding depolarized pulse, as the longer the channels are open the greater the likelihood is of a drug blocking the channel. There is also a delay in channel deactivation following repolarization, as the drug has to unbind before the channel can close. The reduction in the magnitude of the tail current, coupled with a slowing of the kinetics on deactivation, results in a "crossover" of the tail currents recorded with and without drug. Finally, the effective electrical distance of a binding site exposed only when the channel is open should be between 0 and 1, depending on the depth of the binding site in the pore. These defining phenomena are sensitive to the exact kinetics of the blocking compound interacting with the channel, so not all of these criteria will be observed for a particular open channel blocker, or a particular channel type. For example, "crossover" currents will only be seen in response to a depolarizing step in channels that have intrinsic inactivation.

2. Trapping Block

Trapping is a form of open channel block in which the open channel is able to close around the drug while it is still bound to the channel [147–149]. Like open channel block, it requires activation to occur before the blocker can bind. However, unlike conventional open channel block, the drug cannot be removed during wash-off without the channel being activated. This can be represented by a simple four-state model:

$$C \quad \Leftrightarrow \quad O$$
$$\Updownarrow$$
$$C_{\text{blocked}} \quad \Leftrightarrow \quad O_{\text{blocked}}$$

3. Closed-State Block

Wild-type voltage-gated potassium channels in the Kv4.x family exhibit intrinsic inactivation. Application of 4-AP reduces the magnitudes of these currents appreciably, and also slows down the time of both activation and inactivation. As a result, there is "crossover" of the currents initiated by membrane depolarization when comparing records with and without 4-AP [49]. This change in kinetics is thought to result from closed-state binding of the 4-AP, which can be represented as a four-state Markovian model:

$$B \underset{k_b}{\overset{k_f}{\rightleftharpoons}} C \underset{\beta_{-1}}{\overset{\alpha_1}{\rightleftharpoons}} O \underset{\beta_{-2}}{\overset{\alpha_2}{\rightleftharpoons}} I$$

When the channel is in the blocked state it cannot reach the open state until the drug dissociates from the channel. If the drug dissociation rate (k_f) is slow compared to the rate for the $C \rightarrow O$ transition (α_1), then the onset and peak of the current are delayed. Even though the kinetics of activation and inactivation are not altered by the drug, there is an *apparent* alteration in kinetics because k_f is the rate-limiting step, and the channel activation, and therefore coupled inactivation, is delayed.

The closed-state 4-AP block of I_{to} in ferret ventricular myocytes displays a marked dependence on frequency of stimulation [150]. The concentration of 4-AP that produces a half-maximal reduction in peak current is called the apparent dissociation constant, K_d:

$$\frac{I}{I_{\max}} = \frac{[\text{B}]}{[\text{B}] + K_d} \tag{44}$$

As can be seen in Fig. 3, K_d is markedly altered by the frequency at which the current is activated: the larger the interpulse interval, the greater the probability that 4-AP will bind. This reverse use-dependent behavior suggests that, in contrast to other channel types, the intracellular vestibule of the I_{to} channel is open and capable of binding, even when the channel is in a nonconducting state.

This open vestibule hypothesis was studied by Tseng et al. [143] in Kv4.2 channels, which are thought to underlie I_{to} in the majority of ventricular myocytes, and they show the same reverse use dependence to 4-AP that native I_{to} currents do. In contrast, Kv1.4 channels with N-type inactivation removed by N-terminal deletion (Kv1.4ΔN) trap 4-AP in the binding site on deactivation [145,151].

The pore-lining region in voltage-gated potassium channels is thought to be formed from the "loop" of amino acids that connect the transmembrane S5 and S6 segments which form an eight-stranded anti-parallel β-barrel [152,153]. The chain joining segments S4 and S5 is thought to undergo a conformational changed during activation from a state which blocks the entrance to the pore to another that allows ions to flow through the channel [153].

4. Other Changes That Occur with Activation and Deactivation

When the membrane potential is repolarized, the voltage sensors in the S4 transmembrane segment move back to the closed state, and the channel no longer conducts current. The movement associated with deactivation has long been associated with changes in drug affinity. Early studies demonstrated that tetrapentyl ammonium showed open channel block that was similar to the process of inactivation in squid giant axon [28]. Interestingly, block of the open channels showed kinetics that were slow, with relatively high affinity and little voltage dependence. However, when the membrane was repolarized, block was removed very rapidly. This rapid removal of block during repolarization was attributed to a "push-off" process by which energy from deactivation was used to destabilize the drug-binding site. The molecular basis for this phenomenon is still under investigation.

One important question is how movement of the voltage sensor might alter binding affinity directly. The external binding of tetraethyl ammonium to potassium channels involves a coordinated and simultaneous interaction with the four symmetric domains of the potassium channel. It is unclear whether drugs that act at the intracellular side of the channel involve similar simultaneous interactions. It has been proposed for drug binding of 4-AP to the intracellular side of Kv1.4 channels that such interactions do occur. Furthermore, the side chains that form this binding site are proposed to move with activation of the subunit of the channel. This means that there are only two symmetric drug-binding conformations, one with all four subunits fully activated and one with all four subunits deactivated. Such a system can generate a unique pattern of block in which block is greatest for a long series of short-duration pulses or at the end of a pulse of long duration. The putative involvement of activation in coordinating binding suggests that the recovery properties of drugs may be strongly determined by the number of subunits with which a drug interacts simultaneously and by the degree of hetero-multimerization of the subunits underlying a particular current.

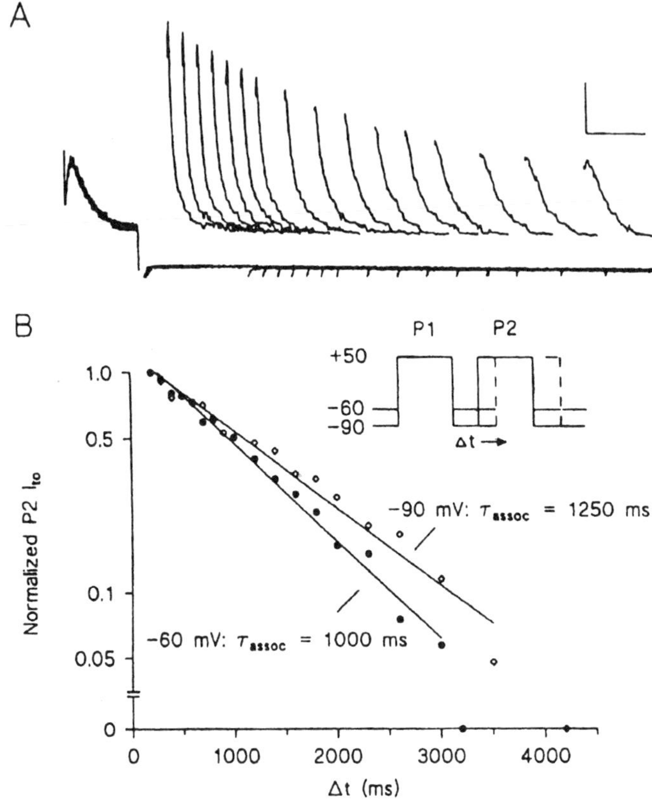

A
B
P1   P2
+50
-60
-90
Δt →
1.0
0.5
0.1
0.05
0
Normalized P2 I_to
-90 mV: τ_assoc = 1250 ms
-60 mV: τ_assoc = 1000 ms
0   1000   2000   3000   4000
Δt (ms)

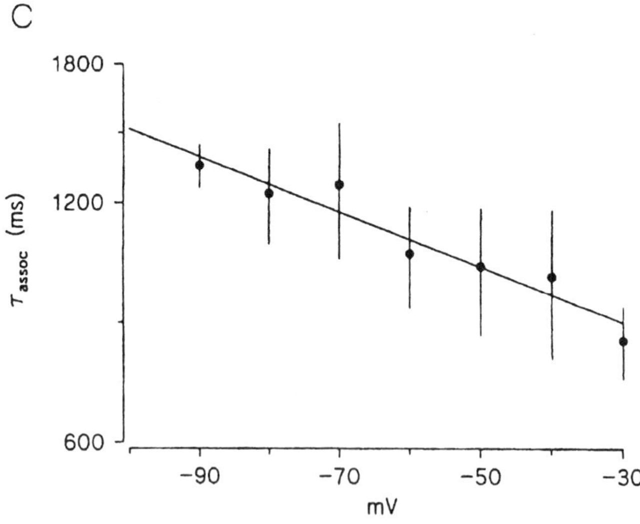

C
1800
1200
600
τ_assoc (ms)
-90   -70   -50   -30
mV

5. Inactivation and Accessibility

N-type inactivation is an important conformational change that is present in many types of potassium channel. Part of the N-terminal domain binds to the intracellular vestibule of the cardiac potassium channel, blocking current flow. The N-terminal binding site is close to drug-binding sites, so competition is expected between N-type inactivation and drug binding. When the N-terminal domain of Kv1.4 is removed, the apparent binding affinity for open channel blockers (measured as a reduction in peak current) such as 4-aminopyridine is increased [145]. Theoretical studies have also demonstrated that the properties (e.g., voltage dependence) of inactivation can determine the use-dependent properties of open channel blockers with slow recovery kinetics [154]. Inactivation is an important process which also strongly modulates the availability of the drug-binding site and can help determine the patterns of use-dependent block.

Drug binding does not always result in just simple blockage of a channel. In some cases, the act of binding to and blocking the channel may initiate additional conformational changes which can have significant effects on activity of the channel.

B. Approximations of Conformation-Dependent Binding and Their Limitations

The intrinsic voltage-dependent inactivation proposed by Hodgkin and Huxley [14] is not a physically accurate description of inactivation for most channels [87,90,111]. The lack of correspondence between the HH formalism and the physical properties of the channel has frequently been outweighed by the mathematical convenience of the HH formalism and the ability of the HH models to reproduce macroscopic current behavior under a wide variety of conditions of interest [155–158]. This ability to reproduce the data arises from the fact that, although inactivation is not voltage gated, it has an indirect voltage dependency because it is coupled to activation, as discussed above.

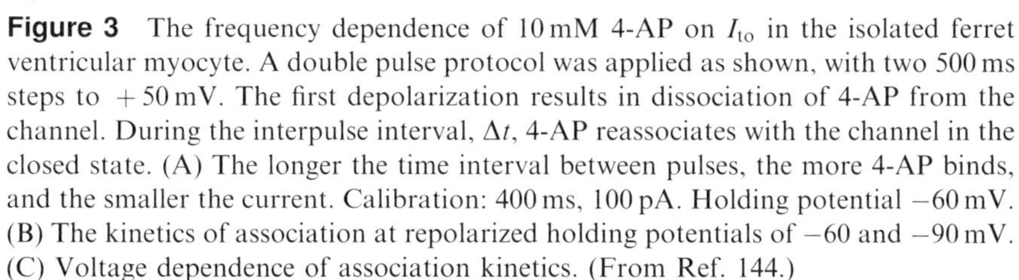

Figure 3 The frequency dependence of 10 mM 4-AP on I_{to} in the isolated ferret ventricular myocyte. A double pulse protocol was applied as shown, with two 500 ms steps to $+50$ mV. The first depolarization results in dissociation of 4-AP from the channel. During the interpulse interval, Δt, 4-AP reassociates with the channel in the closed state. (A) The longer the time interval between pulses, the more 4-AP binds, and the smaller the current. Calibration: 400 ms, 100 pA. Holding potential -60 mV. (B) The kinetics of association at repolarized holding potentials of -60 and -90 mV. (C) Voltage dependence of association kinetics. (From Ref. 144.)

The modeled current during certain simple protocols, such as reproducing an action potential, is insensitive to the whether the voltage dependence of inactivation is direct or indirect. However, when predicting the current due to conformation-specific drug binding, the exact description of the energetics of a particular state may be crucial. Regardless of the similarity of predicted currents, HH-like models with intrinsically voltage-dependent inactivation can lead to significant differences in the predicted drug-binding behavior when compared with voltage-insensitive partially coupled (PC) models of inactivation. Liu and Rasmusson [154] demonstrated this by comparing of two models of ferret I_{to} which reproduced the same macroscopic ion channel behavior. They were both Markovian models with one open state, three closed states, and, in the case of the HH-like model, four inactivated states (Model 1) and, in the PC model, three inactivated states (Model 2). The models have three closed channels, as this gives the best fit of experimental activation data from the ferret I_{to} [49]. The two models are shown in Fig. 4.

The fundamental difference between the models is the voltage sensitivity of the rate constants: α_i and β_i are voltage sensitive, whereas K_f and K_b are not. K_{on} and K_{off}, which govern drug binding and unbinding, have the same voltage-insensitive kinetics in both models.

Although the models produced similar macroscopic current data for voltage clamp simulations, substantial differences arise in the predicted time course and fraction of channels blocked (P_{bound}) at different potentials (Fig. 5A). In the presence of a drug, both models showed only minor changes in P_{open} in response to a single depolarization (Fig. 5B). However, the difference in the value of P_{bound} between Models 1 and 2 predict that recovery from drug binding and, as a result, P_{open} during subsequent pulses will be strongly influenced by model choice (Fig. 5C).

C. Ad-Hoc Modifications to the HH Formalism

The simplicity of the HH equations are appealing, but the independence and voltage dependence of activation and inactivation limit the ability to reproduce certain data. HH-type models effectively reproduce voltage clamp data, but the simulations and calculations of Rasmusson and Liu [154] demonstrate that this type of model may be inappropriate for predicting state-dependent block. PC models are computationally complex, so a full PC model may not be appropriate in all situations. Instead, a computationally simple formulation, which may not be correct from a mechanistic standpoint but which reproduces certain desired features of PC-type block, can be used.

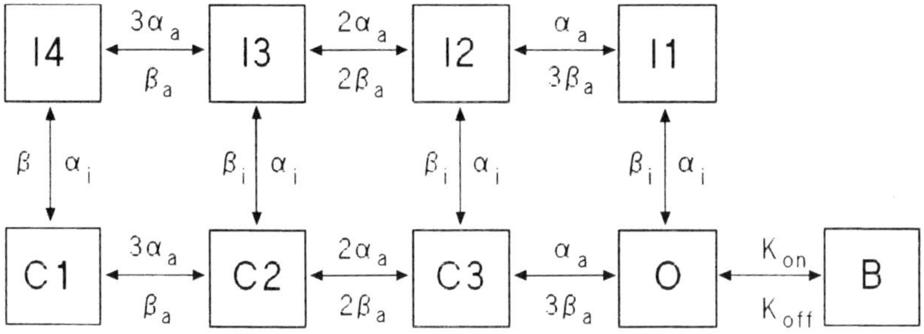

Model 1

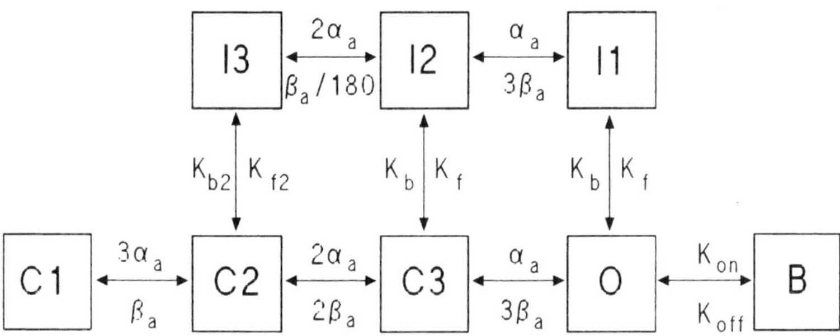

Model 2

Figure 4 Model to reproduce ferret I_{to} current in the presence of an ideal open channel blocker. Model 1 is a Hodgkin-Huxley-type model with three closed states, four inactivated states, and one blocked state. Model 2 is a partially coupled model, with three closed states, three inactivated states, and one blocked state. The rate constants α and β are voltage dependent, but k_f and k_b are voltage insensitive. For both models, k_{on} and k_{off} are the same. (From Ref. 154.)

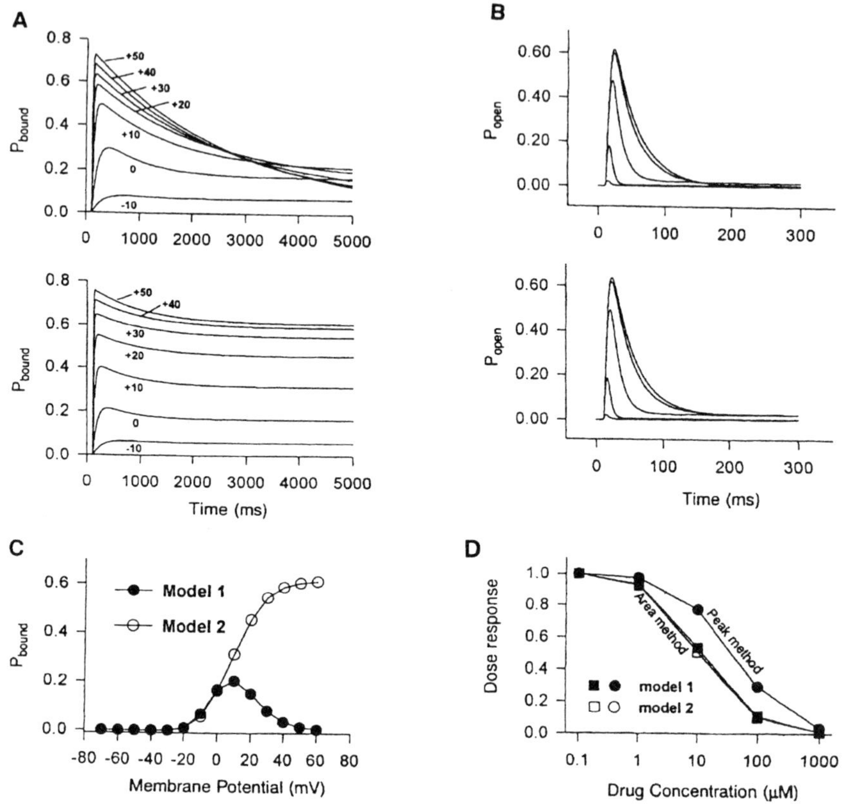

Figure 5 The HH-like and PC models give slightly different results. (A) The fraction of channels in the bond state, P_{bound}, during depolarizations in the presence of 50 mM drug in Model 1 (upper panel) and Model 2 (lower panel). (B) Simulated open-channel probability (P_{open}) in the presence of varying drug concentrations (0, 1, 10, 100, 1000 mM) in response to a step depolarization from -70 to $+50$ mV. (C) Voltage dependence of the steady-state P_{bound} in both models in the presence of 50 mM drug, calculated directly from the coefficient matrix of states. (D) Dose–response curve for Models 1 and 2 for a simulated depolarization from -70 to $+50$ mV. Peak method: normalized P_{open} as a function of drug concentration. Area method: the area under P_{open} as a function of drug concentration. (From Ref. 154.)

1. Ad-Hoc Model 1

The most important feature of PC block is the Boltzmann-like voltage dependence of steady-state binding, with characteristics that parallel steady-state activation. The fraction of channels blocked (P_{bound}) is well approximated by a modified Boltzmann equation:

$$P_{\text{bound}} = \frac{P_{\text{bound, max}}}{1 + \exp[-(V_m - V_{1/2})/K]} \tag{45}$$

where $V_{1/2}$ and K are constants, and $P_{\text{bound, max}}$ is a concentration-dependent term which describes the affinity of the drug for the fully activated channel that can calculated directly from transitions between the final inactivated state, the open state, and the drug bound state:

$$P_{\text{bound, max}} = \frac{1}{1 + (K_{d,\text{apparent}}/[\text{D}])} \tag{46}$$

where [D] is the concentration of the drug, and $K_{d,\text{apparent}}$ is the apparent binding affinity of the drug, which is given by

$$K_{d,\text{apparent}} = \frac{K_{\text{off}}}{k_{\text{off}}[1 - K_f/(K_b + K_f)]} \tag{47}$$

The dependence of $P_{\text{bound, max}}$ on K_f and K_b, the inactivation rate constants, reflects the competition between inactivation and the drug for access to the binding sites on the open channel.

At very negative potentials the time constant for recovery from drug block can be approximated by

$$\tau_{b,\text{negative}} = \frac{1}{K_{\text{off}}} \tag{48}$$

The time constant for initial development of block at very positive potentials and relatively high drug concentrations can be approximated by

$$\tau_{b,\text{positive}} = \frac{1}{K_{\text{off}} + K_{\text{on}}[\text{D}]} \tag{49}$$

The rate at which the channel enters the blocked state is sensitive to P_{open}, but the rate constant from blocked back to the open state is not. One potential approach to modeling voltage-gated blocking behavior is, therefore, to approximate binding by an independent inactivation-like gating variable with a voltage-dependent forward rate constant $\alpha_{\text{block,apparent}}(V)$ and a voltage-independent backward rate constant $\beta_{\text{block,apparent}}$, where

$$\beta_{\text{block,apparent}} = k_{\text{off}} \tag{50}$$

and $\alpha_{\text{block,apparent}}(V)$ is calculated from the relationship

$$P_{\text{bound}} = \frac{P_{\text{bound,max}}}{1 + \exp[-(V_m - V_{1/2})/K]}$$

$$= \frac{\alpha_{\text{block,apparent}}(V)}{\alpha_{\text{block,apparent}}(V) + \beta_{\text{block,apparent}}} \tag{51}$$

Rearranging:

$$\alpha_{\text{block,apparent}}(V) = \beta_{\text{block,apparent}} \left\{ \frac{\frac{P_{\text{bound,max}}}{1+\exp[-(V_m-V_{1/2})/K]}}{1 - \frac{P_{\text{bound}}}{1+\exp[-(V_m-V_{1/2})/K]}} \right\} \tag{52}$$

These equations define an HH-like gating variable that reproduces the equilibrium drug binding. Unfortunately, this approach produces unrealistically slow on rates for block at very positive potentials, and in an apparently reduced binding affinity as the membrane currents are blocked in a time-dependent way. These problems are the result of being constrained in describing equilibrium binding and recovery from inactivation with a single voltage-dependent rate constant, α.

2. Ad-Hoc Model 2

One way to resolve the problem of ad-hoc Model 1 is to develop artificial voltage-sensitive and insensitive components for both α and β. In practice, however, it is simpler to introduce a blocking gating variable, B, where

$$\frac{dB}{dt} = (1 - B)\alpha_{\text{block,apparent}} - \beta_{\text{block,apparent}} B \tag{53}$$

This can be expressed in terms of the experimentally observable quantities $\tau_b(V, [\text{D}])$ and $P_{\text{bound}}(V, [\text{D}])$:

$$\frac{dB}{dt} = (1 - B)\frac{P_{\text{bound}}(V, [\text{D}])}{\tau_b(V, [\text{D}])} - \frac{1 - P_{\text{bound}}(V, [\text{D}])}{\tau_b(V, [\text{D}])} B \tag{54}$$

Equation (54) enables the time constant for the development of block to be calculated independently of the block equilibrium. The initial rate of development of block is roughly proportional to the activation state of the channel, and the asymptotic values for positive and negative potentials are given by Eqs. (48) and (49), so $\tau_b(V, [\text{D}])$ can be written as

$$\tau_b(V, [\text{D}]) = \frac{1}{k_{\text{off}} + k_{\text{on}}[\text{D}][\alpha_a(V)/\alpha_a(V) - \beta_a(V)]^3} \tag{55}$$

where $\alpha_a(V)$ and $\beta_a(V)$ are the voltage-dependent rate constants for activation. The overall equation for describing the channel in the presence of drug is

$$I_{to} = \gamma(V_m - E_K)a^3 i(1 - B) \tag{56}$$

This formulation combines the simplicity of the HH formalism with some of the general characteristics of the PC model of open channel block. It reproduces the concentration-dependent reduction in peak current predicted by the PC model (Model 2). However, this formulation fails to reproduce the dose–response relationship as measured using the current area method.

This difference reflects the different effects that the two formulations have on the inactivation time course. As can be seen in Fig. 6A, the apparent rate of inactivation in the presence of drug increases with increasing drug concentration. This is similar to experimentally observed phenomena [154]. In contrast, ad-hoc Model 2 shows little change in the time course of apparent channel inactivation (Fig. 6C). Despite a very fast development of drug block for a 1 mM concentration of drug, apparent inactivation proceeds with the same time course as normal inactivation, because the fraction of current remaining is already at equilibrium levels of block. The remaining decay represents normal inactivation of the nonblocked fraction [154].

This finite equilibrium value is required to reproduce the biphasic current recovery from block or inactivation, but can do so only ad-hoc Model 1. Ad-hoc Model 2 must be at equilibrium by the peak of the current in order to reproduce the dose–response curve. Therefore, ad-hoc Model 2 cannot reproduce biphasic recovery due to its rapid kinetics. In contrast, ad-hoc Model 1 can reproduce biphasic recovery (Fig. 7) and apparent increase in inactivation rate; however, ad-hoc model 1 cannot reproduce the dose-response curve due to its slow kinetics.

The discrepancy between apparent inactivation rate and affinity is a major limitation of the approach of using a single independent gating variable to simulate macroscopic current. In order to reproduce the time course of decay of the net macroscopic current while preserving the HH formalism, further modifications would be needed. There are at least two alternatives: a voltage- and concentration-dependent variable could be introduced, or the voltage dependence of the inactivation gating variable could be modified to reproduce the inactivation behavior while preserving both steady-state inactivation and recovery characteristics.

Such additional modifications are likely to be quite complex, since the time and voltage dependencies of the apparent inactivation rate in the presence of drugs result from a complex interaction of the drug "on" rate,

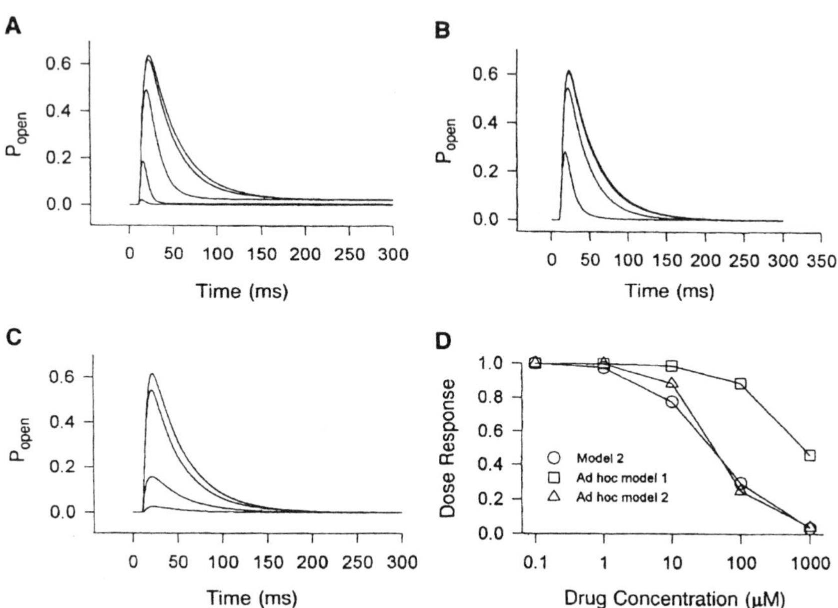

Figure 6 Simulated P_{open} in response to a depolarization from -70 to $+50\,\text{mV}$ for Model 2 (A), ad-hoc Model 1(B), and ad-hoc Model 2 (C) in the presence of 0, 1, 10, 100, and 1000 mM drug. All three models show a reduction in the peak current with increasing drug concentration. Model 2 and ad-hoc Model 1 show an increase in the rate of inactivation with increasing drug concentration, but ad-hoc Model 2 does not. (D) Dose–response curves for all three models, calculated as peak P_{open} against drug concentration. (From Ref. 154.)

the delivery of channels to the open state by channel activation, and the competition between drug binding and inactivation. Therefore, although the ad-hoc HH-approximation formulations are much simpler, they are limitations in their ability to reproduce the exact behavior of the PC model. Depending on the situation, the additional computational difficulty associated with reproducing various aspects of conformation-dependent binding behavior may or may not be worth the additional gain in accuracy.

D. Ligand Binding: Calcium Channel Inactivation by Calcium

Hodgkin and Huxley were fortunate that their initial experiments were performed on the squid giant axon, where only sodium and potassium currents dominate the action potential and there is no complication from the

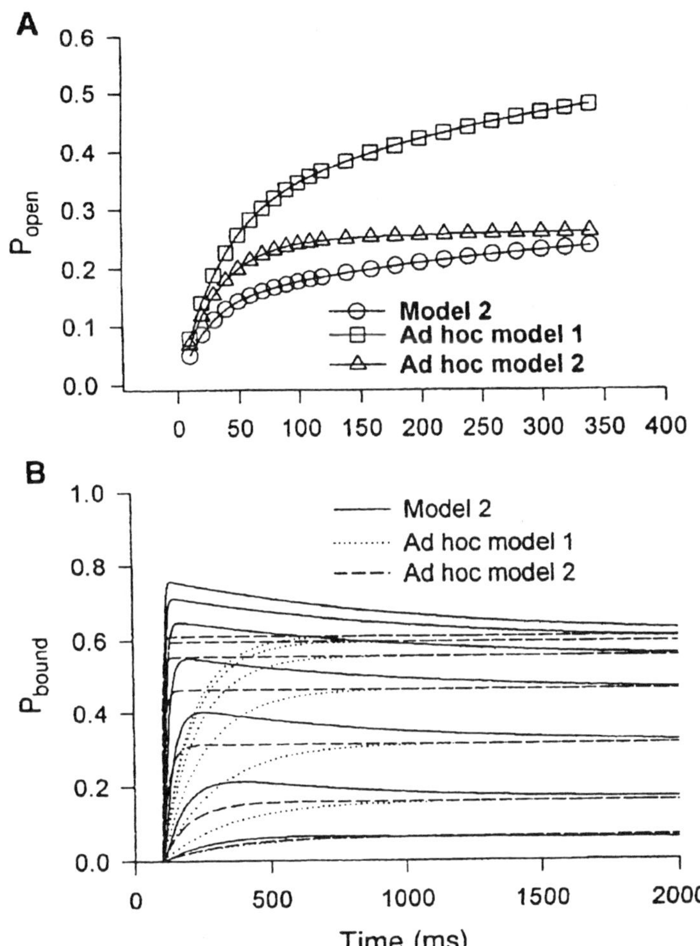

Figure 7 (A) Recovery from drug block in Model 2, ad-hoc Model 1, and ad-hoc Model 2. Equilibrium binding for each model being held at $+30\,mV$ in the presence of 50 mM drug was calculated as the starting value for each model. The membrane was then repolarized to $-70\,mV$ for various durations (Δt), then depolarized to $+50\,mV$. Peak P_{open} values are plotted as a function of Δt. Ad-hoc Model 1 has a biphasic recovery process, but an incorrect magnitude. Ad-hoc Model 2 has a correct magnitude, but does not show biphasic recovery. (B) Voltage dependence of the development of drug blockade. Simulated P_{bound} during depolarization from $-70\,mV$ to between -10 and $+50\,mV$, in 10-mV intervals. Model 2 exhibits a time-dependent decay of P_{bound} after an initial peak, whereas both ad-hoc Models 1 and 2 show a monotonic increase to an equilibrium value. (From Ref. 154.)

calcium current. Even though axons do not have calcium currents, changes in intracellular calcium concentration mediate a wide variety of cellular processes (inducing calcium release from intracellular stores, excitation–contraction coupling, synaptic vesicle release, etc.), so developing a good model of the calcium channel is of vital importance.

The calcium current was first called the "second inward" or "slow inward" calcium-dependent current, I_{si}, which was activated subsequent to the rapid inward sodium current at the beginning of the action potential [160,161]. Initial models of I_{si} represented it as a classic Hodgkin-Huxley-type current with voltage-dependent gating particles:

$$I_{si} = \gamma_s df(V_m - E_s) \tag{57}$$

where d and f are the activation and inactivation parameters, respectively [162,163].

However, what was originally called I_{si} is now known to include current through two types of calcium channel (the long-lasting L type and the transient T type [164]), the sodium–calcium exchanger current, and perhaps some other miscellaneous background currents. Obviously enough, the kinetic properties of the composite I_{si} current varied greatly from cell to cell.

Once the calcium current was studied in isolation, it was apparent that it was not a simple current to model, because of its complex current flow and unusual inactivation. The calcium channel reversal potential calculated from the Nernst equation [see Eq. (4)] is much more positive than that recorded experimentally, suggesting that flow of ions other than calcium may contribute to the reversal potential [165–168]. The Goldman-Hodgkin-Katz constant field current must be therefore be used to represent the current (cf. Eq. (10)):

$$I_{Ca} = P_{Ca} \frac{z_{Ca}^2 EF^2}{RT} \left[\frac{[Ca]_i - [Ca]_o \exp(z_{Ca}EF/RT)}{1 - \exp(z_{Ca}EF/RT)} \right]$$
$$+ \sum P_x \frac{z_x^2 EF^2}{RT} \left[\frac{[X]_i - [X]_o \exp(z_x EF/RT)}{1 - \exp(z_x EF/RT)} \right] \tag{58}$$

where the second term on the right-hand side of the equation represents the sum of the contribution from sodium and potassium ions through the channel.

Inactivation of the L-type calcium channel is both time (i.e., apparently voltage dependent because of the coupling to activation) and calcium dependent [169–174]. Calcium inactivation of the calcium current appears to be modulated by calmodulin [174–177], which may bind to the calcium-binding motif (EF hand) on the carboxyl tail on the the main α_{1C}-subunit [178], thus transducing calmodulin binding into channel inactivation [179].

The dual inactivation mechanisms of calcium current results in a characteristic steady-state inactivation that displays the usual increase in inactivation as the membrane potential becomes more positive, but it then reduces at higher potentials near to the calcium reversal potential, as shown in Fig. 8 [180]. As the membrane potential approaches the reversal potential, fewer and fewer calcium ions enter the cell. Because of this, the contribution to inactivation from calcium binding is reduced, so there is a "tip up" of the steady-state inactivation curve. In the absence of calcium ions, the L-type calcium current does inactivate, but the kinetics are dependent on the ion used to replace Ca^{2+} in solution.

A basic HH model cannot reproduce this type of inactivation, so it must be modified with a calcium-dependent term. Standen and Stanfield [181] developed a model in which calcium bound to an intracellular binding

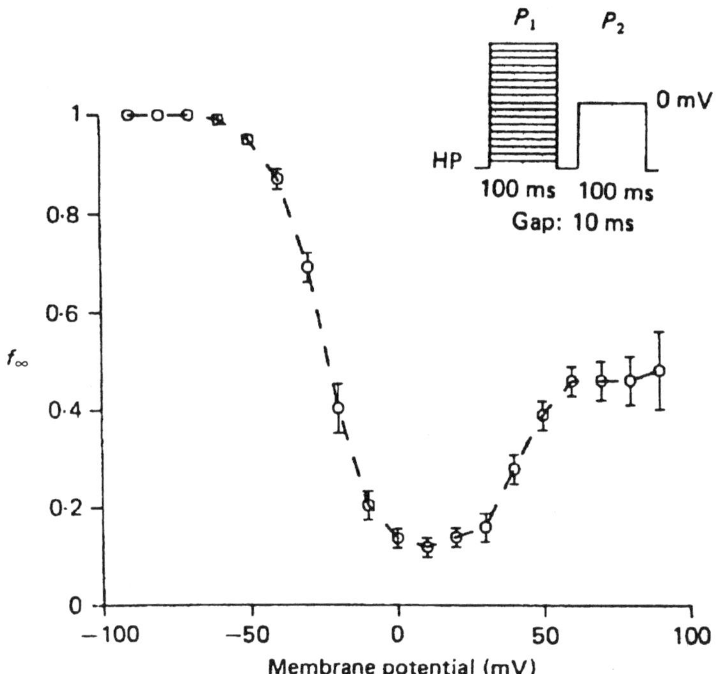

Figure 8 Steady-state inactivation relationship of I_{Ca} in bull frog atrial myocytes. Results from the double-pulse protocol shown were used to calculated f_∞ at various potentials. For potentials less than 0 mV, inactivation increases with depolarization. At potentials positive to +20 mV, inactivation is reduced until it reaches a limiting value of about 0.45. (From Ref. 180.)

site which then resulted in channel inactivation. This can be represented in a state diagram with calcium and the receptor R:

$$\text{Ca} + \text{R} \underset{\beta_{fCa}}{\overset{\alpha_{fCa}}{\rightleftharpoons}} \text{CaR}$$

where α_{fCa} and β_{fCa} are voltage-independent rate constants for the calcium-dependent inactivation gating particle. If the fraction of channels that are *not* inactivated is f_{Ca}, then the rate of change of f_{Ca} is given by

$$\frac{df_{Ca}}{dt} = \beta_{fCa}(1 - f_{Ca}) - \alpha_{fCa} f_{Ca}[\text{Ca}^{2+}]_i \tag{59}$$

In the steady state there will be no net change in inactivation, so Eq. (59) can be set to zero:

$$0 = \beta_{fCa}(1 - f_{Ca}) - \alpha_{fCa} f_{Ca}[\text{Ca}^{2+}]_i$$
$$f_{Ca} = \frac{\beta_{fCa}}{\beta_{fCa} + \alpha_{fCa}[\text{Ca}^{2+}]_i} \tag{60}$$

therefore the steady-state inactivation, $1 - f_{Ca}$, is given by

$$1 - f_{Ca} = \frac{[\text{Ca}^{2+}]}{[\text{Ca}^{2+}] + K_m} \tag{61}$$

where $K_m = \beta_{fCa}/\alpha_{fCa}$.

DiFrancesco and Noble [155] and Hilgemann and Noble [182] used the Standen and Stanfield [181] formulation as the basis of their calcium inactivation term, as did Luo and Rudy [156]. Luo and Rudy increased the steepness of the relationship between intracellular calcium and inactivation by squaring the concentration term:

$$f_{Ca(\text{Luo-Rudy})} = \frac{1}{1 + ([\text{Ca}^{2+}]_i/K_m)^2} \tag{62}$$

where K_m is the concentration of calcium that produces half-maximal calcium inactivation. Luo and Rudy used f_{Ca} in addition to the other voltage-dependent gating particles, f and d.

Hagiwara et al. [164] used a slightly different construction, defining a modified HH-like model with a fast voltage-dependent gating variable, d_L, a slower voltage-dependent gating inactivation variable, f_L, and a modulated conductance, $\gamma_{Ca,L}$, which was sensitive to the *extracellular* calcium concentration:

$$\gamma_{Ca,L} = \frac{\gamma_{Ca,L,max}}{1 + (K_m/[Ca^{2+}]_o)} \tag{63}$$

where $\gamma_{Ca,L,max}$ is the maximum current that flows and K_m is the concentration at which the conductance is half-maximal. The similarity to Eq. (61) is apparent.

Rasmusson et al. [158] did not include a specific term for modulating the current with calcium concentration, but instead modified the inactivation variable to produce a voltage-dependent U shape characteristic of inactivation. A further problem with the characterization of the inactivation gating is the degree of completeness and voltage dependence of the non-calcium-dependent component. Rasmusson et al. [158,183] modeled the bullfrog atrial cell, which does not have calcium release from the sarcoplasmic reticulum (SR) as part of excitation–contraction coupling. In this tissue, inactivation still shows a U-shaped characteristic, which remains even when barium is substituted for calcium. The voltage-sensitive inactivation mechanism may, therefore, be relatively complex. This complex voltage-dependent behavior, which may be critical to reproducing calcium currents during the plateau phase of the action potential, was investigated more thoroughly by Luo and Rudy [156].

In a completely different approach, Jafri et al. [184] developed mode-switching Markov model for calcium inactivation of the L-type calcium current,

$$I_{Ca} = \overline{P_{Ca}} f \, \mathrm{Prob}\{O + O_{Ca}\} I_{Ca} \tag{64}$$

where P_{Ca} is the maximum L-type calcium current conductance, f is the voltage-dependent inactivation particle, and $\mathrm{Prob}\{O + O_{Ca}\}$ is the probability that the channel is in the open state, based on a mode-switching model. The model assumes the channel has four independent subunits that can close the channel, represented by states C_0–C_4 in the normal mode and, C_{Ca0}–C_{Ca4} in the Ca mode.

$$\mathrm{Normal\ mode:}\quad C_0 \underset{\beta}{\overset{4\alpha}{\rightleftharpoons}} C_1 \underset{2\beta}{\overset{3\alpha}{\rightleftharpoons}} C_2 \underset{3\beta}{\overset{2\alpha}{\rightleftharpoons}} C_3 \underset{4\beta}{\overset{1\alpha}{\rightleftharpoons}} C_4 \underset{g}{\overset{f}{\rightleftharpoons}} O$$

$$\gamma \updownarrow \omega \qquad \gamma a \updownarrow \tfrac{\omega}{b} \qquad \gamma a^2 \updownarrow \tfrac{\omega}{b^2} \qquad \gamma a^3 \updownarrow \tfrac{\omega}{b^3} \qquad \gamma a^4 \updownarrow \tfrac{\omega}{b^4}$$

$$\mathrm{Ca\ mode:}\quad C_{Ca0} \underset{\beta'}{\overset{4a'}{\rightleftharpoons}} C_{Ca1} \underset{2\beta'}{\overset{3\alpha'}{\rightleftharpoons}} C_{Ca2} \underset{3\beta'}{\overset{2\alpha'}{\rightleftharpoons}} C_{Ca3} \underset{4\beta'}{\overset{1\alpha'}{\rightleftharpoons}} C_{Ca4} \underset{\delta'}{\overset{f'}{\rightleftharpoons}} O_{Ca}$$

The transitions to the open states, O and O_{Ca}, are controlled by the voltage-independent rate constants f and f'. The transitions between the two modes are controlled by γ, which is calcium dependent. The probability of entering the Ca mode is enhanced at higher voltages, as is reflected in the increased transition rates from left to right in the state diagram.

No matter how well the calcium channel is modeled, it cannot be considered in isolation from its surroundings. In ventricular muscle, L-type calcium channels are found mostly in clusters in the T-tubules [185–188] directly opposed to calcium-induced calcium-release (CICR) channels on the sarcoplasmic reticulum [189]. Calcium that enters the cell through the transmembrane channel initiates CICR from the SR, which will then contribute to the calcium-dependent inactivation of the calcium channel. A faithful model of the L-type calcium channel must therefore include an appropriate representation of the SR release channels, and the local calcium concentration near the membrane. The most common way of dealing with this is by the use of subcellular compartments that represent an average of the calcium concentration in any part of the cell.

E.   A Word of Caution!

All models of voltage-dependent ionic currents are derived at some level from voltage clamp data. In order to study the biophysical characteristics of a current, the experimental electrophysiologist must first isolate it from other overlapping cellular currents. A large number of "tricks" are employed to perform this separation. This means that voltage clamp experiments on cardiac potassium channels are almost always obtained under nonphysiological conditions and must be modified in some way to be incorporated into models of electrical behavior. This section describes a few of the more common manipulations frequently encountered in the experimental literature and provides some cautions on interpreting such data.

1.   Divalent Ion Concentration and "Surface Charge"

Many calcium channel blockers also block potassium currents, therefore block of the calcium channel is frequently achieved using divalent ions such as cadmium, nickel, or cobalt, depending on the channel type and experimental preparation. Unfortunately, such ions tend to bind to fixed negative charges associated with proteins and phospholipids in the bilayer. Charges on or near the surface of the membrane can alter the observed kinetics of a channel by screening the electric field [190]. If the magnitude of the field is changed, the kinetics of the voltage-dependent channel will be correspondingly altered. For example, if the concentration of positively charged ions bound on the outside of the membrane is increased, the membrane will act as though it was being held at a more negative potential.

2. Permeant Ion Concentration

The sodium current in cardiac myocytes can be very large, making it impossible to clamp reliably. In order to obtain a smaller, more manageable current for kinetic analysis, the extracellular sodium concentration is reduced, thus decreasing the driving force. Conversely, the permeant ion concentration can be manipulated to increase the magnitude of relatively small currents, such as I_{kr} [191]. Unfortunately, permeant ions bind to the channel. In general, higher concentrations tend to stabilize the open states and destabilize the nonconducting states such as the resting state and the inactivated state.

3. Temperature Dependence and Q_{10}

For convenience, many electrophysiological experiments are performed at room temperature. Temperature affects many physiological processes, which is quantified by Q_{10}: the amount by which the process changes in response a 10-degree increase in temperature. Gating kinetics generally have a Q_{10} of 2–4, whereas conductance is relatively insensitive with a Q_{10} of only 1.2–1.5 (see [6]). Temperature can also have an indirect effect on ion channel kinetics by changing the rate of enzyme activities within the cell; for example, temperature-driven changes in regulatory processes such as phosphorylation can result in an even stronger temperature dependence of gating. The temperature dependence of the kinetics of a specific channel is not often available, and temperature correction factors used in models are frequently just estimated.

4. Intracellular Dialysis and Calcium Chelators

Most electrophysiological experiments on cardiac myocytes are performed using the whole-cell ruptured patch technique. The advantage of this technique us that it establishes a low-resistance access to the intracellular space, which enables good electrical control of the transmembrane potential. However, this method also creates a contiguous diffusion space reaching from the electrode pipette to the cytosol. The pipette solution can exchange freely with the cytosol, providing a useful method of delivering channel blockers to the inside of the cell. However, this exchange can also lead to dilution of co-factors which are essential for channel function. This dilution process is responsible for the phenomenon referred to in laboratory jargon as "rundown," in which ion channel behavior can change slowly during the course of an experiment.

Maintaining the intracellular-like milieu when experimenting in the ruptured patch configuration therefore requires that the pipette solution contain some solutes present in the intracellular solution. This generally includes potassium, magnesium, and chloride ions, and energy sources such as ATP and creatine phosphate. Perhaps the most important intracellular component is calcium. The cytosol normally has very low levels of calcium. Relatively small fluctuations in cytosolic free calcium regulates many enzymes, and contraction in myocytes. Trace contaminant amounts of calcium found in reagent-grade sodium, potassium, and magnesium salts may be enough to activate these processes. Therefore, a means of reducing calcium to a low level is needed. This is achieved using a calcium chelator, which tightly binds free calcium. The most common is EGTA (ethylene glycol bis(β-aminoethyl ether)-N,N,N$'$N$'$-tetraacetic acid). EGTA has very slow on and off rates, allowing some rapid calcium concentration changes to occur (e.g., calcium-induced calcium release), but maintaining low resting levels. When suppression of faster events is required, the rapid-binding calcium buffer, BAPTA (1,2-bis(2-aminophenoxy)ethane-N,N,N$'$,N$'$-tetraacetic acid) is used. It has been demonstrated that the time dependence of these buffers can profoundly influence the calcium transient and hence the kinetics of calcium channel inactivation. Although difficult to quantify, this often overlooked difference in experimental conditions can help to explain at least some of the variation among experimental results.

VII. CONCLUSION

Advances in molecular biology and biophysics have yielded a startling quantity of detailed knowledge about the properties of the individual component currents underlying the cardiac action potential. At the same time, the explosion of computer speed and availability has provided the tools to analyze and create models of the complex systems based on this new molecular information. Reconciling the genetic and molecular basis of arrhythmias with cellular and whole-organ pathology, and designing more efficacious treatments, will represent the next major challenges for the biomedical engineer.

ACKNOWLEDGMENTS

This work was supported in part by grants from the American Heart Association (9940185N), NSF (KDI Grant DBI-9873173), and NIH (R01 HL-59526-01).

REFERENCES

1. H. Davson The permeability of the erythrocyte to cations. Cold Spring Harbor Symp Quant Biol 8:255, 1940.

2. A Leaf, Maintenance of concentration gradients and regulation of cell volume. Ann NY Acad Sci 72:396–404, 1959.

3. DC Tosteson. JF Hoffman, Regulation of cell volume by active cation transport in high and low potassium sheep red cells. J Gen Physiol 44:169–194, 1960.

4. G Isenberg, W Trautwein. The effect of dihydro-ouabain and lithium-ions on the outward current in cardiac Purkinje fibers. Evidence for electrogenicity of active transport. Pflugers Arch—Eur J Physiol 350:41–54, 1974.

5. W Nernst. Zur Kinetic der in Lösung befindlichen Körper: Theoriw der Diffusion. Z. Phys. Chem. 613–637, 1888.

6. B Hille. *Ion Channels of Excitable Membranes.* 2nd ed. Sunderland, MA: Sinauer Associates, 1992.

7. OP Hamil, A Marty, E Neher, B Sakmann, FJ Sigworth. Improved patch-clamp techniques for high-resolution current recording from cells and cell-free membrane patches. Pfluger Arch—Eur J Physiol 391:85–100, 1981.

8. FJ Sigworth, E Neher. Single Na^+ channel currents observed in cultured rat muscle cells. Nature 287:447–449, 1980.

9. DE Goldman. Potential, impedance, and rectification in membranes. J Gen Physiol 27:37–60, 1943.

10. AL Hodgkin, B Katz. The effect of sodium ions on the electrical activity of the giant axon of the squid. J Physiol 108:37–77, 1949.

11. AL Hodgkin, AF Huxley. Currents carried by sodium and potassium ions through the membrane of the giant axon of *Loligo.* J Physiol 116:449–472, 1952.

12. AL Hodgkin, AF Huxley. The components of membrane conductance in the giant axon of *Loligo.* J Physiol 116:473–496, 1952.

13. AL Hodgkin, AF Huxley. The dual effect if membrane potential on sodium conductance in the giant axon of *Loligo.* J Physiol 116:497–506, 1952.

14. AL Hodgkin, AF Huxley. A quantitative description of membrane current and its application to conduction and excitation in nerve. J Physiol 117:500–544, 1952.

15. A Finkelstein, OS Anderson. The gramicidin A channel: a review of its permeability characteristics with special reference to the single-file aspect of transport. J Membrane Biol 59:155–171, 1981.

16. M Kohlhardt. A quantitative analysis of the Na^+-dependence of V_{max} of the fast action potential in mammalian ventricular myocardium. Saturation characteristics and the modulation of a drug-induced I_{Na} blockade by $[Na^+]_o$. Pflugers Arch—Eur J Physiol 392:379–387, 1982.

17. TB Begenisich, MD Cahalan. Sodium channel permeation in squid axons. II: Non-independence and current-voltage relations. J Physiol 307:243–257, 1980.

18. R Coronado, RL Rosenberg, C Miller. Ionic selectivity, saturation, and block in a K^+-selective channel from sarcoplasmic reticulum. J Gen Physiol 76:425–446, 1980.

19. PG Kostyuk, OA Krishtal. Effects of calcium and calcium-chelating agents on the inward and outward current in the membrane of mollusc neurones. J Physiol 270:569–580, 1977.

20. EW McCleskey, W Almers. The Ca channel in skeletal muscle in a large pore. Proc Natl Acad Sci USA 82:7149–7153, 1985.

21. W Almers, EW McCleskey. Non-selective conductance in calcium channels of frog muscle: calcium selectivity in a single-file pore. J Physiol 353:585–608, 1984.

22. W Almers, EW McCleskey, PT Palade. A non-selective cation conductance in frog muscle membrane blocked by micromolar external calcium ions. J Physiol 353:565–583, 1984.

23. B Hille. Charges and potentials at the nerve surface: divalent ions and pH. J Gen Physiol 51:221–236, 1968.

24. H Drouin, R The. The effect of reducing extracellular pH on the membrane currents of the Ranvier node. Pflugers Arch—Eur J Physiol 313:80–88, 1969.

25. S Hagiwara, K Takahashi. The anomalous rectification and cation selectivity of the membrane of a starfish egg cell. J Membrane Biol 18:61–80, 1974.

26. S Hagiwara, LA Jaffe. Electrical properties of egg cell membranes. Annu Rev Biophys Bioeng 8:385–416, 1979.

27. CM Armstrong. Time course of TEA^+-induced anomalous rectification in squid giant axons. J Gen Physiol 50:491–503, 1966.

28. CM Armstrong. Inactivation of the potassium conductance and related phenomena caused by quaternary ammonium ion injection in squid axons. J Gen Physiol 54:553–575, 1969.

29. CA Vandenberg. Inward rectification of a potassium channel in cardiac ventricular cells depends on internal magnesium ions. Proc Natl Acad Sci USA 84:2560–2564, 1987.

30. H Matsuda. Open-state substructure of inwardly rectifying potassium channels revealed by magnesium block in guinea-pig heart cells. J Physiol 397:237–258, 1998.

31. E Solessio, K Rapp, I Perlman, EM Lasater. Spermine mediates inward rectification in potassium channels of turtle retinal cells. J Neurophysiol 85:1357–1367, 2001.

32. D Oliver, T Baukrowitz, B Fakler. Polyamines as gating molecules of inward-rectifier K^+ channels. Eur J Biochem 267:5824–5829, 2000.

33. SL Shyng, Q Sha, T Ferigni, AN Lopatin, CG Nichols. Depletion of intracellular polyamines relieves inward rectification of potassium channels. Proc Nat Acad Sci USA 93:12014–12019, 1996.

34. AN Lopatin, CG Nichols. $[K^+]$ dependence of polyamine-induced rectification in inward rectifier potassium channels (IRK1, Kir2.1). J Gen Physiol 108:105–113, 1996.

35. AN Lopatin, EN Makhina, CG Nichols. The mechanism of inward rectification of potassium channels: "long-pore plugging" by cytoplasmic polyamines. J Gen Physiol 106:923–955, 1995.
36. AN Lopatin, EN Makhina, CG Nichols. Potassium channel block by cytoplasmic polyamines as the mechanism of intrinsic rectification. Nature 372:366–369, 1994.
37. S Wang, S Liu, MJ Morales, HC Strauss, RL Rasmusson. A quantitative analysis of the activation and inactivation kinetics of HERG expressed in *Xenopus* oocytes. J Physiol 502:45–60, 1997.
38. J Kiehn, AE Lacerda, AM Brown. Pathways of HERG inactivation. Am J Physiol 277:H199–H210, 1999.
39. PL Smith, T Baukrowitz, G Yellen. The inward rectification mechanism of the HERG cardiac potassium channel. Nature 379:833–836, 1996.
40. B Hille. Pharmacological modifications of the sodium channels of frog nerve. J Gen Physiol 51:199–219, 1968.
41. B Neumcke, JM Fox, H Drouin, W Schwarz. Kinetics of the slow variation of peak sodium current in the membrane of myelinted nerve following changes of holding potential or extracellular pH. Biochim Biophys Acta 426:245–257, 1976.
42. AM Woodhull. Ionic blockage of sodium channels in nerve. J Gen Physiol 61:687–708, 1973.
43. GN Mozhayeva, AP Naumov, ED Nosyreva. A study on the potential-dependence of proton block of sodium channels. Biochim Biophys Acta 775:435–440, 1984.
44. B Hille, AM Woodhull, BI Shapiro. Negative surface charge near sodium channels of nerve: divalent ions, monovalent ions, and pH. Phil Trans R Soc Lond B: Biol Sci 270:301–318, 1975.
45. B Hille, W Schwarz. Potassium channels as multi-ion single-file pores. J Gen Physiol 72:409–442, 1978.
46. P Daumas, OS Andersen. Proton block of rat brain sodium channels. Evidence for two proton binding sites and multiple occupancy. J Gen Physiol 101:27–43, 1993.
47. H Eyring. Viscosity, plasticity, and diffusion as examples of absolute reaction rates. J Chem Phys 4:283–291, 1936.
48. H Eyring, R Lumry, JW Woodbury. Some applications of modern rate theory to physiological systems. Rec Chem Prog 10:100–114, 1949.
49. DL Campbell, RL Rasmusson, Y Qu, HC Strauss. The calcium-independent transient outward potassium current in isolated ferret right ventricular myocytes. I. Basic characterization and kinetic analysis. J Gen Physiol 101:571–601, 1993.
50. R MacKinnon. Determination of the subunit stoichiometry of a voltage-activated potassium channel. Nature 350:232–235, 1991.
51. ER Liman, J Tytgat, P Hess. Subunit stoichiometry of a mammalian K^+ channel determined by construction of multimeric cDNAs. Neuron 9:861–871, 1992.

52. K Ho, CG Nichols, WJ Lederer, J Lytton, PM Vassilev, MV Kanazirska, SC Hebert. Cloning and expression of an inwardly rectifying ATP-regulated potassium channel. Nature 362:31–38, 1993.
53. Y Kubo, TJ Baldwin, YN Jan, LY Jan. Primary structure and functional expression of a mouse inward rectifier potassium channel. Nature 362:127–133, 1993.
54. M Li, N Unwin, KA Stauffer, YN Jan, LY Jan. Images of purified Shaker potassium channels. Curr Biol 4:110–115, 1994.
55. M Noda, S Shimizu, T Tanabe, T Takai, T Kayano, T Ikeda, H Takahasi, H Nakayama, Y Kanaoka, N Minamino. Primary structure of *Electrophorus electricus* sodium channel deduced from cDNA sequence. Nature 312:121–127, 1984.
56. ME Gellens, AL George Jr, LQ Chen, M Chahine, R Horn, RL Barchi, RG Kallen. Primary structure and functional expression of the human cardiac tetrodotoxin-insensitive voltage-dependent sodium channel. Proc Nat Acad Sci USA 89:554–558, 1992.
57. B Hille. Ionic channels in excitable membranes. Current problems and biophysical approaches. Biophys J 22:283–294, 1978.
58. MF Schneider, WK Chandler. Voltage dependent charge movement of skeletal muscle: a possible step in excitation-contraction coupling. Nature 242:244–246, 1973.
59. CM Armstrong, F Bezanilla. Currents related to movement of the gating particles of the sodium channels. Nature 242:459–461, 1973.
60. SK Aggarwal, R MacKinnon. Contribution of the S4 segment to gating charge in the Shaker K^+ channel. Neuron 16:1169–1177, 1996.
61. SA Seoh, D Sigg, DM Papazian, F Bezanilla. Voltage-sensing residues in the S2 and S4 segments of the Shaker K^+ channel. Neuron 16:1159–1167, 1996.
62. LM Mannuzzu, MM Moronne, EY Isacoff. Direct physical measure of conformational rearrangement underlying potassium channel gating. Science 271:213–216, 1996.
63. W Stuhmer, F Conti, H Suzuki, XD Wang, M Noda, N Yahagi, H Kubo, S Numa. Structural parts involved in activation and inactivation of the sodium channel. Nature 339:597–603, 1989.
64. N Yang, R Horn. Evidence for voltage-dependent S4 movement in sodium channels. Neuron 15:213–218, 1995.
65. N Yang, AL George Jr, R Horn. Molecular basis of charge movement in voltage-gated sodium channels. Neuron 16:113–122, 1996.
66. WA Catterall. Molecular properties of voltage-sensitive sodium channels. Annu Rev Biochem 55:953–985, 1986.
67. FJ Sigworth. Voltage gating of ion channels. Quart Rev Biophys 27:1–40, 1994.
68. RD Keynes. The kinetics of voltage-gated ion channels. Quart Rev Biophys 27:339–434, 1994.
69. RD Keynes, F Elinder. The screw-helical voltage gating of ion channels. Proc R Soc Lond B: Biol Sci 266:843–852, 1999.

70. G Yellen. The moving parts of voltage-gated ion channels. Quart Rev Biophys 31:239–295, 1998.

71. MF Sheets, JW Kyle, RG Kallen, DA Hanck. The Na channel voltage sensor associated with inactivation is localized to the external charged residues of domain IV, S4. Biophys J 77:747–757, 1999.

72. DM Papazian, LC Timpe, YN Jan, LY Jan. Alteration of voltage-dependence of Shaker potassium channel by mutations in the S4 sequence. Nature 349:305–310, 1991.

73. F Bezanilla. The voltage sensor in voltage-dependent ion channels. Physiol Rev 80:555–592, 2000.

74. LM Mannuzzu, EY Isacoff. Independence and cooperativity in rearrangements of a potassium channel voltage sensor revealed by single subunit fluorescence. J Gen Physiol 115:257–268, 2000.

75. SC Crouzy, FJ Sigworth. Fluctuations in ion channel gating currents. Analysis of nonstationary shot noise. Biophys J 64:68–76, 1993.

76. F Bezanilla, E Stefani. Gating currents. Meth Enzymol 293:331–352, 1998.

77. S Liu, RL Rasmusson, DL Campbell, S Wang, HC Strauss. Activation and inactivation kinetics of an E-4031-sensitive current from single ferret atrial myocytes. Biophys J 70:2704–2715, 1996.

78. WN Zagotta, T Hoshi, RW Aldrich. Shaker potassium channel gating. III: Evaluation of kinetic models for activation. J Gen Physiol 103:321–362, 1994.

79. WN Zagotta, T Hoshi, J Dittman, RW Aldrich. Shaker potassium channel gating. II: Transitions in the activation pathway. J Gen Physiol 103:279–319, 1994.

80. T Hoshi, WN Zagotta, RW Aldrich. Shaker potassium channel gating. I: Transitions near the open state. J Gen Physiol 103:249–278, 1994.

81. K Benndorf. Patch clamp analysis of Na channel gating in mammalian myocardium: reconstruction of double pulse inactivation and voltage dependence of Na currents. Gen Physiol Biophys 7:353–377, 1988.

82. MF Berman, JS Camardo, RB Robinson, SA Siegelbaum. Single sodium channels from canine ventricular myocytes: voltage dependence in relative rates of activation and inactivation. J Physiol 415:503–531, 1989.

83. BE Scanley, DA Hanck, T Chay, HA Fozzard. Kinetic analysis of single sodium channels from canine Purkinje cells. J Gen Physiol 95:411–437, 1990.

84. K Manivannan, RT Mathias, E Gudowska-Nowak. Description of interacting channel gating using a stochastic Markovian model. Bull Math Biol 58:141–174, 1996.

85. S Herzig, P Patil, J Neumann, CM Staschen, DT Yue. Mechanisms of beta-adrenergic stimulation of cardiac Ca^{2+} channels revealed by discrete-time Markov analysis of slow gating. Biophys J 65:1599–1612, 1993.

86. LA Irvine, MS Jafri, RL Winslow. Cardiac sodium channel Markov model with temperature dependence and recovery from inactivation. Biophys J 76:1868–1885, 1999.

87. CM Armstrong, F Bezanilla. Inactivation of the sodium channel II. Gating current experiments. J Gen Physiol 70:567–590, 1977.

88. W Nonner. Relations between the inactivation of sodium channels and the immobilization of gating charge in from myelinated nerve. J Physiol 299:573–603, 1980.

89. R Horn, S Ding, HJ Gruber. Immobilizing the moving parts of voltage-gated ion channels. J Gen Physiol 116:461–476, 2000.

90. WN Zagotta, RW Aldrich. Voltage-dependent gating of Shaker A-type potassium channels in *Drosophila* muscle. J Gen Physiol 95:29–60, 1990.

91. T Hoshi, WN Zagotta, RW Aldrich. Biophysical and molecular mechanisms of Shaker potassium channel inactivation. Science 250:533–538, 1990.

92. GA Lopez, YN Jan, LY Jan. Evidence that the S6 segment of the Shaker voltage-gated K^+ channel comprises part of the pore. Nature 367:179–182, 1994.

93. EY Isacoff, YN Jan, LY Jan. Putative receptor for the cytoplasmic inactivation gate in the Shaker K^+ channel. Nature 353:86–90, 1991.

94. M Holmgren, ME Jurman, G Yellen. N-type inactivation and the S4–S5 region of the Shaker K^+ channel. J Gen Physiol 108:195–206, 1996.

95. WN Zagotta, T Hoshi, RW Aldrich. Restoration of inactivation in mutants of Shaker potassium channels by a peptide derived from ShB. Science 250:568–571, 1990.

96. JP Ruppersberg, R Frank, O Pongs, M Stocker. Cloned neuronal IK(A) channels reopen during recovery from inactivation. Nature 353:657–660, 1991.

97. J Tseng-Crank, JA Yao, MF Berman, GN Tseng. Functional role of the NH_2-terminal cytoplasmic domain of a mammalian A-type K channel. J Gen Physiol 102:1057–1083, 1993.

98. LS Liebovitch, LY Selector, RP Kline. Statistical properties predicted by the ball and chain model of channel inactivation. Biophys J 63:1579–1585, 1992.

99. KL Choi, RW Aldrich, G Yellen. Tetraethylammonium blockade distinguishes two inactivation mechanisms in voltage-activated K^+ channels. Proc Nat Acad Sci USA 88:5092–5095, 1991.

100. J Lopez-Barneo, T Hoshi, SH Heinermann, RW Aldrich. Effects of external cations and mutations in the pore region on C-type inactivation of Shaker potassium channels. Receptors & Channels 1:61–71, 1993.

101. RL Rasmusson, MJ Morales, RC Castellino, Y Zhang, DL Campbell, HC Strauss. C-type inactivation controls recovery in a fast inactivating cardiac K^+ channel (Kv1.4) expressed in *Xenopus* oocytes. J Physiol 489:709–721, 1995.

102. SD Demo, G Yellen. The inactivation gate of the Shaker K^+ channel behaves like an open-channel blocker. Neuron 7:743–753, 1991.

103. L Goldman, CL Schauf. Inactivation of the sodium current in *Myxicola* giant axons. Evidence for coupling to the activation process. J Gen Physiol 59:659–675, 1972.

104. F Bezanilla, CM Armstrong. Inactivation of the sodium channel. I. Sodium current experiments. J Gen Physiol 70:549–566, 1977.

105. CM Armstrong, WF Gilly. Fast and slow steps in the activation of sodium channels. J Gen Physiol 74:691–711, 1979.
106. BP Bean. Sodium channel inactivation in the crayfish giant axon. Must channels open before inactivating? Biophys J 35:595–614, 1981.
107. GS Oxford. Some kinetic and steady-state properties of sodium channels after removal of inactivation. J Gen Physiol 77:1–22, 1981.
108. L Goldman, JL Kenyon. Delays in inactivation development and activation kinetics in myxicola giant axons. J Gen Physi 80:83–102, 1982.
109. J Zimmerberg, F Bezanilla, VA Parsegian. Solute inaccessible aqueous volume changes during opening of the potassium channel of the squid giant axon. Biophys J 57:1049–1064, 1990.
110. MD Rayner, JG Starkus, PC Ruben, DA Alicata. Voltage-sensitive and solvent-sensitive processes in ion channel gating. Kinetic effects of hyperosmolar media on activation and deactivation of sodium channels. Biophys J 61:96–108, 1992.
111. RW Aldrich, DP Corey, CF Stevens. A reinterpretation of mammalian sodium channel gating based on single channel recording. Nature 306:436–441, 1983.
112. Y Liu, ME Jurman, G Yellen. Dynamic rearrangement of the outer mouth of a K^+ channel during gating. Neuron 16:859–867, 1996.
113. O Pongs. Molecular biology of voltage-dependent potassium channels. Physiol Rev 72:S69–S88, 1992.
114. T Hoshi, WN Zagotta, RW Aldrich. Two types of inactivation in Shaker K^+ channels: effects of alterations in the carboxy-terminal region. Neuron 7:547–556, 1991.
115. T Baukrowitz, G Yellen. Modulation of K^+ current by frequency and external $[K^+]$: a tale of two inactivation mechanisms. Neuron 15:951–960, 1998.
116. RL Rasmusson, MJ Morales, S Wang, S Liu, DL Campbell, MV Brahmajothi, HC Strauss. Inactivation of voltage-gated cardiac K^+ channels. Circ Res 82:739–750, 1998.
117. W Nonner, DP Chen, B Eisenberg. Anomalous mole fraction effect, electrostatics, and binding in ionic channels. Biophys J 74:2327–2334, 1998.
118. DP Tieleman, HJ Berendsen. A molecular dynamics study of the pores formed by *Escherichia coli* OmpF porin in a fully hydrated palmitoyloleoylphosphatidylcholine bilayer. Biophys J 74:2789–2801, 1998.
119. S Bek, E Jakobsson. Brownian dynamics study of a multiply-occupied cation channel: application to understanding permeation in potassium channels. Biophys J 66:1028–1038, 1994.
120. SC Li, M Hoyles, S Kuyucak, SH Chung. Brownian dynamics study of ion transport in the vestibule of membrane channels. Biophys J 74:37–47, 1998.
121. G Moy, B Corry, S Kuyucak, SH Chung. Tests of continuum theories as models of ion channels. I. Poisson-Boltzmann theory versus Brownian dynamics. Biophys J 78:2349–2363, 2000.

122. B Corry, S Kuyucak, SH Chung. Tests of continuum theories as models of ion channels. II. Poisson-Nernst-Planck theory versus brownian dynamics. Biophys J 78:2364–2381, 2000b.

123. MF Schumaker, R Pomes, B Roux. Framework model for single proton conduction through gramicidin. Biophys J 80:12–30, 2001.

124. SW Chiu, S Subramaniam, E Jakobsson. Simulation study of a gramicidin/lipid bilayer system in excess water and lipid. I. Structure of the molecular complex. Biophys J 76:1929–1938, 1999.

125. SW Chiu, S Subramaniam, E Jakobsson. Simulation study of a gramicidin/lipid bilayer system in excess water and lipid. II. Rates and mechanisms of water transport. Biophys J 76:1939–1950, 1999.

126. B Roux, TB Woolf. The binding site of sodium in the gramicidin A channel. Novartis Found Symp 225:113–124, 1999.

127. B Roux. Valence selectivity of the gramicidin channel: a molecular dynamics free energy perturbation study. Biophys J 71:3177–3185, 1996.

128. AE Cardenas, RD Coalson, MG Kurnikova. Three-dimensional Poisson-Nernst-Planck theory studies: influence of membrane electrostatics on gramicidin A channel conductance. Biophys J 79:80–93, 2000.

129. DA Doyle, CJ Morais, RA Pfuetzner, A Kuo, JM Gulbis, SL Cohen, BT Chait, R MacKinnon. The structure of the potassium channel: molecular basis of K^+ conduction and selectivity. Science 280:69–77, 1998.

130. SH Chung, M Hoyles, T Allen, S Kuyucak. Study of ionic currents across a model membrane channel using Brownian dynamics. Biophys J 75:793–809, 1998.

131. C Singh, R Sankararamakrishnan, S Subramaniam, E Jakobsson. Solvation, water permeation, and ionic selectivity of a putative model for the pore region of the voltage-gated sodium channel. Biophys J 71:2276–2288, 1996.

132. TW Allen, S Kuyucak, SH Chung. Molecular dynamics estimates of ion diffusion in model hydrophobic and KcsA potassium channels. Biophys Chem 86:1–14, 2000.

133. B Corry, TW Allen, S Kuyucak, SH Chung. Mechanisms of permeation and selectivity in calcium channels. Biophys J 80:195–214, 2001.

134. B Corry, TW Allen, S Kuyucak, SH Chung. A model of calcium channels. Biochim Biophys Acta 1509:1–6, 2000.

135. W Nonner, B Eisenberg. Ion permeation and glutamate residues linked by Poisson-Nernst-Planck theory in L-type calcium channels. Biophys J 75:1287–1305, 1998.

136. J Gullingsrud, D Kosztin, K Schulten. Structural determinants of MscL gating studied by molecular dynamics simulations. Biophys J 80:2074–2081, 2001.

137. IH Shrivastava, MS Sansom. Simulations of ion permeation through a potassium channel: molecular dynamics of KcsA in a phospholipid bilayer. Biophys J 78:557–570, 2000.

138. WA Catterall. Inhibition of voltage-sensitive sodium channels in neuroblastoma cells by antiarrhythmic drugs. Mol Pharmacol 20:356–362, 1981.

139. DJ Snyders, SW Yeola. Determinants of antiarrhythmic drug action. Electrostatic and hydrophobic components of block of the human cardiac hKv1.5 channel. Circ Res 77:575–583, 1995.

140. OS Baker, HP Larson, LM Mannuzzu, EY Isacoff. Three transmembrane conformations and sequence-dependent displacement of the S4 domain in shaker K^+ channel gating. Neuron 20:1283–1294, 1998.

141. JA Yao, GN Tseng. Modulation of 4-AP block of a mammalian A-type K channel clone by channel gating and membrane voltage. Biophys J 67:130–140, 1994.

142. JZ Yeh, GS Oxford, CH Wu, T Narahashi. Dynamics of aminopyridine block of potassium channels in squid axon membrane. J Gen Physiol 68:519–535, 1976.

143. SJ Kehl. 4-Aminopyridine causes a voltage-dependent block of the transient outward K^+ current in rat melanotrophs. J Physiol 431:515–528, 1990.

144. GN Tseng, M Jiang, JA Yao. Reverse use dependence of Kv4.2 blockade by 4-aminopyridine. J Pharmacol Exp Ther 279:865–876, 1996.

145. DL Campbell, Y Qu, RL Rasmusson, HC Strauss. The calcium-independent transient outward potassium current in isolated ferret right ventricular myocytes. II. Closed state reverse use-dependent block by 4-aminopyridine. J Gen Physiol 101:603–626, 1993.

146. RL Rasmusson, Y Zhang, DL Campbell, MB Comer, RC Castellino, S Liu, HC Strauss. Bi-stable block by 4-aminopyridine of a transient K^+ channel (Kv1.4) cloned from ferret ventricle and expressed in *Xenopus* oocytes. J Physiol 485:59–71, 1995.

147. CM Armstrong. Interaction of tetraethylammonium ion derivatives with the potassium channels of giant axons. J Gen physiol 58:413–437, 1971.

148. JR Moorman, R Yee, T Bjornsson, CF Starmer, AO Grant, HC Strauss. pKa does not predict pH potentiation of sodium channel blockade by lidocaine and W6211 in guinea pig ventricular myocardium. J Pharmacol Exp Ther 238:159–166, 1986.

149. CF Starmer, JZ Yeh, J Tanguy. A quantitative description of QX222 blockade of sodium channels in squid axons. Biophys J 49:913–920, 1986.

150. JZ Yeh, J Tanguy. Na channel activation gate modulates slow recovery from use-dependent block by local anesthetics in squid giant axons. Biophys J 47:685–694, 1985.

151. DL Campbell, Y Qu, RL Rasmusson, HC Strauss. "Reverse use-dependent" effects of 4-aminopyridine on the transient outward potassium current in ferret right ventricular myocytes. Adv Exp Med Biol 311:357–358, 1992.

152. RL Rasmusson, Y Zhang, DL Campbell, MB Comer, RC Castellino, S Liu, MJ Morales, HC Strauss. Molecular mechanisms of K^+ channel blockade: 4-aminopyridine interaction with a cloned cardiac transient K^+ (Kv1.4) channel. Adv Exp Med Biol 382:11–22, 1995.

153. S Bogusz, D Busath. Is a beta-barrel model of the K^+ channel energetically feasible? Biophys J 62:19–21, 1992.

154. SR Durell, HR Guy. Atomic scale structure and functional models of voltage-gated potassium channels. Biophys J 62:238–247, 1992.

155. S Liu, RL Rasmusson. Hodgkin-Huxley and partially coupled inactivation models yield different voltage dependence of block. Am J Physiol 272:H2013–H2022, 1997.

156. D DiFrancesco, D Noble. A model of cardiac electrical activity incorporating ionic pumps and concentration changes. Phil Trans R Soc Lond B: Biol Sci 307:353–398, 1985.

157. CH Luo, Y Rudy. A dynamic model of the cardiac ventricular action potential. I. Simulations of ionic currents and concentration changes. Circ Res 74:1071–1096, 1994.

158. CH Luo, Y Rudy. A dynamic model of the cardiac ventricular action potential. II. Afterdepolarizations, triggered activity, and potentiation. Circ Res 74:1097–1113, 1994.

159. RL Rasmusson, JW Clark, WR Giles, K Robinson, RB Clark, EF Shibata, DL Campbell. A mathematical model of electrophysiological activity in a bullfrog atrial cell. Am J Physiol 259:H370–H389, 1990.

160. MT Slawsky, NA Castle. K^+ channel blocking actions of flecainide compared with those of propafenone and quinidine in adult rat ventricular myocytes. J Pharmacol Exp Ther 269:66–74, 1994.

161. W Trautwein. The slow inward current in mammalian myocardium. Its relation to contraction. Eur J Cardiol 1:169–175, 1973.

162. H Reuter. Divalent cations as charge carriers in excitable membranes. Prog Biophys Mol Biol 26:1–43, 1973.

163. GW Beeler, H Reuter. Reconstruction of the action potential of ventricular myocardial fibres. J Physiol 268:177–210, 1977.

164. DG Bristow, JW Clark. A mathematical model of primary pacemaking cell in SA node of the heart. Am J Physiol 243:H207–H218, 1982.

165. N Hagiwara, H Irisawa, M Kameyama. Contribution of two types of calcium currents to the pacemaker potentials of rabbit sino-atrial node cells. J Physiol 395:233–253, 1988.

166. GW Beeler Jr, H Reuter. Membrane calcium current in ventricular myocardial fibres. J Physiol 207:191–209, 1970.

167. M Vitek, W Trautwein. Slow inward current and action potential in cardiac Purkinje fibres. The effect on Mn plus,plus-ions. Pflugers Arch—Eur J Physiol 323:204–218, 1971.

168. W New, W Trautwein. The ionic nature of slow inward current and its relation to contraction. Pflugers Archiv—Eur J Physiol 334:24–38, 1972.

169. H Reuter, H Scholz. A study of the ion selectivity and the kinetic properties of the calcium dependent slow inward current in mammalian cardiac muscle. J Physiol 264:17–47, 1977.

170. P Brehm, R Eckert. Calcium entry leads to inactivation of calcium channel in *Paramecium.* Science 202:1203–1206, 1978.

171. IR Josephson, J Sanchez-Chapula, AM Brown. A comparison of calcium currents in rat and guinea pig single ventricular cells. Circ Res 54:144–156, 1984.

172. RS Kass, MC Sanguinetti. Inactivation of calcium channel current in the calf cardiac Purkinje fiber. Evidence for voltage- and calcium-mediated mechanisms. J Gen Physiol 84:705–726, 1984.

173. KS Lee, E Marban, RW Tsien. Inactivation of calcium channels in mammalian heart cells: joint dependence on membrane potential and intracellular calcium. J Physiol 364:395–411, 1985.

174. P Hess, JB Lansman, RW Tsien. Calcium channel selectivity for divalent and monovalent cations. Voltage and concentration dependence of single channel current in ventricular heart cells. J Gen Physiol 88:293–319, 1986.

175. RW Hadley, JR Hume. An intrinsic potential-dependent inactivation mechanism associated with calcium channels in guinea-pig myocytes. J Physiol 389:205–222, 1987.

176. BZ Peterson, CD DeMaria, JP Adelman, DT Yue. Calmodulin is the Ca^{2+} sensor for Ca^{2+}-dependent inactivation of L-type calcium channels. [Erratum appears in Neuron 1999 Apr;22(4):following 893.] Neuron 22:549–558, 1999.

177. N Qin, R Olcese, M Bransby, T Lin, L Birnbaumer. Ca^{2+}-induced inhibition of the cardiac Ca^{2+} channel depends on calmodulin. Proc Nat Acad Sci USA 96:2435–2438, 1999.

178. RD Zuhlke, GS Pitt, K Deisseroth, RW Tsien, H Reuter. Calmodulin supports both inactivation and facilitation of L-type calcium channels. Nature 399:159–162, 1999.

179. M de Leon, Y Wang, L Jones, E Perez-Reyes, X Wei, TW Soong, TP Snutch, DT Yue. Essential Ca^{2+}-binding motif for Ca^{2+}-sensitive inactivation of L-type Ca^{2+} channels. Science 270:1502–1506, 1995.

180. BZ Peterson, JS Lee, JG Mulle, Y Wang, M de Leon, DT Yue. Critical determinants of Ca^{2+}-dependent inactivation within an EF-hand motif of L-type Ca^{2+} channels. Biophys J 78:1906–1920, 2000.

181. DL Campbell, WR Giles, JR Hume, EF Shibata. Inactivation of calcium current in bull-frog atrial myocytes. J Physiol 403:287–315, 1988.

182. NB Standen, PR Stanfield. A binding-site model for calcium channel inactivation that depends on calcium entry. Proc R Soc Lond B: Biol Sci 217:101–110, 1982.

183. DW Hilgemann, D Noble. Excitation-contraction coupling and extracellular calcium transients in rabbit atrium: reconstruction of basic cellular mechanisms. Proc R Soc Lond B: Biol Sci 230:163–205, 1987.

184. RL Rasmusson, JW Clark, WR Giles, EF Shibata, DL Campbell. A mathematical model of a bullfrog cardiac pacemaker cell. Am J Physiol 259:H352–H369, 1990.

185. MS Jafri, JJ Rice, RL Winslow. Cardiac Ca^{2+} dynamics: the roles of ryanodine receptor adaptation and sarcoplasmic reticulum load. Biophys J 74:1149–1168, 1998.

186. M Wibo, G Bravo, T Godfraind. Postnatal maturation of excitation-contraction coupling in rat ventricle in relation to the subcellular localization and surface density of 1,4-dihydropyridine and ryanodine receptors. Circ Res 68:662–673, 1991.

187. AO Jorgensen, AC Shen, W Arnold, AT Leung, KP Campbell. Subcellular distribution of the 1,4-dihydropyridine receptor in rabbit skeletal muscle in situ: an immunofluorescence and immunocolloidal gold-labeling study. J Cell Biol 109:135–147, 1989.

188. SL Carl, K Felix, AH Caswell, NR Brandt, WJ Ball, PL Vaghy, G Meissner, DG Ferguson. Immunolocalization of sarcolemmal dihydropyridine receptor and sarcoplasmic reticular triadin and ryanodine receptor in rabbit ventricle and atrium. J Cell Biol 129:672–682, 1995.

189. NR Brandt, RM Kawamoto, AH Caswell. Dihydropyridine binding sites on transverse tubules isolated from triads of rabbit skeletal muscle. J Receptor Res 5:155–170, 1985.

190. MD Stern. Theory of excitation-contraction coupling in cardiac muscle. Biophys J 63:497–517, 1992.

191. G Gouy. Sur la constitution de la charge electrique a la surface d'un electrolyte. J Physiol 9, 457–468, 1910.

192. T Shibasaki. Conductance and kinetics of delayed rectifier potassium channels in nodal cells of the rabbit heart. J Physiol 387:227–250, 1987.

193. CM Armstrong. Sodium channels and gating currents. Physiol Rev 61:644–683, 1981.

2
Computation of the Action Potential of a Cardiac Cell

Candido Cabo
*Columbia University and City University of New York,
New York, New York, U.S.A.*

I. INTRODUCTION

Cardiac cells respond characteristically to applied electrical currents. If an electrical stimulus has an intensity that is above a certain threshold, ionic channels are activated and the cell generates an active response called an action potential, and the cell is excited. If the electrical stimulus is below threshold, cells respond passively because no ionic channels are activated. The action potential is a measurement of the variation over time of the electrical potential across the cell membrane after the cell has been excited. Changes in membrane potential are a result of the flow of ions through proteins that are embedded in the cell membrane and form ionic channels with conductivity that is, in general, voltage and time dependent.

Since the first experimental measurements of the action potential in nerve cells it became clear that a mathematical description of the action potential would be useful not only to interpret experimental data but also to generate hypotheses that could later be tested experimentally. Several approaches that differ in the amount of biophysical detail that is incorporated in the model have been used to compute (or simulate) the cardiac action potential. Ionic models, which were pioneered by Hodgkin and Huxley [1], intend to formulate mathematically the cellular processes that lead to the generation of the action potential and that have been measured experimentally. The action potential is then the result of the interaction of all those processes. In other models the goal is not to simulate the cellular processes that lead to the generation of an action potential, but just the dynamics of the action potential [2–4]. In those models, the action potential is the result of predefined rules or mathematical functions that do not correlate precisely with membrane or intracellular processes. In this chapter we will discuss the computation of the action potential using ionic models.

Action potential models have been very useful in investigating different aspects of cardiac electrophysiology, from action potential generation in a single cell to action potential propagation in a multidimensional structure of cardiac cells. Action potentials result from the interaction of many components, including the dynamics of the different ionic channels embedded in the cell membrane, changes in concentrations of ions inside and outside the cell, and how cells are connected. Computer models of the action potential can provide a link between the behavior of those components and the electrical behavior of the whole cell. For example, computer models can be used to predict how changes in ionic channel function caused by an acquired or genetic disease would affect the action potential. It is not usually possible to establish such a direct relationship experimentally. Therefore, action potential models can be used to generate hypotheses that can later be tested experimentally as well as to analyze and understand possible mechanisms of phenomena observed experimentally. Since the action potential is the result of the dynamic activation, inactivation or deactivation of all ionic channels as a function of time and transmembrane voltage that depends on the particular channel, computer models are a unique tool to investigate how each ionic channel contributes to the characteristics of the action potential under different circumstances.

II. ACTIVE RESPONSE OF THE CELL MEMBRANE

A. Excitability of Cardiac Cells: The All-or-None Law

The idea that cardiac cells could produce action potentials in response to an electrical stimulus was known before action potentials could be measured. In 1871, Henry Bowditch established that tissues respond to stimuli in an all-or-none manner, based on his studies on contraction of heart muscle [5]. (For a historical overview of seminal experiments and theories on the electrophysiology of excitable cells, the reader is referred to Ref. 6). Bowditch [5] observed that when an electrical stimulus was of sufficient strength to cause contraction of a frog heart, increasing the stimulus strength did not strengthen the response from the muscle. In Bowditch's words, "An induction shock produces a contraction or fails to do so according to its strength; if it does so, it produces the greatest contraction that can be produced by any strength of stimulus in the condition of the muscle at the time" [5]. Implied in Bowditch's statement is the concept of threshold for stimulation, and that the response of excitable tissues to stimuli above threshold does not depend on the intensity of the stimuli, which are the two properties that define excitable tissues. Also important to determine the response to stimuli above threshold is "the condition of the muscle at the time," that is, tissue viability or refractory state, as we will discuss later. Even though the all-or-none law was first proposed for cardiac tissue, it also applies to other excitable tissues such as skeletal muscle and nerve tissues.

B.   Measured Action Potentials: The Squid Giant Axon

Even though the all-or-none law was proposed for cardiac tissue, the first action potentials were not measured in cardiac cells. Most of the modern electrophysiological concepts and methods were developed in a nerve cell, the giant axon of the squid, because of its large size (its diameter is about 1 mm). Hodgkin and Huxley [7,8] and Curtis and Cole [9,10] were the first to measure an action potential in the squid giant axon using an intracellular micropipette (Fig. 1A). When the micropipette was in contact with the extracellular solution superfusing the axon (essentially sea water), the potential measured in the pipette was zeroed. As the micropipette was inserted into the cell membrane, the measured potential dropped to about $-60\,\mathrm{mV}$. This potential was defined as the resting membrane potential. When an electrical stimulus was applied, the membrane potential rapidly became less negative, until eventually it became positive and then returned to the resting membrane potential after a few milliseconds. This change in potential difference across the cell membrane over time is the action potential.

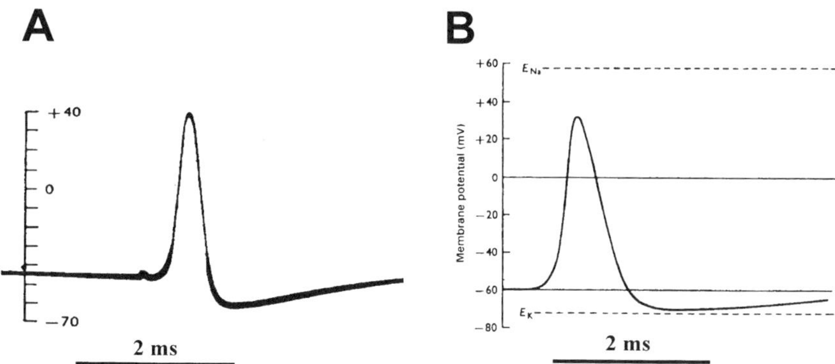

Figure 1   Action potential in a nerve cell (squid giant axon). (A) Experimental measurement. (B) Computer model.

Early on, Hodgkin and Huxley [1] realized that a quantitative understanding of action potential generation and propagation required the separation of the total membrane current causing the action potential into its different ionic components (and therefore the characterization of the different ionic channels). Their measurements of the dynamics of activation, inactivation and deactivation of sodium and potassium channels [1], led to the first quantitative description (model) of the action potential of an excitable cell (Fig. 1B). As we will discuss later in the chapter, the Hodgkin-Huxley formulation on how the voltage and time dependence of the different gates involved in the opening and closing of ionic channels should be modeled is still widely used, 50 years after it was first proposed.

It is important at this point to define some nomenclature that relates to the action potential and the membrane potential. The potential difference across the cell membrane is usually referred to as the membrane (or transmembrane) potential, and it is usually represented as V_m (sometimes also as E_m) Even though the polarity assigned to this potential is a matter of convention, in cellular electrophysiology the membrane potential is defined as the difference between the potential inside the cell and the potential outside the cell. A number of terms have been defined over the years in electrophysiology to describe the directions of changes in membrane potential. Hyperpolarization describes a change toward a more negative membrane potential (i.e., a more negative intracellular potential and V_m if the extracellular potential is constant). Depolarization describes a change toward a less negative membrane potential (i.e., a less negative intracellular potential and V_m if the extracellular potential is constant). Repolarization describes a change toward a more negative membrane potential (i.e., a more negative intracellular potential and V_m if the extracellular potential is constant). Hyperpolarization and repolarization describe changes in membrane potential in the same direction. However, typically hyperpolarization refers to changes in membrane potential of a resting (unexcited) cell, and depolarization refers to changes in membrane potential after a cell has been excited (i.e., depolarized) and the membrane potential is returning to its resting value.

C. Cardiac Action Potential

Cardiac cells are much smaller than the squid giant axon (approximately 100 μm long and with a diameter of 20 μm). Measurement of action potentials in skeletal muscle and cardiac cells had to wait until microelectrode pipettes with a tip external diameter small enough (<0.5 μm) to penetrate the membrane without causing damage could be constructed. Advances in microelectrode construction by Ling and Gerard [11], and by Nastuk and

Hodgkin [12], who were able to record the resting membrane potential and the action potential in skeletal muscle cells, paved the way to similar measurements in cardiac tissues. Caroboeuf and Weidman [13] and Draper and Weidmann [14] were the first to measure the resting membrane potential and the action potential in mammalian cardiac muscle, and Woodbury et al. [15] were the first to measure it in frog heart. Figure 2A shows an action potential measured in a Purkinje fiber (see later) of a dog heart [14]. The morphology of the cardiac action potential is more complex than that of skeletal or nerve cells, and consists of five different phases. In phase 0 (Fig. 2B), the action potential is initiated by a rapid upstroke lasting less than 0.5 msec and with a maximum rate of rise of about 500 mV/msec, which causes the membrane potential to depolarize (and reverse polarity) to about 30 mV (relative to the external solution). During phase 1 there is a brief rapid repolarization followed by a long plateau (phase 2), which is responsible for the long duration of the action potentials over 300 msec). During phase 3, the action potential rapidly repolarizes, and returns to the resting membrane potential (phase 4). Many of the characteristics of the cardiac action potential, such as resting membrane potential and rapid upstroke, are similar to that in skeletal and nerve cells. However, in contrast to skeletal and nerve cell potentials that have duration of less than 5 msec, the cardiac action potential duration has a duration of 300–500 msec.

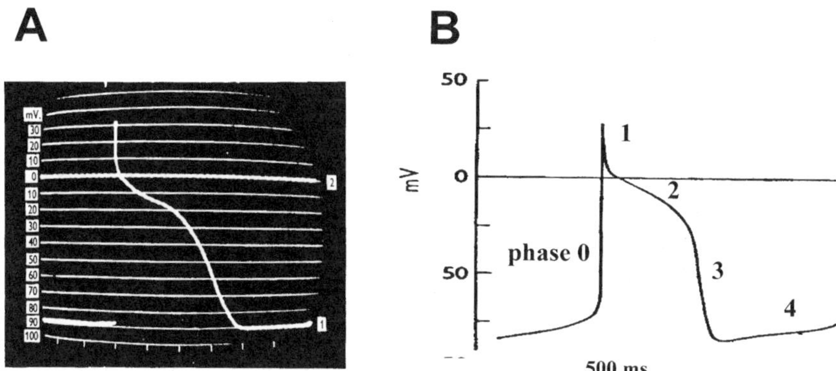

Figure 2 Action potential in a cardiac cell from a dog heart (Purkinje fiber). (A) Experimental measurement. 1 indicates the action potential and 2 is the zero potential recorded when the microelectrode is touching the superfusing bath. The ticks of the time scale at the bottom are spaced 100 msec. (Reproduced with permission from Ref 14.) (B) Computer model of a Purkinje fiber. (Reproduced with permission from Ref. 16.)

Figure 2B shows one of the first computer models of a cardiac action potential, which follows the same formulism used by Hodgkin and Huxley in their model, the squid axon [16].

III. MODELING THE PASSIVE RESPONSE OF THE CELL MEMBRANE

A. Ionic Basis of the Resting Membrane Potential

Before resting membrane potentials could be measured with microelectrodes, it was known that the concentration of potassium inside cardiac cells was greater than that outside cells, and that, at rest, cells are permeable to potassium ions [17,18]. Because inward rectifier potassium channels (a type of potassium channel) are open at the resting membrane potential, the gradient in potassium concentration results in the diffusion of potassium ions from inside the cell to outside. Since potassium ions are positively charged, diffusion results in the accumulation of positive charge outside the cell, and an excess of negative charge inside the cell. The result is an electric field directed from outside to inside the cell, which increases in magnitude as more potassium ions leave the cell. Electric field forces oppose diffusion forces and tend to move potassium ions from outside to inside the cell. Since potassium ions are subject to both diffusion and electric field forces, the growing electric field will eventually prevent the efflux of more potassium ions and a situation of equilibrium will be reached. The equilibrium potential (also called the Nernst potential, because Nernst was the first to show that diffusion of electrolytes in solution creates electrical potentials [19]) is given by the equation

$$ V_m = \left(\frac{RT}{F} \right) \ln \left(\frac{[K]_o}{[K]_i} \right) $$

where V_m is the membrane potential, R is the gas constant, T is the absolute temperature, F is the Faraday constant, $[K]_o$ is the potassium concentration outside the cell, and $[K]_i$ is the potassium concentration inside the cell. The Nernst potential can be interpreted as the potential at which an ion is in equilibrium with its diffusional force. If we consider that in cardiac tissue $[K]_o$ is $\sim 5.4\,\text{mM}$ and $[K]_i$ is $\sim 120\,\text{mM}$, the Nernst potential is $-86\,\text{mV}$, which is remarkably close to measured resting membrane potentials. Therefore, if the cell membrane at rest were permeable only to potassium ions, the ionic current flowing through the membrane at rest would be zero, and the resting membrane potential would be exactly the potassium Nernst potential.

However, measured resting membrane potentials are not always identical to the potassium equilibrium potential. While increasing $[K]_o$ results in a more depolarized resting membrane potential that is almost identical to the calculated Nernst potential, a reduction in $[K]_o$ results in less hyperpolarization than that predicted by the Nernst equation. This indicates that the membrane is also permeable to ions other than potassium at negative membrane potentials. A more accurate estimation of the resting membrane potential can be obtained with Goldman's equation, which takes into account other ions such as sodium and chloride, which can contribute to the resting membrane potential:

$$V_m = (RT/F)\ln\left(\frac{P_K[K]_o + P_{Na}[Na]_o + P_{Cl}[Cl]_i}{P_K[K]_i + P_{Na}[Na]_i + P_{Cl}[Cl]_o}\right)$$

where P_K, P_{Na}, and P_{Cl} are the permeabilities of the cell membrane to potassium, sodium, and chloride, respectively. $[K]_o$, $[Na]_o$, and $[Cl]_o$ are the external concentrations, and $[K]_i$, $[Na]_i$, and $[Cl]_i$ are the internal concentrations. In general, the equilibrium (Nernst) potentials of different ions will be different, and therefore no membrane potential can equilibrate all ions. If a membrane at rest is permeable to several ions (as is the case in cardiac cells), the resting membrane potential represents a dynamic equilibrium in which the total ionic current is zero but the individual ionic currents through the different ionic channels are not zero (because for each ion the equilibrium potential is different from its respective Nernst potential). Still, since the permeability of the membrane to potassium channels at rest is many orders of magnitude larger than the permeability to other ions, the Nernst potential for potassium is a good approximation of the resting membrane potential of cardiac cells.

B.   Passive Cell Membrane: Cole-Curtis Model

The nature of the cell membrane was elucidated by impedance measurements in intact cells before transmembrane potentials could be measured. Following Hermann's suggestion that the cell membrane could be represented under subthreshold conditions by a resistance in parallel with a capacitance [20], Fricke [21], using sinusoidal current analysis to measure the impedance of red blood cell membranes in suspension, showed that the cell membrane could be represented by a capacitance with a specific values of $1\,\mu F/cm^2$. Assuming a value of the dielectric constant of the membrane of 3, Fricke estimated a membrane thickness of about $33\,\text{Å}$, remarkably close to the membrane thickness of $50\,\text{Å}$ measured later with electron microscopy and X-ray techniques. Curtis and Cole [22] measured cell membrane resistance and capacitance in nerve cells and showed that the electrical properties of the membrane are well represented by an *RC* circuit (Fig. 3A). Their careful experiments showed that cells have a high-conductance cytoplasm that is surrounded by a high-resistance membrane with an electrical capacitance of about $1 \, \mu F/cm^2$ (similar to the value measured by Fricke in red blood cells). The capacitor represents the capacitance of the lipid bilayer that forms the cell membrane, and the resistor represents the conductance of the ionic channels that are open at the resting membrane potential. Figure 3B shows the response of a crab axon membrane to hyperpolarizing and depolarizing subthreshold stimuli of different strengths [23]. The 0 value on the abscissa represents the resting membrane potential. After the stimulus is turned off, the membrane potential returns to its resting value. The responses of the cell membrane to subthreshold depolarizing or hyperpolarizing electrical currents (passive responses in Fig. 3B) are well represented by the Cole-Curtis model of the membrane. Eventually, when the strength of a depolarizing stimulus reaches a certain threshold, the stimulus produces an action potential (active responses in Fig. 3B) and not a subthreshold response.

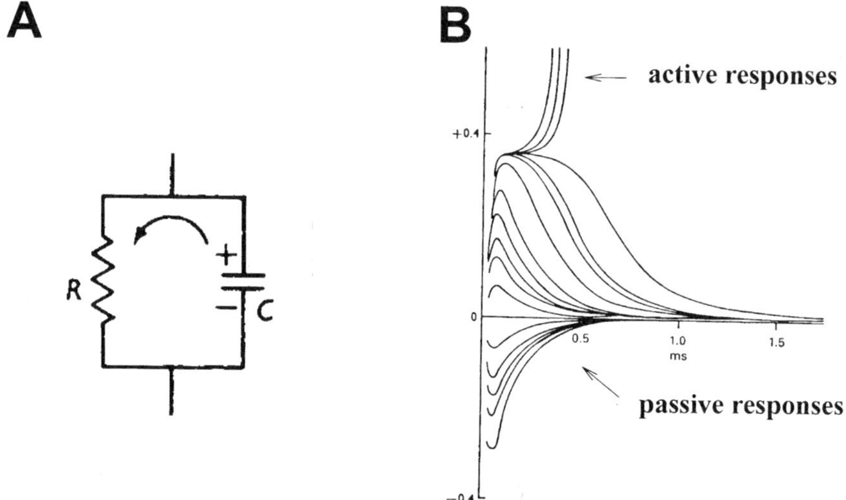

Figure 3 Passive membrane. (A) The Cole-Curtis model that represents the electrical response of a cell membrane by a resistor *R* in parallel with a capacitor *C*. (B) Responses of the membrane to depolarizing and hyperpolarizing subthreshold stimuli (passive responses) and to suprathreshold stimuli (active responses). (Reproduced with permission from Ref. 7.)

IV. MODELING THE ACTIVE RESPONSE OF THE CELL MEMBRANE

A. Parallel Conductance Model

Action potential generation is the result of ionic current flowing through many ionic channels that are embedded in the cell membrane (Fig. 4). Those channels are permeable to different ions (sodium, potassium, calcium) and open and close at different voltage levels with different time constants. The ionic current flowing through a channel is determined not only by the biophysical characteristics of the channel but also by the intracellular and extracellular environment that surrounds the cell membrane. For ionic channels to perform their physiological function, there has to be a gradient in ionic concentrations on both sides of the membrane. Sodium and calcium concentrations are higher outside than inside the cell; potassium concentration is higher on the inside. In maintaining those gradients, ionic pumps transform metabolic energy into potential electrochemical energy that is used by the ionic channels.

Hodgkin and Huxley [1] realized that a quantitative understanding of action potential generation and propagation required the separation of the total membrane current causing the action potential into its different ionic

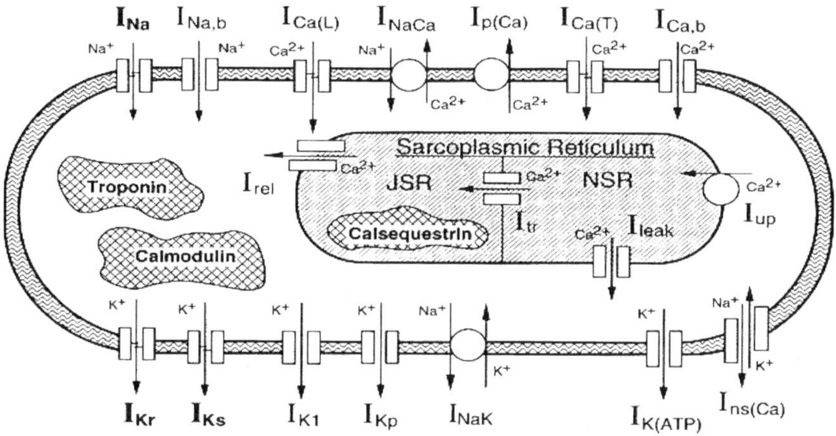

Figure 4 Diagram of a cardiac cell representing the main ionic channels embedded in the cell membrane. The arrows indicate whether the current is inward or outward. The channels are permeable to specific ions (K^+, Na^+, Ca^{2+}) that are also represented. Calcium fluxes in and out the sarcoplasmic reticulum are also indicated, as well as troponin and calmodulin, intracellular proteins that bind to calcium ions (see text for details). (Adapted from Ref. 62.)

components (and therefore the characterization of the different ionic channels). Since then, characterization of the individual currents that contribute to the action potential in cardiac cells (and other excitable tissues) has been an active area of research. Still, after all currents have been characterized, a quantitative understanding of the cell action potential is possible only when all ionic currents are integrated. To integrate the different ionic currents to reconstruct the action potential, Hodgkin and Huxley proposed the parallel conductance model (Fig. 5). The capacitor represents the membrane capacitance (as in the Cole-Curtis model). The branches of the circuit represent the different ways in which ions move between the intracellular and extracellular spaces through the membrane and originate an ionic current. Ions can move as a result of concentration gradients through channels whose conductance is time and voltage dependent, or through channels with constant conductance (leak or background). Ions can also be transported across cell membranes by pumps and exchangers that are necessary to restore concentration gradients. Movement of ions by pumps and exchangers also results in ionic currents that contribute

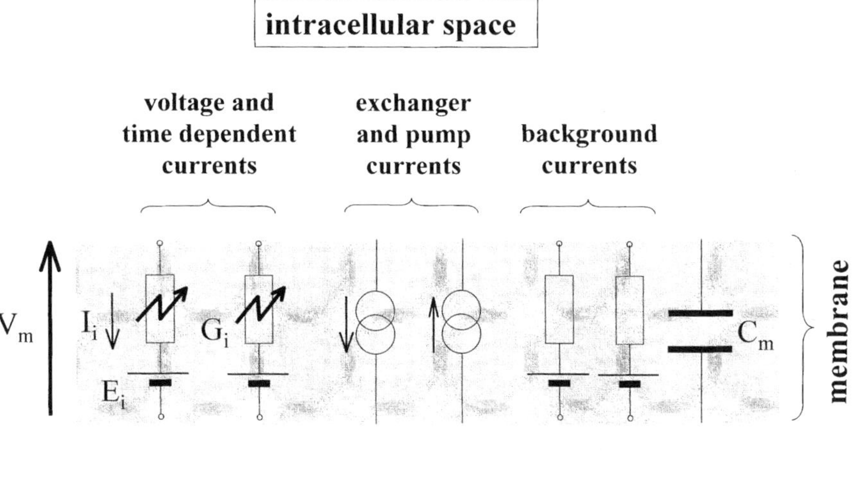

Figure 5 Parallel conductance model to model an active membrane. The shaded area represents the cell membrane that separates the intracellular and extracellular space. Different branches of the circuit represent different types of currents that can flow through the membrane (see text for details).

to the action potential. In the parallel conductance model, it is assumed that each ionic current flows independently from the others (the independence principle that was proposed by Hodgkin and Huxley in 1952). Using modern conventions, positive membrane current flows from the inside to the outside of the cell. The battery on a particular branch represents the equilibrium (or Nernst) potential for that ion, and the variable resistance represents that the resistance (or conductance) of the channel changes as a function of membrane voltage and time.

The expression of the total transmembrane current in the parallel conductance model is the sum of the capacitive and ionic currents. Therefore,

$$I_m = C_m \left(\frac{\partial V_m}{\partial t} \right) + I_{\text{ion}} = C_m \left(\frac{\partial V_m}{\partial t} \right) + \sum I_i$$

where I_m is the total membrane current, C_m is the specific capacitance, V_m is the transmembrane voltage, I_{ion} is the total ionic current, and I_i is the ionic current thorough channel i. Note that, in the above equation, for computations of an action potential in a single cell where there are no spatial changes in transmembrane potential, the total membrane current, I_m, is zero because there is no axial current.

When Hodgkin and Huxley proposed the parallel conductance model, they included three branches in the model to represent the flow of sodium, potassium, and chloride (leakage or background) currents through the membrane of squid axon. The number of branches of the parallel conductance model should represent the state of the art of knowledge of ionic currents that are involved in the generation of the action potential of a particular tissue that we want to model. The number of branches and the formulation of the conductance of those branches (G_i) have changed over time because of the discovery of new currents or a more accurate reformulation of old currents. The number of branches and the formulation of the currents depend on the tissue that we want to model. Therefore, the first step in formulating an ionic model of the action potential is to decide which currents should be part of the model. The second step is to model the individual ionic channels.

B. Models of Ionic Channels

The main currents that originate the action potential in excitable cells flow through ionic channels which are proteins spanning the cell membrane. Ionic channels are structures that can open and close under specific conditions (voltage-gated channels) or can be continuously open (background or leak channels). It is not possible to study the biophysics of an ionic channel during the action potential because the membrane voltage is constantly changing, and the membrane current reflects the contributions of many channels. To characterize a particular ionic channel, the current flowing through the channel is measured while the membrane is kept at constant voltage levels (voltage clamp). There are many techniques to control the membrane voltage that are discussed in detail in Hille [24]. Additionally, intracellular and extracellular bathing solutions could be designed to minimize the contamination by other channels in the measured currents. Voltage- and time-dependent currents can be expressed by applying Ohm's law as

$$I_i = G_i(V_m, t)(V_m - E_i)$$

where I_i is the ionic current through channel i, G_i is the conductance of the channel (which could be a function of transmembrane voltage and time), V_m is the transmembrane potential, and E_i is the equilibrium potential for ion i. Then, modeling an ionic channel requires the derivation of mathematical expressions for the conductance of the channel, including voltage and time dependencies, such that the model reproduces the voltage clamp results obtained experimentally (when subjected to the same protocols). To be able to fit their voltage clamp data, Hodgkin and Huxley [1] expressed the channel conductance as the product of a number of gating variables with first- or higher-order activation and/or inactivation kinetics,

$$G_i = G_{\max}a^m b^n \cdots.$$

where $G_{\max}$ is the maximum conductance of the ionic channel, $0 \leq a, b \leq 1$, and m, n are integers ≥ 1. The gating variables were governed by differential equations of the type

$$\frac{da}{dt} = \frac{a_\infty - a}{\tau_a}$$

where a_∞ is the steady-state value of the gating variable a, and τ_a is its time constant (a_∞ and τ_a are functions of the transmembrane potential only). The solution of the first-order differential equation for each gating variable is

$$a = a_\infty - (a_\infty - a_0)e^{(-t/\tau_a)}$$

where a_∞ is the steady-state value of variable a, and a_0 is the value at the beginning of the integration period.

To obtain the experimental data necessary to model a specific ionic channel (maximum conductance, gating variables, equilibrium potential), a number of voltage clamp protocols and experimental procedures are applied to the cell membrane. The voltage dependence of the steady-state value of the gating variable is usually fitted by a Boltzmann function: $y = 1/(1+\exp[\pm(V_m - V_{1/2})/k])$ to experimental activation and inactivation curves, where V_m is the transmembrane potential, $V_{1/2}$ is the transmembrane voltage at which 50% of the channels are activated, and k is the slope factor. Activation and inactivation kinetics can be fitted by using single- or multiexponential functions, depending on whether the data are best fitted by first- or higher-order kinetics. Fitting of mathematical functions to experimental data can be done using, for example, the Marquardt-Levenberg algorithm for nonlinear regression [25].

Even though the Hodgkin-Huxley formulism to model ionic channels is commonly used and it is appropriate to reproduce the macroscopic response of the channel, it does not provide an accurate description of its structure. When the structure of the channel is important, a generalization of the Hodgkin-Huxley formulism based on Markovian model is often preferred [26]. In response to changes in transmembrane voltage, ionic channels undergo conformational changes that change the functional state of the channel. In general, a channel can be in a closed, open, or inactivated state. A Markovian model represents the structure of ionic channels by describing the states that a channel can occupy and by specifying the transition rates between the different states. The transition rates are in general dependent on the transmembrane potential and can be expressed as $Ae^{(Vm)}$. A channel can be modeled as having several open, closed, or inactivated states. Each state has an associated probability that changes as a function of time and transmembrane potential. The summation of the probabilities associated with each state should be 1 (i.e., the channel has to be in some state). The probability that the channel is in state i, P_i, changes over time according to the expression

$$\frac{dP_i}{dt} = \sum_j (\alpha_{ji} P_j - \alpha_{ij} P_j)$$

where i represents a generic state of the channel, and j represents other states that can be reached by a direct transition from state i; α_{ij} represents the transition rate from state i to state j, and α_{ji} represents the transition rate from state j to i. Formulation of similar expressions for all states of the channel leads to a system of differential equations that the probabilities that the channel is in a given state.

At any given time, the current through the channel can be calculated as:

$$I_i = G_{\max} P_o (V_m - E_i)$$

where P_o is the probability that the channel is in the open state (one or more states defined in the Markovian model). Note that P_o is equivalent to the product of the gating variables in the Hodgkin-Huxley formulism [26].

Sometimes experimental measurements of ionic currents are recorded at room temperature ($22°C$), which is below the physiological temperature ($37°C$) and can thus lead to an underestimation of the rate constants of transitions between states of ionic channels. When necessary, the rate constants need to be adjusted to $37°C$ using the Q_{10} adjustment factor (the factor is usually between 2 and 3 for rate constants between different states in ionic channels). For example, $a_{t1} = a_{t0}(Q_{10})^{(t1-t0)/10}$, where a_{t0} is the measurement of a rate constant at temperature t_0 and a_{t1} is the estimate of that same rate constant at temperature t_1. The rate of different reactions involved in gating might change with temperature with different Q_{10} values. In that case, the steady-state curve might also be a function of temperature. However, given that data on Q_{10} for different reactions is not usually available, it is commonly assumed that all reactions have the same Q_{10}, and therefore the steady-state curve will not be a function of temperature.

C. Modeling Pumps and Exchangers

When a channel opens, ions move by passive diffusion down their concentration gradients. However, if this ionic movement were not reversed, gradients in ionic concentration would eventually disappear. To maintain those gradients, cells use organic molecules that are soluble in the cell membrane (carriers) to transport ions up their concentration gradients. Carriers bind ions in one side of the membrane and deliver them to the other side. There are two mechanisms of carrier-mediated transport. The first mechanism, used by ionic pumps, requires metabolic energy (ATP) to move ions up their gradients. Since those pumps move ions, they also contribute to the overall membrane ionic current. Pump transport is usually described as a binding process and it is formulated by variations of the Michaelis-Menten equation (see Section IV.E). The second mechanism, co-transport, does not depend on metabolic energy to maintain ionic gradients. Instead, it uses the energy available in a given ion's electro-chemical gradient (for example, Na) to transport another ion uphill (for example, Ca). Exchangers use this second mechanism of carrier-mediated transport.

Gradients in sodium and potassium ions are maintained by the sodium/potassium pump, I_{NaK}. This pump extrudes from the cell 3 sodium ions for each 2 potassium ions that it brings in. Several formulations have been proposed for the I_{NaK} pump [27–29]. They all have the same general form:

$$I_{NaK} = f(V_m, [\text{Na}]_o) \left(\frac{[\text{Na}]_i^n}{[\text{Na}]_i^n + [\text{K}_{m,\text{Na}i}]^n} \right) \left(\frac{[\text{K}]_o^m}{[\text{K}]_o^m + [\text{K}_{m,\text{K}o}]^m} \right)$$

where f is a function that represents the dependence of I_{NaK} on the trans-membrane potential and the extracellular concentration of sodium. $[\text{Na}]_i$ is the intracellular concentration of sodium and $[\text{K}]_o$ is the extracellular concentration of potassium. $\text{K}_{m,\text{Na}i}$ and $\text{K}_{m,\text{K}o}$ are the concentrations of sodium and potassium for half-activation. Different models use different values for f, m, and n. DiFrancesco and Noble [27] use $n = m = 1$, and f is independent of the transmembrane voltage and extracellular sodium. Rasmusson et al. [28] use $n = 3$, $m = 2$, and f depends only on the transmembrane voltage. Luo and Rudy [29] use $n = 1.5$, $m = 1$, and f depends on both the trans-membrane potential and extracellular sodium.

Another pump that is usually incorporated in ionic models is the calcium pump. This pump helps the sodium/calcium exchanger (Section V) to extrude calcium ions from the cell [28,29]. Rasmusson et al. [28] and Luo and Rudy [29] describe the calcium pump as a first-order binding process:

$$I_{p,\text{Ca}} = I_{p,\text{Ca,max}} \left(\frac{[\text{Ca}]_i}{[\text{Ca}]_i + [\text{K}_{m,\text{Ca}i}]} \right)$$

where $I_{p,\text{Ca,max}}$ is the current at the maximum pump rate ($\mu\text{A}/\text{cm}^2$ or pA/pF), $[\text{Ca}]_i$ is the intracellular calcium concentration and $[\text{K}_{m,\text{Ca}i}]$ is the half-activation concentration (the $[\text{Ca}]_i$ at which the pump rate is half-maximum).

The main function of the Na/Ca exchanger is to maintain the gradient of calcium ions between the outside and the inside of the cell. The Na/Ca exchanger moves in 3 Na^+ ions for each Ca^{2+} ion that is extruded. The exchanger was first formulated by Mullins [30], and subsequently simplified by DiFrancesco and Noble [27]. That formulation has been subsequently updated by Luo and Rudy [29].

D. Cell Geometry

Cardiac cells have an approximately cylindrical shape with a length of about 0.1 mm, and a radius of about 0.01 mm (Fig. 6). The intracellular volume, V_i, can be easily calculated as $\pi r^2 l$. However, because of the presence of the sarcoplasmic reticulum and mitochondria, only about 65% of the intracellular

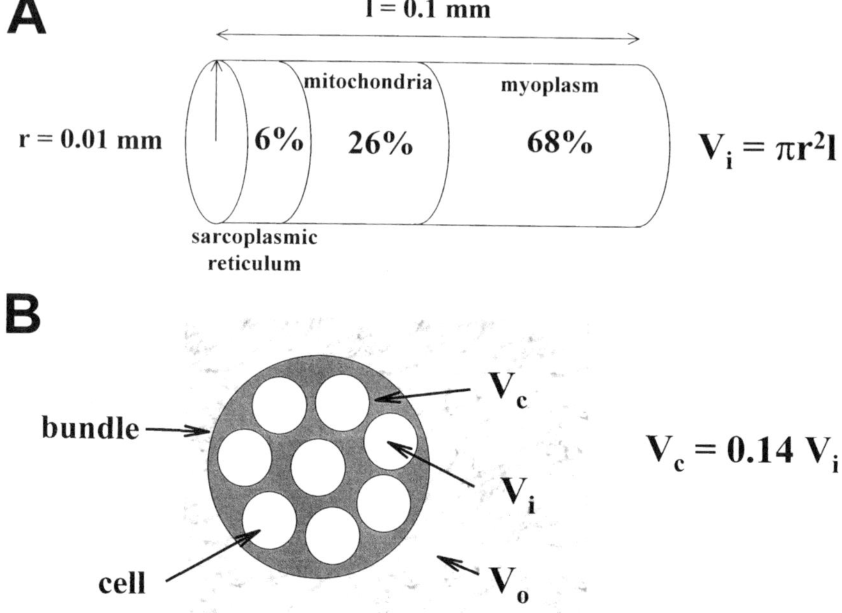

Figure 6 Cell geometry. (A) Diagram representing the dimensions of a cardiac cell with volume V_i, length l and radius r. Relative size of the volume of the myoplasm, mitochondria, and sarcoplasmic reticulum in a cardiac cell. (B) Transverse section of a bundle of cardiac cells to illustrate the differences between the bulk and cleft extracellular spaces and the intracellular space.

space is available for the free movement of ions (Fig. 6A). This volume will be important in the calculation of the changes in ionic concentrations.

Cardiac cells lie very close and parallel to one another in bundles that might be surrounded by a layer of endothelial cells (Fig. 6B). The bundles of cardiac cells are loosely coupled by collagen fibers [31]. As a result of this anatomical arrangement the extracellular space that a cell "sees" inside the bundle (V_c in Fig. 6B) is not well perfused (because the anatomy impedes free diffusion of ions) and the ionic concentrations inside the bundle might be different from the ionic concentrations in the bulk extracellular space (V_o in Fig. 6B). In a bundle, the volume of the restricted extracellular space is a fraction of the volume of the intracellular space (estimated experimentally as $V_c = 0.14 V_i$).

In computer models, the physiological environment of a cell membrane is usually represented by three compartments or spaces that are accessible to the different ions: the intracellular space, a restricted cleft space, and the bulk extracellular space (see Fig. 7) [27,28]. The movement of ions between the intracellular space and the restricted extracellular space as the result of ionic currents flowing through the cell membrane. The movement of an ion between the restricted extracellular space and the bulk extracellular space is the result of the restricted diffusion of that ion. It is important to note that there are situations when the inclusion in the model of a restricted extracellular space is not necessary. For example, simulations of an isolated cardiac cell in a superfused bath do not require a restricted extracellular space because ions flow directly between the intracellular space and the bulk extracellular space. On the other hand, to simulate a piece of tissue in which cells are packed together, simulation of a restricted extracellular space might be necessary.

E. Modeling Intracellular Calcium Dynamics

The major function of the ventricles is to produce a strong contraction to cause blood to circulate. In myocardial ventricular cells, contraction is triggered after an action potential by a process called excitation–contraction coupling. A gradient in calcium ions between inside (~ 0.00012 mM) and outside (~ 2 mM) the cell makes possible the development of a calcium current as a result of the inflow of calcium ions when the calcium channels

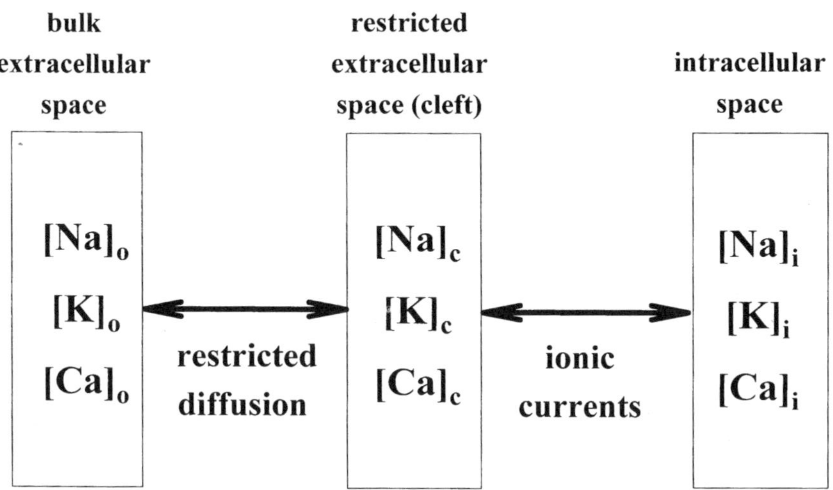

Figure 7 Different spaces and ion concentrations that need to be taken into account in a computer model (see text for explanation).

are open, and this leads to the characteristic long plateau of the cardiac action potential. Calcium ions that enter the cell through the calcium channel ($I_{Ca,L}$ in Fig. 8) induce the release of more calcium ions from the sarcoplasmic reticulum by a process called calcium-induced calcium release (CICR) (J_{rel} in Fig. 8), resulting in an elevation of intracellular calcium. This elevation of intracellular calcium is transient because calcium ions are up-taken by the sarcoplasmic reticulum (J_{up}), and are extruded outside the cell by the sodium/calcium exchanger and a calcium pump (I_{NaCa} and $I_{p(ca)}$ in Fig. 8). In addition to its role in excitation–contraction coupling, intracellular calcium ions regulate other membrane channels (the L-type calcium channel and the slow delayed rectifier (see Section V), and intracellular processes. Therefore, the amount of free intracellular calcium in the myoplasm is determined by many processes (Fig. 8): the calcium that enters/exits the cell through ionic channels, the calcium that leaves the cell through the Na/Ca exchanger, the calcium that is uptaken and released from the sarcoplasmic reticulum, and the calcium that is bound to proteins such as troponin and calmodulin.

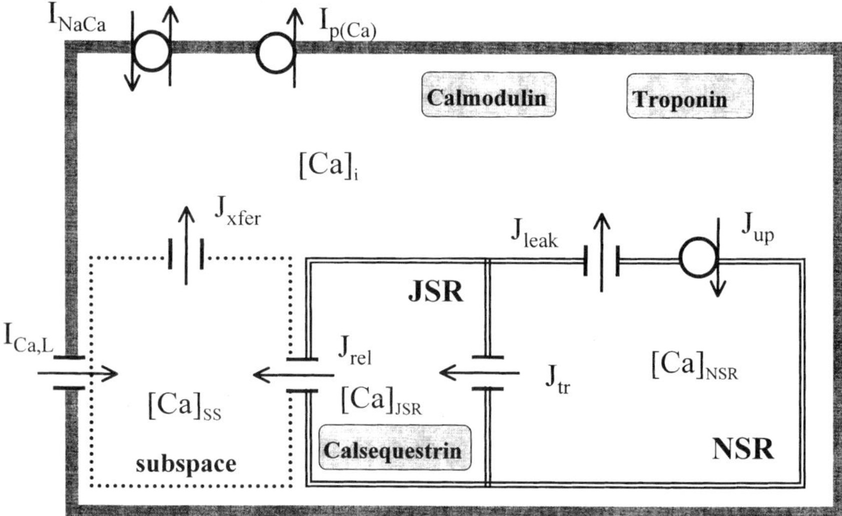

Figure 8 Currents and fluxes that contribute to changes in the concentration of intracellular calcium. The gray rectangle represents the cell membrane. The sarcoplasmic reticulum is divided into two compartments, the NSR and JSR. The dotted rectangle represents a restricted subspace where calcium is released from the JSR. $[Ca]_x$ represents the calcium concentration in compartment X. See text for explanation.

1. Calcium Fluxes[*] in the Sarcoplasmic Reticulum

Several models describing the uptake and release of calcium by the sarcoplasmic reticulum (SR) can be found in the literature [27,29,32–37]. The sarcoplasmic reticulum occupies about 6% of the cell volume (Fig. 6A) and consists of two compartments: the network sarcoplasmic reticulum (NSR), where calcium is uptaken in the SR, and the junctional sarcoplasmic reticulum (JSR), from which calcium is released to the cytoplasm (Fig. 8). The NSR occupies about 90% the SR volume, and the JSR the rest.

We can model the fluxes in the sarcoplasmic reticulum by formulating the uptake (J_{up}) and leakage J_{leak} from the NSR, the translocation from the NSR to the JSR (J_{tr}), and release from the JSR (J_{rel}). The uptake of calcium ions by the NSR is performed by a metabolic pump that can be formulated as (see above the discussion on pumps)

$$J_{up} = J_{upmax} \left(\frac{[Ca]_i^n}{[Ca]_i^n + [K_{m,up}]^n} \right)$$

where J_{upmax} is the maximum pump rate (mmol/L per millisecond), $[Ca]_i$ is the intracellular calcium concentration and $[K_{m,up}]$ is the half-activation concentration (the $[Ca]_i$ at which the pump rate is half-maximum), and n is the order of the binding process. Luo and Rudy [29] use a value of $n = 1$, whereas Jafri et al. [37] use a value of $n = 2$.

Different formulations for the leakage from the NSR to the myoplasm can be found in the literature. Luo and Rudy [29] use the expression

$$J_{leak} = \frac{[Ca]_{NSR}}{\tau_{leak}}$$

where $[Ca]_{NSR}$ is the calcium concentration in the NSR and τ_{leak} is the time constant of the leakage (which is the reciprocal of the leakage rate). On the other hand, Jafri et al. [37] use the expression

$$J_{leak} = \frac{[Ca]_{NSR} - [Ca]_i}{\tau_{leak}}$$

where $[Ca]_{NSR}$ and $[Ca]_i$ are the calcium concentration in the NSR and the myoplasm, respectively, and τ_{leak} is the time constant of the leakage. Both

---

[*]Ionic currents, represented by I, are expressed as current densities indicating the amount of current per unit surface of membrane and have units of, $\mu A/cm^2$ or pA/pF. Ionic fluxes, represented by J, represent changes in ionic concentrations per unit time and have units of mmol/L per millisecond.

formulations are practically equivalent because $[\mathrm{Ca}]_{\mathrm{NSR}} \gg [\mathrm{Ca}]_i$ and therefore $([\mathrm{Ca}]_{\mathrm{NSR}} - [\mathrm{Ca}]_i) \sim [\mathrm{Ca}]_{\mathrm{NSR}}$.

The flux of calcium ions from the NSR to the JSR, J_{tr}, can be formulated as

$$J_{\mathrm{tr}} = \frac{[\mathrm{Ca}]_{\mathrm{NSR}} - [\mathrm{Ca}]_{\mathrm{JSR}}}{\tau_{\mathrm{tr}}}$$

where $[\mathrm{Ca}]_{\mathrm{NSR}}$ and $[\mathrm{Ca}]_{\mathrm{JSR}}$ are the calcium concentrations in the NSR and the JSR, respectively. τ_{tr} is the time constant of the translocation.

The release of calcium from the JSR is a process that is more complex than other calcium fluxes discussed so far, because the rate of release from the JSR depends on the concentration of intracellular calcium by the process of calcium-induced calcium release. Therefore in its more general form J_{rel} can be expressed as

$$J_{\mathrm{rel}} = A_{\mathrm{rel}}([\mathrm{Ca}]_i, t)([\mathrm{Ca}]_{\mathrm{JSR}} - [\mathrm{Ca}]_i)$$

where A_{rel} is the rate of calcium release that in general could be a function of intracellular calcium and time, and $[\mathrm{Ca}]_{\mathrm{JSR}}$ and $[\mathrm{Ca}]_i$ are the calcium concentrations in the JSR and myoplasm, respectively.

DiFrancesco and Noble [27] modeled the basic assumption that release of calcium from the JSR is induced by calcium by formulating A_{rel} as

$$A_{\mathrm{rel}} = (1/\tau_{\mathrm{rel}})\left(\frac{[\mathrm{Ca}]_i^r}{[\mathrm{Ca}]_i^r + \mathrm{K}_{m,\mathrm{rel}}}\right)$$

where the maximum rate of release $1/\tau_{\mathrm{rel}}$ is modulated by the amount of intracellular calcium bound to release sites, and r is the amount of calcium ions assumed to bind to a release site (set in the DiFrancesco-Noble model to 2). The time dependence of A_{rel} results from the time variation of $[\mathrm{Ca}]_i$.

Luo and Rudy [29] incorporated in the formulation of calcium release from the JSR experimental data that were not available at the time when the DiFrancesco-Noble model was formulated. In the Luo-Rudy model the calcium-dependent term is separated from the time-dependent term of the release. The calcium-dependent term is determined by the difference between the amount of calcium entering the cell in the first 2 msec after depolarization ($[\mathrm{Ca}]_{i,2}$) and a fixed threshold ($[\mathrm{Ca}]_{i,\mathrm{th}}$). The time-dependent has two exponents that represent the activation (time constant τ_{on}) and deactivation (time constant τ_{off}) of the release process. They formulated A_{rel} as follows:

$$A_{\mathrm{rel}} = A_{\mathrm{rel,max}}\left(\frac{[\mathrm{Ca}]_{i,2} - [\mathrm{Ca}]_{i,\mathrm{th}}}{[\mathrm{Ca}]_{i,2} - [\mathrm{Ca}]_{i,\mathrm{th}} + \mathrm{K}_{m,\mathrm{rel}}}\right)(1 - e^{-t/\tau_{\mathrm{on}}})e^{-t/\tau_{\mathrm{off}}}$$

where $A_{\mathrm{rel,max}}$ is the maximum rate of calcium release.

The DiFrancesco-Noble and Luo-Rudy models assume that calcium from the JSR is released directly into the myoplasm. However, while this assumption is supported by experimental measurements of calcium transients [38], other investigators have suggested that calcium from the JSR is released in a restricted subspace (Fig. 8) before diffusing to the myoplasm [39]. Jafri et al. [37] modified the Luo-Rudy formulation of calcium release from the JSR by incorporating a restricted subspace and a Markovian model of the calcium release channel (ryanodine receptor) that was originally proposed by Keizer and Levine [40]. The formulation of the calcium release used in their model is

$$J_{\text{rel}} = A_{\text{rel,max}} P_o ([\text{Ca}]_{\text{JSR}} - [\text{Ca}]_{\text{ss}})$$

where $[\text{Ca}]_{\text{ss}}$ is the calcium concentration in the restricted subspace, $A_{\text{rel,max}}$ is the maximum rate of calcium release, and P_o is the probability that the release channel is open (see above Markovian models). The intracellular calcium and time dependence is introduced by P_o.

In the model by Jafri et al. [37] an additional calcium flux between the restricted subspace where the calcium is released from the JSR and the myoplasm has to be introduced. That flux is the result of passive diffusion and can be formulated as

$$J_{\text{xfer}} = \frac{[\text{Ca}]_{\text{ss}} - [\text{Ca}]_i}{\tau_{\text{xfer}}}$$

where $[\text{Ca}]_{\text{ss}}$ and $[\text{Ca}]_i$ are the calcium concentrations in the restricted space and the myoplasm, respectively, and τ_{xfer} is the time constant of the transfer.

2. Calcium Buffering

In addition to its role in cardiac excitation, intracellular calcium is an important second messenger that regulates many intracellular processes by binding to cytosolic proteins. For example, troponin is a contractile protein that binds intracellular calcium, and detects a rise in intracellular calcium as a signal to initiate the interactions between other contractile proteins (actin and myosin) that activate the process of muscle contraction. Intracellular calcium also binds to other cytosolic proteins such as calmodulin. Similarly to troponin, calmodulin responds to a rise in intracellular calcium. Calmodulin is involved in the regulation of metabolic pathways of energy production, muscle contraction, and neurotransmitter release. It is important to note that even if we are not interested in modeling cell processes regulated by calcium, the fact that calcium binds to cell proteins (such as troponin and calmodulin) modulates the amount of free calcium in the cell and, as a result, cell excitation. Binding of calcium to cell proteins has a buffering effect that keeps intracellular calcium at a low level and its variations small.

In its most simplified form, calcium buffering can be modeled using the same formalism used in enzymatic reactions or drug/receptor binding. If the concentration of the buffer is [B] and the concentration of intracellular calcium is $[Ca]_i$, then the binding of the buffer to calcium can be modeled as a chemical reaction as follows:

$$[B]_{\text{free}} + [Ca]_i \underset{K_r}{\overset{K_f}{\longleftrightarrow}} [BCa]_i$$

where $[B]_{\text{free}}$ is the concentration of free buffer, $[Ca]_i$ is the concentration of free intracellular calcium, and $[BCa]_i$ is the concentration of buffer bound to calcium. The forward rate of the reaction is K_f, and the reverse rate is K_r. When equilibrium is reached, the following relation has to be satisfied:

$$\frac{K_r}{K_f} = \frac{[B]_{\text{free}}[Ca]_i}{[BCa]_i} = K_d$$

The ratio K_r/K_f is known as the dissociation constant, K_d, of the reaction. The percent of buffer proteins bound to calcium can be expressed as

$$\frac{[BCa]_i}{[BCa]_i + [B]_{\text{free}}} = \frac{[Ca]_i}{[Ca]_i + K_d}$$

or

$$[BCa]_i = [B]\frac{[Ca]_i}{[Ca]_i + K_d}$$

where $[B](= [B]_{\text{free}} + [BCa]_i)$ is the total concentration of the buffer in the myoplasm.

The equations above quantify free and buffered calcium where equilibrium is reached. While this situation occurs in cells at rest, during an action potential $[Ca]_i$ is constantly changing and the equilibrium equations are not valid. However, when binding of calcium by the buffer occurs much faster than other processes (i.e., calcium release or uptake), equilibrium is reached "instantly" and the equilibrium equations can still be used. In the Luo-Rudy model [29] of the guinea pig ventricular action potential, the fast buffer approximation is used and calcium buffering by troponin and calmodulin in the cytoplasm, and calcium buffering by calsequestrin in the sarcoplasmic reticulum is formulated using the equilibrium expressions derived above. Jafri et al. [37] also use the fast buffer approximation to calculate buffering by calmodulin and calsequestrin. However, buffering by troponin is not considered fast, and therefore reaction kinetics must be taken into account [37].

Modeling calcium buffering is useful not only to simulate binding of intracellular calcium to troponin and calmodulin under normal physiological conditions but also to simulate conditions in voltage clamp experiments in single cells in which calcium buffers such as Bapta or EGTA are included in the pipette solution that determines the composition of the cell myoplasm. A more detailed discussion on how to model calcium buffering can be found in Ref. 41.

3. Changes in Calcium Concentration

Now that we have formulated the calcium fluxes in the sarcoplasmic reticulum and calcium buffering, we have to integrate those two processes to calculate the changes in free intracellular calcium concentration. At any given time, the total concentration of calcium in the myoplasm ($[\text{Ca}]_T$) is the sum of free calcium ($[\text{Ca}]_i$) and buffered calcium $[\text{BCa}]_i$:

$$[\text{Ca}]_T = [\text{Ca}]_i + [\text{BCa}]_i$$

or

$$[\text{Ca}]_T = [\text{Ca}]_i + \left(\frac{[\text{B}][\text{Ca}_i]}{[\text{Ca}_i] + K_d} \right)$$

where [B] is the buffer concentration and K_d is the dissociation constant (see above).

By differentiating with respect to time and rearranging terms [41] we obtain

$$\frac{d[\text{Ca}]_i}{dt} = \left(1 + \frac{[\text{B}]K_d}{([\text{Ca}]_i + K_d)^2} \right)^{-1} \frac{d[\text{Ca}]_T}{dt}$$

or

$$\frac{d[\text{Ca}]_i}{dt} = \beta_{\text{myo}} J_{\text{total,myo}}$$

where $d[\text{Ca}]_i/dt$ is the change in free calcium concentration and $d[\text{Ca}]_T/dt$ is the total flux of calcium in the myoplasm ($J_{\text{total,myo}}$). The term

$$\beta_{\text{myo}} = \left(1 + \frac{[\text{B}]K_d}{([\text{Ca}]_i + K_d)^2} \right)^{-1}$$

scales the total flux as a result of buffering.

Considering the fluxes and currents diagrammed in Fig. 8 and assuming that the subspace compartment does not exist ($J_{xfer} = 0$ and $[Ca]_{ss} = [Ca]_i$), the total flux can be calculated as

$$J_{total,myo} = J_{leak} + J_{rel}\left(\frac{V_{JSR}}{V_i}\right) - J_{up} - \left[\frac{(I_{p(Ca)} - 2I_{NaCa} + I_{Ca,L})S}{2V_iF}\right]$$

where V_{JSR} is the volume of the JSR, V_i is the volume of the intracellular space (myoplasm), S is total membrane area, 2 is the valence of calcium, and F is the Faraday constant. Note that J_{total} consists of terms that are ion fluxes (J_{leak}, J_{rel}, J_{up}) and terms that are current densities ($I_{p(Ca)}$, I_{NaCa}, $I_{Ca,L}$) that need to be converted to ion fluxes (with the proper sign). Also note that the J_{rel} flux is relative to the JSR volume and therefore has to be sealed to the volume of the myoplasm. J_{leak} and J_{up} fluxes are relative to the myoplasm, and scaling is not necessary.

In addition to calculating changes in the concentration of free calcium in the myoplasm, we need to calculate changes of calcium concentrations in the NSR and JSR because those are used to calculate calcium fluxes, which in turn are needed to calculate changes in intracellular free calcium (see above). Changes in $[Ca]_{JSR}$ can be calculated by

$$\frac{d[Ca]_{JSR}}{dt} = \beta_{JSR}\left(J_{tr} - J_{rel}\right)$$

where β_{JSR} is the scaling factor that models the buffering of calcium by calsequestrin in the JSR. Similar to the expression for β_{myo}, β_{JSR} is equal to $(1 + \{[B]K_d/([Ca]_{JSR} + K_d)^2\})^{-1}$, where $[B]$ is the concentration of calsequestrin and K_d is the dissociation constant. Note that fluxes J_{tr} and J_{rel} are relative to the JSR volume and no volume scaling is necessary. Changes in $[Ca]_{NSR}$ can be calculated by

$$\frac{d[Ca]_{NSR}}{dt} = J_{up}\left(\frac{V_i}{V_{NSR}}\right) - J_{leak}\left(\frac{V_i}{V_{NSR}}\right) - J_{tr}\left(\frac{V_{JSR}}{V_{NSR}}\right)$$

where V_i, V_{NSR}, and V_{JSR} are the volumes of the myoplasm, NSR, and JSR, respectively.

So far we have assumed that there are no spatial changes in the distribution of the sarcoplasmic reticulum inside the cell or in the intracellular concentration of calcium. However, the demonstration of calcium waves in single cells [42] indicates that calcium release and uptake (which results in a calcium transient) does not occur simultaneously everywhere in the myoplasm, and therefore there is a spatial distribution of $[Ca]_i$. In studies where calcium wave propagation is important, spatial changes in intracellular calcium concentration should be incorporated in the model.

F.   Modeling the Cell Membrane Environment

Ionic currents result from the movement of sodium, potassium, and calcium ions in and out of the cell. As a result of the current flow, the concentrations of ions inside and outside the cell change during the action potential. Since the pioneering experiments conducted by Ringer between 1881 and 1887, who observed that to keep a frog heart beating it had to be perfused with a solution containing sodium, potassium, and calcium ions mixed in fixed concentrations, it is well known that cell membrane environment is an important determinant of the action potential characteristics. In consequence, in many situations accurate modeling of the changes in ionic concentrations during the electrical activity of the cell is necessary.

1.   Changes in Intracellular Ionic Concentrations

The changes in ionic concentrations in the intracellular space can be calculated with the following expression:

$$\frac{d[X]_i}{dt} = \frac{-(I_X S)}{V_i z_X F}$$

where $[X]_i$ is the concentrations of ion X in the intracellular space, I_X is the sum of all ionic currents carrying ion X, S is total membrane area, V_i is the volume of the intracellular space, z_X is the valence of ion X, and F is the Faraday constant. It is important to notice that because of the high degree of membrane folding, the actual membrane surface S is much larger than the surface that would be predicted from the cylindrical geometry. Luo and Rudy [29] use the expression $S = 2(2\pi r^2 + 2\pi r l)$, where r and l are the radius and the length of the cylindrical representation of the cell. Changes in the intracellular concentrations of potassium and sodium using the previous expression were calculated in Refs. 27–29 and Ref. 43. Changes in the intracellular concentration of calcium are more complicated, as we described earlier.

2.   Changes in Extracellular Ionic Concentrations: Cleft Spaces

In the three-compartment model (Fig. 7), there are two extracellular spaces: the restricted extracellular space and the bulk extracellular space. Changes in the ionic concentrations in the bulk extracellular space are negligible because of the fact that that space is well perfused and big. The changes in ionic concentrations in the restricted extracellular space can be calculated with the following expression:

$$\frac{d[\text{X}]_c}{dt} = \frac{[\text{X}]_o - [\text{X}]_c}{\tau_p} + \frac{I_X S}{V_c z_X F}$$

where $[\text{X}]_c$ is the concentration of ion X in the restricted extracellular space, $[\text{X}]_o$ is the concentration of ion X in the bulk extracellular space, τ_p is the constant for diffusion between the restricted and bulk extracellular spaces, I_X is the sum of all ionic currents carrying ion X, S is total membrane area (see earlier), V_c is the volume of the restricted extracellular space, z_X is the valence of ion X, and F is the Faraday constant. Changes in the concentrations of potassium, sodium, and calcium in the restricted extracellular space using the previous expression were calculated in Refs. 27 and 28.

G. Reconstruction of the Action Potential

Once all ionic currents and changes in ionic concentrations have been formulated, they need to be integrated in the parallel conductance model to produce an action potential. It is not feasible to record all experimental measurements needed for the implementation of the model from the same cell. In general, we have to rely on experimental measurements in different cells, from different laboratories, and sometimes from different species. This means that because the action potential results from an exquisite equilibrium between depolarizing and repolarizing currents, small variations in the formulation of individual currents may lead to big differences between computed and experimental action potentials. Therefore, an important piece of information necessary for the successful completion of the model, in addition to measurements of ionic currents, is experimental measurements of the action potential that we want to simulate.

Often, to reproduce measured action potentials with a computer model, the magnitudes of the ionic currents that have been formulated need to be adjusted. The maximum conductance of the channel is typically the parameter used to adjust the magnitude of each ionic current in a way that important characteristics of the experimental action potential such as maximum rate of depolarization during phase 0, action potential amplitude, action potential duration, and the general shape of the action potential are reproduced by the computer model. The selection of the total magnitude of the current as the parameter to adjust is not capricious. The total magnitude of the currents varies naturally between cells in the same heart, between heart cells of different species, and under different recording conditions, and usually decrease (run down) over time in a particular cell after the cell is patched. This step in the development of an ionic model requires in many cases judicious decisions to resolve conflicting experimental evidence.

H. Numerical Methods: Integration of the Action Potential

To compute the action potential, the integration of the governing differential equation,

$$C_m \left(\frac{dV_m}{dt} \right) = -I_{\text{ion}}$$

where C_m is the specific capacitance (in $\mu F/cm^2$), V_m is the transmembrane voltage (in mV), and I_{ion} is the summation of all ionic currents (in $\mu A/cm^2$ or pA/pF), is necessary. The simplest way to integrate the equation is to use the forward Euler method [25]:

$$\frac{V_m^{t+\Delta t} - V_m^t}{\Delta t} = -\frac{I_{\text{ion}}^t}{C_m}$$

and

$$V_m^{t+\Delta t} = V_m^t - \Delta t \left(\frac{I_{\text{ion}}^t}{C_m} \right)$$

With this expression, from the transmembrane potential (V_m^t) and the total ionic current (I_{ion}^t) at a given time (t) we can calculate the transmembrane potential at a later time ($V_m^{t+\Delta t}$). Δt is the time discretization step. The value of the time discretization step has to be small enough to calculate V_m accurately during rapid changes in the transmembrane potential (i.e., the depolarization phase of the action potential). A value of Δt of 5 or 10 μsec is typically used.

Even though the forward Euler method is a simple and accurate integration method when a small integration time step is used, it is not very efficient. The reason is that even though a small Δt is needed during the depolarization phase of the action potential because V_m changes very fast (phase 0 in Fig. 2B), during the rest of the action potential (phases 1, 2, and 3) a larger Δt could be used because V_m does not change that fast. A number of efficient integration algorithms take advantage of this by using a variable discretization time step [44–46].

V. IONIC CURRENTS DURING THE ACTION POTENTIAL

We discussed earlier that the number of branches in the parallel conductance model depends on the tissue that we want to model, and that, consequently, the first step in formulating an ionic model of the action potential is to decide which currents should be part of the model. In this section we describe the currents that are thought to be important and how they contribute to action potential depolarization and repolarization. Even though the relative contribution of those currents to the action potential may be different for different types of cells (see later), the currents described below are a good starting point in the selection of the ionic currents that are necessary to implement a model of the cardiac action potential.

Figure 9 illustrates how the time course of an action potential (Fig. 9A) relates to the time course of the total ionic membrane current (Fig. 9B) and to the different macroscopic ionic components of the total current (Fig. 9C). The tracings were generated with the Luo-Rudy model [29] of a single cell that is based on experimental measurements of ionic currents in guinea pig myocytes. The negative currents (I_{Na}, $I_{Ca,L}$, $I_{Ca,T}$) are inward currents that depolarize the membrane (i.e., move the membrane potential away from the resting value). The positive currents (I_{Kr}, I_{Ks}, I_{Kl}, I_{Kp}) are outward currents that repolarize the membrane (i.e., move the membrane potential toward the resting value). The Na/Ca exchanger (I_{NaCa}) is positive during the early part of the action potential and negative during the later part. If the cell membrane is depolarized briefly—for example, by using an external current source—to a threshold value (between -50 and -60 mV), the probability of the opening of sodium channels increases. As a result, extracellular sodium ions rush into the cell down their concentration gradient (extracellular sodium ~ 140 mM; intracellular sodium ~ 10 mM) which further depolarizes the membrane and opens more sodium channels, initiating a regenerative process. The sodium current drives the membrane potential to about $+50$ mV The maximum rate of rise of the action potential upstroke (phase 0 in Fig. 9) is related to the strength of the Na current. Sodium channels inactivate very quickly, and when the action potential reaches its maximum amplitude most of the sodium channels are already inactivated, which decreases the sodium current to zero (Fig. 9C).

During the plateau of the action potential (phases 1 and 2 in Fig. 9A), the total membrane ionic current is small but positive (Fig. 9B), indicating a delicate balance between inward (depolarizing) and outward (repolarizing) currents favoring the repolarizing direction. The main depolarizing currents during the plateau are the calcium currents, which cross the membrane through two different types of calcium channels. The long-lasting calcium current (or L-type calcium current), $I_{Ca,L}$, is activated at potentials more positive than -40 mV and inactivates slowly (~ 100 msec). The primary role of this current is to allow the entry of Ca into the cell that is a signal to the sarcoplasmic reticulum to release its Ca stores (see earlier). The increase in intracellular Ca triggers cell contraction. Two mechanisms are in place to restore the low intracellular Ca concentration which normally occurs in heart cells: an ATP-dependent Ca pump moves Ca ions back to the

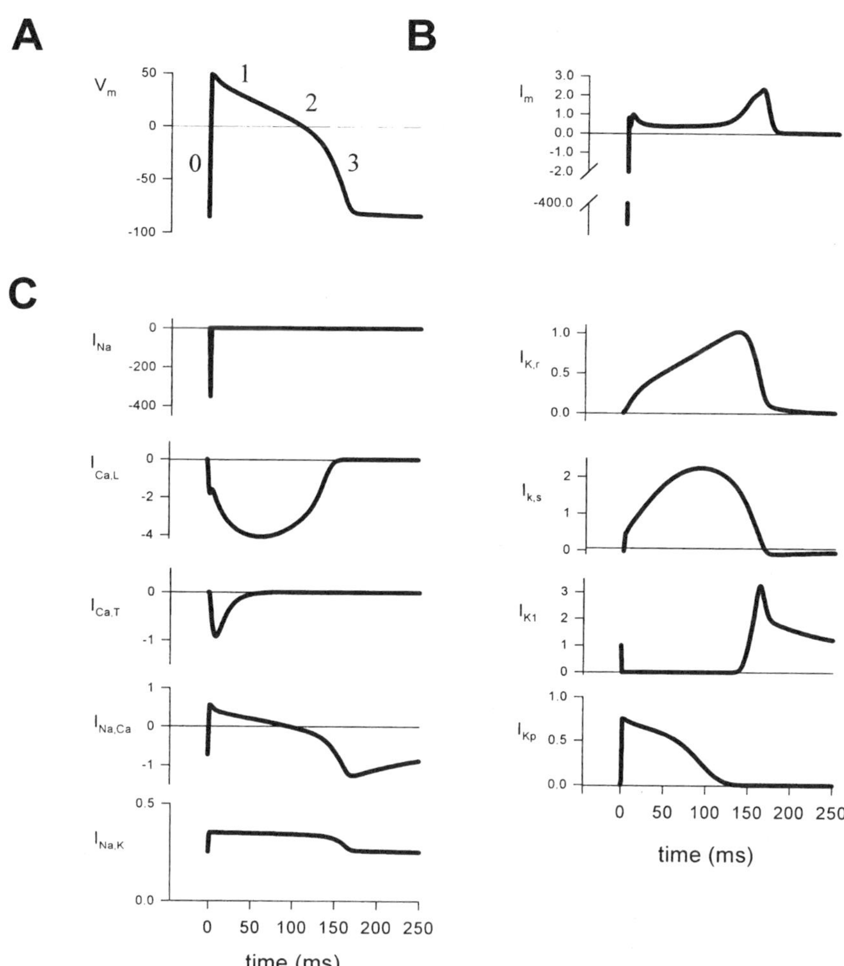

Figure 9 Ionic currents during the action potential. (A) Action potential calculated with the Luo-Rudy model. (B) Total transmembrane current during the action potential in pA/pF. (C) Depolarizing (left) and repolarizing (right) ionic currents. All currents are in pA/pF. Time scale at the bottom.

sarcoplasmic reticulum, and an electrogenic Na–Ca exchange mechanism on the cell membrane. The Na–Ca exchange mechanism moves into the cell three Na ions for each Ca ion extruded from the cell. This results in a net current which is indicated as I_{NaCa} in Fig. 9C. The transient calcium current (or T-type calcium current), $I_{Ca,T}$, activates at potentials more positive than

-70 mV and inactivates rapidly. A potassium current (I_{Kp}) which is rapidly activating and which does not inactivate is important for the rapid repolarization at the beginning of the plateau of the action potential (phase 1). The main repolarizing currents, the delayed rectifier potassium currents, are activated slowly during the plateau of the action potential. In several species the delayed rectifier potassium current consist of two components: I_{Kr} (rapidly activating) and I_{Ks} (slowly activating).

A progressive decrease of the calcium currents with time in combination with a progressive increase of the delayed rectifier potassium currents results in a more positive total ionic membrane current (phase 3 in Fig. 9B), which terminates the plateau and initiates the repolarization of the action potential (phase 3 in Fig. 9A). As repolarization progresses, the inward rectifier potassium current (I_{K1}), which was closed during the plateau phase, is activated and further accelerates the rate of repolarization of the membrane back to its resting membrane potential. Because of the activation of the sodium and potassium currents with each action potential, the cell gains Na ions and loses K ions. To maintain ion gradients between the cell interior and exterior, an ATP-dependent Na–K exchange mechanism extrudes three Na ions for every two K ions moved into the cell; this results in a net outward current which also contributes to the final repolarization (I_{NaK} in Fig. 9C).

VI. CHARACTERISTICS OF THE ACTION POTENTIAL: VALIDATION OF A COMPUTER MODEL

The action potential in Fig. 9 was calculated for a single cell. However, in many situations the model of the action potential will be used to study propagation of cardiac waves or the interaction of electrical stimulus with propagating waves. Therefore, it is important to assess how accurate is the model of the action potential that results from the integration of the different ionic currents during propagation. In what follows, we describe well-established characteristics of the cardiac action potential and how we can validate the computer model by comparing the results of computer simulations to experimental results under similar conditions. As an example, we will use the DiFrancesco-Noble model of a Purkinje fiber [27], but similar validations could be done for any model of the action potential. The action potential characteristics described later relate to the response of the membrane to an electrical stimulus at different phases of the action potential. Studies on electrical excitability often involve decisions on whether a stimulus causes a propagated action potential or not. Therefore, to compare the membrane model to experiments we used a propagation model. To learn how to connect several cardiac cells to implement a propagation model, the reader is referred to later chapters. Validating a membrane model in a context of action potential propagation adds robustness to the test because a propagating action potential results from time and space dependencies that are not directly a part of ionic current measurements (i.e., voltage clamp) from which the membrane model is formulated.

A. Strength–Duration Curve

Whether the application of an electrical stimulus results in an action potential or not depends on the combination of two parameters: strength and duration. Strength–duration curves describe graphically which combination of parameters result in the excitation of the cell. Figure 10 shows the strength–duration curve of a membrane modeled with the DiFrancesco-Noble model [27]. The combination of strength–duration values above the curve result in the excitation of the cell, and the combination of values below the curve do not. Clearly, shorter pulses require higher strengths for stimulation. However, long pulse duration does not guarantee cell activation. A pulse with strength less than a critical value known as rheobase and an infinite duration would not result in cell activation. Another important parameter of the strength–duration curve is the chronaxie, defined as the minimum duration of a stimulus with a strength twice rheobase to excite the cell.

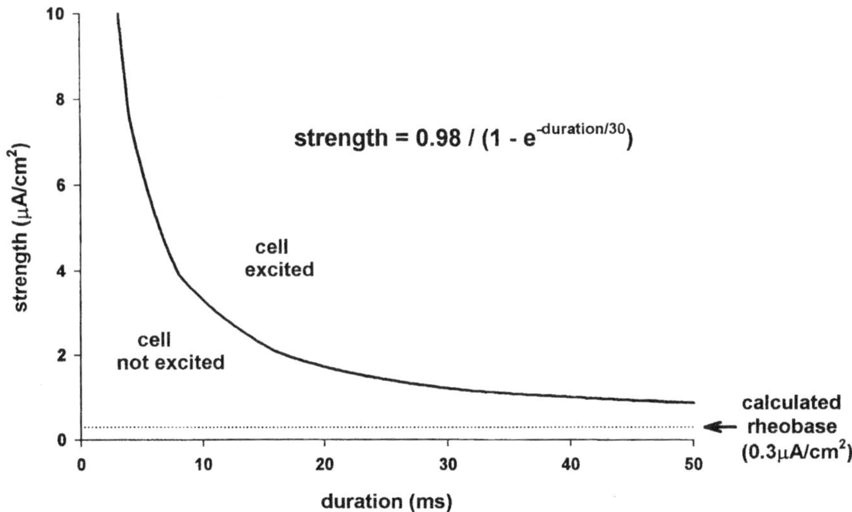

Figure 10 Strength–duration curve of the DiFranscesco-Noble model. The combination of the strength and the duration of a stimulus determines whether the cell is excited or not.

It is possible to find mathematical expressions for the strength–duration curve, rheobase, and chronaxie. At the resting membrane potential, the equivalent electrical circuit of a cell membrane is a resistor in parallel with a capacitance (see Section III.B). If the value of the membrane resistance were R_m and the membrane capacitance C_m, the change in membrane voltage (V_m) resulting from the application of a stimulus current of strength I and infinite duration would be:

$$\Delta V_m = I R_m (1 - e^{-t/\tau})$$

where $\tau = R_m C_m$.

To be able to excite a cell, a current stimulus of strength I and duration T should be able to change the transmembrane voltage to the threshold for excitation, V_{thr}, at the end of its duration T. Therefore the equation for the strength–duration curve is

$$I = \frac{V_{\text{thr}}/R_m}{1 - e^{-T/\tau}}$$

The rheobase would be the value of I when $T \to \infty$, that is, (V_{thr}/R_m), and the chronaxie Tc could be derived from the equation

$$2\left(\frac{V_{\text{thr}}}{R_m}\right) = \frac{V_{\text{thr}}/R_m}{1 - e^{-\text{Tc}/\tau}}$$

Therefore $e^{-\text{Tc}/\tau} = 0.5$, from which we can obtain Tc = 0.693τ.

The strength–duration curve computed with the DiFrancesco-Noble model is shown in Fig. 10. The time constant of the membrane can be estimated by curve-fitting the calculated strength–duration curve ($\tau = 30$ msec, Fig. 10). Since $C_m = 1\,\mu\text{F}/\text{cm}^2$, the resistance of the membrane can be estimated as $R_m = 30\,\text{k}\Omega\,\text{cm}^2$. Once we obtain the membrane resistance, it is possible estimate the excitation threshold, V_{thr}, which is about 30 mV ($0.98\,\mu\text{A}/\text{cm}^2$ times $30\,\text{k}\Omega\,\text{cm}^2$). It is important to note, though, that the rheobase estimated from the strength–duration curve ($0.98\,\mu\text{A}/\text{cm}^2$) is larger than the rheobase obtained directly by using a stimulus with a very long duration ($0.3\,\mu\text{A}/\text{cm}^2$).

B. Strength–Interval Curve: Refractory Period and Supernormal Period

We discussed earlier that the strength and duration of an electrical stimulus are important factors that determine whether a cell will be excited and generate an action potential or not. However, there is another factor to consider: cell excitability. Cell excitability reflects the current condition of the cell and it is a measure of how easy it is to generate an action potential. The excitability of the cell changes during the action potential: when a cell is excited and depolarizes, the excitability decreases and then gradually recovers as the cell repolarizes.

A way to measure changes in excitability is by constructing strength–interval curves like the ones illustrated in Fig. 11 [47]. The figure shows an action potential with the corresponding strength–interval curve in an experiment (Fig. 11A), and in a computer model (Fig. 11B). Strength–interval curves represent the stimulus threshold (i.e., the strength of a stimulus with a fixed duration that is necessary to generate an action potential) at different time intervals after a previous action potential has been generated. That the excitability of the cell changes during the action potential is indicated by the "L" shape of the strength–interval curves. After the cell has repolarized, the strength–interval curve is almost flat, indicating that the stimulation threshold (and the excitability) is almost constant for example, between times 400 and 500 msec in the figure). As the time interval is decreased, the stimulation threshold first decreases during the supernormal period, and then sharply increases. The sharp increase in stimulation threshold indicates a sharp decrease in cell excitability (an action potential is more difficult to

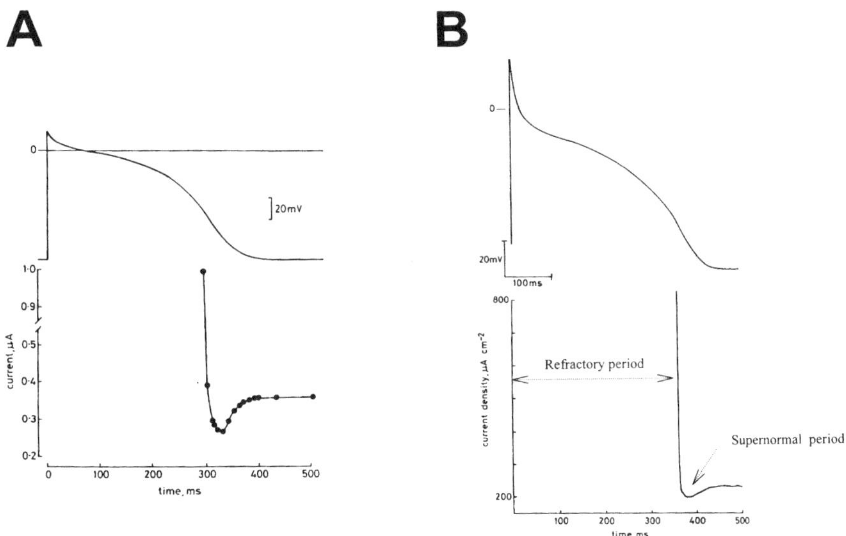

Figure 11   Action potential (top) and corresponding strength–interval curve of a Purkinje fiber. (A) Experiment. (B) Computer model.

generate). At even shorter time intervals the stimulus threshold is so high that an action potential cannot be generated with any stimulus strength. Therefore, after a cardiac cell has depolarized, here is a period in which the cell cannot be reexcited. This time period is called the refractory period.

To generate an action potential, the cell membrane potential has to be depolarized to a threshold voltage at which the sodium channels activate (see earlier). The supernormal period occurs when the cell is repolarizing but not fully repolarized. Cell excitability is improved during this period because the difference between the membrane voltage and the threshold voltage is smaller (and therefore less current is necessary to achieve the voltage threshold) than when the cell is fully repolarized. The period of super-normality is not present in all types of cardiac cells.

C. Graded Response to Current Strength

Figure 12 shows the response of the membrane when a depolarizing current is applied at a fixed time interval during late repolarization, with fixed duration (1 msec) but with different strengths in an experiment (Fig. 12A) and in a computer model (Fig. 12B) of a Purkinje fiber [47]. In the experiment the action potentials were measured at the stimulation site (Fig. 12A,

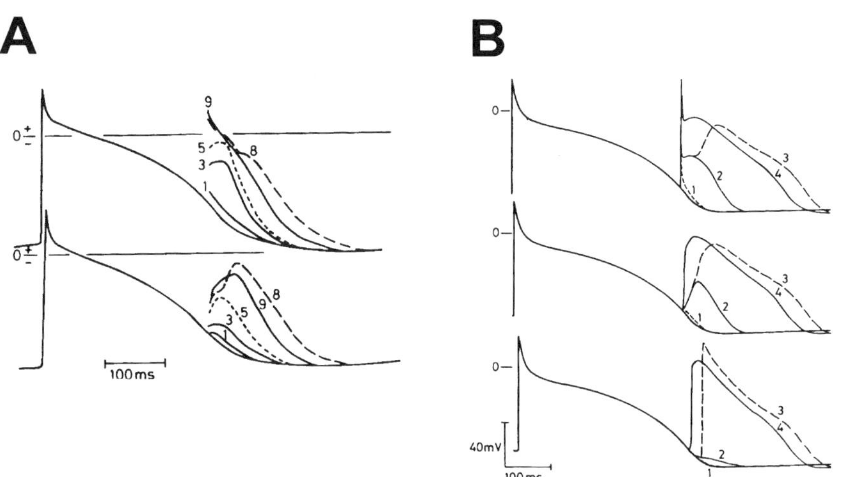

Figure 12 Response of the membrane as depolarizing intracellular current pulses of fixed duration and different strength are applied at a fixed time during the re-polarization of the action potential. (A) Experiment. (B) Computer model. See text for explanation. (Reproduced with permission from Ref. 47.)

top) and 2 mm away from the stimulation site (Fig. 12A, bottom). In the computer model the action potentials were calculated at the stimulation site (Fig. 12B, top), 2 mm (Fig. 12B, middle), and 7 mm away from the stimulation site (Fig. 12B, bottom). The weaker stimuli (1, Fig. 12A; 1 and 2 in Fig. 12B) cause local responses at the stimulation site which appear decremented at distant sites (decremental conduction, no propagated action potential). The stronger stimuli (8 and 9 in Fig. 12A; 3 and 4 in Fig. 12B) cause local responses at the site of the stimulus that appear incremented at distant sites (incremental conduction, there is a propagated action potential). When the strength of the stimulus is just sufficient to cause a propagated response (8 in Fig. 12A, 3 in Fig. 12B), there is a long interval between the stimulus and the response at the distant sites. That latency allows extra time for nodes far away from the stimulus to recover its excitability. For stronger stimuli (9 in Fig. 12A; 4 in Fig. 12B), there is less latency between the stimulus and the response at distant sites. As a result, response 9 precedes response 8 in Fig. 12A, and response 4 precedes response 3 in Fig. 12B. On the other hand, for stronger stimuli, because there is less latency, the tissue has less time to recover and at distant sites the response has lower amplitude and shorter duration than responses to weaker (but still suprathreshold) stimuli (compare responses 3 and 4 at the bottom of Fig. 12B).

Note that the peak in response 3 at the site of stimulation (top, Fig. 12B) and far away from the site of stimulation (bottom, Fig. 12B) happen later in time than response 3 at a site closer to the site of stimulation (middle, Fig. 12B). In this example, the propagated response was originated 2.9 mm from the stimulation site. This phenomenon of a propagated response originating some distance away from the stimulation site has also been observed experimentally.

D. Stimulation Frequency and the Action Potential Duration and Amplitude

The amplitude of the action potential is the difference between the resting membrane potential and the maximum voltage reached during depolarization. The action potential duration is usually defined as the difference between the time the cell depolarizes and the time the cell repolarizes from its peak to 90% of its amplitude. The duration and the amplitude of the action potential are not constant but decrease as the stimulation frequency increases. Figure 13 shows the relationship between action potential amplitude and duration in an experiment (Fig. 13A) and in a computer model (Fig. 13B) [47]. The action potentials shown in the inset in Fig. 13B are calculated with frequencies of 1, 3.3, and 6 Hz, which is the maximum frequency at which the fiber could be stimulated. Even though the relationships shown in

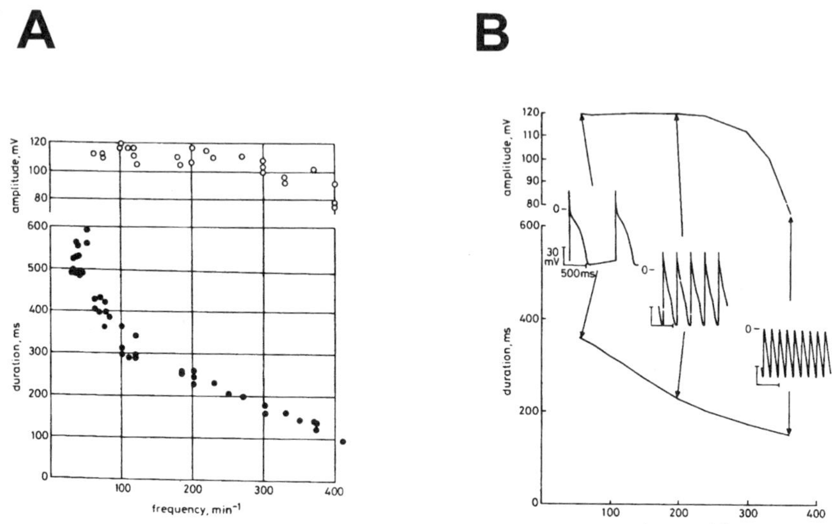

Figure 13 Changes in amplitude and duration of the action potential of a Purkinje fiber as a function of the stimulation frequency. (A) Experiment. (B) Computer model. (Reproduced with permission from Ref. 47.)

Fig. 13 are from Purkinje fibers, similar relationships have been measured (and computed) in other types of cardiac cells. The cardiac restitution curve is commonly used to describe the relationship between action potential duration and frequency of stimulation. In the cardiac restitution curve, the action potential duration is plotted against the basic cycle length of the train of stimuli, which is the reciprocal of the stimulation frequency. Accurate simulation of the response of cells to stimulation frequency is important because parameters extracted from that relationship (slope of the restitution curve) has been correlated with the initiation of arrhythmias [48].

E. Stimulation During the Plateau of the Action Potential

Stimulation during the refractory period of a cell does not generate an action potential (see earlier). However, this does not mean that stimulation during the refractory period does not have an effect on the action potential. Figure 14 shows the effects of cathodal (depolarizing) and anodal (hyperpolarizing) current stimulation during the plateau phase of the action potential in experiments (Fig. 14A) and in a computer model (Fig. 14B) [47]. In the computer model, depolarizing currents shortened the action potential,

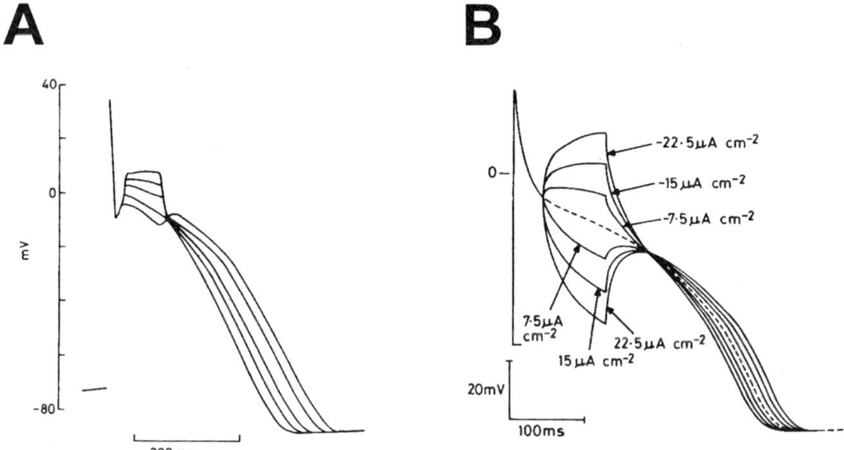

Figure 14   Effect of stimulating a Purkinje fiber during the plateau phase (phase 2) of the action potential with depolarizing and hyperpolarizing intracellular current pulses. (A) Experiment. (B) Computer model. The broken line corresponds to an action potential in the absence of a stimulating current after depolarization. (Reproduced with permission from Ref. 47.)

and hyperpolarizing currents lengthened the action potential as it occurred in the experimental preparations.

F.   All-or-None Repolarization

Stimulation currents that hyperpolarize the cell membrane during the plateau phase of the action potential cause repolarization of the cell during the time that the stimulus is applied. After cessation of the stimulus, the action potential returns to its original course, except for a lengthening, as we discussed earlier. If the current that causes hyperpolarization is strong enough, the action potential does not return to its original course and the cell becomes fully repolarized. This phenomenon is called all-or-none repolarization and is illustrated in Fig. 15 for an experimental preparation of a Purkinje fiber (Fig. 15A), and in a computer model (Fig. 15B) [47]. All-or-none repolarization has also been demonstrated in ventricular muscle cells. Accurate simulation of changes of the time course of repolarization (and therefore of the refractory period) by an electrical stimulus are important in studies of the interaction of electrical stimulus with reentrant arrhythmias or fibrillation because at any given time it is possible to find cells in all phases of the action potential.

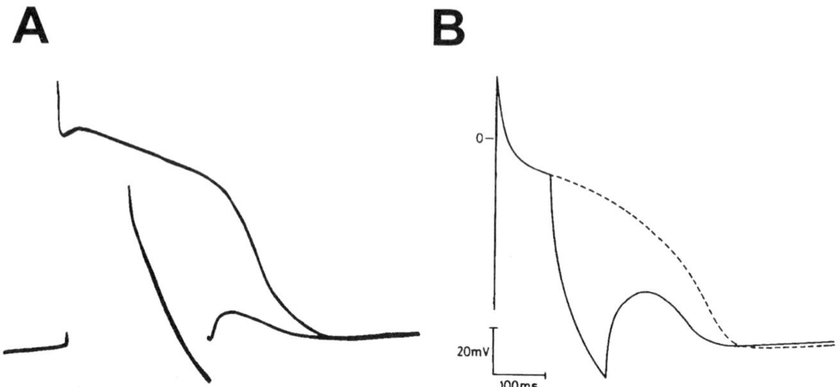

Figure 15   All-or-none repolarization. (A) Experiment. (B) Computer model. The broken line corresponds to an action potential in the absence of a stimulating current after depolarization. (Reproduced with permission from Ref. 47.)

VII.   IONIC MODELS OF THE ACTION POTENTIAL

In this chapter we have used two computer ionic models, the Luo-Rudy model of the ventricular action potential [29] and the DiFrancesco-Noble model [27] of the Purkinje fiber. But before those models were available, the classical ionic models of a Purkinje fiber by Noble [16] and McAllister et al. [49], and the model of a ventricular cell by Beeler and Reuter [50] had been used extensively to increase our understanding of cardiac electrophysiology and had proven the usefulness of computer ionic models. More recently, ionic models of the sinus node [51–54], atrial cells [34,55–58], and ventricular muscle cells [43,59] have been published. The abundance of ionic models reflects the fact that the strength and relative contribution of the different ionic currents to the action potential varies with species and cell type in a normal heart. For example, the sodium current in cells from the sinus or atrioventricular nodes is much smaller than in cells from the atria, the Purkinje, system, or the ventricle. On the other hand, cells in the nodes as well as in the Purkinje system have the property of generating action potentials in the absence of an external stimulus that results from the acti-vation of an ionic (pacemaker) current at diastolic potentials which is not normally active in ventricular cells. Within the ventricle, the relative contribution of the ionic currents to the action potential vary between the endocardium, mid-myocardium, and epicardium [60], sub- and mid-myo-cardial cells have a longer action potential duration than epicardial cells.

Therefore depending on the tissue we want to simulate, the appropriate model should be selected.

The function of a number of ionic channels is modified by acquired diseases such as heart failure, myocardial infarction, and atrial fibrillation, and by genetic diseases such as long QT syndrome. To simulate disease the formulation of one or more ionic channels needs to be modified to reflect the changes in function that have been measured experimentally. Ionic models of heart failure [43,59], and of atrial cells after being remodeled by atrial fibrillation [61], have been published. Sodium and potassium channels have been modified in the Luo-Rudy model [29] to study the effects on the action potential of genetic mutations that cause long QT syndrome [62,63].

VIII.   SIMPLIFIED MODELS OF THE ACTION POTENTIAL

Ionic models of the action potential intend to formulate mathematically the cellular processes that lead to the generation of the action potential and that have been measured experimentally. The action potential is then the result of the interaction of all those processes. While modeling those processes is desirable, integration of all differential equations describing those processes is computationally expensive. Simplified models of the action potential whose goal is to simulate not the cellular processes but the dynamics of the action potential [2–4] have been used extensively in the study of the dynamics of propagation of cardiac waves.

The first model of an action potential, proposed by Wiener and Rosenblueth in their 1946 classic paper [2], was one of those simplified models. In their model, a cell could be in one of three states: active, refractory, or resting. Those states represent the different phases that excitable cells go through during the course of the action potential: depolarization (active state) and subsequent repolarization (refractory state) to the resting membrane potential (resting state). The transitions between those states were governed by a set of rules (laws of conduction) such as: (1) if a cell is at its resting state, it will remain in that state until a wave front passes by; (2) if a cell in the resting state is in contact with a cell in an active state, there is a wave front moving from the active cell to the resting cell with a certain velocity, and the resting cell will become active; (3) after the cell is active, it will remain in the refractory state for a constant time interval (the refractory period). The Wiener-Rosenblueth model did not incorporate biophysical detail on the ionic mechanisms underlying the action potential. However, despite its simplicity, it provided tremendous insight on how electrical waves propagate in cardiac muscle. This model was the first in a class of action potential models known as cellular automata models.

A different approach to simplify a model the action potential was pioneered by FitzHugh [3] and Nagumo et al. [4]. The FitzHugh-Nagumo model could be thought of as having two currents: one depolarizing current and one repolarizing current. Despite the simplified representation of the action potential, this model has been a useful tool in our understanding of the dynamics of propagation of cardiac waves and has been used extensively.

IX. SUMMARY AND CONCLUSIONS

In this chapter we have discussed how ionic models of the action potential are developed and how they can be evaluated by comparing the results of computer simulations to experimental results. We have also described a number of characteristic responses of the action potential to electrical stimulation. When selecting a model to address a specific question, it is important to understand that there is no perfect model that can be used in all situations because the ionic currents that contribute to the action potential vary with cell type and with disease. Finally, it is important to remember that a computer model is as good as the experimental data that has been used for its formulation. And regardless of how good a computer model may be today, it will be obsolete tomorrow. Therefore, computer ionic models have to be modified often to incorporate new experimental results.

REFERENCES

1. Hodgkin AL, Huxley AF. A quantitative description of membrane current and its application to conduction and excitation in nerve. J Physiol (Lond) 117:500–544, 1952.
2. Wiener N, Rosenblueth A. The mathematical formulation of the problem of conduction of impulses in a network of connected excitable elements, specifically in cardiac muscle. Arch Inst Cardiol Med 16:205–265, 1946.
3. FitzHugh R. Impulses and physiological states in theoretical models of nerve membrane. Biophys J 1:445–466, 1961.
4. Nagumo J, Arimoto S, Yoshizawa S. An active pulse transmission line simulating nerve axon. Proc. IRE 50:2061–2070, 1962.
5. Bowditch, H. P. Über die Eigenthümlichkeiten der Reizbarkeit welche die Muskelfasern der Herzens zeigen. In: Arbeiten aus der Physiol. Anstalt in Leipzig, mitgeilt durch Ludwig. C. 6. Jg. 1871 [Ber Sachs Ges (Akad) Wiss 23:625–689, 1871].

6. Piccolino M. Animal electricity and the birth of electrophysiology: the legacy of Luigi Galvani. Brain Res Bull 46(5):381–407, 1998.

7. Hodgkin AL, Huxley AF. Action potentials recorded from inside a nerve fiber. Nature (Lond) 144:710–711, 1939.

8. Hodgkin AL, Huxley AF. Resting and action potentials in nerve fibers. J Physiol (Lond) 104:176–195, 1945.

9. Curtis HJ, Cole KS. Membrane action potentials from the squid axon. J Cell Comp Physiol 15:147–157, 1940.

10. Curtis HJ, Cole KS. Membrane resting potential and action potentials from the squid axon. J Cell Comp Physiol 19:135–144, 1942.

11. Ling G., Gerard RW. The normal membrane potential of frog sartorius fibers. J Cell Comp Physiol 34:383–396, 1949.

12. Nastuk WL, Hodgkin AL. The electrical activity of single muscle fibers. J Cell Comp Physiol 35:39, 1950.

13. Coraboeuf E, Weidmann S. C R Soc Biol (Paris) 143:1329, 1949.

14. Draper MH, Weidmann S. Cardiac resting and action potentials recorded with an intracellular electrode. J Physiol 115:74–94, 1951.

15. Woodbury LA, Woodbury JW, Hecht HH. Membrane resting and action potential of single cardiac muscle fibers. Circulation 1:264–266, 1950.

16. Noble D. A modification of the Hodgkin-Huxley equations applicable to Purkinje fiber action and pacemaker potentials. J Physiol (Lond) 160:317–352, 1962.

17. Bernstein J. Untersuchungen zur Termodynamik der bioelektrischen Strome. Pflugers Arch ges Physiol 92:521–562, 1902.

18. Bernstein J. Elektrobiologie. Viewag, Braunschweig, 1912.

19. Nernst W. Zur kinetic der in Losung befindlichen Korper: Theorie der Diffusion. Z Phys Chem 2:613–637, 1888.

20. Hermann L. Beitrage zur Physiologie und Physik des Nerven. Pflugers Arch Ges Physiol 109:95–144, 1905.

21. Fricke H. The electrical capacity of suspensions with special reference to blood. J Gen Physiol 9:137–152, 1925.

22. Curtis HJ, Cole KS. Transverse electrical impedance of the squid giant axon. J Gen Physiol 21:757–765, 1938.

23. Hodgkin AL. The subthreshold potentials in a crustacean nerve fiber. Proc R Soc Lond B 126:87, 1939.

24. Hille B. Ionic Channels of Excitable Membranes. 2nd ed. Sunderland, MA: Sinauer Associates, 1992.

25. Press WH, Flannery BP, Teukolski SA, Vetterling WT. Numerical Recipes: The Art of Scientific Computing. Cambridge University Press, 1986.

26. Patlak J. Molecular kinetics of voltage-dependent Na^+ channels. Physiol Rev 71:1047–1080, 1991.

27. DiFrancesco D, Noble D. A model of cardiac electrical activity incorporating ionic pumps and concentration changes. Phil Trans R Soc Lond B Biol S 307:353–398, 1985.

28. Rasmusson RL, Clark JW, Giles WR, Shibata EF, Campbell DL. A mathematical model of a bullfrog cardiac pacemaker cell. Am J Physiol 259:H352–H369, 1990.

29. Luo CH, Rudy Y. A dynamic model of the cardiac ventricular action potential. I. Simulations of ionic currents and concentrations changes. Circ Res 74:1071–1096, 1994.

30. Mullins LJ. Ion Transport in the Heart. New York: Raven, 1981.

31. LeGrice IJ, Smaill BH, Chai LZ, Edgar SG, Gavin JB, Hunter PJ. Laminar structure of the heart: ventricular myocyte arrangement and connective tissue architecture in the dog. Am J Physiol 269:H571–H583, 1995.

32. Schouten VJA, van Deen JK, de Tombe P, Verveen AA. Force-interval relationship in heart muscle of mammals: a calcium compartment model. Biophys J 51:13–26, 1987.

33. Nordin C. Computer model of membrane current and intracellular Ca^{2+} flux in the isolated guinea pig ventricular myocyte. Am J Physiol 265:H2117–H2136, 1993.

34. Tang Y, Othmer HG. A model of calcium dynamics in cardiac myocytes based on the kinetics of ryanodine-sensitive calcium channels. Biophys J 67:2223–2235, 1994.

35. Dupont G, Pontes J, Goldbeter A. Modeling spiral Ca^{2+} waves in single cardiac cells: role of spatial heterogeneity created by the nucleus. Am J Physiol 271:C1390–C1399, 1996.

36. Lindblad DS, Murphey CR, Clark JW, Giles WR. A model of the action potential and underlying membrane currents in a rabbit atrial cell. Am J Physiol 271:H1666–H1696.

37. Jafri S, Rice JJ, Winslow RL. Cardiac Ca^{2+} dynamics: The roles of ryanodine receptor adaptation and sarcoplasmic reticulum load. Biophys J 74:1149–1168, 1998.

38. Beuckelmann DJ, Wier WG. Mechanism of release of calcium from sarcoplasmic reticulum of guinea pig cardiac cells. J Physiol (Lond) 405:233–255, 1988.

39. Isenberg G. Cardiac excitation-contraction coupling: from global to microscopic models. In: Physiology and Pathology of the Heart. 3rd ed. Developments in Cardiovascular Medicine, Vol. 151. Boston: Kluwer, 1995, pp 289–307.

40. Keizer J, Levine L. Ryanodine receptor adaptation and Ca^{2+} induced Ca^{2+} release-dependent Ca^{2+} oscillations. Biophys J 71:3447–3487, 1996.

41. Wagner J, Keizer J. Effects of rapid buffers on Ca^{2+} diffusion and Ca^{2+} oscillations. Biophys J 67:447–456, 1994.

42. Wier WG, Cannell MB, Berlin JR, Marban E, Lederer WJ. Cellular and subcellular heterogeneity of $[Ca]_i$ in single heart cells received by fura-2. Science 235:325–328, 1987.

43. Winslow RL, Rice J, Jafri S, Marban E, O'Rourke B. Mechanisms of altered excitation-contraction coupling in canine tachycardia-induced heart failure. II. Models studies. Circ Res 84:571–586, 1999.

44. Moore JW, Ramon F. On numerical integration of the Hodgkin and Huxley equations for a membrane action potential. J Theor Biol 45:249–273, 1974.

45. Rush S, Larsen H. A particular algorithm for solving dynamic membrane equations. IEEE Trans Biomed. Eng 25:389–392, 1978.

46. Vitorri B, Vinet A, Roberge FA, Drouhard JP. Numerical integration in the reconstruction of cardiac action potentials using Hodgkin-Huxley type models. Comput Biomed Res 18:10–23, 1985.

47. Cabo C, Barr RC. Propagation model using the DiFrancesco-Noble equations: comparison to reported experimental results. Med Biol Eng Comput 30(3):292–302, 1992.

48. Garfinkel A, Kim YH, Voroshilovsky O, Qu Z, Kil JR, Lee MH, Karagueuzian HS, Weiss JN, Chen PS. Preventing ventricular fibrillation by flattening cardiac restitution. Proc Natl Acad Sci USA 97(11):6061–6066, 2000.

49. McAllister RE, Noble D, Tsien RW. Reconstruction of the electrical activity of cardiac Purkinje fibers. J Physiol (Lond) 251:1–59, 1975.

50. Beeler GW, Reuter H. Reconstruction of the action potential of ventricular myocardial fibers. J Physiol (Lond) 268:177–2l0, 1977.

51. Yanagihara K, Noma A, Irisawa H. Reconstruction of sino-atrial node pacemaker potential based on the voltage clamp experiments. Jpn J Physiol 30:841–857, 1980.

52. Bristow D, Clark JW. A mathematical model of primary pacemaking cell in S-A node of the heart. Am J Physiol 243:H207–H218, 1982.

53. Bristow D, Clark JW. A mathematical model of the vagally driven primary pacemaker. Am J Physiol 244:H150–161, 1983.

54. Noble D, Noble S. A model of S-A node electrical activity using a modification of the DiFranceseo-Noble (1984) equations. Proc R Soc Lond B Biol Sci 222:295–304, 1984.

55. Rasmusson RI, Clark JW, Giles WR, Robinson K, Clark RB, Shibata EF, Campbell DL. A mathematical model of electrophysiological activity in a bullfrog atrial cell. Am J Physiol 259:H370–H389, 1990.

56. Nygren A, Fiset C, Firek U, Clark JW, Linblad DS, Clark RB, Giles WR. Mathematical model of an adult human atrial cell. The role of K^+ currents in repolarization. Circ Res 82:63–81, 1998.

57. Courtemanehe M, Ramirez RJ, Nattel S. Ionic mechanisms underlying human atrial action potential properties: insights from a mathematical model. Am J Physiol 275:H301–321, 1998.

58. Ramirez RJ, Nattel S, Courtemanche M. Mathematical analysis of canine atrial action potentials: rate, regional factors, and electrical remodeling. Am J Physiol 279:H1767–H1785, 2000.

59. Priebe L, Beuckelmann DJ. Simulation study of cellular electric properties in heart failure. Circ Res 82:1206–1223, 1998.

60. Liu DW, Antzelevitch C. Characteristics of the delayed rectifier current (IKr and IKS) in canine ventricular epicardial, midmyocardial, and endocardial myocytes. A weaker IKs contributes to the longer action potential of the M cell. Circ Res 76:351–365, 1995.

61. Courtemanehe M, Ramirez RJ, Nattel S. Ionic targets for drug therapy and atrial fibrillation-induced electrical remodeling: insights from a mathematical model. Cardiovasc Res 42:477–489, 1999.
62. Viswanathan PC, Rudy Y. Pause induced early afterdepolarizations in the long QT syndrome: a simulation study. Cardiovasc Res 42:530–542, 1999.
63. Wehrens XH, Abriel H, Cabo C, Benhorin J, Kass KS. Arrhythmogenic mechanism of an LQT-3 mutation of the human heart Na(+) channel alpha-subunit: a computational analysis. Circulation 102:584–590, 2000.

3

Modeling the Impact of Cardiac Tissue Structure on Current Flow and Wavefront Propagation

Craig S. Henriquez and Joseph V. Tranquillo
Duke University, Durham, North Carolina, U.S.A.

I. INTRODUCTION

For a multicellular organism to function there must be communication between the cells. The more complex the organism, the greater is the need for specialized tissue to mediate this communication. The communication of electrical information is achieved through the propagation of impulses along the cell membrane of excitable cells. Cardiac electrophysiologists are generally interested in understanding the underlying mechanisms of excitability of cardiac muscle under both normal and abnormal conditions and how the cell-to-cell communication can be affected or modulated through some external means. For example, coronary artery disease may result in ischemia in which regions of myocytes may be transiently or permanently deprived of oxygen or nutrients, degrading both the excitability (the ability to create impulses) and the cellular coupling (the ability for current to flow from cell to cell). Effective treatments of such a condition need be considerate of the consequences of ischemia on both the membrane and the tissue structure.

The basic mechanisms of cellular excitability are typically investigated from a single cell or a small patch of membrane. An excitable cell, however, rarely exists in isolation. Instead, the cell operates as part of a network of similar cells that determine the function of the tissue. The inherent complexity of these networks makes interpretation of isolated cell activities very difficult. Coupling affects the membrane and the membrane alters the impact of coupling. The goal of this chapter is to introduce electrophysiological fundamentals that form the basis of different models of multicellular cardiac tissue and to explore the relationship of the intrinsic membrane activity to the potential fields and current flows that are generated inside and outside the tissue.

II. MULTICELLULAR CARDIAC TISSUE

Cardiac muscle cells, myocytes, communicate primarily via electrical impulses. The cells are typically 30–100 μm long and 8–20 μm wide, with an irregular staircase shape. The bounding membranes of adjacent cells are separated by narrow clefts of interstitial space, except at the point called the *nexus*, where the two membranes join. The nexus connects the intracellular compartments of the cells via connexon protein channels and nexa occur predominantly at the ends of cells and to a lesser extent along the length of cells [1]. Saffitz et al. showed that approximately 70% of the gap junctions embedded in the cell membrane were at the intercalated disks in the longitudinal direction [2]. The remaining 30% formed the transverse connections [2]. Furthermore, it has been shown that this alignment with the intercalated disks occurs late in the maturation process—as late as sexual maturity in rats and early childhood in humans [3].

A typical myocyte is connected to 10 neighboring cells [1] via gap junctions. The membranes of the gap junction or nexus are separated by 2–3 nm. Diffraction and biochemical analysis show that gap junctions are composed of connexons. Each connexon has a central diameter of 2–3 nm and is a hexameric structure with six subunits consisting of a single connexin molecule [4,5]. The type of connexin varies throughout the heart, with connexin43 (Cx43) being most prevalent in the ventricle, Cx45 and Cx40 in the sino-atrial (SA) and atrio-ventral (AV) nodes, and atrial and bundle branch junctions containing all three major cardiac connexins [6,7].

Because of the low conductivity of the cell membrane and the spatial organization of the connexons, current flows more readily in the direction parallel to the longer cell axis (the *longitudinal* direction) than it does in the perpendicular (*transverse*) direction. This organization gives rise to anisotropic electrical properties in which the average conductivity is greater longitudinally than transversely [8]. Under pathological conditions, increases in pH or cytoplasmic Ca^{2+} lead to closure of the gap junctions and thus an increased resistance. Following an infarct, gap junction expression is decreased and the normal, highly ordered organization at the intercalated disk becomes chaotic [9]. Several lipophilic compounds, such as octanol and heptanol, act to block electrical conduction by altering the gap junction resistance.

The myocyte shape, cellular connections, and supporting structures give cardiac tissue the appearance of being composed of fibers. While actual fibers do not exist, it is possible to define an average myocyte direction at each point. This average direction can be interpreted as the local fiber orientation. The work of pioneering anatomists showed that the heart ventricle is an assembly of discrete muscle layers, arranged in nested layers or discrete fiber bundles. The first quantitative analysis of fiber orientation by Streeter [10] and Bassett [11] revealed that there is a smooth transmural variation in fiber angle and that the myocardium is well connected or syncytial throughout. In a more recent quantitative analysis, Le Grice et al. showed that a true syncytial myocardium is not an accurate characterization [12]. Rather, the heart is composed of discrete layers of fibers called sheets. Sheets are approximately four to five cells thick, with neighboring layers of sheets branching into each other. The sheets are surrounded by collagenous connective tissue and the arrangement varies as a function of position in the ventricle. Le Grice et al. found that the sheets lie radial to the ventricular surfaces, though they become almost tangential to the epicardial surface. In this discrete view, the edges formed by cutting the wall tangential to the epicardium define the fiber orientation. Across the ventricular wall, the fibers rotate 120° and the axis of preferred current flow is subject to a corresponding variation [12].

III. ELECTROPHYSIOLOGICAL SOURCES OF THE EXCITABLE CELL

The intrinsic membrane activity produces potentials and currents that can be detected by both intracellular and extracellular electrodes. A useful way to think about current flow in a volume conductor is to consider that it arises from electrophysiological sources. The idea of identifying a physical quantity described as a "source" originates in the discipline of electricity and magnetism.

This concept can be illustrated in electrostatics, beginning with Coulomb's law. In an infinite homogeneous dielectric medium with permittivity κ, the force F exerted by a point charge Q_i on a second point charge Q was found to be given by

$$\mathbf{F} = \frac{Q_i Q}{4\pi\kappa\epsilon_0 r^2} \hat{\mathbf{r}} \tag{1}$$

where ϵ_0 is the permittivity of free space, and r is the radius from Q_i to Q with $\hat{\mathbf{r}}$ as the unit vector in this direction. Using superposition, this result can be generalized to the case where there are N point sources in addition to Q. The force directed on Q is then

$$\mathbf{F} = \sum_{i=1}^{N} \frac{Q_i Q}{4\pi\kappa\epsilon_0 r_i^2} \hat{\mathbf{r}}_i \tag{2}$$

where r_i is the distance from Q_i to Q and $\hat{\mathbf{r}}_i$, is the unit vector in this direction.

It is convenient to imagine that the force on Q arises in two steps. The first is that all the other charges establish an electric field $\mathbf{E}$, namely,

$$\mathbf{E} = \sum_{i=1}^{N} \frac{Q_i}{4\pi\kappa\epsilon_0 r_i^2} \hat{\mathbf{r}}_i \tag{3}$$

where $\mathbf{E}$ is a function of position (of Q), and (the second step) that the force on Q is given by the action of this field on Q, through

$$\mathbf{F} = Q\mathbf{E} \tag{4}$$

Note that the combination of (3) and (4) correctly yields (2), hence supporting the source-field idea.

The field concept is useful in separating the evaluation of the field apart from any reference to the specifics of its action (e.g., the generation of a force on Q). This is also true in electrophysiology, where, for example, we can evaluate the field generated by a fiber carrying a propagating action potential apart from, say, an examination of an electrode system that has been devised to measure a difference in potential within the field. The assumption is often made that the electrode used for evaluating the field does not modify the sources which are present prior to its introduction, but such an effect may be present and requires evaluation.

The electrostatic analog is appropriate to electrophysiology since it turns out that at any instant the fields of physiological origin are quasi-static; that is, they satisfy static equations at that instant. They do, however, involve the flow of currents and therefore can only be maintained by processes that supply energy [13]. The basic source element in electrophysiology is therefore not a point current source but a dipole source (i.e., a battery).

A. Primary Source

If we consider an excitable fiber in an unbounded uniform conducting extracellular space carrying a propagating action potential, then clearly there will be extracellular currents and, therefore, an associated potential field. This field must be arising from sources just as was true in the aforementioned electrostatic problem (though, actually, quasi-static relationships are valid here, as noted above). Because the extracellular and intracellular spaces are normally assumed to be passive conductors (hence, source free), the sources must necessarily be localized to the membrane (which is the only remaining region). This conclusion is not changed if the excitable tissue is multicellular; the sources are confined to the entire system of cellular membranes. Since the membranes are also the location of ionic channels whose behavior underlies the excitable nature of the membranes, one may expect that the membrane is the site of a transfer of energy from the electrochemical field (derived from the difference in extracellular and intracellular ionic composition) to the current flow field (through a process that is mediated by the ionic channels). The sources arising in this process are designated *primary* sources and are characterized by their close association with a supply of nonelectrical energy (ultimately, the ATP from metabolism which powers the processes that transport ions between intracellular and extracellular spaces and is responsible, consequently, for the unequal ionic concentrations of these regions).

The elementary form of source is a double layer lying in the membrane of each excitable cell. For a multicellular tissue it may be possible to space-average these sources to form a continuum (a dipole moment density). Although this is an approximation, it is useful conceptually and may simplify computations. We designate such a source here by $\tau(x, y, z)$. Since the scalar potential field generated by a discrete dipole $\mathbf{p}$ equals $\mathbf{p} \cdot \hat{\mathbf{r}}/4\pi\sigma r^2$, where r is the distance from the dipole to the field point, $\hat{\mathbf{r}}$ is a unit vector in that direction, and σ is the conductivity of the extracellular medium (assumed to be uniform), then the extracellular field, Φ_o, generated by the aforementioned source is

$$\Phi_o = \int_{\text{tissue}} \frac{\tau \cdot \hat{\mathbf{r}}}{4\pi\sigma r^2} \, dV \tag{5}$$

In the extracellular region, by taking the Laplacian of both sides of Eq. (5) one can confirm that Φ_o satisfies Laplace's equation, namely,

$$\nabla^2 \Phi_o = 0 \tag{6}$$

Since this result has been obtained quite generally, one expects this behavior to be valid for any electrophysiologically generated electric field in a source-free region.

B. Secondary Sources

In the above we assumed that the excitable tissue lies in a uniform, unbounded extracellular medium. Except in some approximate sense, this can never actually be the case since any preparation is bounded at some point by the air and there are usually inhomogeneities within any volume conductor. If one assumes that the volume conductor can be approximated by regions, each of which is a uniform conductor, then a discontinuity of normal derivative of potential must occur at the interface between each such adjoining region since the normal component of current density is necessarily continuous (conservation of charge). An examination of this situation shows that charge accumulates at the conductivity interfaces to ensure this boundary condition. The field from this source enhances the electric field on the low-conductivity side and reduces the electric field on the high-conductivity side (to bring about a continuity of current density passing across the interface in the normal direction). This charge density is a field source, and while it is set up to bring about the proper boundary conditions, its effect is felt everywhere. Since this source would not arise without the original field set up by the membrane, it is referred to as a *secondary source*. So, at least in principle, the formulations in the previous sections continue to be valid for inhomogeneous media, so long as additional secondary sources are included which take account of the inhomogeneities. A general expression for the field arising from primary and secondary sources may be given as [14]

$$\Phi_o = \frac{1}{4\pi\sigma}\left[\int_{S_m}(\sigma_e\Phi_e - \sigma_i\Phi_i)\frac{\hat{\mathbf{r}}\cdot d\mathbf{S}_m}{r^2} + \int_{S_k}\Phi_k(\sigma_k'' - \sigma_k')\frac{\hat{\mathbf{r}}\cdot d\mathbf{S}_k}{r^2}\right] \qquad (7)$$

In (7), $\mathbf{S}_m$ denotes the sum of all membrane surfaces and this integral generates the primary field (i.e., the source consisting of double layers, in the membranes, whose strength is the discontinuity of $\sigma\Phi$ across the membrane). On the other hand, $\mathbf{S}_k$ denotes all interface between regions of different conductivity. In the latter case, the sources are double layers lying in all aforementioned interface whose strength is also given by the discontinuity of $\sigma\Phi$ across the interface. The unsubscripted σ in (7) takes on the conductivity at the field point so that the secondary source is an *equivalent* double (dipole) and not a single layer.

C. Equivalent Sources

If the excitable tissue region is considered to be bounded by a surface S_t and the tissue lies in an unbounded volume conductor, then a source-free region

V is identified that is bounded by S_t and the surface at infinity, S_∞. One can apply Green's theorem to the potential field in V, namely,

$$\int_V (\phi \nabla^2 \psi - \psi \nabla^2 \phi)\, dV = \int_S (\phi \nabla \psi - \psi \nabla \phi) \cdot d\mathbf{S} \tag{8}$$

where ϕ and ψ are any twice-differentiable scalar potential functions. In (8), S is the surface that bounds V and therefore $S = S_t + S_\infty$. The surface at infinity, however, can be shown to contribute nothing to the integral if ϕ and ψ are physically realizable potential fields. If we choose

$$\phi = \Phi \qquad \text{and} \qquad \psi = \frac{1}{r} \tag{9}$$

where Φ is the actual potential field within V and r is the distance from an integration element (dV or dS) to an arbitrary field point P, then (8) reduces to

$$\Phi(P) = -\left(\frac{1}{4\pi}\right)\left(\int_{S_t} \frac{\nabla\Phi \cdot \hat{\mathbf{n}}\, dS_t}{r} + \int_{S_t} \Phi \frac{\hat{\mathbf{r}} \cdot \hat{\mathbf{n}}\, dS_t}{r^2}\right) \tag{10}$$

where $\hat{\mathbf{n}}$ is the surface normal. The expression in (10) can be interpreted as evaluating the potential field at P from surface sources on S_t. In the first integral of (10) the quantity $(\nabla\Phi \cdot \hat{\mathbf{n}} = \partial\Phi/\partial n)$ behaves like a single-layer source while the quantity $\Phi\hat{\mathbf{n}}$ behaves like a double layer. [In different words, the field within V (at an arbitrary P) can be evaluated from the field and its normal derivative on S_t.] The aforementioned sources are not real, of course, but are designated as *equivalent* sources. They generate the correct fields in the source-free region, V, but do not generate the correct fields outside V (i.e., in the excitable tissue region), where, in fact, a null field results.

IV. TISSUE MODELS

A. Single Fiber

To model the behavior of excitable tissue, we need descriptions of both the intrinsic membrane properties and the tissue structure (i.e., how the cells are arranged or connected). The model of the membrane current could involve a detailed description of the individual channel kinetics or could include some macroscopic approximation to the average channel behavior.

Hodgkin and Huxley pioneered this approach when they introduced their macroscopic model of the membrane ionic currents into a relatively simple structure (i.e., a circular cylindrical fiber), to better understand the mechanism for impulse propagation in a single unmyelinated nerve fiber. Because of its simplicity, the single cylindrical fiber forms the basis of much electrophysiology theory. One reason for this is that expressions like (7) can be written in more tractable forms if axial symmetry is assumed. For example, applying Gauss's theorem to the first term of (7) we obtain

$$\Phi_o = -\frac{1}{4\pi\sigma} \int_V \nabla \cdot \left[V_f(z)\nabla\left(\frac{1}{r}\right) \right] dV \tag{11}$$

where $V_f(z) = [\sigma_i\Phi_i(z) - \sigma_o\Phi_o(z)]$ and V is the volume occupied by the fiber. Since V_f is independent of angle around the fiber and for points outside the fiber $\nabla^2(1/r) = 0$, (11) simplifies to

$$\Phi_o = -\frac{1}{4\pi\sigma} \int_V \nabla\left(\frac{1}{r}\right) \cdot \hat{\mathbf{a}}_z \frac{\partial V_f(z)}{\partial z} dV \tag{12}$$

where $\hat{\mathbf{a}}_z$ is the unit vector in the z direction. Equation (12) also can be expressed as

$$\Phi_o = -\frac{1}{4\pi\sigma} \int_A dA \int_{-\infty}^{\infty} \frac{\partial V_f(z)}{\partial z} \frac{\partial(1/r)}{\partial z} dz \tag{13}$$

The sources may be regarded as equivalent double-layer disks of strength $\partial V_f(z)/\partial z$, since the remaining mathematical expression describes an axial dipole field. If (13) is integrated by parts, we obtain

$$\Phi_o = -\frac{1}{4\pi\sigma} \int_A dA \int_{-\infty}^{\infty} \frac{\partial^2 V_f(z)}{\partial z^2} \frac{1}{r} dz \tag{14}$$

Here the sources are equivalent single-layer disks, assumed to lie within the intracellular volume of the cylindrical fiber [15].

B. Transmembrane Current Density

Using (14) to evaluate the extracellular potential requires separate knowledge of Φ_i and Φ_o. In many models, however, we are likely to compute membrane quantities such as transmembrane potential V_m, defined as the difference between the intracellular and extracellular potentials, namely,

$$V_m = \Phi_i - \Phi_o \tag{15}$$

and transmembrane current density I_m. As shown below, the transmembrane potential can sometimes be expressed directly in terms of Φ_i.

Assuming axial symmetry, the intracellular current density, $\mathbf{J}_i$ is defined as

$$\mathbf{J}_i = -\nabla\Phi_i = -\sigma_i\left(\frac{\partial\Phi_i}{\partial\rho}\,\hat{\mathbf{a}}_\rho + \frac{\partial\Phi_i}{\partial z}\,\hat{\mathbf{a}}_z\right) \tag{16}$$

while the extracellular current density, $\mathbf{J}_o$, is given by

$$\mathbf{J}_o = -\nabla\Phi_o = -\sigma_o\left(\frac{\partial\Phi_o}{\partial\rho}\,\hat{\mathbf{a}}_\rho + \frac{\partial\Phi_o}{\partial z}\,\hat{\mathbf{a}}_z\right) \tag{17}$$

where $\hat{\mathbf{a}}_z$ and $\hat{\mathbf{a}}_\rho$ are unit vectors in the axial and radial directions, respectively. The total axial intracellular current $I_a{}^i$ is given by Clark and Plonsey [16]:

$$I_a^i = \sigma_i \int_0^a 2\pi\rho\,\frac{\partial\Phi_i}{\partial z}\,d\rho \tag{18}$$

We can evaluate the transmembrane current per unit length, i_m, since by virtue of conservation of charge, it is the change in the total axial intracellular current and can be expressed as

$$i_m = \frac{\partial I_a^i}{\partial z} = \sigma_i \int_0^a 2\pi\rho\,\frac{\partial^2\Phi_i}{\partial z^2}\,d\rho \tag{19}$$

Expanding $\nabla^2\Phi_i = 0$ (Laplace's equation) yields

$$\frac{\partial^2\Phi_i}{\partial z^2} = -\frac{1}{\rho}\frac{\partial}{\partial\rho}\left(\rho\frac{\partial\Phi_i}{\partial\rho}\right) \tag{20}$$

Substituting (20) into (19) gives

$$I_m = \frac{i_m}{2\pi a} = -\sigma_i\left.\frac{\partial\Phi_i^s}{\partial\rho}\right|_{\rho=a} \tag{21}$$

which directly evaluates i_m, by applying Ohm's law at the inside membrane surface.

The intracellular potential can be expressed as a function of $\Phi_i(a,z)$ at the surface by solving Laplace's equation in the intracellular space. Using the method of separation of variables, the result is

$$\Phi_i(\rho,z) = \mathcal{F}^{-1}\left[\Phi_i^s(k)\frac{I_0(|k|\rho)}{I_0(|k|a)}\right] \tag{22}$$

where $\Phi_i^s(k)$ is the Fourier transform of the intracellular potential at the inner surface, defined as

$$\mathcal{F}[\Phi_i^s(z)] = \Phi_i^s(k) = \int_{-\infty}^{\infty} \Phi_i^s(z)e^{jkz}\, dz \tag{23}$$

and the inverse Fourier transform $(\mathcal{F}^{-1})$ is

$$\mathcal{F}^{-1}[\Phi_i^s(k)] = \Phi_i^s(z) = \frac{1}{2\pi}\int_{-\infty}^{\infty} \Phi_i^s(k)e^{-jkz}\, dk \tag{24}$$

and I_0 and I_1 are modified Bessel functions of the first kind (zero and first order).

Substituting (22) into (21) gives

$$I_m = -\sigma_i \left.\frac{\partial \Phi_i^s}{\partial \rho}\right|_{\rho=a} = \sigma_i \mathcal{F}^{-1}\left[\Phi_i^s(k)|k|\frac{I_1(|k|a)}{I_0(|k|a)}\right] \tag{25}$$

For small values of $|k|a$,

$$\frac{1}{k}\frac{I_1(|k|a)}{I_0(|k|a)} = \frac{a}{2} \tag{26}$$

Thus, for action potentials whose axial extent is large compared to the fiber radius (i.e., $|k|a$ is small), I_m is a function of only the axial variation in Φ_i, namely,

$$I_m = \frac{a}{2\varrho_i}\frac{\partial^2 \Phi_i^s}{\partial z^2} \tag{27}$$

C. Core Conductor Model

When modeling propagation in cardiac tissue, it is necessary to make assumptions about the amount of structural detail that should be included. The early models considered the discrete cellular structure to be homogenized into a uniformly continuous domain. The validity of this approach was supported by the experimental work of Weidmann [17] and of Chapman and Fry [18], which showed cardiac tissue to exhibit syncytial electrical behavior.

The core conductor model is a simple circuit analog for a single fiber (Fig. 1). Axial symmetry is assumed and the current flow inside the fiber is effectively one-dimensional. In the circuit analog, the current flow outside the fiber is likewise one-dimensional. This is accomplished by requiring that the fiber lie in a restricted cylindrical-volume conductor. Using the circuit analog of a continuous resistor, R, we can describe the intracellular and extracellular current flow. The total value R of this resistor can be derived

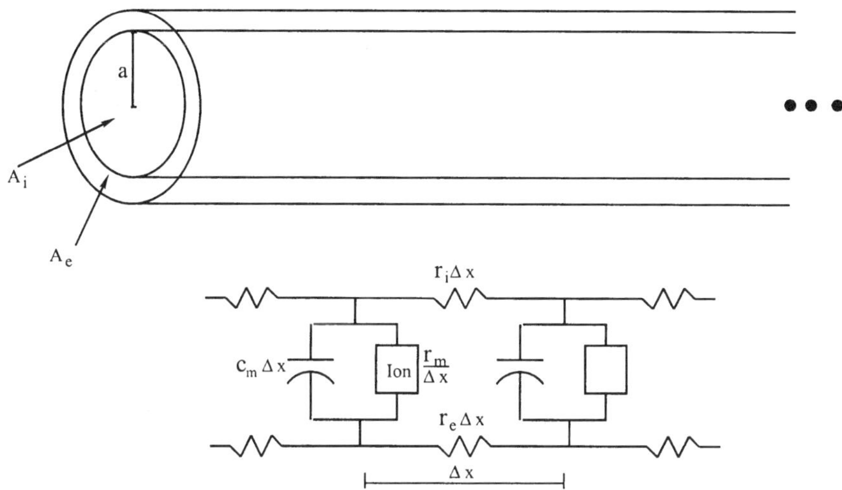

Figure 1   (Top) Schematic of a fiber of radius a in a restricted bath where extracellular and intracellular areas are given by A_e and A_e, respectively. (Bottom) Segment of the resistor analog to the fiber above. Parameters are as follows: r_i is the intracellular resistance per unit length (Ω/cm), r_e is the extracellular resistance per unit length (Ω/cm), c_m is the membrane capacitance ($\mu\text{F}/\text{cm}$), and r_m is the membrane resistance (Ω/cm). As $\Delta x \to 0$, the discrete model approaches a continuous representation.

from the intrinsic material resistivities, ϱ, of the intracellular and extracellular regions using the relationship

$$R = \frac{\varrho \ell}{A} \tag{28}$$

where ℓ is the length of a segment of the fiber and A is the cross-sectional area of the region of interest. For uniform fibers, the continuous resistor can better be described as a resistance on a unit length basis. If we divide both sides of (28) by ℓ, we define a resistance per unit length $r(\Omega/\text{cm})$, given by

$$r = \frac{\varrho}{A} \tag{29}$$

The intracellular and extracellular regions are physically separated by the cell membrane, which is so thin that we can think of it as a surface (interface). The membrane has porelike structures that give resistance to current flow and separate charges in the two spaces, giving capacitance. The total resistance of the membrane to transmembrane current decreases with increasing surface area. In contrast, the membrane capacitance increases as the surface area increases. We therefore often characterize the membrane resistance and membrane capacitance on a per-unit-area basis, that is, as a specific resistance $R_m(\Omega\text{-cm}^2)$ and a specific capacitance C_m (F/cm^2). If we define the membrane properties on a unit length basis, then we obtain the relationship

$$r_m = \frac{R_m}{2\pi a} \tag{30}$$

and

$$c_m = 2\pi a C_m \tag{31}$$

where r_m is the membrane resistance times length (Ω-cm) and c_m is the membrane capacitance per unit length (F/cm).

The electrical model describes this continuous structure as the limit of an infinite number of resistor and capacitor elements found by subdividing the continuum into segments of length Δx. As $\Delta x \to 0$, the discrete representation approaches a continuous representation. Because the assumed fiber is uniform and cylindrical and the intracellular and interstitial currents flow only in the axial direction, the intracellular current density, J_i, is given by

$$J_i = -\frac{1}{\varrho_i}\frac{\partial \Phi_i}{\partial x} \tag{32}$$

and the interstitial current density (outside the fiber), J_e, is

$$J_e = -\frac{1}{\varrho_e}\frac{\partial \Phi_e}{\partial x} \tag{33}$$

Note that the subscript e is used to distinguish interstitial currents (or potentials) that are outside the membrane and inside the tissue from extracellular currents (or potentials) that are outside the membrane *and* outside the tissue (i.e., in the bath).

From conservation of current, any change in longitudinal current must either leave as membrane current or must come or leave via an external electrode. For an electrode in the interstitial space,

$$\frac{\partial J_e}{\partial x} = I_v = \beta_e(I_m + I_p) \tag{34}$$

where I_v is the membrane current *per unit volume* and I_p is an externally applied stimulus current per unit area. The current per unit volume is related to the currents per unit area by the surface-to-volume ratio, β_e, the ratio of the total surface membrane area to the volume of interest. In the interstitial region of a uniform cable,

$$\beta_e = \frac{2\pi a}{A_e} \tag{35}$$

The change in intracellular current is given by

$$\frac{\partial J_i}{\partial x} = -\beta_i I_m \tag{36}$$

with

$$\beta_i = \frac{2\pi a}{A_i} \tag{37}$$

If we apply (15), (32), and (33) to (34) and (36), it can be shown that

$$\frac{1}{(r_i + r_e)} \frac{\partial^2 V_m}{\partial x^2} = (2\pi a) \left[I_m + I_p \frac{r_e}{(r_i + r_e)} \right] = i_m + \frac{r_e}{(r_i + r_e)} i_p \tag{38}$$

where r_i and r_e are the intracellular and interstitial resistances per unit length, respectively, and i_m and i_p are the transmembrane and external (stimulus) currents per unit length, respectively.

Following the derivation by Henriquez, we can use the expressions above to give the potentials inside and outside the fiber as the action potential propagates along the fiber after the stimulus is turned off (i.e., $I_p = 0$) [19]. Using (32), (33), (34), and (36), we obtain

$$\frac{1}{\varrho_i} \frac{\partial^2 \Phi_i}{\partial x^2} = \beta_i (I_m) = \frac{2\pi a}{A_i} I_m \tag{39}$$

and

$$\frac{1}{\varrho_e} \frac{\partial^2 \Phi_e}{\partial x^2} = -\beta_e (I_m) = \frac{-2\pi a}{A_e} I_m \tag{40}$$

Again note the use of the subscript e to denote interstitial quantities [see note after Eq. (33)]. Multiplying both sides of (39) by A_i and both sides of (40) by A_e, using (15), and adding the two resulting expressions, we obtain

$$\left(\frac{A_i}{\varrho_i} + \frac{A_e}{\varrho_e} \right) \frac{\partial^2 \Phi_i}{\partial x^2} = \frac{A_e}{\varrho_e} \frac{\partial^2 V_m}{\partial x^2} \tag{41}$$

and

$$\left(\frac{A_i}{\varrho_i} + \frac{A_e}{\varrho_e}\right)\frac{\partial^2 \Phi_e}{\partial x^2} = -\frac{A_i}{\varrho_i}\frac{\partial^2 V_m}{\partial x^2} \tag{42}$$

From inspection, the intracellular and interstitial potentials arising from conduction along a core conductor are given by

$$\Phi_i = \frac{(\varrho_i/A_i)}{(\varrho_i/A_i) + (\varrho_e/A_e)}\, V_m = \frac{r_i}{(r_i + r_e)}\, V_m \tag{43}$$

$$\Phi_e = -\frac{(\varrho_e/A_e)}{(\varrho_i/A_i) + (\varrho_e/A_e)}\, V_m = -\frac{r_e}{(r_i + r_e)}\, V_m \tag{44}$$

To investigate the effect of the extracellular or interstitial potential on conduction, we rewrite (27) as

$$I_m = \frac{a}{2\varrho_i}\frac{\partial^2 \Phi_i}{\partial x^2} = C_m \frac{\partial V_m}{\partial t} + \sum I_{\text{ion}} \tag{45}$$

where C_m is the membrane capacitance and $\sum I_{\text{ion}}$ is the total membrane ionic current.

Applying (15), (45) is also given by

$$I_m = \frac{a}{2\varrho_i}\left[\frac{\partial^2 V_m}{\partial x^2} + \frac{\partial^2 \Phi_o}{\partial x^2}\right] = C_m \frac{\partial V_m}{\partial t} + \sum I_{\text{ion}} \tag{46}$$

Assigning a zero extracellular resistance per unit length is equivalent to neglecting the contribution of $\partial^2 \Phi_o / \partial x^2$ in (46). For the classical form of (46) ($\partial^2 \Phi_o / \partial x^2 = 0$), Hodgkin argued that the time course of a uniformly propagating transmembrane potential is unique for a given set of membrane properties [20]. For uniform propagation with constant velocity, θ, the temporal and spatial variations in the transmembrane potential [$V_m(t)$ and $V_m(x)$] satisfy the wave equation and are related by θ. We can write (45) as

$$I_m = \frac{a}{2\varrho_i\theta^2}\frac{\partial^2 V_m}{\partial t^2} = C_m \frac{\partial V_m}{\partial t} + \sum I_{\text{ion}} \tag{47}$$

According to Hodgkin, $I_m(t)$ and $V_m(t)$ are single-valued functions and are linked such that

$$I_m = k \frac{\partial^2 V_m}{\partial t^2} \tag{48}$$

where k is a proportionality constant uniquely determined by the properties of the membrane (i.e., C_m and I_{ion} [20]. That is, if (48) is substituted into (46), then the solution $V_m(t)$ is unchanged by changing a and ϱ_i as long as k remains constant. From (47), $k = a/2\varrho_i\theta^2$. Thus, for constant and uniform membrane properties, changes in the spatial distribution of currents resulting from changes in the passive, time-independent properties a or ϱ_i act only to scale the spatial distribution of $V_m(x)$. A change in a for constant ϱ_i or a change in ϱ_i (for constant a) is compensated by a change in θ to keep k constant and $V_m(t)$ the same.

Another important relationship that is derived from (48) is that the time course of the early-rising phase ("foot") of the action potential (where the membrane behavior is linear) is exponential. As an action potential conducts along a fiber, local currents flow away from the site of activity and charge the membrane capacitance of distal sites that are still at rest. The time course of this distal charging can be obtained by assuming that the distal region is passive, so

$$I_m = C_m \frac{\partial V_m}{\partial t} + \frac{V_m}{R_m} \tag{49}$$

The solution to (49) [using (47)] reveals that the initial rising phase of an uniformly propagating action potential along a single fiber where the intracellular and extracellular currents are axial (core conductor fiber), or where $\partial^2 \Phi_o/\partial x^2 = 0$, is exponential with a time constant τ_{foot} given by

$$\tau_{\text{foot}} = \left(\frac{K}{2} + \frac{K}{2} \sqrt{1 + \frac{4}{KR_mC_m}} \right)^{-1} \tag{50}$$

where $K = 2\varrho_i C_m \theta^2 / a$. Usually, $4/KR_mC_m \ll 1$ and $\tau_{\text{foot}} = 1/K$.

D. Transmembrane Potential and Fiber Properties

If the currents are not axial or $\Phi_o \neq 0$ or, more appropriately, $\partial^2 \Phi_o/\partial z^2 \neq 0$, then the dependence of $V_m(t)$ on a or ϱ_i and the time course of the foot are more complicated than that described above. Under steady conditions, we expect that $\Phi_o(x)$ and $\Phi_o(t)$ also satisfy the wave equation and are likewise related by θ. We can then write (46) as

$$\frac{a}{2\varrho_i\theta^2} \left(\frac{\partial^2 V_m}{\partial t^2} + \frac{\partial^2 \Phi_o}{\partial t^2} \right) = C_m \frac{\partial V_m}{\partial t} + \sum I_{\text{ion}} \tag{51}$$

In contrast to (46) with $\partial^2 \Phi_o/\partial x^2 = 0$, the solution of (51) for the steady state $V_m(t)$ will not be uniquely determined by the membrane properties; instead, $V_m(t)$ will be additionally constrained by the contribution of $\partial^2 \Phi_o/\partial t^2$. As before, $V_m(t)$ will be unaffected by changes in the passive properties of the fiber and the surrounding volume conductor only if such changes can be compensated by a change in θ or $\partial^2 \Phi_o/\partial t^2$ to leave the left side of (51) unchanged.

For a fiber in a restricted conductor, the core conductor equations apply and Φ_o (or Φ_e) is given by (44). Applying (44) to (51), we obtain

$$\frac{a}{2[\varrho_i + (f_i/f_e)\varrho_o]\theta^2}\left(\frac{\partial^2 V_m}{\partial t^2}\right) = C_m \frac{\partial V_m}{\partial t} + \sum I_{ion} \tag{52}$$

where f_i and f_e are the fraction of intracellular and interstitial spaces, respectively. Equations (47) and (52) have the same form and thus $I_m(t)$ and $\partial^2 V_m/\partial t^2$ are related by a constant $k = a/2[\varrho_i + (f_i/f_e)\varrho_o]\theta^2$ when the volume conductor constrains the extracellular current to be axial.

If the fiber lies in an unbounded medium, then $\partial^2\Phi_o/\partial t^2$ is not proportional to $\partial^2 V_m/\partial t^2$. For the case where the axial extent of $V_m(x)$ is large compared to the fiber radius a, the line source model can be used to obtain Φ_o. Defining the axial coordinate of the source and field points as

$$x = \theta t \qquad x' = \theta t' \tag{53}$$

Φ_o at $\rho = a$ can be expressed as a function of time, t, and (14) becomes

$$\Phi_o(t) = -\frac{a^2\sigma_i}{4\sigma\theta^2}\int_{-\infty}^{\infty}\frac{\partial^2 V_m}{\partial t^2}\frac{1}{\sqrt{(a/\theta)^2 + (t - t')^2}}\, dt \tag{54}$$

Clearly the time course of Φ_o is not independent of θ. Hence a change in the passive properties cannot be offset by a change in θ and still keep the left-hand side of (51) the same; consequently, $V_m(t)$ will change. It is important to note that for the conditions under which (54) applies, the magnitude of $\partial^2\Phi_o/\partial t^2$ due to a single fiber in isolation is typically small compared to $\partial^2 V_m/\partial t^2$, and the effect of the extracellular potential distribution is negligible. This analysis, however, suggests that if the field outside the fiber is set up from sources other or in addition to the fiber itself and $\partial^2\Phi_o/\partial t^2$ is sufficiently different from $\partial^2 V_m/\partial t^2$, then the influence of Φ_o on the shape of $V_m(t)$ could be significant. In other words, if there exists a radially varying potential outside each bundle in the preparation, the nature of the transmembrane potential of a bundle of fibers may differ from that of a single fiber even if the membrane properties are the same.

V. MULTIDIMENSIONAL TISSUE

Equation (46) has been used for representing the current flow in the intracellular space of a one-dimensional fiber, under the tacit assumption that the effects of the extracellular potential are negligible. The use of a single differential equation (the monodomain model) can be extended to two and three dimensions, in which the effects of anisotropy on conduction can be considered.

The general form of the monodomain model is

$$\nabla \cdot (\mathbf{D}_i \nabla V_m) = \beta I_m - I_{sm} \tag{55}$$

where $\mathbf{D}_i$ is the intracellular conductivity tensor (S/m), β is the surface-to-volume ratio, and I_{sm} is an applied transmembrane current source ($\mu A/cm^3$).

The conductivity tensor $\mathbf{D}_i$ represents the space-averaged electrical properties of the discrete cellular interconnections. Conductivity is the inverse of resistivity. Many multidimensional propagation models are based on an assumption that cardiac tissue is comprised of a collection of parallel fibers and are formulated in a spatial coordinate system whose axes align with the principal axes of $\mathbf{D}_i$. Under this assumption, for a Cartesian coordinate system, $\mathbf{D}_i$ is a diagonal matrix, and the structure may be interpreted as a regular lattice of resistor elements as shown in Fig. 2A. When modeling tissue regions with curved or rotated fibers, $\mathbf{D}_i$ is not necessarily constant throughout the domain and generally cannot be diagonalized everywhere by a single choice of axes.

A. Bidomain Model

The typical interpretation of the monodomain model is that the preparation lies in a large-volume conductor so that the extracellular potential field is small enough to ignore. For a thin sheet of tissue, this assumption is likely valid. For the interior of thick tissue or for tissue in a restricted-volume conductor, however, the extracellular resistance and potentials can be large and, depending on properties, can affect both propagation wave speed and action potential shape. The most appropriate circuit analog is one that includes both intracellular and extracellular resistances such as the linear core conductor model. For multicellular tissue, one can extend the analog by viewing cardiac tissue as comprised of a single intracellular region of convoluted geometry separated by a membrane from a similarly complex extracellular region within the volume of the tissue. This representation is known as the bidomain model [21]. The resistance to intracellular and interstitial current flow depend on the geometry of each region and is generally homogenized to have uniform, continuous properties. For convenience, each space is defined over the entire tissue volume. Figure 2B shows the resistor equivalent network of a bidomain model for the case of parallel fibers. As with the monodomain model described above, a conductivity tensor can be assigned at each point to capture local fiber variation, but in this case this must be done in both spaces [22].

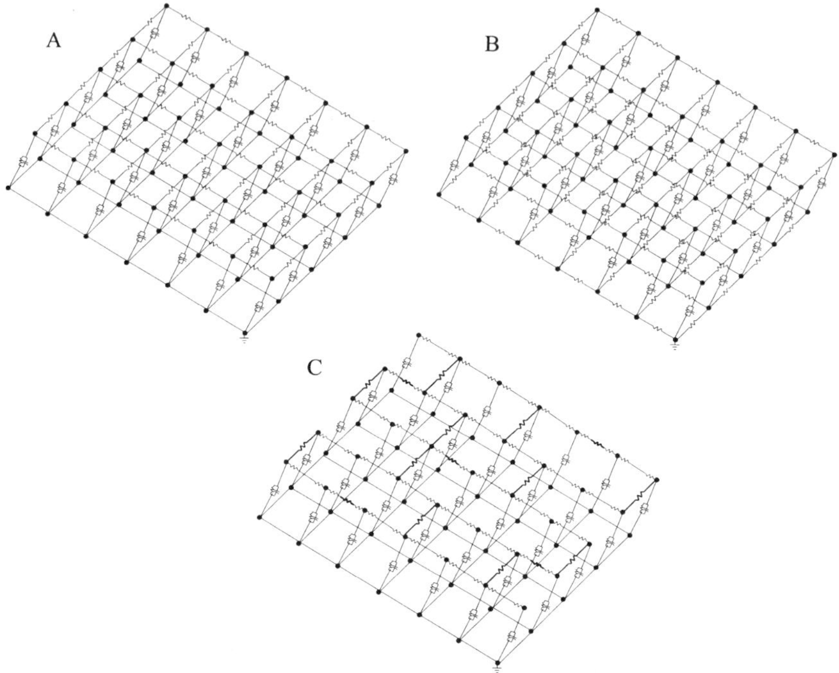

Figure 2 (A) A representative section of a monodomain circuit analog. The extracellular space in an unbounded conductor is considered to have negligible resistance (compared to the membrane and intracellular resistances), and therefore is effectively grounded. Note that passive elements connect the intracellular and extracellular space. In excitable tissue, a more complicated ionic model is substituted. (B) A segment of a bidomain circuit model. Both intracellular and extracellular spaces are well connected by a uniform lattice of resistors. (C) A section of a randomized discrete network. Junctions are represented by darkened resistors. In this depiction, longitudinal connections are composed of both intracellular and junctional resistances while transverse connections are limited to junctional resistances. In many discrete models, however, the transverse connections are composed of both intracellular and junctional resistances.

Converting intrinsic conductivities to bidomain conductivities usually involves a scaling by the fraction of space occupied by each domain. For example, the bulk resistance of the intracellular domain of a single fiber in a uniform bundle is $R_i = \varrho_i \ell / A_i$ (28). If this resistance is spread over the entire tissue volume (the bundle) which comprises the intracellular and interstitial regions and has a cross-sectional area of A_{tot}, and there are n fibers, the effective bidomain conductivity is

$$g_i = \frac{nA_i}{\varrho_i A_{\text{tot}}} = \frac{A_{\text{in}}}{\varrho_i A_{\text{tot}}} \tag{56}$$

where $A_{\text{in}} = nA_i$ is the total intracellular area. If we define the fraction of intracellular area as $f_i = A_i n / A_{\text{tot}}$, and use $\sigma_i = 1/\varrho_i$, then

$$g_i = f_i \sigma_i \tag{57}$$

and similarly,

$$g_e = f_e \sigma_e \tag{58}$$

where the fraction of interstitial space is $f_e = A_e / A_{\text{tot}}$.

The governing equations are continuity of current,

$$\nabla \cdot (\mathbf{J}_e + \mathbf{J}_i) = 0 \tag{59}$$

and Ohm's law,

$$\mathbf{J}_i = -\mathbf{D}_i \nabla \Phi_i \tag{60}$$

$$\mathbf{J}_e = -\mathbf{D}_e \nabla \Phi_e \tag{61}$$

In the bidomain model, any current leaving one domain must enter the other as transmembrane current. In three dimensions, (34) and (36) are generalized to

$$I_v = \nabla \cdot \mathbf{D}_i \nabla \Phi_i \tag{62}$$

and

$$-I_v = \nabla \cdot \mathbf{D}_e \nabla \Phi_e \tag{63}$$

For the special case of tissue made up of parallel fibers, where the tensors are everywhere diagonal, Eqs. (62) and (63) become

$$I_v = g_{ix} \frac{\partial^2 \Phi_i}{\partial x^2} + g_{iy} \frac{\partial^2 \Phi_i}{\partial y^2} + g_{iz} \frac{\partial^2 \Phi_i}{\partial z^2} \tag{64}$$

$$-I_v = g_{ex} \frac{\partial^2 \Phi_e}{\partial x^2} + g_{ey} \frac{\partial^2 \Phi_e}{\partial y^2} + g_{ez} \frac{\partial^2 \Phi_e}{\partial z^2} \tag{65}$$

Using (15), and combining (64) and (65) to eliminate $\partial^2 \Phi_e / \partial z^2$, we obtain

$$I_v = \frac{g_{ez}g_{ix}}{g_{iz}+g_{ez}}\frac{\partial^2 V_m}{\partial x^2} + \frac{g_{ez}g_{iy}}{g_{iz}+g_{ez}}\frac{\partial^2 V_m}{\partial y^2} + \frac{g_{ez}g_{iz}}{g_{iz}+g_{ez}}\frac{\partial^2 V_m}{\partial z^2}$$

$$+ \frac{g_{ez}g_{ix}-g_{iz}g_{ex}}{g_{ez}+g_{iz}}\frac{\partial^2 \Phi_e}{\partial x^2} + \frac{g_{ez}g_{iy}-g_{iz}g_{ey}}{g_{ez}+g_{iz}}\frac{\partial^2 \Phi_e}{\partial y^2}$$

$$= \beta\left(C_m \frac{\partial V_m}{\partial t} + I_{\text{ion}}\right) \tag{66}$$

where $\beta = 2f_i/a$.

For a special set of conductivity values that satisfy

$$\frac{g_{ix}}{g_{ex}} = \frac{g_{iy}}{g_{ey}} = \frac{g_{iz}}{g_{ez}} \tag{67}$$

the coefficients of the terms involving Φ_e in (66) are zero, and the resulting expression reduces to the monodomain equation (55). A bidomain with conductivities that satisfy (67) is said to be equally anisotropic. For a plane wave propagating in an infinite, equally anisotropic bidomain, the intracellular and extracellular potentials are given by (43) and (44), respectively, where, in general, the values of r_i and r_e depend on direction [23]. Equal anisotropy is not expected to be satisfied for real tissue, and thus classical one-dimensional cable theory cannot be extended to three dimensions, in general.

Because real tissue is not infinite in extent, we must account for the presence of an adjoining volume conductor, by defining the interface conditions between tissue and bath. We assume that the interstitial space is the direct link with the extracellular bath. At the interface the intracellular and extracellular potentials are continuous,

$$\Phi_e = \Phi_o \tag{68}$$

as is the normal current,

$$g_{\text{en}}\frac{\partial \Phi_e}{\partial n} = \sigma_o \frac{\partial \Phi_o}{\partial n} \tag{69}$$

where n denotes the direction normal to the tissue surface. We ensure continuity of current through the interstitial space by requiring that the intracellular current vanish at the surface, i.e.,

$$g_{\text{in}}\frac{\partial \Phi_i}{\partial n} = 0 \tag{70}$$

The bounding bath will affect the nature of the current flow inside the tissue. Surface fibers have the extensive extracellular volume with which to exchange currents, while deeper fibers are enveloped by a restricted conducting space created by the presence of other fibers.

B.   The Discrete Monodomain Model

A variation of the monodomain model is one that explicitly includes the gap junctions between cells. In this model, the intracellular regions of the individual cells are discretized into a set of resistors and the junctions are represented by a local change in the resistor value [24–26]. Figure 3C shows a resistor grid representation of the discrete cellular structure. In this model, the values of the resistors are largest at the site of gap junctions and individual cells are connected by either a structured or a random set of resistive connections [27]. More detailed models of realistic cell geometries and interconnections have been proposed, in which a network of resistors defines the irregular two-dimensional cross sections of the cells and resistive junctions of various magnitudes connect the cells, densely at the ends and more sparsely along the lateral edges [28]. Although challenging to implement, the discrete model is extendable to three dimensions.

C.   Electrical Properties

In order to apply the models described above, one needs the components of the conductivity tensors $\mathbf{D}_i$ and $\mathbf{D}_e$ described in (60) and (61). In general, these values are specific to the particular model used to interpret an experimental measurement [8,17,18,30–32]. The most relevant measurements available at this time are summarized in Tables 1 and 2 [29]. The data in Table 1 assume uniform axial propagation (or electrotonus) along a uniform fiber bundle, while those in Table 2 are for propagation in the transverse direction. These tables give the microscopic intracellular and interstitial conductivities, σ_i and σ_e, and the bidomain resistivities along the fiber direction, g_i^{ℓ} and g_e^{ℓ}, and transverse to the axis, g_i^t and g_e^t. The bulk conductivities, g^{ℓ} and g^t are also given. The cellular surface-to-volume ratio is denoted by β, the relative intracellular to total volume is denoted by f_i, the transmembrane action potential magnitude is ΔV_m, and the extracellular difference in potential across the propagating wave is V_{wave}.

An examination of these values shows that there is no clear consistency. The best one can do at this point is to introduce a range of values into the model and note the effect. One possible test is to see which parameters lead to results most consistent with a set of macroscopic experiments. For example, under longitudinal propagation the voltage across the wave measured by Kleber and Riegger [32] in perfused papillary muscle preparation is 51 mV, and this appears to be a dependable experimental value.

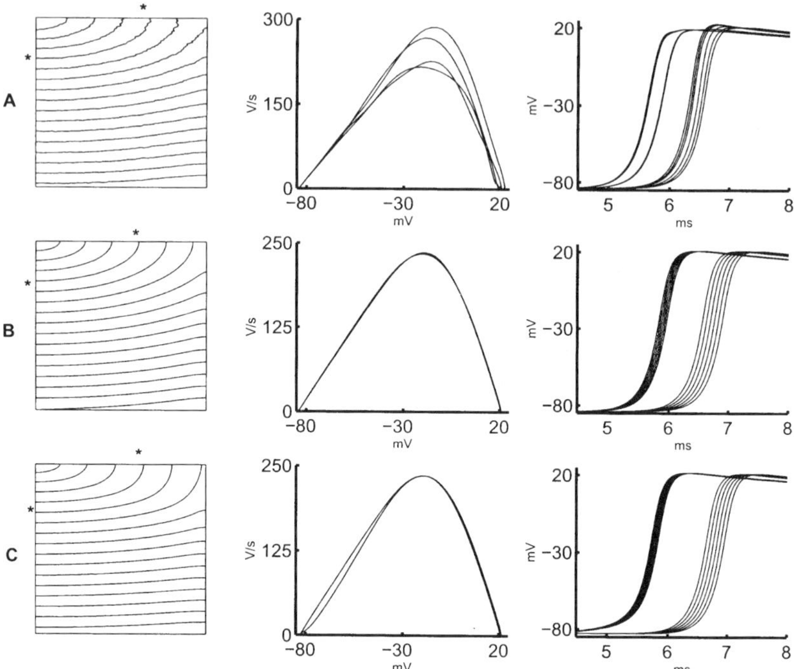

Figure 3 (A, B, C) A comparison of discrete, monodomain, and bidomain propagation. A 0.5-cm × 0.5-cm domain was created with $\Delta x = \Delta y = 15\,\mu m$. Conductivities (for monodomain and bidomain cases) and gap junctions (for the discrete case) were chosen to yield propagation velocities of ~ 60 cm/sec. (longitudinal) and ~ 20 cm/sec. (transverse). Left panels show activation times (spaced 1.5 msec apart). Monodomain and bidomain propagation are uniform, while the discrete isochrones are nonsmooth. Middle panels are of V_m versus $\dot{V}_m$. V_m was taken from two neighboring nodes on the transverse and logitudinal axes (denoted by $*$ in the left panel). In the uniform shape of V_m versus $\dot{V}_m$ in the monodomain and bidomain cases as opposed to the directional variation in the discrete case. Right panels show voltage plots for the longitudinal (left traces) and transverse direction (right traces) from the points denoted by $*$ in the left panels. Note the large variation in node-to-node delays in the discrete case and the nonuniformity in action potential shape. Propagation velocity in the monodomain and bidomain simulations is constant from node to node.

This suggests that the conductivity values of Roberts et al. may be the most reliable [31]. On the other hand, when one considers the many differences among preparations and procedures, and the variety of species used, even this conclusion cannot be strongly supported. The best remedy would be to

Table 1 Measured Linear Cable Parameters for Longitudinal Propagation

	σ_i (mS/cm)	σ_e (mS/cm)	f_i	g_i^{ℓ} (mS/cm)	g_e^{ℓ} (mS/cm)	σ^l (mS/cm)	ΔV_m (mV)	V_{wave} (mV)	β (cm^{-1})
Weidmann [17]	2.13	19.6	0.75	1.60	5.32	6.68	96	24	330
Clerc [8]	2.49	20.8	0.70	1.74	6.25	8.0	100	21.5	330
Roberts [30]	—	—	—	2.78	2.94	5.03	100	46	860
Roberts [31]	—	—	—	3.43	1.17	4.69	99	74	2920
Kleber [32]	6.02	15.9	0.75	4.52	3.97	8.62	98.7	51.5	1080

The data for longitudinal conduction, reported by various authors, are for: σ_i (the intrinsic intracellular conductivity), σ_e (the intrinsic interstitial conductivity), f_i (the fraction of intracellular space), g^{ℓ}_i (the bidomain intracellular conductivity), g^{ℓ}_e (the bidomain interstitial conductivity), g^{ℓ} (the bulk conductivity), ΔV_m (the magnitude of the transmembrane potential), V_{wave} (the difference in extracellular potential across the propagating wave), and β (the cell surface-to-volume ratio).

Table 2 Measured Linear Cable Parameters for Transverse Propagation

	σ_i (mS/cm)	σ_e (mS/cm)	f_i	g^t_i (mS/cm)	g^t_e (mS/cm)	g^t (mS/cm)	ΔV_m (mV)	V_{wave} (mV)	β (cm^{-1})
Clerc [8]	0.28	7.87	0.70	0.19	2.36	2.55	98	8	90
Roberts [30]	—	—	—	0.26	1.33	1.60	100	16	190
Roberts [31]	—	—	—	0.60	0.80	1.40	100	43	430

The data for transverse conduction, reported by various authors, are for: σ_i (the intrinsic intracellular conductivity), σ_e (the intrinsic interstitial conductivity), f_i (the fraction of intracellular space), g^t_i (the bidomain transverse intracellular conductivity), g^t_e (the bidomain transverse interstitial conductivity), g^t (the bulk conductivity), ΔV_m (the magnitude of the transmembrane potential), V_{wave} (the difference in extracellular potential across the propagating wave), and β (the cell surface-to-volume ratio).

make new measurements that take into account the present understanding of cardiac structure and electrophysiology.

The measured values for gap junction resistance also vary widely in the literature, ranging from 0.5 to 40 MΩ [18,33,34]. Most of these experiments were performed on isolated cell pairs, where environmental factors may influence the state of the gap junction. Simulation studies show that values of gap junction resistance in the range of 0.5 to 2.0 MΩ do not lead to significant distortion in the action potential upstroke (e.g., notches and humps), suggesting that this range is nominally normal. Abnormal conditions (ischemia, increased pH, etc), however, are likely to increase the resistance by an order of magnitude or greater.

VI.   CHOOSING A MODEL OF TISSUE STRUCTURE

A.   Discrete Models

The ultimate choice of tissue model often depends on the type of questions being asked and on practical considerations of computational tractability. Despite its simplicity, the one-dimensional fiber continues to play an important role in cardiac biophysics. The primary advantage of this simple virtual preparation is that it is straightforward to couple cells, since the connections are only at the ends. Discrete fiber models are particularly valuable when using more advanced ionic models, because they are formulated for a specific cell size. Discrete models also provide more realistic behavior under conditions of cell decoupling. Increasing the average resistivity in a continuous single fiber will have no impact on the action potential shape, only its conduction velocity. It has been demonstrated in numerous experimental preparations that decoupling of cells leads to marked changes in the action potential upstroke, including notches and humps [35]. In addition, decoupled cells can block action potential propagation [36]. Such propagation failure is not theoretically possible in a uniform continuous fiber except at infinite resistivity.

The extension of the discrete fiber model to higher dimensions is straightforward but requires some strategy for assigning interconnections. Leon and Roberge developed the two-dimensional discrete network in which the preparation was composed of laterally coupled fibers [37]. In this model, the spacing of the transverse resistances was periodic and chosen to give a realistic ratio of transverse to longitudinal conduction velocities. With the regular staggered topology, some conduction produces local collisions increasing the maximum rate of rise of the upstroke while other conduction is branched, leading to a lower $\dot{V}_{max}$ at those sites. Thus the tissue gives rise to a periodic fluctuation in action potential shape and upstroke rate of rise $\dot{V}_{max}$.

Periodic connections are clearly a simplification of the actual cellular topology. Muller-Borer et al. [38] investigated conduction in a two-dimensional tissue model in which the cells' sizes and resistive connections were random. As expected, the random structure produces a nonuniform activation pattern that varies spatially as a function of coupling strength. Such spatially nonuniform activation is consistent with the experimental findings of Spach et al. [35], who hypothesize that the observed variation in $\dot{V}_{max}$ arise because the irregular cell shapes and nonuniform distribution of junctions produce a nonhomogeneous distribution of electrical loading. Spach and Heidlage [28] tested this hypothesis by developing a computer model in which the irregular cell geometry and junction distribution were represented. The simulations showed the variability seen experimentally and emphasized the role of the microstructure on the fine details of high-resolution recordings.

To illustrate discrete propagation, a model of a $0.5\,cm \times 0.5\,cm$ sheet of tissue was created in which cells of random lengths were arranged in a parallel array of fibers. The ionic currents of the cell membrane were modeled using the Luo-Rudy I model. Cell sizes ranged from 5 to 11 nodes ($\Delta x = 15\,\mu m$) per cell (75–$165\,\mu m$) and a fixed diameter of $15\,\mu m$. The intracellular region of each space had a resistivity of $\rho = 100\,\Omega$-cm. The cells are coupled longitudinally at their ends with gap junctions with a resistance of $Rjunc_x = 0.5\,M\Omega$. In the transverse direction, nodes were uncoupled, except at gap junctions, $Rjunc_y = 5.0\,m\Omega$, spaced every 1–4 nodes. Initiation of propagation was induced by a $2000\text{-}\mu A/cm^3$ stimulus in a $60\,\mu m \times 60\,\mu m$ block in the corner of the simulated tissue.

Figure 3A (left panel) shows an activation map with isochrones at every. The jagged isolines are a result of small local changes in conduction velocity, due to the random discrete structure. On average, the propagation velocity is $59.5\,cm/s$ in the longitudinal direction and $20.7\,cm/s$ in the transverse direction. Figure 3A (middle panel) shows a phase plane plot of V_m versus $\dot{V}_m$. This type of plot helps to reveal the variation in the time course of the upstroke resulting from discrete propagation. As has been described previously, the upstroke corresponding to action potentials propagating along cell lengths (longitudinal direction) have a slower rate of rise than those propagating across fibers.

A characteristic feature of discrete propagation is fast propagation within a cell and a delay across the junction. These delays are illustrated in Fig. 3A (right panel). The leftmost traces correspond to 6 consecutive nodes spanning two cells in the longitudinal direction. The transjunction delay is approximately $0.2\,ms$, while the intercellular node-to-node delay is $0.01\,ms$. The rightmost trace shows 6 consecutive nodes and spanning two cells in the transverse direction. The transjunctional delay is of the order of $0.1\,ms$, while intercellular node-to-node delays are approximately 0.04 ms. Wave front conduction is more uniform across cells, as expected.

B. Continuous Monodomain and Bidomain Models

Although directional variability in the action potential shape has been observed, the changes are generally small under normal coupling conditions. In addition, the macroscopic patterns of activation are, on average, consistent with the view that cardiac tissue is syncytial. As noted above, both the monodomain and bidomain models are justified under the assumption that the discrete cellular effects are small. Continuous models of cardiac tissue have several advantages over discrete models. First, there is no need to account explicitly for each cell and its coupling to neighboring cells, simplifying grid generation. Second, current flow in both intracellular and interstitial spaces can be described straightforwardly in three dimensions. Finally, a continuous formulation permits traditional numerical schemes such as finite differencing and finite elements to be applied to the governing equation. Such methods enable spatially varying properties and complex geometries to be considered.

To illustrate continuous propagation, a model of a 0.5-cm $\times$ 0.5-cm sheet of tissue was created with average properties such that the conduction velocities along and across fibers were similar to those obtained in the discrete model shown in Fig. 3B (left panel). In this case, $\sigma_x = 2.1\,\text{mS/cm}$ and $\sigma_y = 0.325\,\text{mS/cm}$. As in the discrete model, initiation of propagation was induced by a 2000-μA/cm^3 stimulus in a 60-μm $\times$ 60-μm block in the corner of the simulated tissue.

Figure 3B (left panel) shows an activation map with isochrones at every 1.5 msec. Unlike the discrete model, the isolines are smooth due to the underlying homogeneous structure. The average propagation velocity is 60.5 cm/sec in the longitudinal direction and 20.5 cm/sec in the transverse direction. Figure 3B (middle panel) shows a phaseplane plot of V_m versus $\dot{V}_m$. Even for tissue with continuous anisotropic properties, the waveshape is independent of direction, since an anisotropic domain is simply a scaled version of an isotropic domain. Figure 3B (right panel) shows that wavefront conduction is uniform in both directions.

Continuous models are often used for three-dimensional preparations where the effects of fiber rotation and curvature more strongly influence conduction. The early eikonal equation-based models (which directly relate wavespeed to continuous tissue properties) showed the speed of propagation to be determined by the speed in the fibers with which the front direction is most closely aligned [39]. In addition, wavefronts arising from below the surface can break through such that the apparent surface velocity is greater than it is for surface stimulation. As the front traverses deeper rotating fibers, the front in a given plane flattens across fibers and eventually begins to dimple. This oblong shape arises because the direction of fastest conduction rotates in depth and distorts the front in adjacent layers. These results have been validated experimentally by Taccardi et al. [40] through a series of pace mapping experiments.

The factors in choosing between using a continuous monodomain model and a continuous bidomain model are manyfold. Muler and Markin were among the first to show that the patterns of activation are affected by the differences in anisotropy in the intracellular and extracellular domains [23]. While the effects are generally small, the activation patterns are not the same. As noted previously, as the ratios of anisotropy become equal, the general bidomain model reduces to a monodomain model. It is important to keep in mind that regardless of the model, wavespeed is affected by both the intracellular and extracellular properties. Hence, when using a monodomain model, the properties used in the simulation should reflect the contribution of both domains.

To illustrate propagation in a bidomain, a model of a 0.5-cm $\times$ 0.5-cm sheet of tissue was again created. In contrast to the mondomain model, conductivities were assigned separately to the intracellular and interstitial domains. To approximately match the propagation velocity of the monodomain case (Fig. 3B), the following conductivities were chosen: $\sigma_{ix} = 3.412\,\mathrm{mS/cm}$, $\sigma_{iy} = 0.4043\,\mathrm{mS/cm}$, $\sigma_{ex} = 10.0\,\mathrm{mS/cm}$, and $\sigma_{ey} = 8.0\,\mathrm{mS/cm}$. For consistency, propagation was initiated by a 2000-μA/cm^3 stimulus in a 60-μm $\times$ 60-μm block in the corner of the simulated tissue.

Figure 3C (left panel) shows an activation map with isochrones at every 1.5 msec. As with the monodomain model, the isolines are smooth due to the underlying homogeneous structure. The front, however, is not elliptical. There is a slight flattening of the front across fibers. The average propagation velocity is 60.1 cm/sec in the longitudinal direction and 20.5 cm/sec in the transverse direction.

Figure 3C (middle panel) shows a phaseplane plot of V_m versus $\dot{V}_m$. For a bidomain tissue with unequal anisotropy, the waveshapes are not the same in both directions (particularly for the early rising phase) since a simple scalar cannot transform both domains into an isotropic region. Note, however, that the shape change is small. Figure 3C (right panel) shows that wavefront conduction is again uniform in both directions.

While the monodomain is generally believed to be a good approximation to a tissue lying in a large-volume conductor, there are still a few subtle effects of the adjoining bath that are not captured. For example, the shunting effects of the bath on the local circuit currents are expected to vary as a function of distance into the tissue (away from the bath). Measurements by Suenson [41] and by Knisley et al. [42] suggested that the bath non-uniformity is in the interstitial resistance. In the monodomain model, this bath effect would have to be introduced through the assignment of a non-uniform conductivity tensor, which would have to be modified every time the bath or tissue properties changed. This bath effect, however, is more easily handled in the bidomain formulation. In the bidomain model, the bath–tissue boundary conditions can be included. A bidomain model with the bath effect can explain the average directional differences in waveshape observed by Spach et al. [35] and Tsuboi et al. [43], which can arise in a continuous model if unequal anisotropy is assumed [44].

Finally, Neu and Krassowska [45] argue that since the bidomain is formulated under the assumption that properties are spatial averages and the membrane is everywhere present, its application to a given problem should be done cautiously. They note that the validity of the bidomain is questionable when (1) the region of interest is near the surface or sources, (2) the phenomena of interest is smaller that a few length constants, (3) the phenomena of interest occurs on a time constant of less than 1 msec, (4) the transmembrane potential magnitude is significantly greater than 100 mV, and (5) the electric field is greater than 70 V/cm and nonuniform. In general, these restrictions are not difficult to overcome when modeling wavefront propagation under normal conditions.

C. Computational Considerations

The sophistication and choice of the model is often limited practically by considerations of computational tractability. With cell dimensions of approximately $10 \times 10 \times 100 \, \mu m$, even small volumes of tissue require a large computational domain. For example, a discrete model of a 1-cm $\times$ 1-cm $\times$ 1-cm region of myocardium, at the resolution of a single cell, would need a computational grid of at least $1000 \times 1000 \times 100$, or 100 million nodes! Using a more syncytial model, such as a monodomain, allows the use of a larger element size that encompasses multiple cells. An element size of $100 \times 100 \times 100 \, \mu m$ reduces the computational-grid 1-cm^3 domain by a factor of 100. This domain size can be significantly reduced if advanced computational techniques such as adaptive mesh refinement are used [46]. These methods have been generally applied to continuous models.

An interesting paradigm is one that would combine spatial adaptation with microstructural models. Using this approach the local region encompassing the narrow wavefront would include the discrete structure at cell resolution, while the region outside the front would be a continuous approximation using significantly larger elements. The obvious advantage is that the benefits of both models can be considered. As of yet, such an approach has not been attempted.

D. Conclusions

Cardiac structure has a clear impact on wavefront conduction, the time course of action potentials, and the extracellular field. Inhomogeneities at the cell or subcellular level can give rise to measurable spatial variation in wavespeed and action potential rate of rise. While on a microscopic scale conduction is discontinuous, effectively jumping from cell to cell, the significant coupling in normal tissue acts to produce relatively smooth activation over the spatial resolution of most measurement techniques. In regions of injury or disease, the coupling is sparse and can easily lead to regions where current-to-load mismatches occur, creating the conditions for impulse conduction to fail. The fine details of interpreting how these mismatches lead to block, however, depend on choice of the model.

As noted, continuous monodomain or bidomain models are often adequate to study macroscopic activation and recovery in normal tissue. There are only subtle differences in the properties of conduction in an equally anisotropic bidomain (a monodomain) and a unequally anisotropic bidomain. Hence, in many practical situations, there is no need to account for the extracellular properties, except in its important contribution to the overall conduction velocity, and the tissue can be treated as a computationally more tractable monodomain. Although the extracellular space can influence propagation and waveshape, the most common reason for choosing a bidomain model over the monodomain models is a desire either to introduce or to sense current in the interstitial space. In most experimental and clinical situations, stimulation and recording are extracellular. Since the interstitial space is accessible in the bidomain model, it is possible to model realistic stimulation and extracellular potential distributions. Several stimulation effects, such as virtual electrodes in which adjacent regions of depolarization and hyperpolarization adjoin at the stimulus site, are a consequence of the unequal anisotropy [47]. Again, while some of these stimulus effects can be mimicked in a monodomain through some nonuniform modification of the tissue properties and transmembrane currents, the steps in the transformation may be problem specific.

Clearly, much has been learned from the models already regarding underlying interrelationships of tissue structure and membrane excitability. Until computational methods improve enough so each cell can be explicitly represented, investigators will continue to make compromises regarding the choice of the model and its size. As a result, understanding the basic assumptions and interrelationship of the different models will remain critically important in the study of the electrical communication in excitable tissue.

REFERENCES

1. Hoy RH, Cohen ML, Saffitz JE. Distribution and three-dimensional structure of inter-cellular junctions in canine myocardium. Circ Res 64:563–574, 1989.
2. Saffitz JE, Hoyt RH, Luke RA, Kanter HL, Beyer EC. Cardiac myocyte interactions at gap junctions: role in normal and abnormal electrical conduction. Trends Cardiovasc Med 2:56–60, 1992.
3. Gourdie RG. A map of the heart: gap junctions, connexin diversity and retroviral studies of conduction myocyte lineage. Clin Sci 88:257–262, 1995.
4. Makowski L. X-ray diffraction studies of gap junction structure. Adv Cell Biol 2:119–158, 1988.
5. Unwin PNT, Zampighi G. Structure of the junction between communicating cells. Nature 283:545–549, 1980.
6. Beyer EC. Gap junctions. Int Rev Cytol 137C:1–37, 1993.
7. Davis LM, Rodefeld ME, Green K, Beyer EC, Saffitz JE. Gap junction protein phenotypes of the human heart and conduction system. J Cardiovasc Electrophysiol 6(10):813–822, 1995.
8. Clerc L. Directional differences of impulse spread in trabecular muscle from mammalian heart. J Physiol 255:335–346, 1976.
9. Huang XD, Sandusky GE, Zipes DP. Heterogeneous loss of connexin43 protein in ischemic dog hearts. J Cardiovasc Electrophysiol 10(1):79–91, 1999.
10. Streeter DD. Gross morphology and fiber geometry of the heart. In: Geiger SR, ed. Handbook of Physiology, Sec 2. The Cardiovascular System American Physiology Society, Bethesda, MD: 1979, pp. 61–112.
11. Streeter DD, Bassett DL. An engineering analysis of myocardial fiber orientation in the pig's left ventricle in systole. Anat Rec 155:503–511, 1966.
12. Le Grice IJ, Smaill BH, Chai LZ, Edgar SG, Gavin JB, Hunter PJ. Laminar structure of the heart: ventricular myocyte arrangement and connective tissue architecture in the dog. Am J Physiol 269:H571–H582, 1995.
13. Plonsey R. Bioelectric Phenomena. New York: McGraw-Hill, 1969.
14. Plonsey R. Fundamentals of electrical processes in the electrophysiology of the heart. In: Ghista DN, ed. Advances in Cardiovascular Physics, 1979, pp. 1–28.
15. Plonsey R. The active fiber in a volume conductor. IEEE Trans Biomed Eng 21:371–381, 1974.
16. Clark J, Plonsey, R. A mathematical evaluation of the core conductor model. Biophys J 6:95–112, 1966.
17. Weidmann S. Electrical constants of trabecular muscle from mammalian heart. J Physiol 210:1041–1054, 1970.
18. Chapman RA, Fry CH. An analysis of the cable properties of frog ventricular myocardium. J Physiol 283:263–282, 1978.

19. Henriquez CS. Simulating the electrical behavior of cardiac tissue using the bidomain model. Crit Rev Biomed Eng 21:1–77, 1993.

20. Hodgkin AL. A note on conduction velocity. J Physiol (Lond) 125:221–224, 1954.

21. Tung L. A bidomain model for describing ischemic myocardial D.C. potentials. Ph D dissertation, Massachusetts Institute of Technology, Cambridge, MA, 1978.

22. Schmitt OH. Biological information processing using the concept of interpenetrating domains. In: Leibovic KN, ed. Information Processing in the Nervous System, Berlin, Heidelberg, New York: Springer-Verlag, 1969, p. 329.

23. Muler AL, Markin VS. Electrical properties of anisotropic nerve-muscle syncytia. II. Spread of flat front of excitation. Biophysics 22:518–522, 1977.

24. Diaz PJ, Rudy Y, Plonsey R. Intercalated discs as a cause for discontinuous propagation in cardiac muscle: a theoretical simulation. Ann Biomed Eng 11:177–189, 1983.

25. Henriquez CS, Plonsey R. Effect of resistive discontinuities on waveshape and velocity in a single cardiac fiber. Med Biol Eng Comput 25:428–438, 1987.

26. Rudy Y, Quan W. A model study of the effects of the discrete cellular structure on electrical propagation in cardiac tissue. Circ Res 61:815–823, 1987.

27. Leon LJ, Roberge FA. Structural complexity effects on transverse propagation in a two-dimensional model of myocardium. IEEE Trans Biomed Eng 38(10):997–1009, 1991.

28. Spach MS, Heidlage JE. The stochastic nature of cardiac propagation at a microscopic level. Electrical description of myocardial architecture and its application to conduction. Circ Res 76(3):366–380, 1995.

29. Plonsey R, van Oosterom A. Implications of macroscopic source strength on cardiac cellular activation models. J Electrocardiol 24:99–112, 1991.

30. Roberts DE, Hersh LT, Scher AM. Influence of cardiac fiber orientation on wavefront voltage, conduction velocity, and tissue resistivity in the dog. Circ Res 44(5):701–712, 1979.

31. Roberts D, Scher AM. Effect of tissue anisotropy on extracellular potential fields in canine myocardium in situ. Circ Res 50:342–351, 1982.

32. Kleber AG, Riegger CB. Electrical constants of arterially perfused rabbit papillary muscle. J Physiol 385:307–324, 1987.

33. Weingart R. Electrical properties of the nexal membrane studied in rat ventricular cell pairs. J Physiol 370:267–284, 1986.

34. Weingart R, Maurer P. Action potential transfer in cell pairs isolated from adult rat and guinea pig ventricles. Circ Res 63(1):72–80, 1988.

35. Spach MS, Miller WT, Geselowitz DB, Barr RC, Kootsey JM, Johnson EA. The discontinuous nature of propagation in normal canine cardiac muscle. Evidence for recurrent discontinuities of intracellular resistance that affect the membrane currents. Circ Res 48(1):39–54, 1981.

36. Wang Y, Rudy Y. Action potential propagation in inhomogeneous cardiac tissue: safety factor considerations and ionic mechanism. Am J Physiol Heart Circ Physiol 278(4):H1019–H1029, 2000.

37. Leon LJ, Roberge FA. Directional characteristics of action potential propagation in cardiac muscle. A model study. Circ Res 69(2):378–395, 1991.

38. Muller-Borer BJ, Erdman DJ, Buchanan JW. Electrical coupling and impulse propagation in anatomically modeled ventricular tissue. IEEE Trans Biomed Eng 41(5):445–454, 1994.

39. Colli Franzone P, Guerri L, Rovida S. Wavefront propagation in an activation model of the anisotropic cardiac tissue: asymptotic analysis and numerical simulations. J Math Biol 28(2):121–176, 1990.

40. Taccardi B, Macchi E, Lux RL, Ershler PR, Spaggiari S, Baruffi S, Vyhmeister Y. Effect of myocardial fiber direction on epicardial potentials. Circulation 90(6):3076–3090, 1994.

41. Suenson M. Interaction between ventricular cells during the early part of excitation in the ferret heart. Acta Physiol Scand 125(1):81–90, 1985.

42. Knisely S, Maruyama T, Buchanan JW Jr. Interstitial potential during propagation in bathed ventricular muscle. Biophys J 59(3):509–515, 1991.

43. Tsuboi N, Kodama I, Toyama J, Yamada K. Anisotropic conduction properties of canine ventricular muscles. Influence of high extracellular K^+ concentration and stimulation frequency. Jpn Circ J 49(5):487–498, 1985.

44. Henriquez CS, Muzikant AL, Smoak CK. Anisotropy, fiber curvature and bath loading effects on activation in thin and thick cardiac tissue preparations: simulations in a three dimensional bidomain model. J Cardio Electrophys 7:424–444, 1996.

45. Neu JC, Krassowska W. Homogenization of syncytial tissues. Crit Rev Biomed Eng 21(2):137–199, 1993.

46. Cherry EM, Greenside HS, Henriquez CS. A space-time adaptive method for simulating complex cardiac dynamics. Phys Rev Lett 84(6):1343–1346, 2000.

47. Sepulveda NG, Roth BJ, Wikswo JP, Jr. Current injection into a two dimensional anisotropic bidomain. Biophys J 55:987–999, 1989.

4
Electrical Stimulation of Cardiac Cells

Wanda Krassowska
Duke University, Durham, North Carolina, U.S.A.

Bradley A. Stone
Medtronic USA, Inc., Columbus, Ohio, U.S.A.

John C. Neu
University of California at Berkeley, Berkeley, California, U.S.A.

I. INTRODUCTION

A single cell in vitro is a common object of electrophysiological investigations. Isolated from its environment and maintained under tightly controlled experimental conditions, a single-cell preparation reveals intrinsic excitability properties of the cell membrane. Following methodology introduced by Hodgkin and Huxley [1], most studies are performed under conditions of spatial clamp, during which the cell membrane or a portion of it is exposed to spatially uniform electrical conditions. This methodology allows measurements of temporal changes in membrane currents and, combined with the use of channel-blocking drugs or specially designed voltage clamp protocols, reveals the underlying kinetics of ionic currents responsible for the membrane excitability. These types of studies lead to quantitative models of the membrane kinetics (see Ch. 1, 2).

Since the 1980s, researchers in cardiac electrophysiology have been increasingly interested in studies of isolated cardiac cells exposed to an external field. When a cell is placed in an electric field, it becomes polarized [2–4]. The transmembrane potential (i.e., the difference between the potential inside and outside the cell) in the part of the cell facing the positive electrode (anode) decreases below resting potential, hyperpolarizing the membrane. The transmembrane potential in the part of the cell facing the negative electrode (cathode) increases above the resting potential, depolarizing the membrane. This field-induced polarization of a single cell (Fig. 1) resembles the "sawtooth potential" that theoretically should arise in a fiber consisting of separate cells connected by gap junctions [5–8]. Hence, just as the polarization of a single cell by the field is known to affect its physiological state (e.g., by causing it excitation), the sawtooth potential may contribute to the field stimulation, electrical induction of rotors, and defibrillation of the heart. Therefore, studying the response of a single cell in an external field may give insight into the processes occurring in the bulk of tissue at a considerable distance from the electrodes.

One obvious similarity between the effects of an electric field on a single cell and on cardiac muscle is that both are characterized by spatial changes in the transmembrane potential. In a cell, these spatial changes one very rapid: during field stimulation, the transmembrane potential is expected to change by at least 50 mV over a distance of 20–200 μm (i.e., between the two opposite ends of the cell). Because the hyperpolarized and depolarized portions of the cell membrane are tightly coupled through a highly conductive cell interior, the activation of the cell by an electric field differs in many respects from the activation caused by the intracellular injection of stimulating current. In this sense, single-cell investigations bridge the gap between fundamental electrophysiological studies conducted

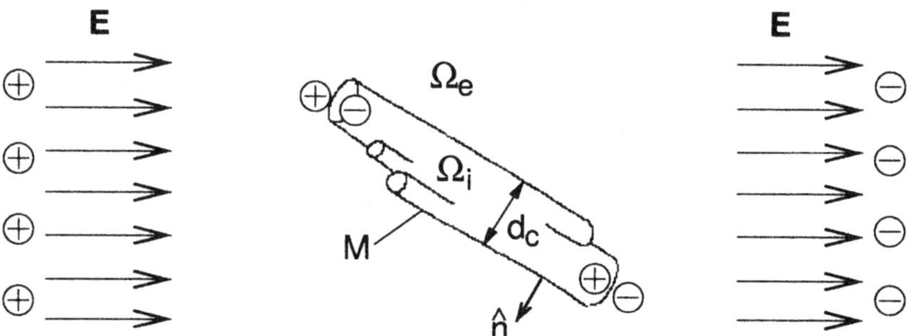

Figure 1 An idealized representation of a single cardiac cell in an external electric field. An excitable membrane is represented by a closed surface M that separates the interior of the cell Ω_i from the surrounding medium Ω_e. The cell has a typical diameter d_c, usually on the order of 10–20 μm. Electric field E, established by external electrodes, is assumed to be uniform in the vicinity of the cell.

on space-clamped preparations and clinical applications occurring under spatially changing electrical conditions.

This chapter summarizes the present understanding of the behavior of a single cell in an external electric field. Section II provides a brief review of experimental studies performed on isolated cardiac cells. Section III is devoted to the theory describing the response of a single cell to an external field. Section IV addresses the issue of whether the theoretical understanding of the single-cell behavior is adequate to reproduce experimental results. Finally, Sec. V discusses implications of the single cell studies for understanding the behavior of multicellular tissues.

II. EXPERIMENTAL STUDIES OF STIMULATION OF CARDIAC CELLS

A. Preparation of Cells

Experimental studies on single cardiac cells typically use cells from the ventricles of mature guinea pigs [4,9–11], dogs [12], frogs [10], and rabbits [3], or from the ventricles of chick embryos [13]. In the case of chick embryos, cardiac myocytes are isolated from 11-day-old embryos by using trypsin [14]. In the case of mature animals, the whole-heart enzymatic dissociation technique is used [9,10], in which the excised heart is placed in a Langerdorff apparatus and perfused for a few minutes with a solution containing collagenase and protease. After rinsing the heart and removing it from the perfusion apparatus, the ventricles are cut into small slices that are placed in a petri dish and gently shaken to release dissociated cells. The cell suspension can be used immediately or incubated overnight at 37°C in a culture solution [4,9,13].

During field stimulation experiments, cells are placed in a HEPES-buffered salt solution: Tyrode's solution for mammalian cells and Ringer's solution for frog cells. The experiments are conducted either at room temperature (23–27°C) [10–12,15] or at body temperature (36–37°C) [4,13], which is maintained by placing the culture dish in a microincubator. To prevent evaporation at 37°C, a thin layer of mineral oil is placed on the surface of the solution.

Ventricular cells of mature animals such as rabbit myocytes, shown in Fig. 2a, have an approximate rod shape and visible cross-striations. Rabbit cells are typically $119 \pm 29\,\mu\text{m}$ long and $22 \pm 4\,\mu\text{m}$ wide [3]. Guinea pig myocytes are $110–270\,\mu\text{m}$ long and $10–40\,\mu\text{m}$ wide [9,10]. Dog cells are somewhat smaller, $50–100\,\mu\text{m}$ long and $10–20\,\mu\text{m}$ wide [12]. Frog cells have the highest aspect ratio, being $200–400\,\mu\text{m}$ long (when stretched) and $5–10\,\mu\text{m}$ wide [10]. Chick embryo cells (Fig. 2b) are spherical, $15–27\,\mu\text{m}$ in

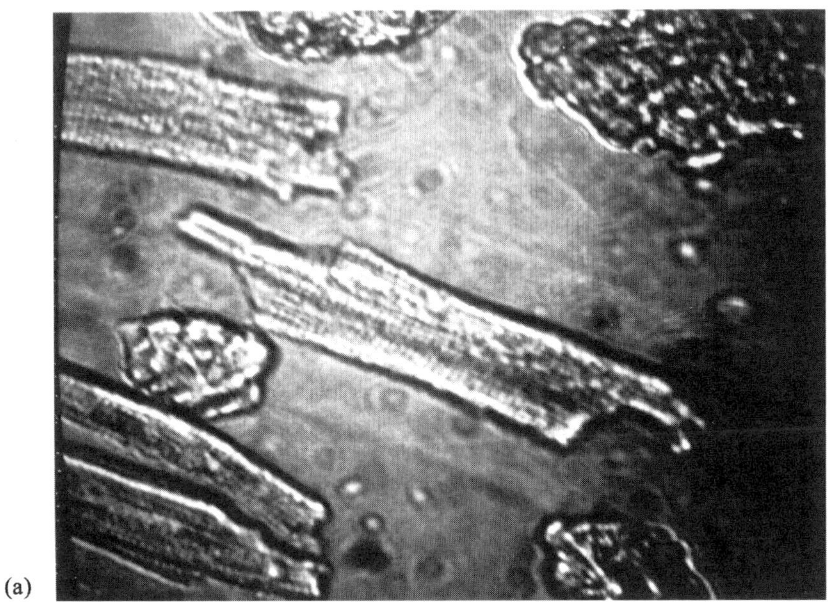

(a)

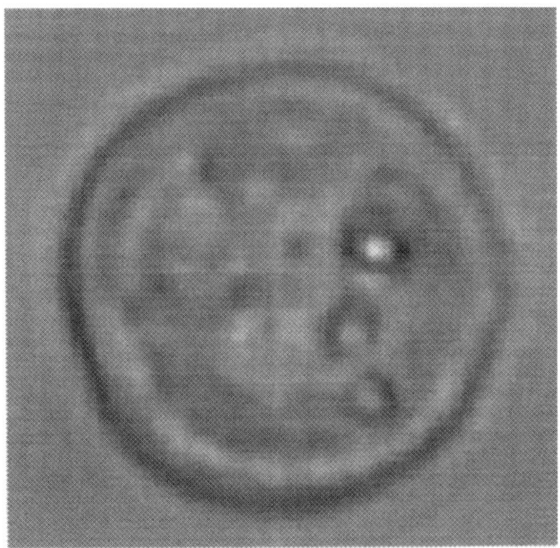

(b)

Figure 2   (a) Enzymatically dissociated cardiac ventricular cells of a mature rabbit. (Courtesy of Dr. Stephen B. Knisley, University of Alabama, Birmingham.) (b) An enzymatically dissociated cardiac ventricular cell of an 11-day-old chick embryo.

diameter [13]. Once dissociated, approximately 250,000–500,000 cells are deposited in a petri dish. Many cells are not suitable for the field stimulation experiment because of damage during the dissociation process or they are near the electrodes and may be exposed to a nonuniform field. Cells selected for the study should be located near the center of the experimental chamber, have no neighboring cells in their immediate proximity, be quiescent at the beginning of the study, and have visible signs of contraction when stimulated. During the study, cells are observed under a microscope. Most studies use a camera mounted on a microscope to display the cell on a video monitor [13], to videotape the cell response [9], or to transfer the cell image to a computer for further processing [12].

B. Field Stimulation

Field stimulation experiments are conducted in a chamber designed to provide a uniform stimulating field. A chamber is constructed from Teflon or Plexiglass and has a rectangular shape. The stimulus is delivered through a pair of parallel platinum-black electrodes attached to the opposite sides of the chamber (Fig. 3a). The electrodes are typically 1 mm high, up to 1.4 cm long, and positioned 0.6–1.1 cm apart [9,11–13]. In a study by Bardou et al. [9], the electrode assembly was mounted on a rotating ring so that the field could be applied at any angle with respect to the cell. Tung et al. [10] used micropaddle electrodes, 0.8–1.2 mm in length and 0.4–0.6 mm apart. With a micromanipulator, the electrodes were lowered into the suspension as a unit to straddle an individual cell.

Electrical shocks are generated by a commercially available stimulator (e.g., a clinical Grass stimulator [10,11,13]) or by a custom-built pulse generator [3,10,12]. The cells are allowed to rest for a few seconds between successive stimuli. The stimulus magnitude (in volts), current (in amperes), and/or duration (in milliseconds) are measured with an oscilloscope. The nominal strength of the stimulating field can be computed by dividing the potential drop across the chamber by the distance between the stimulating electrodes. Realizing that this calculation could contain errors caused by edge effects and losses on the electrode–solution interface (see Ch. 7), some researchers performed direct measurements of the field strength [3,10,13]. Tung et al. [10] measured potentials at fixed distances between micropaddles with a Teflon-coated stainless steel wire. These measurements provided a calibration to compute the field from the current delivered by the stimulator. Stone et al. [13] used a dual-wire probe (Fig. 3b) to measure potentials in several locations within the chamber. The field strength determined by the probe measurements was independent of the position of the probe and differed by 3–10% from the nominal field computed from the voltage drop across the chamber.

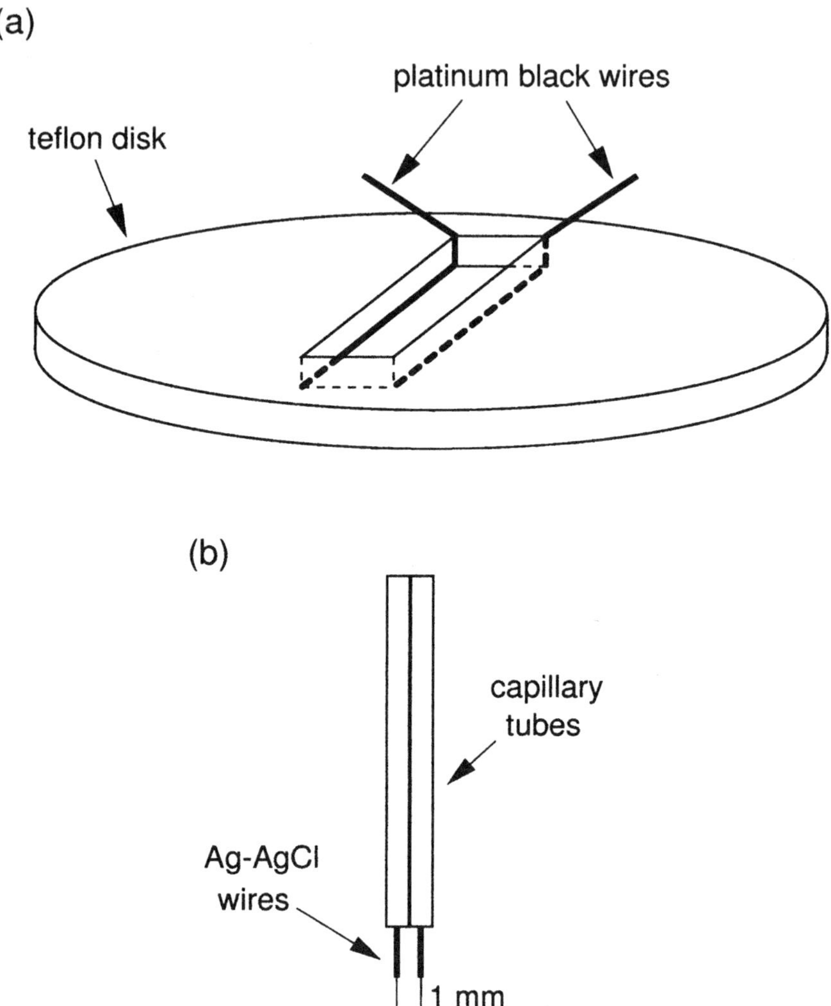

Figure 3 (a) An example of an experimental chamber used for stimulating cells with a uniform electric field. The chamber consists of a circular Teflon plate with a 6-mm × 14-mm rectangular window in the center. The stimulus is delivered through two platinum-black wires flattened to a height of approximately 1 mm and placed along the bottom edge of the long sides of the window. The diameter of the chamber, 35 mm, allows it to fit directly into the culture dish. (b) A dual-wire probe used to measure the strength of a field. The probe consists of two leads made of Ag–AgCl wires, 0.3 mm in diameter, aligned 1 mm apart by capillary tubes that are glued together with epoxy. A micromanipulator is used to lower the probe to the bottom of the chamber at several locations within the window. (From Ref. 13.)

C. Measurements of the Thresholds for Field Stimulation

The earliest studies on the stimulation of single cardiac cells concentrated on measurements of the cell activation threshold, its dependence on pulse duration, and the orientation of the field with respect to the cell. As microelectrode measurements are technically difficult in the presence of large electric fields, most of these studies relied on the contraction of the cell as a criterion of activation [9,10,13]. An exception is the study by Ranjan and Thakor [12], who used a perforated patch clamp technique to measure the intracellular potential of a field-stimulated cell. The stimulus artifact was eliminated by using a feedback circuit to apply the field symmetrically with respect to the reference electrode.

In most studies, thresholds are expressed in terms of the applied electric field (E_{th}), either measured directly or estimated from the inter-electrode distance (Sec. II.B). However, thresholds expressed in terms of the field strength are dependent on the cell dimension, with small cells having higher thresholds than large ones [13]. In order to compare thresholds from different cells, some researchers converted E_{th} to the transmembrane potential V_m by using a theoretical estimate of the maximum magnitude of V_m induced in a cell by a uniform field of strength E [11,13]. For a spherical cell (Sec. III.B.3),

$$V_m = \frac{3}{4} d_c E \tag{1}$$

where d_c is the diameter of the cell. For a cylindrical cell in the transverse field (Sec. III.B.2),

$$V_m = d_c E \tag{2}$$

There is no exact formula for a cylindrical cell in the longitudinal field; the often-used approximate formula is [16]

$$V_m = \frac{1}{2} l_c E \tag{3}$$

where l_c is the length of the cell.

Equations (2) and (3) imply that for mature cardiac myocytes (Fig. 2a), the activation threshold should depend on the orientation of the cell with respect to the field. Such a dependence was indeed observed experimentally [9,10,12]. Tung et al. [10] showed that the field strength required for activation was lower when the cell was placed with its long axis parallel to the stimulating field. For a 2-ms pulse, the longitudinal and transverse thresholds were $2.4 \pm 0.6\,\text{V/cm}$ and $13.8 \pm 5.8\,\text{V/cm}$ for frog cells, and

$2.8 \pm 1.5\,\mathrm{V/cm}$ and $7.3 \pm 1.5\,\mathrm{V/cm}$ for guinea pig cells. Bardou et al. [9] observed that the threshold increased gradually as the angle between the cell axis and the field increased from $0°$ to $90°$ in increments of $10°$. A later study by Ranjan and Thakor [12] added more details to this picture: while the activation thresholds for canine cells were $182 \pm 83\%$ higher when the field was transverse to the cell, the thresholds could also change by $98.9 \pm 71\%$ when the field direction was reversed. This result implies a difference in the magnitude of the membrane potential on the opposite ends of the cell. This issue is discussed further in Sec. IV.C.3. Two studies, by Tung et al. [10] and by Stone et al. [13], investigated the dependence of the cell activation thresholds on the duration of the pulse. As an example, Fig. 4 shows the activation thresholds measured for 10 chick heart cells with stimulus durations ranging from 0.2 to 40 ms [13]. Such strength—duration curves are usually represented by the Weiss-Lapicque relationship [17,18],

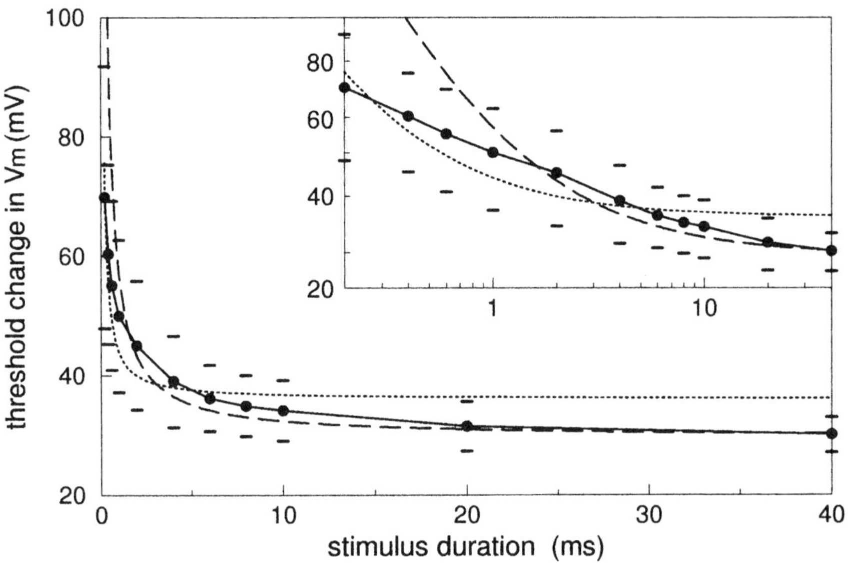

Figure 4   Activation thresholds measured experimentally and expressed in terms of the maximum change of the transmembrane potential from rest. The experimental data points are shown with filled circles connected by a solid line. The standard deviation for each point is shown with short lines above and below the circle. The inset shows the same data as the main panel but on a logarithmic scale. In each panel, the measured thresholds are compared to the Weiss-Lapicque formula, which is fitted by using the Weiss method (dashed line) and nonlinear regression (dotted line). (From Ref. 13.)

$$V_{\text{th}}(d) = V_{\text{rh}}\left(1 + \frac{c}{d}\right) \tag{4}$$

where V_{th} is the stimulation threshold expressed as a maximum change of the transmembrane potential from rest, V_{rh} is the rheobase, c is the chronaxie, and d is the stimulus duration. As described by Mouchawar et al. [19], a nonlinear least-square regression (e.g., the SAS NLIN procedure [20]) can be used to fit Eq. (4) to the experimental data. Alternatively, in the so-called Weiss curve-fitting method, Eq. (4) is multiplied by d, making the right-hand side of the resulting equation a linear function of d, with a slope equal to V_{rh} and an intercept equal to the product $V_{\text{rh}}c$. Hence, V_{rh} and $V_{\text{rh}}c$ can be found by a linear regression. For the data of Fig. 4, the rheobase and chronaxie were 29.82 mV and 0.90 ms (Weiss method) and 36.03 mV and 0.22 ms (nonlinear regression).

Quantitatively, the thresholds of Fig. 4 are consistent with the threshold voltage of the space-clamped chick heart membrane. For a 10-ms stimulus, activation occurred when the depolarized end of the cell was 34.17 mV above rest. With V_{rest} of -83 mV, the depolarized end was at approximately -48.8 mV, above the -51 mV value given as the threshold voltage for the chick heart membrane by Clay et al. [21]. This level of depolarization is also consistent with the range of V_m at which fast sodium current starts flowing, -60 to -40 mV [22–24]. However, Fig. 4 shows that the shape of the experimental strength–duration curve cannot be reproduced correctly by the Weiss-Lapicque relationship (4). When parameters V_{rh} and c are determined by the Weiss method, Eq. (4) overestimates the thresholds for short stimulus durations. For parameters determined by nonlinear regression, Eq. (4) overestimates thresholds for long durations while underestimating thresholds for intermediate durations. Similar problems were also reported by Tung et al. [10] for strength–duration curves of the guinea pig cells. In contrast, the Weiss-Lapicque relationship approximates reasonably well strength–duration curves obtained by stimulating the myocardium with a unipolar electrode [25]. Hence, as argued further in Sec. IV.C.1, the field stimulation of isolated cells may be qualitatively different from the stimulation by current injection.

Field stimulation thresholds are sensitive to temperature, as demonstrated by Stone et al. [15,26], who measured activation thresholds in chick heart cells at both 37 and 27°C. Lowering the temperature reduced the thresholds for all stimulus durations (Fig. 5). For stimulus durations greater than or equal to 1 ms, Student's t test has shown that the thresholds measured at 37°C were significantly higher than the thresholds measured at 27°C, with a confidence level of 0.95. When determined by the Weiss method, the rheobase decreased from 29.82 to 20.8 mV, and the chronaxie

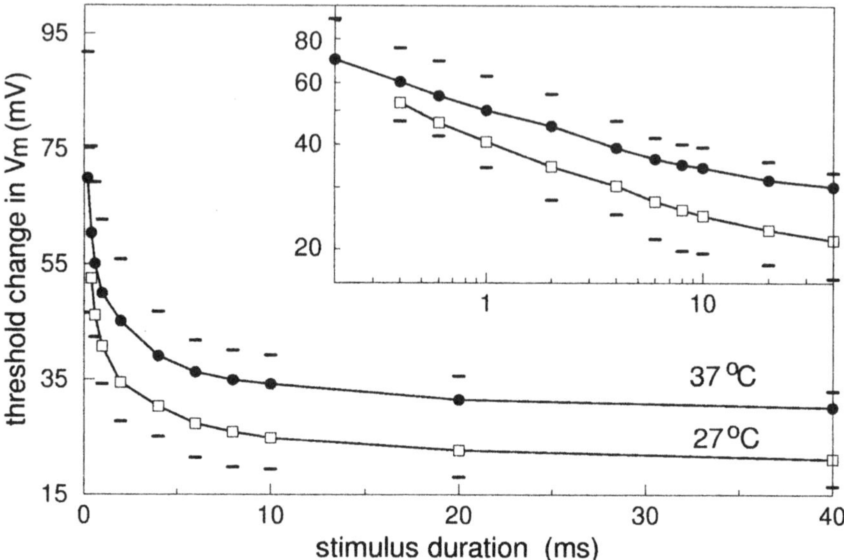

Figure 5 Activation thresholds measured at 37 and 27°C. The experimental data points are shown with filled circles (37°C) and open squares (27°C) connected by solid lines. The standard deviation for each point is shown with short lines above (37°C) or below (27°C) the symbol. The inset shows the same data as the main panel but on a logarithmic scale.

increased from 0.90 to 1.4 ms. For nonlinear regression, the rheobase and chronaxie changed from 36.03 to 25.41 mV and from 0.22 to 0.45 ms, respectively. With an average cell diameter of 20 μm, a rheobase of 20.8 mV corresponds to an electric field of approximately 14 V/cm. This value is larger than those reported for other species. For a 10-ms stimulus, Ranjan and Thakor [12] reported thresholds of 2–10 V/cm for fields parallel to isolated canine cells. Even smaller threshold fields were measured in guinea pig myocytes by Tung et al. [10]: average thresholds were 1.4 and 3.5 V/cm for fields applied parallel and perpendicular to the cell. The chronaxie measured by Tung et al. was 2.2–2.6 ms, a twofold increase compared to the embryonic chick heart cells. The reasons for these differences are discussed in Sec. IV.C.2.

The temperature dependence of field stimulation thresholds for single cells differs from that observed for cardiac fibers. Evans [27], comparing field stimulation thresholds for rabbit papillary muscle at 24 and 37°C, demonstrated that the thresholds at 24°C were slightly lower for stimulus durations greater than 6 ms, but higher for stimulus durations less than 6 ms. This behavior is qualitatively similar to the behavior of thresholds for current injection in a giant squid axon, in the sense that for long stimulus durations (100 ms) thresholds measured at a low temperature are below thresholds measured at a high temperature [28], while for short stimuli (1 ms) the relationship is reversed [29]. A theoretical explanation of this crossover was given by FitzHugh [30] on the basis of the Hodgkin-Huxley nerve model. It is unclear whether such a crossover also exists for the field stimulations of single cells. For chick heart cells, Fig. 5 shows that the thresholds at 27 and 37°C become closer as the stimulus duration decreases; for durations below 1 ms, the difference is not statistically significant. Hence, it is possible that these thresholds might eventually cross, but for a duration much smaller than observed in cardiac fibers and giant squid axons.

D. Measurements of the Field-Induced Transmembrane Potential

The introduction of optical detection methods in the 1980s made it possible to measure the magnitude and distribution of the transmembrane potential induced by the electric shock (Chaps. 14–17). At first, the direct observations of the field-induced transmembrane potential were performed in nonexcitable cells and vesicles [2,31,32]. For cardiac cells, this type of optical measurement was performed on rabbit cells [3] and guinea pig ventricular cells [4,11]. Cells are dissociated as described in Sec. II.A and stained with a voltage-sensitive fluorescent dye, such as di-4-ANEPPS, dissolved in ethanol and diluted in Tyrode's solution to produce a final concentration of 5–12 μm. A calcium-sensitive dye, fura-2, is also used to monitor shock-induced changes in the intracellular calcium concentration [33]. During measurements, a 14–22-μm spot on a cell is illuminated by laser light (argon or He–Ne laser [3,4,11]), which excites the dye to emit fluorescence. The fluorescence signal is measured with a photodiode coupled to a high-gain current-to-voltage amplifier [3,11]. In order to minimize phototoxic side effects, illumination of the cell is limited to 20–500 ms. Simultaneous recordings of the electrical and optical signals demonstrated that the fractional change in fluorescence $(-\Delta F/F)$ is proportional to the change in the transmembrane potential. Typically, fluorescence changes by 8–9% per 100 mV [4], and the response of the dye is linear in the range of −280 to 140 mV [11].

To obtain measurements of the transmembrane potential induced by the field, without any contamination from the upstroke of an action potential, the cell is first excited by a small S_1 stimulus, and the S_2 shock is applied 30–50 ms later, during the plateau. The measurements confirmed the existence of depolarization at the end of the cell facing the cathode and hyperpolarization at the end facing the anode. A study of Knisley et al. [3], in which 20- and 40-V/cm shocks were used, also confirmed that the magnitude of V_m correlates with the cell length and agrees approximately with the theoretical estimate of Eq. (3). However, the magnitude of V_m was larger at the hyperpolarized end: -93 versus 79 mV for the field of 20 V/cm, and -214 versus 177 mV for 40 V/cm. A more pronounced asymmetry in the shock-induced V_m was observed by Cheng et al. in the guinea pig cells [11]. This study, using a wider range of field strengths, revealed that for V_m below -100 mV or above 50 mV, the magnitude of V_m no longer changed linearly with the field strength. In general, V_m at the depolarized end was lower than predicted by Eq. (3), and V_m at the hyperpolarized end was higher. The ratio of V_m at the hyperpolarized and depolarized ends increased with the field strength: from 1.54 when the nominal V_m [i.e., computed from Eq. (3)] was 70 mV to 3.29 for 180 mV.

E. Mapping the Excitation Process in a Single Cell

The studies summarized above used optical fluorescence measurements to monitor the temporal behavior of the transmembrane potential at only one spot on a cell. In contrast, an optical mapping system developed by Windisch et al. [4,34,35] monitors the transmembrane potential at several spots on the cell simultaneously, giving a more complete picture of the cell response during field stimulation and the ensuing activation. In these experiments, the fluorescence emitted by the cell was collected by a 10×10 photodiode array connected to 100 amplifiers; the output of 24 amplifiers was transferred to a digital transient recorder. The field of an individual photodiode was 15×15 or $22 \times 22\,\mu$m, and the interelectrode spacing was 15 or 23 μm. Hence, each cell yielded about 12 individual simultaneous signals.

An example of the results is shown in Fig. 6. The optical signals collected from 14 sites on a cell show depolarization at the end of the cell facing the cathode, hyperpolarization at the opposite end, and essentially no shock-induced V_m at the center (Fig. 6A). Both the onset and termination of the shock-induced potential are fairly rapid (Fig. 6B). After the stimulus is turned off, the cell membrane is at a spatially uniform potential; after a short delay, the upstroke of an action potential is seen. Despite different transmembrane potentials during the shock, the activation process in a single cell during field stimulation is tightly coupled: the delay between the activation of the first and the last part of the cell membrane is approximately 20 μs (which is also the temporal resolution of the system). However, the effect of

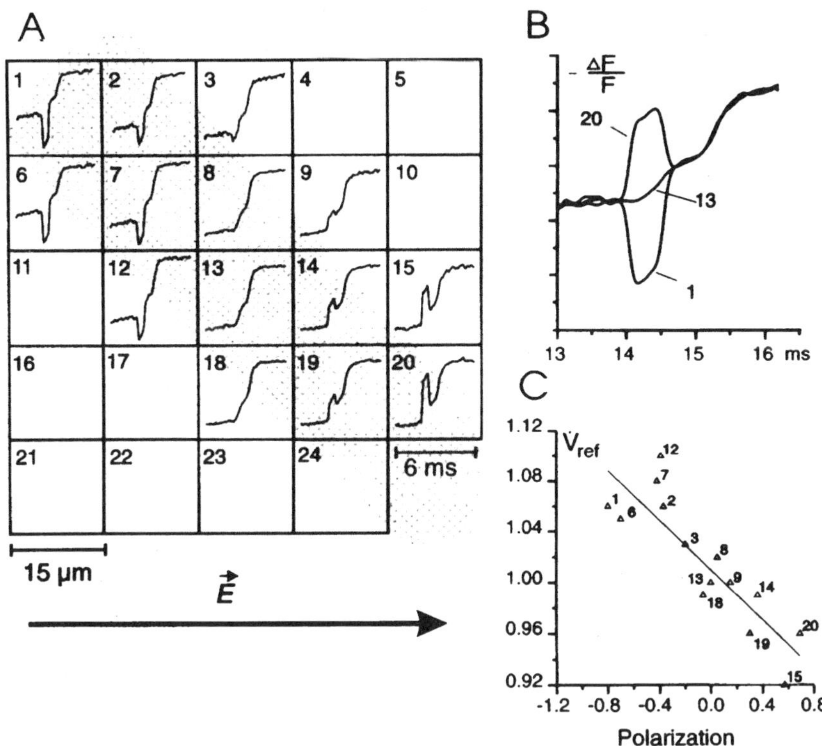

Figure 6 Optical mapping of the field stimulation process in an isolated guinea pig cell. (A) Outline of a cell overlaid on a photodiode array. The subpanels show optical signals measured at 14 numbered sites on a cell. (B) Optical signals $(-\Delta F/F)$ versus time from sites 1, 13, and 20 normalized to the same amplitude. The pulse duration was 0.5 ms, and the field strength was slightly above the threshold. (C) Maximum upstroke velocity $\dot{V}_{\text{ref}}$ plotted as a function of membrane polarization. $\dot{V}_{\text{ref}}$ is a maximum derivative of the optical signal $-\Delta F/F$ measured during the upstroke and expressed relative to its value at site 13. Membrane polarization is the maximum change of optical signal $-\Delta F/F$ during the stimulus, expressed relative to the action potential amplitude. Positive and negative polarization values indicate depolarization and hyperpolarization, respectively. (From Ref. 4.)

different shock-induced V_m is reflected in $\dot{V}_{\text{ref}}$, the maximum slope of the optical signal during the upstroke. As seen is Fig. 6C, $\dot{V}_{\text{ref}}$ is up to 10% higher in the parts of the cell that have been hyperpolarized by the shock than in the parts that have been depolarized. Section IV.B compares these results with the predictions of the model.

III. MODELING STIMULATION OF A SINGLE CARDIAC CELL

A. Statement of the Problem

Consider an idealized model of a single cell immersed in an extracellular bath and exposed to an electric field (Fig. 1). Both the cell and the bath are usually assumed to be source-free, purely resistive regions, with conductivities σ_i and σ_e (Table 1). Hence, intra- and extracellular potentials Φ_i and Φ_e satisfy Laplace's equations,

$$\nabla^2 \Phi_i = 0 \quad \text{in } \Omega_i \tag{5}$$

$$\nabla^2 \Phi_e = 0 \quad \text{in } \Omega_e \tag{6}$$

where Ω_i denotes the interior of the cell, and Ω_e denotes the extracellular bath. The boundary conditions on the external surface $\partial\Omega_e$ are determined by the experimental setup. However, if the cell is small compared to the extracellular region, and if it is located away from the electrodes, the electric field is approximately uniform in the vicinity of the cell. Therefore, the details of the experimental setup can be ignored. In such a case, the extracellular potential far away from the cell corresponds to a uniform electric field $\mathbf{E}$,

$$\Phi_e(\mathbf{x}, t) \sim -\mathbf{E} \cdot \mathbf{x} \qquad \text{as } |\mathbf{x}| \to \infty \tag{7}$$

where $\mathbf{x}$ is the position vector, t is time, and $\sim$ means that $\mathbf{E} \cdot \mathbf{x}$ is a leading-order approximation of Φ_e as $\mathbf{x} \to \infty$. On the membrane M, the potential is discontinuous, and the difference between intra- and extracellular potentials defines the transmembrane potential V_m,

$$V_m = \Phi_i - \Phi_e \qquad \text{on } M \tag{8}$$

Table 1 Typical Parameters of a Cardiac Cell

Symbol	Value	Definition
d_c	15 μm	Cell radius
l_c	100 μm	Cell length
σ_i	4 mS/cm	Intracellular conductivity
σ_e	20 mS/cm	Extracellular conductivity
C_m	1 μF/cm^2	Specific membrane capacitance
R_m	6 kΩ cm^2	Specific membrane resistance
V_{rest}	−84 mV	Rest potential

The current through the membrane is continuous. In both regions Ω_i and Ω_e, normal current densities are equal to the membrane current density I_m,

$$-\hat{n} \cdot \{\sigma_i \nabla \Phi_i\} = -\hat{n} \cdot \{\sigma_e \nabla \Phi_e\} = I_m \quad \text{on } M \tag{9}$$

where $\hat{n}$ is the outward unit normal vector on M. I_m consists of capacitive and ionic components:

$$I_m = C_m \frac{\partial V_m}{\partial t} + I_{\text{ion}} \tag{10}$$

where C_m is the membrane surface capacitance and I_{ion} is the combined ionic current through channel proteins, pumps, and exchangers, representing the complex dynamic response of the excitable membrane. The initial condition at time $t = 0^+$ (i.e., immediately after the field has been turned on) is usually assumed in the form

$$V_m(\mathbf{x}, 0^+) = V_o \quad \text{on } M \tag{11}$$

i.e., the transmembrane potential is independent of the position on the membrane, and its value V_o depends on the cell's physiological state at the time the field is applied. Equations (5)–(11) determine solutions for the potentials up to an arbitrary function of time added to both Φ_i and Φ_e. To obtain a unique solution, Eqs. (5)–(11) must be supplemented by a normalization condition. It is convenient to assume it in the form

$$\int_M \Phi_e \, da = 0 \tag{12}$$

The initial-boundary-value problem (5)–(12) does not, in general, have a closed-form solution. For certain cell geometries and under the assumption of a passive membrane, separation-of-variables solutions were obtained (reviewed in Secs. III.B.2 and III.B.3). For an arbitrary cell geometry or when the excitability of the membrane is included, problem (5)–(12) was solved numerically (Sec. III.C). However, the best insight into the behavior of a single cell is obtained via singular perturbation analysis. In the past, this type of analysis was performed for a membrane patch and for myelinated and unmyelinated axons by FitzHugh [36]; for a single cell stimulated by an intracellular source and an extracellular sink by Barcilon et al. [37] and Peskoff et al. [38,39]; and for a field stimulation of single cells by Neu and Krassowska [40]. Therefore, the remainder of this section will present a singular perturbation analysis for a single cardiac cell in an external field, drawing on and expanding the results of Krassowska and Neu [40].

B. Singular Perturbation Analysis

The first step in a singular perturbation analysis is to choose a suitable system of units (Table 2) that will be used to convert the single cell problem (5)–(12) to a nondimensional form. Note the existence of two different time scales in the single-cell problem. The cellular time constant τ_c, typically less than 1 µs, describes charging and discharging of the cell with currents flowing in the intra- and extracellular space. The membrane time constant, typically several milliseconds, describes charging and discharging of the cell with currents flowing through the membrane. These two time scales differ by approximately four orders of magnitude, indicating that the single cell responds to an external field in two separate stages. The first stage, "initial polarization" of the cell, takes place immediately after the external field is turned on. The second stage, which takes place after the cell is fully charged, is the actual change of the physiological state of the cell caused by the field. For example, during excitation, the second stage takes the cell from the rest state to the excited state. In the language of singular perturbation theory, the initial polarization that proceeds with the cellular time constant τ_c constitutes an initial layer [41]. This initial layer determines the effective initial conditions for the second stage, which proceeds with a much longer membrane time constant τ_m. The advantages of singular perturbation analysis is that these two stages can be formally separated and that equations describing each stage are simpler than the original single-cell problem (5)–(12). The next section will derive the set of equations for the initial polarization, and examples of its solutions will be given in Secs. III.B.2 and III.B.3. Section III.B.4 will derive the equation that governs the change of the cell's physiological state, with examples following in Secs. III.B.5 and III.B.6.

Derivations presented in Secs. III.B.1 and III.B.4 rely on comparing orders of magnitude. Hence, all variables and equations in these sections are

Table 2 Scaling Units for the Single-Cell Problem

Quantity	Unit	Definition
Potential	ΔV	Amplitude of action potential
Conductivity	σ_i	Intracellular conductivity
Distance	d_c	Cell diameter
Membrane current	$\Delta V / R_m$	
Electric field	$\Delta V / d_c$	
Time	$\tau_c = d_c C_m / \sigma_i$	Cellular time constant
	$\tau_m = R_m C_m$	Membrane time constant

nondimensional, scaled with the units given in Table 2. Section III.B.5, which deals with FitzHugh-Nagumo model, also uses nondimensional variables. However, in order to emphasize the role of cell parameters and use a realistic model of membrane dynamics, the examples in Secs. III.B.2, III.B.3, and III.B.6 return to dimensional quantities.

1. Initial Polarization

The problem describing initial polarization is obtained by scaling Eqs. (5)–(12) with the system of units given in Table 2 and with a cellular time constant τ_c as a unit of time. Laplace's Eqs. (5)–(6), boundary conditions (7)–(8), and normalization condition (12) remain unchanged. Initial condition (11) retains the same form, with V_o measured in units of ΔV, as indicated in Table 1. The boundary conditions on currents (9)–(10) become

$$-\hat{n} \cdot \nabla\Phi_i = -\mu\hat{n} \cdot \nabla\Phi_e = \frac{\partial V_m}{\partial t} + \varepsilon I_{\text{ion}} \quad \text{on } M \tag{13}$$

Here, $\mu \equiv \sigma_e/\sigma_i$, I_{ion} is the nondimensional ionic current, and ε is a small, dimensionless parameter equal to the ratio of the two time scales appearing in Table 2,

$$\varepsilon \equiv \frac{\tau_c}{\tau_m} = \frac{d_c}{\sigma_i R_m} \tag{14}$$

With cell parameters given in Table 1, the cellular time constant τ_c is $0.375\,\mu\text{s}$, and the membrane time constant τ_m is $6\,\text{ms}$, making ε equal to 6.3×10^{-5}. In boundary conditions (13), ε multiplies I_{ion}, indicating that during initial polarization the ionic current is orders of magnitude smaller than the capacitive current or currents flowing in the intra- and extracellular space. To formally eliminate I_{ion} from the initial polarization problem, potentials Φ_i, Φ_e, and V_m are expressed as expansions in powers of ε:

$$\Phi_i(\mathbf{x}, t, \varepsilon) = \phi_i^0 + \varepsilon\phi_i^1 + \cdots \quad \text{in } \Omega_i \tag{15}$$

$$\Phi_e(\mathbf{x}, t, \varepsilon) = \phi_e^0 + \varepsilon\phi_e^1 + \cdots \quad \text{in } \Omega_e \tag{16}$$

$$V_m(\mathbf{x}, t, \varepsilon) = v_m^0 + \varepsilon v_m^1 + \cdots \quad \text{on } M \tag{17}$$

These expansions are introduced into Eqs. (5)–(8) and (13). Taking the limit $\varepsilon \to 0$, the leading-order potentials ϕ_i^0, ϕ_e^0, and v_m^0 are shown to satisfy Laplace's equations,

$$\nabla^2\phi_i^0 = 0 \text{ in } \Omega_i \quad \text{and} \quad \nabla^2\phi_e^0 = 0 \text{ in } \Omega_e \tag{18}$$

with the following boundary conditions:

$$\phi_e^0(\mathbf{x}, t) \sim -\mathbf{E} \cdot \mathbf{x} \quad \text{as } |\mathbf{x}| \to \infty \tag{19}$$

$$v_m^0 = \phi_i^0 - \phi_e^0 \quad \text{and} \quad -\hat{n} \cdot \nabla \phi_i^0 = -\mu \hat{n} \cdot \nabla \phi_e^0 = \frac{\partial v_m^0}{\partial t} \quad \text{on } M \tag{20}$$

Note that problem (18)–(20) no longer contains ionic current, indicating that during initial polarization the membrane behaves as a pure capacitance.

The initial polarization problem (18)–(20) requires initial conditions on potentials at time $t = 0^+$. These are determined, in general, by solving a boundary-value problem determined by Eqs. (18)–(19), subject to the boundary conditions on the membrane:

$$-\hat{n} \cdot \nabla \phi_i^0 = -\mu \hat{n} \cdot \nabla \phi_e^0 \quad \text{and} \quad v_m^0(\mathbf{x}, 0^+) = V_o \quad \text{on } M \tag{21}$$

Equation (21) indicates that at $t = 0^+$, the current across the membrane is continuous, and the transmembrane potential remains unchanged because the finite external current cannot instantaneously change the charge on the membrane capacitance.

Of particular interest are the steady-state solutions to problem (18)–(20) because they serve as effective initial conditions to the problem governing the subsequent evolution of potentials that proceeds with the time constant τ_m (to be derived in Sec. III.B.4). These solutions are determined by the following boundary-value problem. In the extracellular space,

$$\nabla^2 \phi_e^0 = 0 \quad \text{in } \Omega_e \tag{22}$$

$$\hat{n} \cdot \nabla \phi_e^0 = 0 \quad \text{on } M \tag{23}$$

$$\phi_e^0(\mathbf{x}) = -\mathbf{E} \cdot \mathbf{x} \quad \text{as } |\mathbf{x}| \to \infty \tag{24}$$

Problem (22)–(24) has a linear dependence on the electric field $\mathbf{E}$. Thus, extracellular potential is sought in the form

$$\phi_e^0(\mathbf{x}, \mathbf{E}) = \mathbf{E} \cdot \mathbf{w}(\mathbf{x}) \tag{25}$$

where $\mathbf{w}$ is a vector of weight functions. Computing w_i involves solving Eqs. (22)–(24) with the ith component of the electric field $\mathbf{E}$ set to one and other components to zero. Weight functions $\mathbf{w}$ must also satisfy $\int_M \mathbf{w} \, da = 0$, so that ϕ_e^0 satisfies normalization condition (12). In the intracellular space,

$$\nabla^2 \phi_i^0 = 0 \quad \text{in } \Omega_i \tag{26}$$

$$\hat{n} \cdot \nabla \phi_i^0 = 0 \quad \text{on } M \tag{27}$$

Equations (26)–(27) indicate that the leading-order intracellular potential will have a uniform value throughout the interior of the cell. Normalization condition (12) together with boundary conditions (27) and the initial condition on transmembrane potential (21) imply that

$$\phi_i^0(\mathbf{x}) = V_o \quad \text{in } \Omega_i \tag{28}$$

i.e., the intracellular potential is determined by the state of the cell at the time the field was applied. Consequently, the leading-order transmembrane potential has the form

$$v_m^0(\mathbf{x}, \mathbf{E}) = V_o - \mathbf{E} \cdot \mathbf{w}(\mathbf{x}) \tag{29}$$

evaluated for $\mathbf{x}$ belonging to M. The form of Eq. (29) shows that the transmembrane potential induced by the field is a linear combination of v_m^0 induced by the three orthogonal components of the field. This is a general rule that applies to cells of any shape in an external field. Hence, the transmembrane potential varies with the position around the cell, and its maximum magnitude occurs at a location determined by the cell shape and the relative magnitudes of the field components. As seen in experiments, v_m^0 is most negative at the end of the cell facing the anode and most positive at the end facing the cathode (Fig. 6A).

2. Analytical Solution for a Cylindrical Cell in a Transverse Field

For simple geometries, the initial polarization problem described by (18)–(20) can be readily solved. This section presents such a solution for a cylindrical cell placed in an electric field transverse to its axis. The cell is assumed to be initially at rest, so V_o, the transmembrane potential prior to the application of the field, is equal to the intrinsic rest potential of a cardiac cell, V_{rest}. However, since this solution focuses on the shock-induced transmembrane potential, the rest potential will be ignored, assuming that the rest state occurs at the zero membrane potential. Hence, in this and the next section, potentials v_m^0, ϕ_i^0 and ϕ_e^0 are understood as deviations from their rest values.

Solutions to problem (18)–(20), obtained by separation of variables [40,42], express potentials in cylindrical coordinates as functions of radius r, angle θ, and time t. Written in dimensional variables, these solutions are

$$\phi_i^0(r, \theta, t) = -\frac{2\sigma_e}{\sigma_i + \sigma_e} Er \cos\theta \, e^{-t/\tau_{ip}} \qquad r \leq a \tag{30}$$

$$\phi_e^0(r, \theta, t) = -Er \cos\theta \left[1 + \frac{a^2}{r^2} \left(1 - \frac{2\sigma_i}{\sigma_i + \sigma_e} e^{-t/\tau_{ip}} \right) \right] \qquad r \geq a \tag{31}$$

$$v_m^0(\theta, t) = 2Ea \cos\theta (1 - e^{-t/\tau_{ip}}) \qquad r = a \tag{32}$$

where $a = d_c/2$ is the cell radius. All three potentials evolve with a time constant of initial polarization,

$$\tau_{ip} = aC_m \left(\frac{1}{\sigma_i} + \frac{1}{\sigma_e} \right) \tag{33}$$

Note that τ_{ip} is not equal to the cellular time constant τ_c of Table 1. The role of τ_c is to determine the correct order of magnitude for scaling time in the original single-cell problem. In contrast to τ_c, τ_{ip} depends on the shape of the cell. For a cylindrical cell and conductivities assumed in Table 1, τ_{ip} is shorter than τ_c, 0.225 µs rather than 0.375 µs.

The behavior of intra- and extracellular potentials is illustrated in Fig. 7. Immediately after the field is turned on, potentials ϕ_i^0 and ϕ_e^0 assume values determined by the solution of the boundary-value problem (18)–(19) with boundary conditions (21) (dashed lines). Intracellular potential ϕ_i^0 changes linearly with the distance and has a slope higher than the strength of the external field. This is because the intracellular conductivity σ_i is assumed to be five times smaller than the extracellular conductivity σ_e (Table 1). On the membrane M, the potential at time 0^+ is continuous, so initially transmembrane potential v_m^0 (understood as a deviation from rest) is zero. Extracellular potential ϕ_e^0 asymptotes with the distance to the straight line corresponding to the applied electric field. If the intra- and extracellular conductivities were equal, the electric fields at $t = 0^+$ would have been equal to E in both intra- and extracellular space and the cell would be totally invisible to the field. When the intra- and extracellular conductivities are different, the field "sees" the interior of the cell, but does not "see" the membrane.

As time increases, the slope of the intracellular potential decreases in magnitude, and ϕ_i^0 asymptotes to a constant value, shown by a solid line. In contrast, the magnitude of the extracellular potential in the proximity of the cell slightly increases in time. At steady state, the slope of ϕ_e^0 at the membrane M is zero. This indicates that after the initial polarization is complete, the membrane is fully charged and behaves as an insulator: it admits no further current from the environment.

As intracellular and extracellular potentials evolve in time, they lose continuity on the membrane M, and the transmembrane potential v_m^0 develops. Figure 8a shows v_m^0 increasing in time to the steady-state value of $2Ea$ at the end of the cell facing the cathode and to $-2Ea$ at the end facing the anode. After the initial polarization is complete, the final values of the potentials are

$$\phi_i^0(r, \theta) = 0 \qquad r \leq a \tag{34}$$

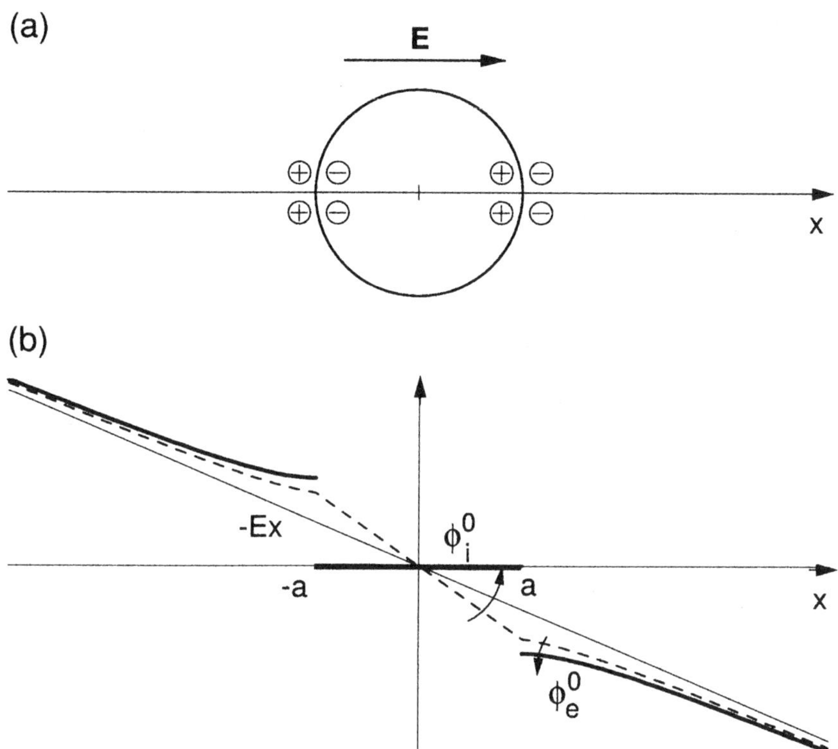

Figure 7 Leading-order intracellular and extracellular potentials during initial polarization. The potentials are plotted along axis x connecting two poles of the cell as shown in (a). In (b), the part of the graph between $-a$ and a corresponds to the intracellular potential ϕ_i^0, and the parts beyond this range correspond to the extracellular potential ϕ_e^0. Dashed lines show intra- and extracellular potentials immediately after the electric field is turned on ($t = 0^+$). As time increases, intra- and extracellular potentials move in the directions indicated by the arrows until they reach the steady state shown by solid lines. A thin line marked $-Ex$ is the potential that corresponds to the uniform external field. (Based on Ref. 40.)

$$\phi_e^0(r, \theta) = -Er\cos\theta\left(1 + \frac{a^2}{r^2}\right) \qquad r \geq a \tag{35}$$

$$v_m^0(\theta) = 2Ea\cos\theta \qquad r = a \tag{36}$$

This co-sinusoidal distribution of the transmembrane potential v_m^0 along the circumference of the cell is shown in Fig. 8b.

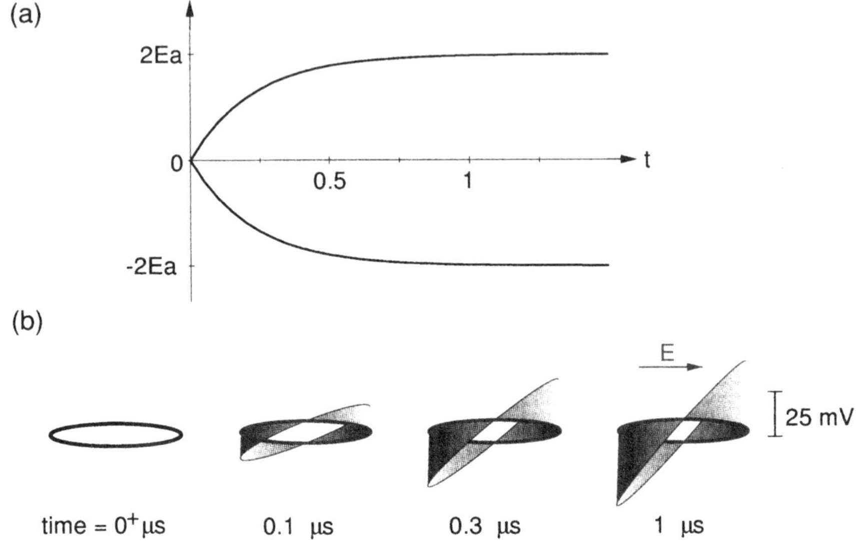

Figure 8 Leading-order transmembrane potential during initial polarization. (a) Time course of v_m^0, which increases exponentially at the end of the cell facing the cathode and decreases at the opposite end. (Based on Ref. 40.) (b) Distribution of v_m^0 along the circumference of the cell at four different instants of time. The thick circle in each of the parts indicates the rest potential.

3. Analytical Solutions for Cells of Other Shapes

Solutions of the initial polarization problem (18)–(20) for cells of other geometries, both steady-state and time-dependent, can be found in the literature. A spherical cell is encountered most often, as it is a reasonable representation for many cell types, especially those grown in vitro. Even embryonic cardiac cells grown in vitro assume a spherical shape (Fig. 2b). Separation-of-variables solutions for a spherical cell were obtained by, among others, Schwan [43,44]:

$$v_m^0(\theta, t) = 1.5 Ea \cos\theta (1 - e^{-t/\tau_{ip}}) \qquad r = a \tag{37}$$

with

$$\tau_{ip} = aC_m \left(\frac{1}{\sigma_i} + \frac{1}{2\sigma_e} \right) \tag{38}$$

This formula differs from the solution (32)–(33) for a transverse cylinder by a shape factor (1.5 instead of 2) and by a slightly shorter time constant (0.206 μs instead of 0.225 μs).

In addition to Eq. (37) for v_m^0, Schwan computed the magnitude of the transmembrane potential induced by an a.c. field,

$$v_m^0 = \frac{1.5Ea\cos\theta}{\sqrt{1 + (2\pi f\tau_{ip})^2}} \tag{39}$$

where f is the frequency of the applied field. Equation (39), known in the literature as the "Schwan equation," allows one to examine the dependence of the field-induced transmembrane potential on frequency. Namely, the cutoff frequency f_c,

$$f_c = \frac{1}{2\pi\tau_{ip}} \tag{40}$$

is for cardiac cells in the range of megahertz. Hence, under typical pulse parameters used for cardiac stimulation, one does not have to account for frequency effects.

For spheroidal cells, analytical solutions for prolate and oblate spheroids were obtained by Klee and Plonsey [45]. Their solutions allowed the electric field to assume an arbitrary direction with respect to the cell axis. The general formula for v_m^0 is

$$v_m^0 = (E_x x + E_y y)A + E_z z C \tag{41}$$

where E_x, E_y, and E_z are three orthogonal components of the applied field, z is the axis of revolution for the spheroid, and A and C are constants. A and C depend only on the dimensions of the cell in the two principal directions; full expressions for the prolate and oblate spheroids can be found in the original paper [45] and will not be repeated here. Formula (41) illustrates a general rule established in Eq. (29): the field-induced transmembrane potential is a linear combination of V_m induced by the three orthogonal components of the field, E_x, E_y, and E_z. In this case, the vector of weight functions $\mathbf{w}$ is (Ax, Ay, Cz).

4. Change of the Cell's Physiological State

Further time evolution of these potentials and, in particular, the possibility of changes in the physiological state of the cell membrane, depend on the ionic current I_{ion}. The problem governing the active response is obtained by scaling problem (5)–(12) with units given in Table 3 and with τ_m as a unit of

Table 3 Ratios of Thresholds for Transverse and Longitudinal Fields

Species	Duration (ms)	Dimensions (μm $\times$ μm)	Aspect ratio	E_t/E_1 (meas.)	E_T/E_1 (est.)	Source
Guinea pig	10	153×18	8.6	3.73	4.3	[9]
Guinea pig	2	135×25	5.4	2.61	2.7	[10]
Dog	10	75×15	5.0	1.82	2.5	[12]
Frog	2	300×10	30	5.75	15	[10]

time. Laplace's Eqs. (5)–(6), boundary conditions (7)–(8), and normalization condition (12) remain unchanged. Boundary conditions on currents (9)–(10) become

$$-\hat{n} \cdot \nabla\Phi_i = \varepsilon \left(\frac{\partial V_m}{\partial t} + I_{\text{ion}} \right) \tag{42}$$

$$-\mu\hat{n} \cdot \nabla\Phi_e = \varepsilon \left(\frac{\partial V_m}{\partial t} + I_{\text{ion}} \right) \tag{43}$$

on M. Note that in Eqs. (42)–(43), all the currents flowing through the membrane, i.e., both the capacitive and the ionic components are multiplied by a small parameter ε. Hence, in a leading-order approximation, the membrane behaves as an insulator. That is precisely the situation that developed at the end of the initial polarization; consequently, the steady-state solutions (34)–(36) of initial polarization serve as effective initial conditions to the problem describing the active response of the cell.

Solutions to this problem can be obtained numerically by discretizing Eqs. (5)–(8) and (42)–(43) and solving them on a computer (Sec. III.C). In essence, the cell is treated as a collection of membrane patches, and the physiological states of those patches will determine the physiological state of the whole cell. An alternative approach is to recognize that the boundary-value problem (5)–(8), (42)–(43) reduces at leading order to an ordinary differential equation that governs the response of a cell treated as a whole. To show this, one must examine the macroscopic balance of current. The net current entering or leaving the cell is computed by integrating Eq. (42) over membrane M. The integral of the intracellular current (left-hand side) is zero because the cell is source-free. Hence, the net capacitive current is balanced by the net ionic current:

$$\int_M \frac{\partial V_m}{\partial t} \, da = - \int_M I_{\text{ion}}(V_m) \, da \tag{44}$$

Integral identity (44) can be approximated by an ordinary differential equation. First, recognize that in the limit $\varepsilon \to 0$, problem (5)–(8), (42)–(43) has no time dependence. Thus, the leading-order potentials will retain the same form as established by the initial polarization [Eqs. (25), (28), and (29)] except for a time-dependent constant. As shown in the previous section, the intracellular potential is uniform throughout the interior of the cell [Eq. (28)]; this uniform value may subsequently evolve with time. The extracellular potential is given by $\mathbf{E} \cdot \mathbf{w}$ and does not depend on time. This result is motivated physically: a small net ionic current crossing the membrane can easily change the potential inside a cell that is only several micrometers in diameter. However, the same current has practically no impact on the extracellular potential, because the extracellular region Ω_e is large, and because the extracellular field is enforced by the electrodes. Thus, in the limit $\varepsilon \to 0$, the leading-order intracellular potential depends only on time t, while the leading-order extracellular potential depends on the position $\mathbf{x}$ and, parametrically, on the electric field $\mathbf{E}$:

$$\Phi_i(\mathbf{x}, t, 0) = \phi_i^0(t) \tag{45}$$

$$\Phi_e(\mathbf{x}, t, 0) = \phi_e^0(\mathbf{x}, \mathbf{E}) = \mathbf{E} \cdot \mathbf{w}(\mathbf{x}) \tag{46}$$

$$V_m(\mathbf{x}, t, 0) = v_m^0(\mathbf{x}, t) = \phi_i^0(t) - \mathbf{E} \cdot \mathbf{w}(\mathbf{x}) \tag{47}$$

Second, define the macroscopic ionic current i_{ion} as an average of the pointwise ionic current I_{ion} over membrane M:

$$i_{\text{ion}}(\phi_i^0, \mathbf{E}) = \frac{1}{|M|} \int_M I_{\text{ion}}(v_m^0) \, da = \frac{1}{|M|} \int_M I_{\text{ion}}(\phi_i^0 - \mathbf{E} \cdot \mathbf{w}) \, da \tag{48}$$

where $|M|$ is the surface area of the membrane. For a given geometry of a cell and the dynamics of the membrane, this macroscopic ionic current is a function of the intracellular potential ϕ_i^0 and depends parametrically on the field $\mathbf{E}$. Introducing Eqs. (45)–(48) allows one to write the macroscopic balance of current (44) as

$$\frac{\partial \phi_i^0}{\partial t} = -i_{\text{ion}}(\phi_i^0, \mathbf{E}) \tag{49}$$

or, in the dimensional variables,

$$C_m \frac{\partial \phi_i^0}{\partial t} = -i_{\text{ion}}(\phi_i^0, \mathbf{E}) \tag{50}$$

Equation (50) demonstrates that the active response of a cell to an electric field is governed by an ordinary differential equation. This equation is similar to the equation governing the response of a space-clamped membrane patch, indicating that a single cell in an external electric field can be treated as a unit. Further similarities and differences in the behavior of a single cell and a membrane patch will be explored in the next two sections.

5. Example: Excitation of a Cylindrical Cell with FitzHugh-Nagumo Membrane Dynamics

The theoretical results presented above are illustrated with an example of a cylindrical cell activated from the rest state by an electric field transverse to its axis. To underscore the similarities and differences between the behavior of a single cell and a membrane, the figures will also show the excitation of a membrane patch. This comparison is clearest when a membrane dynamics is described by the FitzHugh-Nagumo (FN) model [46,47], in which the excitability is represented by a simple cubic function. For a space-clamped membrane, the governing equations are

$$\frac{dV_m}{dt} = -I_{\text{ion}}(V_m, u) + I_{\text{stim}} \tag{51}$$

$$\frac{du}{dt} = b(cV_m - u) \tag{52}$$

where V_m denotes the fast activation variable (i.e., transmembrane potential), u denotes the slow activation variable responsible for refractoriness and recovery, I_{stim} is an external stimulating current, and a, b, and c are constants. For the FN model, the pointwise ionic current I_{ion} is

$$I_{\text{ion}} = V_m(a - V_m)(1 - V_m) + u \tag{53}$$

The macroscopic ionic current can be computed from Eq. (48), using Eq. (53) and expression (36) for the transmembrane potential for a cylindrical cell in a transverse field:

$$\begin{aligned}
i_{\text{ion}}(\phi_i^0, \mathbf{E}) &= \frac{1}{2\pi a} \int_0^{2\pi} I_{\text{ion}}(\phi_i^0 + 2Ea\cos\theta) a\, d\theta \\
&= \frac{1}{2\pi} \int_0^{2\pi} [(\phi_i^0 + 2Ea\cos\theta)(a - \phi_i^0 + 2Ea\cos\theta) \\
&\qquad \times (1 - \phi_i^0 + 2Ea\cos\theta)]\, d\theta \\
&= \phi_i^0(a - \phi_i^0)(1 - \phi_i^0) + u^0 + 0.5(2Ea)^2(3\phi_i^0 - a - 1) \tag{54}
\end{aligned}$$

Here, u^0 is the macroscopic slow variable defined as

$$u^0(t) = \frac{1}{2\pi a} \int_M u(\mathbf{x}, t) a\, d\theta \tag{55}$$

i.e., as a surface average of the pointwise slow variable u, which as a function of $\mathbf{x}$ may vary with the position around the cell. The governing equation for u^0 has the same form as Eq. (52) for u of a space-clamped membrane. Hence, the ordinary differential equations governing the change of physiological state in a single cell are

$$\frac{d\phi_i^0}{dt} = -[I_{\text{ion}}(\phi_i^0, u^0) + 0.5(2Ea)^2(3\phi_i^0 - a - 1)] \tag{56}$$

$$\frac{du^0}{dt} = b(c\phi_i^0 - u^0) \tag{57}$$

Comparing Eqs. (51)–(52) governing the physiological state of the space-clamped membrane and Eqs. (56)–(57) for the single cell, one recognizes that their form is very similar, with intracellular potential ϕ_i^0 replacing transmembrane potential V_m, and macroscopic slow variable u^0 replacing u. The difference lies in the term that is responsible for external stimulation; stimulating current I_{stim} is in the single cell replaced by term $0.5\,(2Ea)^2\,(3\phi_i^0 - a - 1)$, which describes the contribution of external field E. In this term, the scaling factor of 0.5 is due to the cylindrical geometry of the cell. $2Ea$ is equal to the maximum magnitude of the field-induced transmembrane potential [Eq. (36)]. The contribution of this term is proportional to the square of the maximum magnitude of v_m^0, and hence it is independent of the polarity of the field. Finally, $(3\phi_i^0 - a - 1)$ describes how the effect of the field depends on the initial state of the cell. When an electric field is applied to a resting cell ($\phi_i^0 = 0$ in FN model), this term is negative, and consequently i_{ion} flows inward, increasing ϕ_i^0 and bringing the cell toward depolarization. When an electric field is applied during plateau ($\phi_i^0 \approx 1$), this term is positive, and i_{ion} flows outward and repolarizes the cell.

Figure 9a compares the time courses of the intracellular potential ϕ_i^0 in a single cell and the transmembrane potential V_m in a membrane patch. Except for a slight delay in time, caused by a small difference in the magnitude of the stimuli with respect to the thresholds, these two courses are essentially identical.

Further similarities can be observed when these two activations are analyzed using methods of nonlinear dynamics. Figures 9b and 9c show phase portraits for a single cell and a membrane patch. As implied by Eqs. (56)–(57) and (51)–(52), the coordinates are the intracellular potential ϕ_i^0 and the macroscopic slow variable u^0 for the cell and the transmembrane potential V_m and pointwise slow variable u for the membrane. In the absence of an external stimulus, cubic nullclines of the fast variable $d\phi_i^0/dt = 0$ ($dVm/dt = 0$ for the membrane) are identical. The intersection of the fast-variable nullcline with the slow-variable nullcline $du^0/dt = 0$ ($du/dt = 0$)

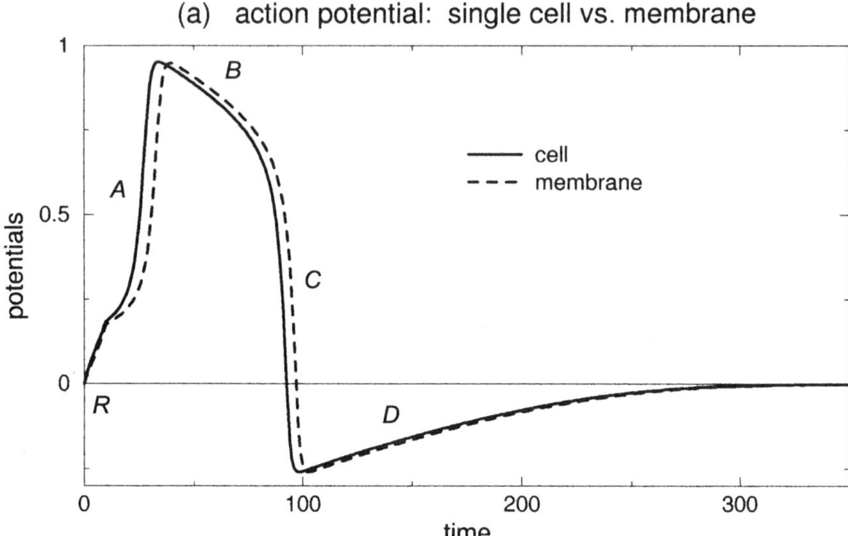

Figure 9   (a) Time course of the intracellular potential ϕ_i^0 in a cell (solid line) and the transmembrane potential V_m in a membrane (dashed line) during an action potential. Membrane dynamics is represented by the FitzHugh-Nagumo model, with parameters $a = 0.13$, $b = 0.094$, and $c = 0.37$. Action potential is elicited by $E = 0.11$ in a cell and $I_{\text{stim}} = 0.02$ in a membrane. (b) and (c) Action potential in a single cell and a membrane patch illustrated on a phase plane. The trajectory of a phase point is shown with a thick solid line; filled arrows show the direction of the point movement; dots indicate positions of the phase point at constant time intervals. Letters A, B, C, and D label consecutive phases of an action potential as seen in (a). Thin solid lines are nullclines for the fast and slow variables, and thin dashed lines show the position of the fast variable nullcline during the stimulus. R and R' indicate the positions of the fixed point in the absence and presence of a stimulus; R corresponds to the rest state. The open arrow indicates the position on the trajectory when the stimulus ends. All quantities in this figure are dimensionless.

defines the fixed point R, which corresponds to the rest state of the cell (membrane). Application of an external stimulus changes the position of the fast-variable nullcline to that shown by a dashed line. In a membrane, current I_{stim} raises the entire V_m nullcline; in a cell, field E deforms the shape of the ϕ_i^0 nullcline, raising its left part and lowering its right one. Yet the end result is the same: the fixed point moves from R to R'. The phase point that corresponds to the state of the system, and which initially resides at the rest

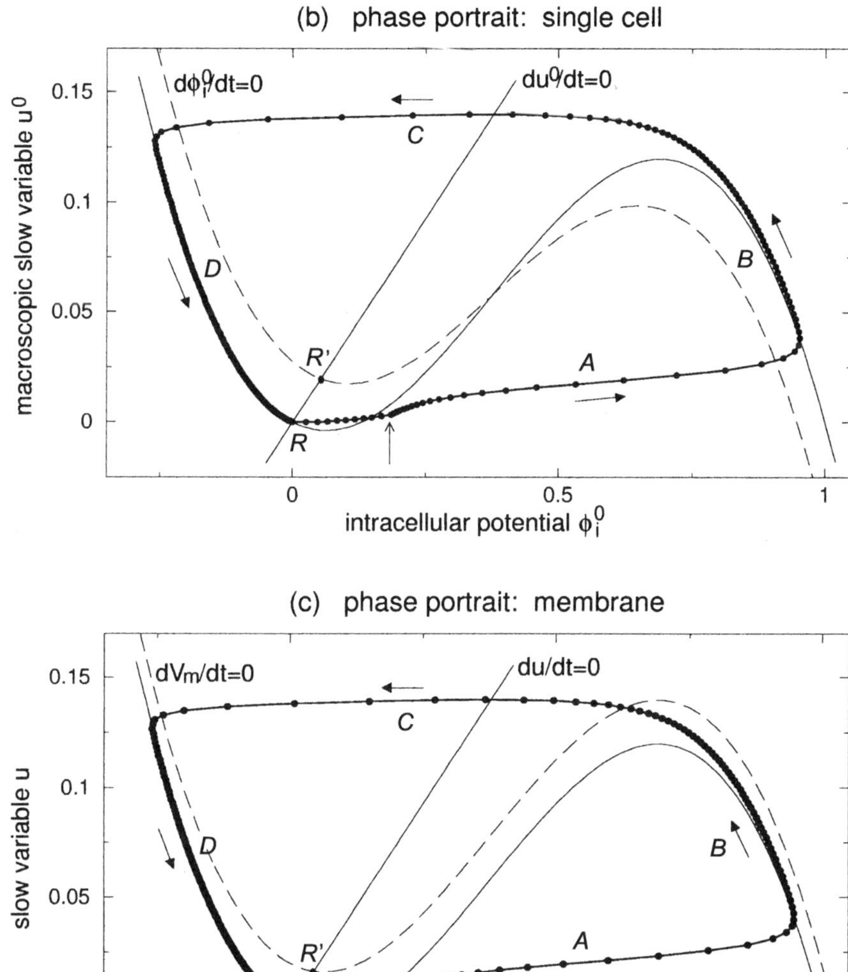

Figure 9 (Continued)

state R, is no longer at a fixed point and must move toward increasing ϕ_i^0 (V_{m}). When the stimulus is turned off (open arrow), the fast-variable null-cline instantaneously returns to its initial position (thin solid line). However, by this time the phase point has already cleared the middle branch of the cubic nullcline, which in this phase plane represents crossing the threshold for excitation. Hence, the phase point continues to move along a trajectory that describes an action potential, through the branches corresponding to the upstroke (A), plateau (B), repolarization (C), and afterpotential (D), before returning to the rest state R.

Figures 9b and 9c demonstrate remarkable similarities between a cell and a membrane in the mechanism of excitation and the resulting trajectory of the phase point during an action potential. Hence, a single cell can indeed be regarded as a unit. Just as a membrane patch has distinct physiological states associated with the transmembrane potential V_m, a single cell can be considered to have its own physiological states, defined in terms of in-tracellular potential ϕ_i^0. The main difference is that for the membrane patch, current I_{stim} merely moves the V_m nullcline up or down; in a cell, E changes the shape of the ϕ_i^0 nullcline. Also, for the membrane patch, a positive I_{stim} raises the nullcline, whereas a negative I_{stim} lowers it; in a cell, the de-formation of the nullcline by the electric field E is independent of the po-larity. Thus, excitation of a single cell can be achieved with either polarity of the external field.

6. Example: Excitation of a Cylindrical Cell with Luo-Rudy Membrane Dynamics

Figures 10 and 11 continue the analysis of the field stimulation of a single cylindrical cell using realistic membrane dynamics, the 1991 Luo-Rudy (LR) model [48]. Numerical solutions for these figures were obtained by in-tegrating Eq. (50) and its counterpart for the membrane patch using the Euler method with a time step of $1\,\mu s$. To compute i_{ion}, the cell's cir-cumference was divided into 64 equipotential patches of equal area. The pointwise ionic current in each patch, I_{ion}, was computed by using the field-induced transmembrane potential given by Eq. (36) and the LR model of membrane dynamics. The macroscopic ionic current i_{ion} was computed from Eq. (48) by using the trapezoidal rule to evaluate the integral of I_{ion} over surface M of the cell.

Figure 10 illustrates the mechanism of field stimulation with a short (0.5-ms) pulse. In this case, activation occurs after the pulse. The process starts with the field being turned on; at time $t = 0^+$, the initial polarization is completed. (Recall that now t is measured on the scale of milliseconds, not microseconds.) This is immediately reflected by the pointwise

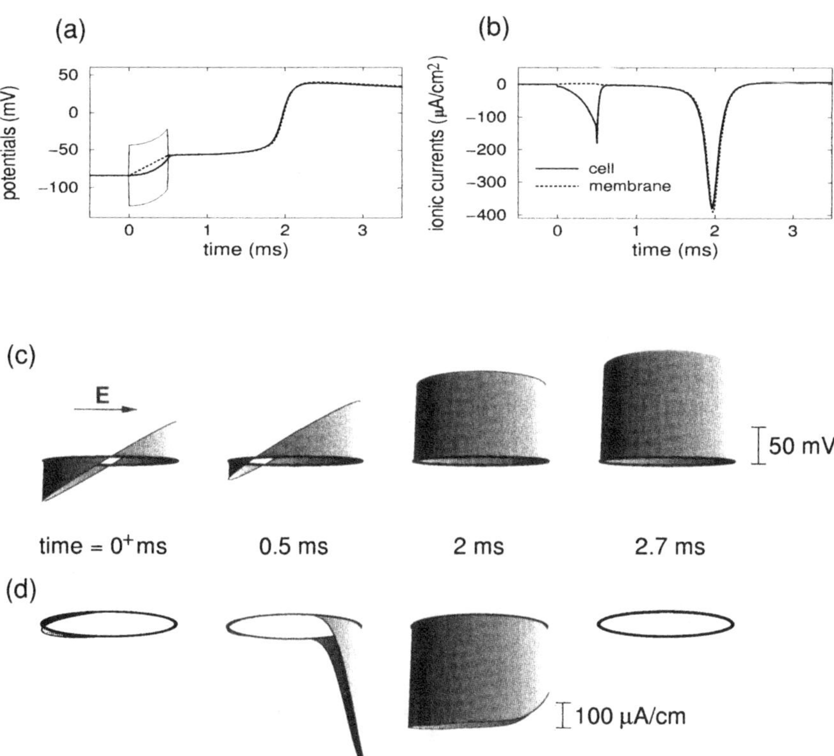

Figure 10 Time course of excitation in a cylindrical cell and in a membrane patch with membrane dynamics represented by the Luo-Rudy model. The stimulus duration is 0.5 ms, and the strength is 27 V/cm (cell) and 55.9 µA/cm^2 (membrane); both stimuli are just above the threshold. (a) Time course of the intracellular potential ϕ_i^0 (cell, solid line) and the transmembrane potential V_m (membrane, dashed line). Thin lines show transmembrane potentials at the depolarized and hyperpolarized poles of the cell. (b) Macroscopic ionic current i_{ion} (cell, solid line) and pointwise ionic current I_{ion} (membrane, dashed line) plotted as functions of time. (c) and (d) Distribution of the transmembrane potential and ionic current along the circumference of a cell at four different time instants. The thick circles in (c) are drawn at rest potential, and in (d) at zero ionic current.

transmembrane potentials that jump to their maximum values (Fig. 10c, "0^+ ms" and thin lines in Fig. 10a), even though the intracellular potential ϕ_i^0 remains at rest (Fig. 10a, thick solid line). The subsequent gradual increase in ϕ_i^0 is due to the fast sodium current that comes from the depolarized part of the cell. The development of this current can be seen as a first

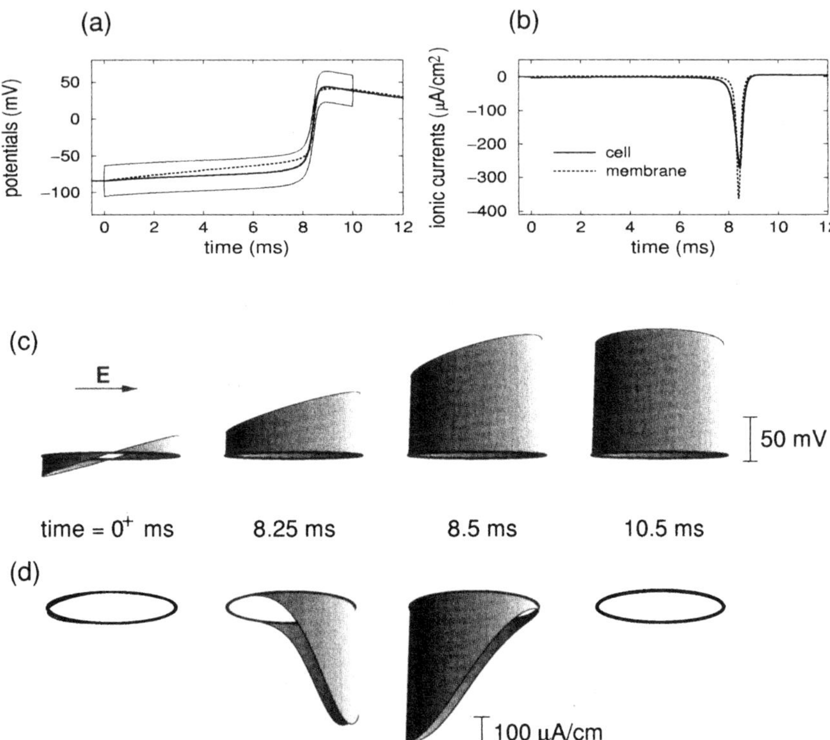

Figure 11 Time course of excitation in a cylindrical cell and in a membrane patch with a long (10-ms) stimulus. The stimulus strength is 14 V/cm (cell) and 5.2 µA/cm² (membrane); both stimuli are just above the threshold. See the caption to Fig. 10 for a description of (a)–(d).

spike in Fig. 10b, coinciding in time with the activation of the depolarized end (Fig. 10d, "0.5 ms"). The magnitude of this spike is smaller in Fig. 10b than in Fig. 10d because to obtain i_{ion} for Fig. 10b, I_{ion} shown in Fig. 10d is averaged over the cell surface. After the field is turned off, the sodium current abruptly ends, but it has managed to bring the intracellular potential ϕ_i^0 just above threshold, so that the activation process can continue. At 2 ms, the cell activates (Fig. 10b, second spike). Since I_{ion} is nearly uniform along the circumference (Fig. 10d, "2 ms"), the cell fires essentially as a unit.

The mechanism of field stimulation with a long (10-ms) pulse is somewhat different (Fig. 11). In this case, the cell fires while the field is still on. The most conspicuous difference is the absence of a separate spike in i_{ion}, which reflects the flow of fast sodium current through the depolarized part of the membrane (Fig. 11b). For short pulses, this spike is instrumental in bringing the cell to the threshold. However, for long pulses just above the threshold, V_m at the depolarized end is too small to immediately activate sodium channels (Fig. 11c, "0^+ ms"). Instead, the slow increase in ϕ_i^0 is due to a very small inward current, which initially comes from the hyperpolarized end of the cell and is carried by potassium ions (Fig. 11b, solid line, and Fig. 11d, "0^+ ms"). Thus, at time $t = 0^+$, the hyperpolarized end of the cell is not a "load" that counteracts excitation. On the contrary, the current entering through the hyperpolarized end is instrumental in initiating the movement of ϕ_i^0 toward the threshold. Potassium current is soon aided by the sodium current from the depolarized part of the cell, which very slowly grows in magnitude and at 8.25 ms is large enough to initiate an upstroke of an action potential (Fig. 11d, "8.25 ms"). As other parts of the cell are being raised above the threshold, their sodium channels open and complete the activation process (Fig. 11d, "8.5 ms"). Hence, for long pulses, the upstroke is not as synchronized as for short ones, with a delay of approximately 0.25 ms between activations of the two opposite ends of the cell.

For comparison, dashed lines in parts a and b of Figs. 10 and 11 show V_m and I_{ion} of a membrane patch activated with 0.5- and 10-ms stimuli. Immediately after the onset of the stimulus, the time courses of both ϕ_i^0 (cell) and V_m (membrane) show only a gradual increase. In a single cell, this increase is due to a sodium current from the depolarized end (0.5-ms stimulus) or to a potassium current from the hyperpolarized end (10-ms stimulus). In contrast, in a membrane patch, this increase is due to charging of the membrane by the external current I_{stim}. The ionic current I_{ion} flows outward and opposes the depolarizing influence of the external current (Figs. 10b and 11b, dashed line; on the scale used in these figures, the positive offset is barely visible). During the upstroke, ϕ_i^0 in a cell closely matches V_m in a membrane. For the short pulse, the shape and magnitude of i_{ion} in a cell is nearly identical to I_{ion} in a membrane. This is the consequence of the cell firing as a unit (Fig. 10d, "2 ms"). For the long pulse, the shape of i_{ion} in a cell is broader but lower in magnitude than I_{ion} in a membrane, a consequence of a 0.25-ms delay between activations of the depolarized and the hyperpolarized ends of the cell (Fig. 11d, "8.25 ms" and "8.5 ms").

Figures 10 and 11 demonstrate that the time course of activation in a single cell, when represented by an intracellular potential, is very similar to the activation of a membrane patch, supporting the use of ϕ_i^0 as a state variable for the cell. These similarities are not limited to activation from rest. Tung and Jain [49] reported that shocks applied during the relatively refractory period cause similar prolongation of an action potential in a cell and in a space-clamped membrane.

C. Numerical Solutions

1. Direct Numerical Solutions

The response of a single cell to an external field has also been studied by using numerical simulations. This method is especially important when the cell membrane is excitable, since in this case the single-cell problem (5)–(12) does not have an analytical solution. A straightforward approach is to discretize the cell interior and the surrounding extracellular space and to solve the single-cell problem (5)–(12) with a finite-difference or a finite-element method. This approach has an advantage of avoiding any simplifications of the original problem, so the only loss of accuracy comes from discretization errors. To our knowledge, this approach has to date been applied only to passive cells. In 1972, Klee and Plonsey [50] developed a finite-difference model for axially symmetric cells with a leaky membrane. The main difference between typical finite-difference problems encountered in the literature and the finite-difference model for a single cell lies in the boundary conditions on the cell membrane, where both the potential and its normal derivative have jumps [Eqs. (8)–(9)]. Klee and Plonsey proposed a suitable discretization scheme for these types of conditions and examined the accuracy and convergence properties of the successive overrelaxation method used to solve the finite-difference equations. This finite-difference model was applied by Klee and Plonsey to compute steady-state solutions for prolate and oblate spheroidal cells [45]. The results of this study confirmed that the intracellular potential is essentially uniform in the cell interior and that it is appropriate to treat the membrane as an insulator.

Two-dimensional finite-element models for single cells with realistic geometries were developed by Ranjan and Thakor [12]. The cell geometries were based on the digitized video images of cardiac cells used in the experimental part of this study. The use of the finite-element method made it possible to represent complex shapes of cardiac myocytes. The steady-state distributions of the potentials inside and outside the cells were computed with a commercial finite-element package, ANSYS (Swanson Analysis, Inc., Houston, PA). Qualitatively, the realistically shaped cells behave similarly to cells with regular shapes. For example, the interiors of the cells were found to be isopotential. However, there were three quantitative differences. First, the simulations revealed discrete "hot spots," i.e., small regions of the membrane that, because of the cell's irregular geometry, had a larger-than-expected magnitude of transmembrane potential. Second, the magnitude of transmembrane potential at the opposite ends of the cell differed by up to 20%. Finally, for one cell of an unusual geometry, the transmembrane potential induced by the transverse field exceeded that induced by the longitudinal field, even though the length-to-width ratio was approximately 4. However, results obtained with a two-dimensional model should be viewed with caution. Some predictions of a two-dimensional model, such as "hot spots," disappear or greatly diminish when the model is extended to three dimensions (B. A. Stone, unpublished results, 1996).

The two models discussed above were limited to the steady-state solution of the single-cell problem. A time-dependent model for a spherical cell was developed by DeBruin and Krassowska [51]. This study investigated the cell with a passive but electroporating membrane; i.e., when the transmembrane potential exceeded a certain critical value, the membrane increased its conductance in a manner consistent with the creation of a large number of conducting pores [52]. The study by DeBruin and Krassowska, although not related directly to the stimulation of single cells, illustrates the problem with applying the "brute-force" numerical approach to the single-cell problem: the computational expense is prohibitive. The reason is that the finite-difference or finite-element method requires discretizing the entire volume under study, resulting in a large number of nodes in intra- and extracellular space. Potentials at these nodes are unknown and have to be computed at each time step by solving a large set of simultaneous equations. Even if inverting the matrix is avoided by transforming the equations using LU decomposition so that each time step would require only a forward and backward substitution [53], the cost is substantial: simulating 2 s of the response of a cell with a passive membrane was estimated to require 180 hr on a Sun Ultra 1 workstation [51]. Therefore, the direct numerical simulations of single-cell problem (5)–(12) are suitable only for steady-state solutions. For time-dependent problems, especially when a sequence of several beats has to be simulated, the original problem must be simplified to make it more manageable. In the past, three different methods of dealing with the simulations of excitable cells have been proposed: a Green's theorem approach, a "cell as a short fiber" approach, and a "cell as a unit" approach. The advantages and limitations of these approaches are discussed in the following sections.

2. Green's Theorem Approach

This approach was proposed by Leon and Roberge in 1990, first as a more realistic representation of a one-dimensional fiber than a cable model [54]. Later, it was extended to a single cell stimulated by an extracellular electrode [55]. The approach takes advantage of Green's theorem, which allows one to express potentials inside isotropic conductive regions in terms of potentials and their derivatives on boundaries. In a single-cell problem, Green's theorem is applied to Laplace's Eqs. (5)–(6), resulting in the following expressions for the intra- and extracellular potentials at point P, just inside and outside the membrane:

$$\Phi_i(P) = \frac{1}{2\pi} \int_M \frac{1}{r} \hat{n} \cdot \nabla\Phi_i \, da - \frac{1}{2\pi} \int_M \Phi_i \frac{\hat{n} \cdot \mathbf{r}}{r^3} \, da \tag{58}$$

$$\Phi_e(P) = -\frac{1}{2\pi} \int_M \frac{1}{r} \hat{n} \cdot \nabla\Phi_e \, da + \frac{1}{2\pi} \int_M \Phi_e \frac{\hat{n} \cdot \mathbf{r}}{r^3} \, da \tag{59}$$

Here, $\mathbf{r}$ is a vector that joins a source point on the membrane to field point P, and r is the length of $\mathbf{r}$. Expression (59) for the extracellular potential has the sign reversed because the outward vector $\hat{n}$ points into the region, contrary to the statement of Green's theorem. Using boundary conditions (9)–(10), normal derivatives of Φ_i and Φ_e can be expressed in terms of capacitive and ionic currents through the membrane, resulting in

$$\Phi_i(P) = -\frac{1}{2\pi\sigma_i} \int_M \frac{1}{r} \left[C_m \frac{d}{dt}(\Phi_i - \Phi_e) + I_{\text{ion}} \right] da - \frac{1}{2\pi\sigma_i} \int_M \Phi_i \frac{\hat{n} \cdot \mathbf{r}}{r^3} \, da \tag{60}$$

$$\Phi_e(P) = \frac{1}{2\pi\sigma_e} \int_M \frac{1}{r} \left[C_m \frac{d}{dt}(\Phi_i - \Phi_e) + I_{\text{ion}} \right] da - \frac{1}{2\pi\sigma_e} \int_M \Phi_e \frac{\hat{n} \cdot \mathbf{r}}{r^3} \, da \tag{61}$$

External stimulation by a point electrode located in the intra- or extracellular region can be represented by including term I_{si}/r or I_{se}/r in Eq. (60) or (61), respectively. The surface of membrane M is then discretized into N patches, each assumed to be equipotential, and the integral identities (60)–(61) can be converted into a system of $2N$ equations with potentials Φ_i and Φ_e as unknowns. Thus, this approach shares many similarities with the boundary-element method [56].

The advantage of the Green's theorem approach is that it does not make any simplifying assumptions. Transformation of Laplace's Eqs. (5)–(6) into integral identities (60)–(61) is purely mathematical, so computing potentials from these identities is equivalent to solving the original problem. Integral identities (60)–(61) still result in a system of equations that must be solved numerically in each time step. However, since only the surface of the cell has to be discretized, the number of equations is much smaller. For example, Leon and Roberge's model of a cylindrical cell [55] contained 96 unknowns compared to 960 unknowns in DeBruin and Krassowska's model of a spherical cell [51]. Hence, even though the Green's theorem approach results in a full matrix, as opposed to a sparse one in the finite-difference and finite-element methods, computational savings can be substantial.

Vigmond and Bardakijan proposed a variation of the above approach based on differentiation of Green's theorem [57]. This alternative formulation can be solved directly for an electric field, resulting in a more accurate solution and in a further decrease of computational expense.

3. The Cell as a Short Fiber

This approach maps the interior of a cell and its membrane onto a one-dimensional cable model [58]. Such mapping was used for a spherical cell by Quan and Cohen [59], and for a prolate spheroidal cell by Tung and Borderies [60] and by Fishier et al. [61]. Typically, the cell is divided into N slices that are perpendicular to the direction of the field. Each slice is assumed to be equipotential and is mapped onto the circuit representation of the cable (Fig. 12). Intracellular resistance per length R_i is determined based on the cross section, length, and shape of the corresponding slice. Except for a cylindrical cell in a longitudinal field, R_i changes with the position along the cell. Likewise, the magnitude of the current flowing through each membrane element M_k $(k = 1, 2, \ldots, N)$ changes with the position, being proportional to the surface area of the slice. Extracellular potentials are assumed to be known and values V_k, which are assigned at each extracellular node, correspond to the steady-state solution of the relevant initial

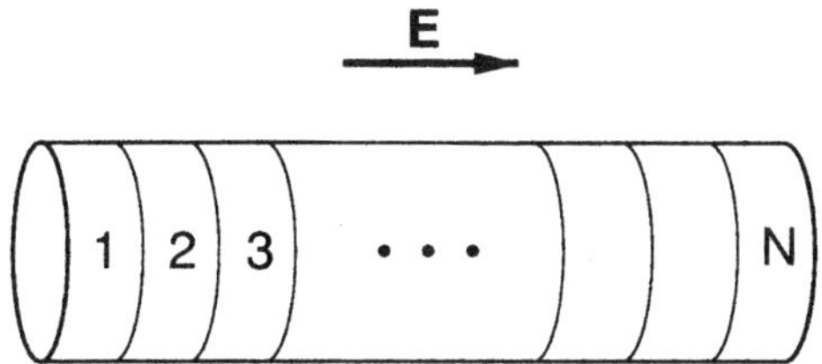

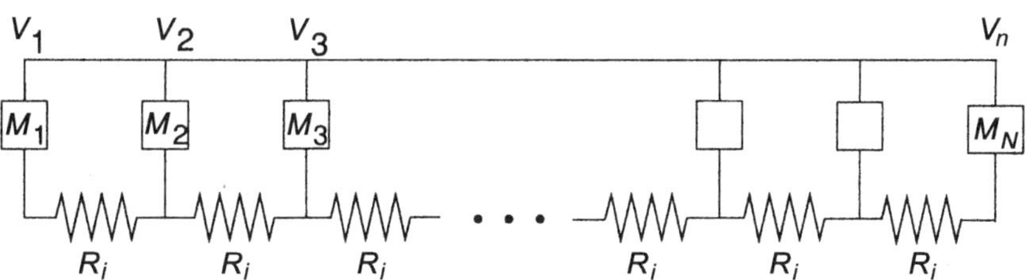

Figure 12 Mapping a cylindrical cell onto a circuit representation of the cable model. $V_1, V_2, \ldots, V_n$ are potentials at extracellular nodes; their values are computed from the steady-state solutions of the initial polarization problem. $M_1, M_2, \ldots, M_N$ represent isopotential membrane patches; resistors R_i represent the intracellular resistance between adjacent nodes.

polarization problem. Intracellular potentials or, equivalently, transmembrane potentials are unknown and must be computed at each time step by solving a system of N equations.

The advantages of this method are that the development of the model is very intuitive, relies on the well-understood formalism of a one-dimensional cable model, and involves solving only a small number of equations. In the literature, the number of slices representing the cell varies from three [60] to eleven [61]. An accurate estimate of the activation threshold calls for at least 21 slices [13]; but even with a larger number of slices, the computational efficiency remains high because the system is diagonal and can be solved without inverting a matrix.

4. The Cell as a Unit

Some studies that used the "cell as a short fiber" approach made an additional assumption that the interior of the cell was equipotential [61,62]. This simplification reduces the number of unknowns to only one, the intracellular potential of the cell, Φ_i. Hence, computing the response of the cell to an external field involves solving only one ordinary differential equation, in which the driving force for the change of Φ_i is a weighted sum of transmembrane currents flowing through all the slices representing the cell membrane. Therefore, with this simplification, the "cell as a short fiber" model reduces to Eq. (50), which governs the change of the physiological state of the cell and was rigorously derived in Sec. III.B.4.

Hence, the "cell as a unit" model consists of one differential equation [Eq. (50)], which governs the time evolution of the only state variable of the system, the leading-order intracellular potential Φ_i. The computational savings are obvious: with only one equation to solve, there is no matrix to invert. In fact, the cost of solving the single-cell problem becomes comparable to solving the equation for a space-clamped membrane. The difference is that the pointwise ionic current I_{ion} must be computed for several membrane patches and integrated numerically to determine the macroscopic ionic current i_{ion} [Eq. (48)]. Another advantage is that the "cell as a unit" model gives better insight into the mechanism of the cell's response to the field. For simple models, such as the FitzHugh-Nagumo model analyzed in Sec. III.B.5, an insightful qualitative analysis can be carried out by using methods of nonlinear dynamics. At the same time, the model does not lose sight of the contribution of different parts of the membrane to the activation process (Sec. III.B.6). If the membrane dynamics is described by a realistic model, such as Luo-Rudy kinetics [48], the "cell as a unit" model will also allow one to examine the spatial and temporal contributions of currents carried by different ions.

Finally, Eq. (50) is a rigorous simplification of single-cell problem (5)–(12). It keeps only the elements that are truly important, discarding those whose contribution to the excitation process is on the order of the small parameter ε or smaller. Thus, ε also serves as a measure of how well the "cell as a unit" model approximates problem (5)–(12). The approximation will be acceptable so long as ε is sufficiently small, i.e., so long as the initial polarization of the cell is much faster than the processes that follow it. By and large, this condition is satisfied during cell stimulation by an external field. The one instance when the model breaks down is when the field is so strong that it causes electroporation of the parts of the cell membrane that experience the highest transmembrane potential [63]. Creation of pores is very fast, on the order of microseconds, so the initial polarization of the cell and the creation of pores cannot be separated. These two processes occur simultaneously: the polarization of the cell initiates the creation of pores, and the development of pores limits and partially reverses the polarization of the cell. This is the reason why DeBruin and Krassowska's study [51] had to use a "brute-force" finite-difference model to simulate the first few microseconds of the electroporation process. Nevertheless, so long as the membrane is intact or its resistance is not altered artificially by the incorporation of a large number of conductive channels [32], the "cell as a unit" model should give an accurate description of the processes occurring during stimulation of a cell by an external field.

IV. CAN MODELS REPRODUCE EXPERIMENTAL RESULTS?

A. Field-Induced Transmembrane Potential

The first question in comparing experimental and modeling results is whether experiments can confirm the characteristics of the first stage of a cell's response to a field, the initial polarization. As seen in Sec. III.B.1, theory predicts that during this stage only the capacitive properties of a cell membrane are important. A cell should undergo a very rapid charging process, which within a microsecond establishes the distribution of the transmembrane potential, which depends only on the cell shape. The experiments that measured V_m optically, summarized in Sec. II.D [3,4,11], have demonstrated that the general pattern of cell polarization agrees with theoretical predictions: the end of the cell facing the cathode is depolarized, and the end of the cell facing the anode is hyperpolarized. These studies also confirmed that the maximum magnitude of the field-induced V_m depends linearly on the electric field and the cell size, as predicted by the steady-state solution to the initial polarization problem (Secs. III.B.2 and III.B.3).

However, the linear dependence on the field strength is limited to fields that induce V_m between -100 and $50\,\text{mV}$ [11]. Beyond this range, the hyperpolarization becomes larger than the depolarization [3,11]. As stated by Cheng et al. [11], the 1994 Luo-Rudy (LRd) model of a cardiac membrane [64,65] cannot reproduce this asymmetry, which suggests that the field-induced V_m is not determined by the ionic currents that are active in the physiological range. To account for the asymmetry, Cheng et al. had to supplement the LRd model by two additional currents: a time-independent outward current I_a, which activates at strongly positive V_m, and an electroporation current I_{ep}, which reflects the disruption of the barrier function of the membrane. Both currents become active at transmembrane potentials that are larger than the threshold for activation (I_a at $160\,\text{mV}$, and I_{ep} at approximately $= 350\,\text{mV}$). Similar changes were also proposed by Platzer et al. [66], who used the 1991 Luo-Rudy model [48]. Nevertheless, in the range of potentials that arise during field stimulation, this asymmetry is not expected to be a significant factor. Note that in the optical recordings shown in Fig. 6b, the positive and negative deflections have nearly the same magnitude.

Theory also predicts that the cell polarizes very rapidly, with the time constant below $1\,\mu\text{s}$ (Secs. III.B.2 and III.B.3). The optical recordings show that charging and discharging of the cell membrane are indeed very fast (Fig. 6b). Platzer and Windisch [62], using a "cell as a unit" model with Beeler-Reuter membrane dynamics [67], successfully reproduced their optical recordings of field-induced transmembrane potential. However, presently the temporal resolution of optical recordings is limited to a value from $1\,\mu\text{s}$ [11] to $20\,\mu\text{s}$ [4], which is not adequate to confirm the theoretically predicted exponential character of the initial polarization [Eqs. (32) and (37)] or to measure its time constant.

B. Time Course of Activation

The second stage of the cell's response, the actual change of its physiological state, is governed by first-order differential Eq. (50), with the intracellular potential ϕ_i^0 as a state variable. This theoretical result is in agreement with experimental measurements. As seen in Fig. 6b, after the field is turned off, transmembrane potentials at all three sites on a cell assume essentially the same value, just as predicted by the model in Fig. 10a. Theory also postulates that the cell should activate as a unit. Indeed, Windisch et al. [4] observed that the upstroke at the different sites on a cell is highly synchronized: the entire cell activates within $20\,\mu\text{s}$. As $20\,\mu\text{s}$ is also the limit of the temporal resolution of Windisch et al.'s optical system, the delay in the activation of the opposite ends of the cell may be even smaller. In general, there is a close correspondence in the spatial and temporal patterns of activation predicted by the model and measured experimentally (Fig. 10a versus Fig. 6b).

The main quantitative difference is the spatial nonuniformity in the maximum rate of rise of the upstroke ($\dot{V}_{ref}$) observed by Windisch et al. [4]: $\dot{V}_{ref}$ is up to 10% larger at the end of the cell that has been hyperpolarized by the shock. This result can be qualitatively explained by the model: the shock partially inactivates sodium channels in the depolarized end of the cell, so during the upstroke this part of the cell has a smaller magnitude of sodium current (Fig. 10d, "2 ms"). However, the model that treats the cell as a unit cannot, by its very nature, quantitatively reproduce this result. This is because in the "cell as a unit" model, the pointwise ionic current I_{ion} does not influence local V_m: all current goes into a common "bag," the macroscopic ionic current i_{ion}, which drives the evolution of the intracellular potential ϕ_i^0. As the temporal evolution of V_m is tied to the evolution of ϕ_i^0 [Eqs. (45)–(47)], the rate of rise of the upstroke in the model is identical everywhere on the membrane (Fig. 10a). In a real cell mapped by Windisch et al., the pointwise I_{ion} does have some influence on local V_m. However, as seen in Fig. 6c, the change in $\dot{V}_{ref}$ caused by this influence is below 5%. This difference, treated as a measure of the accuracy of the "cell as a unit" model, indicates that this simple model provides a reasonable approximation of the time course of activation in a single cell stimulated by an external field.

C. Activation Thresholds for Field Stimulation

1. Strength–Duration Curves

The final issue in assessing the accuracy of the theoretical models of a single cell is whether field stimulation thresholds predicted by models are in quantitative agreement with thresholds measured experimentally. This question was addressed by Stone et al. [13], who compared measured and computed field stimulation thresholds for embryonic chick heart cells. Chick cells are ideal for this type of study because, when isolated, they assume a spherical shape [22], allowing a straightforward analytical computation of field-induced transmembrane potential (Sec. III.B.3). Experimental thresholds, in the form of a strength–duration curve, are shown in Fig. 4. Theoretical thresholds were computed from the "cell as a unit" model: the initial polarization of a spherical cell was determined by Eq. (37) (in a limit $t \to \infty$), and the evolution of the intracellular potential was computed numerically from Eq. (50), using the procedure described in Sec. III.B.6.

The quantitative agreement in predicting activation thresholds critically depends on choosing a suitable representation of the ionic current I_{ion} as well as other parameters of the model. For chick heart cells, there exist two models of membrane dynamics: by Ebihara and Johnson [68] and by Shrier and Clay [69,70]. The former model is better suited for 11-day-old chick ventricular cells, but it describes only the fast sodium current I_{Na}. Therefore, Stone et al. [13] combined the fast sodium current of the Ebihara-Johnson (EJ) model with the non-sodium currents of the 1991 Luo-Rudy (LR) model, forming an EJLR model. Other parameters, such as membrane capacitance ($1.3\,\mu F/cm^2$), rest potential ($-83\,mV$), maximum sodium conductance ($23\,mS/cm^2$), and sodium reversal potential ($29\,mV$), were also based on measurements in embryonic thick heart cells. The simulations were performed with a cell radius of $20\,\mu m$ and the thresholds were expressed in terms of a maximum change of the transmembrane potential, using Eq. (1).

Activation thresholds predicted by the cell model with EJLR membrane dynamics (dotted line) are shown in Fig. 13 together with thresholds measured experimentally (solid line). For long stimulus durations, the theoretical thresholds were within one standard deviation of experimental ones. However, for short stimulus durations, the theoretical thresholds increased much more rapidly than the experimental ones. Moreover, for stimuli below

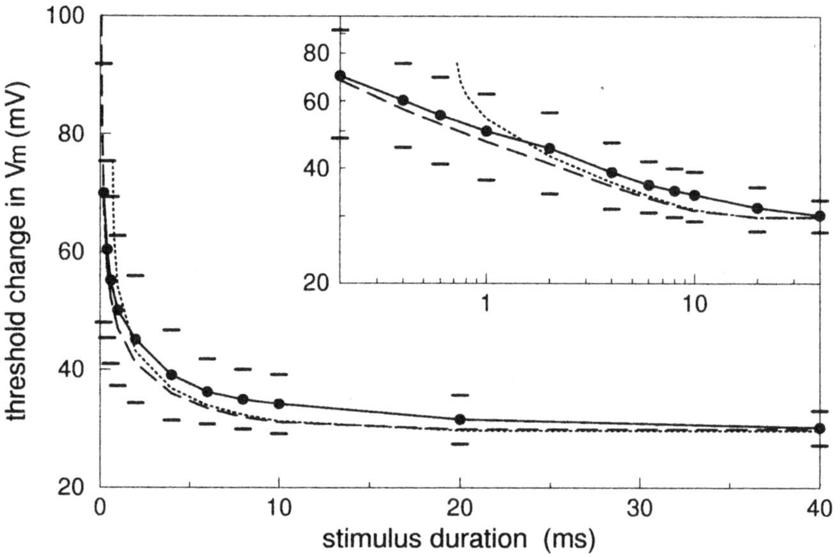

Figure 13 Activation thresholds measured experimentally at 37°C (filled circles connected by a solid line) and predicted by cell models with an EJLR membrane. The dotted line corresponds to the EJLR model with the original dynamics, and the dashed line corresponds to the EJLR modified to account for the slow deactivation of the fast sodium current. (From Ref. 13.)

0.8 ms, the cell model was inexcitable. One possibility is that the model was correct and that the excitability of cells in the experiment was due to the applied field being different from the assumed rectangular pulse– for example, because of the current flow after the nominal termination of the stimulus [4,10]. Direct measurements of the electric field disproved this hypothesis: the shape of the electric field was very close to rectangular, and its strength dropped to zero essentially instantaneously (Fig. 2 in Ref. [13]). Hence, the problem lay with the model, and an analysis was performed to explain whether the inexcitability resulted from simplifications made when the cell model was created or from the choice of model parameters. None of the tested mechanisms, alone or in combination, could fully explain the inexcitability of the cell models for short stimulus durations, leading Stone et al. to a conclusion that the existing membrane models do not contain the element responsible for the excitation of single cells by short stimuli. A search of the literature suggested the hypothesis that this missing element is a "tail current," i.e., a flow of fast sodium current occurring after the termination of the stimulus. In the EJ model, the tail current is practically nonexistent because of an extremely fast (15 µs) closing of the m gates. Similarly fast deactivation is also observed in the Luo-Rudy 1991 [48] and the Luo-Rudy 1994 [64,65] membrane models. These models are all based on voltage clamp experiments performed by Ebihara and Johnson [24,68], in which the time constant of the m gates for V_m just above rest was not measured directly but instead was extrapolated from measurements at higher V_m. Thus, all these models miss the phenomenon of the sodium tail current.

To test the hypothesis that the tail current is responsible for the activation of cells exposed to short stimuli, Stone et al. modified the kinetics of the sodium current in the EJ model by allowing the m gates to close with a voltage-independent time constant of 1.4 ms. This value was chosen because it provided the best fit to the experimental strength–duration curve. With this modification, the cell model was able to predict thresholds for stimulus durations down to and below 0.2 ms (Fig. 13, dashed line). For all stimulus durations, the thresholds predicted by the model were within one standard deviation of the experimental data, the correlation coefficient between the experimental and predicted thresholds was 0.9972, and the overall root-mean-square error was 6.14%. However, the assumed time constant for closing m gates, 1.4 ms, may be too large. The time constant of deactivation of the sodium channels, measured by Murray et al. [71] in guinea pig myocytes at 37°C, is 0.061 ms, which corresponds to closing of the m gates with a time constant of 0.18 ms. This value is almost eight times lower than the 1.4 ms used by Stone et al. This discrepancy can be explained in part by the fact that the dynamics of the guinea pig membrane may be faster than that of the chick heart membrane, as evidenced by a faster action potential upstroke [48]. In addition, trypsin used by Stone et al. to dissociate the chick cells has been reported to cause an up-to-threefold increase of τ_m [72].

An indirect argument in favor of the tail current hypothesis comes from comparing the ability of the Weiss-Lapicque relationship to approximate activation thresholds for the stimulation of cardiac muscle with a small unipolar electrode and the stimulation of single cells with an external field. Theoretically, the Weiss-Lapicque relationship [Eq. (4)] approximates thresholds for a space-clamped membrane charged by an external current [17]. A study by Pearce et al. [25] demonstrated that Eq. (4) is an accurate representation of thresholds for the cardiac muscle stimulated with a unipolar electrode, even for stimuli as short as 1 µs. In contrast, the strength–duration curves for single cells are flatter than the Weiss-Lapicque relationship (Fig. 4). The tail current hypothesis offers the following explanation. Assume that for a successful activation, a critical amount of charge must be forced into a cell. To activate the myocardium with a unipolar electrode, the entire charge must be delivered by the stimulating current. To activate a single cell, the field only needs to deliver a fraction of this charge; the rest enters the cell after the stimulus as a sodium tail current. As the duration of a stimulus decreases, the tail current plays an increasingly important role in the activation of a cell, and consequently thresholds for single cells increase less rapidly than thresholds for the myocardium. Nevertheless, the mechanism of field stimulation with short pulses is not yet fully understood and needs further study.

2. Effects of Temperature

As seen in Sec. II.C on an example of embryonic chick heart cells (Fig. 5), lowering the temperature from 37 to 27°C reduces the measured thresholds for all stimulus durations. To reproduce this change in the model, Stone et al. [15,26] made several modifications to the model described in Sec. IV.C.1. The first modification followed Cooley and Dodge [73], who reproduced threshold dependence on temperature for intracellular stimulation of the giant squid axon by multiplying the rate coefficients for the ionic channels by a Q_{10} factor. In the EJLR model of Stone et al., the Q_{10} factor equal to 3 was used. However, the rate coefficients alone could not fully explain the change in thresholds at the different temperatures. The rheobase, 25.8 mV, was still 24% higher than the experimental rheobase of 20.8 mV. More important, thresholds for short stimulus durations increased instead of decreasing. For example, the threshold for a 0.5-ms stimulus increased from 53.9 to 62.3 mV. The second factor affected by the temperature is the rest potential, which for cardiac preparations tends to become less negative as the temperature decreases [74,75]. For chick cells, the rest potential at 22°C was measured by Josephson et al. [76] as $-70\,$mV and by Sada et al. [77] as $-73.5\,$mV. Hence, the rest potential in the model was increased from -83 to $-73.5\,$mV, which for long pulse durations brought the theoretical thresholds within a standard deviation of the experimental ones. Finally, predicting thresholds for short stimulus durations required modifications of the deactivation time constant of the fast sodium current. As in the case of the model for 37°C, the model for 27°C assumed that m gates close with a voltage-independent time constant. The value assumed for the 27°C model was $4.5\,$ms, i.e., three times larger than the $1.5\,$ms assumed for 37°C, which is consistent with the Q_{10} factor of 3. With these three modifications, the model was able to predict activation thresholds at 27°C within one standard deviation of the experimental measurements (Fig. 14). The correlation coefficient between measured and predicted thresholds was 0.9976, and the root-mean-square error was 5.86%.

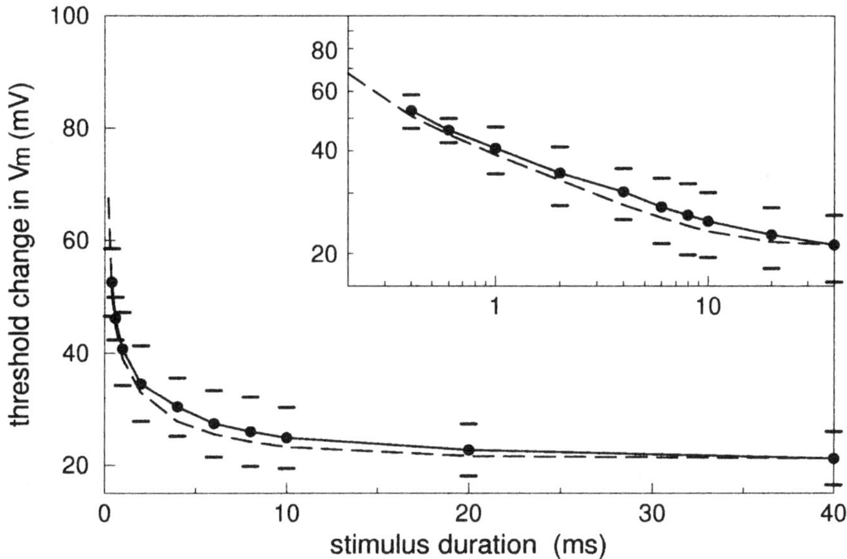

Figure 14 Activation thresholds measured experimentally at 27°C (filled circles connected by a solid line) and predicted by the cells model with the modified EJLR membrane (dashed line). To account for the temperature, the rate coefficients and the deactivation time constant are multiplied by a Q_{10} factor of 3, and the rest potential is decreased to $-73.5\,$mV.

The decrease of the thresholds for field stimulation with decreasing temperature may explain the low values of thresholds for guinea pig myocytes measured by Tung et al. [10]. For the stimulus duration of 10 ms, Tung et al. reported thresholds of only 1.4 V/cm (longitudinal) and 3.5 V/cm (transverse). Taking into account that the typical cell dimensions for adult guinea pig heart cells are $25 \times 135\,\mu m$, the longitudinal threshold of 1.4 V/cm implies a field-induced transmembrane potential of 9.5 mV [Eq. (3)] and the transverse threshold of 3.5 V/cm implies V_m of 8.8 mV [Eq. (2)]. Assuming rest potential of -84 mV, the transmembrane potential at the depolarized end is expected to be approximately -75 mV, still 10–15 mV short of the threshold voltage for the guinea pig membrane, which has been reported to lie between -65 and -60 mV [48,78]. However, the study by Tung et al. was conducted at room temperature, which increases rest potential to approximately -70 mV. If, as indicated by some studies [79,80], the voltage at which sodium channels open does not change appreciably with temperature, then 8.8–9.5 mV may be sufficient to reach the threshold.

3. Effects of a Cell's Orientation and Shape

Since mature cardiac cells have an elongated, rodlike shape (Fig. 2a), an electric field induces a larger transmembrane potential when it is longitudinal to the cell axis rather than transverse to it. Hence, thresholds for field stimulation are expected to have a strong directional dependence. Using equations for the maximum transmembrane potential induced in a cylindrical cell by transverse [Eq. (2)] and longitudinal [Eq. (3)] fields, the ratio of thresholds is predicted to be

$$\frac{E_t}{E_L} = \frac{1}{2}\frac{l_c}{d_c} \tag{62}$$

where E_t and E_l are the transverse and longitudinal thresholds, respectively. Formula (62) indicates that the threshold ratio is proportional but not equal to the aspect ratio of the cell, l_c/d_c. Experimental measurements discussed in Sec. II.C and summarized in Table 3 confirm that in all cases the threshold ratio is lower than the aspect ratio. The best agreement between measured and predicted ratios is for guinea pig cells, for which the ratio computed from Eq. (62) is probably within a standard deviation of the experimental data. The worst agreement is for frog cells, for which measured and computed ratios differ by a factor of almost three. This poor performance is most likely due to frog cells resembling tortuous threads (as seen in Fig. 1 of Tung et al. [10]) more than the short rods assumed by Eq. (62). In particular, the effective length of a frog cell is shorter than its stretched length l_c.

Tung et al. [10] also measured threshold ratios for stimuli of different durations and observed that for guinea pig cells, E_t/E_l, increases from approximately 2 for a 0.02-ms duration to 5.5 for durations above 0.2 ms. The same trend occurred for frog myocytes. This change in threshold ratios is most likely caused by membrane dynamics and cannot be reproduced using Eq. (62), which is based on the steady-state solution to the initial polarization problem. A model of a cylindrical or ellipsoidal cell with an excitable membrane is needed to understand and reproduce this decrease in threshold ratios for short stimulus durations.

However, models based on cells with a regular geometry, such as a cylinder or an ellipsoid, have a definite limitation as far as the quantitative predictions of thresholds are concerned. This limitation is seen in the study by Ranjan and Thakor [12], who observed that the threshold can change by $98.9 \pm 70.1\%$ with the reversal of the field's polarity. Since field-induced transmembrane potential in cells with a regular geometry is symmetric, this result indicates a possible role played by the irregularities in the cell shape. As discussed in Sec. III.C.1, Ranjan and Thakor tested this hypothesis with two-dimensional models of cells with realistic shapes. While these cells had an asymmetric field-induced V_m, this asymmetry (about 20%) was smaller than the observed change in thresholds (98.9%). It is unclear whether the nonlinearities associated with the excitability of the membrane can amplify this modest difference in the transmembrane potential to an almost twofold difference in thresholds. Other factors may be involved, such as a nonuniform distribution of sodium channels on the cell surface [81]. To resolve this issue, one needs a three-dimensional model of a cell with a realistic shape and membrane dynamics that are suitable for the preparation in question. The construction of such a model is one of the remaining tasks in theoretical studies of field stimulation of single cardiac cells.

V. DISCUSSION

This review implies, through a comparison of experimental and theoretical results published in the literature, that the existing theory provides a reasonably accurate representation of processes occurring during field stimulation of isolated cardiac cells. More precisely, there is good qualitative agreement between the models and the experiments; a good quantitative agreement can also be reached if one is willing to put sufficient effort into the proper representation of cell geometry, membrane dynamics, and model parameters. For cell geometry, the most important factors are the aspect ratio and, in some cases, irregularities of the cell shape [12]. For membrane dynamics, the crucial factor is the fast sodium current, since models developed for different species do not match experimental data as well as the model developed for the same species and the same type of preparation [13]. Among model parameters, the most important one is V_{rest}, which should be adjusted for temperature [13]. There still remain some details of the cell activation process that have to be examined more closely. One example is the effect of a realistic cell shape on the ratio of transverse to longitudinal activation thresholds and the reason for the decrease of this ratio for short stimuli. Another example is the possible role of the sodium tail current; the resolution of this issue is a prerequisite for successful modeling of field stimulation with short-duration pulses or high-frequency waveforms. Nevertheless, there is reason to believe that the response of a single cardiac cell to an external electric field is already fairly well understood.

The situation is quite different when one considers the effect of electric field on multicellular cardiac preparations. Despite extensive experimental and theoretical efforts during the last two decades, this problem is, at best, only partially understood (see Ch. 5 and 6). Experimental measurements using optical dyes and double-barelled microelectrodes revealed complex patterns of the shock-induced transmembrane potential (see Ch. 16). These patterns can be predicted or reproduced by the model only for a few rather simple preparations, e.g., virtual anodes and cathodes near a point source [82,83] or a papillary muscle subject to defibrillation-strength fields [84,85]. Also, while the mechanisms of activation and induction of arrhythmias by a unipolar electrode have been successfully reproduced by modeling and computer simulations [86–88], the processes that involve the myocardium away from electrodes, such as field stimulation, induction of rotors by cross-field stimulation, or defibrillation, remain elusive [89].

As mentioned in the Introduction of this chapter, several studies have pointed out that the polarization of individual cells of cardiac tissue (i.e., the sawtooth potential) provides an elegant theoretical explanation of these processes [5–8,61,90,91]. There are at least five arguments supporting the role of the sawtooth. First, just as in a single cell, the magnitude of the sawtooth is proportional to the electric field [16,92,93], which explains the experimental finding that the success or failure of stimulating and defibrillating shocks depends on the field's strength [94–96]. Second, in contrast to the macroscopic transmembrane potential, which decays to zero away from electrodes, the electric field retains an appreciable magnitude throughout most of the heart [97–99]. Hence, the sawtooth appears to be capable of having a direct effect on the entire heart. Third, the threshold for field stimulation of the myocardium depends on the orientation of the field with respect to cardiac fibers in the same way it does for a single cell: the threshold for the longitudinal field is lower than that for the transverse one [94]. Moreover, there is experimental evidence that the defibrillation threshold decreases if shocks are applied along two orthogonal pathways [100,110]. This behavior is consistent with that expected from the sawtooth, since with two orthogonal paths more cells are exposed to the longitudinal field. Fourth, the sawtooth potential affects each cell in the same way, depolarizing one half and hyperpolarizing the other. This can therefore explain the relative insensitivity of the thresholds for field stimulation and defibrillation to the polarity of the electrodes [102,103]. Finally, the response of a cell polarized by a sawtooth potential to biphasic waveforms [55,59,61] can qualitatively explain decrease of the defibrillation threshold for biphasic waveforms observed in experiments [104–106].

However, to date the sawtooth potential has been observed in isolated single cells [3,4,11] and cell pairs [107] but not in multicellular tissue in vitro, even when the potential is measured with sufficient spatial resolution to detect cellular events [108,109]. This is a very puzzling result: the absence of a potential drop across the membrane, whose resistivity is eight orders of magnitude higher than resistivity of intra- and extracellular space, seems to defy Ohm's law. Of course, there is a possibility that the sawtooth exists but at present cannot be measured. Just like the initial polarization of a single cell, the sawtooth arises in a fraction of a microsecond and may be immediately followed by a synchronized response of the entire cell. Optical systems, with a temporal resolution of 1–20 μs and spatial averaging over a 5–22 μm spot [3,4,11] may miss very localized and transient manifestations of the sawtooth and measure instead the second, spatially synchronized stage of the cell response. Also, preparations used in experimental studies of the sawtooth may not be appropriate: a small cardiac strand or a monolayer surrounded by a large volume of extracellular bath tends to polarize as a unit, producing essentially no sawtooth potential. Finally, the sawtooth is unlikely to appear as an interspersed sequence of depolarization and hyperpolarization. This pattern would occur in a multicellular preparation only if all the cells were of the same length and were connected by the same junctional resistance. In a real cardiac muscle, because of its structural inhomogeneities, the sawtooth potential will have a positive or negative offset. Hence, the sawtooth will manifest itself as an abrupt change of potential across the cell boundary, but not necessarily as a polarity reversal of the transmembrane potential [107].

However, there is a *theoretical* explanation for the absence of the sawtooth in cardiac muscle. For a single cell in a conductive bath, the transmembrane potential induced by the field is given by Eq. (29). Ignoring an offset V_0, $v_m^0 = -\mathbf{E} \cdot \mathbf{w}$, that is, v_m^0 is a product of an electric field $\mathbf{E}$ and a weight function $\mathbf{w}$ that reflects the drop of potential across the membrane. However, when a cell is surrounded by other cells, computing its polarization has to account for the presence of neighbors. Using a rigorous homogenization method, Neu and Krassowska [93] determined that the sawtooth potential arising in a multicellular tissue is given by the following formula:

$$v_m^0 = \mathbf{E} \cdot (\mathbf{w_i} - \mathbf{w_e}) \tag{63}$$

This formula shows that the sawtooth potential is determined not by one but by two weight functions: $\mathbf{w_i}$, which reflects discontinuities present in the intracellular space, and $\mathbf{w_e}$, which reflects discontinuities present in the extracellular space. In computer simulations conducted to date, extracellular conductivity has been assumed to be spatially uniform and $\mathbf{w}_e$ has been ignored. However, in real cardiac muscle, both $\mathbf{w}_i$ and $\mathbf{w}_e$ may exist. If their shapes and magnitudes approximately match, then according to Eq. (63) a cancellation will occur and the magnitude of the sawtooth potential will be very close or equal to zero. One scenario when such a cancellation may occur is illustrated in Fig. 15. Let us assume that the transmembrane potential is being measured at discrete spots within the two abutting cells in the direction of the applied field. Note that at the same location where the

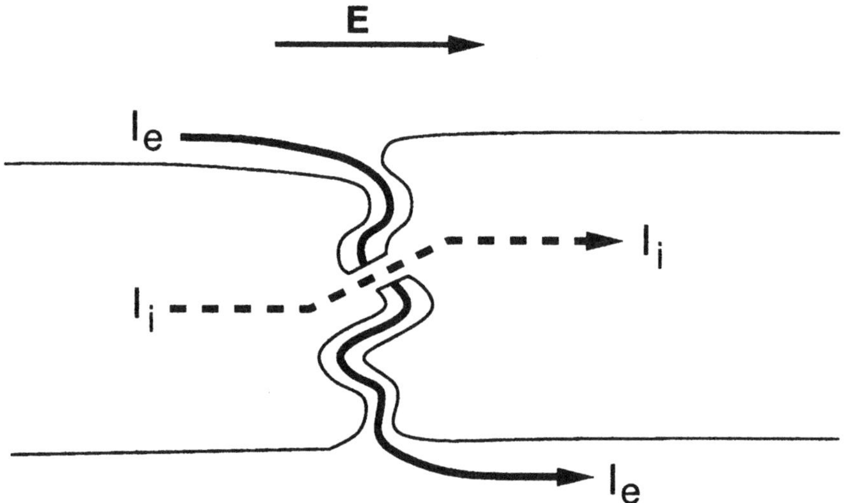

Figure 15 Example of a situation in which the sawtooth potential may not arise. The path of intracellular current I_i is indicated by a thick dashed line; the path of extracellular current I_e is indicated by a thick solid line. The intracellular current flows through a gap junction, shown schematically as a channel connecting the interiors of the two adjacent cells. The direction of the electric field is indicated by arrow **E**.

intracellular current negotiates the gap junction, extracellular current flows through a narrow and tortuous path formed by the end processes of the cells. If the potential drop along this part of the extracellular path has a magnitude similar to the potential drop across the gap junction, then the sawtooth potential can be very small.

Additional arguments against the sawtooth come from a closer examination of the qualitative similarities between the theoretical and experimental results listed above. This examination reveals two important quantitative discrepancies. First, the dependence on the electric field of the thresholds for field stimulation, rotor induction, and defibrillation may be an artifact of the limited range of the shock strengths tested in the experiments. This view is supported by a recent study by Idriss [110], who induced rotors in the heart by using a larger range of the S_2 shock strengths. As the shock strength increased and the site of a rotor moved away from the electrode, the field strength measured at the earliest sites of activation (which Idriss took as an approximate measure of the field at a critical point) increased from below $2\,\mathrm{V/cm}$ at $60\,\mathrm{V}$ to about $10\,\mathrm{V/cm}$ at $460\,\mathrm{V}$. This observation cannot be explained by sawtooth alone, for which the shock-induced transmembrane potential depends on the cell dimensions and the magnitude of the field but not on the distance of the cell from the electrode. The second discrepancy is that the thresholds for field stimulation that are measured in multicellular preparations are too small to establish a sawtooth potential of a physiologically meaningful magnitude. For example, field stimulation thresholds reported by Frazier et al. [94] for canine myocardium are $0.64\,\mathrm{V/cm}$ (along fibers) and $1.84\,\mathrm{V/cm}$ (across fibers). Assuming that the dimensions of cardiac cells are $100 \times 20\,\mu\mathrm{m}$ (12), sawtooth potentials induced by these fields are 3.2 and $1.84\,\mathrm{mV}$, respectively. Hence, the maximum magnitude of the sawtooth is less than one-eighth of the value needed to reach the excitation threshold of the cardiac membrane, -65 to $-60\,\mathrm{mV}$ [48,78]. Since Frazier's experiment was conducted in vivo, low threshold fields cannot be explained by the effect of the room temperature (Sec. IV.C.2).

One way of reconciling the qualitative arguments for the sawtooth with quantitative arguments against it is to postulate that the sawtooth potential works together with the macroscopic transmembrane potentials. Such macroscopic potentials are due to the activation function, the unequal anisotropy ratios of intra- and extracellular conductivities, the curvature of cardiac fibers, tissue heterogeneities, and other factors, all of which are reviewed in this volume by Fishler, Trayanova et al., and Efimov and Cheng (see Chs. 5, 6, 16). A recent model that combined the sawtooth with the activation function succeeded in qualitatively and quantitatively reproducing the initiation of a rotor by a cross-field stimulation [111,112]. It has also been postulated that units larger than single cells are involved in the mechanisms of field stimulation, induction of arrhythmias, and defibrillation [84,89,113]. The existence of such macroscopic units was confirmed experimentally by Gillis et al. [109], who observed regions of depolarization and hyperpolarization on the opposite sides of extracellular clefts that separate groups of cells. Moreover, these macroscopic units appear to play a role in field stimulation, as demonstrated by White et al. [114], who reported that the creation of a lesion decreases the local threshold for field stimulation. A counterexample comes from Zhou et al. [115], who mapped V_m over the heart surface with a spatial resolution ranging from 30 μm to 3 mm and found no evidence of oppositely polarized units. Therefore, the existence of the sawtooth potential in cardiac tissue, its spatial scale, and its possible role in the heart's response to an external field requires further studies, both experimental and theoretical.

REFERENCES

1. AL Hodgkin, AF Huxley, B Katz. Measurement of current-voltage relations in the membrane of the giant axon of *Loligo*. J Physiol 116:424–448, 1952.
2. D Gross, LM Loew, WW Webb. Optical imaging of cell membrane potential changes induced by applied electric fields. Biophys J 50:339–348, 1986.
3. SB Knisley, TF Blitchington, BC Hill, AO Grant, WM Smith, TC Pilkington, RE Ideker. Optical measurements of transmembrane potential changes during electric field stimulation of ventricular cells. Circ Res 72:255–270, 1993.
4. H Windisch, H Ahammer, P Schaffer, W Müller, D Platzer. Optical multisite monitoring of cell excitation phenomena in isolated cardiomyocytes. Pflugers Archiv—Eur J Physiol 430:508–518, 1995.
5. R Plonsey, RC Barr. Effect of microscopic and macroscopic discontinuities on the response of cardiac tissue to defibrillating (stimulating) currents. Med Biol Eng Comput 24:130–136, 1986.
6. W Krassowska, TC Pilkington, RE Ideker. Periodic conductivity as a mechanism for cardiac stimulation and defibrillation. IEEE Trans Biomed Eng 34:555–560, 1987.
7. AM Chernysh, VY Tabak, MS Bogushevich. Mechanisms of electrical defibrillation of the heart. Resuscitation 16:169–178, 1988.
8. SM Dillon. Optical recordings in the rabbit heart show that defibrillation strength shocks prolong the duration of depolarization and the refractory period. Circ Res 69:842–856, 1991.
9. AL Bardou, J-M Chesnais, PJ Birkui, M-C Govaere, PM Auger, D Von Euw, J Degonde. Directional variability of stimulation threshold measurements in isolated guinea pig cardiomyocytes: relationship with orthogonal sequential defibrillating pulses. PACE 13:1590–1595, 1990.

10. L Tung, N Sliz, MR Mulligan. Influence of electrical axis of stimulation on excitation of cardiac muscle cells. Circ Res 69:722–730, 1991.

11. DK -L Cheng, L Tung, EA Sobie. Nonuniform responses of transmembrane potential during electric field stimulation of single cardiac cells. Am J Physiol 277:H351–H362, 1999.

12. R Ranjan, NV Thakor. Electrical stimulation of cardiac myocytes. Ann Biomed Eng 23:812–821, 1995.

13. BA Stone, M Lieberman, W Krassowska. Field stimulation of isolated chick heart cells: comparison of experimental and theoretical activation thresholds. J Cardiovasc Electrophysiol 10:92–107, 1999.

14. CR Horres, M Lieberman, JE Purdy. Growth orientation of heart cells on nylon monofilament. Determination of the volume-to-surface area ratio and intracellular potassium concentration. J Membrane Biol 34:313–329, 1977.

15. BA Stone, M Lieberman, W Krassowska. Field stimulation of heart cells at 37°C and 27°C. Ann Biomed Eng 25:S62, 1997.

16. W Krassowska. Modeling the interaction of cardiac muscle with strong fields. In: HG Othmer, FR Adler, MA Lewis, JC Dallon, eds. Case Studies in Mathematical Modeling—Ecology, Physiology and Biofluids. Upper Saddle River, NJ: Prentice Hall, 1997, pp. 277–308.

17. LA Geddes, JD Bourland. Tissue stimulation: theoretical considerations and practical applications. Med Biol Eng Comput 23:131–137, 1985.

18. W Irnich. The fundamental law of electrostimulation and its application to defibrillation. PACE 13:1433–1447, 1990.

19. GA Mouchawar, LA Geddes, JD Bourland, LA Pearce. Ability of the Lapicque and Blair strength-duration curves to fit experimentally obtained data from the dog heart. IEEE Trans Biomed Eng 36:971–974, 1989.

20. SAS User's Guide: Statistics, Version 5 Edition. Cary, NC: SAS Institute, 1985, pp. 575–606.

21. JR Clay, RM Brochu, A Shrier. Phase resetting of embryonic chick atrial heart cell aggregates. Experiment and theory. Biophys J 58:609–621, 1990.

22. S Fujii, R Ayer Jr, R DeHaan. Development of the fast sodium current in early embryonic chick heart cells. J Membrane Biol 101:209–223, 1988.

23. H Sada, M Kojima, N Sperelakis. Use of single heart cells from chick embryos for the Na^+ current measurements. Mol Cell Biochem 80:9–19, 1988.

24. L Ebihara, N Shigeto, M Lieberman, EA Johnson. The initial inward current in spherical clusters of chick embryonic heart cells. J Gen Physiol 75:437–456, 1980.

25. JA Pearce, JD Bourland, W Neilsen, LA Geddes, M Voeltz. Myocardial stimulation with ultrashort duration current pulses. PACE 5:52–58, 1982.

26. BA Stone. Activation Thresholds for Chick Heart Cells During Field Stimulation, Modeled and Measured. MS thesis, Duke University, Durham, NC, 1997.

27. F Evans. Factors Affecting Field Stimulation of Cardiac Tissue. PhD thesis, Duke University, Durham, NC, 1996.

28. R Guttman, B Sandler. Effect of temperature on the potential and current thresholds of the axon membrane. J Gen Physiol 46:257–266, 1962.

29. RA Sjodin, LJ Mullins. Oscillatory behavior of the squid axon membrane potential. J Gen Physiol 42:39–47, 1958.

30. R FitzHugh. Theoretical effect of temperature on threshold in the Hodgkin-Huxley nerve model. J Gen Physiol 49:989–1005, 1966.

31. B Ehrenberg, DL Farkas, EN Fluhler, Z Lojewska, LM Loew. Membrane potential induced by external electric field pulses can be followed with a potentiometric dye. Biophys J 51:833–837, 1987.

32. Z Lojewska, DL Farkas, B Ehrenberg, LM Loew. Analysis of the effect of medium and membrane conductance on the amplitude and kinetics of membrane potentials induced by externally applied electric fields. Biophys J 56:121–128, 1989.

33. V Krauthamer, JL Jones. Calcium dynamics in cultured heart cells exposed to defibrillator type electric shocks. Life Sci 60:1977–1985, 1997.

34. H Windisch, H Ahammer, P Schaffer, W Müller, D Platzer. Fast optical potential mapping in single cardiomyocytes during field stimulation. Proc 14th Annu Int Conf of the IEEE Engineering in Medicine and Biology Society, 1992, pp. 634–635.

35. H Windisch, W Müller, H Ahammer, P Schaffer, D Dapra, M Hartbauer. Optical potential mapping helps to reveal discrete-natural-phenomena in cardiac muscle. Int J Bifurcation Chaos 6:1925–1933, 1996.

36. R FitzHugh. Dimensional analysis of nerve models. J Theor Biol 40:517–541, 1973.

37. V Barcilon, JD Cole, RS Eisenberg. A singular perturbation analysis of induced electric fields in nerve cells. SIAM J Appl Math 21:339–354, 1971.

38. A Peskoff, RS Eisenberg. The time-dependent potential in a spherical cell using matched asymptotic expansions. J Math Biol 2:277–300, 1975.

39. A Peskoff, RS Eisenberg, JD Cole. Matched asymptotic expansions of the Green's function for the electric potential in an infinite cylindrical cell. SIAM J Appl Math 30:222–239, 1976.

40. W Krassowska, JC Neu. Response of a single cell to an external electric field. Biophys J 66:1768–1776, 1994.

41. CM Bender, S Orszag. Advanced Mathematical Methods for Scientists and Engineers. New York: McGraw-Hill, 1978, pp. 417–430.

42. C Dixon. Applied Mathematics of Science and Engineering. New York: Wiley, 1971, pp. 281–324.

43. HP Schwan. Biophysics of the interaction of electromagnetic energy with cells and membranes. In: M Grandolfo, SM Michaelson, A Rindi, eds. Biological Effects and Dosimetry of Nonionizing Radiation. New York: Plenum, 1983, pp. 213–231.

44. HP Schwan. Dielectrophoresis and rotation of cells. In: E Neumann, AE Sowers, CA Jordan, eds. Electroporation and Electrofusion in Cell Biology. New York: Plenum, 1989, pp. 3–21.

45. M Klee, R Plonsey. Stimulation of spheroidal cells—The role of cell shape. IEEE Trans Biomed Eng BME-23:347–354, 1976.

46. B. FitzHugh. Impulses and physiological states in theoretical models of nerve membrane. Biophys J 1:445–466, 1961.

47. J Nagumo, S Arimoto, S Yoshizawa. An active pulse transmission line simulating nerve axon. Proc IRE 50:2061–2070, 1962.

48. C-H Luo, Y Rudy. A model of the ventricular cardiac action potential: depolarization, repolarization, and their interaction. Circ Res 68:1501–1526, 1991.

49. L Tung, SK Jain. Simulations of electrical field stimulation of cardiac myocytes during the relative refractory period. Proc 14th Annu Int Conf of the IEEE Engineering in Medicine and Biology Society, 1992, pp. 644–645.

50. M Klee, R Plonsey. Finite difference solution for biopotentials of axially symmetric cells. Biophys J 12:1661–1875, 1972.

51. KA DeBruin, W Krassowska. Modeling electroporation in a single cell. I: Effects of field strength and rest potential. Biophys J 77:1213–1224, 1999.

52. RW Glaser, SL Leikin, LV Chernomordik, VF Pastushenko, AI Sokirko. Reversible electrical breakdown of lipid bilayers: formation and evolution of pores. Biochim Biophys Acta 940:275–287, 1988.

53. G Dahlquist, A Björck. Numerical Methods. Englewood Cliffs, NJ: Prentice-Hall, 1974, pp. 146–157.

54. LJ Leon, FA Roberge. A new cable model formulation based on Green's theorem. Ann Biomed Eng 18:1–17, 1990.

55. LJ Leon, FA Roberge. A model study of extracellular stimulation of cardiac cells. IEEE Trans Biomed Eng 40:1307–1319, 1993.

56. CA Brebbia, JCF Telles, LC Wrobel. Boundary Element Techniques: Theory and Applications in Engineering. Tokyo: Springer-Verlag, 1984.

57. EJ Vigmond, BB Bardakjian. Efficient and accurate computation of the electric fields of excitable cells. Ann Biomed Eng 24:168–179, 1996.

58. AL Hodgkin, WAH Rushton. The electrical constants of a crustacean nerve fibre. Proc Roy Soc B 133:444–479, 1946.

59. W-L Quan, TJ Cohen. Field stimulation of single cardiac cell—The dependency of membrane excitation threshold on waveform shape and cellular refractoriness. Proc 15th Annu Int Conf of the IEEE Engineering in Medicine and Biology Society, 1993, pp. 869–870.

60. L Tung, J-R Borderies. Analysis of electric field stimulation of cardiac muscle cells. Biophys J 63:1–16, 1992.

61. MG Fishler, EA Sobie, NV Thakor, L Tung. Mechanisms of cardiac cell excitation with premature monophasic and biphasic field stimuli: a model study. Biophys J 70:1347–1362, 1996.

62. D Platzer, H Windisch. Simulation of excitation of single cardiomyocytes under field stimulation. Proc 14th Annu Int Conf of the IEEE Engineering in Medicine and Biology Society, 1992, pp. 642–643.

63. K Kinosita, I Ashikawa, N Saita, H Yoshimura, H Itoh, K Nagayama, A Ikegami. Electroporation of cell membrane visualized under a pulsed-laser fluorescence microscope. Biophys J 53:1015–1019, 1988.

64. C-H Luo, Y Rudy. A dynamic model of the cardiac ventricular action potential. I. Simulations of ionic currents and concentration changes. Circ Res 74:1071–1096, 1994.

65. J Zeng, KR Laurita, DS Rosenbaum, Y Rudy. Two components of the delayed rectifier K^+ current in ventricular myocytes of the guinea pig type: theoretical formulation and their role in repolarization. Circ Res 77:140–152, 1995.

66. D Platzer, D Dapra, C Günter, H Windisch. Model-augmented investigations on field-stimulated cardiomyocytes. Proc 1st Joint BMES/EMBS Conf, 1999, p. 140.

67. GW Beeler, H Reuter. Reconstruction of the action potential of ventricular myocardial fibres. J Physiol 268:177–210, 1977.

68. L Ebihara, EA Johnson. Fast sodium current in cardiac muscle. A quantitative description. Biophys J 32:779–790, 1980.

69. A Shrier, JR Clay. Repolarization currents in embryonic chick atrial cell aggregates. Biophys J 50:861–874, 1986.

70. VC Kowtha, A Kunysz, JR Clay, L Glass, A Shrier. Ionic mechanisms and nonlinear dynamics of embryonic chick heart cell aggregates. Prog Biophys Mol Biol 61:255–281, 1994.

71. KA Murray, T Anno, PB Bennett, LM Hondeghem. Voltage clamp of the cardiac sodium current at 37°C in physiologic solutions. Biophys J 57:607–612, 1990.

72. CA Vandenberg, R Horn. Inactivation viewed thorough single sodium channels. J Gen Physiol 84:535–564, 1984.

73. JW Cooley, FA Dodge. Digital computer solutions for excitation and propagation of the nerve impulse. Biophys J 6:583–599, 1966.

74. E Coraboeuf, S Weidmann. Temperature effects on the electrical activity of Purkinje fibers. Helv Physiol Acta 12:32–41, 1954.

75. MS Suleiman, RA Chapman. Effect of the temperature on the rise in intracellular sodium caused by calcium depletion in ferret ventricular muscle and the mechanism of the alleviation of the calcium paradox by hypothermia. Circ Res 67:1238–1246, 1990.

76. JR Josephson, N Sperelakis. Developmental increases in the inwardly-rectifying K^+ current of embryonic chick ventricular myocytes. Biochim Biophys Acta 1052:123–127, 1990.

77. H Sada, M Kojima, N Sperelakis. Fast inward current properties of voltage-clamped ventricular cells of embryonic chick heart. Am J Physiol 255:H540–H552, 1988.

78. H Kishida, B Surawicz, LT Fu. Effects of K^+ and K^+-induced polarization on $(dV/dt)_{max}$, threshold potential, and membrane input resistance in guinea pig and cat ventricular myocardium. Circ Res 44:800–814, 1979.

79. A Portela, MI Guardado, H Jenerick, PA Stewart, RJ Perez, C Rodriguez, JR de Xamar Oro, E Zothner, TC Rozell, AL Gimeno. Temperature dependence on the passive and dynamic electrical parameters of muscle cells. Acta Physiol Latinoam 29:15–43, 1979.

80. L. Ebihara. Inward currents in spherical clusters of chick embryonic heart cells. PhD thesis, Duke University, Durham, NC, 1980.

81. B Hille. Ionic Channels of Excitable Membranes. 2nd ed. Sunderland, MA: Sinauer Associates, 1992, pp. 514–519.

82. NG Sepulveda, BJ Roth, JP Wikswo Jr. Current injection into a two-dimensional anisotropic bidomain. Biophys J 55:987–999, 1989.

83. BJ Roth. Approximate analytical solutions to the bidomain equations with unequal anisotropy ratios. Phys Rev E 55:1819–1826, 1997.

84. W Krassowska, MS Kumar. The role of spatial interactions in creating the dispersion of transmembrane potential by premature electric shocks. Ann Biomed Eng 25:949–963, 1997.

85. KA DeBruin, W Krassowska. Electroporation and shock-induced transmembrane potential in a cardiac fiber during defibrillation strength shocks. Ann Biomed Eng 26:584–596, 1998.

86. BJ Roth. Strength-interval curves for cardiac tissue predicted using the bidomain model. J Cardiovasc Electrophysiol 7:722–737, 1996.

87. JM Saypol, BJ Roth. A mechanism for anisotropic reentry in electrically active tissue. J Cardiovasc Electrophysiol 3:558–566, 1992.

88. BJ Roth. Nonsustained reentry following successive stimulation of cardiac tissue through a unipolar electrode. J Cardiovasc Electrophysiol 8:768–778, 1997.

89. BJ Roth, W Krassowska. The induction of reentry in cardiac tissue. The missing link; how electric fields alter transmembrane potential. Chaos 8:204–220, 1998.

90. NA Trayanova, TC Pilkington. A bidomain model with periodic intracellular Junctions: a one-dimensional analysis. IEEE Trans Biomed Eng 40:424–433, 1993.

91. JP Keener. Direct activation and defibrillation of cardiac tissue. J Theor Biol 178:313–324, 1996.

92. W Krassowska, TC Pilkington, RE Ideker. The closed form solution to the periodic core-conductor model using asymptotic analysis. IEEE Trans Biomed Eng 34:519–531, 1987.

93. JC Neu, W Krassowska. Homogenization of syncytial tissues. CRC Crit Rev Biomed Eng 21:137–199, 1993.

94. DW Frazier, W Krassowska, P-S Chen, PD Wolf, EG Dixon, WM Smith, RE Ideker. Extra-cellular field required for excitation in three-dimensional anisotropic canine myocardium. Circ Res 63:147–164, 1988.

95. JM Wharton, PD Wolf, WM Smith, P-S Chen, DW Frazier, S Yabe, N Danieley, RE Ideker. Cardiac potential and potential gradient fields generated by single, combined, and sequential shocks during ventricular defibrillation. Circulation 85:1510–1523, 1992.

96. X Zhou, JP Daubert, PD Wolf, WM Smith, RE Ideker. Epicardial mapping of ventricular defibrillation with monophasic and biphasic shocks in dogs. Circ Res 72:145–160, 1993.

97. E Lepeschkin, HC Herrlich, S Rush, JL Jones, RE Jones. Cardiac potential gradients between defibrillation electrodes. Med Instrum 14:57, 1980.

98. P-S Chen, PD Wolf, FJ Claydon III, EG Dixon, HJ Vidaillet Jr, ND Danieley, TC Pilkington, RE Ideker. The potential gradient field created by epicardial defibrillation electrodes in dogs. Circulation 74:626–636, 1986.

99. ASL Tang, PD Wolf, Y Afework, WM Smith, RE Ideker. Three-dimensional potential gradient fields generated by intracardiac catheter and cutaneous patch electrodes. Circulation 85:1857–1864, 1992.

100. DL Jones, GJ Klein, GM Guiraudon, AD Sharma, MJ Kallok, WA Tacker, JD Bourland. Sequential pulse defibrillation in man: comparison of thresholds in normal subjects and those with cardiac disease. Med Instrum 21:166–169, 1987.

101. AL Bardou, J Degonde, PJ Birkui, P Auger, J-M Chesnais, M Duriez. Reduction of energy required for defibrillation by delivering shocks in orthogonal directions in the dog. PACE 11:1990–1995, 1988.

102. JC Schuder, H Stoeckle, WC McDaniel, M Dbeis. Is the effectiveness of cardiac ventricular defibrillation dependent upon polarity? Med Instrum 21:262–265, 1987.

103. PG O'Neill, KA Boahene, GM Lawrie, LF Harvill, A Pacifico. The automatic implantable cardioverter-defibrillator: effect of patch polarity on defibrillation threshold. J Am Coll Cardiol 17:707–711, 1991.

104. JL Jones, RE Jones. Improved defibrillator waveform safety factor with biphasic waveforms. Am J Physiol 245:H60–H65, 1983.

105. SA Feeser, ASL Tang, KM Kavanagh, DL Rollins, WM Smith, PD Wolf, RE Ideker. Strength-duration and probability of success curves for defibrillation with biphasic waveforms. Circulation 82:2128–2141, 1990.

106. JL Jones, OH Tovar. Threshold reduction with biphasic defibrillator waveforms. Role of charge balance. J Electrocardiol 28(suppl):25–30, 1995.

107. V Sharma, L Tung. Theoretical and experimental study of intercellular junction-induced sawtooth effect in cardiac cell-pairs. Proc 1st Joint BMES/EMBS Conf, 1999, p. 138.

108. X Zhou, WM Smith, DL Rollins, RE Ideker. Spatial changes in transmembrane potential during a shock. PACE 18:935, 1995.

109. AM Gillis, VG Fast, S Rohr, AG Kleber. Spatial changes in transmembrane potential during extracellular electrical shocks in cultured monolayers of neonatal rat ventricular myocytes. Circ Res 79:676–690, 1996.

110. S Idriss. Characterization of the upper limit of ventricular vulnerability. PhD thesis, Duke University, Durham, NC, 1995.

111. J Wall, NA Trayanova, K Skouibine, W Krassowska. Modeling induction of reentry with realistic S2 stimulus. Proc 1st Joint BMES/EMBS Conf, 1999, p. 154.

112. K. Skouibine, J. Wall, W. Krassowska, NA Trayanova. Modeling induction of a rotor in cardiac muscle by perpendicular electric shocks. Med Biol Eng Comput, 2002 (in press).

113. W Krassowska, DW Frazier, TC Pilkington, RE Ideker. Potential distribution in three-dimensional periodic myocardium: part II. Application to extracellular stimulation. IEEE Trans Biomed Eng 37:267–284, 1990.
114. JB White, GP Walcott, RE Ideker. Myocardial discontinuities: a substrate for producing virtual electrodes to increase directly excited areas of the myocardium by shocks. PACE 20:1234, 1997.
115. X Zhou, RE Ideker, TF Blitchington, WM Smith, SB Knisley. Optical transmembrane potential measurements during defibrillation-strength shocks in perfused rabbit hearts. Circ Res 77:593–602, 1995.

5

Computer Modeling of Defibrillation I: The Role of Cardiac Tissue Structure

Matthew G. Fishler
St. Jude Medical, Inc., Sunnyvale, California, U.S.A.

I. INTRODUCTION

Cardiac defibrillation has become an indispensable clinical interventional tool for the acute treatment of otherwise fatal episodes of sudden cardiac arrest. For decades (if not centuries [1,2]) it has been recognized that the delivery of a strong electric discharge across a fibrillating heart could successfully resuscitate an individual from almost certain death. Consequently, ever since the commercialization of the first defibrillator, the availability and use of both internal and external defibrillators has continued to accelerate, propelled in recent years by the ever-expanding list of clinical indications for implantable cardioverter-defibrillators [3] and the improved accessibility, functionality, and educational efforts with respect to public access of automatic external defibrillators [4–6].

Obviously, defibrillation works. But how? What are the fundamental mechanisms underlying the defibrillation process? Certainly most individuals in need of immediate defibrillation are probably not particularly concerned with *how* an electric shock delivered by their defibrillator will interact with their myocardium to terminate the offending arrhythmia; they just hope it works! However, other individuals *not* in need of immediate defibrillation have been inspired by these questions and have dedicated much of their careers to help answer them. Over the years, these research endeavors have accumulated a wealth of knowledge, significantly elevated our overall level of understanding, and successfully elucidated many important facets about the defibrillation process. Nevertheless, the "how" of defibrillation is still not yet fully characterized—details of some underlying mechanisms still remain incompletely described, resolved, and/or confirmed.

The methods by which these research endeavors have progressed have included clinical, experimental, computational, and theoretical approaches. Each of these approaches provides a set of advantages—balanced by a corresponding set of limitations—for mechanistic investigation, hypothesis testing, and process discovery. Computer modeling has proven to be particularly valuable for its ability to provide exceptionally exquisite control over the entire "experimental" design, enabling one to produce conditions *in silico* that might be impractical or impossible to replicate in vitro. Such control is invaluable when, for example, one wants to create "idealized" conditions, or when one wants to control, minimize, or eliminate potentially confounding factors that might otherwise obscure or adversely influence the mechanism under study. Furthermore, computer models have also provided the best (and sometimes the only) windows into many underlying processes not readily accessible or measurable using experimental techniques. For example (and of relevance to the topic covered here), modeling has provided an important avenue for investigating the role and impact of myocardial tissue structure on intrinsic and extrinsic cardiac electrophysiological processes. Defibrillation is one such process of extrinsic origin.

Most generally, defibrillation can be dissected into three major subprocesses: (1) the shock establishes an electric field between the anodal and cathodal electrodes; (2) this electric field induces changes in transmembrane voltages (V_m) throughout the myocardium; and (3) these induced changes in V_m alter the electrophysiological kinetics of the affected tissues from their preshock dynamics, and thereby disrupt (and hopefully terminate) the underlying arrhythmic behavior. How the electric field is established (subprocess 1) is generally well understood, with modeling being used primarily for helping to illustrate and quantify the spatial distributions of current densities, extracellular potentials, and field strengths throughout the heart and thorax for various electrode configurations [7–12]. In contrast, how subprocesses 2 and 3 transpire during defibrillation is still not fully elucidated, and remains the focus of active research. Computer modeling has made substantial contributions to these research endeavors, both as a medium for hypothesis testing and corroboration and as a platform for mechanistic discovery and prediction.

The next two chapters describe and review how computer modeling has been used in defibrillation research and how it has been an invaluable quantitative vehicle with which to extend and expand our level of understanding of the underlying processes. In particular, this chapter concentrates on the contributions and insights that computer modeling has provided into how the electric field of a defibrillation shock interacts with cardiac tissue to induce V_m changes throughout the entire heart (i.e., subprocess 2). The next chapter then explores how computer modeling has contributed to the investigation of how these shock-induced changes in V_m result in subsequent arrhythmia termination (i.e., subprocess 3).

II. THEORETICAL TOOLS FOR INVESTIGATING SHOCK–TISSUE INTERACTIONS

The previous chapter discussed how an isolated cardiomyocyte can be excited by electrical stimulation. Such investigations have been extremely important to the overall understanding of cardiac stimulation and defibrillation. However, intact cardiac tissue is not a simple conglomerate of isolated cells; rather, cardiomyocytes within the myocardium are strongly interconnected via relatively low-resistance gap junctions. The presence of these intercellular couplings thus causes the tissue to behave electrophysiologically as a functional "syncytium," meaning that the characteristic length over which electrophysiological influences extend is significantly longer than the underlying cells themselves. Consequently, defibrillation cannot be described or analyzed as a simple spatial superposition of the excitation dynamics of isolated cells. Instead, the influences of those syncytial properties must also be considered.

A. The Foundation of Cardiac Stimulation Analysis: One-Dimensional Cable Theory

Although cardiac tissue is constructed from discrete cardiomyocytes and is thus anatomically discontinuous, the conductive strength of the intercellular junctional connections enable the tissue to be well approximated macroscopically as a functionally continuous excitable medium. Such an approximation is quite advantageous, since the electrophysiological characteristics of the syncytium (i.e., its macroscopic behaviors) can then be analyzed using significantly simplified quantitative methods. One-dimensional cable theory represents the foundational approach to the quantitative inquiry and analysis of cardiac tissue stimulation, excitation, and propagation [13–15]. This method—originally borrowed from the toolbelt of neuroscientists studying stimulation of and propagation along neuronal axons—provides a robust mathematical framework for investigating the electrophysiological behaviors of an idealized one-dimensional strand of cardiac tissue suspended within a restricted interstitial space.

Figure 1A illustrates an idealized representation of this one-dimensional cardiac "cable." In this cylindrical geometry, a cardiac fiber of finite length L is centrally suspended within a restricted interstitial space. A sarcolemmal membrane ensheathes the intracellular space and separates it from the surrounding interstitial space. Currents (capacitive or ionic) can pass across this membrane from one space to the other, but currents within each space are constrained by definition to flow only along the long axis of the fiber itself (i.e., axially). The membrane and both spaces are all spatially uniform and continuous—any discontinuities or heterogeneities have been removed through homogenization of the associated structural properties.

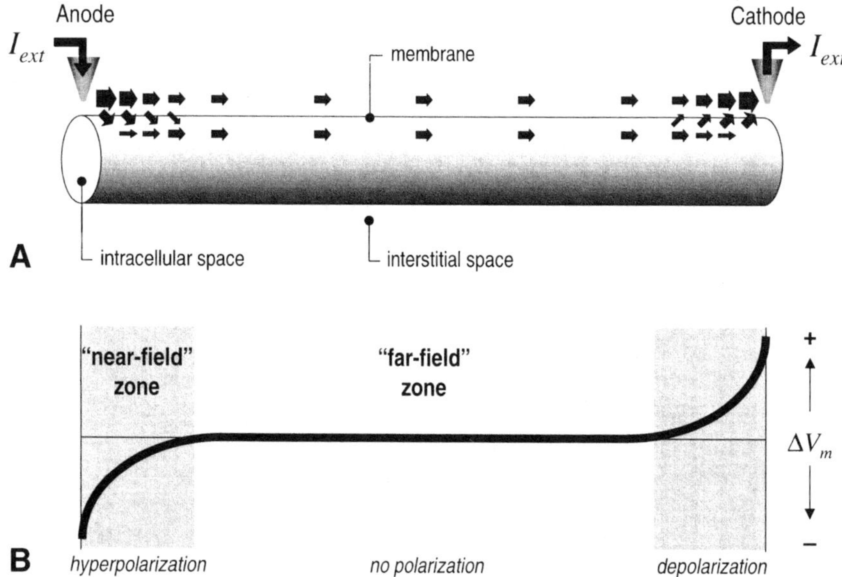

Figure 1 (A) Schematic representation of a finite-length one-dimensional cardiac "cable" undergoing external stimulation. A continuous sarcolemmal membrane separates the intracellular and interstitial spaces of the fiber. Arrows superimposed along the fiber illustrate the relative magnitudes and directions of the passive axial and transmembrane currents as induced by this stimulus. (B) Steady-state transmembrane potential, V_m, profile corresponding to the stimulus delivered to the fiber in (A). Zones of relative hyperpolarization and depolarization develop proximal to the anode and cathode, respectively, with the magnitudes of these induced polarizations diminishing exponentially with increasing distance from those electrodes. Outside of these "near-field" zones, a "far-field" region exists in which the fiber remains relatively unperturbed by the stimulus.

Two differential equations [Eqs. (1) and (2)], coupled through a third [Eq. (3)], can be used to fully describe the spatiotemporal dynamics of the intracellular and interstitial potentials (ϕ_i and ϕ_e, respectively) manifested along this fiber:

$$g_i \frac{\partial^2 \phi_i}{\partial x^2} = +\beta I_m \tag{1}$$

$$g_e \frac{\partial^2 \phi_e}{\partial x^2} = -\beta I_m \tag{2}$$

$$I_m = C_m \frac{\partial V_m}{\partial t} + I_{\text{ion}} \tag{3}$$

where g_i and g_e are the effective intracellular and interstitial axial conductivities (mS/cm), β is the ratio of membrane surface area to tissue volume (cm^{-1}), I_m and I_{ion} are the total (i.e., capacitive plus ionic) and ionic-only transmembrane current densities (μA/cm^2), C_m is the specific membrane capacitance (μF/cm^2), $V_m = \phi_i - \phi_e$ is the transmembrane potential (mV), x is distance along the fiber (cm), and t is time (msec). Furthermore, by scaling Eq. (1) by g_e and Eq. (2) by g_i, Eqs. (1)–(3) can be combined into a single equation in terms of V_m only:

$$\frac{g_i g_e}{g_i + g_e} \cdot \frac{\partial^2 V_m}{\partial x^2} = \beta \left(C_m \frac{\partial V_m}{\partial t} + I_{\text{ion}} \right) \tag{4}$$

In the example of Fig. 1A, extracellular electrodes are positioned at either end of the fiber, providing a source of stimulating current (I_{ext}, in μA) that flows from anode to cathode. While intracellular electrodes could be (and have been [15]) modeled as well, this extracellular configuration idealizes the actual electrode arrangement found in practice (clinically and experimentally), and thus is more appropriate and relevant to the study of the myocardium's electrophysiological responses to typical pacing or defibrillation stimuli. Mathematically, this extracellular stimulation arrangement can be represented by the enforcement of boundary conditions on Eqs. (1) and (2):

$$-g_i \cdot \left. \frac{\partial \phi_i}{\partial x} \right|_{\pm L/2} = 0 \qquad -g_e \cdot \left. \frac{\partial \phi_e}{\partial x} \right|_{\pm L/2} = J_{\text{ext}} \tag{5}$$

where $J_{\text{ext}} = I_{\text{ext}}/A_{\text{tot}}$ is the equivalent current density of the stimulating current, I_{ext}, as computed relative to the total cross-sectional area, A_{tot}, of the combined intracellular and extracellular spaces. These boundary conditions ensure that the ends of the cardiac fiber are appropriately "sealed"—that the stimulating current cannot communicate directly with the intracellular space, but instead must approach the fiber via the interstitial space [16]. With appropriate scaling, the boundary conditions in Eq. (5) can also be combined to define the associated boundary conditions on V_m:

$$\left.\frac{\partial V_m}{\partial x}\right|_{\pm L/2} = \frac{J_{\text{ext}}}{g_e} \tag{6}$$

Although the cardiac membrane has a significantly nonlinear current–voltage relationship overall (see Chap. 2), this relationship is quite linear when transmembrane voltage excursions from resting potential are relatively small. In this zone, the nonlinear transmembrane ionic currents have not yet been activated, so the membrane responds mostly passively. Under these conditions, the current–voltage relationship can be approximated as $I_{\text{ion}} = V_m/R_m$, where R_m is the passive specific resistance of the membrane ($k\Omega\cdot cm^2$) and V_m is specified relative to its resting value. Utilizing this definition for I_{ion} conjunction with Eqs. (4) and (6), the steady-state spatial solution for V_m along this fiber ($-L/2 \leq x \leq L/2$) is given by

$$V_m(x) = \frac{J_{\text{ext}}\lambda}{g_e} \cdot \frac{\sinh(x/\lambda)}{\cosh(L/2\lambda)} \tag{7}$$

where

$$\lambda = \sqrt{\frac{R_m g_i g_e}{\beta(g_i + g_e)}} \tag{8}$$

is the length constant for the fiber, and defines the distance over which local electrophysiological responses attenuate in magnitude by $1/e$ ($\sim 63\%$). For reference, experimental measurements of the myocardial length constant along the fiber direction vary between approximately 0.3 and 1.0 mm [17–19].

This steady-state solution for V_m [Eq. (7)] is graphed schematically in Fig. 1B in registration with the fiber above. Note that the membrane is hyperpolarized only in the vicinity of the anode, and depolarized only in the vicinity of the cathode. Furthermore, the magnitudes of these polarizations decay exponentially with distance from these electrodes, such that beyond a few length constants from either electrode, the membrane remains essentially unperturbed by the stimulus. To note, while Fig. 1A depicts the extracellular electrodes located immediately adjacent to the fiber ends, these electrodes may in fact be distal to these ends and yet still have the same impact on the fiber. That is because the polarized regions at the fiber ends develop as a consequence of the interaction of the stimulus current with the sealed boundary conditions at those ends, not because of the proximity of the electrodes per se.

Additional insight into these results can be realized by observing the associated stimulus-induced current distributions along this passive fiber. The relative magnitudes and directions of the axial and transmembrane currents induced by this stimulus are indicated by arrows superimposed on the fiber in Fig. 1A. At the anode, all of the stimulus current is delivered to the interstitial space. Over the distance of a few space constants, some fraction of this current (determined by the relative magnitudes of the axial conductivities within the two spaces) crosses the membrane and redistributes into the intracellular space. As the distance from the anode increases, the magnitude of these transmembrane currents diminishes exponentially to essentially zero (coincident with the decay in V_m) as the redistribution reaches an effective equilibrium. Thus, along the central section of the fiber, only axial currents are present. Of course, all of the current that crossed into the intracellular space near the anode must eventually return to the interstitial space for removal by the extracellularly located cathode. This return of current to the interstitial space occurs within a few space constants of the cathode, with the magnitudes of the associated transmembrane currents increasing exponentially with proximity to the cathode itself.

One-dimensional cable theory has proven to be a very constructive and insightful tool for the basic study of propagation through and point stimulation of cardiac tissue [17,20–22]. However, this theory appears insufficient to explain many documented cardiac responses to defibrillation shocks. Perhaps most significantly, this theory cannot adequately explain how a defibrillation shock induces direct excitation throughout the entire heart. Experiments have demonstrated that defibrillation shocks can induce direct excitation of the entire heart, including midmyocardial regions far from any stimulating electrodes or heart surfaces [23–25]. Yet, as illustrated in Fig. 1B, traditional cable theory predicts that direct polarization will occur only in regions of tissue proximal to the electrodes (sometimes referred to as the "near-field" zone), while essentially no polarization will develop in regions distal to those electrodes (the "far-field" zone). How then are these far-field regions stimulated by the shock? The most obvious conclusion from these results is that one or more features of cardiac tissue critical to the development of a far-field response are not embodied in this simplified one-dimensional cable model. Identifying such features and characterizing the associated mechanisms involved has been the focus of a substantial amount of subsequent theoretical and computational research.

B. Bidomain Theory

In order to quantitatively explore behaviors of cardiac tissue beyond a single dimension, cable theory had to be extended and generalized. The result is bidomain theory [26–28]. As with traditional cable theory, bidomain theory is designed to provide a macroscopic representation of cardiac tissue. This macroscopic perspective is achieved by using volume-averaged quantities for all tissue properties, thereby abstracting any underlying microscopic features [29]. Furthermore, consistent with this volume-averaged approach, the intracellular and interstitial domains are conceptualized to be completely interpenetrating—and thus coexistent spatially—separated everywhere by a continuous sarcolemmal membrane. The governing equations of bidomain theory are two coupled parabolic reaction-diffusion differential equations given as*

$$\nabla \cdot (\tilde{g}_i \nabla \phi_i) = +\beta I_m \tag{9}$$
$$\nabla \cdot (\tilde{g}_e \nabla \phi_e) = -\beta I_m \tag{10}$$

where ∇ and $\nabla\cdot$ are gradient and divergence operators, respectively, I_m is total transmembrane current as defined in Eq. (3), and $\tilde{g}_i$ and $\tilde{g}_e$ are now tensors describing the effective intracellular and interstitial bidomain conductivities for this multidimensional tissue. Tensors are required to embody the fact that the tissue can have different conductivity magnitudes along its different principal axes—a property known as anisotropy.

Since $\tilde{g}_i$ and $\tilde{g}_e$ are mathematically independent quantities, each domain can foster individual—and hence potentially different—extents of anisotropy. If each domain is anisotropic to the same degree as the other (i.e., equal anisotropy ratios), $\tilde{g}_i$ and $\tilde{g}_e$ are related through a simple scalar factor (i.e., $\tilde{g}_i = k \cdot \tilde{g}_e$), thereby enabling Eqs. (9) and (10) to be combined into a single effective governing equation in terms of V_m only. In this way, while this equally anisotropic system is described physically as a bidomain with separate intracellular and interstitial spaces, it actually behaves mathematically as though it has only one effective combined domain. Under such equivalent monodomain conditions, the tissue responds analogous to the one-dimensional cable, in that stimulus-induced polarizations occur only in the vicinity of the electrodes (and/or the sealed ends of the tissue) and decay spatially at a hyperexponential rate [14,18]. Thus, equally anisotropic conditions are also insufficient to explain far-field excitation of bulk myocardium.

---

*In fact, there exists an entire family of equivalent representations for these governing equations [30].

However, real cardiac tissue is not equally anisotropic, but rather demonstrates decisively unequally anisotropic conductivity ratios between the two domains. In the intracellular space, the ratio between conductivity magnitudes along versus across the tissue fiber direction has been estimated experimentally at about 10, while in the interstitial space it is about 2.5 [31]. With unequal anisotropy ratios, there is no scalar relationship between $\tilde{g}_i$ and $\tilde{g}_e$, and thus Eqs. (9) and (10) cannot be combined into a single equation in terms of V_m alone. Thus, a true mathematical bidomain is preserved. As will be described in more detail below, the presence of unequal anisotropic ratios within the bidomain creates a medium in which many interesting and unexpected stimulus-induced responses are possible.

C. Insights from the Generalized Activating Function

Some qualitative hints into what factors could contribute to the far-field bulk stimulation of cardiac tissue can be revealed through some simple manipulations of the original bidomain equations [32]. Rearrangement of Eq. (9) after substitution of the definition $\phi_i = V_m + \phi_e$ gives

$$\beta\left(I_{\text{ion}} + C_m \frac{\partial V_m}{\partial t}\right) - \nabla \cdot (\tilde{g}_i \nabla V_m) = \nabla \cdot (\tilde{g}_i \nabla \phi_e) \tag{11}$$

For a stimulus applied when the entire tissue is at rest, V_m is initially constant everywhere and the net transmembrane ionic current (I_{ion}) is zero everywhere. Thus, the first and third terms on the left-hand side of Eq. (11) drop out, and the relationship simplifies to

$$\frac{\partial V_m}{\partial t} \propto \nabla \cdot (\tilde{g}_i \nabla \phi_e) \tag{12}$$

Equation (12) suggests that the initial change in V_m will be driven by the term given on its right-hand side. This term is known as the generalized activating function [32], and represents an effective source term for V_m.

Further insight can be obtained by expanding the divergence operation in this generalized activating function:

$$\frac{\partial V_m}{\partial t} \propto [\tilde{g}_i : \nabla(\nabla \phi_e)] + [(\nabla \cdot \tilde{g}_i) \cdot \nabla \phi_e] \tag{13}$$

where the colon indicates a tensor inner product. The expansion elucidates the conditions necessary for inducing changes in transmembrane potentials anywhere within the tissue. The first term on the right-hand side of Eq. (13)

indicates that a change in V_m can be induced even in the presence of a spatially uniform tissue conductivity profile (i.e., $\tilde{g}_i = $ constant, thus $\nabla \cdot \tilde{g}_i = 0$) as long as the gradients in extracellular potentials are not also spatially uniform [i.e., $\nabla \phi_e \neq$ constant, thus $\nabla(\nabla \phi_e) \neq 0$]. Alternatively, from the second term in Eq. (13), a change in V_m can be induced even in the presence of a spatially uniform gradient in extracellular potentials [i.e., $\nabla \phi_e = $ constant, thus $\nabla(\nabla \phi_e) = 0$] as long as the underlying tissue conductivity profile is not also spatially uniform (i.e., $\tilde{g}_i \neq$ constant, thus $\nabla \cdot \tilde{g}_i \neq 0$). In other words, this generalized activating function reveals that stimulus-induced polarization of cardiac tissue can develop only when either the extracellular potential gradients or the underlying tissue conductivity profile (or both) is spatially nonuniform. The next section describes several predicted and/or proven elements of cardiac tissue structure that have been hypothesized to provide the source(s) of nonuniformity underlying the generation of stimulus-induced membrane polarization within the bulk myocardium.

III. MECHANISMS FOR SHOCK–TISSUE INTERACTION

To date, four distinct features of cardiac tissue structure have been hypothesized to interact directly with an applied shock such that polarizations might be induced within the bulk tissue. These four structurally based mechanisms—described in detail below—are not mutually exclusive, and, in fact, are all likely to coexist and contribute to the total shock-induced response. However, not all of these predicted mechanisms have yet been observed experimentally, and the relative weights with which they might contribute to a total response are still largely unknown.

A. Virtual Electrodes

Before the introduction of the bidomain representation of myocardium, theoretical investigations of the responses of multidimensional cardiac tissue utilized equations that were essentially direct extensions of one-dimensional cable theory. Using modern terminology, these previous models effectively provided only a monodomain representation of the tissue, with or without (equal) anisotropy. Consequently, the predicted responses of the tissue to point stimuli were also analogous to those predicted for the fiber: point stimulation would induce transmembrane polarizations that diminish monotonically in magnitude with increasing distance from that stimulating electrode. Moreover, the rate at which these polarizations would decay toward zero within this multi-dimensional environment were predicted to be even faster than the exponential rate for a one-dimensional fiber [14,18].

Sepulveda et al. [33] were first to predict that *unequally* anisotropic cardiac tissue does *not* respond to point stimulation with monotonically decaying potentials, but rather generates an unexpectedly complicated and nonintuitive spatial distribution of polarized regions, including areas of both depolarization and hyperpolarization. These researchers used a finite-element method to solve the passive steady-state bidomain equations numerically and determine the responses of a two-dimensional sheet of anisotropic cardiac tissue to unipolar point current injection. (Because of the inherent symmetry of the problem, only one quadrant of the tissue was actually analyzed.) Three sets of anisotropy ratios were investigated: equal (equivalent to a monodomain condition), nominally unequal (corresponding to experimentally measured values within heart muscle), and reciprocally unequal (an extreme case in which the directions of highest conductivity within the two domains are perpendicular). Figure 2 presents isopotential contour plots of induced transmembrane potentials obtained for cathodal point stimulation of this tissue model with (A) equal or (B) nominally unequal anisotropy ratios. In both cases, the extracellular cathodal current

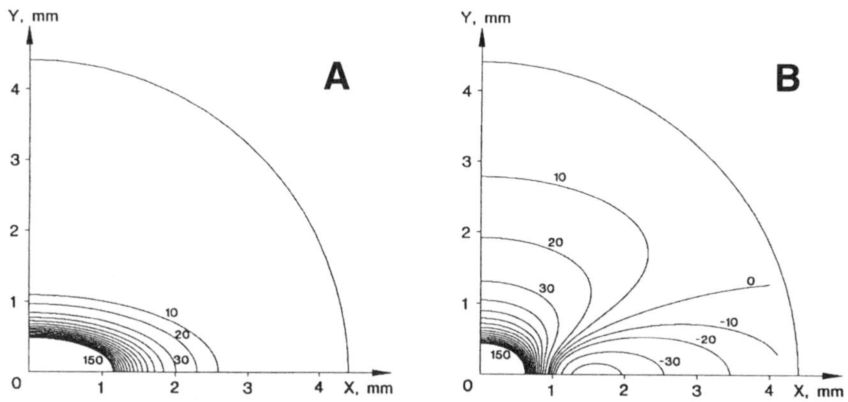

Figure 2 Isopotential contour plots of transmembrane potentials (in millivolts) across one quadrant of a two-dimensional bidomain tissue as induced by a steady-state extracellular cathodal point stimulus (located at the origin). The tissue was modeled with either (A) equal or (B) nominally unequal anisotropy ratios. In (A), potentials decay monotonically in a elliptic pattern with increasing distance from the electrode. In (B), a "dog-bone"-shaped depolarized zone is flanked by two symmetrically located hyperpolarized "virtual anodes." (From Ref. 33.)

source was located at the origin, while the indifferent electrode was at infinity. Figure 2A illustrates that the associated depolarizations induced across the equally anisotropic myocardium create an elliptical pattern of isopotential lines that—as described above—decay quickly and monotonically with increasing distance from the electrode. In contrast, Fig. 2B reveals that the associated depolarizations induced across the unequally anisotropic myocardium create a "dog-bone" pattern of isopotential lines that is symmetrically flanked by two smaller regions of induced hyperpolarizations. Thus, in addition to the expected—but oddly shaped—depolarized response under the cathode, this stimulus also induced two weaker "virtual anodes" symmetrically located on the longitudinal axis of the tissue approximately 1–2 mm on either side of the origin. Reversing the polarity of the stimulus to that of an anode source likewise reverses the polarity of the induced responses, thereby resulting in a strong, centrally located, hyperpolarized dog-bone region flanked symmetrically by two weaker "virtual cathodes."

Direct experimental confirmation of these dog-bone and associated virtual electrode polarization responses was obtained several years after their theoretical existence was originally predicted. Using optical imaging techniques to record from tissues stained with voltage-sensitive dyes, three groups [34–36] successfully mapped the transmembrane potentials induced around an extracellular point electrode. Their results provided irrefutable evidence for the generation of these dog-bone responses and their associated virtual electrodes, and thereby strongly supported the unequally anisotropic bidomain representation of cardiac tissue.

If cardiac tissue is indeed unequally anisotropic, then the generation of virtual electrodes should not be specific to only point-source stimulation, but rather should develop wherever there are nonuniform electric fields and a concomitant divergence of source currents. These conditions also occur during defibrillation, and can be especially pronounced when the shock is applied transvenously. Indeed, Efimov and colleagues [37–39] have recently recorded and analyzed the establishment of large-scale virtual electrodes during the administration of defibrillation-level shocks delivered through such a transvenous electrode system. These results are thus suggestive that the generation of virtual electrodes throughout the entire heart volume—formed specifically because of the unequally anisotropic structure of the myocardium—might represent an important underlying mechanism of defibrillation.

B. Resistive Discontinuities

The concept of resistive discontinuities was one of the first mechanisms proposed to explain how an applied *uniform* electric field could induce polarizations within the bulk myocardium. As originally hypothesized, resistive discontinuities refer specifically to the boundaries between individual cells within cardiac tissue [40–43], although the term has since been generalized to include supercellular discontinuities as well [44–48]. In traditional cable and bidomain theories, tissue discontinuities are purposely removed from the tissue via a volume-averaging homogenization process [28,29]. The resultant continuous syncytium is therefore only appropriate for studying macroscopic electrophysiological behaviors. For processes such as propagation and point stimulation, this macroscopic perspective has been extremely valuable and insightful. Yet, as discussed above, this same traditional approach seems to be insufficient when applied to the investigation of bulk defibrillation, especially in regions where electric fields are nearly uniform. It was thus speculated that perhaps this discrete nature of the tissue was an important structural detail that should not be ignored when modeling the interaction of a strong electric field stimulus with cardiac tissue.

To test this hypothesis, the pioneering studies by Plonsey and Barr [40,41] and Krassowska et al. [42,43] used theoretical [41–43] and computational [40] techniques (described below) to evaluate the steady-state stimulus-induced transmembrane responses along a passive one-dimensional cardiac fiber in which periodic resistive intracellular discontinuities associated with intercellular gap junctions were retained. An example of such an idealized discontinuous fiber is illustrated schematically in Fig. 3A. As with the idealized continuous fiber of traditional cable theory (see Fig. 1A), this cylindrical fiber is centrally located within a restricted interstitial space, with anodal and cathodal extracellular electrodes positioned at its ends. However, the intracellular space (and only the intracellular space) of this fiber is interrupted at periodic intervals by discrete resistive boundaries. These boundaries are a consequence of the implied construction of the fiber from individual cardiac cells which are interconnected end to end via gap junctions (Fig. 3B).

The steady-state transmembrane voltage response (ΔV_m) of such a passive discontinuous fiber to an applied electric field is plotted in Fig. 3D. Similar to the response seen for the continuous fiber (see Fig. 1B), the response of this fiber demonstrates significant membrane hyperpolarization proximal to the anode and significant membrane depolarization proximal to the cathode, both of which decay exponentially with increasing distance from the corresponding electrode. Superimposed on this large-scale cable response, however, is an additional sawtoothlike polarization pattern that extends undiminished along the entire length of the fiber. Figures 3B and 3C illustrate in closer detail the current and voltage responses of four central cells of this fiber. The arrows in Fig. 3B trace the paths of axial and transmembrane currents along and between these cells, and reveal that each

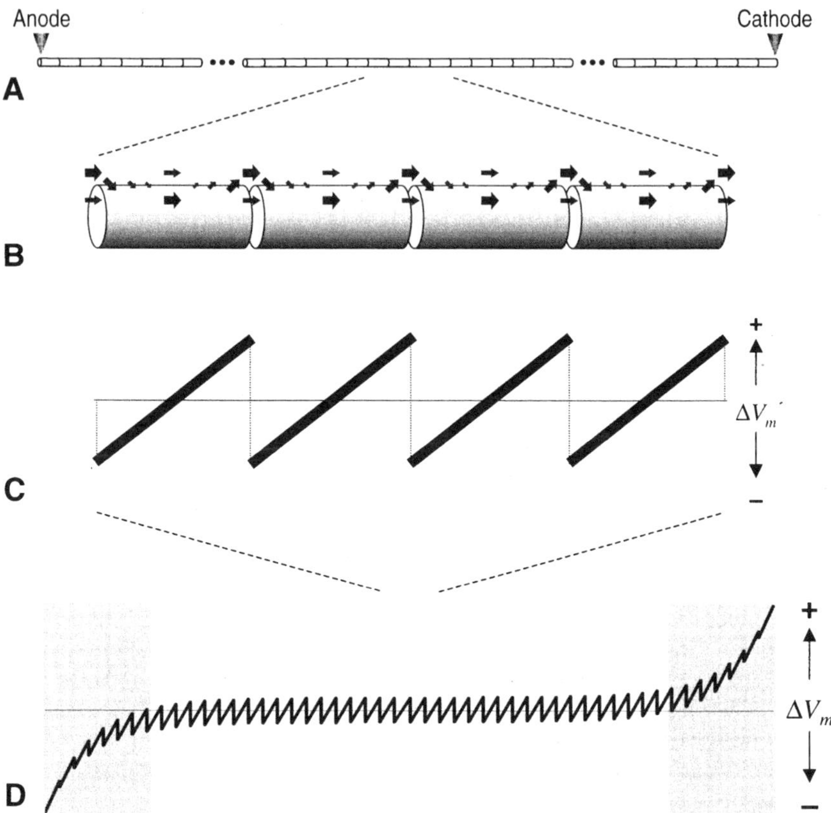

Figure 3 (A) Schematic representation of a one-dimensional discontinuous cardiac fiber undergoing external stimulation. This fiber is similar to the continuous fiber of Fig. 1A, except that the intracellular space of this fiber is interrupted at periodic intervals by discrete resistive boundaries. (B) Magnified view of four central cells and their associated stimulus-induced axial and transmembrane current distributions. Because the intercellular junctions present strong resistive impediments to the otherwise unobstructed flow of intracellular current, some of this current bypasses these junctions by redistributing locally into the interstitial space. (C) Spatial profile of the corresponding transmembrane voltage responses (ΔV_m) of these four cells from (B). Within each cell, ΔV_m varies almost linearly from most hyperpolarized at the cell end nearest the anode to most depolarized at the cell end nearest the cathode. (D) The spatial profile of ΔV_m along this entire passive discontinuous fiber. Note the sawtooth pattern of polarization superimposed on the larger-scale continuous-cable response (compare to Fig. 1B).

intercellular junction presents a strong resistive impediment to the flow of intracellular axial current. To compensate, some of this current bypasses these junctions by redistributing locally into the interstitial space. The resulting transmembrane currents are accompanied by corresponding spatial changes in transmembrane voltages (Fig. 3C) such that, within each cell, ΔV_m varies almost linearly, from most hyperpolarized at the cell end nearest the anode to most depolarized at the cell end nearest the cathode. When viewed over several cells, this spatially repeating development of polarizations forms a sawtoothlike pattern—thus, this hypothesis is also commonly known as the sawtooth mechanism for far-field excitation.

This sawtooth hypothesis generated significant interest because it seemed to offer explanations to many unresolved questions about defibrillation. Most obviously, it provided an elegant theoretical mechanism to explain how the bulk myocardium could be directly polarized by a defibrillation shock. Furthermore, several general predictions based on this hypothesis were at least indirectly consistent with experimental and clinical observations. For example, the predicted direct proportionality between the magnitudes of the induced sawtooth polarizations and the strength of the surrounding electric field (see below) was consistent with the experimental findings that defibrillation success likewise depended directly on the strength of the applied shock [24,49,50]. Additionally, since a reversal of the anode and cathode electrodes simply reverses the direction of the induced sawtooth response (i.e., it reverses the halves of each cell which are depolarized and hyperpolarized) but does not otherwise distort the overall cellular responses, it supported some experimental and clinical evidence that defibrillation thresholds were relatively insensitive to the polarity of the applied shock [51,52].

Nevertheless, enthusiasm for this mechanism (at least with respect to cellular-level discontinuities) has since wavered significantly. Recent modeling efforts which have explored this mechanism in more detail have revealed that, while junctional discontinuities can indeed induce excitation in these idealized models of cardiac tissue, the predicted field strengths necessary to reach these excitation thresholds were significantly larger in magnitude than experimental evidence would suggest [53]. Moreover, direct experimental explorations for this mechanism have been even less supportive. While recent experiments have recorded reciprocal polarization of a single isolated cell within an electric field [54–56], other in-vitro experiments in which the cardiomyocytes were part of a multicellular tissue preparation have thus far failed to detect any evidence of similar cellular-level sawtooth responses [45,57,58]. However, in contrast to the lack of observed responses to such cellular-level discontinuities, larger-scale supercellular discontinuities within cardiac tissue have been shown in vitro to induce measurable transmembrane polarizations responses [45,47,48].

The following subsections describe several theoretical and computational methods that have been developed and employed to specifically investigate the influences of resistive discontinuities during field stimulation.

1. Secondary Sources

Within the intracellular space, each junctional site presents an abrupt resistance to the flow of intercellular axial current. As such, by Ohm's law the potential drop across that junction must be discontinuous with magnitude $V_j = I_j \cdot R_j$, where R_j is the magnitude of one junctional resistance and I_j is the total axial intracellular current through that junction. Plonsey and Barr recognized that this junctional effect was equivalent to the effect expected from a secondary (dipole) current source located at each junction [41]. They utilized this equivalence to develop an exact solution procedure which combined the analytic steady-state solutions of the continuous finite-length passive fiber [e.g., Eq. (7)] with the analytic solutions derived for these secondary-source junctional effects. Advantageously, because dipoles are linear elements, the total influence of all of the dipole sources within the fiber can be determined as a superposition from the separate influences of each individual dipole. However, these junctional responses cannot be considered in complete isolation, since each secondary current source will contribute to the total axial current passing through every other junction within the fiber. Consequently, a system of linear equations—one equation for each junction within the fiber—is required to solve this problem exactly.

Cartee and Plonsey [59] extended this secondary source technique so as to be able to determine the stimulus-induced *transient* evolution of responses anywhere within a passive finite-length discontinuous fiber. Their solution technique was analogous to that utilized for the steady-state case [41], except that all of the time-dependent governing equations were first Laplace-transformed into the frequency domain, thereby reducing the partial differential equations into more tractable ordinary differential equations. However, since the resulting system of linear equations is now frequency-dependent, the system must be solved repeatedly across the complete complex frequency spectrum. Once accomplished, a final solution in the time domain can be obtained by computing the numerical inverse transform of the associated frequency-domain results.

Fishler [60] subsequently derived a significantly simplified analytic closed-form solution for describing the stimulus-induced transient evolution of responses within the subset of cells of the fiber located far from the stimulating electrodes. This simplification was achieved by assuming that the fiber approached infinite length, and thus every cell and every junction in the fiber responded identically. With this assumption, the total axial current passing through each junction could be determined analytically and in closed form, and thus no longer required a system of simultaneous equations to compute.

Interpreting and analyzing intercellular junctions as secondary sources has provided important intuitive insights into the effects of these resistive discontinuities on the stimulus-induced changes in transmembrane potentials and axial currents. However, as a solution method, it is extremely limited, since it is only amenable to analyzing one-dimensional fibers with passive membrane characteristics.

2. Spectral Techniques

Trayanova and Pilkington devised an alternative analytic approach employing spectral techniques that can also be extended to multidimensional preparations [61–63]. Their approach takes specific advantage of the idealized periodicity with which these intercellular junctions repeat within the intracellular domain, enabling them to use a combination of Fourier transforms and Fourier series to convert the coupled differential bidomain equations [Eqs. (9) and (10)] into an alternative system of algebraic equations in terms of the transform of the intracellular and interstitial potentials. Exact analytic equations for the potentials themselves can then be obtained by applying a Fourier series inverse transform to those results.

For example, consider the application of these spectral techniques to determining the steady-state solution of a one-dimensional passive fiber of length L with periodic intercellular junctions [61]. In this case, the governing bidomain equations reduce to

$$\frac{d}{dx}\left[g_i(x)\cdot\frac{d\phi_i}{dx}\right] = +\alpha(\phi_i - \phi_e) \tag{14}$$

$$g_e\frac{d^2\phi_e}{dx^2} = -\alpha(\phi_i - \phi_e) \tag{15}$$

where $\alpha = \beta/R_m$. Since it is assumed that intracellular bidomain conductivity profile, $g_i(x)$, is spatially periodic and also symmetric within a judicially chosen unit cell, $g_i(x)$ can be equivalently represented by an infinite Fourier cosine series:

$$g_i(x) = \sum_{m=0}^{\infty} S_m \cos\left(\frac{2m\pi x}{x_o}\right) \tag{16}$$

where x_o is the unit cell length and the S_m are the associated Fourier series coefficients. With $g_i(x)$ so defined, each of the periodic bidomain equations [Eqs. (14) and (15)] can themselves then be restated in the frequency domain by using finite Fourier cosine transforms [61]:

$$\frac{-k_n}{2} \sum_{m=0}^{\infty} \{S_m[(k_n + 2pk_m) \cdot F_i(n + 2pm) + (k_n - 2pk_m)$$

$$\times F_i(|n - 2pm|)]\} = +\alpha[F_i(n) - F_e(n)] \tag{17}$$

$$-k_n^2 g_e F_e(n) + [(-1)^n - 1] \cdot J_{\text{ext}} = -\alpha[F_i(n) - F_e(n)] \tag{18}$$

where $n = 0, 1, 2, \ldots$, is the Fourier transform index; $k_n = n\pi/L$; $p = L/x_o$ is the number of cells in the fiber; and $F_i(n)$ and $F_e(n)$ are the finite Fourier cosine transforms of the potentials $\phi_i(x)$ and $\phi_e(x)$, respectively. The cosine transform is chosen so that the boundary conditions based on potential gradients [e.g., Eq. (5)] can be incorporated explicitly into the resulting transformed equations. A single equation in terms of F_i only can be readily obtained by substituting the expression for $F_e(n)$ determined from Eq. (18) into the right-hand side of Eq. (17). However, this one equation must then be solved for all $n = 0, 1, 2, \ldots$, in order to fully characterize the spectral details of the intracellular potential. Furthermore, as is apparent in Eq. (17), the solution for $F_i(n)$ also depends on the values of $F_i(|n + 2pm|)$ and $F_i(|n - 2pm|)$. Thus, an infinite system of interdependent linear algebraic equations (one per n) is required to solve for the complete transform of intracellular potential, $F_i(n)$. In matrix form, this linear system can be expressed as $\mathbf{AF}_i = \mathbf{B}$, where $\mathbf{F}_i$ is a vector of the unknown $F_i(n)$s, $\mathbf{B}$ is a vector incorporating factors related to fixed tissue parameters and the applied stimulus, and $\mathbf{A}$ is an operator matrix. For practical purposes, this infinite system needs to be truncated to a finite dimension which can then be solved using standard numerical techniques. Fortuitously, the matrix $\mathbf{A}$ is structurally sparse; therefore, an efficient sparse matrix solver (e.g., an iterative conjugate gradient method) can be employed for this task. Once these $F_i(n)$'s are evaluated, the corresponding $F_e(n)$'s can be computed directly from Eq. (18). Finally, analytic expressions describing the intracellular and interstitial potentials as functions of position, $\phi_i(x)$ and $\phi_e(x)$, can be determined from the inverse Fourier cosine transform of the corresponding set of $F_i(n)$'s and $F_e(n)$'s. An expression for the induced transmembrane potentials then follow naturally as $V_m(x) = \phi_i(x) - \phi_e(x)$.

As with the secondary source approach described above, this spectral technique provides a procedure with which to derive exact analytical solutions for the stimulus-induced responses of discontinuous tissues. However, while this technique has been extended to analyze the responses of multidimensional tissues [62,63], it can only be used under highly idealized conditions where the intracellular axial conductivity profile is spatially periodic in all principal directions. Moreover, this method is thus far not appropriate for analyzing responses of tissues with nonconstant (i.e., nonpassive) membrane resistances, or for determining any responses as functions of time (i.e., non-steady-state).

3. Two-Scale Asymptotic Analysis (Homogenization Method)

Krassowska et al. [42,43] introduced the formal homogenization method to the study of stimulus-induced responses in periodic cardiac tissues. This rigorous mathematical approach takes advantage of the dichotomy between the relative spatial scales over which the aperiodic and periodic components of the resulting stimulus-induced responses develop. The aperiodic component reflects the macroscopic response of the fiber, while the periodic component reflects the microscopic responses within each cell (or equivalent unit cell). A technique known as two-scale asymptotic analysis [64] provides a mechanism by which to decompose the original problem into two simpler problems—one each for the macroscopic and microscopic components— which can be solved separately.

Once again, consider a one-dimensional discontinuous passive fiber with cells of length ε [42]. All quantities associated with this fiber are functions of a spatial variable z. However, to facilitate the separation of macroscopic and microscopic phenomena, z is treated as consisting of two separate variables: a conventional macroscopic spatial variable x, and a supplemental spatial variable $y = x/\varepsilon$ which is introduced to specifically reference the microscopic phenomena. Moreover, since these microscopic phenomena are assumed to be periodic, it is sufficient to consider y over only one such periodic interval $-\frac{1}{2} \le y \le +\frac{1}{2}$ (the unit cell). As such, the periodic cable equations of Eqs. (14) and (15) can be rewritten as

$$\frac{d}{dz}\left[g_i(y) \cdot \frac{d\phi_i(z)}{dz} \right] = +\alpha V_m(z) \tag{19}$$

$$g_e \frac{d^2\phi_e(z)}{dz^2} = -\alpha V_m(z) \tag{20}$$

where $V_m(z) = \phi_i(z) - \phi_e(z)$. Notice that since intracellular conductivity is strictly periodic on the unit cell, g_i depends only on y. The remaining quantities are functions of z; that is, they depend on both x and y. To separate the x and y dependence, potentials are expressed as "two-scale" power series expansions of the parameter ε:

$$\phi_i(z) = \phi_i^0(x) + \varepsilon \cdot \phi_i^1(x, y) + \varepsilon^2 \cdot \phi_i^2(x, y) + \cdots$$

$$\phi_e(z) = \phi_e^0(x) + \varepsilon \cdot \phi_e^1(x, y) + \varepsilon^2 \cdot \phi_e^2(x, y) + \cdots \tag{21}$$

$$V_m(z) = V_m^0(x) + \varepsilon \cdot V_m^1(x, y) + \varepsilon^2 \cdot V_m^2(x, y) + \cdots$$

In this representation, the zero-order terms are the macroscopic potentials which are obtained by solving the equivalent continuous cable model. Since there is no periodicity involved in these macroscopic potentials, these terms appropriately do not depend on the periodic variable y. In contrast, the first- and higher-order terms describe the microscopic potentials that change primarily as functions of position within the unit cell. However, because their magnitudes also depend on the position along the fiber, these higher-order terms are functions of both x and y.

The differential equations that determine the zero- and first-order potentials are derived by introducing the power-series expansions [Eq. (21)] into the fiber equations [Eqs. (19) and (20)]. In this step, the total derivative with respect to z must account for both x and y:

$$\frac{d}{dz} = \frac{\partial}{\partial x} + \frac{1}{\varepsilon} \cdot \frac{\partial}{\partial y} \tag{22}$$

After using Eq. (22) to perform differentiations in Eqs. (19) and (20), terms on the left- and right-hand sides of the resulting equations are grouped based on powers of ε. Since these equations must be satisfied for all possible values of ε, the respective coefficients of the successive powers of ε on each side of the equations must themselves be equal. Specifically, terms multiplied by ε^{-1} determine the differential equations governing microscopic potentials ϕ_i^1 and ϕ_e^1, while terms multiplied by ε^0 determine the differential equations governing macroscopic potentials ϕ_i^0 and ϕ_e^0. These two sets of equations can then be solved independently. The total potentials are then re- constructed from these macroscopic and microscopic components using Eq. (21). The complete derivation, resulting equations, and solutions for this one-dimensional problem can be found in the original work by Krassowska et al. [42].

The analysis reveals that the solution for the aperiodic macroscopic component of the total induced transmembrane potential profile [that is, $V_m^0(x)$] is identical to that for the homogenized equivalent of the periodic fiber, in which the periodically varying intracellular conductivity, $g_i(y)$, is replaced by its harmonic mean value, $\bar{g}_i$, averaged across a unit cell (thereby eliminating its dependence on y). Additionally, the solution for the first- order periodic microscopic component of the total induced transmembrane potential profile [that is, $V_m^1(x, y))$] is given by

$$V_m^1(x, y) = \frac{\partial \phi_i^0}{\partial x} \cdot w(y) \tag{23}$$

where $w(y)$ is a weight function equivalent to the distribution of the transmembrane potentials within the unit cell as induced by a unit gradient of the zero-order (aperiodic) term of intracellular potential. The exact profile of the weight function $w(y)$ is itself determined by the profile of the axial conductivities. For the idealized case where each cell has a uniform cytoplasmic conductivity, g_c, and junctional resistances are lumped at the intercellular boundaries, $w(y) = (\bar{g}_i/g_c - 1) \cdot y$. Furthermore, in the far-field regions of the fiber, $\partial \phi_i^0/\partial x \approx -E_x$, where E_x is the strength of the applied uniform electric field. Thus, with these insights, Eq. (23) effectively embodies to a leading order the periodic sawtooth response expected under these conditions. Advantageously, it only takes two terms of the power-series expansions of Eq. (21) to give an excellent approximation of the total potentials in a discrete fiber; the contributions from higher-order terms are relatively insignificant.

Unlike the various limitations restricting the expanded applications of the techniques of secondary sources and spectral analysis, this homogenization technique has been successfully extended and adapted for use with dynamic nonlinear membrane models [29,65–68] and with multidimensional tissues [29,44,69–70] with embedded fiber curvature [44,66,67,69] and/or nonuniform electric fields [66,67]. However, a potential limitation of this method is that it implicitly presumes strict periodicity of the unit cells—that is, the dimensions and axial conductivity profile of every unit cell must be identical. This restriction can be relaxed somewhat by constructing the unit cell from many cardiac cells, at the cost of increased computational complexity for the microscopic potentials [70].

4. Resistive Networks

The three previous techniques provide mathematically rigorous analytic approaches to determining the transmembrane potential profiles induced by stimuli applied to cardiac tissue with resistive intracellular discontinuities. A less analytical—but generally more flexible—approach entails effectively representing the cardiac tissue as an interconnected network of discrete resistive elements, and then solving this system for the corresponding potentials, currents, etc., at each distinct node of the network. In this way, each node describes the average lumped behavior of a finite subsection of cardiac tissue, while the magnitude of each discrete internodal resistance reflects the lumped equivalent of the cumulative resistance associated with that finite internodal pathway. Refining the discretization of the tissue improves the spatial resolution of the network's approximation to the original system and thereby can result in more accurate results, but will also require proportionately more computational effort to solve.

Figure 4 illustrates schematically a resistive network representation of the one-dimensional discontinuous cardiac fiber of Fig. 3. In this simple example, each cell has been discretized into four subunits. Each subunit is defined by a discrete membrane element (gray square) across which the discrete intracellular and interstitial domain nodes (black dots) communicate. Spatial interactions between adjacent subunits occur through the discrete axial resistors whose magnitudes are determined by the total cumulative resistance between nodes. For example, if the interstitial domain is homogeneous, each of the interstitial axial resistors can be computed (in ohms) as $R_e = \rho_e \cdot l/A_e$, where ρ_e is the effective interstitial resistivity (Ω-cm), l is the distance (cm) between adjacent nodes, and A_e is the cross-sectional area (cm^2) of the interstitial domain. Note that in Fig. 4 each of the intracellular axial resistors associated with each intercellular junction is drawn larger than the other intracellular resistors, since this resistor incorporates resistance due to both the distributed cytoplasmic resistivity as well as the junction itself.

Since the first description of this sawtooth mechanism [40] this method has been used extensively for computing and exploring shock-induced membrane responses in idealized discontinuous tissues. The earliest studies determined just the passive (subthreshold) responses of a one-dimensional fiber [40,59]. Soon thereafter, active membrane kinetics were incorporated

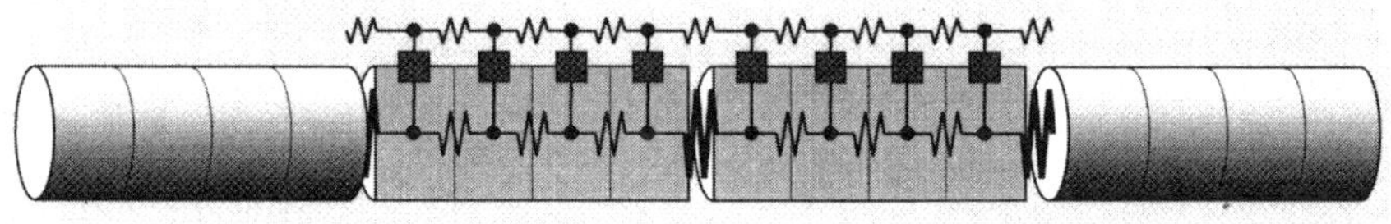

Figure 4 A resistive network representation of a one-dimensional discontinuous cardiac fiber. Four cells from this fiber are visible, with the central two cells "opened" to reveal the associated resistive network. Each cell has been discretized into four subunits. The intracellular and interstitial nodes of each subunit (black dots) interact via discrete membrane elements (gray squares). Spatial interactions between adjacent subunits occur through discrete axial resistors. As a visual reminder that the intercellular junctions represent significant resistive impediments to axial current, those resistors associated with these junctions are drawn correspondingly larger than the other intracellular resistors.

into the models so as to be able to explore the time-dependent dynamics of these responses along the fibers [53,71–73]. More recently, two-dimensional discontinuous tissues have been investigated for analogous shock-induced responses [74].

While the use of resistive networks has been introduced here with reference only to solving problems involving resistive discontinuities, this method has much broader appeal and is in fact the most common general-purpose method for adapting electrophysiological problems [e.g., Eqs. (9) and (10)] into forms appropriate for numerical solutions via computer simulations. Descriptions of all of the numerical methods by which resistive network problems can be solved are well beyond the scope of this chapter. However, see Chap. 3 for some discussions and examples of how resistive networks can be applied to wave propagation problems. Also, see Refs. [75–80].

C. Syncytial Heterogeneities

Recently, Fishler [81,82] proposed that inhomogeneities of the tissue structure need not be discrete (such as gap junctions, detailed above) in order to induce redistributions of axial currents and generate corresponding membrane polarizations. Even spatially continuous variations in the conductivity profile of the tissue could theoretically induce these polarizations. Thus, this structural mechanism might in fact be relevant during defibrillation, since such "syncytial heterogeneities" [81] are inherent within and throughout cardiac tissue. Examples of potential sources of syncytial heterogeneities that are manifested at different spatial scales include cell-to-cell variations in myocyte shape [83]; spatial variations in cellular packing efficiency [84]; spatial variations in the sizes, shapes, and extents of fiber bundles [84,85]; spatial variations in capillary densities [86]; etc. Evidence from one histological study—although not specifically designed to determine syncytial heterogeneity measures—has reported both long-distance, well-correlated trends in syncytial properties as well as short-distance, seemingly uncorrelated fluctuations with standard deviations of approximately 5% [84]. In general, however, histological studies designed specifically to quantify these heterogeneities are lacking; thus, accurate measurements of the spatial scales and magnitudes over which syncytial heterogeneities are manifested are still essentially unknown. Computational investigations have provided important fundamental insights into many qualitative aspects of this predicted mechanism [81,82], but until such quantitative estimates are determined, the actual relevance and impact of syncytial heterogeneities during defibrillation cannot be specifically or accurately assessed.

As described earlier, traditional bidomain theory uses volume averaging specifically to remove tissue heterogeneities. Thus, previous computational studies that investigated shock–tissue interactions had no ability to discern the potential influences of syncytial heterogeneities. Fishler resolved this shortcoming by reintroducing syncytial heterogeneities back into a modified two-dimensional bidomain model of cardiac tissue. These heterogeneities were incorporated over multiple length scales simultaneously through spatial variations in local intracellular volume fractions, f_i. As suggested by its name, this variable represents the fraction of a unit volume of bidomain tissue that is composed of intracellular (versus interstitial) space [28]. In turn, variations in f_i cause corresponding spatial variations in the magnitudes of local effective axial conductivities (components of $\tilde{g}_i$ and $\tilde{g}_e$) and membrane surface-to-volume ratios (β) [28,87,88]. Because it was unknown a priori which spatial scale(s) of heterogeneity might prove most sensitive to an applied electric field, compound heterogeneities in f_i with spatial correlations persisting over several different length scales simultaneously were incorporated into the model. These compound multiscale heterogeneities were created from the superposition of multiple independent distributions in f_i, each manifesting spatial heterogeneities from just a single underlying correlation scale. To construct each of these simple distributions, random values chosen from a normal distribution were first assigned to all points of the tissue that were orthogonally separated by the desired correlation length (from 40 to 2560 μm). Next, bilinear interpolation was performed between these positions so as to assign correspondingly correlated values to the intervening regions. After superposition of several of these independent simple distributions with correlation lengths spanning a desired range of values, the resulting compound distribution was translated and scaled as necessary to produce a final heterogeneous profile with a specified mean and standard deviation. An example of one such compound distribution of f_i—with overall mean of 0.80 and standard deviation of 0.05—is given in the two-dimensional intensity profile of Fig. 5A. Evidence of the multiple simultaneous correlation scales can be identified across this spatial profile— for example, seemingly uncorrelated pixilated "static" is superimposed onto more coherent variations that one might envision would be representative of some larger underlying structural influences.

Using these two-dimensional syncytial heterogeneities in f_i, Fishler solved the bidomain equations [Eqs. (9) and (10)] to compute the corresponding passive steady-state [81] and active time-dependent [82] shock-induced transmembrane responses. Since only the far-field responses to these applied uniform electric fields were of interest in these investigations, the boundary conditions associated with the applied shocks were modified

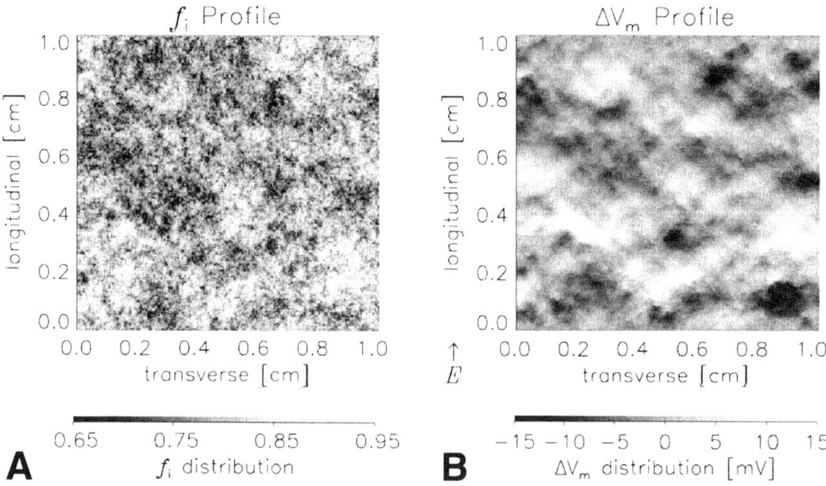

Figure 5 (A)   Intensity profile of the compound distribution of syncytial heterogeneities manifested across a representative two-dimensional region of cardiac tissue. In this example, syncytial heterogeneities were incorporated over multiple length scales simultaneously through superimposed spatial variations in local intracellular volume fractions (f_i). The overall mean and standard deviation of this distribution is 0.80 ± 0.05. (B) Corresponding intensity profile of steady-state transmembrane potentials, V_m, as induced by the far-field stimulation of the heterogeneous syncytium in (A). This macroscopically uniform 1-V/cm field stimulus was directed along the longitudinal axis of the tissue (indicated by "E" with arrow), and resulted in many interspersed and variously shaped and sized regions of relative depolarization (gray→white) and hyperpolarization (gray→black).

so as to "short-circuit" the strong near-field polarization effects typically generated within tissue regions proximal to the stimulating electrodes (e.g., Fig. 1B). Ordinarily, boundary conditions for an externally applied shock would be similar to those described in Eq. (5): all stimulus current enters and exits via the interstitial domain, and the intracellular axial current at the boundary is zero. The near-field zone then defines the region proximal to each electrode (or tissue boundary) over which this applied stimulus current redistributes between the interstitial and intracellular domains. By precalculating estimates of local redistribution ratios, the boundary conditions can be modified accordingly so that the applied stimulating currents are already distributed appropriately between the interstitial and intracellular domains. In this way, the near-field effects are effectively eliminated, thereby enabling a relatively uncontaminated study of the far-field responses induced across the entire syncytium.

Figure 5B presents an intensity map of the steady-state transmembrane potentials, V_m, across the passive syncytium of Fig. 5A as induced by a longitudinal far-field 1-V/cm electric field stimulus. This uniform shock engages the heterogeneous tissue by producing many interspersed and variously shaped and sized regions of relative depolarization (gray→white) and hyperpolarization (gray→black). The locations, extents, and magnitudes of these polarizations are correlated primarily to the spatial gradients of the underlying heterogeneities, not the heterogeneities themselves—a result that is consistent with the expectations elucidated via the generalized activating function [see Eq. (13)] [32]. When the dynamic responses were evaluated using an active membrane model, the computed diastolic thresholds were sufficiently low so as to suggest that even modest syncytial heterogeneities could contribute at least a substantial fraction of the membrane polarization responsible for suprathreshold excitation [82]. If these predictions are confirmed, syncytial heterogeneities could in fact be a significant structural mechanism contributing to the far-field excitation process.

D. Fiber Curvature

Cardiac tissue has an underlying fiber direction that defines the principal axis of both contraction and conduction. For simplicity, many models assume that this fiber direction is uniformly straight throughout the region being studied. In reality, however, the heart's fiber geometry is quite complex and nonuniform, with fibers curving in conformance with the shape of the ventricular walls as well as rotating with depth through the walls [84,89]. Consequently, a defibrillation shock actually interacts with myocardial tissue such that the relative angles between the local electric field vectors and local fiber directions vary significantly throughout the heart volume. Several modeling and experimental studies have recently demonstrated that this changing relative fiber angle represents another important structurally-based mechanism for the induction of far-field polarizations within the myocardial bulk.

Fiber curvature is incorporated into these computational models via manipulations of the conductivity tensors, $\tilde{g}_i$ and $\tilde{g}_e$, to adjust for the local fiber direction relative to a fixed global coordinate system (x, y, z). At every point within the tissue, a variable local coordinate system (l, t, u) can be established and aligned with the underlying local fiber direction. Because the axes of this local coordinate system are now aligned with the axes of local anisotropy, the corresponding conductivity tensors are diagonal, and have the general form

$$\tilde{g}_{(f)} = \begin{bmatrix} g_l & 0 & 0 \\ 0 & g_t & 0 \\ 0 & 0 & g_u \end{bmatrix} \tag{24}$$

where the subscript (f) indicates that this conductivity tensor is defined with respect to the local fiber coordinate system, g_l is the local conductivity magnitude in the longitudinal direction along the fiber axis, and g_t and g_u are the local conductivity magnitudes in the two principal (transverse) directions orthogonal to the fiber axis. This fiber conductivity tensor, $\tilde{g}_{(f)}$, is then transformed into its equivalent form for the global coordinate system, $\tilde{g}_{(g)}$, by applying a rotation tensor R and its transpose R^T as follows:

$$\tilde{g}_{(g)} = \begin{bmatrix} g_x & g_{xy} & g_{xz} \\ g_{xy} & g_y & g_{yz} \\ g_{xz} & g_{yz} & g_z \end{bmatrix} = R \cdot \tilde{g}_{(f)} \cdot R^T \tag{25}$$

The elements of R are determined by the angles through which the axes need to be rotated to align with the global coordinate system. These transformed conductivity tensors can then be used as normal in the governing bidomain equations defined and implemented within the global coordinate system [e.g., Eqs. (9) and (10)].

In the first such study to investigate the potential impact of fiber curvature, Trayanova et al. used an idealized spherical-shell model of the heart in the presence of a uniform electric field [90]. With this passive steady-state model, they demonstrated that fiber curvature could indeed induce polarizations throughout the midmyocardial (far-field) mass of the heart, but only when the myocardium was assumed to manifest unequally anisotropic conductivity ratios within the interstitial and intracellular domains. Under conditions of equal anisotropy, the bulk myocardium remained unperturbed by the shock, and only boundary (surface) polarizations were observed. These fundamental insights were recently extended to include and explore the relative additional influences of transmural fiber rotation through and elliptical eccentricity of this idealized heart [39]. These results reaffirmed the predicted ability of fiber curvatures to induce bulk polarizations—with smaller radii of curvature resulting in stronger responses—and also revealed that transmural fiber rotations noticeably modulate this response profile along the radial direction.

The field-induced responses from more realistic cardiac anatomies and their associated fiber geometries have also been computed. Trayanova et al. [91] determined the passive steady-state transmembrane responses of a two-dimensional slice of myocardium to various applied shocks. All of the structural information for this two-dimensional slice was extracted from a mathematical model of the three-dimensional geometry and fibrous architecture of canine ventricles, developed at Auckland University from detailed measurements taken from actual preserved canine hearts [92]. Figure 6A illustrates the vector field of fiber directions and their relative magnitudes across the transverse slice of ventricular myocardium used by Trayanova for their simulations [91], and Fig. 6B presents the associated distribution of transmembrane polarizations as induced within this slice by two epicardial patch electrodes (locations of which are indicated in the inset). In addition to the expected surface polarizations, large interspersed regions of depolarization and hyperpolarization attributable to the underlying fiber curvature were observed throughout the bulk of the myocardium. Analogous simulations recently performed using the entire three-dimensional Auckland heart and realistic defibrillation electrode locations likewise reported similar distributions of fiber-induced bulk polarizations throughout the entire myocardial volume [93].

Finally, some recent studies have performed parallel computational and experimental investigations to assess the predicted role of fiber curvature and/or transmural fiber rotation during cardiac stimulation [38,94]. The qualitative correlations observed between their model predictions and

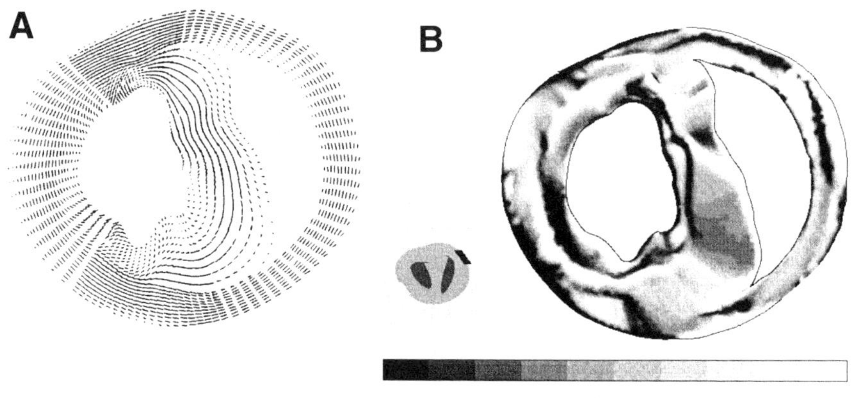

Figure 6 (A) Vector field of fiber directions and their relative magnitudes across a transverse slice of ventricular myocardium. (B) Corresponding intensity map of the distribution of transmembrane polarizations as induced within this slice by two epicardial patch electrodes (locations of which are indicated in the inset). In addition to the expected strong surface polarizations, large interspersed regions of depolarization and hyperpolarization attributable to the underlying fiber curvature were observed throughout the bulk of the myocardium. (From Ref. 91.)

experimental results provide valuable confirming evidence that fiber curvatures might indeed be an important structural mechanism underlying the induction of polarizations within and throughout the bulk myocardium.

IV. CONCLUSIONS

As described above, mathematical and computer models have been invaluable vehicles for investigating in detail the underlying mechanisms of the shock—tissue interactions that are the basis of defibrillation. Indeed, it is a testament to the power and insight of these models that many of their predictions still cannot be confirmed or even explored because of the inherent limitations of the experimental techniques available today. Thus, these models have filled—and continue to fill—an important role in helping to understand and unravel the complexities of the entire defibrillation process.

With respect to this process, the perspective and insights provided in this chapter describe only the current hypotheses of how the shock and tissue interact to induce transmembrane polarization responses directly. However, equally as important to this defibrillation process are the spatiotemporal dynamics that transpire throughout the myocardium *after* the shock has been delivered. This aspect of defibrillation—and how computer models provide valuable insights therein as well—is the focus of the next chapter.

REFERENCES

1. JL Prevost, F Batelli. Some effects of electric discharge on the hearts of mammals. CR Acad Sci 129:1267–1268, 1899.
2. RA DeSilva, TB Graboys, PJ Podrid, B Lown. Cardioversion and defibrillation. Am Heart J 100:881–895, 1980.
3. G Gregoratos, MD Cheitlin, A Conill, AE Epstein, C Fellows, TB Ferguson, Jr., RA Freedman, MA Hlatky, GV Naccarelli, S Saksena, RC Schlant, MJ Silka. ACC/AHA guidelines for implantation of cardiac pacemakers and antiarrhythmia devices: a report of the American College of Cardiology/American Heart Association Task Force on Practice Guidelines (Committee on Pacemaker Implantation). J Am Coll Cardiol 31:1175–1209, 1998.
4. G Nichol, AP Halistrom, R Kerber, AJ Moss, JP Ornato, D Palmer, B Riegel, S Smith Jr, ML Weisfeldt. American Heart Association report on the second public access defibrillation conference, April 17–19, 1997. Circulation 92:1309–1314, 1998.
5. SC Smith, Jr., RS Hamburg. Automated external defibrillators: time for federal and state advocacy and broader utilization. Circulation 97:1321–1324, 1998.

6. JW Gundry, KA Comess, FA DeRook, D Jorgenson, GH Bardy. Comparison of naive sixth-grade children with trained professionals in the use of an automated external defibrillator. Circulation 100:1703–1707, 1999.

7. NG Sepulveda, JP Wikswo, Jr., DS Echt. Finite element analysis of cardiac defibrillation current distributions. IEEE Trans Biomed Eng 37:354–365, 1990.

8. WJ Karlon, SR Eisenberg, JL Lehr. Effects of paddle placement and size on defibrillation current distribution: a three-dimensional finite element model. IEEE Trans Biomed Eng 40:246–255, 1993.

9. DB Jorgenson, DR Haynor, GH Bardy, Y Kim. Computational studies of transthoracic and transvenous defibrillation in a detailed 3-D human thorax model. IEEE Trans Biomed Eng 42:172–184, 1995.

10. TF Kinst, MO Sweeney, JL Lehr, SR Eisenberg. Simulated internal defibrillation in humans using an anatomically realistic three-dimensional finite element model of the thorax. J Cardiovasc Electrophysiol 8:537–547, 1997.

11. X Min, R Mehra. Finite element analysis of defibrillation fields in a human torso model for ventricular defibrillation. Prog Biophys Mol Biol 69:353–386, 1998.

12. AL de Jongh, EG Entcheva, JA Replogle, RS Booker, III, BH KenKnight, FJ Claydon. Defibrillation efficacy of different electrode placements in a human thorax model. PACE 22(pt II):152–157, 1999.

13. AL Hodgkin, WAH Rushton. The electrical constants of a crustacean nerve fibre. Proc R Soc B133:444–479, 1946.

14. JJB Jack, D Noble, RW Tsien. Electric Current Flow in Excitable Cells. Oxford: Clarendon Press, 1975.

15. R Plonsey, RC Barr. Bioelectricity: A Quantitative Approach. New York: Plenum, 1988.

16. W Krassowska, JC Neu. Effective boundary conditions for syncytial tissues. IEEE Trans Biomed Eng 41:143–150, 1994.

17. S Weidmann. Electrical constants of trabecular muscle from mammalian heart. J Physiol (Lond) 210:1041–1054, 1970.

18. RA Chapman, CH Fry. An analysis of the cable properties of frog ventricular myocardium. J Physiol (Lond) 283:263–282, 1978.

19. AG Kléber, CB Riegger. Electrical constants of arterially perfused rabbit papillary muscle. J Physiol 385:307–324, 1987.

20. S Weidmann. The electrical constants of Purkinje fibers. J Physiol (Lond) 118:348, 1952.

21. RW Joyner, J Picone, R Veenstra, D Rawling. Propagation through electrically coupled cells. Effects of regional changes in membrane properties. Circ Res 53:526–534, 1983.

22. Y Rudy. Reentry: insights from theoretical simulations in a fixed pathway. J Cardiovasc Electrophysiol 6:294–312, 1995.

23. PG Colavita, P Wolf, WM Smith, FR Bartram, M Hardage, RE Ideker. Determination of effects of internal countershock by direct cardiac recordings during normal rhythm. Am J Physiol 250:H736–H740, 1986.

24. DW Frazier, W Krassowska, P-S Chen, PD Wolf, EG Dixon, WM Smith, RE Ideker. Extracellular field required for excitation in three-dimensional anisotropic canine myocardium. Circ Res 63:147–164, 1988.

25. RJ Sweeney, RG Gill, MI Steinberg, PR Reid. Ventricular refractory period extension caused by defibrillation shocks. Circulation 82:965–972, 1990.

26. L Tung. A bi-domain model for describing ischemic myocardial D-C potentials. PhD thesis, Massachusetts Institute of Technology, Cambridge, MA, 1978.

27. WT Miller, III, DB Geselowitz. Simulation studies of the electrocardiogram. I. The normal heart. Circ Res 43:301–315, 1978.

28. CS Henriquez. Simulating the electrical behavior of cardiac tissue using the bidomain model. Crit Rev Biomed Eng 21:1–77, 1993.

29. JC Neu, W Krassowska. Homogenization of syncytial tissues. Crit Rev Biomed Eng 21:137–199, 1993.

30. N Hooke, CS Henriquez, P Lanzkron, D Rose. Linear algebraic transformations of the bidomain equations: implications for numerical methods. Math Biosci 120:127–145, 1994.

31. BJ Roth. Electrical conductivity values used with the bidomain model of cardiac tissue. IEEE Trans Biomed Eng 44:326–328, 1997.

32. EA Sobie, RC Susil, L Tung. A generalized activating function for predicting virtual electrodes in cardiac tissue. Biophys J 73:1410–1423, 1997.

33. NG Sepulveda, BJ Roth, JP Wikswo Jr. Current injection into a two-dimensional anisotropic bidomain. Biophys J 55:987–999, 1989.

34. M Neunlist, L Tung. Spatial distribution of cardiac transmembrane potentials around an extracellular electrode: dependence on fiber orientation. Biophys J 68:2310–2322, 1995.

35. SB Knisley. Transmembrane voltage changes during unipolar stimulation of rabbit ventricle. Circ Res 77:1229–1239, 1995.

36. JP Wikswo, Jr., S-F Lin, RA Abbas. Virtual electrodes in cardiac tissue: a common mechanism for anodal and cathodal stimulation. Biophys J 69:2195–2210, 1995.

37. IR Efimov, YN Cheng, M Biermann, DR Van Wagoner, TN Mazgalev, PJ Tchou. Transmembrane voltage changes produced by real and virtual electrodes during monophasic defibrillation shock delivered by an implantable electrode. J Cardiovasc Electrophysiol 8:1031–1045, 1997.

38. E Entcheva, J Eason, IR Efimov, Y Cheng, R Malkin, F Claydon. Virtual electrode effects in transvenous defibrillation—modulation by structure and interface: evidence from bidomain simulations and optical mapping. J Cardiovasc Electrophysiol 9:949–961, 1998.

39. E Entcheva, NA Trayanova, FJ Claydon. Patterns of and mechanisms for shock-induced polarization in the heart: a bidomain analysis. IEEE Eng Med Biol Mag 46:260–270, 1999.

40. R Plonsey, RC Barr. Effect of microscopic and macroscopic discontinuities on the response of cardiac tissue to defibrillating (stimulating) currents. Med Biol Eng Comput 24:130–136, 1986.

41. R Plonsey, RC Barr. Inclusion of junction elements in a linear cardiac model through secondary sources: application to defibrillation. Med Biol Eng Comput 24:137–144, 1986.
42. W Krassowska, TC Pilkington, RE Ideker. The closed form solution to the periodic core-conductor model using asymptotic analysis. IEEE Trans Biomed Eng BME-34:5l9–53l, 1987.
43. W Krassowska, TC Pilkington, RE Ideker. Periodic conductivity as a mechanism for cardiac stimulation and defibrillation. IEEE Trans Biomed Eng BME-34:555–560, 1987.
44. W Krassowska, DW Frazier, TC Pilkington, RE Ideker. Potential distribution in three-dimensional periodic myocardium—part II: application to extracellular stimulation. IEEE Trans Biomed Eng 37:267–284, 1990.
45. AM Gillis, VG Fast, S Rohr, AG Kléber. Spatial changes in transmembrane potential during extracellular electrical shocks in cultured monolayers of neonatal rat ventricular myocytes. Circ Res 79:676–690, 1996.
46. W Krassowska, MS Kumar. The role of spatial interactions in creating the dispersion of transmembrane potential by premature electric shocks. Ann Biomed Eng 25:949–963, 1997.
47. VG Fast, S Rohr, AM Gillis, AG Kléber. Activation of cardiac tissue by extracellular electrical shocks: formation of "secondary sources" at intercellular clefts in monolayers of cultured myocytes. Circ Res 82:375–385, 1998.
48. JB White, GP Walcott, AE Pollard, RE Ideker. Myocardial discontinuities: a substrate for producing virtual electrodes that directly excite the myocardium by shocks. Circulation 97:1738–1745, 1998.
49. JM Wharton, PD Wolf, WM Smith, P-S Chen, DW Frazier, S Yabe, N Danieley, RE Ideker. Cardiac potential and potential gradient fields generated by single, combined, and sequential shocks during ventricular defibrillation. Circulation 85:1510–1523, 1992.
50. X Zhou, JP Daubert, PD Wolf, WM Smith, RE Ideker. Epicardial mapping of ventricular defibrillation with monophasic and biphasic shocks in dogs. Circ Res 72:145–160, 1993.
51. JC Schuder, H Stoeckle, WC McDaniel, M Dbeis. Is the effectiveness of cardiac ventricular defibrillation dependent upon polarity? Med Instrum 21:262–265, 1987.
52. WD Weaver, JS Martin, MJ Wirkus, S Morud, S Vincent, PE Litwin, C Morgan. Influence of external defibrillator electrode polarity on cardiac resuscitation. PACE 16:285–290, 1993.
53. MG Fishler, BA Sobie, L Tung, NV Thakor. Modeling the interaction between propagating cardiac waves and monophasic and biphasic field stimuli: the importance of the induced spatial excitatory response. J Cardiovasc Electrophysiol 7:1183–1196, 1996.
54. SB Knisley, TF Blitchington, BC Hill, AO Grant, WM Smith, TC Pilkington, RE Ideker. Optical measurements of transmembrane potential changes during electric field stimulation of ventricular cells. Circ Res 72:255–270, 1993.

55. H Windisch, H Ahammer, P Schaffer, W Müller, D Platzer. Optical multisite monitoring of cell excitation phenomena in isolated cardiomyocytes. J Physiol (Lond) 430:508–518, 1995.

56. DK-L Cheng, L Tung, EA Sobie. Nonuniform responses of transmembrane potential during electric field stimulation of single cardiac cells. Am J Physiol 277:H351–H362, 1999.

57. X Zhou, WM Smith, DL Rollins, RE Ideker. Spatial changes in transmembrane potential during a shock. PACE 18 (pt. II):935, 1995.

58. X Zhou, SB Knisley, WM Smith, D Rollins, AE Pollard, RE Ideker. Spatial changes in the transmembrane potential during extracellular electric stimulation. Circ Res 83:l003–1014, 1998.

59. LA Cartee, R Plonsey. The effect of cellular discontinuities on the transient subthreshold response of a one-dimensional cardiac model. IEEE Trans Biomed Eng 39:260–270, 1992.

60. MG Fishler. The transient far-field response of a discontinuous one-dimensional cardiac fiber to subthreshold stimuli. IEEE Trans Biomed Eng 44:10–18, 1997.

61. N Trayanova, TC Pilkington. A bidomain model with periodic intracellular junctions: a one-dimensional analysis. IEEE Trans Biomed Eng 40:424–433, 1993.

62. N Trayanova. Discrete versus syncytial tissue behavior in a model of cardiac stimulation—I: mathematical formulation. IEEE Trans Biomed Eng 43:1129–1141, 1996.

63. N Trayanova. Discrete versus syncytial tissue behavior in a model of cardiac stimulation—II: results of simulation. IEEE Trans Biomed Eng 43:1141–1151, 1996.

64. CM Bender, SA Orszag. Advanced Mathematical Methods for Scientists and Engineers. New York: McGraw-Hill, 1978.

65. W Krassowska, TC Pilkington, RE Ideker. Two-scale asymptotic analysis for modeling activation of periodic cardiac strand. Math Comput Model 16:121–130, 1992.

66. JP Keener, AV Panfilov. A biophysical model for defibrillation of cardiac tissue. Biophys J 71:1335–1345, 1996.

67. JP Keener. The effect of gap junctional distribution on defibrillation. Chaos 8:175–187, 1998.

68. JP Keener, TJ Lewis. The biphasic mystery: why a biphasic shock is more effective than a monophasic shock for defibrillation. J Theor Biol 200:1–17, 1999.

69. W Krassowska, TC Pilkington, RE Ideker. Potential distribution in three-dimensional periodic myocardium—part I: solution with two-scale asymptotic analysis. IEEE Trans Biomed Eng 37:252–266, 1990.

70. SP Juhlin, JB Pormann. Dimensional comparison of the sawtooth pattern in transmembrane potential. Comput Cardiol 21:413–416, 1994.

71. LA Cartee, R Plonsey. Active response of a one-dimensional cardiac model with gap junctions to extracellular stimulation. Med Biol Eng Comput 30:389–398, 1992.

72. LJ Leon, FA Roberge. A model study of extracellular stimulation of cardiac cells. IEEE Trans Biomed Eng 40:1307–1319, 1993.
73. MG Fishler, EA Sobie, L Tung, NV Thakor. Cardiac responses to premature monophasic and biphasic field stimuli: results from cell and tissue modeling studies. J Electrocardiol 28:174–179, 1996.
74. MG Fishler. Mechanisms of cardiac cell excitation and wave front termination via field stimulation. PhD thesis, The Johns Hopkins University School of Medicine, Baltimore, MD, 1995.
75. RW Joyner, M Westerfield, JW Moore, N Stockbridge. A numerical method to model excitable cells. Biophys J 22:155–170, 1978.
76. L Lapidus, GF Pinder. Numerical Solution of Partial Differential Equations in Science and Engineering. New York: Wiley, 1982.
77. DM Harrild, CS Henriquez. A finite volume model of cardiac propagation. Ann Biomed Eng 25:315–334, 1997.
78. W Quan, SJ Evans, HM Hastings. Efficient integration of a realistic two-dimensional cardiac tissue model by domain decomposition. IEEE Trans Biomed Eng 45:372–385, 1998.
79. JP Keener, K Bogar. A numerical method for the solution of the bidomain equations in cardiac tissue. Chaos 8:234–241, 1998.
80. Z Qu, A Garfinkel. An advanced algorithm for solving partial differential equation in cardiac conduction. IEEE Trans Biomed Eng 46:1166–1168, 1999.
81. MG Fishler. Syncytial heterogeneity as a mechanism underlying cardiac far-field stimulation during defibrillation-level shocks. J Cardiovasc Electrophysiol 9:384–394, 1998.
82. MG Fishler, K Vepa. Spatiotemporal effects of syncytial heterogeneities on cardiac far-field excitations during monophasic and biphasic shocks. J Cardiovasc Electrophysiol 9:1310–1324, 1998.
83. AL Sorenson, D Tepper, EH Sonnenblick, TF Robinson, JM Capasso. Size and shape of enzymatically isolated ventricular myocytes from rats and cardiomyopathic hamsters. Cardiovasc Res 19:793–799, 1985.
84. IJ LeGrice, BH Smaill, LZ Chai, SG Edgar, JB Gavin, PJ Hunter. Laminar structure of the heart: ventricular myocyte arrangement and connective tissue architecture in the dog. Am J Physiol 269:H571–H582, 1995.
85. JR Sommer, PC Dolber. Cardiac muscle: ultrastructure of its cells and bundles. In: A Paes de Carvalho, BF Hoffman, M Lieberman, eds. Normal and Abnormal Conduction of the Heart Beat. Mount Kisco, NY: Futura, 1982, pp. 1–28.
86. AM Gerdes, FH Kasten. Morphometric study of endomyocardium and epimyocardium of the left ventricle in adult dogs. Am J Anat 159:389–394, 1980.
87. KS Cole, HJ Curtis. Bioelectricity: electric physiology. In: O Glasser, ed. Medical Physics. Chicago: Year Book, 1950, pp. 82–90.
88. BJ Roth. The electrical properties of tissue. In: JD Bronzino, ed. The Biomedical Engineering Handbook. Boca Raton, FL: CRC Press, 1995, pp. 126–138.

89. DD Streeter, Jr. Gross morphology and fiber geometry of the heart. In: RM Berne, N Sperelakis, SR Geiger, eds. Handbook of Physiology. Section 2: The Cardiovascular System. Baltimore, MD: American Physiological Society, 1979, pp. 61–112.

90. NA Trayanova, BJ Roth, LJ Malden. The response of a spherical heart to a uniform electric field: a bidomain analysis of cardiac stimulation. IEEE Trans Biomed Eng 40:899–908, 1993.

91. N Trayanova, J Eason, CS Henriquez. Electrode polarity effects on the shock-induced transmembrane potential distribution in the canine heart. Proc IEEE Eng Med Biol Soc 17:317–318, 1995.

92. PM Nielsen, IJ Le Grice, BH Smaill, PJ Hunter. Mathematical model of the geometry and fibrous structure of the heart. Am J Physiol 260:H1365–H1378, 1991.

93. F Aguel, NA Trayanova, G Siekas, JC Eason, MG Fishler, AM Street. Virtual electrodes induced throughout bulk myocardium by ICD defibrillation. Proc BMES/EMBS 1:289, 1999.

94. SB Knisley, N Trayanova, F Aguel. Roles of electric field and fiber structure in cardiac electric stimulation. Biophys J 77:1404–1417, 1999.

6
Computer Modeling of Defibrillation II: Why Does the Shock Fail?

Natalia Trayanova, James Eason, Cory Anderson, and Felipe Aguel
Tulane University, New Orleans, Louisiana, U.S.A.

I. INTRODUCTION

An electric shock delivered to the heart attempts to terminate fibrillation by changing the transmembrane potential of cardiac cells. The success or failure of the shock is a function of both the level of shock-induced change in transmembrane potential as well as the preshock electrical activity in the myocardium. The series of two chapters on defibrillation included in this book dissect these underlying processes: The preceding chapter [1] examined the interaction between the applied electric field and the tissue structure in establishing a shock-induced change in transmembrane potential. The present chapter explores how this induced transmembrane potential alters the preshock electrical state of the tissue to result in a successful or failed defibrillation attempt. Both chapters describe insight achieved through computer simulations of the defibrillation process, underscoring the role of modeling and simulation as an indispensable research tool.

This chapter presents and summarizes recent research from our group demonstrating the involvement of the shock-induced virtual electrodes (VEs) in defibrillation and arrhythmogenesis. VEs refer to the change in transmembrane potential that extends beyond the immediate vicinity of the defibrillation electrodes. Typically, these are large-scale (i.e., of the order of a space constant and above) regions of shock-induced membrane depolarization (the transmembrane potential is increased) and hyperpolarization (the transmembrane potential is lowered). We believe it is the VEs that determine to a large degree the outcome of defibrillation therapy. The VE concepts is a relatively new hypothesis that has taken a prominent place among the existing theories of defibrillation [2] and is steadily becoming a commonplace in cardiac electrophysiology with "major implications for clinical electrophysiologic work and for defibrillator design" [3].

II. SHOCK-INDUCED VIRTUAL ELECTRODES

To better understand the role of the VEs in the defibrillation process, research first focused on uncovering the mechanisms by which the VEs are formed in *passive* cardiac tissue following field stimulation. Although this is one of the main topics of the preceding chapter, here we briefly review these concepts since they have major consequences for understanding why the shock fails or succeeds. Several studies by our group [4–6] as well as other investigators [7–9] (see also the preceding chapter in this book) have made contributions to this inquiry. Passive tissue is a "computational trick": it is a model representation of cardiac tissue whose (nonlinear) excitability properties have been eliminated. Thus, the response of the tissue to any external stimulus is governed by the simple laws of current passing through linear networks of resistors (the intra- and extracellular spaces) and RC circutis (the cell membrane). The benefit of using such a representation of the myocardium is in that it allows the researchers to "divide and conquer" the processes underlying the defibrillation process: the VE distribution during the shock is determined without "contamination" from preshock or postshock electrical activity. Being well acquainted with the VE distribution, one can easily then distinguish electrical activations arising from the shock itself from activations that are continuation of preshock activity. Experimentally, the passive distribution of VEs can be observed by delivering the defibrillation shocks during the plateau of a paced activation, which ensures that active membrane responses are (partially) suppressed and passive tissue behavior is revealed [10]. Here we include a short summary of the basic mechanisms responsible for the formation of VEs in passive tissue and provide an example demonstrating a computer simulation of these in slice through the dog heart.

The bidomain model [11] has emerged as the most appropriate modeling tool in examining the electrical behavior of cardiac tissue, particularly its response to external electrical stimuli. The bidomain model is based on known electrophysiological evidence of the syncytial nature of cardiac tissue and the recognition of tissue structure as consisting of two distinct domains, intra- and extracellular, separated by the cell membrane. Employing the passive bidomain representation of the myocardium, recent studies [5–7] have found that the shock-induced VE pattern is determined by the interaction of the applied electric field with tissue geometry and fibrous structure of the myocardium. Two types of VEs are now recognized:

> *Surface VEs*, which penetrate the ventricular wall over several cell layers [5,7], due to secondary sources arising at the boundaries separating myocardium from blood cavity or surrounding bath (more specifically, over a distance of about three length constants into the tissue with the length constant at each point on the myocardial surface being dependent on the fiber orientation around that point [6])
>
> *Bulk VEs* throughout the ventricular wall, due to either spatial nonuniformity in applied electric field, or nonuniformity in tissue architecture, such as fiber curvature [4,5,7], fiber rotation [6], fiber branching and anastomosis [4], and local changes in tissue conductivity [9].

Examples of VEs following field stimulation in passive cardiac tissue are shown in Figs. 1b–1e (Fig. 1b–e; see color plate). Figure 1a shows a slice through the ventricles extracted from the Auckland canine heart model [12,13]. The slice is characterized by a set of fiber directions shown as dots and dashes. Long dashes indicate fibers in the plane of the slice, while dots represent fiber directions perpendicular to the slice. The rotation of the fibers through the ventricular wall is evident from the change in dash length between endocardium and epicardium. The region within the ventricles represents the blood. The slice is immersed in a square conductive bath (see insets in Figs. 1b–1e).

The VE distributions presented in Fig. 1 correspond to various locations of the shock electrodes (locations are shown in the insets). Electrode locations both in the bath (Figs. 1b and 1c) and at the tissue surface (Figs. 1d and 1e) are considered. In all cases the field strength is 4.55 V/cm. For all electrode locations examined, the distribution of shock-induced transmembrane potential is highly nonuniform. Surface VEs can be clearly outlined in each slice. Stripes of high-magnitude surface depolarization (hyperpolarization) are seen at tissue borders facing, or in contact with, the cathode (anode). Similar VEs are observed at the cavity–tissue borders. VEs are also present throughout the myocardium ("bulk" VEs). Since this two-dimensional tissue slice incorporates only fiber curvature and not other elements of bulk tissue structure as specified above, the formation of the VEs is due only to the change in fiber spatial orientation in the tissue and/or spatial nonuniformity of the applied field [5].

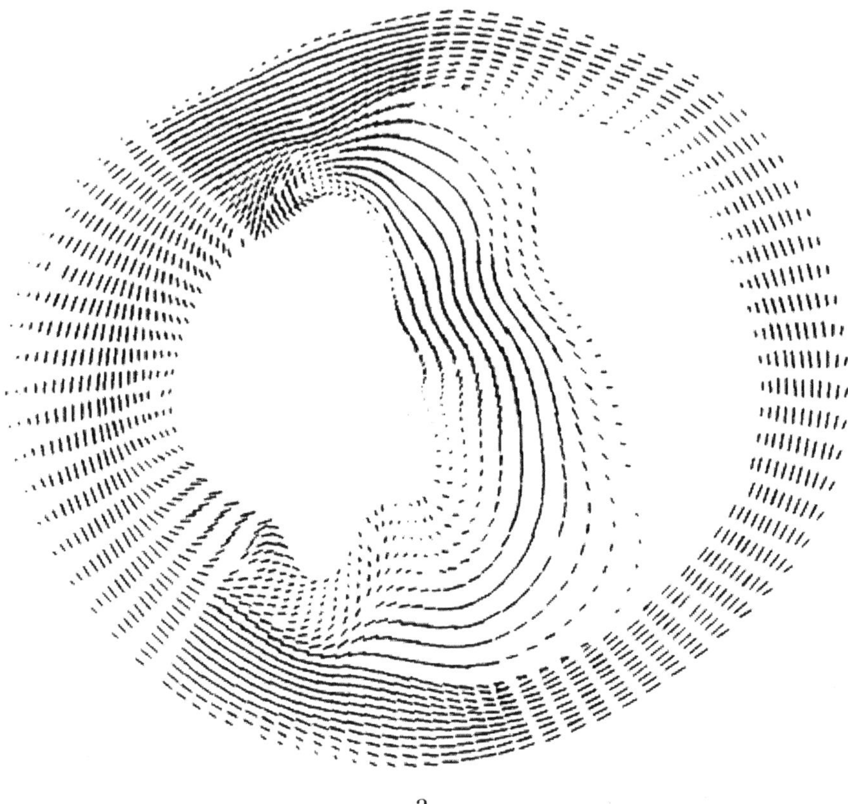

a

Figure 1a   Geometry of the slice. Bars indicate fiber orientations. The longest bars represent fibers that lie within the plane of the slice. The dots denote fibers oriented perpendicularly to the slice.

Examining all four transmembrane potential maps in Fig. 1, one notices that the bulk VEs do not change dramatically with the change in electrode location. Perhaps the most significant differences are observed in the slice where the line electrodes are perpendicular to each other (Fig. 1b). The reason for the similarity between Fig. 1b, in which the slice is field-stimulated by large parallel line electrodes, i.e., uniform electric field, and Figs. 1d and 1e, in which the electrodes represent "patches," is that although the patch electrodes create a spatially nonuniform field, they are far apart so that the majority of the myocardium between them is, in effect, subjected to a fairly uniform field. Thus, it is the fiber curvature that mostly determines the shape and location of the bulk VEs in this example. Other simulation and experimental studies confirm the role of the fiber curvature in the generation of the bulk VEs [10].

It is worth noting that in the simulation presented here significant portions of the myocardium experience a change in their transmembrane potential as a result of the shock. Both positively and negatively polarized VEs are dispersed through the septum and the ventricular wall. The range of their magnitude is approximately ±100 mV. Although two-dimensional, this example clearly demonstrates the far-field effects of the shock within the myocardium. Using the passive tissue representation, we are able to evaluate the sole effect of the shock. What follows after the shock is a function of both magnitude and spatial location of the shock-induced VEs as well as of the preshock activity in the myocardium. Thus, the effect of the same shock-induced depolarization or hyperpolarization can be different depending on the local electrical state of the tissue. To better understand the complex electrical phenomena that take place after the defibrillation shock, and in accordance with our "divide and conquer" approach to understanding defibrillation and arrhythmogenesis, we need first to elucidate the response of a single cell (membrane patch) to shocks for various electrical states of the cell. The electrical events that follow the delivery of the shock to a single isopotential cell represent the local shock response that is not "contaminated" by contributions from surrounding cells.

III. EFFECTS OF SHOCKS ON SINGLE CELLS: EXTENSION OF REFRACTORINESS AND DEEXCITATION

Shock-induced depolarization results in activation of a single isopotential cell if the cell is excitable, or in prolongation of action potential duration if the cell is refractory (this is often referred to as "extension of refractoriness") [14–18]. Shock-induced hyperpolarization brings the cell membrane below rest if the cell is quiescent. It causes repolarization and abbreviates the action potential if the cell is refractory [19–24]. If the negative change in transmembrane potential is large enough, it completely abolishes the action potential and fully restores local excitability [25,26]. The term "deexcitation" [2,26] is typically used to describe the latter phenomenon.

To illustrate these effects, Fig.. 2a shows the effect of monophasic current-injection pulses (S2) on the single-cell action potential duration (APD). The cell membrane kinetics are represented by the Drouhard-Roberge modification [27,28] of the Beeler-Reuter model [24]. Cell responses to both hyper- and depolarizing pulses of various strengths are examined.

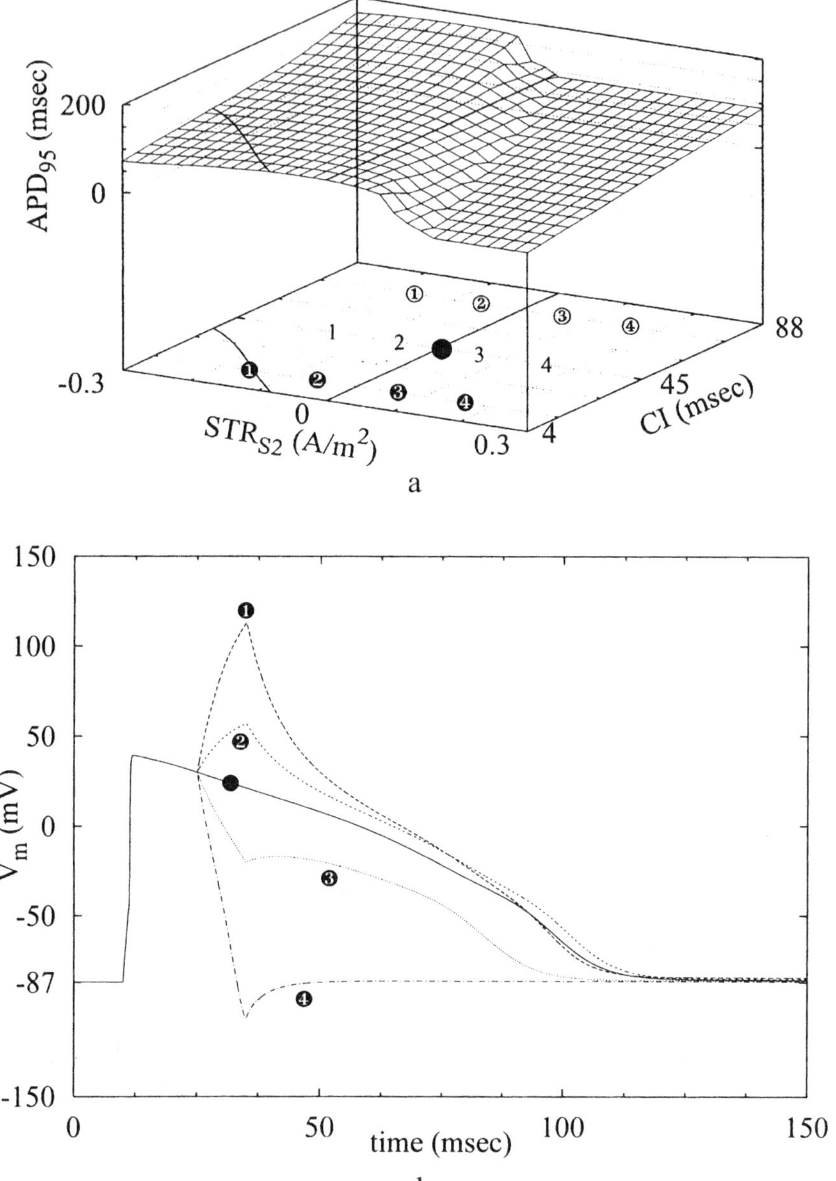

Figure 2   Response of a membrane patch to defibrillation shocks. (a) 3-D surface showing the change in action potential duration (APD) at 95% repolarization as a function of coupling interval (CI) and stimulus (S2) strength. Thick lines indicate no change in APD. (b–d) Individual traces of action potentials in time. The filled black dot corresponds to the unperturbed action potential (zero shock strength and zero CI on the 3-D plot). Numbered traces (both black and white, circled and not-circled numbers) indicate responses to shocks of strength and CI as shown in the 3-D plot.

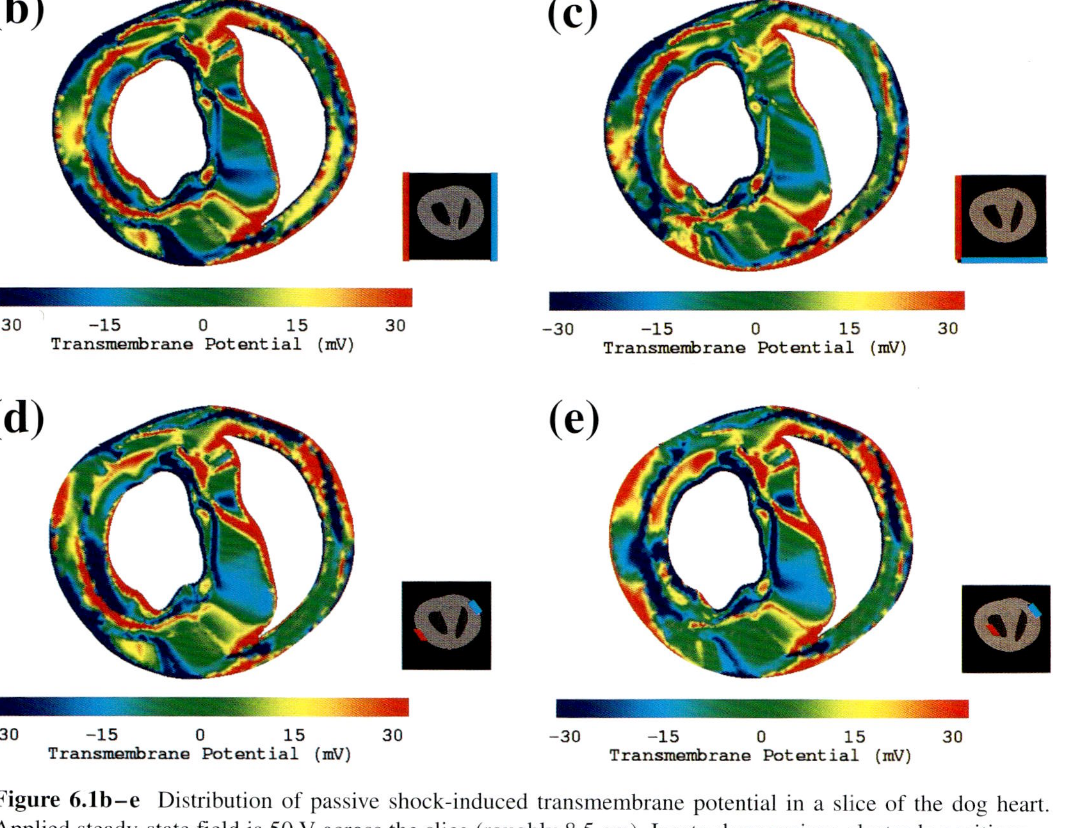

Figure 6.1b–e Distribution of passive shock-induced transmembrane potential in a slice of the dog heart. Applied steady-state field is 50 V across the slice (roughly 8.5 cm). Insets show various electrode positions.

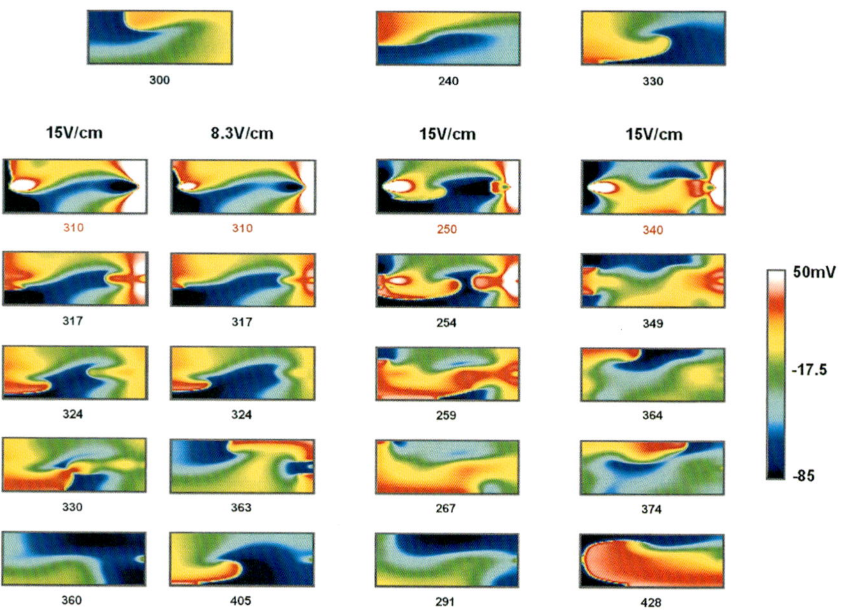

Figure 6.3 Transmembrane isopotential maps following four 10-msec monophasic defibrillation shocks delivered at three different instants of time to a signal spiral wave. The sheet measures 14×5 mm and consists of straight fibers running horizontally. The shock is administered via two small ("wire"-like) electrodes located in the middle of opposite vertical tissue boundaries. Anode is on the left, cathode is on the right. Top panels show the spiral wave at the moment of shock delivery. Time at which the map is calculated is displayed under it (red numbers indicate instants during which the shock is on). Time is measured from the onset of the S1 stimulus used to initiate the spiral wave. Shock strength is shown above each column of transmembrane potential maps.

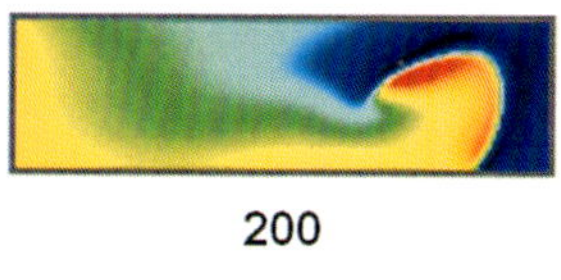

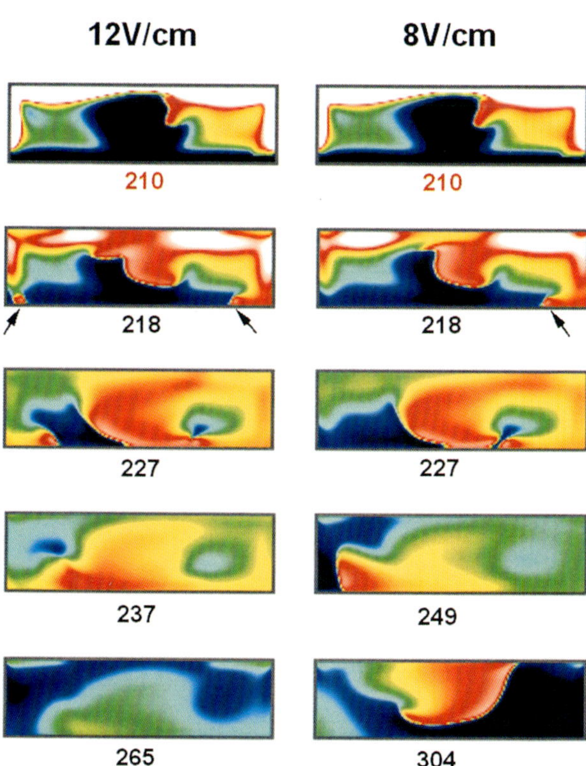

Figure 6.4 Transmembrane isopotential maps following two 10-msec monophasic defibrillation shocks of strengths 12 and 8 V/cm, respectively, delivered to a myocardial sheet 200 msec after the S1 stimulus used to initiate the spiral wave shown in the top panel. The sheet size is 22×6 mm. The defibrillation shock is delivered via two line electrodes occupying the top and bottom tissue borders. The cathode is on the top, the anode is on the bottom. Fibers in the sheet form parallel arcs; the radius of fiber curvature is 11.5 cm. The fibers are uniformly convex with respect to the anode and concave with respect to the cathode. Time at which the map is calculated is displayed under it (red numbers indicate instants during which the shock is on). Arrows indicate break excitations along the bottom tissue boundary.

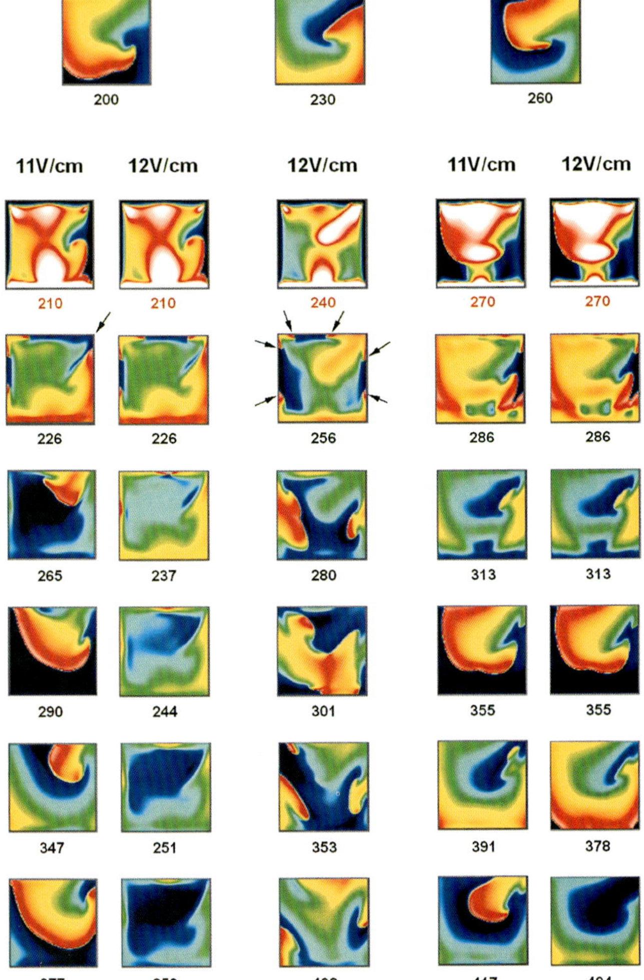

Figure 6.5 Transmembrane isopotential maps following five 10-msec monophasic defibrillation shocks delivered at three different instants of time. The sheet measures 2×2 cm. The defibrillation shock is delivered via two line electrodes occupying the top and bottom tissue borders. The cathode is on the bottom, the anode is on the top. Fibers in the sheet form parallel parabolas, i.e., they have a nonuniform curvature; an expression for the fiber pathways can be found in Ref. 38. The fibers are convex with respect to the cathode and concave with respect to the anode (i.e., shock polarity is reversed as compared to Fig. 4). Top panels show the spiral wave at the moment of shock delivery. Time at which the map is calculated is displayed under it (red numbers indicate instants during which the shock is on). Time is measured from the onset of the S1 stimulus used to initiate the spiral wave. Shock strength is shown above each column of transmembrane potential maps. Arrows indicate break excitations or the lack thereof (see text for details).

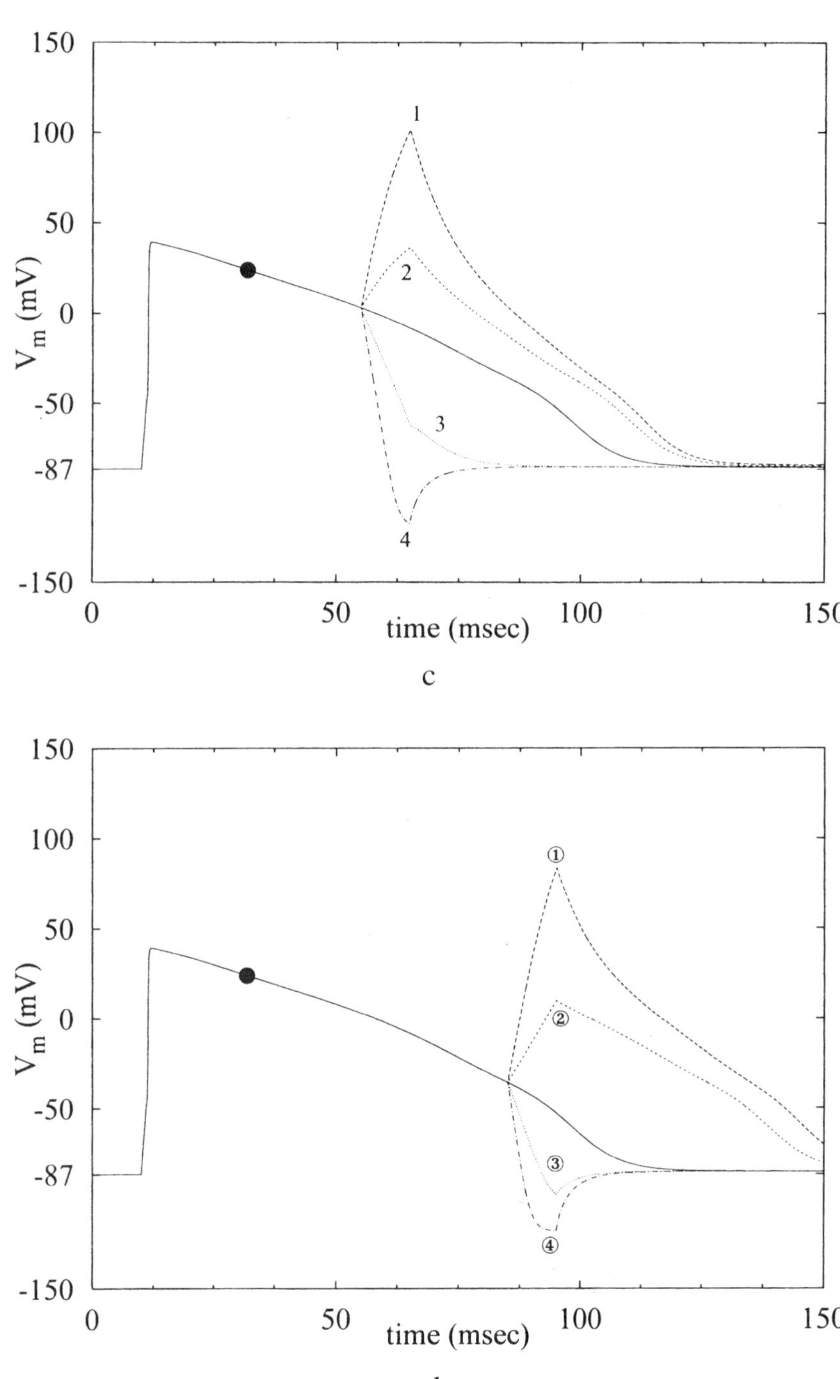

c

d

Figure 2 (*Continued*)

The shocks are delivered at increasing coupling intervals (CIs) and the APD is measured at 95% repolarization. Accompanying the three-dimensional plot of APD as a function of CI and shock strength are plots of action potential time course affected by the shock (Figs. 2b–2d). The presented action potentials correspond to shock strengths and CIs as marked on the 3-D plot.

The figure demonstrates that stimulus-induced hyperpolarization (right side of the 3-D plot) shortens the APD. Indeed, all action potential traces marked by the numbers 3 and 4 in Figs. 2b–2d are of duration shorter than normal. The stronger hyperpolarizing shocks (all traces marked 4) abolish the action potential: the cell is "deexcited" by the shock and becomes fully excitable. Strong hyperpolarizing shocks quickly "send" the transmembrane potential to a level below rest; the APD thus follows the CI (seen on the right side of the 3-D plot as a linear increase in APD as CI grows). The shock strength that is needed for such a response decreases as the CI increases (compare traces 3 and 4 in Fig. 2d with corresponding traces in Fig. 2c). Weak hyperpolarizing shocks shorten APD without immediately abolishing the action potential (traces marked 3, Figs. 2b and 2c). The weaker the shock, the smaller the disturbance in the action potential it causes and the closer the postshock APD to the original.

The depolarizing pulses in Fig. 2 delivered at long CIs cause an extension of the APD, and thus of the refractory period of the cell (left side of Fig. 2a, beyond the 0-level APD line). Extension of APD is demonstrated by all five traces that belong to this region in Fig. 2a; these are traces 1 and 2 in Figs. 2c and 2d and trace 2 in Fig. 2c. As these traces indicate, stronger shocks cause a larger extension in APD. For the same shock strength, the shock-induced extension in APD increases with the increase in CI (compare traces marked 2 in Figs. 2b–2d).

A strong depolarizing shock delivered at a very short CI shortens the APD as seen in trace 1, Fig. 2b. The stronger the shock, the larger the effect. However, this shortening is relatively small.

Overall, the response of cells shocked while undergoing an action potential varies, depending on the shock strength as well as on the timing of the stimulus. Armed with this knowledge about the cellular response to shocks, the postshock electrical events in the myocardium and their mechanisms can be better understood. The events that take place during and after the defibrillation shock incorporate both the local cellular response to the shock (as examined above) as well as the spatial interaction between cells in the myocardium. In the sections that follow we present our model of the defibrillation process, series of simulations of termination of a spiral wave with a defibrillation shock, as well as insights into the mechanisms by which the shock fail or succeeds that we have obtained from the computer simulations.

IV. COMPUTER MODELING OF THE DEFIBRILLATION PROCESS

Modeling the process of defibrillation on the tissue (and possibly organ) level incorporates several key components. *First*, the tissue is represented by the active bidomain model. The following coupled differential equations govern the potentials in this model representation of cardiac tissue:

$$\nabla \cdot (\hat{\sigma}_i \nabla \Phi_i) = i_m \tag{1}$$

$$\nabla \cdot (\hat{\sigma}_e \nabla \Phi_e) = -i_m \tag{2}$$

$$i_m = \beta \left[C_m \frac{\partial V_m}{\partial t} + I_{\text{ion}}(V_m) - I_{\text{stim}}(t) + G(V_m, t) V_m \right] \tag{3}$$

where $\hat{\sigma}_i$ (mS/cm) and $\hat{\sigma}_e$ are conductivity tensors in the intra- and extracellular domains, respectively, i_m (μA/cm^3) is the volume density of the transmembrane current, β(cm^{-1}) is the surface-to-volume ratio of the membrane, C_m (μF/cm^2) is the specific membrane capacitance, I_{stim} (μA/cm^2) represents the transmembrane stimulation current density, and G (mS/cm^2) is the variable membrane conductance that incorporates membrane electroporation.

Second, we incorporate in the model the additional variable membrane conductance $G(V_m, t)$, which accounts for the pore generation in the membrane during strong electric shocks [29].

Third, in the simulations presented in this chapter the ionic current [the term I_{ion} in Eq. (1)] is represented by the Drouhard-Roberge modification [27] of Beeler-Reuter kinetics [24] (BRDR model). We alter the original BRDR model in order to accommodate strong electric fields. The summary of rate coefficient revisions for the various gates can be found in reference [28]. The modifications in the membrane model take effect only under the defibrillation electrodes where the range of shock-induced transmembrane potentials exceed the range for which the original membrane model has been derived. Using this approach, we are able to achieve physiologically meaningful results while maintaining numerical stability in our transmembrane potential solutions under the defibrillation electrodes. Further, to account for the fact that APD in a fibrillating ventricle is considerably shorter than a normal action potential, we decrease the APD of a single cell to approximately 100 msec [30].

Fourth, in the simulations included in this chapter the homogeneous Neumann boundary conditions (no current crossing) are used to represent the fact that the preparation is surrounded by an insulator. If the tissue boundaries are in contact with a volume conductor (bath) such as blood, then the boundary conditions incorporate continuity of interstitial and bath potentials as well as currents.

Fifth, we account for change in fiber orientation in the myocardium. In a model preparation that incorporates changing fiber orientation, the intrinsic longitudinal and transverse conductivities of a fiber (and its associated extracellular space) remain constant. As the fiber bends, however, the global conductivities (with respect to a fixed coordinate system) in the tissue change from point to point. It is the latter conductivities (rather, conductivity tensors) that participate in the equations above. The relationship between the global tissue conductivities and the ones intrinsic to a fiber is expressed via a matrix reflecting the local change in fiber orientation as shown in reference [31]. The conductivity values intrinsic to a fiber used here are the same as in a paper by Roth [32].

Sixth, we represent arrhythmia-like electrical activity by initiating a single spiral wave in the tissue. An S1-S2 cross-stimulation protocol described in detail in reference [31] is employed to start the reentry. The spiral waves shown in this chapter are stable for hundreds of milliseconds and have periods of rotation between 70 and 80 msec.

Seventh, we use variety of shock electrode locations as well as shock waveforms. The simulations included in this chapter represent the responses to monophasic, 10-msec duration shocks of varying strengths only; the shocks are delivered at various times during the spiral wave cycle.

Finally, the numerical implementation of our model for the simulations presented here is based on the method of lines. A detailed description of the numerical aspect of our studies can be found in a previous publication of ours [28]. In brief, we first replace the spatial differential operators by finite differences and then solve the resulting system of nonlinear differential-algebraic equations. The ordinary differential equations are solved using a predictor-corrector method (PECE) and, after each time step, Φ_e is updated by solving the algebraic equations. We employ an iterative technique, GMRES (Generalized Minimal Residual Method), which is a variation of the conjugate gradient method. We use a banded, diagonally preconditioned version of GMRES. The semi-implicit PECE method used here is a two-step Adams-Bashforth predictor with a two-step Adams-Moulton corrector. It is of higher order than the explicit solvers, e.g., the forward Euler method, and avoids the computational expense of fully implicit methods.

V. SIMULATIONS OF DEFIBRILLATION

A. General Aspects

While we are making progress in simulating the effect of the defibrillation shock in three dimensions for realistic heart geometry and fiber orientation, simulation of defibrillation in two dimensions is far more advanced. Here we present several examples of defibrillation and postshock activity in tissue slices undergoing spiral wave reentry. Two groups of simulations studies are presented based on the mechanisms by which the *bulk* VEs underlying the defibrillation process are generated. In the first group of simulations the VEs are the result of a nonuniform external field applied to a homogeneous tissue consisting of straight fibers. In other words, the nonuniformity needed for the formation of VEs is in the applied field. In the second group of simulations the applied electric field is uniform, while the fiber orientation in the tissue is not. These examples incorporate fiber curvature that leads to the formation of bulk VEs. In both groups of simulations the VEs play an important role in the postshock activity in the myocardium.

During the shock cardiac tissue is simultaneously positively and negatively polarized in different areas. As a result, the cells in these regions will behave in a fashion similar to the single isopotential cell experiencing a depolarizing or hyperpolarizing stimulus as shown in Fig. 2: the final effect will be a function of the preshock electrical state in this region as well as of the sign and magnitude of the polarization (the VEs). The difference in behavior between a single cell as shown in Fig. 2 and multicellular cardiac tissue consists in the fact that VEs of opposite polarity in the tissue are often in immediate contact with each other. The electrotonic spatial interactions between regions thus play a very important part in the defibrillation process. Deexcitation induced by a negative (anodal) VE creates an area of excitable tissue. Electrotonic currents from adjacent depolarized areas can easily excite the deexcited area and create new wavefronts, a process known as break excitation [32]. As the results presented below indicate, these new wavefronts determine, to a large extent, the success or failure of the defibrillation shock.

The simulations presented in the next section describe the termination of a single spiral wave by 10-msec monophasic defibrillation shocks of different strengths delivered at various times during the spiral wave cycle. Although shocks might have the same strength and thus induce the same polarization in the tissue, the postshock distribution of depolarization and deexcitation could be very different depending on the preshock distribution of transmembrane potential. It is this issue that is also addressed by the simulations presented below.

B.   Example 1: Nonuniform External Field and Straight Fibers

The results presented here demonstrate extinguishing of spiral wave reentry by a defibrillation shock delivered via two small "wire" electrodes at the opposite tissue boundaries. The anode is located on the left, while the cathode is on the right. Point-source electrodes provide the highest degree of field nonuniformity in the vicinity of the electrodes. All simulation results refer to the same tissue size, $14 \times 5\,mm$, the smallest sheet of myocardium that could maintain reentrant activity.

In Fig. 3 (Fig. 3; see color plate), the leftmost column displays the tissue behavior following a defibrillation shock of strength $15\,V/cm$ administered at $300\,msec$ after the onset of the S1 stimulus that was used to elicit the spiral wave. At the time of shock delivery, the spiral wavefront is approaching the left vertical tissue border in a counterclockwise direction (see top panel on left); the tissue in front of the wave is excitable. The shock results in VEs (310-msec panel): membrane depolarization in the vicinity of cathode of a half-"dog-bone" shape and a hyperpolarization adjacent to it [34,35]. Analogous opposite membrane polarization is observed in the vicinity of the anode. Note that before the shock delivery the tissue in the vicinity of the cathode is depolarized by the propagating wave. Nonetheless, the shock-induced region of hyperpolarization on the right is strong enough to deexcite the tissue there.

After the shock is turned off, the cathodal depolarization is able to "invade" the adjacent island of hyperpolarization; this is cathode-break excitation. The new activation front formed at the right hyperpolarized region further attains a "bullet-like" shape and rapidly propagates across the middle of the sheet (see 324-msec panel). At this time the central portion of the sheet is excitable (the excitable core of the spiral wave) and can sustain this propagation. Simultaneously, the reentrant wave propagates through the anodal deexcited area on the left virtually undisturbed by the small island of anodal depolarization. As it makes its way in the longitudinal direction at the bottom of the sheet, the reentrant wave collides with the new cathode-break excitation (330-msec panel). The resultant wavefront following the collision quickly dies out since it has nowhere to go — the tissue is refractory to the point of being incapable of maintaining any activation. The time interval between administering the shock and decrease of the transmembrane potential throughout the tissue to a maximum of $-30\,mV$ is $66\,msec$. In this case the reentrant wave is extinguished by the defibrillation shock over an interval comparable to a single spiral wave revolution. The remaining small island of depolarization adjacent to the cathode corresponds to the area of electroporation in the tissue. There the transmembrane potential is at a zero level, since no potential difference exists across a "hole" in the membrane.

The next column in Fig. 3 represents a decrease in the shock strength with the other parameters in the simulation remaining unchanged. The smaller shock strength results in smaller-in-size VEs. The propagation of the spiral wave through the VE region on the left is the same as in the previous simulation; the only difference is that the wavefront velocity is smaller for the weaker shock. The reason for this is that after the weaker shock the wave propagates through a less deexcited area under the anode (compare the two 317-msec panels). On the right, a cathode-break activation takes place again; it invades the small virtual anode, but wavefront propagation farther into the tissue fails since the curvature of this new wavefront is too small to maintain propagation [36] (smaller than the one following the stronger shock). As a result, the initial spiral wave continues to recirculate through the tissue only mildly disturbed by the area of electroporation that takes place under the cathode. The shock fails to terminate the reentrant activity.

The third and fourth columns in Fig. 3 demonstrate the effect of shock timing on postshock events. Two different timings of the shock, 240 and 330 msec, respectively, are used; the top panels depict the spiral wave at the moment of the shock. In both cases the shock strength is 15 V/cm. In the "240-msec" simulation, a new cathode-break activation wavefront is formed at the right hyperpolarized region (250-msec panel). This new wavefront propagates freely toward the center of the tissue. Simultaneously, anode-break excitation occurs on the left. It is combined with the remnant of the initial reentrant activity into a broad wavefront propagating down and to the right. The collision of this wavefront with the postshock activation on the right renders the tissue too refractory to maintain the reentry. The spiral wave is extinguished within 60 msec from the delivery of the shock. In contrast, the shock applied 330 msec after the onset of the S1 stimulus does not result in a collision of wavefronts that annihilate each other. During the time the shock is applied, the spiral wave propagates to the right and reaches the VEs on the right. Although tissue becomes deexcited under the virtual anode there, the newly formed excitable gap is immediately invaded by the spiral wave. The spiral wave cannot propagate further, since it is surrounded by tissue rendered refractory by the shock-induced cathodal depolarization. However, an anode-break activation is generated under the anode (348-msec panel), which results in two propagating waves, one traveling upward and the other downward. The front directly below the physical anode dies out. The other front propagates along the top border toward the cathode (364-msec panel) and makes a revolution *in a direction opposite to the rotation of the original spiral wave*. This new activity is not sustained, however, and represents type II defibrillation. The total time interval between administering the shock and decrease of the transmembrane potential throughout the tissue to a maximum of $-30\,\text{mV}$ in this case is 160 msec.

C. Example 2: Uniform Applied Field and Curved Fibers

This section examines simulation results in which the defibrillation shock was delivered via two large line electrodes located at the opposite (top and bottom) boundaries of the tissue slice, thus simulating a uniform applied field. The fiber orientation in the sheet, however, is changing in space. Figure 4 (Fig. 4; see color plate) depicts tissue slices in which the fibers have uniform curvature, i.e., they form parallel arcs with convexity toward the bottom tissue border. Tissue size in this case is $22 \times 6\,\text{mm}$, and the radius of fiber curvature is 11.5 mm. Figure 5 (Fig. 5; see color plate) demonstrates the postshock activity in a larger sheet, $2 \times 2\,\text{cm}$, in which the fibers have a nonuniform curvature: they are of parabolic shape. The curvature of the fibers is highest along the central vertical axis; all of the fibers remain parallel to each other. The VEs in both cases are a result of the fiber curvature in the sheet; the shape of the VEs has been described previously in a series of publications from our group [5,37,38].

In Fig. 4 the shock is delivered to a stable spiral wave 200 msec after the initiation of reentry. The top panel of Fig. 4 shows the location of the activation at the moment of shock delivery. The postshock activity following two defibrillation episodes is presented in the panels below: a successful 12 V/cm shock on the left and an unsuccessful 11 V/cm shock on the right. The cathode is located on the top and the anode is on the bottom border of the sheet. Both shocks result in an increase in transmembrane potential under the cathode and decrease of it under the anode. Further, a large central area of shock-induced deexcitation is formed, in addition to two regions of depolarization along the vertical tissue border. Despite the difference in shock strengths, the induced VEs are visually quite similar (also due to the saturation of the color bar). Careful examination of the transmembrane potential values indicates, however, that the stronger shock is associated with larger values of the shock-induced depolarization and hyperpolarization.

As indicated previously, the close proximity of depolarized and deexcited areas could result in break excitations. The regions in the vicinity of the bottom left and right corners of the tissue seem to fulfill this requirement. Indeed, two break excitations take place following the stronger shock (see arrows, 218-msec panel, left). They result in two wavefronts propagating along the bottom tissue border. However, only one break excitation (at the right corner) occurs after the weak shock (see arrow, 218-msec panel, right). The current resulting from the transmembrane potential gradient between positively and negatively polarized areas around the left corner in this case is insufficient to elicit a new (break) activation there. The transmembrane potential maps at 227 msec show that the remnant of the spiral wave has managed to propagate through the central deexcited area and reach the bottom tissue border, where it has collided with the activation front coming from the right. In the case of the stronger shock, further propagation resulting from this collision is halted by the wavefront approaching from the left (237-ms panel, left); activations die out since they are surrounded by (partially) refractory tissue. In the case of the 8-V/cm shock, a significant excitable gap exists at 227 msec that further supports the propagation of a wavefront formed after the collision. The spiral wave is reinstated and defibrillation fails.

Similar events take place in the simulations presented in Fig. 5. However, the polarity of the electrodes there is reversed: the cathode is on the bottom and the anode is on the top. The figure also demonstrates the dependence of postshock activity on the timing of the shock. Two defibrillation shocks, 11 and 12 V/cm, are delivered to the spiral wave 200 msec after its initiation, a 12-V/cm shock is given at 230 msec, and two shocks, again 11 and 12 V/cm, are applied at 260 msec. The top panel of Fig. 4 depicts the spiral wave at the instants of shock delivery. In all cases, the VEs induced by the shocks are similar, since the shocks are roughly of the same strength and the electrodes are in the same position. The VE patterns differ only by the fact that the underlying activity is different: different portions of the tissue are excitable or refractory depending on the location of the preshock spiral wave. The shock induces depolarization under the cathode and hyperpolarization under the anode. In addition, a central region of the sheet plus two small areas near the top corners of the tissue are depolarized. Hyperpolarization is induced along the vertical tissue borders.

The two shocks delivered at 200 msec have a different outcome: the stronger (right) succeeds, while the weaker (left) fails. Similar to the events depicted in Fig. 4, the lack of a single break excitation after the weaker shock is what makes all the difference. Indeed, after the 10-msec shock is turned off (210-msec panels), break excitations ensue in all regions where a strong depolarization is in contact with deexcited tissue (226-msec panels) except for the top right corner of the tissue subjected to the weaker shock. The tissue around this corner remains excitable (shown with arrow), and by the time the adjacent activations propagate through it, central portions of the myocardium recover enough to be able to sustain propagation (265-msec panel, left). The wavefront invades the newly recovered areas and the spiral wave activity is reestablished (panels 290–377 msec, left); the shock thus fails. In contrast, the numerous break activations elicited after the stronger shock collide with each other and the remnant of the spiral wave, thus erasing any postshock excitable gap in the sheet; the myocardium returns to rest (panels 237–259 msec, right) and the shock succeeds.

The shock delivered at 230 msec results in an interesting postshock activity. It is strong enough to elicit break activations in any region that contains a shock-induced depolarization next to deexcited tissue (see arrows in 256-msec panel). One would expect that these excitations would quickly propagate, collide with each other, and leave no excitable gap in the tissue. However, the combination between VEs and preshock activity is such that large areas of the myocardium experience small depolarizations from which they quickly recover. Activations invade these areas; note the two merged break activations invading the tissue from the left, and the merged wave-front propagating from the right (280-msec panel). By the time these wavefronts propagate through excitable tissue, the rest of the myocardium recovers, making it possible to maintain propagation (301-msec panel). As a result, two to three wandering wavelets traverse the myocardium at any time and maintain the arrhythmia (353- and 408-msec panels). Despite its relative strength, the defibrillation shock fails, providing a clear indication that the outcome of the shock is a function of preshock activity for a given strength of the shock.

The two leftmost columns in Fig. 5 show the activity resulting from shocks delivered at 260 msec. Again, due to the similar shock stregths, the VEs are visually nearly identical (270-msec panels); so are the first 85 msec of postshock activity (355-msec panels). After the shock, in both cases the activity does not cease—the wavefront that invaded the excitable gap at 313 msec manages to propagate through recovered tissue and reestablish the reentrant circuit. However, in the case of the stronger shock the central portion of the tissue is less recovered (378-msec panel) and the wavefront dies after a single rotation (type II defibrillation).

D. Effect of Shock Timing

The collection of simulation results presented here offers also a unique opportunity to explore the role of shock timing on the outcome of a defibrillation attempt. Figures 3 and 5 include results regarding shocks of the same strength delivered at different times during the spiral wave cycle. Figure 3 compares the postshock activity for three shocks, all of strength 15 V/cm, given at 240, 300, and 330 msec after the S1 used to initiate the spiral wave. The first two shocks succeed in extinguishing the spiral wave without any postshock activity, while in the third simulation the new spiral wavefront makes a single rotation before it runs out of excitable tissue. In

Fig. 5, however, success and failure are both attributed to the 12-V/cm shock: the shocks delivered at 200 and 260 msec, respectively, succeed, while the shock administered at 230 msec fails. Comparison between postshock transmembrane potential distribution for successful and unsuccessful shocks in Fig. 5 indicates that the postshock excitable gap at the "230-msec" shock was much larger than for the other two shocks, thus directly relating the amount of postshock excitable tissue to the probability of success of the shock.

VI. THE NEW INSIGHTS INTO DEFIBRILLATION

The simulation results presented in this chapter offer a new aspect of our understanding of the defibrillation process. They clearly demonstrate the importance of both depolarization and deexcitation by the shock. Depolarization by the shock and its consequences, excitation or extension of refractoriness, have been long considered the only effect that the shock exerts on the myocardium. Indeed, the extension of refractoriness [14,17,18,39,40], the progressive depolarization [41] and the upper limit of vulnerability [42] hypotheses maintain that the success of the shock is due to its ability to depolarize cardiac tissue. *Reestablishment of fibrillation is then believed to be caused by proliferation of fibrillatory wavelets that were *not terminated* by the depolarizing effect of the shock [43,44]. These wavefronts typically emanate from the low shock-gradient zone of the heart [45,46], where the shock-induced depolarization is expected to be of low magnitude (thus assuming a direct relationship between shock gradient and induced transmembrane potential [47]).

The simulations presented here, other recent simulations of ours published elsewhere [48], as well as experimental evidence by other researchers [2,10,49–53] underscore the major contribution of the negative shock-induced polarization in defibrillation and arrhythmogenesis. Negative polarization creates a new excitable gap—thus, even if all preexisting activity is erased by the VEs *new* wavefronts can still arise at the border between positive and negative VEs (break excitation) and quickly spread through deexcited areas [54,55]. This postshock activity can lead to a success or failure of the shock depending on the dynamics of both break activations and remnants of preshock activity, and their interaction with the adjacent areas of depolarization.

---

*For a detailed comparison of the VEs hypothesis for defibrillation with other existing theories, refer to a recent publication by Efimov et al. [2].

When does the shock fail and when does it succeed according to the VEs hypothesis? First, our results indicate that a stronger shock extends refractoriness to a larger degree and over larger areas than a weaker shock. This ensures that new and/or old propagating wavefronts are blocked when they encounter these areas of shock-induced depolarization. *Therefore, the crucial factor in determining the outcome of the shock is the time interval needed for the wavefronts to traverse the postshock excitable gap.* If this time interval is short, the wavefronts will propagate through deexcited areas before the adjacent areas of depolarization (or extended refractoriness) have recovered; further propagation will be blocked. If it takes longer for the excitable gap to become fully excited so that the wavefronts cannot erase it by the time the adjacent areas recover from shock-induced depolarization, then the wavefronts will invade these adjacent areas and wavefront propagation will continue.

Our results presented here demonstrate that slow or incomplete eradication of the excitable gap is due to one or more several reasons (note: detailed presentation of these arguments can be found in a previous publication of ours [38]): *First, some break excitations do not take place for weaker shocks.* Certain break excitations occur for the stronger shock but are absent in the case of a weaker shock. This is evident from all cases of defibrillation presented here: in Fig. 3, 300-msec timing of the shock, the stronger shock results is a successful break excitation on the right; in Fig. 4 there are two break excitations associated with the strong shock and only one following the weak shock; in Fig. 5, 200-msec timing, the arrow indicates the absence of a break excitation for the weaker shock. Weaker shocks generate smaller transmembrane potential gradients and thus stimulating currents between depolarized and deexcited areas that might fail to produce break excitations. Our simulations demonstrate that the break excitations combine with preexisting wavefronts to excite the postshock excitable gap—when many excitations invade the excitable gap simultaneously from all sides, the probability of it being fully consumed while the adjacent areas are still refractory is much larger. *Second, break excitations that follow a weaker shock, if any, take longer to develop.* Careful examination of the data in the present simulations indicate that the same break excitation can occur earlier or later after the shock, depending on the strength of the shock. The same behavior was observed in another simulation study of ours [38]. Again, the reason is that the transmembrane potential gradient between adjacent VEs of opposite polarity is larger for the stronger shock, resulting in a larger current at the border between the regions. This stimulating current elicits a break excitation—the larger the stimulus, the smaller the latency of the new activation (consistent with the strength–duration relationship). An excitation occuring in a deexcited region early after the shock will likely manage to traverse this region before the adjacent depolarized areas recover from refractoriness; such an activation will be blocked. In contrast, activations occurring late after the shock are more likely to reach and propagate through areas recovered from refractoriness and become wavefronts of refibrillation. *Third, break excitations propagate slower through weakly deexcited areas.* Weaker shocks are associated with a slower propagation through the (weakly) deexcited areas, resulting in an extended time interval for wavefront propagation. For instance, the anode break excitation front on the left in Fig. 3, 300-msec timing, is propagating faster through the area deexcited by the stronger shock, as already described in Sec. V.B. Similar observations regarding conduction velocity have been made recently in rabbit heart experiments [26] and in another simulation study of ours [28]. Stronger shocks are associated with high propagation velocities in deexcited areas due to the fact that these areas are more negatively polarized, and thus have a lower threshold for propagation. Therefore, faster propagation through a deexcited region ensures arrival of the excitation at the borders of the region before its surroundings have recovered from depolarization and refractoriness. The three reasons outlined above provide a clear indication why stronger shocks are likely to succeed and weaker to fail.

The arguments presented above reveal the mechanisms behind the higher success rate of strong shocks as compared to weak shocks. However, one issue is not addressed by the above agruments: the role of shock timing. This issue has not received much attention in the recent experimental studies of virtual electrode arrhythmogenesis, since these studies have focused mostly on VE-induced phase singularities in predominantly refractory tissue [53]. Here we demonstrate that for shocks of the same strength, i.e., the same VEs, the outcome of the shock is also dependent on preshock electrical activity. For certain combinations of VEs and preshock distribution of transmembrane potential there is a larger postshock excitable gap (compare 12-V/cm shocks in Fig. 5). These shocks are more likely to fail, since it takes longer for the postshock activations to traverse the larger excitable gap (consistent with the arguments above). We believe that for shocks of the same strength, it is the extent of the postshock excitable areas (both VE-induced and remnants of preshock excitable gaps, if any) that underlies the probabilistic nature of defibrillation.

The results presented in this chapter underscore the importance of computer simulations in unraveling the mechanisms for defibrillation and postshock arrhythmogenesis. Modeling and simulation provide means to examine the contribution of VEs, shock strength and timing in a controlled environment: we are able to clearly delineate electrical phenomena induced by VEs from preshock activity, and thus to dissect the various aspect of the spatial interactions in the success or failure of the shock. Bidomain simulations have guided the design of experiments [54] and have been used to analyze numerous experimental results [8,10,56]. It is our expectation that this symbiosis between simulation and experiment will continue to flourish.

ACKNOWLEDGMENTS

This work was supported by National Science Foundation grants BES-9809132 and DMF-9709754, National Institutes of Health grants HL63195, and HL67322.

REFERENCES

1. Fishler MG. Computer modeling of defibrillation I: the role of cardiac tissue structure. In Cabo C, Rosenbaum D., eds. Quantitative Cardiac Electrophysiology, New York, NY: Marcel Dekker, Inc., 2002, pp. 199–233.
2. Efimov IR, Gray RA, Roth BJ. Virtual electrodes and de-excitation: new insights into fibrillation induction and defibrillation. J Cardivasc Electrophysiol 11:339–353, 2000.
3. Winfree A. Various ways to make phase singularities by electric shock. J Cardiovasc Electrophysiol 11:286–289, 2000.
4. Trayanova NA, Roth BJ, Malden LJ. The response of a spherical heart to a uniform electric field: a bidomain analysis of cardiac stimulation. IEEE Trans Biomed Eng 40:899–908, 1993.
5. Trayanova NA, Skouibine K, Aguel F. The role of cardiac tissue structure in defibrillation. Chaos 8:221–233, 1998.
6. Entcheva E, Trayanova NA, Claydon F. Patterns of and mechanisms for shock-induced polarization in the heart: a bidomain analysis. IEEE Trans Biomed Eng 46:260–270, 1999.
7. Sobie EA, Susil RC, Tung I. A generalized activating function for predicting virtual electrodes in cardiac tissue. Biophys J 73:1410–1423, 1997.
8. Entcheva E, Eason J, Efimov I, Cheng Y, Malkin R, Claydon F. Virtual electrode effects in transvenous defibrillation-modulation by structure and interface: evidence from bidomain simulations and optical mapping. J Cardiovasc Electrophysiol 9:949–961, 1998.
9. Fishler MG. Syncytial heterogeneity as a mechanism underlying cardiac far-field stimulation during defibrillation-level shock. J Cardiovasc Electrophysiol 9:384–394, 1998.
10. Knisley SB, Trayanova NA, Aguel F. Roles of electric field and fiber structure in cardiac electric stimulation. Biophys J 77:1404–1417, 1999.
11. Henriquez CS. Simulating the electrical behavior of cardiac muscle using the bidomain model. Crit Rev Biomed Eng 21:1–77, 1993.

12. Nielsen PMF, LeGrice IJ, Smaill BH, Hunter PJ. Mathematical model of geometry and fibrous structure of the heart. Am J Physiol 260:H1365–H1378, 1991.

13. Hunter PJ, Nielsen PMF, Smaill BH, LeGrice IJ, Hunter IW: An anatomical model with applications to myocardial activation and ventricular mechanics. In: Pilkington TC, et al., eds. High Performance Computing in Biomedical Research. Boca Raton, FL: CRC press, 1993, pp. 3–26.

14. Sweeney RJ, Gill RM, Reid PR. Characterization of refractory period extension by transcardiac shock. Circulation 83:2057–2066, 1991.

15. Knisley SB, Smith WM, Ideker RE. Prolongation and shortening of action potentials by electrical shocks in frog ventricular muscle. Am J Physiol 266:H2348–H2358, 1994.

16. Knisley SB, Hill BC. Optical recordings of the effect of electrical stimulation on action potential repolarization and the induction of reentry in two-dimensional perfused rabbit epicardium. Circulation 88:2402–2414, 1993.

17. Jones JL, Jones RE, Milne KB. Refractory period prolongation by biphasic defibrillator waveforms is associated with enhanced sodium current in a computer model of the ventricular action potential. IEEE Trans Biomed Eng 41:60–68, 1994.

18. Jones JL, Tovar OH. Threshold reduction with biphasic defibrillator waveforms. J Electrocardiol 28(suppl):25–30, 1995.

19. Weidmann S. Effect of current flow on the membrane potential of cardiac muscle. J Physiol 115:227–236, 1951.

20. Hall AE, Noble D. Transient responses of Purkinje fibers to non-uniform currents. Nature 199:1294–1295, 1963.

21. Vassale M. Analysis of cardiac pacemaker potential using a "voltage clamp" technique. Am J Physiol 210:1335–1341, 1966.

22. Noble D. The Initiation of the Heartbeat. Oxford: Clarendon, 1975.

23. Goldman Y, Morad M. Regenerative repolarization of the frog ventricular action potential: a time and voltage-dependent phenomenon. J. Physiol (Lond) 268:575–611, 1977.

24. Beeler GW, Reuter H. Reconstruction of the action potential of ventricular myocardial fibers. J Physiol 268:177–210, 1977.

25. Pumir A, Romey G, Krinsky V. Deexcitation of cardiac cells. Biophys J 74:2850–2861, 1998.

26. Cheng Y, Mowrey KA, Wagoner DRV, Tchou PJ, Efimov IR. Virtual electrode induced re-excitation: a mechanism of defibrillation. Circ Res 85:1056–1066, 1999.

27. Drouhard JP, Roberge FA. A simulation study of the ventricular myocardial action potential. IEEE Trans Biomed Eng 29:494–502, 1982.

28. Skouibine K, Trayanova N, Moore P. A numerically efficient model for simulation of defibrillation in an active bidomain sheet of myocardium. Math Biosci 166:85–100, 2000.

29. Krassowaska W. Effects of electroporation on transmembrane potential induced by defibrillation shocks. PACE 18:1644–1660, 1995.

30. Skouibine K, Trayanova NA, Moore PK. Anode/cathode make and break phenomena in a model of defibrillation. IEEE Trans Biomed Eng 46:769–777, 1999.

31. Trayanova NA, Aguel F, Skouibine K. Extension of refractoriness in a model of cardiac defibrillation. In: Proc Pacific Symp on Biocomputing, Altman RB, Dunker AK, Hunter L, Klein TE, Lauderdale K, eds. Singapore: World Scientific, 1999, pp. 240–251.

32. Roth BJ. A mathematical model of make and break electrical stimulation of cardiac tissue by a unipolar anode or cathode. IEEE Trans Biomed Eng 42:1174–1184, 1995.

33. Trayanova NA, Skouibine K, Moore P. Virtual electrode effects in defibrillation. Prog Biophys Molec Biol 69:387–403, 1998.

34. Sepulveda NG, Roth BJ, Wikswo JP Jr. Current injection into a two-dimensional anistropic bidomain. Biophys J 55:987–999, 1989.

35. Trayanova NA, Pilkington TC. The use of spectral methods in bidomain studies. In: Pilkington TC, et al., eds. High-Performance Computing in Biomedical Research. Boca Raton, FL: CRC Press, 1993, pp. 403–425.

36. Cabo C, Pertsov A, Baxter W, Davidenko J, Gray R, Jalife J. Wave-front curvature as a cause of slow conduction and block in isolated cardiac muscle. Circ Res 75:1014–1028, 1994.

37. Trayanova NA, Skouibine K. Modeling defibrillation: effects of fiber curvature. J Cardiol 31(suppl):23–29, 1998.

38. Skouibine K, Trayanova N, Moore P. Success and failure of the defibrillation shock: insights from a simulation study. J Cardiovasc Electrophysiol, 11: 785–796, 2000.

39. Dillon SM. Optical recordings in the rabbit heart show that defibrillation strength shocks prolong the duration of depolarization and the refractory period. Circ Res 69:842–856, 1991.

40. Dillon SM. Synchronized depolarization after defibrillation shocks: a possible component of the defibrillation process demonstrated by optical recordings in rabbit heart. Circulation 85:1865–1878, 1992.

41. Dillon SM, Kwaku KF. Progressive depolarization: a unified hypothesis for defibrillation and fibrillation induction by shocks. J Cardiovasc Electrophysiol 9:529–552, 1998.

42. Chen PS, Swerdlow C, Hwang C, Karagueuzian HS. Current concepts of ventricular defibrillation. J Cardiovasc Electrophysiol 9:553–562, 1998.

43. Chen PS, Shibata N, Dixon EG, Wolf PD, Danieley ND, Sweeney MB, Smith WM, Ideker RE. Activation during ventricular defibrillation in open-chest dogs: evidence of complete cessation and regeneration of ventricular fibrillation after unsuccessful shocks. J Clin Invest 77:810–823, 1986.

44. Walcott GP, Knisley SB, Zhou X, Newton JC, Ideker RE. On the mechanism of ventricular defibrillation. PACE 20(part 2):422–431, 1997.

45. Chen PS, Ideker RE, Smith WM, Danieley ND, Melnick SD, Wolf PD. Comparison of activation during ventricular fibrillation and following unsuccessful defibrillation shocks in open-chest dogs. Circ Res 66:1544–1560, 1990.

46. Blanchard SM, Ideker RE. The process of defibrillation. In: Estes N III, et al., eds. Implantable Cardioverter- Defibrillators, New York: Dekker, 1994, pp. 1–27.
47. Krassowaka W, Pilkington TC, Ideker RE. Periodic conductivity as a mechanism for cardiac stimulation and defibrillation. IEEE Trans Biomed Eng 34:555–559, 1987.
48. Lindblom A, Roth B, Trayanova N. Role of virtual electrodes in arrhythmogenesis: pinwheel experiment revisited. J Cardiovasc Electrophysiol 11:274–285, 2000.
49. Knisley SB, Hill BC, Ideker RE. Virtual electrode effects in myocardial fibers. Biophys J 66:719–728, 1994.
50. Knisley SB. Transmembrane voltage changes during unipolar stimulation of rabbit ventricle. Cir Res 77:1229–1239, 1995.
51. Efimov IR, Cheng YN, Mowrey K, Van Wagoner DR, Mazglev T, Tchou PI. High resolution fluorescent imaging reveals "virtual electrode" phenomenon during application of monophasic shock from implantable cardiac defibrillator lead in isolated rabbit heart. PACE 20:1080, 1997.
52. Efimov IR, Cheng YN, Biermann M, Van Wagoner DR, Mazagalev TN, Tchou PI. Transmembrane voltrage changes produced by real and virtual electrodes during monophasic defibrillation shock delivered by an implantable electrode. J Cardiovasc Electrophysiol 8:1031–1045, 1997.
53. Efimov IR, Cheng YN, Van Wagoner DR, Mazgalev TN, Tchou PI. Virtual electrode-induced phase singularity: a basic mechanism of defibrillation failure. Circ Res 82:918–925, 1998.
54. Wikswo JP Jr, Lin SF, Abbas RA. Virtual electrodes in cardiac tissue: a common mechanism for anodal and cathodal stimulation. Biophys J 69:2195–2210, 1995.
55. Roth BJ, Wikswo JP Jr. The effect of externally applied electrical fields on myocardial tissue. Proc IEEE 84:379–391, 1996.
56. Efimov IR, Aguel F, Cheng YN, Wollenzier B, Trayanova NA. Virtual electrode polarization in the far field: implications for external defibrillation. Am J Physiol 279: H1055–H1070, 2000.

7
Theoretical and Practical Considerations for Cardiac Recording and Stimulating Electrodes

Robert A. Malkin
The University of Memphis, Memphis, Tennessee, U.S.A.

Bradford D. Pendley
Rhodes College, Memphis, Tennessee, U.S.A.

I. INTRODUCTION

Electronic equipment that displays electrocardiograms (ECGs), delivers pacing pulses, or defibrillates carries current in the form of electrons. The body, on the other hand, carries current in the form of ions. Electrodes—physically in contact with both the body and the instrument—translate, or transduce, between the body's ionic current and the instrument's electronic current. Transduction is best described as a chemical reaction between the material carrying the electronic current, usually a metal, and the material carrying the ionic current, a bodily fluid. For this reason, most of our understanding of how electrodes work has been borrowed from the field of chemistry. About one-half of this chapter is devoted to understanding the chemistry, hydrodynamics, and thermodynamics of the electrode–body interface. However, a second area of concern is that the body reacts to the presence of the electrode, a foreign substance. Depending on the electrode's composition, shape, placement, and other factors, the body's reaction to the electrode could overwhelm the signal, impede the stimulating current, or mechanically destroy the electrode. Finally, the last section of this chapter describes several of the most common electrodes for stimulation and recording in electrocardiology.

II. THE ELECTRODE/TISSUE INTERFACE

A. The Generation of a Double Layer

To illustrate how the translation of ionic information into electronic information occurs, consider a metallic electrode comprised of a pure metal, M, placed in contact with an aqueous solution containing a cation, C^+, and an anion, A^-. When the metallic electrode comes into contact with the solution, events are set into motion because of the dissimilar conducting phases. The nature of these events can be very complicated and are specific to the metal and the solution's constituents. The properties of the interface between the electrode and the solution are governed by the excesses and deficiencies in the quantities of the pertinent components, and this system will reach a state of balance or equilibrium after some time. In order to reach this equilibrium state, chemical reactions occur and the charge distribution around the electrode/solution interface is altered. For example, if C^+ is not the cation of the electrode metal, a minute amount of metal will dissolve in the solution adjacent to the electrode, creating additional cations, M^+, in the solution and excess electrons on the electrode. Thus, in this example, the electrode tends to become more negatively charged with respect to the solution.

These events result in the region near the electrode (on the order of nanometers) adopting a charge distribution different from that of the bulk of the solution (where cations balance anions). Helmholtz [1] was the first to model the consequences of this charge distribution, and proposed that the ions in the solution reside at the surface of the electrode, forming a double layer of charge similar to a parallel-plate capacitor (Fig. 1). However, this early model of charge distribution around the electrode could not accurately and completely account for the properties of the interface. While it is true that any charge on the electrode must reside on its surface, the ions in the solution need not be confined to the electrode surface. Gouy and Chapman [1] postulated that the interplay between the attraction/repulsion of cations/anions to the electrode, and the thermal processes that tend to randomize them, leads to a diffuse layer of charge adjacent to the electrode (Fig. 1). Later modifications (Stern, etc.) further refined our modeling of this charge distribution.

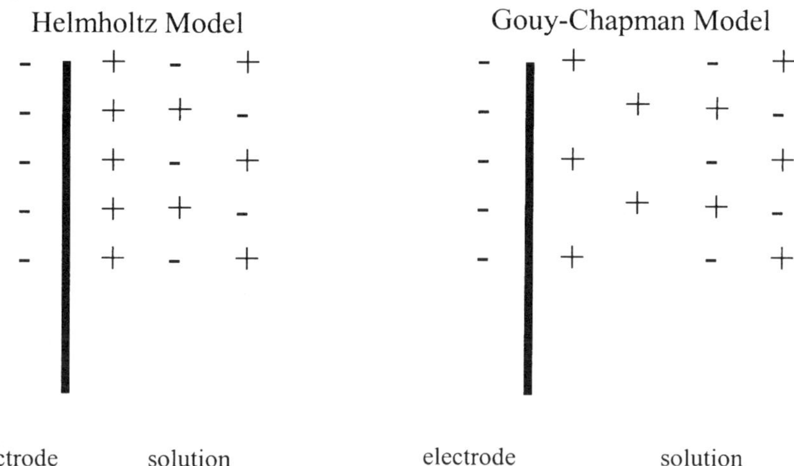

Figure 1 Two models of the charge distribution around an electrode. Notice the difference in the positive charge layer nearest to the electrode.

The net effect of the processes described leads the electrode to adopt a potential that depends on the electrode's composition, the solution's ionic constituents, and the temperature, since temperature influences the equilibrium state. In reaching an equilibrium state, the electrode has converted, via chemical processes, ionic information into electronic information. This is precisely what was sought from the electrode and provides a means to measure potentials within the body.

B. The Electrode's Ideal Potential

All potential measuring devices require two inputs; that is, they measure the difference in potential between two points. Thus, two electrodes are necessary to make a measurement. As a convenience for comparing the potentials of different metal electrodes, scientists often refer their measurements back to the standard, or normal, hydrogen electrode (SHE or NHE). The SHE is comprised of a platinum electrode immersed in a solution of an acid over which hydrogen gas is bubbled. The platinum catalyzes the following reaction at its surface;

$$2H^+ + 2 \text{ electrons} \rightleftharpoons H_2(g)$$

As with any electrode, the potential of the SHE depends on the amounts of hydrogen ion and hydrogen gas. Consequently, very specific constraints are placed on this "reference" electrode to facilitate comparison; the hydrogen gas must have a pressure of 1 atm, and the activity (a quantity related to the concentration of the species) of H^+ must likewise equal 1. When these conditions are met, the potential of the SHE is *defined* as 0.000 V *at all temperatures*. This allows the potential of other electrodes immersed in a solution to be measured versus the SHE, and the potential measured is referred to as the standard reduction potential, E^0. In practical applications, one typically uses a more convenient reference electrode with a known offset relative to the SHE.

Table 1 lists the standard reduction potential of several metals commonly used in cardiac measurements [2]. For comparison purposes, the standard reduction potential for electrode processes assumes the activity of all species is equal to 1. It is clear from Table 1 that the potential of the electrode depends on the electrode's composition and can vary widely. However, the data listed in Table 1 are deceptively simple. These values of the standard reduction potential are measured under very specific conditions, conditions that do not emulate physiological conditions. Since the potential developed at an electrode also depends on the solution's ionic constituents and their concentrations, the potential of an electrode will almost always vary from the value listed in Table 1. Unfortunately, we are unaware of any comprehensive measurements of electrode potentials under physiological conditions.

The situation is even more complicated when electrode materials are made from alloys of metals such as stainless steel 316L or titanium/aluminum/vanadium. For these alloys, the potential can vary several hundred millivolts and depends on the pertinent metallic species and the physiological conditions [3,4]. For example, with the commonly employed

Table 1 Standard Reduction Potentials at 25°C for Several Common Electrode Materials

Metal and reaction	Potential versus SHE V
$Pt^{2+} + 2e^- \rightarrow Pt$	1.2
$Ag^+ + e^- \rightarrow Ag$	0.7996
$Cu^{2+} + 2e^- \rightarrow Cu$	0.3402
$AgCl + e^- \rightarrow Ag + Cl^-$	0.2223
$2H^+ + 2e^- \rightarrow H_2$	0.0000 by definition
$Fe^{3+} + 3e^- \rightarrow Fe$	−0.036
$Ti^{2+} + 2e^- \rightarrow Ti$	−1.63
$Al^{3+} + 3e^- \rightarrow Al$	−1.706

Source: Ref. 2.

Ti_6Al_4V alloy, the pertinent electrode material is the passifying layer of TiO_2 formed at the electrode surface, whereas for stainless steel 316L it is a chromium species [3,4].

C. Deviations from the Ideal Potential

1. Nonpolarizable Electrodes

The capacity of an electrode to faithfully transduce depends on its ability to instantaneously respond to changes in potential without altering the potential being measured. Ideally, this means that as ionic distributions change, charge transfer (infinitesimal as it may be) must occur infinitely quickly across the electrode/solution interface and that the electrode must return to its equilibrium state instantly. In other words, the electrode should return to equilibrium instantly, no matter how much current it passes. An electrode which possess these qualities is called an ideally nonpolarizable electrode. Of course, no such electrode exists, but several real electrodes approach it. This concept is shown schematically in Fig. 2.

The chemical reaction that occurs at such a nonpolarized electrode can be given by the following general chemical equation:

$$aA(s) + n \text{ electrons} \rightleftharpoons bB(s) + cC(aq) \tag{1}$$

where A represents the reactant, B and C the products, a, b, c are the stoichiometric coefficients of the reaction, the symbol (s) means solid phase and the symbol (aq) means aqueous phase. The potential of the non-polarizable electrode in this example is given by the Nernst equation,

$$E = E^0 - \frac{RT}{nE} \ln \frac{a_B^b a_C^c}{a_A^a} \tag{2}$$

where R is the ideal gas law constant (8.3145 J/K mol), T is the temperature in kelvin, F is the Faraday constant (96,485 C), and a_A, a_B, and a_C are the activities of the reactant and products, respectively. The activity of a pure solid or liquid is 1. The activity of an ionic species in solution approaches its molar concentration at low concentrations but is markedly lower than its molar concentration at higher concentrations. Hence, Eq. (2) simplifies to

$$E = E^0 - \frac{RT}{nF} \ln a_C^c \tag{3}$$

The Nernst equation shows the variables which affect the ideally nonpolarizable electrode: the electrode material (E^0), the solution's ionic constituents (activity of C), and the temperature.

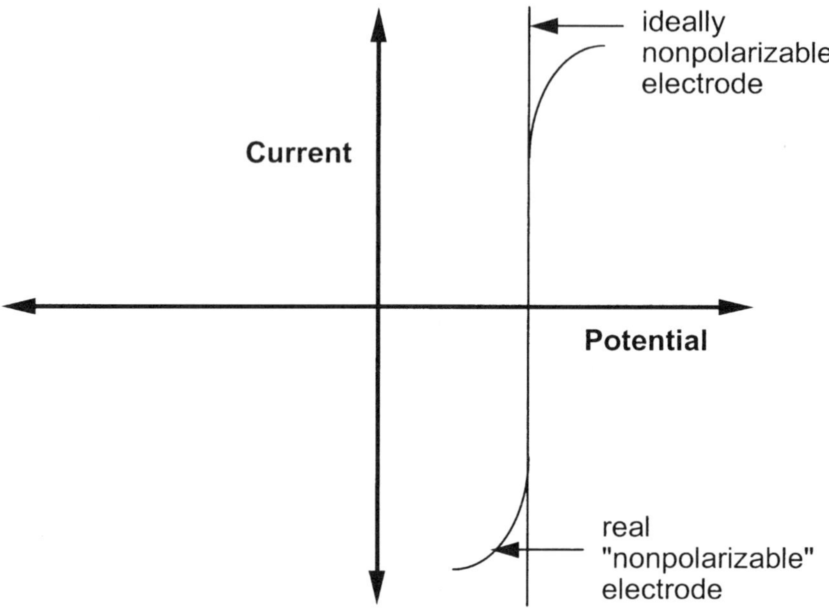

Figure 2 Variation in the potential of a nonpolarizable electrode when current flows through it. Ideally nonpolarizable electrodes do not exist, but are approximated by several common electrodes.

A real electrode that approaches the ideally nonpolarizable electrode is the silver–silver chloride electrode. For this electrode, the pertinent net chemical equation is

$$AgCl(s) + 1 \text{ electron} \rightleftharpoons Ag(s) + Cl^-(aq) \tag{4}$$

The Ag/AgCl electrode is often constructed of a silver electrode that has been electrolytically coated with a layer of silver chloride. However, sintered AgCl electrodes are available. The potential and stability of the silver–silver chloride electrode has been studied extensively and has been found to closely approximate an ideally nonpolarizable electrode [5,6].

2. Potential Stability—Chemical Considerations

The potential of any real electrode in contact with tissue depends on at least those variables that affect the ideal electrode: the composition of the electrode, the composition of the ionic constituents near the electrode, and the temperature. Thus, stability in the measured potential must likewise depend on the constancy of these factors. In addition, for real electrodes, variations in potential occur due to the transduction process itself, as well as the nature of the electrode/tissue interface.

As mentioned previously, two electrodes are required for any application. The two electrodes can be made of the same or different materials. If two electrodes of differing materials are used to record bioelectric events, several problems can result. The magnitude of the problems depends on the choice of materials. From Table 1, it can be seen that the difference in the potential of the two electrode materials will lead to a DC offset voltage. For example, if one were to select a silver electrode and a silver–silver chloride electrode, the difference between the standard reduction potentials is 577 mV, a large offset. Selecting Pt and Al would result in a DC potential of nearly 3 V, possibly more than the voltage of the power supply, leading to unexpected interference with the electronics. In addition, the potential difference could force current to flow. In the short term, this current may be sufficient to stimulate tissue. Over time, this current flow changes the potential difference between the electrodes for reasons that will be discussed shortly. Despite all of these problems, it is relatively common to use dissimilar materials in clinical electrodes, where other factors are of primary concern.

However, many researchers choose to use two electrodes of the same material to record bioelectric events. Ideally, if all other factors were the same, the potential difference between these electrodes would be zero. However, surface impurities normally preclude the electrode composition of such electrode pairs from being identical, and this leads to a small potential offset between the electrodes as shown in Table 2 [7]. While the magnitude of this offset can be small, large (ca. 100 mV) offsets can occur for some metals. One way to minimize these offsets is to immerse the shorted electrodes in an appropriate solution, e.g., 0.9% NaCl solution. A galvanic reaction occurs that minimizes the offset voltage and, after a suitable time period has passed, the electrode pair will have nearly the same potential [8]. However, in use, the electrodes are exposed to different levels and/or different impurities, again resulting in an increasing offset over time.

The ability of an electrode to return quickly to its equilibrium potential after passing current can be important for some recordings. For example, Witkowski and co-workers [9] have shown that a silver electrode is a poor selection compared with a silver–silver chloride electrode to record events following defibrillation. They placed the electrodes in a saline bath and simulated a defibrillatory waveform superimposed on a myocardial electrocardiogram by using a 20 mV triangular waveform mixed with a large defibrillatory shock (Fig. 3). The small triangular waveform was observed shortly after the defibrillatory shock when recorded by a silver–silver

Table 2 Fluctuations in Potential Between Electrodes

Electrode metal	Electrolyte type	Potential difference between electrodes
PbHg	$PbCl_2$ in chamois on human skin	0–600 μV (basal) 1.3–6.8 μV (Fluctuations)
Calomel	Saline	1–20 μV
Zn-ZnSO$_4$	Saline	180 μV
Zn	Saline	450 μV
Stainless Steel	Saline	10 mV
Zn	Saline	100 mV
ZnHg	Saline	82 mV
Ag	Saline	94 mV
AgHg	Saline	90 mV
Ag-AgCl	Saline	2.5 mV
Pb	Saline	1 mV
PbHg	Saline	1 mV
Pt	Saline	320 mV
Ag, AgCl sponge	ECG paste	0.2 mV 0.07 mV drift in 1 hr
Ag, AgCl (11-mm disk)	ECG paste	0.47 mV 1.88 mV drift in 1 hr
Pb (11-mm disk)	ECG paste	4.9 mV 3.70 mV drift in 1 hr
Zn, ZnCl$_2$ (11-mm disk)	ECG paste	15.3 mV 11.25 mV drift in 1 hr

Source: Ref. 7.

chloride electrode, whereas the silver electrode required a much longer time before the triangular waveform could be observed. Mayer et al. have measured the "electrode recovery potential" for a number a electrode materials [10].

There are two main factors that determine the time it takes an electrode to return to its equilibrium potential after current has passed through it. Which of these factors dominates depends on the electrode and the current density [11,12]. If a relatively small amount of current is passed, very little chemical transformation occurs and the time is determined by the impedance characteristics of the electrode–tissue interface. As was described previously, a voltage is developed at the tissue–electrode interface, but as this interface has physical dimensions, it also creates a resistance and a capacitance. Most cardiac stimulation and measurements can be understood with a model that places the voltage, resistance, and

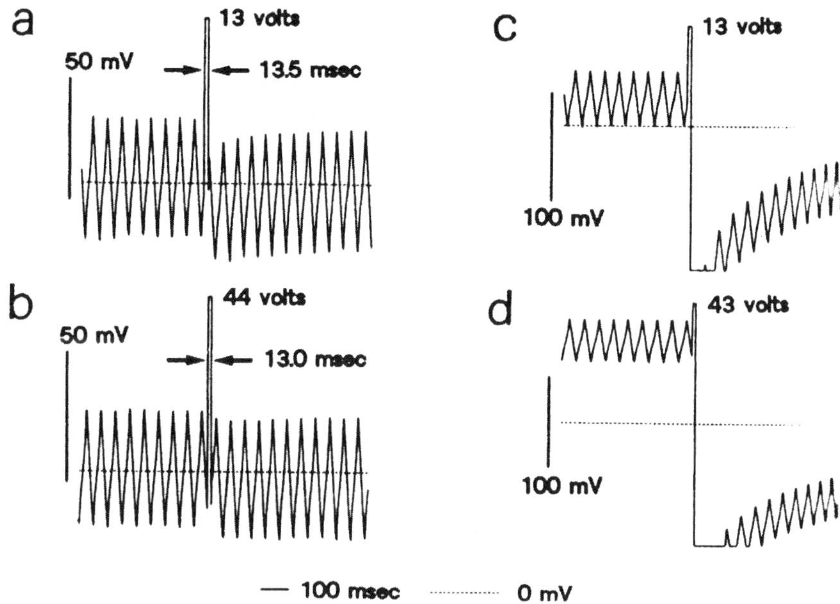

Figure 3 The effect of electrode composition on its ability to transduce electrical signals after current polarization. Results obtained by Witkowski et al. in vitro on two sets of electrodes: one pair of nonpolarizable Ag–AgCl electrodes and on set of polarizable Ag electrodes. The triangular waveform in the Ag–AgCl electrodes (panels a and b) is clearly visualized after the defibrillating shock, but the identical defibrillating field creates a much larger offset potential in the Ag electrodes (panel c) with the subsequent inability to visualize the known triangular waveform shortly thereafter. This behavior becomes more profound as the shock field is increased, as shown in panel d. (From Ref. 9.)

capacitance in series as shown in Fig. 4 (see below for more details). Therefore, when a small current is passed through this equivalent circuit, it returns to its initial state with an *RC* time constant.

If a relatively large current is passed, e.g., a defibrillatory pulse is applied, chemical transformations occur. These transformations can alter the electrode's potential. The return to the equilibrium potential after a large current requires that the electrode–tissue interface be restored to its initial condition. This requires the transport of pertinent material to and/ or away from the electrode–tissue interface. It is these transport processes which dictate the time it takes for the electrode to return to its equilibrium potential.

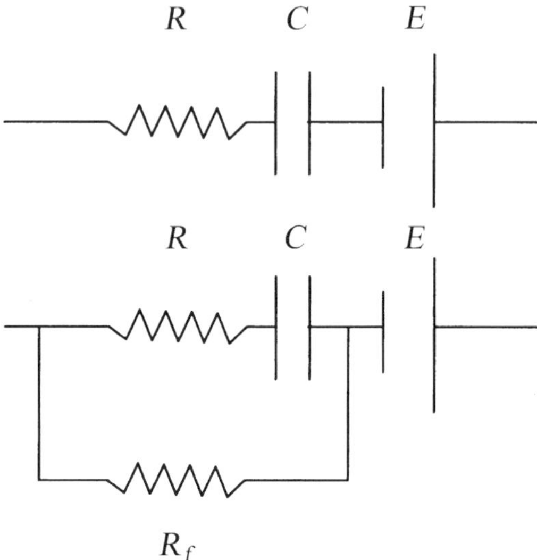

Figure 4 The series model of R, E, and C is often sufficient to model the electrode body interface for cardiology. For very low frequency work, the added resistor, R_f, gives better agreement with measured results.

The designer of an electrode must consider more than the material properties of the electrode. Variations in potential can occur as a result of the transduction process itself. Consider the net reaction occurring at a silver–silver chloride electrode, as shown in Eq. (4). In fact, two reactions are occurring simultaneously: silver chloride is being reduced, while silver is being oxidized. Under equilibrium conditions, no current flows through the circuit because both reactions occur at the same rate. Consequently, the potential remains constant. However, if either the forward or reverse reaction dominates, i.e., current flows through the circuit, changes in the concentration of chloride near the electrode occur, changing the potential of the electrode.

3. Potential Stability—Other Factors

A potential difference can also exist between identical electrodes in contact with different ionic constituents even without current flow. One source of variation involves the establishment of a junction potential at the interface between two solutions having different ionic constituents and/or concentrations. This can occur anywhere between the electrodes, even far from the electrodes. The differences in the ionic mobilities of each species, a transient junction potential, typically of the order of several millivolts, develops. For example, it is common to bath the heart in 0.9% saline (NaCl) solution while recording bioelectric events. Since the composition and concentration of the saline solution is not identical to that in the epicardium, a small potential difference develops at the interface where the saline solution meets the extracellular fluid. This potential difference is small (less than 1 mV) and dissipates as the saline solution mixes with the extracellular fluid.

For external use, bathing solutions are sometimes replaced with recording gels. However, dry electrodes can be used [13] for short-term recordings of stationary subjects (as during an experimental animal procedure). A drop of water or alcohol improves the signal fidelity. In fact, electrolyte jelly is not required to make most clinical ECG recordings. In an interesting study, Lewes replaced electrode gel in 4000 ECGs on humans with substances ranging from KY Jelly to mayonnaise and toothpaste [14]. All of the substances produced ECGs indistinguishable from the standard recordings.

However, whenever the subject is in motion, the shifting interface of the dry electrode generates a potential change which appears as a large motion artifact in the ECG recording. The most direct way of reducing this artifact in acute animal work is to puncture the skin [15,16]. Needles or alligators clips can be used to penetrate the outer layers of the skin. However, sandpaper [15] and pointed electrodes [16] have been successfully employed. These techniques have seen some limited clinical use where extreme conditions make the use of gels impractical.

For human work and chronic recordings where motion is a factor, a wet, or gel electrode is preferred. The contents of the gels varies tremendously [17,18]. However, many follow the AAMI standard [19]. In each case, the gel acts to reduce the motion artifact by creating a stable electrode–tissue interface (and therefore interface potential). As the patient moves, the gel must be present in sufficient quantity to shift with the motion, maintaining a constant environment for the electrode. For this reason, recessed electrodes are used (Fig. 5). Recessed electrodes for recording probably date to 1921 [20], when they were used on elephants. The most common current version uses a chlorided silver disk with a electrolyte-soaked sponge between the metal and the body. In this case, the sponge acts like the gel, shifting with the patient to maintain the electrode's environment.

There is a second source of motion artifact. The skin itself generates a potential across its various layers. As these layers are compressed or expanded, their potential shifts. Unfortunately, electrode gel does not affect this second source of motion artifact. Techniques which depend on abrading the skin or puncturing the skin do largely eliminate this source of noise.

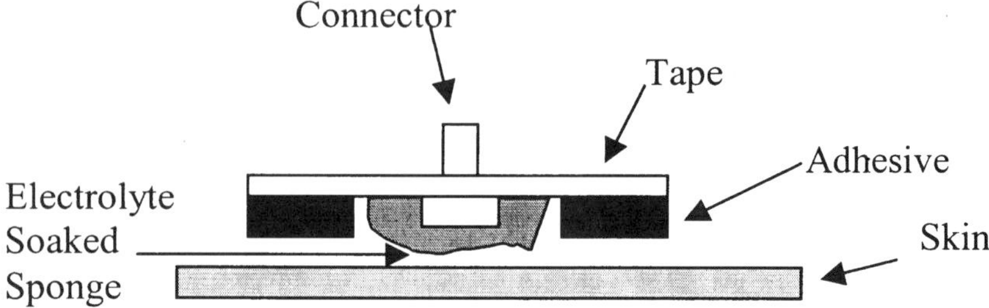

Figure 5   The most common electrode for clinical recordings is the recessed, disposable electrode. In this electrode, an electrolyte-soaked sponge acts as the body–electrode interface and reduces the effects of skin irritation and motion artifact.

When stimulating, either the electrodes must penetrate the skin or an electrode gel must be used, since the impedance of the dry skin–electrode interface would drop too much voltage for efficient stimulation and variations in skin–electrode impedance would cause areas of potentially damaging high current density. However, care must be used in selecting a gel for stimulation. Because the input impedance of the amplifiers is typically very large in modern ECG machines, the resistivity for solutions labeled "electrode paste" ranges tremendously [21]. All of these are adequate for recording, but the larger-resistivity gels are not suitable for stimulation.

III. ELECTRICAL MODELS OF THE ELECTRODE/TISSUE INTERFACE

A. Recording Electrodes

As was described previously, a voltage is developed at the tissue–electrode interface, but as this interface has physical dimensions, it also creates a resistance and a capacitance. In many cases, it is the capacitance and resistance which interfere, more than the voltage, in cardiac measurements and stimulation. Most cardiac stimulation and measurement applications can be understood with a model which places the voltage, resistance, and capacitance in series (Fig. 4) [22].

Unfortunately, just as the voltage developed at the tissue–electrode interface has many dependencies, so do the series resistance and capacitance. The values of the capacitance and resistance depend on many factors, including the material of the electrode and electrolyte, frequency, electrode area, and current density. In general, the capacitance drops with frequency. This is often captured as

$$C = Kf\alpha$$

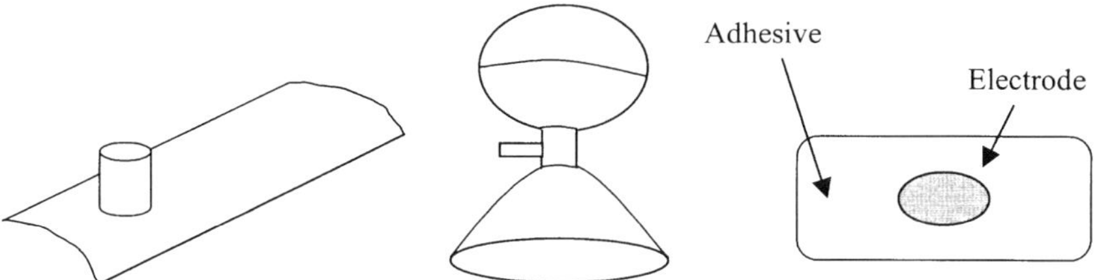

Figure 6 Clinical electrodes for recording the body surface electrocardiogram can be found in many forms. For short-term recordings of motionless subjects, the suction-cup electrode is both common and effective.

where C is the series capacitance; K is a factor which depends on the metal, electrolyte, electrode area, and temperature; f is frequency; and α varies from 0.2 to 0.8, but is generally close to 0.5 [23]. For example, a platinum black (platinum deposited on platinum) electrode in 0.9% NaCl gives $11759/f^{0.221}\,\mu\text{F}/\text{cm}^2$ of electrode area. This number is given in terms of electrode area because K depends on the electrode area. Steel in 0.9% NaCl yields $161/f^{0.525}\,\mu\text{F}/\text{cm}^2$ [23,24], but steel in muscle gives $179/f^{0.387}$. The latter change is due to the change in electrolyte, another key component of K.

The series resistance of the electrode–electrolyte interface is approximately equal to the reactance of the capacitance:

$$R = \frac{1}{2}\pi f C$$

where R is the series resistance [25]. However, this relationship is not exact [26]. For example, for stainless steel in 0.9% NaCl with an area of $0.157\,\text{cm}^2$ and a current density of $0.025\,\text{mA}/\text{cm}^2$, Geddes et al. reported that $R = 7269/f^{0.504}\,\Omega$ (the value depends on frequency because the capacitance depends on frequency) [26]. However, the reactance was equal to $6963/f^{0.489}\,\Omega$, or an equivalent capacitance of $25.2/f^{0.525}\,\mu\text{F}$—a reasonable but not exact match. In addition, using the reactance of the capacitance as the resistance is valid only for low current densities ($<1\,\text{mA}/\text{cm}^2$). There is a sharp drop in resistance at low frequencies for measurements using current densities beyond $1\,\text{mA}/\text{cm}^2$.

As the frequency approaches DC, in fact, the series RC approximation cannot be used. The model cannot pass DC, but electrodes can. Some investigators compensate for this weakness by adding a resistance across the RC, called the faradic leakage (Fig. 4). However, in cardiology, the DC behavior of the electrodes is rarely of interest.

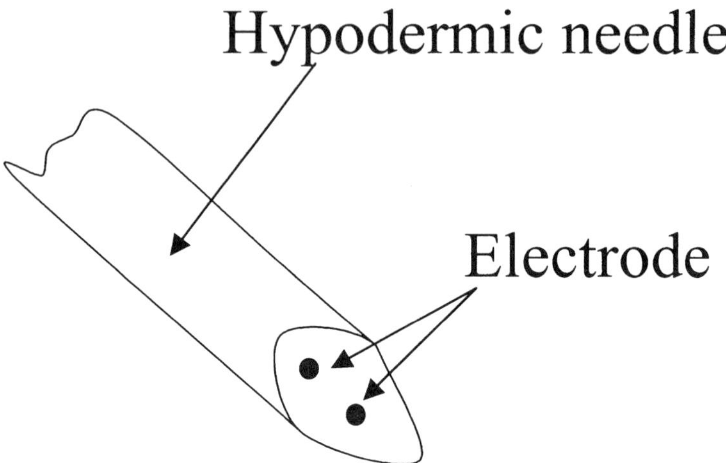

Figure 7   For experimental, subepicardial recordings, the plunge needle electrode has seen frequent use. The electrodes either protrude from the shaft or bevel of the hypodermic needle. Construction of these electrodes can be tedious.

B.   Stimulating Electrodes

Unlike the recording electrodes, the impedance of stimulating electrodes cannot be modeled as a series resistance and capacitance during cardiac stimulation. In fact, the number and variety of effects present during stimulation have prevented the creation of a single model useful for all cardiac stimulation applications. In general, individual models are used for each specialty area, e.g., defibrillation, ablation. In this section, we can only list some of the most common considerations.

For most stimulation applications, the stimulating current does not mirror the voltage. In other words, the body/electrode exhibits a strong reactive component. This cannot be attributed completely to the capacitive component of the stimulating electrode, since the total reactivity has a strong dependence on the size of the stimulating current. For example, a rectangular current pulse of 1 mA invokes an apparent impedance (ratio of voltage to current) of about $500\,\Omega$ across the heart, while a current of 1 A invokes an apparent impedance of about $50\,\Omega$ across the same electrodes [27]. In addition, these values differ depending on whether a sinusoidal, rectangular, or other waveform is used. Thus, the body/electrode presents a highly nonlinear impedance to stimulation. Linear assumptions are nevertheless used, but must be confined to a single application under limited conditions.

Large stimulating currents can affect not only the electrode (as described above), but the surrounding tissues. For example, a defibrillation pulse of 1 A can evolve bubbles [28], which may be damaging. Also, strong stimuli can break down the cells [29] or cause excessive heating [30] either of which can cause necrosis. Both electroporation [29] and heating [30] may be therapeutic or toxic. Successive stimuli of the same strength may encounter different tissue impedances, due to several possible mechanisms [31]. Each stimulation protocol, and even each stimulus within the protocol, can be assumed to encounter unique chemical and physiological processes.

IV. BIOCOMPATIBILITY

Biocompatibility is the study of the body's acceptance of the electrode, both by the contacting tissue and by the body as a whole. Any foreign substances placed in the body will provoke a response. The response can be inflammatory, allergic, immunological, or even carcinogenic. In addition, the body may attack the material, causing mechanical degradation or electrical isolation of an electrode. Stimulation complicates any prediction of the body's response, since the stimulation current can promote or inhibit certain responses. Usually, the primary issue for electrocardiology is whether the body's response interferes with the intended operation of the electrode. Fortunately, with the proper selection of materials and careful use of the electrodes, biocompatibility interference can be kept to a minimum.

Most of our understanding of the selection of recording electrode materials for biocompatibility comes from empirical studies. Fischer et al. compared silver, copper, and stainless steel, showing that the necrotic and edematous region around silver and copper ranged between 1.5 and 7 mm [32]. Only stainless steel offered a minimal amount of tissue response. Collias and Maneulidis described the body's reaction to stainless steel as acute (within the first 24 hr) [33]. Within 2 weeks, the necrotic region seen acutely was revascularized. After 4 months, a thick capsule completely surrounded the electrode. Robinson and Johnson compared gold, platinum, silver, stainless steel, tantalum, and tungsten in cat brains [34]. Gold and stainless steel provoked the least response, followed by tantalum, platinum, and tungsten. Silver provoked a vigorous response, though all were encapsulated after 2 weeks. For most applications, encapsulation, if not excessive, does not interfere with the operation of the recording electrode, since the conductivity of the encapsulation material does not vary significantly from other body tissues. Longer-term studies must also consider chemical corrosion of the material [35].

Unlike recording electrodes, milliampers of current or more can regularly pass through stimulating electrodes. Such large current densities cause a net chemical reaction at the electrode–electrolyte interface. This chemical reaction can release substances from the surface of the electrode, provoking a response from an otherwise biocompatible material. Also, the reaction can deposit substances from the body onto the electrode or change the chemistry of the electrode. Both deposited substances and the altered electrode chemistry may evoke a different response from the body than the original material. For all of these reasons, stimulating electrodes are typically constructed of noble metals, such as platinum and titanium. Stainless steel, which has many of the properties of the noble metals, is also used for stimulation.

Stainless steels used for implantation are recommended by the American Society of Materials Testing (1992, F139-86, p 61) to be of type 316L, which means the steel must contain iron with 0.03% carbon, 2% manganese, 2–4% molybdenum, 12–14% nickel, and 17–20% chromium. This is also called austenitic stainless steel. Titanium is also a common electrode. It provokes as little response as stainless steel, but has approximately one-half the density. Neither show significant impact on recording or stimulating from corrosion, but titanium is superior to stainless steel for long-term implantation. The most common titanium alloy for implantation is Ti_6Al_4V, which contains titanium, 6% aluminum, and 4% vanadium. Platinum is extremely resistant to corrosion, but has poor mechanical properties. Nevertheless, for electrodes which will be implanted for a very long time, such as pacemaker tips, platinum is an excellent and common choice. Another is a platinum/iridium alloy containing 10% iridium.

For external recordings the biocompatibility issues are typically minimized due to the short duration of the application. Stainless steel, copper, silver, and other materials, including silver–silver chloride, can be used as an effective external electrode with little short-term response from the body. However, when stimulating, an electrode gel is required, and the gel can be an irritant. In addition, for chronic use, the gel can provoke allergic reactions, erythema, and discolorations [36]. A more common problem with external electrodes arises when a gel is used for recording. The gel itself can produce a response in the skin which appears as a signal in the ECG. One component of this response is the galvanic skin response [37]. Certain gels, such as those containing calcium chloride, increase the galvanic skin response three fold, which can distort the ECG, appearing most likely as a baseline wander.

V. COMMON ELECTRODES FOR RECORDING

Hundreds of different electrode types have been developed. Most of these are used only in specialized applications. In this section, we will review the most common types. Many are based on the Ag/AgCl electrode presented previously because it is approximately nonpolarizable, it is relatively easy to make, it is largely nontoxic (external use), and it has fairly low noise characteristics. For example, a simple process for making Ag–AgCl electrodes [38] is to chloridize a pure silver wire (anode) using 1.5 V in series with a few hundred ohms for 30 min (another, larger Ag wire is the cathode). Other construction methods have also been suggested [39].

A. Electrocardiogram

Probably the most common electrical recording of the heart is the electrocardiogram. It is not surprising that elegant electrodes have been developed for the most common recording applications. For short-term veterinary research electrocardiograms, an excellent electrode choice is stainless steel needles penetrating the skin. A similar effect can be achieved with stainless steel alligator clips applied directly to the skin. The sharp teeth of the alligator clips penetrate the skin. Wetting of the skin near the clips, such as with alcohol or water, also reduces the impedance and artifact generation [13]. Impedances of 10–100 Ω have been reported for dry electrodes and subjects. However, even well-controlled studies demonstrate tremendous variability [40], including impedances up to megaohms. All reports indicate that the impedance falls with frequencies above 100 Hz, dropping as much as eightfold up to 500 Hz, a typical upper limit for ECGs [41]. Despite the variability and relatively high values of the impedances, since modern amplifier impedances are routinely $10^{12}\,\Omega$ or higher, either set of dry electrodes—needles or clips—are quite common and deliver excellent performance where little motion is encountered.

Clinical electrocardiogram recording electrodes come in many forms (Fig. 6). One simple electrode is a variation of the dry metal electrode described for veterinary use. These can be in the form of plates, circles, or suction cups, which combine both electrical and mechanical connection in one application. When the subject is motionless, any of these can be used without gels. However, to reduce motion artifact, the dry plate electrodes can also be used with electrode recording gel.

A popular electrode for clinical use is the recessed, disposable electrode (Fig. 5). This device uses a Ag–AgCl button in contact with a sponge which is soaked in an electrolyte. A plastic cover is removed which reveals the sponge and an adhesive surrounding structure that attaches to the skin. Disposable, recessed electrodes come in a large variety of shapes and sizes, including pediatric, flexible, and radiotransparent [42] variants.

Internal (chronic) electrogram electrodes for veterinary use that will not be used for stimulation are often stainless steel. For research purposes, stranded or coiled stainless steel wire insulated to the within about 1 mm of the tip with polyvinyl chloride or polyethene is adequate [43]. It is relatively rare to see chronic recording electrodes in clinical use which are not also associated with stimulation. (Electrodes used for both stimulation and recording are discussed below.) A notable exception is the internal Holter "loop recording" monitor [44]. These are implanted ECG recorders for the diagnosis of syncope and do not deliver stimuli.

B. Other Noninvasive Electrodes

The standard ECG gives only a limited view of the heart. A more complete picture would outline the complete activation pattern of the heart. There have been many attempts to derive this activation pattern from body surface recordings—the so-called inverse problem of cardiology [45]. In some cases, solving the inverse problem has motivated the development of new electrodes configurations [46].

Several groups [47] have developed vests which carry 100–200 electrodes. Each electrode is typically of the active, dry type [47]. Because they offer a more complete view of the electrical activity on the body surface, they are more suited to calculation of the heart surface electrical activity [48]. This approach has seen some clinical use [48], but has not been widely accepted.

Another electrode configuration which promises to improve the quality of solutions to the inverse problem is the Laplacian electrode [49]. In this configuration, three standard ECG electrodes, placed in close proximity, are electronically combined to approximate the surface Laplacian. Alternatively, a concentric ring electrode can give the Laplacian [50]. Theoretically [49], this electrode configuration should offer a more penetrating view of the electrical activity at a distance, such as at the heart's surface. However, the Laplacian electrode has also seen only limited clinical use.

One noninvasive electrode that deserves special note is the esophageal pill electrode [51]. This is a small, bipolar electrode approximately the size of a pill, with a lead which is externalized through the mouth. The patient swallows the pill. Then the clinician advances the wire until the electrodes are adjacent to the atrium. The proximity of the esophagus to the atrium is responsible for the excellent atrial electrogram recordings possible with the pill electrode. Atrial stimulation has also been reported [52]. The pill recording electrode has clinical utility, particularly in situations where the ECG does not clearly indicate the P waves or for recording the pathology of the posterior heart.

C. Epicardial and Transmural Electrodes

Just as there is a large variety of body surface recording electrodes, many investigators have chosen to construct custom electrode arrays for epicardial and transmural recordings. These epicardial electrode arrays generally fall into two categories. Electrodes are either placed on a flexible superstructure (electrode socks), which can conform to nonplanar surfaces but are limited to low spatial densities [53], or offer high spatial densities but restrict the electrodes to a single plane (electrode plaques) [54,55]. Hybrid designs have used small plaques affixed to the flexible structure [56]. Electrodes as small as 50 μm in diameter, spaced 200 μm apart on rigid plaques, have been used [54]. However, there is a constant trend toward larger arrays with ever finer-spaced electrodes, presumably to enhance our understanding of the fine structure of arrhythmias.

One of the limits on the spacing of electrodes may be that the spacing between the electrodes must be much larger than the electrodes themselves [57]. Electrode arrays violating this paradigm, it is conjectured, will interface with conduction [58] and will produce signals that can not be analyzed [57]. However, recent theoretical work challenges this view [59]. Using a finite-element approach, Eason and Malkin found that high-density arrays do not significantly affect the recorded activation times. Center-to-center spacings as little as two times the electrode width exhibit reduced signal strength, but no significant change in the timings of the signals. As newer electrode construction techniques become available, such as inexpensive photolithographic approaches [55], experimental challenges are sure to follow.

Steel, titatnium [60], platinum, and silver [61] have seen use for recording electrode arrays. The electrodes are prepared and placed dry, or moistened in saline, onto the surface of the heart. Though the heart is often beating vigorously, since the skin potential is absent, and the myocardium provides a continuously moist electrode–electrolyte interface, motion artifact is negligible. When the electrodes will be exposed to large potentials, which could lead to relatively large currents through the electrodes (such as during defibrillation), and recordings are required soon after the current pulse, then the only material which has been successfully used is silver–silver chloride [9], owing to is superior approximation to a nonpolarizable electrode.

For transmural recordings, needle electrodes are common. In this approach, standard hypodermic needles are modified to allow several electrodes to penetrate the shaft of the needle (Fig. 7). The modified needles are plunged directly into the myocardium. More recently, plunge electrodes made from metal films on polymers have been used [62] (Fig. 8). These are produced photolithographically with a very repeatable electrode geometry.

D. Endocardial

The transmural recording electrodes described in the previous paragraph are sometimes used to obtain endocardial recordings in experimental

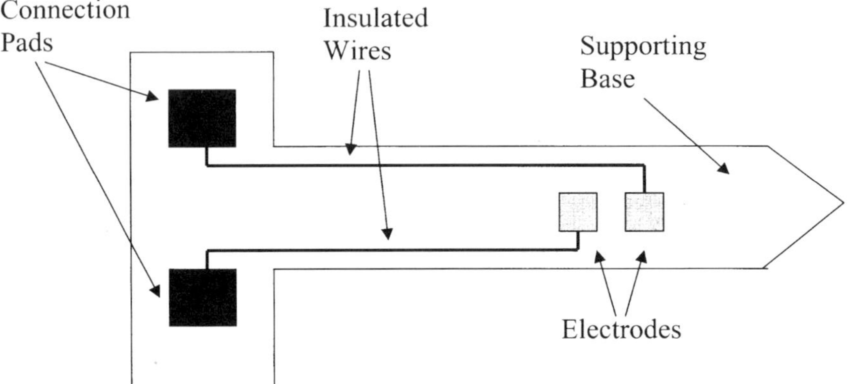

Figure 8 Plunge needles can be replaced with mass-produced electrodes. Photolithographic techniques are used to define the shape and position of the electrodes. Both thin and thick-film techniques have been used. In this example, a thick-film electrode is suggested with two small electrodes on a planar needle.

preparations. However, this is not an acceptable clinical approach. For clinical use, endocardial recordings are usually gathered from catheter-borne, stainless steel or platinum electrodes.

One such catheter has an olive-shaped tip. The olive was introduced by Taccardi in 1987 [63]. This multielectrode catheter tip is made from an epoxy resin with dozens or hundreds of, typically, Ag or AgCl recording wires exposed on the surface. Once inserted, generally into the right ventricle, the electrodes record the spacial distribution of the potentials in the cavity. However, since the electrodes do not necessarily contact the myocardium, the potentials in the muscle must be derived from the recorded signals, a problem-plagued step [64]. This problem is actually a variation of the inverse problem of cardiography, mentioned above. Tremendous strides have been made in the solution of this inverse problem, which may lead to widespread clinical use [65].

A variation on the olive is the expandable or basket catheter [66]. Once inserted into the ventricle, the basket can be expanded to contact the walls of the endocardium. Since the electrodes contact the endocardium, baskets of this type bypass the problem of having to calculate the endocardial potentials. Baskets have seen frequent clinical use, though primarily for research purposes.

Far more common than the basket for clinical recording from the endocardium is the multielectrode catheter. The multielectrode catheter is essentially a tube with 8, 10, or more, often stainless steel or platinum, rings spaced several millimeters apart starting from the distal end of the catheter. The catheter is advanced into, typically, the right ventricle. In some cases, the electrodes are placed under fluoroscopy in the desired location. In some varieties of this device, the distal end of the catheter can be flexed—or steered—for precise movement of the electrodes. Once in place, the clinician can record the electrogram from multiple sites along the catheter, and therefore in the heart, simultaneously. One of the reasons for the popularity of this approach is that, in many cases, therapy can be delivered from the same catheter, including pacing and ablation.

However, one major disadvantage of the multielectrode approach is that the catheter must be moved in order to view the heart's activity at another site that is not along the catheter. If the arrhythmia under consideration is regular and repeatable, then theoretically, recordings from several sites can be spatially aligned by knowing the relative location of the catheter for each recording. Since the rhythm is regular, the signals can be temporally aligned using one fixed electrode, located on the body surface or the coronary sinus, for example. However, determining the relative location of two such sites in the heart has proven to be a difficult problem [67]. Biplane fluoroscopy, a standard approach, gives only a limited view of the three-dimensional nature of the problem. More elaborate approaches have been proposed, but only recently have techniques based on technologies such as magnetics, impedance, or ultrasound become commercially available.

A special category of recording electrodes is the MAP (monophasic action potential) recording electrodes [68]. These are frequently used on the endocardium, but are also used on the epicardium. Unlike typical epicardial electrodes, MAP electrodes have very-small diameter tips which are brought in close contact with the cells, typically by suction (Fig. 9). MAP recordings

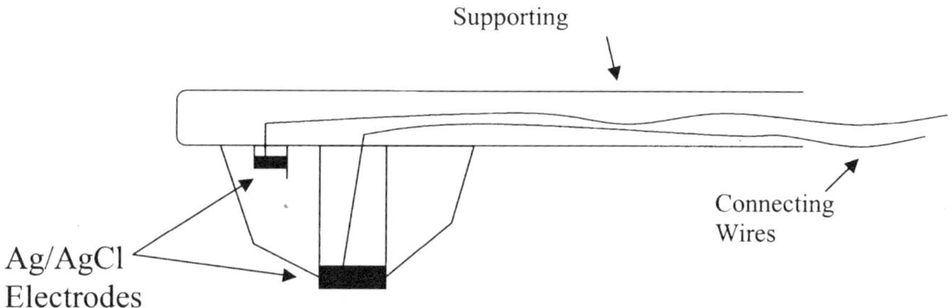

Figure 9 Monophasic action potential (MAP) electrodes puts one small electrode in direct contact with the tissue, either using suction or pressure. The result is an approximation to a recording of the transmembrane potential.

are unique because they give a good approximation to the action potential of the cell. Normally, action potentials can only be recorded by penetrating the cell, which has limited experimental and clinical application on a beating heart. However, the MAP electrode is rapidly applied and reliably used, both clinically and experimentally, without penetrating the cell. In experimental preparations, arrays of MAP electrodes have also proven useful [69].

E. Intracellular Electrodes

MAP electrodes cannot make precise measurements of the action potential. So, sometimes it is desirable to measure the electrical activity across the membrane of an individual cell. When such measurements are made, the tip of a miniature electrode whose size is small compared to the size of the cell, e.g., micrometers, is placed inside the cell while another electrode (a reference electrode) is placed outside the cell.

Two types of intracellular electrodes are used: metallic and electrolyte-filled. Metallic electrodes consist of a metal or carbon fiber sheathed in an insulating material, typically glass or quartz. Several authors have described procedures for constructing these electrodes [23,70]. The transduction process for these electrodes is identical to that described earlier in this chapter, and they are subject to the same stability considerations described. Three potentials exist in this measurement scheme: the potential of the metallic electrode inserted inside the cell, the potential of the reference electrode positioned outside the cell (any associated junction potential if applicable), and the potential across the cell membrane. For situations where the measuring electrode and reference electrode are of the same material, the first two are assumed to be constant (and are often assumed to cancel), allowing the measurement of the cell membrane potential.

In contrast, electrolyte-filled intracellular electrodes are typically glass pulled pipettes that are filled with an electrolyte, e.g., KCl solution. Several methods exist for their construction [70], but typically a glass tube is heated while being drawn during the softening of the glass. This creates two tapered glass pipettes that are broken apart and filled with an electrolyte solution and into which a metallic electrode, e.g., a Ag/AgCl electrode, is placed. In this case, the transduction process for the metallic electrode inserted into the pulled pipette is again identical to that described earlier in the chapter. However, these electrodes possess much greater impedance than metallic intracellular electrodes. Further, in addition to measuring the potential difference between the inner metallic electrode, the membrane potential, and the external reference electrode, two additional potentials are encountered: a junction potential at the tip of the pipette and what is termed the tip potential. Since it is common to fill the pipette electrode with an electrolyte solution that is different from the intracellular fluid, a transient junction potential exists at the tip due to differences in the ionic concentration and mobilities of species within the pipette and in the cell. The potential dissipates with time as the intracellular and pipette fluids mix. In addition, a tip potential exists because the thin glass wall at the tip of the pipette behaves like a glass membrane, developing a potential that responds in a fashion similar to a glass pH electrode. This potential does not dissipate, but can be relatively small [70].

VI. COMMON ELECTRODES FOR STIMULATING

There are currently only three common applications for stimulating electrodes in cardiology: pacing, ablation, and defibrillation/cardioversion. Nevertheless, these applications have had a tremendous impact on the field of cardiology.

A. Pacing

Probably the most commonly used cardiac stimulating electrode is the catheter-borne pacing electrode (Fig. 10). Several electrodes are mounted on a 1–2-mm-diameter catheter which is threaded through a vein into the right ventricle. In most applications, the distal tip of the catheter is the cathode and the anode is either the titanium case of the pacemaker (monopolar pacing) or a second electrode on the catheter (bipolar pacing).

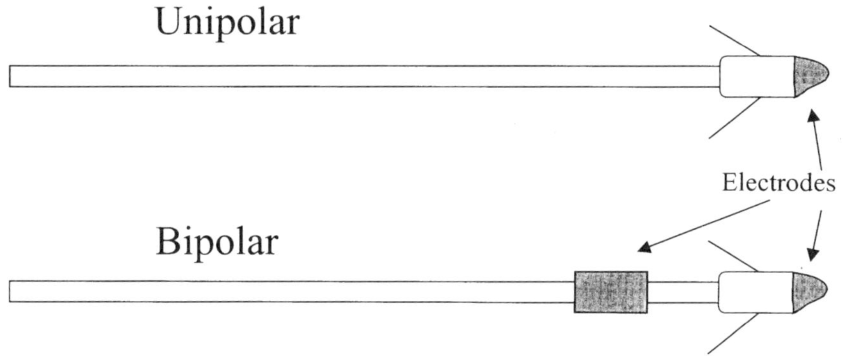

Figure 10 The most widely used electrode is probably the catheter-borne pacing electrode. In most cases, two electrodes are carried on a long flexible wire approximately 2–3 mm in diameter.

The insulating material, while not strictly part of the electrode, is often the source of failure of a chronically implanted electrode. Very few materials are used for the insulation. Silicone now accounts for nearly 100% of the insulation materials used in pacemaker leads. This is partially due to some well-publicized insulation failures and partially due to the fact that the approval of new materials for chronic use in the body can be costly and time consuming

There are also very few materials used for the pacing tip itself, due to the demanding nature of the application. Unlike the pacemaker, the pacing catheter is expected to last for more than five years. The need for a very low pacing threshold (to conserve battery power) and the demand for a long life has lead pacing electrodes to be made of titanium, platinum, or platinum-iridium (Fig. 11). Activated surfaces, such as platinum black, have also been used [71,72]. Temporary internal pacing leads can also be made of stainless steel, since they are intended for short-term use.

Typical impedances between electrodes for pacing are 200–1000 Ω. However, this changes as the body encapsulates the electrodes. It typically drops during the first week post-implant, then increases with time to a value about 10–20% below the impedance at implant.

One special case to consider is when the same electrode will be used for both recording and stimulating. This arises most frequently when the pacemaker is intended to deliver stimuli only when the heart fails (demand pacing). Furthermore, since most pacing leads have two electrodes at the distal end, it arises only when bipolar pacing is used. Under these conditions, the stimulating current leaves a charge at the electrode–electrolyte interface, which develops a potential, typically so large that no signal can be sensed using that electrode. One solution is to remove this charge with a

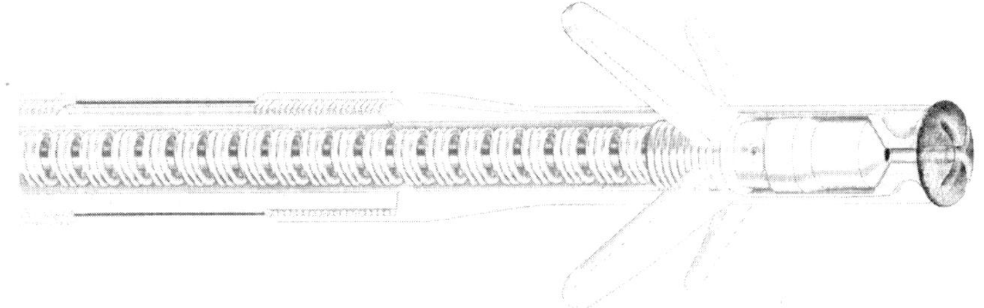

Figure 11 The tip of the pacing electrode is often made of platinum-iridium. The shape of the tip may be constructed to improve the chances that the tip will become lodged in the endocardium, as shown here with small tines sticking out of the tips. Active fixation, with an auger tip, is also used. (Image courtesy of St. Jude Medical.)

longer, lower-strength stimulus of the opposite polarity and equal charge. This is called charge dumping or interface depolarization [73].

The vast majority of pacing is delivered using transvenous catheters, but there are several other configurations. Transcutaneous pacing [74] is used where emergency pacing is required. Unfortunately, transcutaneous pacing can be painful and ineffective. Transesophageal pacing [75] offers the advantages of lower thresholds and usually less skeletal muscle contractions, one source of patient pain. In this configuration, one electrode is advanced into the esophagus to the level of sphincter or into the stomach. A second electrode is placed on the thorax.

B. Ablation

Ablation is the technique whereby large amounts of energy are dissipated through cardiac tissue to cause necrosis, probably by heating. Intentional, focused necroses can be therapeutic, if the targeted tissue was arrhythmogenic. The energy is most often delivered in bursts of microwave current lasting a few seconds. The catheter-borne ablation electrode is advanced into the heart under fluoroscopic guidance. As with endocardial recording electrodes, locating the ablation electrode is still problematic. Not only must the electrode be located with respect to anatomical features, and with respect to its previous locations, but it must also be located with respect to the target tissue. Typically, the target tissue is identified using endocardial electrogram recordings, from the same or a separate catheter. So, the ablation catheter must also be located with respect to the other catheters. Ablation catheters can be stainless steel, titanium, or platinum. The ablation electrode may be coupled with thermocouples to monitor the heating near the electrode. More details about ablation and ablation catheters can be found in later chapters in this book.

C. Defibrillation

Defibrillation is the process of delivering a very large amount of stimulating current to terminate ventricular fibrillation. As much as 4000 V may be delivered to the patient for 5 or 10 msec during external defibrillation. There are three types of defibrillators, acute external, acute internal, and implanted. Acute external defibrillators are now found in police cars and even airplanes. They generally use stainless steel electrodes with a large surface area and a highly conductive electrode gel. Unlike the case of the ECG, the gel must be carefully selected, since it forms the flexible, conductive interface between the body and the rigid steel electrodes. The gel must also be carefully applied, since it can form a low-resistance pathway between the electrodes, shunting current away from the heart.

Figure 12 Defibrillation catheters carry one or two electrodes of a relatively large surface area. Typically these electrodes are implanted such that one of them is in the right ventricular apex and the other is in the superior vena cava. This illustration shows the distal electrode, with the colors altered to highlight the internal structure. (Image courtesy of St. Jude Medical.)

Spoon-shaped, typically stainless steel, paddles are also used for internal acute defibrillation, such as in the operating room. Voltages below 1000 V are required for internal defibrillation. However, since the heart is bathed in a conducting solution, no interface gel is needed.

For chronic use, a transvenous defibrillating electrode is most common. In this configuration a coiled electrode (2.5–3.5 mm in diameter, 3–5 cm in length) made of titanium or platinum is advanced into the right ventricle (Fig. 12). The insulating materials are typically the same as used in pacing, specifically silicone and polyurethane. Overall catheter diameters are continuously decreasing, but are currently about 2 mm. In most cases, the titanium case of the implanted defibrillator is the other electrode. Either electrode may act as the anode. However, biphasic stimulation, where the polarity is reversed during stimulation, is now nearly universal, because of its increased efficacy.

Impedances vary considerably across defibrillation applications. External defibrillation typically sees about 500 Ω, while internal defibrillation sees 30–60 Ω. Interestingly, this impedance drops during the course of an experimental study, and drops markedly after the first defibrillation shock [31]. Most investigators do not report a significant change in impedance over the life of the implanted electrodes.

REFERENCES

1. AJ Bard, LR Faulkner. Electrochemical Methods: Fundamentals and Applications. New York: Wiley, 1980, pp. 488–549.
2. RC Weast, ed. CRC Handbook of Chemistry and Physics. 63rd ed. Boca Raton, FL: CRC Press, 1982, pp. D162–D167.
3. P Kovacs, JA Davidson. Chemical and electrochemical aspects of the biocompatibility of titanium and its alloys. In: SA Brown and JE Lemons, eds. Medical Applications of Titanium and Its Alloys: The Material and Biological Issues. West Conshohocken, PA: ASTM, 1996, pp. 163–177.
4. P Kovacs. Electrochemical techniques for studying the corrosion behavior of metallic implant materials. Proc Corrosion/92 Symp. Techniques for Corrosion Measurement, Houston, TX, 1992, pp. 5-1 to 5-14.

5. GJ Janz. Silver-silver halide electrodes. In: DJG Ives, GJ Janz, eds. Reference Electrodes: Theory and Practice. New York: Academic Press, 1961, pp. 179–230.

6. GJ Janz, H Taniguchi. The silver-silver halide electrodes: preparation, stability, reproducibility and standard potentials in aqueous and nonaqueous media. Chem Rev 53:397–437, 1953.

7. LA Geddes. Electrodes and the Measurement of Bioelectric Events. New York: Wiley Interscience, 1972, chap. 8.

8. LA Geddes. Electrodes and the Measurement of Bioelectric Events. New York: Wiley Interscience, 1972, p. 364.

9. FX Witkowski, PA Penkoske, R Plonsey. Mechanism of cardiac defibrillation in open-chest dogs with unipolar DC-coupled simultaneous activation and shock potential recordings. Circulation 82(1):244–260, 1990.

10. S Mayer, LA Geddes, JD Bourland, L Ogborn. Electrode recovery potential. Ann Biomed Eng 20:385–394, 1992.

11. KJ Vetter. Electrochemical Kinetics: Theoretical and Expermental Aspects. New York: Academic Press, 1967.

12. AJ Bard, LR Faulkner. Electrochemical Methods: Fundamentals and Applications. New York: Wiley, 1980, pp. 249–277.

13. JD Bourland, LA Geddes, G Sewell, R Baker, J Kruer. Active cables for use with dry electrodes for electrocardiography. J. Electrocardiol 11(1):71–74, 1978.

14. D Lewes. Electrode jelly in electrocardiography. Br Heart J 27:105–115, 1965.

15. HW Tam, JG Webster. Minimizing electrode motion artifact by skin abrasion. IEEE Trans Biomed Eng 24(2):134–139, 1977.

16. D Lewes. Multipoint electrocardiography without skin preparation. Lancet 1965.

17. MM Asa, AH Crews, EL Rothfield, ES Lewis, IR Zucker, A Berstein. High fidelity radioelectrocardiography. Am J Cardiol 14:530–532, 1964.

18. FW Fascenelli, C Cordova, DG Simons, J Johnson, L Pratt, LE Lamb. Biomedical monitoring during dynamic stress testing. Aerosp Med 37:911–922, 1966.

19. Disposable ECG Electrodes, ANSI/AAMI EC12:1991. In: Monitoring and Diagnostic Equipment. Arlington, VA: AAMI, 1995.

20. A Forbes, S Cobb, M Cattell. An electrocardiogram and an electromyogram in an elephant. Am J Physiol 55:385–389, 1921.

21. A Searle, LA Kirkup. Direct comparison of wet, dry and insulating bioelectric recording electrodes. Physical Meas 21(2):271–283, 2000.

22. E Warburg. Uber die Polarizationscapacitat des Platins. Ann Phys (Leipzig) (Ser. 4) 6:125–135, 1901.

23. LA Geddes. Electrodes and the Measurement of Bioelectric Events. New York: Wiley Interscience, 1972, chap. 4.

24. T Ragheb, LA Geddes. The polarization impedance of common electrode metals operated at low current density. Ann Biomed Eng 19:151–163, 1991.

25. LA Geddes. Measurement of electrolytic resistivity and electrode-electrolyte impedance with a variable-length conductivity cell. Chem Instrum 4:147–168, 1973.

26. LA Geddes, CP Da Costa, G Wise. The impedance of stainless-steel electrodes. Med Biol Eng 9(5):511–521, 1971.

27. LA Geddes, W A Tacker, W Schoenlein, M Minton, S Grubbs, P Wilcox. The prediction of the impedance of the thorax to defibrillating current. Med Instrum 10(3):159–162, 1976.

28. R Pendekanti, C Henriquez, G Tomassoni, W Miner, E Fain, D Hoffmann D, P Wolf. Surface coverage effects on defibrillation impedance for transvenous electrodes. Ann Biomed Eng 25(4):739–746, 1997.

29. JL Jones, CC Proskauer, WK Paull, E Lepeschkin, RE Jones. Ultrastructural injury to chick myocardial cells in vitro following "electric countershock." Circ Res 46(3):387–397, 1980.

30. Z Gu, CM Rappaport, PJ Wang, BA VanderBrink. A 2 1/4-turn spiral antenna for catheter cardiac ablation. IEEE Trans Biomed Eng 46(12):1480–1482, 1999.

31. SJ Sirna, RA Kieso, KJ Fox-Eastham, J Seabold, F Charbonnier, RE Kerber. Mechanisms responsible for decline in transthoracic impedance after DC shocks. Am J Physiol 257:H1180–H1183, 1989.

32. G Fischer, GP Sayre, RG Bickford. Histologic changes in the cat's brain after introduction of metallic and plastic coated wire used in electro-encephalography. Proc Staff Meet Mayo Clin 32:14–22, 1957.

33. JC Collias, EE Manuelidis. Histopathological changes produced by implanted electrodes in cat brains. J Neurosurg 14:302–328, 1957.

34. FR Robinson, MT Johnson. Histopathological studies of tissue reactions to various metals planted in cat brains. ASD Tech Rept 61-397, Wright-Patterson AFB, Ohio: USAF, 1961.

35. J Black. Biological Performance of Materials. 2nd ed. New York: Dekker, 1992.

36. RJ Cochran, T Rosen. Contact dermatitis caused by ECG electrode paste. S Med J 73(12):1667–1668, Dec 1980.

37. R Edelberg, NR Burch. Skin resistance and galvanic skin response. Arch Gen Psych 7:163–169, 1962.

38. GJ Janz, DJC Ives. Silver silver-chloride electrodes. Ann NY Acad Sci 148:210–221, 1968.

39. MR Neuman. Bipotential electrodes. In: JG Webster, ed. Medical Instrumentation Application and Design. 3rd ed. New York: Wiley, 1998, pp. 191–193.

40. OH Schmitt, M Okajima, M Blaug. Skin preparation and electrocardiographic lead impedance. Dig IRE Int Conf Med Electron, 1961.

41. Geddes, Baker, Principles of Applied Biomedical Instrumentation, 3rd ed. New York: Wiley, 1989, pp. 693–702.

42. HJ Marriott, HT Castillo, F LaCamera, EJ Swanick. Improved ECG monitoring during cardiac catheterization using radiotransparent electrodes and chest leads. J Electrocardiol 10(2):119–122, 1977.

43. CF Opitz, GF Mitchell, MA Pfeffer, JM Pfeffer. Arrhythmias and death after coronary artery occlusion in the rat: continuous telemetric ECG monitoring in conscious, untethered rats. Circulation 92(2):253–261, 1995.

44. SR Cumbee, RE Pryor, M Linzer. Cardiac loop ECG recording: a new non-invasive diagnostic test in recurrent syncope. S Med J 83(1):39–43, 1990.

45. F Greensite, G Huiskamp. An improved method for estimating epicardial potentials from the body surface. IEEE Trans Biomed Eng 45(1):98–104, 1998.

46. RL Lux, RS MacLeod, M Fuller, LS Green, F Kornreich F. Estimating ECG distributions from small numbers of leads. J Electrocardiol 28(suppl):92–98, 1995.

47. J Leibman, CW Thomas, Y Rudy, R Plonsey. Electrocardiographic body surface potential maps of the QRS of normal children. J Electrocardiol 14(3):249–260, 1981.

48. LE Widman, J Liebman, C Thomas, R Fraenkel, Y Rudy. Electrocardiographic body surface potential maps of the QRS and T of normal young men. Qualitative description and selected quantifications. J Electrocardiol 21(2):121–136, 1988.

49. DB Geselowitz, JE Ferrara. Is accurate recording of the ECG surface Laplacian feasible? IEEE Trans Biomed Eng 46(4):377–381, 1999.

50. PR Johnson, D Kilpatrick. An asymptotic estimate for the effective radius of a concentric bipolar electrode. Math Biosci 161(1–2):65–82, 1999.

51. R Arzbaecher. A pill electrode for the study of cardiac arrhythmia. Med Instrum 12(5):277–281, 1978.

52. JM Jenkins, M Dick, S Collins, W O'Neill, RM Campbell, DJ Wilber. Use of the pill electrode for transesophageal atrial pacing. Pacing Clin Electrophysiol 8(4):512–527, 1985.

53. L Harrison, RE Ideker, WM Smith, GJ Klein, AG Wallace, JJ Gallagher. The sock electrode array: a tool for determining global epicardial activation during unstable arrhythmias. Pacing Clin Electrophysiol 3(5):531–540, 1980.

54. EE Johnson, SF Idriss, C Cabo, SB Melnick, WM Smith, RE Ideker. Evidence that organization increases during the first minute of ventricular fibrillation in pigs mapped with closely spaced electrodes (abstr). J Am Coll Cardiol 19:90A, 1992.

55. RA Malkin, BD Pendley. Construction of a very high density extracellular electrode array. Am J Physiol Heart Circ Physiol 279:H437–H442, 2000.

56. SF Idriss, PD Wolf, WM Smith, RE Ideker. Effect of pacing site on ventricular fibrillation initiation by shocks during the vulnerable period. Am J Physiol 277(5, pt 2):H2065–2082, 1999.

57. FX Witkowski, KM Kavanagh, PA Penkoske, R Plonsey. In vivo estimation of cardiac transmembrane current. Circ Res 72(2):424–439, 1993.

58. R Plonsey, CS Henriquez, N Trayanova. Extracellular (volume conductor) effect on adjoining cardiac muscle electrophysiology. Med Biol Eng Comput 25:126–129, 1988.

59. J Eason, RA Malkin. A stimulation study evaluating the performance of high density electrode arrays on myocardial tissue, IEEE Trans Biomed Eng 47(7), 2000, 893–901.

60. M Cohen, R Hoy, J Saffitz, P Corr, J Loslo. A high density in vitro extracellular electrode array: description and implementation. Am J Physiol 89:H681–H689, 1989.

61. P-S Chen, N Shibata, EG Dixon, PD Wolf, ND Danieley, MB Sweeney, WM Smith, RE Ideker. Activation during ventricular defibrillation in open-chest

dogs: Evidence of complete cessation and regeneration of ventricular fibrillation after unsuccessful shocks. J Clin Invest 77:810–823, 1986.

62. JJ Mastrotatoro, HZ Massoud, TC Pilkington, RE Ideker. Rigid and flexible thin-film multielectrode arrays for transmural cardiac recording. IEEE Trans Biomed Eng 39(3):271–279, 1992.

63. B Taccardi, G Arisi, E Macchi, S Baruffi, S Spaggiari. A new intracavitary probe for detecting the site of origin of ectopic ventricular beats during one cardiac cycle. Circulation 75(1):272–281, Jan 1987.

64. DL Derfus, TC Pilkington, EW Simpson, RE Ideker. A comparison of measured and calculated intracavitary potentials for electrical stimuli in the exposed dog heart. IEEE Trans Biomed Eng 39:1192–1206, 1992.

65. ZW Liu, P Jia, PR Ershler, B Taccardi, RL Lux, DS Khoury,Y Rudy. Noncontact endocardial mapping: reconstruction of electrograms and isochrones from intracavitary probe potentials. J Cardiovasc Electrophysiol 8(4):415–431, Apr 1997.

66. C Schmitt, B Zrenner, M Schneider, M Karch, G Ndrepepa, I Deisenhofer, S Weyerbrock J Schreieck, A Schomig. Clinical experience with a novel multielectrode basket catheter in right atrial tachycardias. Circulation 99(18):2414–2422, May 1999.

67. R Fenici, J Nenonen, K Pesola, P Korhonen, J Lotjonen, M Makijarvi, L Toivonen, VP Poutanen, P Keto, T Katila. Nonfluoroscopic localization of an amagnetic stimulation catheter by multichannel magnetocardiography. Pacing Clin Electrophysiol 22(8):1210–1220, Aug 1999.

68. MR Franz. Long term recording of monophasic action potentials from human endocardium. Am J Cardiol 51:1629, 1983.

69. ML Koller, ML Riccio, RF Gilmour. Dynamic restitution of action potential duration during electrical alternans and ventricular fibrillation. Am J Physiol 275(5):H1635–1639, 1998.

70. M Fleischmann, S Pons, DR Rolison, PP Schmidt. eds. Ultramicroelectrodes; Datatech Systems, Inc. Morgantown, NC, 1987.

71. R Onaral, HP Schwan. Linear and nonlinear properties of platinum electrode polarization. Part I, Med Biol Eng Comput 20:299–306, 1982.

72. R Onaral, HP Schwan. Linear and nonlinear properties of platinum electrode polarization. Part II, med boil end comput 21:210–216, 1983.

73. BS Manley, EM Chong, C Walton, AP Economides, An animal model for the chronic study of ventricular repolarisation and refractory period. Cardiovasc Res v23(1):16–20, 1989.

74. RO Cummins, JR Graves, MP Larsen, AP Hallstorm, TR Hearne, J Ciliberti, RM Nicola, S Horan. Out-of-hospital transcutaneous pacing by emergency medical technicians in patients with asystolic cardiac arrest. N Engl J Med 328(19):1377–1382, May 1993.

75. DJ McEneaney, DJ Cochrane, JA Anderson, AA Adgey. Ventricular pacing with a novel gastroesophageal electrode: a comparison with external pacing. Am Heart J 133(6):674–680, Jun 1997.

8
Impedance Measurements in Cardiac Tissue

Tamara C. Baynham
The University of Georgia, Athens, Georgia, U.S.A.

Wayne E. Cascio and Stephen B. Knisley
University of North Carolina, Chapel Hill, North Carolina, U.S.A.

I. INTRODUCTION

A. Definitions of Resistance and Impedance

Resistance is the property of a conductor, which determines the flow of electrical current. We calculate the resistance of a conductor by applying a potential difference between two points on the conductor, measuring the current, and dividing:

$$R = \frac{V}{I} \tag{1}$$

where R is resistance, V is the potential difference, and I is current. This expression is known as Ohm's law. The SI unit for resistance is ohms, abbreviated as Ω.

A quantity related to resistance is resistivity, ρ. Resistivity is a property of a material rather than of a specimen of material and is defined for isotropic materials as

$$\rho = \frac{E}{J} \tag{2}$$

where E is the applied electric field and J is the current density. The SI units of resistivity are ohm-meters, abbreviated as Ω-m. We often refer to the conductivity, σ, of a material rather than its resistivity. Conductivity is the reciprocal of resistivity,

$$\sigma = \frac{1}{\rho} \tag{3}$$

and has SI units $(\Omega\text{-m})^{-1}$.

The relationship between a material's resistivity and its resistance is shown below. Consider a cylindrical conductor of cross-sectional area A and length l, carrying a constant current, I. If a potential difference, V, is applied between its ends and the circular cross sections at each end are equipotential, then the electric field and current density will be constant for all points in the cylinder, thus

$$E = \frac{V}{l} \tag{4}$$

and

$$J = \frac{I}{A} \tag{5}$$

Resistivity is written as

$$\rho = \frac{E}{J} = \frac{V/l}{I/A} \tag{6}$$

Resistance equals V/I, thus

$$R = \rho\frac{l}{A} \tag{7}$$

In determining the electrical properties of biological tissues, one often assumes that the tissue is purely resistive, i.e., obeys Ohm's law, thus measurements of tissue resistance and resistivity are sufficient to describe the tissue of electrical properties. However, this assumption is not true for most biological tissues. For instance, the electrical behavior of cell membranes has been shown to fit an electrical circuit containing an electrical resistance and capacitance [1–3]. Thus, a complete description of the electrical properties includes impedance, Z, which is the opposition to flow of steady-state alternating current and which can possess resistive, capacitive, and inductive elements. The following relation describes impedance:

$$Z = \sqrt{R^2 + (X_L - X_C)^2} \tag{8}$$

where R is resistance, X_C is capacitive reactance, and X_L is inductive reactance.

B. Elements of Biological Impedance

Impedance, as defined above, includes resistance. In biological tissues there are at least two types of resistances: time-invariant and time-variant [4]. The time-invariant resistances are what are commonly called ohmic resistances, i.e., they obey Ohm's law. In cardiac cells, the resistances of the interstitial and intracellular spaces are largely ohmic resistances.

Rectification occurs when a resistance is altered by a change in voltage. In an electrical circuit a rectifier is a device, usually used to convert alternating current to direct current, that passes current preferentially in one direction or another [5]. Rectifiers can be classified as time-variant resistances in which voltage changes over time. When referring to rectification in the heart, the most prominent example is the rectification of the potassium channels. Outward rectification occurs when the membrane passes current more readily in the outward direction, i.e., from inside to outside the cell, which occurs when the membrane is in a depolarized state. This outward current, I_k, is carried by potassium and contributes to repolarization at the end of the cardiac action potential. There is also a current carried by potassium in which the membrane passes current more readily in the inward direction. This is termed the anomalous or inward rectifier current, I_{K1}. This current is inactive for depolarized membrane so it does not contribute much to repolarization. I_{K1} is dominant during the resting state and is termed anomalous because its closure favors depolarization of the action potential [5]. Another potassium rectifier current is the transient outward current, I_{to}. I_{to} helps the heart repolarize during phase 1 repolarization.

Impedance also includes capacitance. Capacitance can be considered the ability to store charge. Capacitive elements are represented by negative reactance [1]. Measurements of capacitance have been associated with cell membrane properties [1]. Several investigators have described the cell membrane as having a parallel resistance and capacitance.

C. Importance of Impedance Measurements for Impulse Propagation

Electrical resistance is important for models of cardiac propagation [6–8]. The very nature of propagation, in which the action potential travels to adjacent cells by electrotonic spread of current, implicates the knowledge of impedance as significant in the process of impulse propagation. The cardiac tissue can be described as a syncytium, in which adjacent cardiac cells and fibers are connected to each other allowing cell-to-cell propagation of wave fronts. Weidmann demonstrated evidence for the electrotonic spread of current reporting measurements of a space constant, λ, greater than the length of one cell [9]. Weidmann also illustrated cell-to-cell communications by monitoring this diffusion of radioactive potassium across the intercalated disc [10]. This increased the evidence for theories concerning the existence of low resistance junctions i.e., gap junctions, between the cells. Electron microscope studies of cardiac tissue revealed a region of close apposition believed to be sites of low resistance [11]. Gap junctions are one of three types of intracellular junctions [11]. The fascia adherens and desmosomes are the two other types, which fulfill mechanical and adhesive roles [11].

The propagation of the cardiac impulse relies on the distribution of the gap junctions [12]. Histological sections of cardiac tissue reveal that gap junctions are more likely to be present near ends of cells (longitudinal) rather than along the lateral borders (transverse) [13]. Changes in coupling due to variations in the number and distribution of gap junctions accounts for anisotropic propagation, i.e., preferential propagation in one direction over another [14,15]. Spach et al. [16] found that the distribution of lateral gap junctions decreased with age. This suggests that slow transverse conduction, further slowed in aged tissue due to decreased lateral gap junctions, can provide a substrate for microreentrant circuits [16]. Other tissue pathologies that have been postulated to alter gap junction distribution are ischemia, hypertrophy, and cardiomyopathy [17,18–20]. An increase in intracellular resistance has been observed during ischemia [21] and hypoxia [22], suggesting that the consequences of metabolic stress decrease the conductance of gap junctions and are responsible for increased intracellular resistance (see below). The resistance of the gap junctions has been measured using diffusion [10], by subtracting cytoplasmic resistance from the total intracellular resistance [23], and by using dual voltage clamp methods [24,25].

D. Importance of Impedance Measurements in Defibrillation

So far, an electrical countershock is the only therapeutic method that effectively terminates ventricular fibrillation. Several investigators emphasize the importance of the electrical field strength and shock duration for reestablishing normal rhythm in fibrillating hearts by interrupting multiple wave fronts [26–30]. The defibrillation shock is produced by electrodes positioned either on the torso for external defibrillation or by a catheter positioned inside the right ventricle for internal defibrillation. In each case, the impedance of the defibrillation system influences both the distribution of the electric field and the temporal characteristics of the shock. The impedance of a defibrillation "system" includes several components: lead wire impedance, electrode/tissue interface impedance, as well as tissue impedance. We will focus solely on the tissue impedance seen through the heart; for a more extensive discussion on all components of the defibrillation system, see the review by KenKnight et al. [31]. Cardiac tissue exhibits behavior inconsistent with a purely ohmic resistance; therefore, it is necessary to consider the capacitance of the cardiac tissue [9,32–35]. However, the assumption that the heart behaves as a purely ohmic resistance has been utilized in the analysis of ideal paddle placement during external defibrillation [36–39].

Some mathematical models evaluating defibrillation use the assumption that the resistance of cardiac tissue is isotropic, i.e., the resistance does not vary as a function of direction, despite evidence of anisotropic resistances in cardiac tissue [40,41]. Eason and colleagues performed a computer simulation utilizing finite-element analysis to study the influence of resistive anisotropy on defibrillation. They found that the inclusion of anisotropic characteristics of the myocardium in a heart–torso model changed the estimated defibrillation threshold for a right ventricular transvenous lead placement by only 4.5% compared to an isotropic model [42]. Thus, in most cases isotropic models are sufficient to characterize the electric field in the heart. Also, during internal defibrillation, it has been determined from a mathematical model that >90% of the applied current traverses the extracellular space of the myocardium, while only 10% flows in the intracellular space [43]. Because the majority of the bulk resistivity is represented by the extracellular space [44], it may be sufficient to neglect the highly anisotropic intracellular resistivity for some purposes. However, several investigators emphasize the importance of transmembrane or intracellular current flow in the defibrillation process [45–48].

II. MEASUREMENTS OF IMPEDANCE

Tissue electrical properties have been described in terms of models or equivalent circuits containing components that correspond to structural features known from histological studies of the tissue. Cardiac tissue contains cells connected to each other through gap junctions located laterally or at the ends of cells. A group of cardiac cells, each of which is approximately 100 μm long and 15 μm wide, forms cardiac fibers [13]. The angle of the ventricular fibers has been found to vary 180° from epicardium to endocardium in canine ventricles [49,50]. This is similar to that of human ventricles. Because of their complexity, structural features of cardiac tissue have not always been fully represented in models. However, the realism that is achievable with cardiac models has increased over the years. In this chapter, we will consider resistance calculations in terms of the simplest models, in which tissue was assumed homogeneous and continuous (isotropic monodomain), up to the more complex anisotropic bidomain, which is widely accepted today. Anisotropy can be defined as a material property that varies as a function of direction. Isotropic tissue can likewise be defined as a material that does not vary as a function of direction. When referring to domains, we refer to a division of the cellular or fiber structure. For instance, the intracellular domain refers to the space inside the cell; the interstitial domain refers to the space between cells; and the extracellular domain refers to any conductive space outside the cell such as a tissue bath. In bidomain representations of tissue, the interstitial and extracellular spaces are combined to describe one domain, while the second domain is the intracellular space. We will divide our discussion based on measurements in which either alternating or direct current was applied to the tissue, and in which properties were described in terms of either one-dimensional or multidimensional models.

A. AC Measurements

1. One-Dimensional AC Measurements

AC (alternating current) measurements have been used extensively to analyze the impedance properties of skeletal muscle, as illustrated in the classic work of Falk and Fatt [51] and reviewed more extensively by Eisenberg [52]. Early measurements in cardiac tissue centered on determination of an appropriate equivalent circuit to describe cable-like structures such as papillary muscle, Purkinje fibers, and trabecular muscles. Cole et al. evaluated the phase angle of various biological cell membranes [1]. Figure 1 was evaluated as a one-time constant electrical equivalent circuit for the cell membrane. Work in nerve fibers by Hodgkin and Huxley [53] also supports a one-time constant electrical equivalent circuit model for nerve membranes. Impedance loci graphs were plotted of data from different biological specimens such as calf blood, rabbit muscle, and frog skin and nerve [1]. Such graphs give a vectorial representations of biological impedance [54] that provide for the identification of time constants. General rules for constructing impedance loci graphs are presented by Schanne and P.-Ceretti [4].

Freygang and Trautwein [34] evaluated electrical constants in strands of sheep Purkinje fibers. The authors expanded on the DC (direct current) analysis of Fozzard et al. [33] (to be discussed later), in which a two-time constant equivalent circuit model of Purkinje fiber was introduced. Freygang and Trautwein were interested in correlating the structure of the Purkinje fiber to the existing equivalent circuit representations of biological

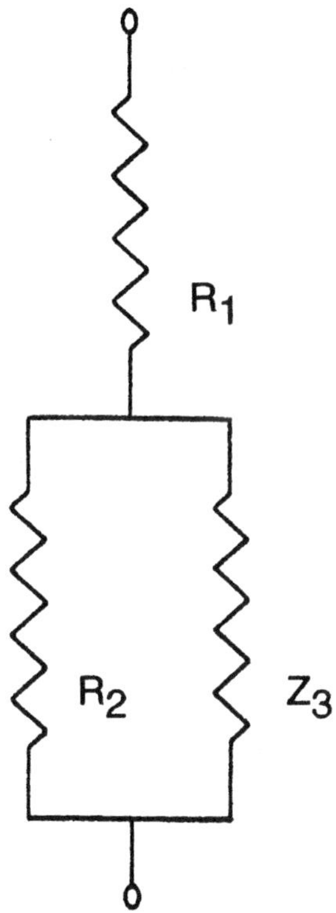

Figure 1 A one-time constant equivalent tissue circuit of biological membrane. The terminals at the top and bottom represent the outside and inside of the cell, respectively. (From Ref. 1.)

tissues. The two-time constant electrical equivalent circuit model, first described by Falk and Fatt for skeletal muscle and investigated by Fozzard in Purkinje fiber, suggests that the second time constant was a result of capacitance introduced by the transverse tubule system [33,51]. Freygang and Trautwein, however, suggested that since the transverse tubule system in Purkinje tissue is now well developed [55], it is not likely the source of the second time constant. Freygang and Trautwein implicated gap junctions as the source of the second time constant. In this investigation, the resistance and capacitance over a range of frequencies was determined with a bridge circuit. Schwan discussed extensively the use of bridge circuits for the measurement of biological impedance [56]. Figure 2 is the electrical equivalent circuit suggested by Freygang and Trautwein containing two-time constants. In the equivalent circuit the Purkinje tissue was modeled as

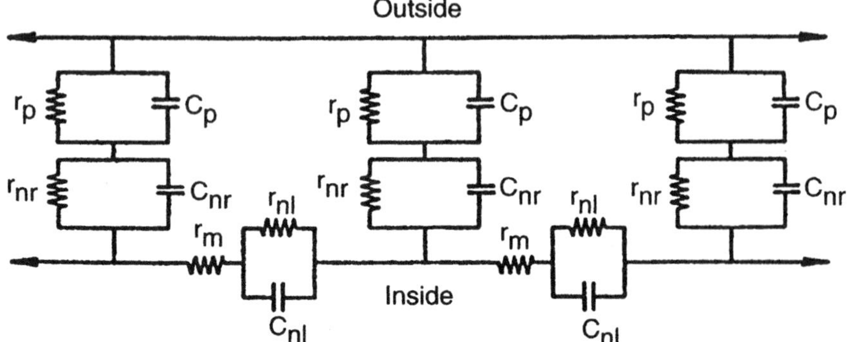

Figure 2   The general circuit. All the parameters are for a unit length of the strand. Thus, r_p and r_{nr} have the dimensions of Ω-cm, C_p and C_{nr} are μF/cm, r_m and r_{nl} are Ω/cm, and C_{nl} is μF-cm. (From Ref. 34. Copyright 1970, The Rockefeller University Press.)

a cylinder. The resistance and capacitance of the plasma membrane are denoted by r_p and c_p, respectively. The nexal, or gap junction, resistance and capacitance are r_{nr} and c_{nr} in the radial direction and r_{nl} and c_{nl} in the longitudinal direction. The myoplasm resistance is denoted by r_m.

Chapman and Fry [58] performed a one-dimensional AC and DC analysis of frog trabeculae. The authors measured and calculated passive properties using the voltage-ratio method introduced by Weidmann [9]. A bridge circuit was used in the AC analysis of longitudinal impedance. Figure 3 is a representation of the fiber as a one-dimensional cable. This study reported the presence of a second time constant in the intracellular pathway.

In an attempt to evaluate the existence of low-resistance gap junctions important for cell-to-cell wavefront propagation, Sperelakis and Hoshiko [35] examined the electrical impedance in strips of feline papillary and trabecular muscle. The muscles were kept in an isotonic sucrose solution. The authors found high cell resistance (5560 Ω-cm) and calculated high gap junction resistance (5106 Ω-cm). Thus, they suggested that the low-resistance gap junctions described by others did not exist. It is likely that the tissue was ischemic, which is now known to increase intracellular calcium and close gap junctions, which might explain the high resistance.

2.   Multidimensional AC Measurements

Sperelakis and MacDonald [57] extended their one-dimensional analysis to account for the multidimensional nature of cardiac tissue by modeling the feline papillary muscle as a bundle of parallel fibers. The preparations were

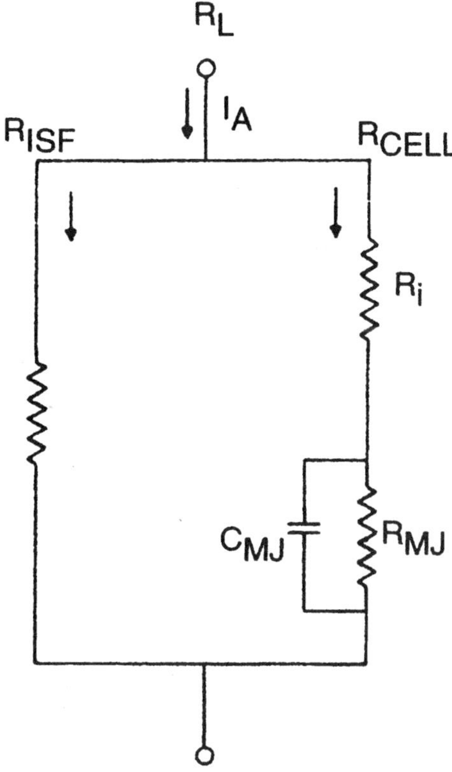

Figure 3 Electrical equivalent circuit for the measured longitudinal resistance (R_L, in ohms) of a cardiac muscle bundle equilibrated in normal Ringer's solution. For simplicity, it was assumed that there are only two parallel current (R_{ISF} and R_{CELL}) pathways through the tissue for applied current (I_a). R_i is the resistance of the myoplasm. R_{MJ} is the resistance of the transverse cell junctions (intercalated discs), and C_{MJ} is the corresponding capacitance of these cell junctions. R_{ISF} is the resistance of the interstitial fluid (ISF). R_{ISF} divided by the fractional volume of the ISF space (Vol_{ISF}) gives the tissue resistivity of the ISF pathway (ρ_{ISF}). R_i divided by $1 - Vol_{ISF}$ gives the tissue resistivity of myoplasm, and R_{MJ} divided by $1 - Vol_{ISF}$ gives the tissue resistivity of the cell junctions; the tissue resistivity of the cell pathway (ρ_{CELL}) is the sum of the myoplasmic and junctional resistivities. (Adapted from Ref. 57.)

kept in oxygenated Ringer's solution or in a Ringer's solution that included isotonic sucrose solutions. The authors analyzed a glass rod model of fibers in which Ringer's solution filled the spaces between the rods. The resistivities of both the interstitial fluid domain and the intracellular domain were

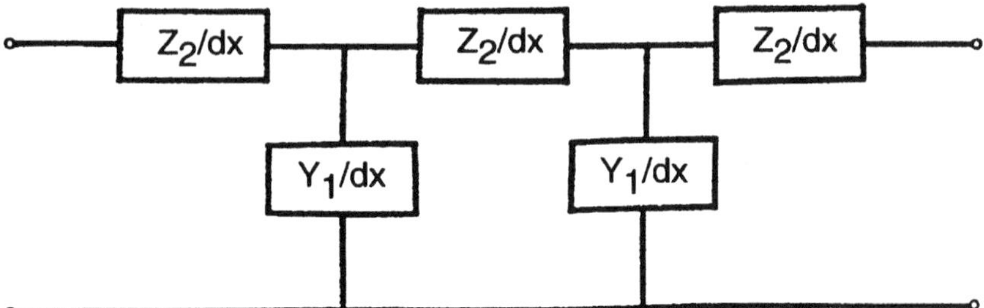

Figure 4 A one-dimensional cable represented by a ladder network of lumped impedances. Y_1/dx represents the membrane admittance per unit length. Admittance is reciprocal of impedance. Z_2/dx represents the intracellular impedance per unit length. The extracellular impedance is assumed to be negligible. One purpose of the analysis is to determine whether Z_2 has purely resistive or some reactive properties. (From Ref. 58.)

considered. The resistivity ratios represented a combination of both domains. The ratio of transverse to longitudinal resistivities for the glass rods was 7. The ratio for cardiac tissue averaged 11. Figure 4 illustrates the one-time constant electrical equivalent circuit for the measured longitudinal resistance, R_L. This study was one of the earlier attempts to calculate the transverse resistivity of isolated parallel fiber muscle.

Schwan and Kay [59,60] calculated resistance in canine heart muscle. In their study, the tissue was assumed to be an isotropic monodomain. Evaluation of resistance in such a medium results in single values for potential gradient and current density. The measurements were taken after the electrical activity of the heart ceased. A bipolar electrode was used to both inject current and measure the corresponding voltage. This configuration may increase the polarization of the electrodes, which could affect measurement. The resistivity of the cardiac tissue was calculated in response to a subthreshold current pulse delivered at rates that varied from 10 Hz to 10 kHz. A potential limitation is that the heart was arrested before measurements were taken. This could artificially increase resistance, since ischemia may cause gap junctions to close [61].

van Oosterom et al. modeled the canine heart as an anisotropic monodomain and used a four-electrode technique to evaluate intramural resistivity in cardiac tissue [62]. Sinusoidal currents of varying frequencies were injected via two outer electrodes. The potential difference that developed was measured between two inner electrodes minimizing effects of electrode polarization. Resistivity was independent of frequency between 10 Hz and 5 kHz and phase in the cardiac cycle. Ischemia was induced by coronary artery occlusion during some experiments. During ischemia, resistivity was found to increase with increasing time of ischemia.

Steendijk et al. [63] performed an analysis of a four-electrode resistivity technique with an anisotropic monodomain model. The authors presented a theoretical evaluation of the technique and of the effect of a disturbing epicardial layer on the measurements. The influence of the layer depended on its thickness compared to the interelectrode distance, and on its resistivity relative to that of the medium of interest. The effect of a thin insulating layer, such as the continuous layer of connective tissue present on the epicardium [5], was negligible if the ratio of the connective tissue thickness to the interelectrode distance was less than 0.1. In this case, the ratio of the layer to epicardial resistivity was greater than 1. A layer of conducting fluid of the same ratio of layer thickness to interelectrode distance (0.1) was found to cause a large effect. In this case, the ratio of the layer to epicardial resistivity was equal to 0.1. Experimental measurements were obtained from anesthetized dogs. The electrode array consisted of two orthogonal linear electrodes. Each electrode was 0.4 mm in diameter, with an interelectrode spacing of 1 mm. The electrode array was positioned to measure along and across the epicardial fiber direction in the perfusion region of the left anterior descending coronary artery. Measurements were made over a frequency range of 5 to 60 Hz. They reported an anisotropy ratio (ratio of transverse to longitudinal resistance) of 2.38:1.

Bidomain models are sometimes considered the most realistic representations of electrical properties of cardiac tissue. In the bidomain model, the cardiac tissue is assumed to consist of two domains: one represents the volume-averaged properties of the intracellular space and the other represents the volume-averaged properties of the interstitial space. The intracellular and interstitial spaces are often assumed to be uniform and continuous. The bidomain model also assumes that the domains coexist and are separated by a cell membrane. Current may flow in each domain and from one domain to the other via the cell membrane [6]. The bidomain representation of the cardiac tissue in three dimensions is given by

$$\bar{J}_e = -\left(g_{ex} \frac{\partial \Phi_e}{\partial x} + g_{ey} \frac{\partial \Phi_e}{\partial y} + g_{ez} \frac{\partial \Phi_e}{\partial z} \right) \tag{9}$$

for the interstitial domain and

$$\bar{J}_i = -\left(g_{ix} \frac{\partial \Phi_i}{\partial x} + g_{iy} \frac{\partial \Phi_i}{\partial y} + g_{iz} \frac{\partial \Phi_i}{\partial z} \right) \tag{10}$$

for the intracellular domain [6]. J_e and J_i represent the interstitial and intracellular current densities, respectively. The bidomain conductivities are represented by g_e and g_i in the x, y, and z directions. Φ_e and Φ_i represent the interstitial and intracellular potential, respectively.

Plonsey and Barr [64,65] developed a bidomain formulation in which the four-electrode method can be used to find the six bidomain conductivities. The major assumption of this formulation was that the ratio of intracellular to interstitial bidomain conductivities along the principal axes was assumed equal. With electrode separation less than or equal to the space constant, current injected into the interstitial space remained essentially in the interstitial space. For electrode separation greater than five times the space constant, the current was distributed in both the intracellular and interstitial spaces, which prevented determination of individual conductivities.

LeGuyader et al. [66] presented a modification of the four-electrode technique developed by Plonsey and Barr [64]. The model added a transmembrane capacitance in parallel with the transmembrane resistance. LeGuyader et al. predicted that at very small interelectrode spacing, the current would be confined to the interstitial space at low frequencies and to the intracellular space at high frequencies. The electrode array consisted of a linear array of four electrodes in each of two perpendicular directions. The electrodes had a diameter of 25 μm with interelectrode spacing of 158 μm. Experiments were performed in isolated canine papillary muscle. The papillary muscle was placed in a tissue bath of oxygenated Tyrode solution. Impedance measurements in both the longitudinal and transverse directions were taken for frequencies between 10 Hz and 10 kHz. The voltage-to-current ratio (in mV/μA) in the longitudinal direction was 5 at 10 Hz and 3.5 at 10 kHz. The voltage-to-current ratio in the transverse direction was 5.5 at 10 Hz and 4.5 at 10 kHz. This formulation was suitable for measuring the conductivity of a tissue preparation in which transmural rotation is not an issue.

LeGuyader et al. [67] also used their modified four-electrode technique to evaluate the bidomain conductivities in isolated arterially perfused canine atrium bathed in oxygenated Tyrode's solution. The electrodes in this case were 50 μm in diameter with interelectrode spacing of 340 μm. Although spacing of the electrodes helped to ensure measurement of the local anisotropic properties, this preparation neglected the effect of structure of the atria in the tissue depth.

Trelles et al. [68] performed a theoretical analysis of a method for measuring conductivities. The major assumption was that the use of a pair of long parallel line electrodes to inject current into the tissue results in a current distribution that varies only in two dimensions along the surface of the preparation (x and y axes). In this analysis, the x direction was considered to be along the fiber axis, and the y direction was considered to be in the direction across fibers. The conductivities in the y and z directions were assumed equal ($g_{ey} = g_{ez}$ and $g_{iy} = g_{iz}$). The authors were able to reproduce results presented from previous analyses with [41,64] and without [66,67] the assumption of equal ratio of intracellular-to-interstitial bidomain conductivities along the principal axes (x, y, and z).

B. DC Measurements

1. One-Dimensional DC Measurements

Several techniques have been used to determine the electrical properties of cardiac tissue including the tissue resistance. One method used to determine tissue resistance, the voltage-ratio method [9,32], was based on a model consisting of a conducting core representing intracellular space, a cell membrane having a parallel resistance and capacitance, and an extracellular conductor [69]. The core-conductor model is essentially a one-dimensional bidomain model. The currents in the intracellular and interstitial spaces are assumed to flow axially, and the resistances in the spaces are assumed to behave as ohmic resistances. Figure 5 is a representation of the core-conductor model. Figure 5A illustrates the geometry in which the conducting core, the intracellular space, is bound by a restricted interstitial space. An electrical circuit representation of the model is also shown (Fig. 5B). The arrows denote axial current flow in which I_e and I_i represent the interstitial and intracellular current, respectively. I_m is the membrane current. By applying Ohm's law to the core-conductor model, one obtains

$$I_i = -\left(\frac{1}{r_i}\right)\left(\frac{\partial \Phi_i}{\partial x}\right) \tag{11}$$

and

$$I_e = -\left(\frac{1}{r_e}\right)\left(\frac{\partial \Phi_e}{\partial x}\right) \tag{12}$$

where r_i and r_e are intracellular and interstitial resistance, Φ_i and Φ_e are intracellular and interstitial potential, and x is the distance along the preparation. A corresponding bidomain representation when tissue width is taken into account is given by

$$J_i = -g_i \frac{\partial \Phi_i}{\partial x} \tag{13}$$

and

$$J_e = -g_e \frac{\partial \Phi_e}{\partial x} \tag{14}$$

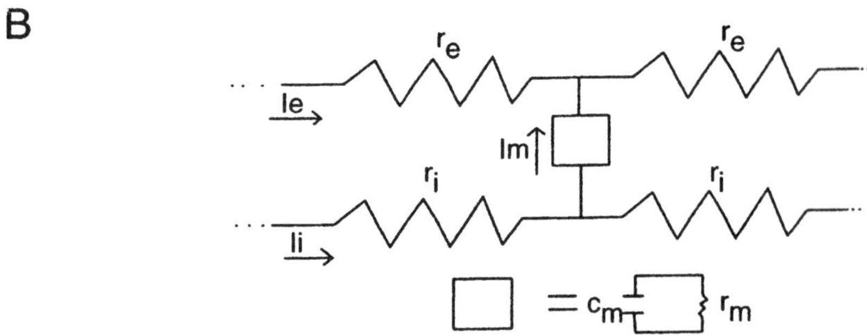

Figure 5 Core-conductor model. (A) A representation of a cardiac cell or fiber as a cable with a conducting core surrounded by a restricted interstitial space. This geometry is used in the core-conductor model of the cardiac tissue. (B) An electrical representation of the core-conductor model. The resistances r_i and r_e represent the intracellular and interstitial spaces, respectively. The intracellular resistance is a combination of the myoplasm and gap junctional resistance. The axial current flow in the interstitial and intracellular spaces is represented by I_e and I_i. The membrane resistance and capacitance are represented by r_m and c_m.

where g_i and g_e represent the intracellular and interstitial bidomain conductivities.

The voltage-ratio method has been used to calculate intracellular and interstitial resistivities. This method uses the assumptions of the core-conductor model, and measurements of the intracellular, transmembrane, and interstitial potentials, and from them, resistances are calculated. The application of Ohm's law yields

$$V_i = I_i r_i \tag{15}$$

$$V_e = I_e r_e \tag{16}$$

where V_i and V_e are longitudinal voltage differences and I_i and I_e are local circuit currents produced at the leading edge of the propagating wavefront. From the local circuit theory of action potential propagation and the law of conservation of charge,

$$I_e + I_i = 0 \tag{17}$$

Using this relationship and dividing Eq. (15) by Eq. (16),

$$\frac{V_i}{V_e} = \frac{-r_i}{r_e} \tag{18}$$

Also, the application of a subthreshold stimulus yields a membrane current of zero in the region midway between stimulation electrodes. Thus, the change in voltage per unit length in this region is the same intra- and extracellularly. From this voltage gradient and the applied current strength (I), the parallel resistance, r_t, is determined:

$$r_t = \frac{r_e r_i}{r_e + r_i} = \frac{dV/dx}{I} \tag{19}$$

With knowledge of V_i, V_e, and r_t, one can calculate r_i and r_e from the last two equations. The intracellular resistance is converted to intracellular resistivity by using the following relationship:

$$R_i = r_i \pi a^2 \tag{20}$$

where a represents the radius of the muscle bundle.

The measurements sometimes require recordings with microelectrodes to gain access to intracellular potentials. Technical considerations for microelectrode measurements include compensation for the microelectrode tip capacitance by use of a negative-input capacitance amplifier [70]. This can "cancel" the effect of capacitance produced by the microelectrode and any lead wire or cables.

Weidmann conducted a study in which the electrical constants of Purkinje fibers were evaluated with the voltage-ratio method [9]. The Purkinje fibers were excised and placed in a tissue bath where they were surrounded by oxygenated Tyrode's solution. The fiber was surrounded by a layer of silicon oil to limit the amount of extracellular solution on the surface of the fiber.

One microelectrode was used to inject a square wave of stimulating current into the fiber. Another microelectrode was connected to an amplifier and used to record the intracellular potential. The Purkinje fiber, described by the core-conductor model, was evaluated. The results agreed with

Hodgkin and Rushton's [53] findings using nerve fibers in that the spread of the potential was found to decrease with distance from the electrode. Passive electrical properties were studied in trabecular muscle bundles obtained from the right ventricle of sheep or calf hearts using the core-conductor model and voltage-ratio method [32]. Silicon oil served as an extracellular insulator. The muscle bundles received oxygenated solution at room temperature.

Clerc [41] used an anisotropic bidomain model and voltage-ratio method to examine longitudinal and transverse resistivities. Smooth cylinders oriented longitudinally represented the individual fibers. In the case of transverse propagation, it was assumed that the intracellular and interstitial volumes, as well as the total membrane surface, were fixed quantities. This idealization was possible because of Clerc's investigation of a uniform sample of tissue, which was placed into a parallel-plane electrode system. This arrangement provided for only a one-dimensional potential and current variation. The electrical constants in the longitudinal and transverse directions were analyzed separately.

Fozzard et al. [33], in an experiment using DC pulses, considered the one-time constant equivalent circuit model discussed by Cole [1] and the two-time constant equivalent circuit model presented by Falk and Fatt [51]. The equivalent circuits are shown in Fig. 6. The cable constants were determined as described by Weidmann [9]. Voltage clamp measurements were also performed. Both analyses were in agreement with the two-time constant electrical equivalent circuit model presented by Falk and Fatt.

While the principles underlying the voltage-ratio method, which were used in the experiments described above, have not changed, the reliability of measurements has been improved by the use of arterially perfused papillary muscles and moist-air insulation surrounding the muscle. This will be described in Sec. III.

2. Multidimensional DC Measurements

Rush et al. performed experiments with anesthetized dogs using a four-electrode configuration to evaluate resistivity [71]. Although this study provided a significant improvement over the two-electrode technique of Schwan et al. [56], it is unclear whether the dimensions of the electrodes and the spacing between the electrodes were sufficient to provide accurate measures of resistivity. Several authors have explored the significance of electrode size, spacing, and configuration [63,64,72,73], and we will discuss the importance of each parameter in Sec. II.C. This approach also neglects possible effects produced by transmural rotation of cardiac fibers [74–76].

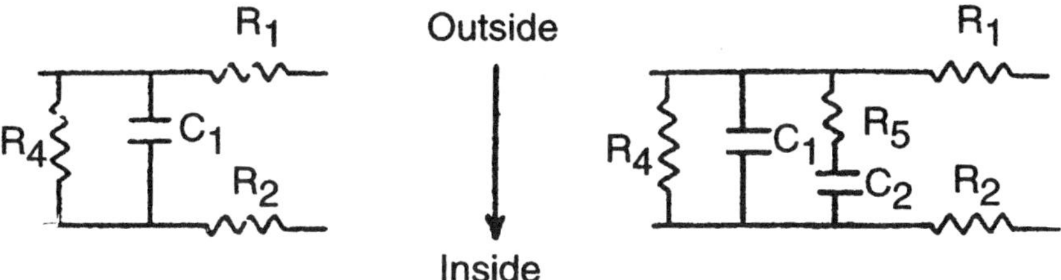

Figure 6 The circuit on the left represents a unit of a simple cable with the membrane resistance (R_4) and the membrane capacitance (C_1) in parallel. R_1 is the resistance of the external fluid and R_2 is the internal (myoplasm) resistance. The circuit on the right is modified by placing part of the capacitance (C_2) in series with a resistance (R_5). (Adapted from Ref. 33.)

Roberts et al. investigated the influence of fiber orientation on wavefront voltage, conduction velocity, and tissue resistivity in anesthetized dogs [77]. The cardiac tissue was modeled as an anisotropic bidomain. Left ventricular epicardial stimulation was applied to the central terminal of an epicardial array covering an area 9×16.5 mm. The array consisted of 84 silver electrodes 0.25 mm in diameter with interelectrode spacing of 1.5 mm. In the determination of tissue resistivity, a map of the potential produced in response to a constant-current stimulus injected into the tissue was produced and resistivity values represented values averaged over the 83 recording electrodes.

Roberts et al. [78] measured resistivities in dogs using the 84-terminal array described above, which was positioned on the left ventricular epicardium near the left anterior descending coronary artery. The return was located on the torso of the dog. Gross resistivities were calculated using maps of stimulus artifact magnitude in response to a suprathreshold constant-current pulse delivered through the central electrode terminal. Regression analysis was used to determine the wavefront voltages used in the calculation of resistances.

Baynham and Knisley [75] used a variation of the four-electrode method, similar to that of Trelles et al. [68], to calculate bulk resistivities from the epicardium of rabbit ventricles. Current was injected from 1-cm-long parallel line electrodes spaced 1 cm apart, and the potential difference was measured from 1-mm-diameter recording electrodes positioned between the line electrodes and spaced 3 mm apart. Epicardial resistance did not vary with direction as expected in anisotropic myocardium. A 3-D monodomain computer model of the ventricle was performed including a 3/1 transverse-to-longitudinal specific resistivity ratio and providing for transmural fiber rotation of 180° [74,75]. Transmural rotation of fibers resulted in shunting of current that depended on electrode orientation with respect to fibers. This finding suggests that fiber rotation in ventricles can cause resistance measured in a sufficiently large region to become effectively isotropic. Complexities of epicardial and three-dimensional multifiber structure may mask resistive anisotropy of localized regions of fibers [65,79].

C. Limitations of Impedance Measurements

Roth described the importance of having accurate measurements of conductivity when modeling stimulation in cardiac tissue [80]. In Table 1, the resistivity values reported by some of the authors mentioned above are summarized.

The reported resistivities vary widely, which may be attributed to differences introduced by the tissue preparation or the electrode configuration. Plonsey and Barr [64,65] discussed factors that can lead to errors in measurement when two- or four-electrode systems are used to inject current and measure voltage into a cardiac tissue sample. The interpretation of these voltages depends on the model assumed for the tissue structure [65]. Of the five listed, three assume an anisotropic bidomain, and two, those of Rush et al. [71] and Steendijk et al. [63], assume an anisotropic monodomain.

Plonsey and Barr [64,65] also discussed the proper separation of electrodes when using the four-electrode configuration. They suggested that by combining measurements with small interelectrode distance ($d \leq 0.1\lambda$) and large interelectrode distance ($d = 1\lambda$), it is possible to calculate all six bidomain conductivities. Steendijk et al. used a four-electrode system to calculate resistance with interelectrode distance of 1 mm. Rush did not report the interelectrode distance. LeGuyader et al. [67] had an interelectrode distance within the range suggested by Plonsey and Barr. Also, Wang et al. [72] presented an error analysis of the four-electrode method in which they evaluated the effect of increasing electrode diameter or length relative to the interelectrode spacing and decreasing tissue size. They found that both of these resulted in a decrease in anisotropy ratio.

Entcheva et al. [73,76] investigated the optimal location of the electrodes in the four-electrode configuration in the presence of transmural fiber rotation. They found that the linear arrangement was not optimal since it did not account for transmural rotation. When rotation was not accounted for, the transverse and longitudinal conductivities were almost indistinguishable [73,76], consistent with the direction-independent resistance found by Baynham and Kinsley [75]. Entcheva et al. [76] suggested that asymmetrical placement or placement of the voltage-measuring electrodes outside the current electrodes may be optimal. Thus, the complex structure of cardiac tissue affects the measurement of impedance in hearts.

Table 1 Measured Resistivity Values (in Ω-cm)

Reference:	Rush et al. [71]	Clerc [41]	Roberts and Scher [78]	Steendijk et al. [63]	LeGuyader et al. [67]
	$R_i = 252 \pm 30\%$	$R_{il} = 402 \pm 30$	$R_l = 213 \pm 25$	$R_l = 210 - 310$	$R_{il} = 354$
	$R_t = 563 \pm 15\%$	$R_{it} = 3620 \pm 280$	$R_t = 705 \pm 80$	$R_t = 380 - 500$	$R_{it} = 2989$
		$R_{ol} = 48 \pm 4$			$R_{el} = 102$
		$R_{ot} = 127 \pm 38$			$R_{et} = 153$

III. RESISTIVE PROPERTIES UNDER PATHOLOGICAL CONDITIONS

A. Historical Perspective

Interest in the electrical resistive properties of muscle tissue during normal and pathological conditions has existed for over a century. In 1901, Kodis [81] sought insight into the physiochemical changes connected with cell death of frog skeletal muscle by measuring the changes in resistivity during the process of death using an alternating current method introduced by Kohlrausch [82]. This and other studies showed that the resistivity of tissue changes as cellular injury progresses, so that the viability of tissue can be assessed by measures of impedance [83]. In 1979, a four-electrode technique was used by van Oosterom et al. [62] to measure tissue resistance in canine heart during normal perfusion and ischemia. After occlusion of the coronary artery, intramural resistance increased from 450 to 1200Ω-cm after 3 hr of ischemia. These characteristic changes in tissue resistive properties during ischemia led to the possibility of using resistance as a means to study the effectiveness of cardiac preservation strategies. For example, Garrido et al. [84] showed that cardioplegic solutions modified favorably the time course of the increase in tissue resistance during ischemia. Ellenby et al. [85] utilized a four-electrode technique to show that a β-adrenergic blocking agent delayed the onset of the rise in resistance. A delayed onset of the rise in lactate concentrations and the concurrent fall in ATP concentration in ischemic myocardium indicates that the change in resistance may be coupled to a metabolic process.

In recent years investigators have characterized changes of electrical resistance and used these changes to better understand the mechanisms underlaying cell-to-cell electrical communication, impulse propagation, and arrhythmia formation in compromised myocardium. These applications proceeded logically from the fact that in an idealized cable, composed of a core conductor and insulator, the speed of impulse propagation is determined by the internal and external resistance. In the case of compact ventricular myocardium, current flow proceeds through intracellular, intercellular, and extracellular spaces, regardless of whether that current arises from a propagating impulse or from a subthreshold current pulse. Thus, the resistive properties of the whole tissue are composed of the specific resistance of the elements of the intracellular/intercellular pathway, i.e., the myoplasmic and gap junctional resistance, and those of the extracellular pathway, i.e., the interstitial space and intravascular space. Any condition that affects the specific resistance of these parameters is expected to influence current flow and impulse propagation. For example, clinically important conditions such as hypoxia, anoxia, ischemia, reoxygenation, and reperfusion will affect both the external and internal resistance through attendant changes in tissue architecture, microvasculature, interstitial fluid content, cell size, and cell-to-cell coupling. These changes in the passive properties of the tissue will directly influence the sources and sinks of excitatory current required for impulse propagation and the flow of injury current across ischemic border zones. Because the previous studies [62,84,85] measured only changes in total tissue resistivity in intact hearts during ischemia, these studies did not provide insight into the effects of ischemia on the individual external or internal resistive components. At that time, however, there was significant interest in assessing the effect of metabolic inhibition on the internal resistance (r_i), which reflects gap junctional conductance during metabolic inhibition caused by hypoxia or anoxia.

The voltage-ratio method developed by Weidmann based on the core-conductor model provided a means to assess the change of internal resistance directly [32]. In 1979 Wojtczak [86] defined changes in passive cable-like properties, namely, the specific internal resistance (R_i), in hypoxic cardiac tissue. Using the voltage-ratio method, Wojtczak measured an R_i of 265 Ω-cm in superfused bovine trabecular muscle. After 1 hr of hypoxia in a glucose-free solution, R_i increased three-fold. Interestingly, these changes were reversed after restoration of O_2 and glucose and were accelerated by pharmacological interventions intended to increase intracellular Ca^{2+} (Ca_i^{2+}), such as an increase in stimulation rate, application of epinephrine, and increasing extracellular Ca^{2+} (Ca_o^{2+}). These were the first experiments to directly relate the decrease in electrical coupling, as measured by the increase in internal resistance, to a decrease in conduction velocity as predicted by the core-conductor model. This study suggested that a rise in Ca_i^{2+} contributed to the changes in internal resistance during metabolic inhibition, analogous to the effects of exogenous Ca_i^{2+} and cardiac glycosides [24].

The Weidmann method required silicon oil as the extracellular insulator and was a difficult method to master. In 1986 Buchanan et al. [87] developed a modification of the Weidmann method in which a cylindrical papillary muscle was superfused with a thin layer of crystalloid solution rather than silicon oil. In this method the volume in the extracellular compartment, and hence the external shunt resistance, could be controlled. In this study the mean R_i was 341 Ω-cm, and was in close agreement with values obtained by Weidman in ungulate Purkinje fibers (181 Ω-cm) [9] and in ventricular muscle (470 Ω-cm [32] and 240 Ω-cm [10]). This method was used to assess rate-dependent effects of hypoxia on R_i in guinea pig papillary muscles [22]. The results of this study were qualitatively similar to those of Wojtczak. For example, after 30 min of hypoxia when stimulated at 0.5 Hz, R_i increased from 252 to 286 Ω-cm, and to 373 Ω-cm when stimulated at 3 Hz. However, the external shunt resistance was dominated by the volume of the superfusion solution, as the maximum extracellular potential during a propagated action potential, V_o, was only 1–12 mV or about 20% of the absolute value of V_o measured in the interstitium of compact ventricular muscle. This proved to be a serious methodological problem because the changes in the external resistance during hypoxia could not be accurately determined.

Although the voltage-ratio method of Weidmann offered a means to measure R_i under a variety of physiological and pathological conditions, it was limited in its ability to detect changes in the external component of tissue resistance. For this reason Kléber [88] developed the arterially perfused papillary muscle preparation in which the cylindrical papillary muscle is insulated by humidified air. In this model the ventricular septum and right ventricular muscles were isolated and perfused through the septal artery. By perfusing the tissue preparation through its normal microvascular system, viability could be maintained indefinitely, and the external shunt resistance of the superfusion solution was eliminated. Because extracellular current flow was restricted to the extracellular compartment proper, this method provided a means to directly and accurately measure changes in the external and internal resistance simultaneously. Based on this model, the specific external resistance (R_o) calculated was 63 Ω-cm, while R_i was 166 Ω-cm. The principal difference between this method and other methods that use the voltage-ratio arises from the marked increase in the extracellular bipolar electrogram, V_o, and the decrease in the intracellular voltage, V_i. This is most apparent in the ratio of V_i/V_o. In perfused tissue the contribution of the changes in r_o and r_i to the changes in total tissue resistance are approximately equal. The ratios and internal resistances reported among the various voltage-ratio methods are summarized in Table 2.

Because we believe that the arterially perfused papillary muscle with humidified air as the extracellular insulator represents the most physiological method for measurement of cardiac resistive properties, the remainder of the discussion will report changes in passive cable-like properties only from studies using perfused tissues.

B. Changes in Resistive Properties in Hypoxic and Ischemic Myocardium

1. Anoxia and Hypoxia

In 1989 Riegger and colleagues [89] described the changes in external (r_o), internal (r_i), and whole tissue (r_t) resistance caused by anoxia and hypoxia in isolated cylindrical rabbit papillary muscles insulated by humidified gas. During arterial perfusion with anoxic solution lacking glucose substrate, r_t

Table 2 Voltage Ratio Among Various Methods

	Weidmann [32]	Wojtczak [86]	Hiramatsu et al. [22]	Buchanan et al. [87]	Kléber and Riegger [88]
V_i (mV)	72	80	110	113	47.2
V_o (mV)	24	30	10.9	10	51.5
V_i/V_o	3	2.69	10.7	11	1*
R_i (Ω-cm)	470	265	244.8	341	166

* Kléber and Riegger used a voltage ratio of V_o/V_i (ratio of extra- to intracellular voltage).

fell during the first 20 min of O_2/substrate deprivation. The fall in r_t was attributed to a decrease in r_o and was associated with a decrease in vascular resistance and an increase in conduction velocity. The experimental findings suggested that alterations in the interstitial compartment caused by anoxia resulted in changes in the r_o. After 12 min of anoxic perfusion, r_o decreased to about 50% of the initial value [89]. In the isolated ischemic papillary muscle, r_i is determined by myoplasmic and gap junctional resistance. Because the gap junctional resistance is much greater than the myoplasmic resistance, changes in the bulk internal resistance reflect primarily changes in cell-to-cell electrical coupling, i.e., changes in gap junctional conductance. Approximately 20 min after the onset of O_2/substrate deprivation, r_t increased and inexcitability quickly followed, findings consistent with a rapid rise of r_i and the onset of cell-to-cell electrical uncoupling [89]. In contrast to anoxia, hypoxic perfusion (PO_2, 20–25 mmHg) did not lead to inexcitability or cell-to-cell electrical uncoupling. As in anoxia, the r_o decreased during hypoxic perfusion. The decrease in r_o was not dependent on the presence or absence of glucose. In this case r_o decreased to 40–50% of the baseline value after 25–30 min of hypoxic perfusion [89]. These findings show that the external resistance is sensitive to changes in the microvasculature and interstitium. Thus, changes in r_o provide an indirect and qualitative measure of microvascular permeability and formation of edema during metabolic inhibition.

Whether the intravascular compartment of perfused cardiac tissue contributes to the external resistance is not well established. In order to address this question, Fleischhauer et al. [90] measured changes in r_o in response to changes in intravascular resistivity and the volume of the interstitial space. The effect of changes of intravascular resistivity was assessed by varying the hematocrit of the perfused blood. Hematocrits of 10%, 40%, and 60% spanned a three-fold increase in intravascular resistivity relative to blood-free perfusate. The volume of the interstitial space was modified by means of adjusting the colloidal osmotic pressure, via alterations of the dextran concentration. As the colloidal osmotic pressure decreased, r_o and the amplitude of the extracellular voltage field decreased, while the diameter of the muscle increased. Conversely, as the colloidal osmotic pressure increased, r_o and the amplitude of the extracellular voltage field increased. In contrast, a three-fold alteration of intravascular electrical resistivity did not induce any significant changes in dimensions of the muscle, r_o, or the amplitude of the extracellular voltage field. These studies indicated that the microvasculature of ventricular myocardium is generally electrically insulated from the interstitial space, and the electrical current flow in the extracellular space during excitation is largely confined to the narrow, anisotropic interstitial space [90].

2. Ischemia

Changes in whole r_t, consisting of r_o and r_i in parallel, and r_o and r_i were measured in isolated cylindrical rabbit papillary muscles insulated by humidified air during no-flow ischemia by Kléber et al. [21]. Total tissue resistance, r_t, increased during the first 30 min after the onset of ischemia with biphasic response. During the first minute after arrest of perfusion, r_t increased by 30–40%. For the next 10–15 min r_t remained constant, until a second rapid and sustained increase occurred. The initial increase in r_t reflects an immediate increase in r_o; see Fig. 7 (open circles). This initial rise in r_o was related to the decrease in perfusion pressure, and is believed to be independent of ischemic metabolism, because the increase in r_o was measured in the presence of sufficient O_2 [88], and occurred 10–15 beats after the arrest of coronary flow (Kiser and Cascio, unpublished observation). This early ischemia-independent rise in r_o was most likely the consequence of the collapse of the intravascular compartment and the narrowing of the interstitial space. This initial increase in r_o was followed by a gradual and slow increase of r_o [88]. Subsequently, Yan et al. [91] utilized choline as an extracellular marker and a choline biosensor to show that the extracellular space decreases 10–20% of control value after 10 min of ischemia. The external resistance, r_o, was found to be an excellent indicator of the change in the extracellular space based on the equation:

$$\Delta ECS(\%) = [(1 - 0.0016 * \Delta[Na^+]_o)/(r_o/r_o') - 1] * 100 \tag{21}$$

where r_o is the external resistance at any time during ischemia, r_o' is the value of the external longitudinal resistance during the control perfusion, and $\Delta[Na^+]_o$ is the change in extracellular Na^+ concentration.

Little data are available regarding the influence of drugs on the time course or magnitude of changes in r_o during normal perfusion or ischemia. It is likely that drugs or metabolites that influence cell volume homeostasis

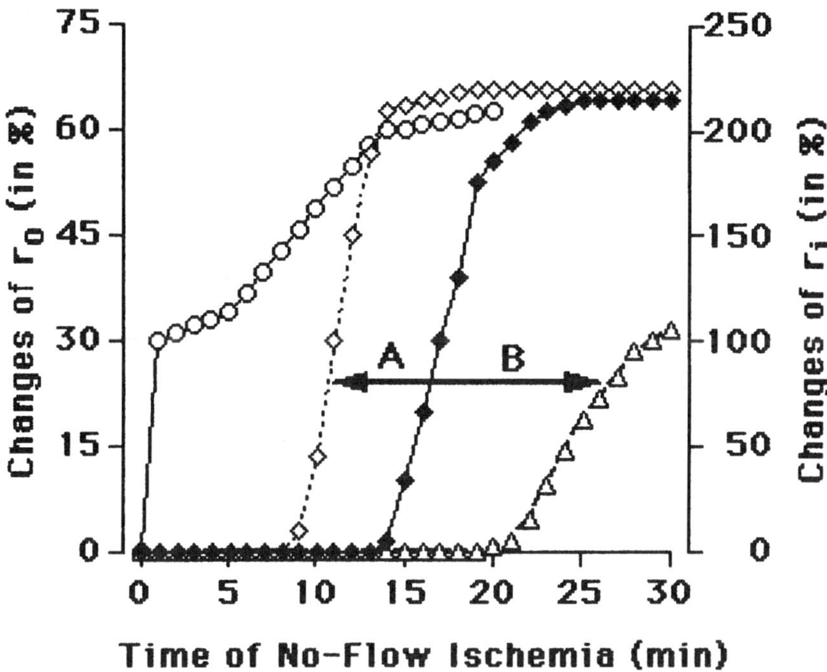

Figure 7 Schematic representation of the change of external longitudinal resistance (r_o, open circles) and internal longitudinal resistance (r_i, solid diamonds) after abrupt and complete arrest of coronary flow. Experimental conditions or drugs that increase intracellular Na^+, Ca^{2+} loading or energy depletion of the ischemic myocardium hasten the onset of cell-to-cell electrical uncoupling. This phenomenon is represented by the shift in the r_i curve to the left, which is represented by the arrow labeled A. By contrast, experimental conditions or drugs that attenuate Na^+, Ca^{2+} loading or energy depletion of the ischemic myocardium delay the onset of the rise in r_i (open triangles), which is represented by the r_i curve marked with the arrow B. In general, factors that delay the onset of r_i also decrease the speed of cell-to-cell electrical uncoupling. This graph represents concepts published in several studies [17,21,96–99,106].

or interstitial fluid accumulation, such as microvascular/capillary permeability, will influence this parameter. For example, the Ca^{2+} channel antagonist and vasodilator, verapamil (0.5 µmol/L), decreased r_o by 15% during 30 min of perfusion [17]. Because r_o contributes to the scaling of the extracellular voltage and excitatory current, modification in the value of r_o will have effects on impulse propagation and may modify clinically important parameters such the electrocardiographic voltages. Further study in this area is needed.

On the other hand, the r_i remained unchanged during the first 10 min of ischemia, then increased abruptly and rapidly—see Fig. 7 (solid diamonds). During the first 10 min r_o/r_i increased and then progressively fell until inexcitability ensued. The change of r_o/r_i is relevent to electrocardiographic changes observed during myocardial ischemia. For example, the early increase in r_o/r_i will increase the TQ-segment depression and ST elevation across an ischemic border zone, whereas the decrease in r_o/r_i caused by the increase in r_i will result in a decrease in the TQ depression and ST elevation. Such an effect was shown in a porcine model of regional ischemia (see below). The time of onset of cell-to-cell electrical uncoupling varied among individual muscles. The variability depended on the rate of stimulation, the diameter of the muscle, the hematocrit, and the buffer capacity of the extracellular compartment. No data exist regarding the time course of changes in transverse cell-to-cell coupling during myocardial ischemia, although changes in transverse impulse propagation in ischemic whole heart follow the same time course as longitudinal impulse propagation [92].

3. Cellular Mechanisms of Changes in Resistive Properties

The ionic and metabolic mechanisms responsible for cell-to-cell electrical uncoupling are controversial. The potential role of Ca^{2+} and the metabolic rate for the onset of cell-to-cell electrical uncoupling is suggested by the observation that conditions or drugs that affect inward excitatory currents or cellular Na^+ and Ca^{2+} loading, such as Ca^{2+} channel agonists [93], cyclic AMP [94], and increased stimulation rate both in the presence [22] and absence [95] of ischemia or hypoxia, hasten the onset of cell-to-cell electrical uncoupling—see Fig. 7 (open diamonds). By contrast, electrical quiescence [96], mechanical inhibition [96], R 56865 [97], Ca^{2+} channel antagonists [17], and preconditioning [17,98–100] delay the onset of the rise in r_i—see Fig. 7 (open triangles). However, the time course of cell-to-cell electrical uncoupling may be influenced by the state of the myocardium. For example, preconditioning delayed the onset of the rise in r_i in normal myocardium [99,100] but accelerated it in hypertrophied myocardium [99].

Using ratiometric epifluorescence imaging of Ca_i^{2+}, Dekker et al. [98] concluded that Ca_i^{2+} was the principal effector of cell-to-cell electrical uncoupling and the increase of r_i in ischemic rabbit papillary muscle. This conclusion was reached because the onset of cellular electrical uncoupling was concordant with an increase in Ca_i^{2+} and independent of changes in external pH, pH_o. In other experimental models, however, Ca_i^{2+} must rise to high concentrations to affect gap junctional conductance [101], the purported determinant of the cell-to-cell coupling. In other experimental models, high concentrations of Ca_i^{2+} are present during the early reversible phase of ischemia without evidence of cell-to-cell electrical uncoupling [102], therefore other factors such as ischemia-induced intracellular acidosis or other metabolic compounds may act synergistically with Ca^{2+} [101,103,104]. Using the same experimental model as Dekker et al. [98], Muller-Borer et al. [105] measured pH_i during ischemia and found that pH_i decreases from 7.1 to 6.8 after 15 min. Thus, the onset of cell-to-cell electrical uncoupling found in their experiments was likely to have occured with a decrease in pH_i of at least 0.3 pH units, suggesting that changes r_i may not be completely independent of pH_i, which was not measured in their experiments.

This view is supported by other data obtained using NMR techniques in isolated ischemic rabbit hearts. Owens et al. [102] showed that a rise in Ca_i^{2+} occurred early in ischemia and preceded the secondary rise of tissue resistance corresponding to cell-to-cell electrical uncoupling. Cell-to-cell electrical uncoupling accompanied a further increase in Ca_i^{2+} and a fall in pH_i to <6.0. These results emphasize that the process by which cell-to-cell electrical uncoupling occurs is likely to be sensitive to the combined effects of Ca^{2+} and H^+ on gap junctions. The different dependence on Ca_i^{2+} is likely to be the consequence of the differences in experimental methods that result in different magnitudes of pH_i change during ischemia. In the isolated perfused papillary muscle surrounded by gas, CO_2 diffusion is favored, whereas in the whole heart CO_2 diffusion is limited. At the surface of the papillary muscle, the onset of cell-to-cell electrical uncoupling is related primarily to Ca_i^{2+} accumulation rather than pH_o or pH_i. By contrast, in whole hearts the onset of cell-to-cell electrical uncoupling is related to the combined increase of Ca_i^{2+} and H_i^+. Other ischemic metabolites that may play a role in cell-to-cell electrical uncoupling during anoxia and ischemia include long chain acyl carnitines [106], arachidonic acid, and free fatty acids.

4. Physiological Consequences of Changes in Resistive Properties During Ischemia

Cellular uncoupling may explain several observations in ischemic boundaries of regionally ischemic whole heart. Cell-to-cell electrical uncoupling contributes to ventricular arrhythmias in regionally ischemic hearts [107, 108] and is believed to signal the onset of irreversible cellular injury. For example, $[K^+]_o$, extracellular electrograms [109], ST-TQ shifts [110], and conduction velocity [92] tend to normalize at ischemic boundaries after 20–30 min of ischemia. In 1995, Smith et al. [108] used DC current and the four-electrode technique in the regionally ischemic in-situ porcine heart to show that loss of cell-to-cell electrical interaction is related to the occurrence of ventricular fibrillation (Figs. 8 and 9), and to changes in the amplitude of the injury current. As shown in Fig. 8, regional ischemia is characterized by two phases of ventricular arrhythmia designated as 1a and 1b. The 1a phase probably relates to alterations in active membrane properties, whereas arrhythmias during the 1b phase are closely linked to the onset of the increase in myocardial tissue resistance [108]. This relationship is best seen in Fig. 9, where the onset of ventricular fibrillation was closely linked to the onset of cell-to-cell electrical uncoupling.

In 1997, Cinca et al. [107] confirmed this observation. However, in contrast to Smith et al., Cinca et al. utilized an AC current source (10 μA, 1110 Hz) and measured phase angle as well as tissue resistivity. In their studies the tissue resistance increased from 237 Ω-cm to 359 Ω-cm during the first 30–35 min of ischemia, and then further increased to 488 Ω-cm at 60 min and 718-Ωcm at 150 min—see Fig. 10. Phase angle was sensitive to changes in the tissue resistance induced by ischemia and appeared to identify the onset of cell-to-cell electrical uncoupling more accurately. Interestingly, changes in phase angle appeared to be independent of changes in the external resistance associated with the collapse of the extracellular compartment.

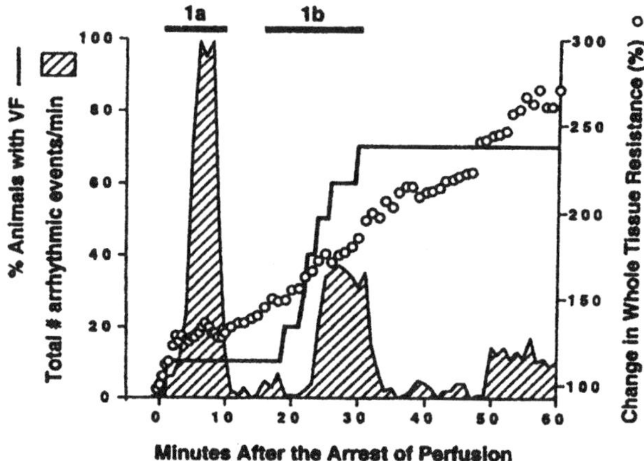

Figure 8 Plot showing the incidence of spontaneous ventricular fibrillation (solid line, left axis) and total ischemic events (PVCs, short runs of nonsustained ventricular tachycardia) per 1-min interval (hatched area) in nine regionally ischemic porcine hearts. The relationship between ventricular events and changes in myocardial resistance during ischemia is shown by the superimposition of the change in whole-tissue resistance, an indirect measure of cell-to-cell electrical coupling (open circles), and the arrhythmic events [108]. 1a and 1b represent the known biphasic distribution of ventricular arrhythmias during ischemia [113]. (From Ref. 108. With Permission of the American Heart Association.)

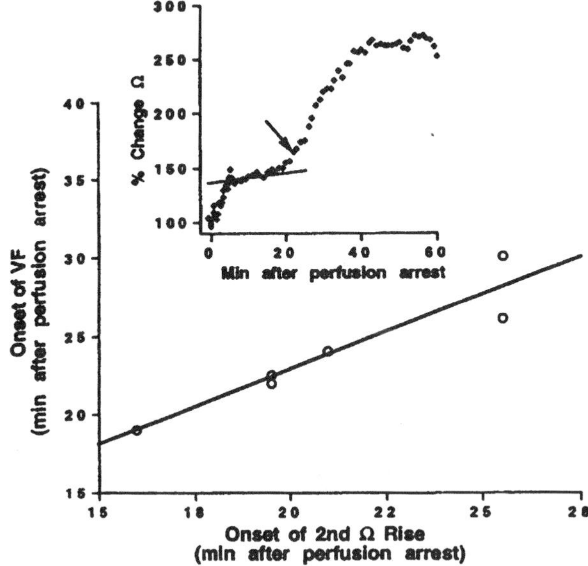

Figure 9 Linear regression plot comparing the time to the onset of ventricular fibrillation to the time of onset of the second rise in resistivity in six experiments. The regression equation is $y = 0.95x + 3.81$; $R_2 = 0.885$; $P = 0.005$. Inset shows the pattern of resistivity change during a single experiment. The rapid rise in r_t during ischemia corresponds to the rapid rise of r_i shown in Fig. 7 and represents the rapid process of cell-to-cell electrical uncoupling. The onset of cell-to-cell electrical uncoupling (second rise time) indicated by an arrow [108]. (From Ref. 108. With Permission of the American Heart Association.)

This is to be expected, given that the membrane capacitance is in parallel with the external resistance and in series with the internal resistance [111]. As shown in Fig. 11, the onset of cell-to-cell electrical uncoupling as measured by either the rise in resistance or phase angle was delayed by preconditioning.

C. Effect of Myocardial Disease on Passive Electrical Parameters

Little is known about the effects of cardiac disease on the changes in passive electrical properties of heart muscle—for example, healed myocardial infarction, ventricular hypertrophy, infiltrative heart disease, or congestive heart failure. An understanding of the changes of passive electrical properties of the heart is important, given that each of these conditions carries

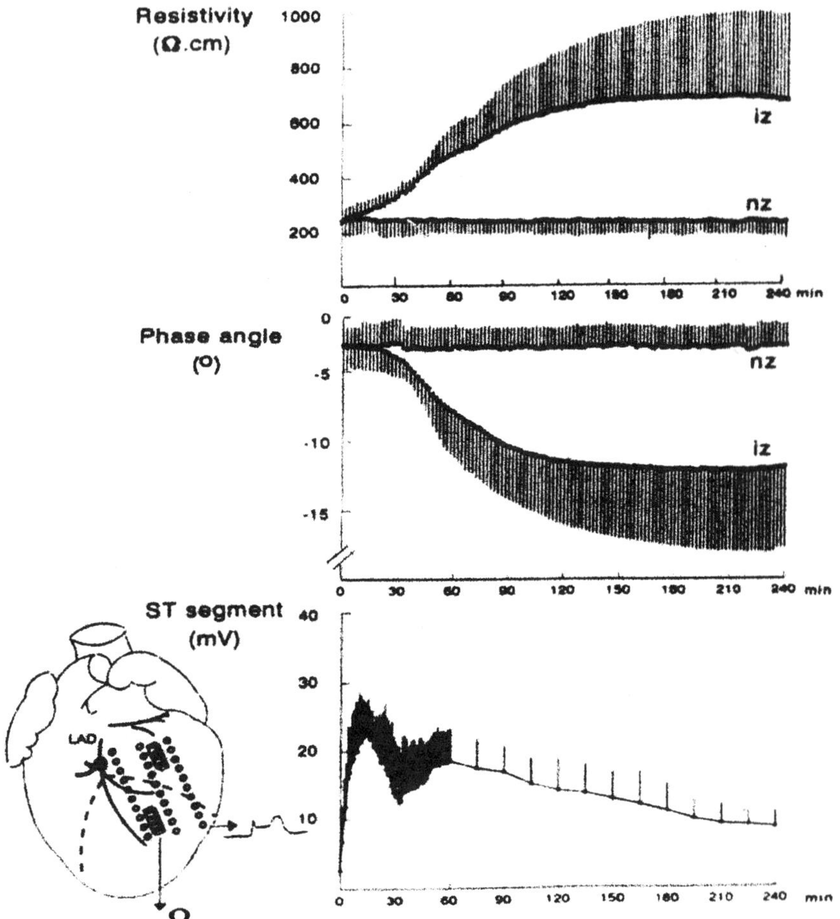

Figure 10 Sequential changes in myocardial resistivity, tissue phase angle, and epicardial ST-segment potential during 4 hr of regional ischemia in the in-situ porcine heart. Dots represent mean value and bars 1 SD of samples from ischemic (iz) and normal (nz) myocardial zones. Note that the resistivity rises early and more gradually than simultaneous measures of phase angle. The inset shows a schematic representation of the experimental method illustrating the spatial relationship between impedance probe and surrounding epicardial electrodes. (From Ref. 107. With Permission of the American Heart Association.)

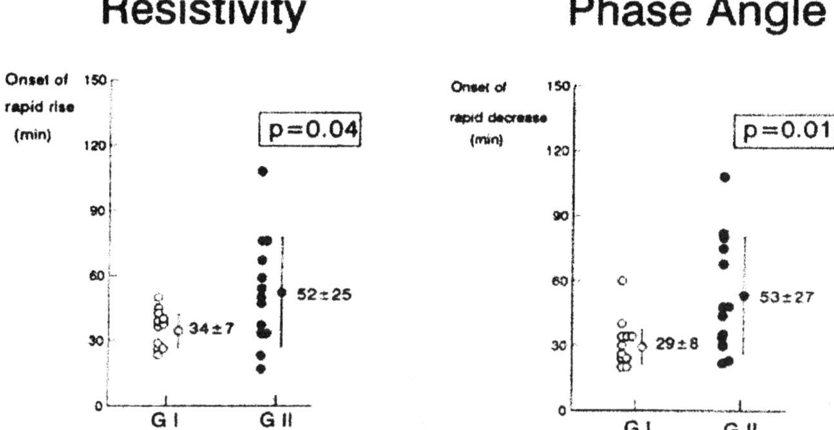

Figure 11   Time to onset of rapid change in myocardial resistivity and phase angle in nonpreconditioned (GI) and preconditioned pigs (GII). Preconditioning delays the onset of the rapid rise in myocardial resistivity and phase angle. (From Ref. 107. With Permission of the American Heart Association.)

with it an increased risk of ventricular arrhythmia and sudden cardiac death. The incidence of ventricular arrhythmias is higher in failing hearts, especially during acute ischemia. Vermeulen et al. [112] studied isolated papillary muscles in a model of volume—and pressure-overloaded rabbit hearts (heart failure model) during no-flow ischemia and found no difference between the onsets of cell-to-cell electrical uncoupling in these hearts compared to controls. However, other factors such as action potential duration and conduction velocity decreased more in preparations from failing than from control hearts, while extracellular K^+ accumulation was greater in failing hearts, indicating that primary changes in r_i do not explain the increased propensity for arrhythmias in this population. However, this study did not address changes in r_o or r_m. In a follow-up study from the same laboratory, Dekker et al. [99] showed that in the absence of ischemic preconditioning, the onset of cell-to-cell electrical uncoupling was not changed in failing hearts; however, the durations of the uncoupling process was influenced. In failing hearts the duration of the uncoupling process was increased as the degree of failure increased. With ischemic preconditioning, the onset of uncoupling was prolonged in normal hearts but was accelerated in failing hearts. In each of these cases a rise Ca_i^{2+} preceded the onset of uncoupling.

REFERENCES

1. KS Cole. Electric phase angle of cell membranes. J Gen Physiol 19:641–649, 1932.
2. KS Cole, HJ Curtis. Electric impedance of nitella during activity. J Gen Physiol 21:37–64, 1938.
3. KS Cole, HJ Curits. Electric impedance of the squid giant axon during activity. J Gen Physiol 22:649–670, 1939.
4. OF Schanne, ER P.-Ceretti. Impedance Measurements in Biological Cells. New York: Wiley, 1978, pp. 359–365.
5. AM Katz. Physiology of the Heart. 2nd ed. New York: Raven Press, 1992, pp. 438–472.
6. CS Henriquez. Simulation of the electrical behavior of cardiac tissue using the bidomain model. CRC Crit Rev Biomed Eng 21:1–77, 1993.
7. AE Pollard, N Hooke, CS Henriquez. Cardiac propagation simulation. CRC Crit Rev Biomed Eng 20:171–210, 1992.
8. AE Pollard, MJ Burgess, KW Spitzer. Computer simulations of three-dimensional propagation in ventricular myocardium: effects of intramural fiber rotation and inhomogeneous conductivity on epicardial activation. Circ Res 72:744–756, 1993.
9. S Weidmann. The electrical constants of purkinje fibers. J Physiol 118:348–360, 1952.
10. S Weidmann. The diffusion of radiopotassium across intercalated disks of mammalian cardiac muscle. J Physiol 187:323–342, 1966.
11. FS Sjostrand, E Anderson. Electron microscopy of the intercalated discs of cardiac muscle tissue. Experientia 10:369–370, 1954.
12. Y Rudy. Cardiac conduction: an interplay between membrane and gap junction. J Electrocardiol 31(suppl):1–5, 1998.
13. JR Sommer, B Scherer. Geometry of cell and bundle appositions in cardiac muscle: light microscopy. Am J Physiol 248:H792–H803, 1985.
14. JW Woodbury, WE Crill. On the problem of impulse conduction in the atrium. In: E. Florey, ed. Nervous Inhibition. New York: Pergamon Press, 1961, pp. 124–135.
15. MS Spach, WT Miller III, DB Geselowitz, RC Barr, JM Kootsey, EA Johnson. The discontinuous nature of propagation in normal canine cardiac muscle. Evidence for recurrent discontinuities of intracellular resistance that affect the membrane currents. Circ Res 48(1):39–54, 1981.
16. MS Spach, PC Dolber. Relating extracellular potentials and their derivatives to anisotropic propagation at a microscopic level in human cardiac muscle: evidence for electrical uncoupling of side-to-side fiber connections with increasing age. Circ Res 58:356–371, 1986.
17. WE Cascio, GX Yan, AG Kleber. Passive electrical properties, mechanical activity, and extracellular potassium in arterially perfused and ischemic ventricular muscle. Effects of calcium entry blockade or hypocalcemia. Circ Res 66(6):1461–1473, 1990.

18. EA Luque, RD Veenstra, EC Beyer, LF Lemanski. Localization and distribution of gap junctions in normal and cardiomyopathic hamster heart. J Morph 222:203–213, 1994.

19. NS Peters, CR Green, PA Poole-Wilson, NJ Severs. Cardiac arrhythmogenesis and the gap junction. J Mol Cell Cardiol 27:37–44, 1995.

20. NS Peters, J Coromilas, NJ Severs, AL Wit. Disturbed connexin43 gap junction distribution correlates with the location of reentrant circuits in the epicardial border zone of healing canine infarcts that cause ventricular tachycardia. Circulation 95:988–996, 1997.

21. AG Kléber, CB Riegger, MJ Janse. Electrical uncoupling and increase of extracellular resistance after induction of ischemia in isolated, arterially perfused rabbit papillary muscle. Circ Res 61(2):271–279, 1987.

22. Y Hiramatsu, JW Buchanan, SB Knisley, LS Gettes. Rate-dependent effects of hypoxia on internal longitudinal resistance in guinea pig papillary muscles. Circ Res 63:923–929, 1988.

23. N Sperelakis, T Hoshiko, RM Berne. Nonsyncytial nature of cardiac muscle: membrane resistance of single cells. Am J Physiol 198(3):531–536, 1960.

24. R Weingart. The actions of ouabain on intercellular coupling and conduction velocity in mammalian ventricular muscle. J Physiol 264:341–365, 1977.

25. R Weingart, P Maurer. Cell-to-cell coupling studied in isolated ventricular cell pairs. Experientia 43:1091–1094, 1987.

26. P-S Chen, N Shibata, EG Dixon, RO Martin, RE Ideker. Comparison of the defibrillation threshold and the upper limit of ventricular vulnerability. Circulation 73:1022–1028, 1986.

27. P-S Chen, N Shibata, EG Dixon, PD Wolf, ND Danieley, MB Sweeney, WM Smith, RE Ideker. Activation during ventricular defibrillation in open-chest dogs. Evidence of complete cessation and regeneration of ventricular fibrillation after unsuccessful shocks. J Clin Invest 77:810–823, 1986.

28. P-S Chen, PD Wolf, SD Melnick, ND Danieley, WM Smith, RE Ideker. Comparision of activation during ventricular fibrillation and following unsuccessful defibrillation shocks in open chest dogs. Circ Res 66(6):1544–1560, 1990.

29. DW Frazier, PD Wolf, JM Wharton, ASL Tang, WM Smith, RE Ideker. Stimulus-induced critical point: mechanism for electrical initiation of reentry in normal canine myocardium. J Clin Invest 83:1039–1052, 1989.

30. N Shibata, P-S Chen, EG Dixon, PD Wolf, ND Danieley, WM Smith, RE Ideker. Epicardial activation following unsuccessful defibrillation shocks in dogs. Am J Physiol 255:H902–H909, 1988.

31. B KenKnight, M Eyüboglu, RE Ideker. Impedance to defibrillation countershock: does an optimal impedance exist? PACE 18(11):2068–2087, 1995.

32. S Weidmann. Electrical constants of trabecular muscle from mammalian heart. J Physiol 210:1041–1054, 1970.

33. HA Fozzard. Membrane capacity of the cardiac Purkinje fiber. J Physiol 182:255–267, 1966.

34. WH Freygang, W Trautwein. The structural implications of linear electrical properties of cardiac Purkinje stands. J Gen Physiol 55:524–547, 1970.
35. N Sperelakis, T Hoshiko. Electrical impedance of cardiac muscle. Circ Res 9:1280–1283, 1961.
36. FJ Claydon, TC Pilkington, AS Tang, MN Morrow, RE Ideker. A volume conductor model of the thorax for the study of defibrillation fields. IEEE Trans Biomed Eng 35:981–992, 1988.
37. BJ Fahy, Y Kim, A Ananthaswamy. Optimal electrode configuration for external cardiac pacing and defibrillation: an inhomogenous study. IEEE Trans Biomed Eng 34(9):743–748, 1987.
38. WJ Karlon, SR Eisenberg, JL Lehr. Effects of paddle placement and size on defibrillation current distribution: a three-dimensional finite element model. IEEE Trans Biomed Eng 40:246–255, 1993.
39. IF Ramirez, SR Eisenberg, JL Lehr, FJ Schoen. Effects of cardiac configuration, paddle placement, and paddle size on defibrillation current distribution: a finite element model. Med Biol Eng Comput 27:587–594, 1989.
40. LA Geddes, LE Baker. The specific resistance of biological material: a compendium of data for the biomedical engineer and physiologist. Med Biol Eng 5:271–293, 1967.
41. L Clerc. Directional differences of impulse spread in trabecular muscle from mammalian heart. J Physiol 255:335–346, 1976.
42. J Eason, J Schmidt, A Dabasinskas, G Siekas, F Aguel, N Trayanova. Influence of anisotropy on local and global measures of potential gradient in computer models of defibrillation. Ann Biomed Eng 26:840–849, 1998.
43. R Plonsey, RC Barr. Effect of microscopic and macroscopic discontinuities on the response of cardiac tissue to defibrillating (stimulating) currents. Med Biol Eng Comput 24:130–136, 1986.
44. JJ Ackmann, MA Seitz. Methods of complex impedance measurements in biologic tissue. Crit Rev Biomed Eng 11(4):281–311, 1984.
45. SM Dillon. Synchronized repolarization after defibrillation shocks: a possible component of the defibrillation process demonstrated by optical recordings in rabbit heart. Circulation 85:1865–1878, 1992.
46. SB Knisley, BC Hill. Optical recordings of the effect of electrical stimulation on action potential repolarization and the induction of reentry in two-dimensional perfused rabbit epicardium. Circulation 88:2402–2414, 1993.
47. SB Knisley, WM Smith, RE Ideker. Prolongation and shortening of action potentials by electrical shocks in frog ventricular muscle. Am J Physiol 266:H2348–H2358, 1994.
48. JL Jones, RE Jones, G Balasky. Improved cardiac cell excitation with symmetrical biphasic defibrillator waveforms. Am J Physiol 253:H1418–H1424, 1987.
49. MA Fernandez-Teran, JM Hurle. Myocardial fiber architecture of the human heart ventricles. Anat Rec 204:134–147, 1982.
50. DD Streeter Jr, HM Spotnitz, DP Patel, J Ross Jr, EH Sonnenblick. Fiber orientation in the canine left ventricle during diastole and systole. Circ Res 24:339–347, 1969.

51. G Falk, P Fatt. Linear electrical properties of striated muscle fibers observed with intracellular electrodes. Proc R Soc Lond B 160:69–123, 1964.

52. RS Eisenberg. Impedance measurement of the electrical structure of skeletal muscle. In: Handbook of Physiology. Bethesda, MD: American Physiological Society, 1983, pp.301–323.

53. AL Hodgkin, WAH Rushton. The electrical constants of a crustacean nerve fibre. Proc R Soc Lond B 133:444–479, 1946.

54. ME Valentinuzzi. Bioelectical impedance techniques in medicine: Part 1. Bioimpedance measurement. First section: General concepts. Crit Rev Biomed Eng 24(4–6):223–255, 1996.

55. JR Sommer, EA Johnson. Cardiac muscle. A comparative study of Purkinje and ventricular fibers. J Cell Biol 36(3):497–526, 1968.

56. HP Schwan. Determination of biological impedance. In: WL Nastuk, ed. Physical Techniques in Biological Research. New York: Academic Press, 1963, pp. 323–406.

57. N Sperelakis, RL MacDonald. Ratio of transverse to longitudinal resistivities of isolated cardiac muscle fiber bundles. J Electrocardiol 7(4):301–314, 1974.

58. RA Chapman, CH Fry. An analysis of the cable properties of frog ventricular myocardium. J Physiol 283:263–282, 1978.

59. HP Schwan, CF Kay. Specific resistance of body tissues. Circ Res 4:664–670, 1956.

60. HP Schwan, CF Kay. The conductivity of living tissues. Ann NY Acad Sci 65:1007–1013, 1957.

61. C Peracchia. Calcium effects on gap junction structure and cell coupling. Nature 271:669–671, 1978.

62. A van Oosterom, RW de Boer, RT van Dam. Intramural resistivity of cardiac tissue. Med Biol Eng Comput 17:337–343, 1979.

63. P Steendijk, G Mur, ET Van Der Velde, J Baan. The four-electrode resistivity technique in anisotropic media:theoretical analysis and application on myocardial tissue in vivo. IEEE Trans Biomed Eng 40:1138–1148, 1993.

64. R Plonsey, RC Barr. The four-electrode resistivity technique as applied to cardiac muscle. IEEE Trans Biomed Eng 29(7):541–546, 1982.

65. R Plonsey, RC Barr. A critique of impedance measurements in cardiac tissue. Ann Biomed Eng 14:307–322, 1986.

66. P LeGuyader, P Savard, R Guardo, L Pouliot, F Trelles, M Meunier. Myocardial impedance measurements with a modified four-electrode technique. Proc 16th Annual IEEE Engineering in Medicine and Biology Society, 1994, pp. 880–881.

67. P LeGuyader, P Savard, F Trelles. Measurement of myocardial conductivities with an eight-electrode technique in the frequency domain. Proc 17th Annual IEEE Engineering in Medicine and Biology Society, 1995, pp. 71–72.

68. F Trelles, P Savard, P LeGuyader. A new method for measuring myocardial conductivities: the parallel electrode technique. Proc 17th Annual IEEE Engineering in Medicine and Biology Society, 1995, pp. 73–74.

69. JJB Jack, D Noble, RW Tsien. Linear cable theory. In: Electric Current Flow in Excitable Cells. Oxford: Clarendon Press, 1975, pp. 25–66.

70. MR Neuman. Biopotential amplifiers. In: JG Webster, ed. Medical Instrumentation: Application and Design. Boston: Houghton Mifflin, 1992, pp. 288–353.

71. S Rush, JA Abildskov, R McFee. Resistivity of body tissues at low frequencies. Circ Res 12:40–50, 1963.

72. Y Wang, PH Schimpf, DR Haynor, Y Kim. Geometric effects on resistivity measurements with four-electrode probes in isotropic and anisotropic tissues. IEEE Trans Biomed Eng 45(7):877–884, 1998.

73. E Entcheva, J Eason, F Claydon. Stimulation of cardiac tissue: sample volume and penetration depth. Ann Biomed Eng. 25(suppl 1):S-62, 1997.

74. TC Baynham, SB Knisley. The role of fiber rotation in effective epicardial resistance of rabbit ventricles. Ann Biomed Eng 26:S-20, 1998.

75. TC Baynham, SB Knisley. Effective epicardial resistance of rabbit ventricles. Ann Biomed Eng 27(1):96–102, 1999.

76. E Entcheva, J Eason, F Claydon, R Malkin. Spatial effects from bipolar current injection in 3D myocardium: implications for conductivity measurements. Proc Computers in Cardiology, 1997, pp. 717–720.

77. DE Roberts, LT Hersh, AM Scher. Influence of cardiac fiber orientation on wavefront voltage, conduction velocity, and tissue resistivity in the dog. Circ Res 44(5):701–712, 1979.

78. DE Roberts, AM Scher. Effect of tissue anisotropy on extracellular potential fields in canine myocardium in situ. Circ Res 50(3):342–351, 1982.

79. AM Pertsov, SF Mironov, O Berenfeld, J Jalife. Significant increase in anisotropy of epicardial activation during first minutes of ischemia. PACE 20(4):1102, 1997.

80. BJ Roth. Electrical conductivity values used with the bidomain model of cardiac tissue. IEEE Trans Biomed Eng 44(4):326–328, 1997.

81. T Kodis. The electrical resistance in dying muscle. Am J Physiol 5:267–273, 1901.

82. F Kohlrausch, L Holborn. Das Leitvermgen der Elektrolyte, insbesondere der Isungen. Methoden, Resultate und chemische Anwendungen. Leipzig: Teubner, 1898.

83. C Fourcade. Contribution a l'etude de la mort cellulaire par mesure de l'impedance tissulaire. Lyon: Universite Claude Bernard, 1973.

84. H Garrido, J Sueiro, J Rivas, J Vilches, J Romero, F Garrido. Bioelectrical tissue resistance during various methods of myocardial preservation. Ann Thorac Surg 36(2):143–151, 1983.

85. MI Ellenby, KW Small, RM Wells, DJ Hoyt, JE Lowe. On-line detection of reversible myocardial ischemic injury by measurement of myocardial electrical impedance. Ann Thorac Surg 44:587–597, 1987.

86. J Wojtczak. Contractures and increase in internal longitudinal resistance of cow ventricular muscle induced by hypoxia. Circ Res 44:88–95, 1979.

87. JW Buchanan Jr, S Oshita, T Fujino, LS Gettes. A method for measurement of internal longitudinal resistance in papillary muscle. Am J Physiol 251:H210–H217, 1986.

88. AG Kléber, CB Riegger. Electrical constants of arterially perfused rabbit papillary muscle. J Physiol 385:307–324, 1987.

89. CB Riegger, G Alperovich, AG Kléber. Effect of oxygen withdrawal on active and passive electrical properties of arterially perfused rabbit ventricular muscle. Circ Res 64(3):532–541, 1989.

90. J Fleischhauer, L Lehmann, AG Kléber. Electrical resistances of intersititial and microvascular space as determinants of the extracellular electrical field and velocity of propagation in ventricular myocardium. Circulation 92(3):587–594, 1995.

91. GX Yan, J Chen, KA Yamada, AG Kléber, PB Corr. Contribution of shrinkage of extracellular space to extracellular K^+ accumulation in myocardial ischaemia of the rabbit. J Physiol (Lond) 490 (pt 1):215–228, 1996.

92. AG Kleber, MJ Janse, FJ Wilms-Schopmann, AA Wilde, R Coronel. Changes in conduction velocity during acute ischemia in ventricular myocardium of the isolated porcine heart. Circulation 73(1):189–198, 1986.

93. T Maruyama, WE Cascio, SB Knisley, J Buchanan, LS Gettes. Effects of ryanodine and BAY K8644 on membrane properties and conduction during simulated ischemia. Am J Physiol 261(6 pt 2):H2008–H2015, 1991.

94. J Wojtczak. Influence of cyclic nucleotides on the internal longitudinal resistance and contractures in the normal and hypoxic mammalian cardiac muscle. J Mol Cell Cardiol 14(5):259–265, 1982.

95. J Bredikis, F Bukauskas, R Veteikis. Decreased intercellular coupling after prolonged rapid stimulation in rabbit atrial muscle. Circ Res 49:815–820, 1981.

96. HL Tan, MJ Janse. Contribution of mechanical activity and electrical activity to cellular electrical uncoupling in ischemic rabbit papillary muscle. J Mol Cell Cardiol 26(6):733–742, 1994.

97. HL Tan, AO Netea, ME Sleeswijk, P Mazon, R Coronel, T Opthof, MJ Janse. R56865 delays cellular electrical uncoupling in ischemic rabbit papillary muscle. J Mol Cell Cardiol 25(9):1059–1066, 1993.

98. LR Dekker, JW Fiolet, E VanBavel, R Coronel, T Opthof, JA Spaan, MJ Janse. Intracellular Ca^{2+}, intercellular electrical coupling, and mechanical activity in ischemic rabbit papillary muscle. Effects of preconditioning and metabolic blockade. Circ Res 79(2):237–246, 1996.

99. LR Dekker, H Rademaker, JT Vermeulen, T Opthof, R Coronel, JA Spaan, MJ Janse. Cellular uncoupling during ischemia in hypertrophied and failing rabbit ventricular myocardium:effects of preconditioning. Circulation 97(17):1724–1730, 1998.

100. HL Tan, P Mazon, HJ Verberne, ME Sleeswijk, R Coronel, T Opthof, MJ Janse. Ischaemic preconditioning delays ischaemia induced cellular electrical uncoupling in rabbit myocardium by activation of ATP sensitive potassium channels. [published erratum appears in Cardiovasc Res 27(7):1385]. Cardiovasc Res 27(4):644–651, 1993.

101. L Firek, R Weingart. Modification of gap junction conductance by divalent cations and protons in neonatal rat heart cells. J Mol Cell Cardiol 27(8):1633–1643, 1995.

102. LM Owens, TA Fralix, E Murphy, WE Cascio, LS Gettes. Correlation of ischemia-induced extracellular and intracellular ion changes to cell-to-cell electrical uncoupling in isolated blood-perfused rabbit hearts. Circulation 94(1):10–13, 1996.

103. JM Burt. Block of intercellular communication:interaction of intracellular H^+ and Ca^{2+}. Am J Physiol 253(4 pt 1):C607–C612, 1987.

104. RL White, JE Doeller, VK Verselis, BA Wittenberg. Gap junctional conductance between pairs of ventricular myocytes is modulated synergistically by H^+ and Ca^{++}. J Gen Physiol 95(6):1061–1075, 1990.

105. B Muller-Borer, H Yang, DR Sandiford, T Johnson, J Lemaster, WE Cascio. Confocal microscopy technique measures subendocardial pH_i and pH_e gradient in perfused and ischemic rabbit papillary muscle. Circulation 94:I-713, 1996.

106. KA Yamada, J McHowat, GX Yan, K Donahue, J Peirick, AG Kléber, PB Corr. Cellular uncoupling induced by accumulation of long-chain acylcarnitine during ischemia. Circ Res 74(1):83–95, 1994.

107. J Cinca, M Warren, A Carreno, M Tresanchez, L Armadans, P Gomez, J Soler-Soler. Changes in myocardial electrical impedance induced by coronary artery occlusion in pigs with and without preconditioning: correlation with local ST-segment potential and ventricular arrhythmias. Circulation 96(9):3079–3086, 1997.

108. WT Smith IV, WF Fleet, TA Johnson, CL Engle, WE Cascio. The Ib phase of ventricular arrhythmias in ischemic in situ porcine heart is related to changes in cell-to-cell electrical coupling. Circulation 92(10):3051–3060, 1995.

109. BJ Scherlag, N el-sherif, R Hope, R Lazzara. Characterization and localization of ventricular arrhythmias resulting from myocardial ischemia and infarction. Circ Res 35(3):372–383, 1974.

110. AG Kleber, MJ Janse, FJ van Cappelle, D Durrer. Mechanism and time course of S-T and T-Q segment changes during acute regional myocardial ischemia in the pig heart determined by extracellular and intracellular recordings. Circ Res 42(5):603–613, 1978.

111. MM Gebhard, E Gersing, CJ Brockhoff, A Schnabel, HJ Bretschneider. Impedance spectroscopy: a method for surveillance of ischemia tolerance of the heart. Thorac Cardiovasc Surg 35:26–32, 1987.

112. JT Vermeulen, HL Tan, H Rademaker, CA Schumacher, P Loh, T Opthof, R Coronel, MJ Janse. Electrophysiologic and extracellular ionic changes during acute ischemia in failing and normal rabbit myocardium. J Mol Cell Cardiol 28(1):123–131, 1996.

113. E Kaplinsky, S Ogawa, CW Balke, LS Dreifus. Two periods of early ventricular arrhythmia in the canine acute myocardial infarction model. Circulation 60(2):397–403, 1979.

9

Electrical Stimulus, Reentry, Fibrillation, and Defibrillation: Insights Gained by the Graded Response and Restitution Hypotheses

Hrayr S. Karagueuzian and Peng-Sheng Chen
Cedars-Sinai Medical Center and the University of California, Los Angeles, Los Angeles, California, U.S.A.

I. INTRODUCTION

Ventricular fibrillation (VF) induced by an electrical stimulus in situ is normal hearts is initiated by a reentrant wave front of activation that is both functional and transient in nature. The characteristics of the induced functional reentry are compatible with the phenomenon of spiral wave of excitation, whose 2-D and 3-D (scroll-wave) dynamics and mechanisms have become the subject of intense recent experimental and simulation studies. The presence of an in-situ protective zone (i.e., termination of the induced in-situ scroll wave that heralds VF) by a single-point electrical stimulus strongly suggests the presence of only one or two counterrotating ("figure-eight") scroll waves in both ventricles at the onset of VF. The graded response hypothesis provides a cellular mechanism of a strong point electrical stimulus-induced reentry in the normal ventricular muscle. The induced reentry, with a relatively fast rotation period (around 100 ms), has a transient lifespan. Within few cycles (1–3), the core of the reentrant wave front meanders (drifts), then breaks up into multiple wavelets, each of which then propagates with its own "independent" regime. The restitution hypothesis provides an adequate explanation of the destabilization of the single scroll wave (meandering and breakup) that signals the transition from the tachysystolic Stage I VF (one or a pair of scroll waves) to the "convulsive incoordination," Stage II VF (multiple independent wavelets). These two sequential dynamic stages of VF were elegantly described by Wiggers more than 60 years ago using electrocardiographic and cinematographic methods [1]. While stage I VF can be terminated by a timed, single-point electrical stimulus ("protective zone"), stage II VF cannot. For an electrical stimulus to terminate the Stage II VF, a critically large portion of both ventricles must to be engaged by the electrical shock. In this chapter we present recent experimental and simulation findings that provide insight into the dynamic scenarios of VF induced by an electrical stimulus. Termination of reentrant and nonreentrant wave fronts by an electrical stimulus ("defibrillation") can also be adequately explained by the graded response (progressive depolarization) hypothesis of vulnerability to reentry. Not surprisingly, this hypothesis of the upper limit of vulnerability (ULV) in humans, which is characteristically identical to the defibrillation threshold (DFT).

II. METHODS AND MATERIALS

The studies were conducted in in-situ anesthetized dogs and in vitro on isolated cardiac tissues. A high-resolution (1.6-mm spatial resolution) plaque electrode, 3.2 cm by 3.8 cm, containing 500 bipolar electrodes, was used for in-vitro [2,3] and in-vivo [4,5] studies. Activation maps were analyzed either by dynamic visualization of the activation wave fronts or by constructing isochronal activation maps [6,7]. We also used conventional glass microelectrodes to simultaneously record single-cell transmembrane potentials (TMP) from two specific sites during the formation of a functional reentrant wave front in isolated tissues [2]. The choice of these two TMP recordings sites was based on an initially determined site of premature stimulus (S2)-induced conduction block and the site of earliest activation after the S2 [2] (see Fig. 1 for recording arrangement). The results that describe the genesis of VF upon electrical stimulus will be presented and discussed in the following order:

1. Induction of functional reentry and VF by an S2 stimulus
2. Cellular graded responses and the ULV hypothesis
3. The protective zone and the case for a single scroll wave during Stage I VF
4. The restitution hypothesis and wave front breakup
5. Defibrillation and the graded response hypothesis of the ULV
6. Virtual electrode effect and defibrillation

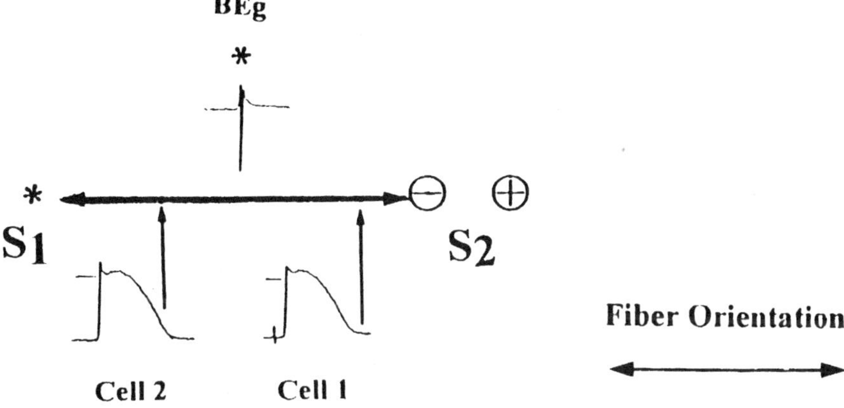

Figure 1 Diagram showing the locations of simultaneous two microelectrode recordings (two upward-pointing arrows) and that of the bipolar electrograms (asterisk). S1 and S2 are the locations of the regular pacing and premature stimulation sites. The circled negative symbol is the cathode side and the positive symbol is the anode side of the bipolar S2 stimulating electrode. The double-headed arrow shows the long axis of the fiber orientation.

III. RESULTS AND DISCUSSION

A. Induction of Functional Reentry and VF by an S2 Stimulus

1. In-Situ Studies

Chen et al. [8] have shown in open-chest anesthetized dogs that VF induced by a critically timed single-point stimulus (S2) is initiated by a figure-eight reentry around the S2 site. In this in-situ model of VF induction he basic S1 and the premature S2 stimuli are applied at different sites (near the outflow tract of the right ventricle), with the line connecting the S1S2 sites being parallel to the long axis of epicardial fiber orientation. With such an arrangement, an S2 of a critical strength (usually above 10 mA) and of critical coupling interval (during the relative refractory period) induces an activation that emerges some 5 mm away from the S2 site. The front then propagates in all directions but blocks near the S2 site. The wave front rotates around the S2 site of block, then reenters through it as it recovers, inscribing a figure-eight reentry (scroll wave) with a period of about 100 msec [8,9]. Subsequent to the first activation, the reentrant wave front drifts away from the mapping site during the second to the fourth cycle, causing loss of organized activation. The drifted reentrant wave front then undergoes breakup, and complete disorganization of activation wave fronts occur that signal the onset of VF.

2. In-Vitro Studies

The cellular basis of reentry formation by a strong electrical point stimulus was investigated in vitro on thin canine epicardial slices isolated from the same right ventricular sites mapped in the in-situ hearts during the S2-induced VF [2]. With the exact same stimulation protocol and similar electrode size and locations (Fig. 1), reentry induction was tested by an S2 stimulus to compare it with the in-situ results. Two transmembrane action potentials were recorded simultaneously from the most superficial (first three) epicardial cell layers with conventional glass capillary microelectrode [10,11]. The locations of these two cellular recordings were based on the results of initial activation map data. Specifically, one recording was made from a cell in the area of S_2-induced conduction block and the second recording was made from the region of the earliest activation after the S_2 stimulus. Cellular recordings were made during regular (S_1–S_1) pacing with a bipolar stimulating electrode (cycle lengths of 600 msec) from the left edge of the tissue and during premature stimulation with a second bipolar stimulating electrode placed in the center of the tissue. The line connecting the two stimulating electrode was aligned along the long axis of the fiber orientation as in the in-situ studies. Figure 2 shows that a critical S2 stimulus, as in the in-situ studies, induces an activation that originates away from the S2 site. The wave front then propagates around the S2 site of block, then reenters from the initial S2 site of block as this site recovers. While this method of stimulation typically induces figure-eight reentry in in-situ model, single-arm reentry is not uncommon in isolated tissues [2]. Subsequent to the first reentrant beat, a second reentrant activity may ensue, which in the in-situ case may degenerate to VF. The analysis of the transmembrane recordings by the two microelectrodes shows that the premature S2 stimulus applied during the relative refractory period induces a nonregenerative graded response that propagates slowly toward the more recovered cells near the S1 site to initiate an activation (Fig. 2) The graded response properties change as the S2 stimulus changes. Since the S2 characteristics (amplitude, coupling interval, and polarity) play a decisive role in reentry induction (vulnerability), a review of the S2-induced graded responses that mediate and lead to reentry formation is in order.

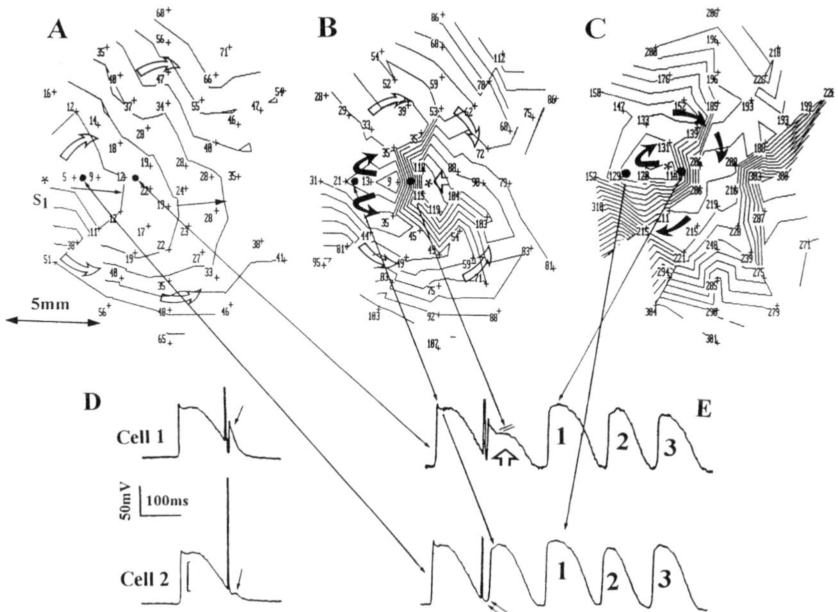

Figure 2 Sequential activation map and two simultaneous action potential recordings. A 56-channel bipolar electrode array was used in this study. (A) Isochronal activation map (10-msec isochrone interval) during regular S_1–S_1 pacing at 600-msec cycle length (asterisk). The crosses represent electrode locations and the numbers give the time of activation, with the onset of S_1 as time zero. The two dots represent the two sites from which subsequent simultaneous action potentials are recorded. The arrows in (A)–(C) point to the direction of wave front propagation. The horizontal double-headed arrow indicates the fiber orientation and also serves as a length scale. (B) Isochronal activation map of an S_2 stimulus (40 mA at 136-msec interval) applied in the center tissue (asterisk pointed by an open arrow). The site of earliest activation is located 3 mm away from the S_2, toward the S_1 site (isochrone encircling 9-msec site). The S_2-initiated wave fronts propagates first toward the S_1 site, then rotates (double curved arrows) around the site of block and reaches proximal to the site of block in 104 msec, forming a figure-eight. (C) Activation continues through the initial site of the block. (D) Two simultaneous action potential recordings from sites indicated in (A). An S_2 stimulus (40 mA, 122-msec interval) induces a graded response in Cell 1 (arrow), which propagates to Cell 2 with decrement in amplitude (35 mV to 5 mV) (single arrows). (E) An S_2 (40 mA and 136-msec interval) initiates a graded response in Cell 1 with an 8-msec delay, and an action potential in Cell 2 with an 18-msec delay that arises from the graded response (double arrows). The action potential initiated in Cell 2 blocks at the site of Cell 1 (large open arrow with double horizontal lines in Cell 1) with an electronic depolarization as in Figs. 8, 9 and 10. The reentrant wave front in (C) excites Cell 1, then Cell 2, as shown in (E), with action potential number 1. Two subsequent reentrant action potentials are also shown (2 and 3). (From Ref. 2. With permission of the American Heart Association.)

B. Cellular Graded Responses and the ULV Hypothesis

1. Time Domain

Tissue vulnerability to reentry is critically dependent on the timing on the S_2 stimulus. The S_1–S_2 coupling interval that initiated reentry ("the vulnerable period") is confined to a specific period that precedes the effective refractory period (ERP) by up to 50 msec (the relative refractory period) (Fig. 3) [2].

2. Current Domain

Reentry is induced only when the current strength of the S_2 is above (known as the lower limit of vulnerability) or below (known as the upper limit of vulnerability of ULV) a specific current amplitude. Current falling outside this region do not induce reentry no matter what stimulus timing is inclusive of the "vulnerable period" (Fig. 3). The mean threshold current strength for reentry induction is 28 ± 13 mA, and the mean strongest S_2 current above which no reentry could be induced is 72 ± 21 mA in canine thin epicardial slices [2]. These results are qualitatively compatible with earlier finding of reentry induction by a premature stimulus in isolated tissues [12], in-situ ventricles at the onset of VF [8], and in humans [13].

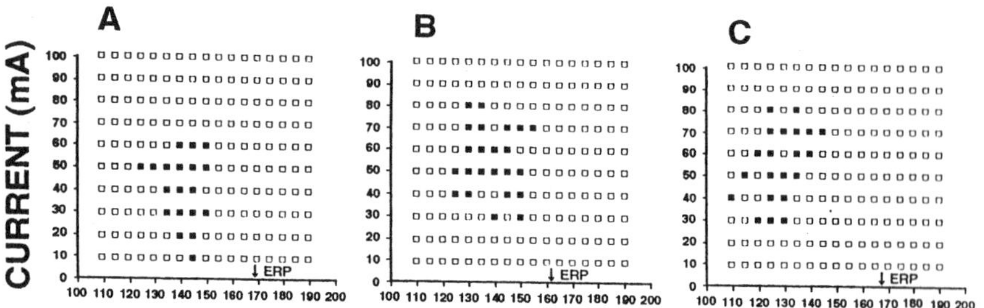

Figure 3 Strength–interval plots for reentry induction in three epicardial tissue slices. The filled squares indicate the S_2 trials that induced reentry and the open squares are the S_2 trails that failed to induce reentry. The ordinate, in milliamperes, is the S_2 current strengths, and the abscissa is the coupling intervals in milliseconds. ERP is the refractory period (downward-pointing arrow) measured with twice the diastolic current threshold during 600-msec cycle length. (From Ref. 2. With permission of the American Heart Association.)

3. The Relation Between the S_2 and the Graded Responses

The properties of ventricular muscle cell graded responses are a function of the S_2 current strength and current timing, transmembrane voltage, and electrode polarity near which the graded response is recorded.

4. Effects of the Current Strength and Transmembrane Voltage

An increase in the strength of the S_2 current from 5 to 100 mA, at a fixed interval (ERP-20 msec), progressively an increases the amplitude and duration of the graded responses. Figure 4 illustrates an example. For a given current strength, the greater the voltage negativity, the higher is the graded response amplitude. These findings [2] are compatible with those of Kao and Hoffman seen on the subendocardial fibers [14].

5. Effects of the Coupling Interval

An increase in the coupling interval of the S_2, at a fixed current strength, progressively increases the amplitude and the duration of the graded responses (Fig. 4) [2].

6. Graded Response Amplitude–Duration Relationship

A regression analysis shows a significant ($P < 0.01, r = 0.79$) positive linear correlation between the graded response amplitude and the graded response duration:

$$\text{Duration (msec)} = 0.98 \, \text{ms/mV} * \text{amplitude (mV)} + 8.41 \, \text{msec}$$

A regression analysis shows a significant positive correlation ($P < 0.01, r = 0.96$) between the ERP and the total APD over 100 msec of APD prolongation (Fig. 5):

$$\text{ERP (msec)} = 1.067 * \text{APD (msec)} - 24.51 \, \text{msec}.$$

7. Graded Responses Near the Anodal Pole

All the above measurements were made near the cathodal pole of the stimulating electrode. The graded response characteristics near the anodal pole of the S2 stimulating electrode are, however, different from those recorded near the cathodal pole (Fig. 6). No graded responses could be induced at distances greater than 3 mm from the anode, nor during the entire plateau range of the action potential for current strengths up to 100 mA. S_2 applied at a slightly later part of the plateau induced small-amplitude (2–4-mV) graded responses with a net shortening of the total APD. However, with relatively late-coupled S_2 stimuli (i.e., >110 msec), a graded response with a net prolongation of the total APD occurs (Fig. 6C), consistent with previous

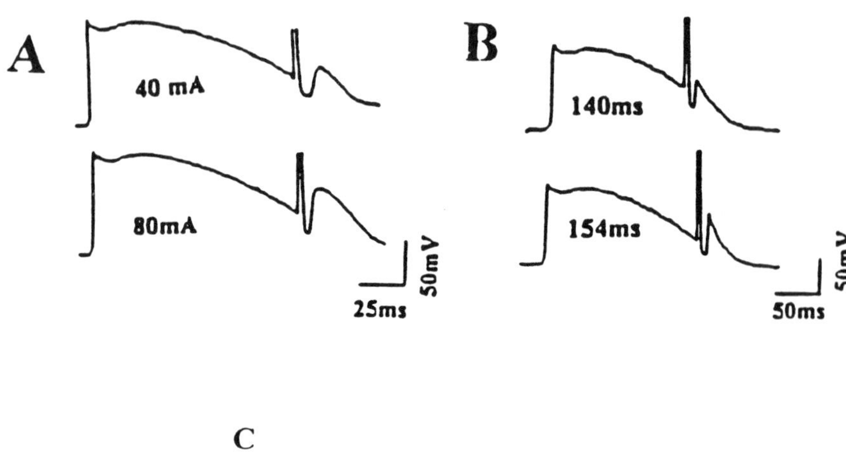

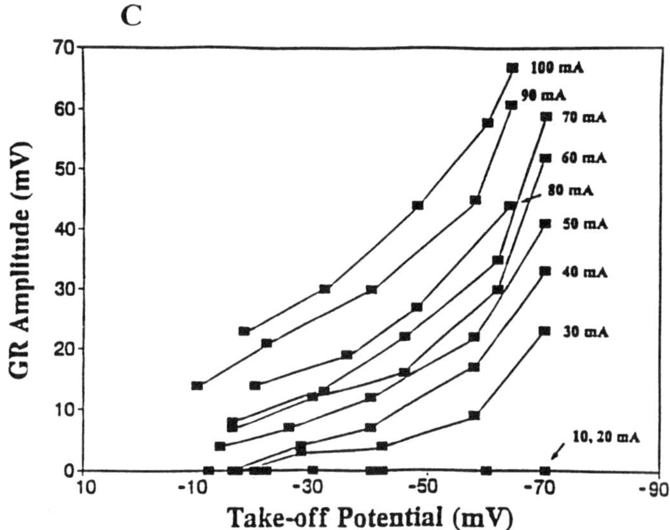

Figure 4 Relation of the graded response (GR) properties to the S_2 stimulus characteristics. (A) Effects of increasing the S_2 current strength from 40 to 80 mA at a fixed coupling interval. (B) Effects of increasing S_2 coupling intervals from 140 to 154 msec on the graded responses in a different tissue. An increase in the amplitude and the duration of the graded response occur in both cases (A and B). (C) Relation between the graded response amplitude and the take-off potential for currents strengths of 10 to 100 mA.

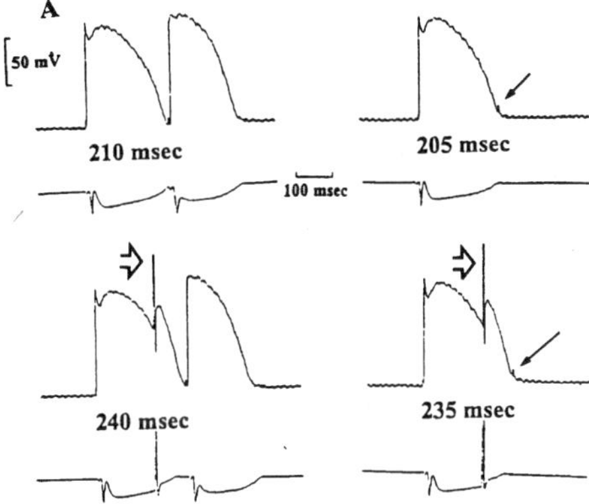

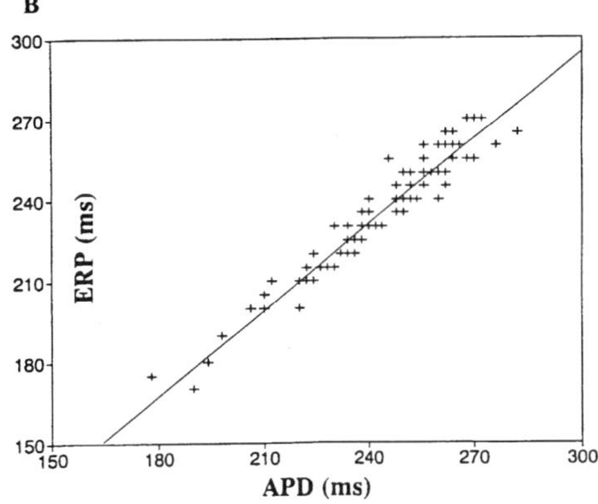

Figure 5   Relation between graded response-induced prolongation of the action potential during (APD) and the effective refractory period (ERP). (A) Top recordings show induction of a the earliest premature stimulus with an S1S2 coupling interval of 210 msec and a block with an S1S2 of 205 msec. After the induction of a graded response (lower left recording), the earliest premature responses is now initiated at an S1S2 coupling interval of 240 msec with block occuring at an S1S2 interval of 235 msec, reflecting a 30-msec increase in the ERP. Action potentials are recorded along the fiber 1 mm away from the S$_2$'s cathodal pole. The lower recording is a bipolar electrogram. (B) Relation between the graded response-induced prolongation of the total APD (abscissa), and the resultant increase in the ERP, ordinate. (From Ref. 2. With permission of the American Heart Association.)

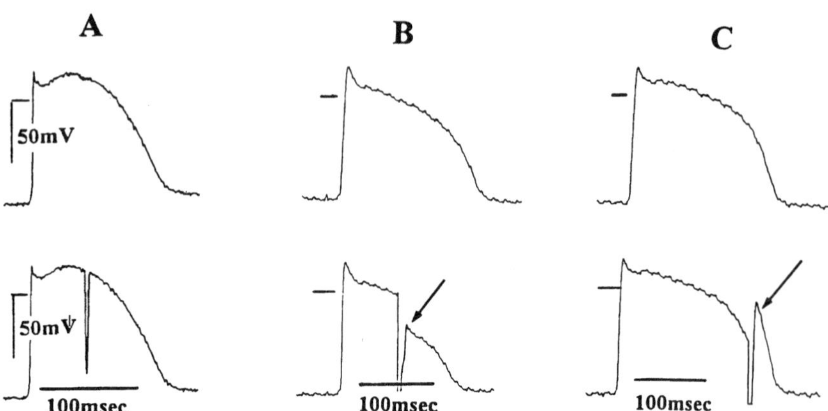

Figure 6 Effects of S$_2$ at the anodal side of the S$_2$ stimulus. Recordings are made 2 mm away from the anode, opposite the S$_1$ site (A–C). In (A–C), recordings of the top row show the last regularly driven action potentials at 600-msec cycle length prior to the S$_2$. Note lack of effect of an "anodal" current of 80 mA when applied during the plateau phase of the action potential (A, bottom). (B, bottom) shows the effect of an S$_2$ of 60 mA applied 80 msec after the upstroke. A small-amplitude graded response is induced (arrow) with a net (26-msec) shortening of the total action potential duration. (C) Effect of a relatively late-coupled S$_2$ (60 mA at a coupling interval of 165 msec), that induces a graded response (arrow) with a net (16-msec) prolongation of the total action potential duration. (From Ref. 2. With permission of the American Heart Association.)

studies on Purkinje fibers [15], endocardial ventricular muscle cells [16], and rabbit ventricular epicardium [17]. At equal distances from the two poles of the S$_2$ along the fiber (analysis done 2 mm away from each pole and with an S$_2$ of 80 mA), the amplitude and the duration of the graded responses were significantly ($P < 0.01$) lower in the cells at the anodal side than in the cells at the cathodal side (6 ± 2 mV) versus 28 ± 9 mV and 16 ± 4 msec versus 36 ± 10 msec, respectively [2].

8.   The Velocity and Extent of Graded Response Propagation

The graded responses propagate in an anisotropic and decremental fashion. The distance traveled along the long axis of the cells extends to about 5 mm and to 2–3 mm in the transverse direction [2]. Propagation is decremental, as the amplitude of the graded response progressively decreases at distances farther away from the S2 source (Figs. 2 and 7). The longitudinal velocity of the propagation is 5–6 times slower than the velocity of

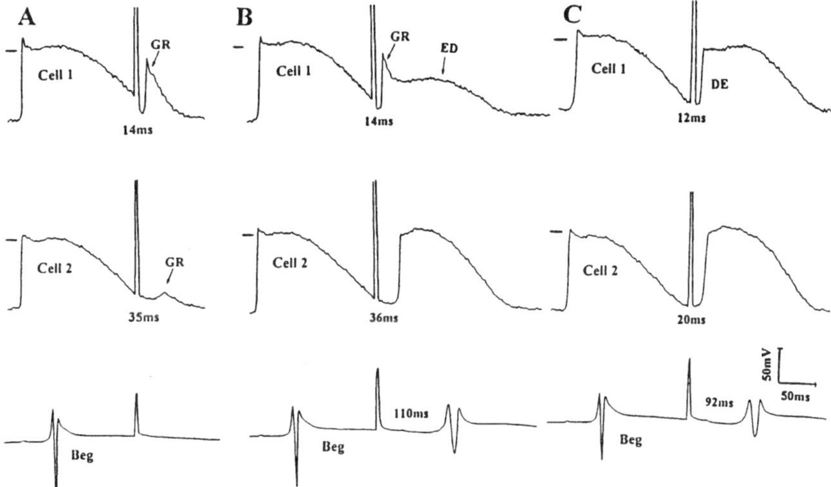

Figure 7   Comparative graded response (A and B) and regenerative response (C) conduction time in an isolated canine epicardial tissue. Cell 1 is 1 mm and Cell 2 is 5 mm away from the cathodal pole of the S_2. (A) shows the propagation of an S_2 (45 mA, 156-msec coupling interval)-induced GR that propagates to the distal Cell 2 with a 21-msec delay. (B) shows that when the coupling interval of the S_2 was increased to 162 msec, an action potential with a 22-msec delay after the GR is initiated in the distal Cell 2. This causes an ED in the proximal Cell 1. (C) shows initiation of a regenerative response by a direct excitation (DE) with an S_2 applied after full recovery (interval 180 msec) that propagates to the distal Cell 2 with an 8-msec delay. (From Ref. 2. With permission of the American Heart Association.)

regenerative responses $(18.2 \pm 3.8\,\mathrm{cm/sec})$ versus $60 \pm 10.9\,\mathrm{cm/sec}$, respectively [2].

9.   Propagation of the Graded Responses and Initiation of Action Potentials

When the amplitude of the propagated graded responses in the recovering cells toward the S_1 site reaches threshold, an action potential is initiated (Figs. 2,7–9).

10.   Graded Responses and Initiation of Reentry

Figure 2 shows one such example of reentry initiation. Figure 2A shows activation during regular pacing. Figure 2B shows that an S_2 of 35 mA strength applied with a coupling interval of 150 msec caused a local block

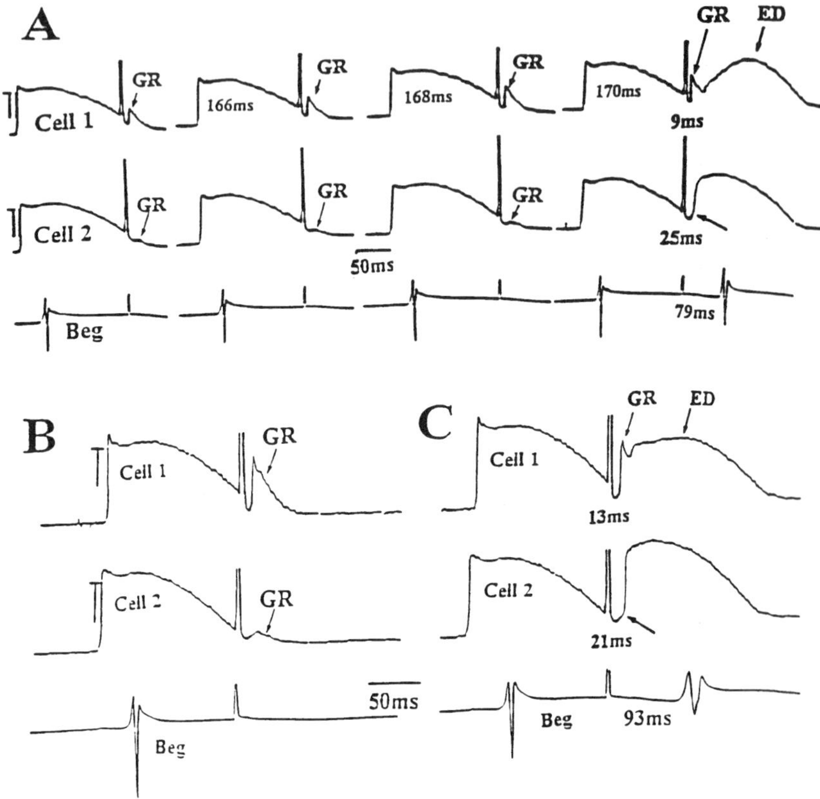

Figure 8 Propagation of the graded responses and initiation of action potential in the distal recovered cells. (A–C) show two simultaneous action potential recordings, Cell 1 is 1 mm and Cell 2 is 3 mm away from the cathode of the S_2. (A) is from one experiment and (B) and (C) from another. The bottom recording is bipolar electrogram (Beg) in the middle, with the two cells toward the top of the tissue. In (A) the effects of increasing S_2 (40 mA) coupling intervals (from 164 to 170 msec) are shown. At 170-msec interval the graded response (GR, single arrows) in the distal cell initiated an action potential with a smooth transition from phase 4 to phase zero of the action potential (single upward-pointing arrow). The proximal Cell 1 shows a slowly rising electronic depolarization (ED) that occurs during the repolarization of the GR 25–30 msec after the upstroke of the Cell 2 action potential. (B) and (C) show that increasing the S_2 current strength from 35 mA (B) to 55 mA (C), with interval fixed at 132 msec, increases the distal cell GR amplitude and initiates an action potential in Cell 2 (single upward-pointing arrow in C). As in (A), Cell 1 shows ED during the falling phase of the GR (arrow). The vertical bar in all panels is 50 mV and the horizontal bar is zero reference potential. (From Ref. 2. With permission of the American Heart Association.)

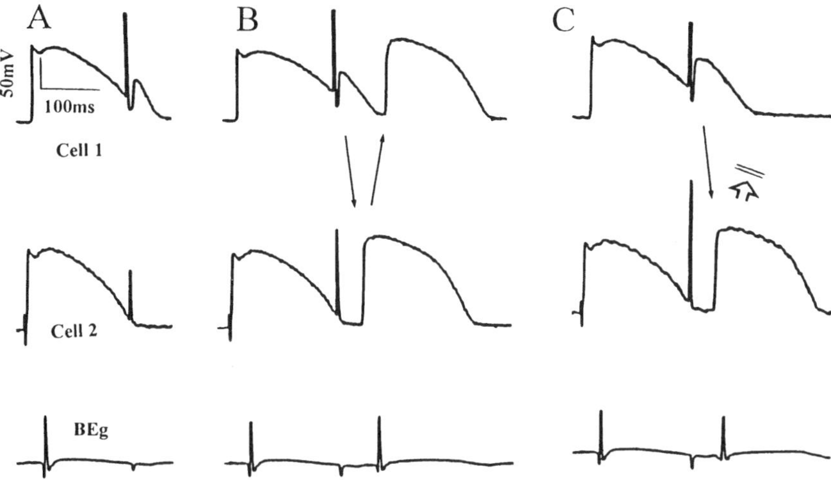

Figure 9   Super-strong S_2 stimulus prevents reentry. Recording arrangements and abbreviations same as in Figure 2–5. Cell 1 was 1 mm and Cell 2 was 6 mm away from the cathode of the S_2. Increasing the S_2 current strength from 40 mA (A) to 50 mA (B) (interval fixed at 170 msec) initiates an action potential in the distal Cell 2 (arrow pointing downward) with delay of 48 msec, which then reenters and excites Cell 1 (upward-pointing arrow in B). (C) shows that when the S_2 becomes 70 mA (10 mA above the ULV), the duration of the GR increases from 58 msec (B) to 102 msec. In this case, distally originated action potential fails to excite Cell 1 (open arrow intercepted by double horizontal lines). The numbers under the recordings are delay times after he onset of the S2. (From Ref. 2. With permission of the American Heart Association.)

and distal early activation (two curved arrows), leading to the first reentrant wave front (Fig. 2C). Subsequent simultaneous recordings of two trans-membrane potentials (site of block and site of earliest activation, 2 dots in (Figs. 2A–2C) are shown in Figs. 2D and 2E. The distally originated front (Fig. 2B) rotates around the site of block, then reenters (Fig. 2C), initiating the first action potential (#1 in Fig. 2E).

11.   Super-Strong S₂ Stimulus Prevents Reentry

As shown in Figure 3, when the strength of the S2 stimulus exceeds a critical limit, reentry can no longer be induced (ULV). This is caused by the excess prolongation of the ERP at the S2 site, which converts unidirectional conduction block to bidirectional block and prevents reentry formation. Figure 9 illustrates one example in which the ERP was progressively increased by the graded responses while progressively increasing the strength of the S_2 current at a fixed interval. Increasing the current strength from 40 mA (Fig. 9A) to 50 mA (Fig. 9B), caused a graded response-mediated origination of action potential in the distal Cell 2, which rotates around the site of block and reenters through this site of block after it recovers it excitability (Fig. 9B). In this case the graded response duration was 58 msec and the ERP was 283 msec. However, when the current strength was raised to 80 mA (Fig. 9C), the graded response duration became 102 msec and the ERP became 288 msec. In this case the distally originated action potential could not reenter [2].

12. Activation Pattern After a Super-Strong S_2 Stimulus That Does Not Induce Reentry

The activation map showed a super-strong S2-induced bidirectional block near the S2 site and prevented graded response-induced origination of activation from reentering from the site of initial block (Fig. 9) [2]. These findings provide a cellular basis for the well-known phenomenon of the ULV.

C. The Protective Zone and the Case of Single Scroll Wave During Stage I VF

A point electrical stimulus (S3) subsequent to a premature S2 stimulus that induces VF can prevent the emergence of VF [9]. For the S3 stimulus to prevent induction of VF, the S3 stimulus must be applied at critical coupling intervals after the S2 stimulus. The time interval(s) during which an S3 stimulus prevents the emergence of VF is known as the "protective zone" [18]. An S3 applied in the protective zone terminates the Stage I reentrant wave front induced by the S2. It has been hypothesized that the stimulus exerts its protective effect by terminating local reentry [19]. The graded response hypothesis of vulnerability to reentry explains the protective zone phenomenon. When the strong stimulus falls before complete recovery, a graded response is induced at the stimulus site that prolongs the refractory period. When the leading edge of the reentrant wave front revisits the area of graded response-induced ERP prolongation, it cannot reenter, resulting in the termination of reentry. This is analogous to the super-strong S2 current (ULV) that can be induce reentry because of conversion of the site of unidirectional block to bidirectional block. Figure 10 is a diagram that illustrates our proposed protective zone hypothesis by which S3 may terminate reentry during Stage I VF. The figure depicts only one of the two reentrant wave fronts of the figure-eight reentry. The duration of the

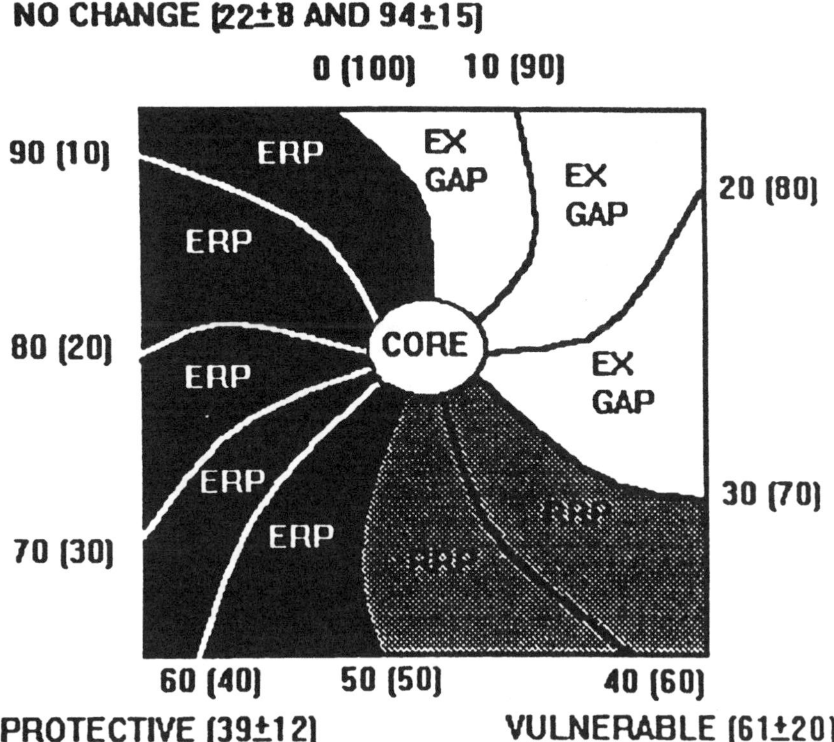

Figure 10 Diagram showing interaction between strong electrical stimulation and reentrant wave fronts. This figure shows only one of the two reentry wave fronts of the figure-eight reentry that is induced by a strong S2 in the in-situ heart. This reentrant wave front moves in a clockwise direction with a cycle length of 100 msec. The numbers 0–90 indicate the times of the isochronal activation lines. The numbers in parentheses indicate the recovery intervals. The black, gray, and white areas indicated the effective refractory period (ERP), the relative refractory period (RRP), and the excitable gap (EG), respectively. These measurements were made at the instant when the activation wave front is at the 0 isochronal line. When an electrical stimulus is given approximately 39 msec after the preceding activation, it falls in the protective zone and terminates reentry. However, if the stimulus is applied 56 msec after the preceding activation (vulnerable zone), it terminates the reentry but initiates a new one. If the shock is given too early (20 msec of recovery interval) or too late (92 msec of recovery time), the reentry continues in the same direction. The figure shows that the effect of electrical stimulation on the reentrant wave front are dependent on the relationship between the time of the stimulus and the recovery interval. (From Ref. 9. With permission of the American Heart Association.)

effective refractory period (ERP) of fibrillating ventricular muscle ranges between 48 and 77 msec [20]. We hypothesize that the first 50 msec constitutes the ERP, followed by 20–30 msec of relative refractory period. The remaining part of the reentrant cycle is therefore the excitable gap. Since Stage I VF may involve more than one cycle of organized stationary reentrant wave front, the protective interaction of the S3 stimulus with the reentrant wave front occurs recurrently [9]. When the S3 is applied about 40 msec after the preceding activation, the reentrant wave front is terminated and the VF is aborted. When the S3 is applied 61 msec after the preceding activation, the existing pattern of activation is disrupted and the S2-induced reentry terminated. Instead, a new activation pattern arises that evolves to VF. We call the interval that follows the protective zone during which the S3 changes the existing pattern of activation by terminating reentry but nevertheless causes the VF to be induced the "vulnerable period" [9]. The recovery interval associated with new patterns of activation averages 61 msec, an interval that corresponds to the RRP in fibrillating ventricular muscle cells [20]. As expected, when the S3 is applied 22 msec after the preceding activation, at a time when the reentrant activation wave front would have just revisited the site, no change of activation pattern is observed [9]. When the S3 is given 94 msec after the preceding activation, when the cells near the S3 are fully recovered, the patterns of activation also remain unchanged and the S2-induced reentry continues unperturbed [9]. These findings support the hypothesis that, depending on the time of the electrical stimulation relative to the reentrant activation, a stimulus may result if termination, reinitiation, or no change of the reentrant wave front occurs during Stage I VF (Fig.10).

D. The Restitution Hypothesis and Wave Front Stability

Earlier in-situ canine mapping studies [8] and subsequent simulation [21] of excitable media strongly suggest that the transition from Stage I VF to Stage to II VF is brought about by the breakup of the single scroll wave (functional reentry in 3-D) present during Stage I VF. Allessie et al. first made the seminal observation that functional reentry can occur in cardiac tissue in the absence of an anatomical obstacle (originally called leading circle reentry) [22]. Subsequently, functional reentry (spiral wave) was documented experimentally in animal [12,23–25] and human [26] ventricle. This raised the possibility that the unstable spiral wave (functional reentry) present during the Stage I VF may undergo breakup into multiple wave fronts evolving to Stage II VF. Spiral waves are a generic property of excitable media; they arise in cardiac tissue because conduction velocity depends on wave-front curvature [27]. The more convexly curved a wave front is, the more slowly it propagates, because the depolarizing current at the leading edge is diluted into a larger sink of resting cells. At a critical curvature, the source of depolarizing current is too small to bring the sink (resting tissue) to threshold, and propagation fails despite the presence of fully excitable cells downstream. A spiral wave is born when a break occurs along a propagating wave front; at the break, the end is highly curved and forms the tip of the spiral wave, which precesses around a core defined by the critical curvature. Along the arm of the spiral wave, curvature progressively decreases, allowing conduction velocity to increase with radial distance from the tip. The landmark study of Davidenko et al. [28] demonstrated spiral wave as a mechanism of functional reentry in normal cardiac tissue. That study also documented that spiral-wave reentry could be unstable, as predicted by computer simulations [21,29,30]. In this case, the core around which the spiral-wave arm rotates is not stationary but meanders through the tissue, producing the ECG appearance of polymorphic tachycardia or even fibrillation [31]. In addition to meandering, computer simulations also demonstrated that spiral waves can break up to form multiple spiral waves similar to wave fronts observed during fibrillation [29,31–33]. Movement of multiple spiral waves and their cores (also called phase singularities) is complex [34], so the classic spiral morphology become distorted, characteristic of disorganized wave fronts in fully developed fibrillation.

Breakup of a single spiral wave (scroll wave in 3-D), more often than not, is preceded by meandering (drift), a phenomenon that promotes wave front breakup. At the present there is no direct experimental proof of a single scroll wave breakup that converts VT to VF (i.e., transition from Stage I VF to Stage II VF) [1]. While single spiral drift may promote fibrillation-like activity, breakup of the meandering wave front eventually develops causing the characteristic multiple-wave-front VF [35]. Recently, studies on the mechanism(s) leading to wave front breakup became the focus of intense research. These studies unequivocally show that the restitution properties of the APD (Fig. 11) and the CV are important determinants of reentrant and nonreentrant wave front stability. When the restitution curve (the relationship between the APD or CV to the preceding DI), has a steep slope (>1), the reentrant and nonreentrant wave fronts manifest complex oscillations in cycle length (CL) and APD [36]. When the magnitudes of these oscillations was critically elevated, the "1-D" reentry around an anatomical obstacle abruptly terminated [36]. Similarly, increasing the magnitude of CL oscillation during AV nodal reentry abruptly terminated the reentry [37]. Figure 12 shows the relationship between the steepness of the APD restitution and spiral-wave stability. For spiral wave rotating at constant CL, APD and DI equilibrate at the intersection of the restitution curve (solid line) with the dotted line defined by relationship

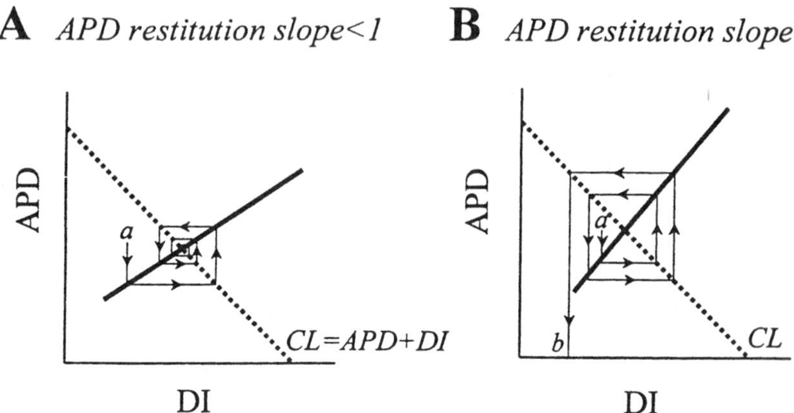

Figure 11 Relationship between steepness of APD restitution and spiral-wave stability. For spiral wave rotating at constant CL, APD and DI equilibrate at intersection of restitution curve (solid line) with dotted line defined by relationship APD + DI = CL. (A) For a shallow APD restitution (slope <1), a small perturbation (a) that shortens DI results in even smaller change in APD, producing smaller change in DI, and so forth, until equilibrium is reestablished. (B) For steep APD restitution (slope >1), a small decrease in DI (a) produces larger change in APD, which produces larger change in DI, etc. The oscillation is amplified rather than damped. When DI becomes too short to generate action potential (i.e., at b), conduction fails, causing wave break along the spiral-wave arm. (From Ref. 5. With permission of the American Heart Association.)

APD + DI = CL. (A) For a shallow APD restitution (slope <1), a small perturbation (a) that shortens DI results in even smaller change in APD, producing smaller change in DI, and so forth, until equilibrium is reestablished. (B) For steep APD restitution (slope <1), a small decrease in DI (a) produces larger change in APD, which produces larger change in DI, etc. The oscillation is amplified rather than damped. When DI becomes too short to generate action potential (i.e., at b), conduction fails, causing wave break along spiral-wave arm. The mechanism by which the slope of the APD restitution acts as a difference amplifier is as follows. The curve provides the next value of APD as a function of the previous DI. For a spiral wave rotating at a constant CL, for example, the equilibrium values of APD and DI occur at the intersection of the APD restitution curve, with the dashed line representing the CL (Fig. 11). With a slight perturbation (change) in the DI, the APD of the next beat will differ according to the APD restitution curve. The new APD in turn will generate a new DI. Whether this difference in DI is greater or smaller is determined by the APD

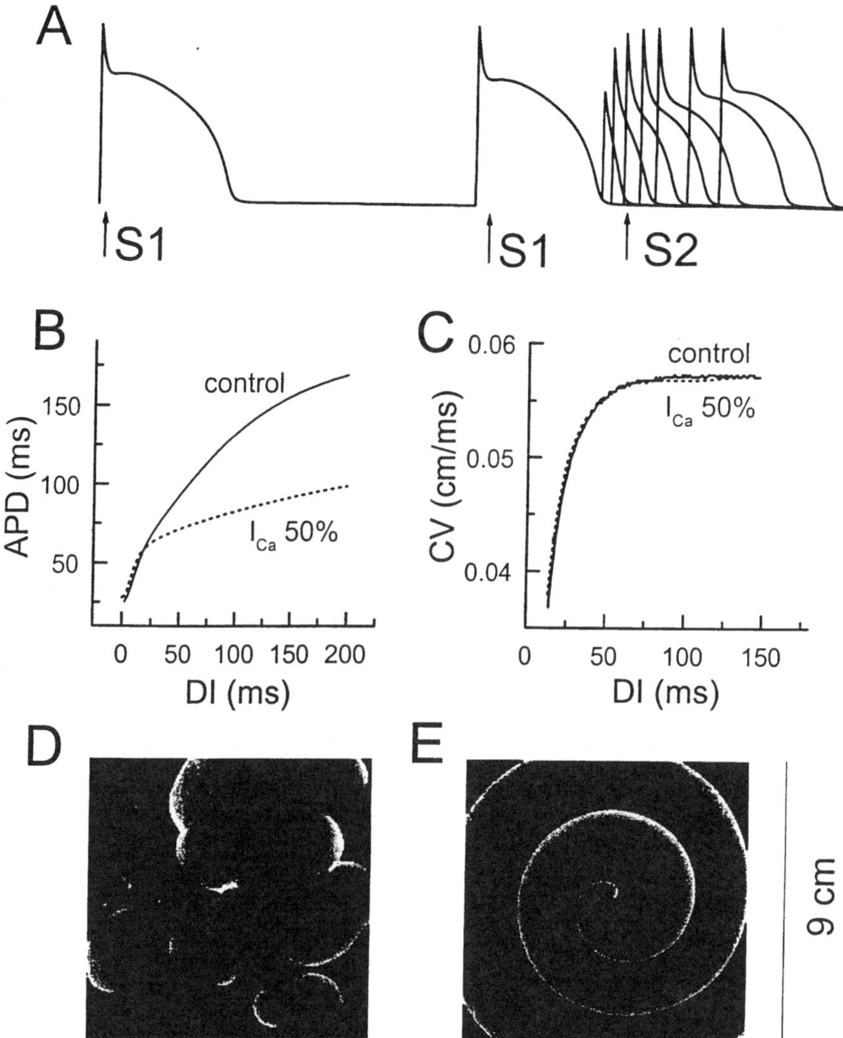

Figure 12 Effects of APD and CV restitution on spiral-wave stability in Luo-Rudy ventricular action potential model. (A) During pacing at fixed CL (S_1–S_1), a premature stimulus (S_2) is introduced at progressively shorter DIs. APD progressively shortens as DI decreases until cell is refractory. (B) APD restitution curve before (solid line) and after (dashed line) Ca current is reduced by 50%. (C) CV restitution curves under identical conditions. (D, E) Snapshots of activation patterns 5 sec after initiation of spiral-wave reentry. Under control conditions (D), spiral wave breaks up into multiple wave fronts resembling fibrillation; after Ca current is reduced to flatten APD restitution (E), the spiral wave remains intact. (From Ref. 5. With permission of the American Heart Association.)

restitution slope. If it is <1 (Fig. 11), the next difference is smaller, and if it is >1 (Fig. 11B), the next difference is larger. In this way, a shallow slope restores APD and DI back to their equilibrium values, whereas a steeply sloped APD restitution curve amplifies the differences so that they progressively diverge. If this oscillation grows large enough, the DI will eventually becomes shorter than the refractory period, causing a wavebreak at some point along the spiral wave arm. APD restitution produces oscillations in the wave back (repolarization phase), whereas CV restitution creates oscillations in the wave front (depolarization phase) by slowing CV in regions with short DIs. This creates a spatial mode of CL oscillation. This interaction between APD and CV restitution creates spatiotemporal oscillations with regime consistent with quasiperiodic dynamics [38].

We now provide evidence acquired from simulation and experimental studies that shows how a steeply sloped APD and conduction velocity (CV) restitution curves in 2-D and 3-D tissues promote wavebreak.

1. Simulation Studies

The influence of steep slope of APD restitution (slope>1) promoting wavebreak was reported by Karma [32], who realized the instability resulting from restitution, if it occurred in a 2-D matrix along the arm of a spiral wave, would produce wave break. Courtemanche et al. [32], however, found that increasing the maximum slope of the APD restitution prevented spiral wave breakup. These finding were different from Karma's findings, which showed increased breakup with increased steepness of the APD restitution curve in a simplified 2-D model of spiral wave [21,32]. More recently, Qu et al. [39], using a relatively realistic cardiac action potential model (Luo-Rudy I), stressed that spiral breakup is closely related to the DI over which the slope of the APD restitution curve remains >1 rather than to the maximum slope of the APD restitution curve. These authors suggested, based on theoretical arguments, that the excitable gap near the tip of the spiral wave is very narrow, a property that is functionally equivalent to a very short DI. Moving away from the tip toward the arm of the spiral, the excitable gap progressively increases, equivalent to relatively longer DIS. Therefore, if the APD restitution slope is steep only at very short DIS, only the spiral tip will be subject to oscillations, causing the tip of the spiral for meander. However, spiral breakup will not occur, because the spiral arm is subject to longer diastolic intervals, at which the slope of the APD restitution is <1. Thus, oscillations in APD and DI along the spiral arm will be damped and wavebreak will not occur. In contrast, however, when the range of DIS over which the slope is >1 extends to wider range to encompass the long DIS experienced by the spiral arm, oscillation in APD and diastolic along the spiral arm becomes amplified, leading to wave break away from the tip [39]. These observations explain the discrepancy between Karma's and Courtemanche's results.

In addition, Qu's simulation studies showed that the CV restitution promotes spiral breakup independent of APD restitution [39].

2.   Experimental Studies

a. In-Situ Studies.   VF was induced in anesthetized open-chest dogs by rapid pacing while mapping epicardial activation pattern using 477 bipolar electrodes [5]. A dynamic restitution curve [40] was constructed by plotting the ERP versus the DI. The DI was estimated by subtracting the ERP from the pacing interval (PI), i.e., the interval between two consecutive S1–S1 inputs. When the PI was between 600 and 260 msec the restitution curve was relatively flat, with a slope of 03 ± 025. However, when the PIs were < 260 msec, the slope became steeper (1.04 ± 1.38, $P < 0.001$) and was associated with the development of significant CL variations, i.e., the interval between two consecutive activations. The slope of the restitution within 50 msec of the CL that induced VF was 2.57 ± 1.68 [5]. As the slope of the restitution curve increased during rapid pacing, temporal and spatial variations in the conduction time also increased. When the PI shortened, both morphology and CL variations developed. When the PI shortened to 200 and 190 msec, transient complicated patterns of CL and morphology variations emerged at the beginning of the pacing train, which then settled into a stable alternans toward the end of the pacing train. As the PI further shortened to 180 msec, VF was initiated at the beginning of the pacing train, accompanied by large variations in the CL. There was a significant positive correlation between the longest PI associated with significant CL variations and the PI that induced VF (the VFT) in all dogs ($n = 5$) that we studied. That is, the longer the PI at which greater than 50% of the sites showed variability, the longer the VFT was. The consequences of larger spatio-temporal CL and CV variations (alternans), typically seen at short PIs, are the creation of wavebreaks [5].

We now discuss the dynamic scenario of wavebreak in light of our in-situ findings. Spatial variations in CL, resulting from CV restitution, will also result in spatial variations in DI, because $CL = APD + DI$. This fact directly links CV restitution to ERP restitution. As a consequence of CV restitution, two nearby cells develop a slight difference in their Dis during rapid pacing. Because of ERP restitution, these differences in DI will cause the ERP of the next beat to differ at the two sites. Those two slightly different ERPs will then generate two different next DIs. Whether this difference will be greater or smaller than the preceding differences is determined by the slope of their ERP restitution curves. If it is >1, the next difference will be larger, and if it is <1, the next difference will be smaller. In this way, a steeply sloped ERP restitution curve becomes a "difference amplifier," the gain of which is the slope. Thus any spatial differences in CL and DI resulting from CV restitution will be amplified on the next paced beat by a steeply sloped ERP restitution curve and further increase the functional dispersion of refractoriness. In this way, a steep ERP restitution amplifies over time the spatial differences in DI and ERP produced by the CV restitution. That is, CV restitution excites a spatial mode of oscillation in DI, while the ERP restitution excites a temporal mode [5]. In a homogeneous and isotropic tissue, for the sake of theoretical argument, growing spatiotemporal oscillations might lead to a DI which may be short to generate and action potential, causing the propagation to fail. Alternatively, initiation of activation at very short DI will produce activation with very short APDs that have low safety margin for propagation and as a result will block, causing wave break [41]. Because the spatial variations in CL and DI resulting from CV restitution are radially symmetrical in homogeneous tissue, propagation failure occurs simultaneously everywhere along the wave front at once, leading to extinction of the target wave induced by the pacing stimulus. In heterogeneous tissue, since no such radial symmetry exists, a short DI with failure of propagation will occur at one point along the wave front (wave break), while the remaining portion of the wave front will continue to propagate. This situation is analogous to the well-known phenomenon of "source-to-sink mismatch." This wavebreak may lead to reentry signaling the onset of VF. Alternatively, if the wavebreak occurs during fully developed VF, the daughter wavelets of a wave break maintain the VF by providing constant source of activation. Wave break occurs because of the intrinsic heterogeneity that exists in the normal canine ventricle. These in-situ canine studies provide evidence for a casual role of alternans, and shows how alternans caused wave break, i.e., long–short couplings create conduction failure which, in 2- and 3-D media yield wave break with the subsequent creation of reentrant spiral wave and scroll waves, respectively. These findings are consistent with the results of Pastore et al. [42], who reported that "discordant alternans" of APD induced by rapid pacing is casually related to the initiation of VF.

b. In-Vitro Experimental Studies.   We tested the restitution hypothesis of wave break using pharmacological agents in isolated in-vitro swine RV preparations. The hypothesis is based on the premise that flattening the slope of the APD restitution curve will prevent wave break. Since VF is maintained by the continuous breakup of activation wave fronts, an agent that flattens the slope of the APD restitution curve should prevent breakup and therefore should terminate the VF. Bretylium is an effective antifibrillatory drug capable of chemical defibrillation. We found that bretylium reduces the slope of the dynamic APD restitution curve while at the same time also increasing the ERP, a property that lengthens the wavelength. Wavelength plays an important role in spiral-wave reentry by setting the minimal space required for a spiral wave to sustain itself. If the wavelength is too long relative to tissue size, tachycardia or fibrillation will self-terminate [43,44]; this is the basis of the "critical mass hypothesis" [45–47]. To separate effects of wavelength from those of restitution we added to the bretylium perfusate cromakalim, an ATP-sensitive potassium channel opener that shortens the ERP [3], and then retested the restitution hypothesis of vulnerability. The combination of drugs caused a net shortening of the ERP and a concomitant flattening of the APD restitution curve from 1.88 ± 0.22 to 0.18 ± 0.32. With such an effect the maximum number of wavelets that supported the VF was reduced from about 5 to 1, an effect that converted the VF to monomorphic VT supported by only one spiral wave [48]. More recently, we found that nicotine increases the slope of the APD restitution curve and the DI over which the slope remained > 1 in the surviving epicardial border zone of dogs with healed myocardial infarction [49]. This effect of nicotine promoted wave break during pacing at a slower pacing rate, which led to reentry formation with subsequent degeneration of the reentry to VF [49].

The restitution hypothesis of vulnerability also supports that VF is maintained by multiple wavelets. While transiently stationary scroll waves can produce short-duration periodic activity during VF, these periodic sources are rapidly destroyed and as such cannot maintain the VF with their high frequencies of activation [50]. Furthermore, the restitution hypothesis of VF also provides indirect evidence of the breakup hypothesis of VF maintenance. As the slope of the APD restitution curve is flattened, the number of VF wavelets progressively decreases and eventually settles to one reentrant (rotor) source with much slower frequency than the frequency of VF. Similarly, our own work using the isolated RV swine preparation showed the progressive tissue size reduction causes reduction in the number of wave fronts, slowing of activation, and eventual termination of VF [43]. The progressive slowing of the VF CL argues against a single source of activation as a cause of fibrillation, slower activation rates would prevent breakup of the presumed single high-frequency rotor source would have maintained the same (presumably fast) activation CLC. However, the opposite happens. The VF is terminated rather than maintained. Similarly, failure to capture the VF by rapid pacing as a surrogate of "high-frequency dominant source" at distance longer than just a few centimeters in the in-situ fibrillating canine ventricle argues against "high-frequency periodic sources" as a "robust" mechanism of VF maintenance [51].

E. Defibrillation and the Graded Response Hypothesis of the ULV

It is well known that successful defibrillation depends on the strength of the electrical shock: the higher the shock strength, the greater is the probability of successful defibrillatior [52]. However, the nature and the influence of the shock on the rapidly activating muscle cells on the outcome of defibrillation have been poorly explored.

The demonstration of the presence of a protective zone at distinct time intervals after the induction of the first reentrant wavefront of the VF shows the importance of the timing of the shock relative to the recovering cells engaged in the reentry. A stimulus not falling in the protective zone may terminate existing reentrant and nonreentrant wave fronts but induces a new wave front that is different from the one present at the time of the stimulus application [9]. The presence of a discrete time interval during which a stimulus induces a different reentrant wave front known as the "vulnerable period," exists both during Stage I VF [9] and Stage II VF [53]. Application of a shock in the vulnerable period during Stage I VF fails to prevent the termination of Stage I as Stage II VF emerges [9]. Similarly, application of a shock during Stage II VF that coincides in time with the vulnerable period terminates all existing activation wave fronts (after a threshold defibrillation shock there is period of nonregenerative activity regardless of the outcome of the shock), but at the same time reinitiates new wave fronts that eventually evolve to VF [53]. In fact, during in-situ canine VF, when the defibrillation shock occurs at a site that was activated 64 ± 11 msec prior to the shock, a new activation wave front arises from that site while other wandering wave fronts having different recovery intervals become extinct. The newly initiated activation, which appears to be the earliest site activation, evolves to VF. Such shocks, falling in regions of the ventricle that had about 60 msec of recovery time prior to the defibrillation shock, are typically seen after unsuccessful defibrillation shocks [54]. Successful defibrillation occurs when all sites of the fibrillating ventricle experience a shock strength that is above the ULV even though such shocks fall during the vulnerable period [55].

The graded response hypothesis can explain the cellular basis of the ULV for the induction of reentry and the protective zone. The presence of the ULV for graded response-induced reentry in in-vitro and in-situ canine hearts [8,56] and in humans [57] suggests a common underlying mechanism of defibrillation. Similarly, Since ULV closely agrees with the defibrillation threshold [57], it is suggested that the graded response hypothesis of vulnerability may have relevance in the understanding the mechanism of defibrillation. Because the values of the ULV and the defibrillation energy requirements are closely related [57], the graded response mechanism of vulnerability to reentry may also have relevance to the mechanism of ventricular defibrillation.

Successful defibrillation occurs when the shock strength is high enough to convert unidirectional block to bidirectional block, preventing reentry [9]. It is not uncommon to observe one or two activation wave fronts after a successful shock that terminates the VF. It is possible that these wave fronts arise as a result of the shock-induced graded responses that propagate slowly and initiate activation wave fronts from recovered areas but subsequently die out because they cannot find recovered areas to reentry. The latter phenomenon occurs because of excess prolongation of the ERP by the graded responses. So far as we know, no single hypothesis of defibrillation provides a unifying hypothesis, at the cellular level, combining vulnerability to defibrillation. Recently, Kwaku and Dillon, using optical mapping in the rabbit ventricle, found that unsuccessful defibrillation failed to show critical points in 236 of 239 episodes of defibrillation that they analyzed [58]. They concluded that "such wave front dynamics [induction of reentry by critical point formation in defibrillation] are not absolutely necessary for defibrillation to fail" [58]. These authors further suggested that a critical degree of ERP prolongation (>60%) was necessary for the cessation of fronts and prevention of reentry formation. This proposed mechanism of defibrillation is compatible with the graded response hypothesis of vulnerability, ULV, and defibrillation. The shock strength should be high enough (ULV) to cause sufficient (i.e., >60% [58]) prolongation of ERP to terminate all fronts without including new ones, even such shocks falling during the vulnerable period. The graded response hypothesis can also explain the greater defibrillation efficacy of biphasic shock compared to monophasic shocks. With the high likelihood that myocardial cell during VF may be in different stages of excitability, the defibrillation shock may not develop graded responses of appreciable amplitude in cells during their plateau phase of the cardiac cycle. As a result, no or only minimal extension of their refractory period will develop during the first depolarizing phase of the biphasic shock. However, during the hyperpolarizing phase of the biphasic shock, an acceleration of repolarization will occur, which upon the break of the anodal phase of the biphasic shock results in a graded response, often causing a net prolongation of the ERP [17].

F.   Virtual Electrode Effect and Defibrillation

The region of an excitable tissue that becomes directly depolarized by the stimulus current and from which propagation proceeds has been termed the "virtual cathode." In the 3-D complex cardiac syncytium, one-dimensional cable theory fails to explain the unexpected patterns of virtual electrode effect because of the complexity of the stimulus current flow in the heart. Simulation studies showed that the use of a bidomain model of cardiac tissue replicated much of the experimental findings [59]. Bidomain models are based on passive properties of the cardiac tissue, and assume that the ratios of electrical conductance along the fiber and perpendicular to it differ (anisotropy) in the intracellular and extracellular spaces (domains). The difference in anisotropy generates a complex myocardial charge distribution, causing complex TMP distribution in response to strong electrical stimuli.

Earlier studies using excitation measurements and subsequent epifluoresent maps provided unexpected 2-D images of virtual electrode (cathode and anode) effects in in-situ hearts that may exert profound influence on the outcome of the defibrillation shocks. Such a potential is appreciated, as cardiac tissue can be electrically stimulated with the onset (make) or termination (break) of an electrical current that is applied with either cathodal (negative) of anodal (positive) stimuli. Wikswo et al. [59], using modified one-dimensional passive cable analysis to account for 3-D anisotropic tissue, found that the size of the virtual cathode (measured during cathodal stimulation with 0.25-mm titanium wire), was dependent on both the strength of the cathodal stimulus and the fiber orientation. With the fastest longitudinal propagation as $0°$ and the slowest as $90°$, the size of the virtual cathode was largest between $45°$ and $60°$ (1–3 mm, depending on the stimulus strength) and smallest in the longitudinal direction (1 mm, at both 1 and 7 mA current strength) [59]. The edge of the virtual cathode was defined as the point at which the transmembrane voltage deflection equaled the membrane threshold potential at the end of the stimulus initiating activation [59]. The 2-D size of the virtual cathode was calculated based on recordings of the arrival times of the wave fronts at several electrode positions away from the stimulating electrode, and then by back-extrapolation to find the position of the wave front at the time of the end of the stimulus. The virtual electrode with cathodal stimuli creates a dogbone-shaped area of depolarization extending in the direction transverse to the epicardial fibres. Two areas of hyperpolarization (virtual anodes) are also induced at the same time along the fiber on both sides of the central depolarized area. This phenomenon is referred to as virtual electrode effect, because cathodal stimulation produces two transient areas of hyperpolarization, called virtual anodes, and, as expected, an area of depolarization near the cathode [60].

Knisley et al. [61] studied the transmembrane potential (TMP) changes induced by extracellular electrical field stimulation using voltage-sensitive epicardial maps in arterially perfused rabbit hearts. These authors emphasized that the TMP changes were not consistent with the classical exponential decay and space constant predicted by one 1-D cable theory [61]. Suprathreshold cathodal stimulation could produce hyperpolarizing effect (virtual anode) and depolarization (virtual cathode) by anodal stimulation away from the stimulating electrode site [60,61]. Recently, using simulation and epiflourescence imaging techniques, the mechanism of excitation during and after stimulation of both refractory and excitable (diastole) tissue were analyzed.

Ranjan et al. [62] identified an active mechanism for anodal break excitation at the cellular level using modified Luo-Rudy I action potential in a 2-D bidomain model. Hyperpolarization-activated I_f current then caused excitation at the site of the anode independent of depolarizing influences exerted by the adjoining two virtual cathode sites created by the anodal stimulation. This direct hyperpolarization-induced excitation at the virtual anode by the I_f activation was apparent at relatively weaker current strengths when the depolarizing amplitude in the adjoining virtual cathode sites failed to reach threshold. It was suggested in this study that with stronger anodal currents of stimulation, the amplitude of the depolarization in the adjoining virtual cathode sites may reach threshold potential at a faster rate and depolarize the hyperpolarized regions before anodal break excitation occurs under the anode [62].

Virtual electrode effect was demonstrated by Efimov et al. during VF in isolated rabbit hearts in response to monophasic [17] and biphasic [63] shocks delivered by internal transvenous cardioverter defibrillator (ICD) lead. These authors explained failure of defibrillation shocks by the creation of phase singularities produced by the shock-induced virtual electrode effects [63]. It was suggested that the dynamic interactions of unsuccessful shocks could induce areas of depolarization and hyperpolarization, creating "points of singularity," i.e., areas surrounded by excitable (hyperpolarized), excited (depolarized), refractory (nonpolarized) tissue that may evolve to reentry after the shock and reinitiate the VF [63]. Successful shocks failed to produce phase singularity, at least in the epicardial mapped region (11.5 mm by 11.5 mm) [63]. While these studies provide an attractive working hypothesis for defibrillation failure, the cellular mechanisms by which a failed shock leads to reentry formation remain undetermined. Each epifluoresent signal was recorded from an area of $710 \times 710\,\mu m$ [63], an area that may encompass up to 1000 myocardial cells. In fact, in only 10.7% of cases (12 of 112) of shock-induced point singularities did the induced reentry sustain itself and the VF continue. For example, in 24 cases, reentry propagated along a line of conduction block, turned around it, and then self-terminated by encountering refractory tissue (i.e., bidirectional block) [63]. These findings suggest that the presence of a phase singularity per se cannot predict the ultimate outcome of the shock. Here, the graded response hypothesis of the ULV, as detailed above, offers a cellular basis for the observed outcome resulting from the dynamic interaction between adjoining depolarized and hyperpolarized areas created by the virtual electrode effect after the shock.

More recently, the role of virtual electrode-induced hyperpolarization (accelerated repolarization) on ventricular vulnerability was evaluated in rabbits [64]. It was shown that a strong electrical shock given during the refractory period prolongs the APD at the cathode side while accelerating repolarization in the adjoining virtual anode side. Under these conditions, these authors found that an activation wave front arises from the virtual anode side, where the repolarization is accelerated, which then undergoes wave break as it approaches the virtual cathode side where the APD is prolonged. Depending on the level of TMP from which activation arises in the virtual anode side, the velocity of conduction velocity can be fast (if activation arises from more negative TMP) or slow (if the activation arises from less negative TNP) [64],.With faster conduction no reentry or arrhythmia occurs. However, when the slower conduction reentry and arrhythmia arise [64], these authors explained that with slower propagation the wave break will encounter an "excitable gap" so to generate reentry and arrhythmia [64], these authors further suggested that the "reexcitation of these gaps through progressive increase in shock strength may provide the basis for the lower and upper limits of vulnerability. The former [LLV] may correspond to the origination of slow wave front of excitation and phase singularities. The latter [ULV] corresponds to fast conduction during which wave breaks no longer produce sustained arrhythmia" [64].

It is all too remarkable that the cellular basis of these interesting findings can be readily explained by the graded response hypothesis of vulnerability. Recall that the graded response hypothesis was tested by applying an S2 shock during the relative refractory period at a site different from the S1 site. In this manner the S1 site depolarizes earlier and thus recovers sooner than the S2 site. Activation then arises at the S1 site by the depolarizing graded responses evoked at the S2 site as they propagate toward the more recovered S1 site. This dynamic scenario is identical by the juxtaposition of virtual cathode and virtual anode in close vicinity, where the S2 shock applied during the refractory period at the virtual cathode side propagates to the side of "forced" or "accelerated" repolarization located just in the adjoining virtual anode side. With very strong S2 shocks, the wave front arising from the recovered side (S1 side in our studies [2] and the virtual anode side, i.e., the area of forced or accelerated repolarization side in the studies of Efimov's group [64]) undergoes block in the area of graded response-induced prolongation of the APD (S2 side in our studies

[2] and the virtual cathode side in the Efimov's [54]. However, the wave break cannot reenter because the very strong current-induced excessive prolongation of the ERP does not recover to allow reentrant excitation (absence of excitable gap as suggested by the Effimov's group [64]). The graded response hypothesis therefore offers a cellular basis for the phenomenon of the upper limit of vulnerability. In contrast, when the shock strength is not too strong, the wave front originating from the accelerated repolarization side (S1 or virtual anode side) undergoes block (wave break) at the S2 site with graded response-induced increase APD. However, the wave break in this case successfully reenters through the initial site of block, as this area recovers its excitability sooner because of the relatively shorter APD prolongation with less strong current strength. The graded response hypothesis therefore also offers a cellular basis for the phenomenon of the lower limit of vulnerability [2].

IV. CONCLUSIONS

We presented two hypotheses that adequately explain functional reentry formation by a strong electrical stimulus and its subsequent breakup to multiple wave fronts in the normal myocardium. Both of these hypotheses provide a mechanism of activation wave front breakup. The graded response hypothesis explains the cellular basis of reentry formation and the ULV phenomenon during stimulation with a strong electrical stimulus. The restitution hypothesis explains the breakup of a reentrant and a non-reentrant wave front. Since the ULV is closely linked to the defibrillation threshold, the graded response hypothesis of the ULV might be a useful working hypothesis to study the mechanism(s) of defibrillation.

ACKNOWLEDGMENTS

This study was supported in part by a National Institutes of Health specialized Center of Research (SCOR) Grant for Sudden Death (HL52319), University of California Tobacco Related Disease Research Program (9RT-0041), American Heart Association National Center Grants-in-Aid (9759623N and 92009820), National Institutes of Health SCOR Grant P50HL5231 and 1R01HL and 389-02), the Electrocardiographic Heartbeat Organization, the Ralph M. Parsons Foundation, Los Angeles, CA, Award, and by Pauline and Harold Price Endowment Fund (P-S.C).

REFERENCES

1. Wiggers CJ, Bell JR, Paine M. Studies of ventricular fibrillation caused by electric shock II. Cinematographic and electrocardiographic observation of the natural process in the dog's heart. Its inhibition by pottassium and the revival of coordinated beats by calcium. Am Heart J 5:351–365, 1930.
2. Gotoh M, Uchida T, Mandel WJ, Fishbein MC, Chen P-S, Karagueuzian HS. Cellular graded responses and ventricular vulnerability to reentry by a premature stimulus in isolated cainine ventricle. Circulation 95:2141–2154, 1997.
3. Uchida T, Yashima M, Gotoh M, Qu Z, Garfinkel A, Weiss JN, Fishbein MC, Mandel WJ, Chen P-S, Karagueuzian HS. Mechanism of acceleration of functional reentry in the ventricle; effects of ATP-sensitive potassium channel opener. Circulation 99:704–712, 1999.
4. Wu T-J, Ong JJ, Hwang C, Lee JJ, Fishbein MC, Czer L, Trento A, Blanche C, Kass RM, Mandel WJ, Karagueuzian HS, Chen P-S. Characteristics of wave fronts during ventricular fibrillation in human hearts with dilated cardiomyopathy: role of increased fibrosis in the generation of reentry. J. Am Coll Cardiol 32:187–196, 1998.
5. Cao J-M, Qu Z, Kim YH, Wu TJ, Garfinkel A, Weiss JN, Karagueuzian HS, Chen PS. Spatiotemporal heterogeneity in the induction of ventricular fibrillation by rapid pacing: importance of cardiac restitution properties. Circ Res 84:1318–1331, 1999.
6. Ikeda T, Wu T-J, Uchida T, Hough D, Fishbein MC, Mandel WJ, Chen P-S, Karagueuzian HS. Meandering and unstable reentrant wave fronts induced by acetylcholine in isolated canine right atrium. Am J Physiol 273:H356–H370, 1997.
7. Kamjoo K, Uchida T, Ikeda T, Fishbein MC, Garfinkel A, Weiss JN, Karagueuzian HS, Chen P-S. The importance of location and timing of electrical stimuli in terminating sustained functional reentry in isolated swine ventricular tissues–evidence in support of a small reentrant circuit. Circulation 96:2048–2060, 1997.
8. Chen P-S, Wolf PD, Dixon EG, Danieley ND, Frazier DW, Smith WM, Ideker RE. Mechanism of ventricular vulnerability to single premature stimuli in open-chest dogs. Circ Res 62:1191–1029, 1988.
9. Bonometti C, Hwang C, Hough D, Lee JJ, Fishbein MC, Karagueuzian HS, Chen P-S. Interaction between strong electrical stimulation and reentrant wavefronts in canine ventricular fibrillation. Circ Res 77:407–416, 1995.
10. Kobayashi Y, Peters W, Khan SS, Mandel WJ, Karagueuzian HS. Cellular mechanisms of differential action potential duration restitution in canine ventricular muscle cells during single versus double premature stimui. Circulation 86:955–967, 1992.
11. Karagueuzian HS, Khan SS, Hong K, Kobayashi Y, Denton T, Mandel WJ, Diamond GA. Action potential alternans and irregular dynamics in quinidine-intoxicated ventricular muscle cells. Implications for ventricular proarrhythmia. Circulation 87:1661–1672, 1993.

12. Davidenko JM, Persow AV, Salomonza R, Baxter W, Jalife J. Sustained vortex-like waves in normal isolated ventricular muscle. Proc Natl Acad Sci USA 355:349–351, 1990.

13. Chen P-S, Feld GK, Kriett JM, Mower MM, Tarazi RY, Fleck RP, Swerdlow CD, Gang ES, Kass RM. Relation between upper limit of vulnerability and defibrillation threshold in humans. Circulation 88:186–192, 1993.

14. Kao CY, Hoffman BF. Graded and decremental response in heart muscle fibers. Am J Physiol 194:187–196, 1958.

15. Weidmann S. Effects of current flow on the membrane potential of cardiac muscle. J Physiol 15:227–236, 1951.

16. Ino T, Karagueuzian HS, Hong K, Meesmann M, Mandel WJ, Peter T. Relation of monophasic action potential recorded with contact electrode to underlying transmembrane action potential properties in isolated cardiac tissues: a systematic microelectrode validation study. Cardiovasc Res 22:255–264, 1988.

17. Efimov IR, Cheng YN, Biermann M, Van WD, Mazgalev TN, Tchou PJ. Transmembrane voltage changes produced by real and virtual electrodes during monophasic defibrillation shock delivered by an implantable electrode. J Cardiovasc Electrophysiol 8:1031–1045, 1997.

18. Verrier RL, Brooks WW, Lown B. Protective zone and the determination of vulnerability to ventricular fibrillation. Am J Physiol 234:H592–H596, 1978.

19. Euler DE, Moore NE. Continuous fractionated electrical activity after stimulation of the ventricles during the vulnerable period: evidence for local reentry. Am J Cardiol 46:783–791, 1980.

20. Cha Y-M, Birgersdotter-Green U, Wolf PL, Peters BB, Chen P-S. The mechanisms of termination of reentrant activity in ventricular fibrillation. Circ Res 74:495–506, 1994.

21. Karma A. Spiral breakup in model equations of action potential propagation in cardiac tissue. Phys Rev Lett 71:1103–1106, 1993.

22. Allessie MA, Bonke FI, Schopman FJ: Circus movement in rabbit atrial muscle as a mechanism of tachycardia. III. The "leading circle" concept: a new model of circus movement in cardiac tissue without involvement of an anatomical obstacle. Circ Res 41:9–18, 1977.

23. Gough WB, Megra R, Restivo M, Zeiler RH, EI-Sherif N. Reentrant ventricular arrhythmias in the late myocardial infarction period in the dog. 13 correlation of activation and refractory maps. Circ Res 57:432–445, 1985.

24. Dillon SM, Allessie MA, Ursell PC, Wit AL. Influences of anisotropic tissue structure on reentrant circuits in the epicardial border zone of subacute canine infarcts. Circ Res 63:182–206, 1988.

25. Schalij MJ, Lammers WJEP, Rensma PL, Allessie MA. Anisotropic conduction and reentry in perfused epicardium of rabbit left ventricle. Am J Physiol 263:H1466–H1478, 1992.

26. Downar E, Kimber S, Harris L, Mickleborough L, Sevaptsidis E, Masse S, Chen TCK, Genga A. Endocardial mapping of ventricular tachycardia in the intact human heart. II. Evidence for multiuse reentry in a functional sheet surviving myocardium. J Am Coll Cardiol 20:869–878, 1992.

27. Cabo C, Pertsov AM, Baxter WT, Davidenko JM,. Gray RA, Jalife J. Wavefront curvature as a cause of slow conduction and block in isolated cardiac muscle. Circ Res 75:1014–1028, 1994.
28. Davidenko JM, Pertsov AM, Salomonsz R, Baxter W, Jalife J. Stationary and drifting spiral waves of excitation in isolated cardiac tissue. Nature 355:349–351, 1992.
29. Courtemanche M, Winfree AT. Re-entrant rotating waves in a Beeler-Reuter based model of two-dimensional cardiac electrical activity. Int J Bifurc Chaos 1:431–444, 1991.
30. Efimov IR, Krinsky VI, Jalife J. Dynamics of rotating vortices in the Beeler-Reuter model of cardiac tissue. Chaos, Solutions & Fractals 5:513–526, 1995.
31. Gray RA, Jalife J, Panfilov AV, Baxter WT, Cabo C, Davidenko JM, Pertsov AM, Hogeweg P. Mechanisms of cardiac fibrillation. Science 270:1222–1223, 1995.
32. Karma A. Electrical alternans and spiral wave breakup in cardiac tissue. Chaos 4:461–472, 1994.
33. Panfilov A, Hogeweg P. Scroll breakup in a three-dimensional excitable medium. Phys Rev E 53:1740–1743, 1996.
34. Gray RA, Pertsov AM, Jalife J. Spatial and temporal organization during cardiac fibrillation. Nature 392:75–78, 1998.
35. Lee JJ, Hough D, Hwang C, Fan W, Fishbein MC, Bonometti C, Karagueuzian HS, Chen P-S. Reentrant wavefronts during Wiggers' stage II Ventricular fibrillation in dogs (abstr). J Am Coll Cardiol 424A, 1995.
36. Frame LH, Simson MB. Oscillations of conduction, action potential duration, and refractoriness. A mechanism for spontaneous termination of reentrant tachycardias. Circulation 78:1277–1287, 1988.
37. Simson MB, Spear JF, Moore EN. Stability of an experimental atrioventricular reentrant tachycardia in dogs. Am J Physiol 240:H947–H953, 1981.
38. Garfinkel A, Chen P-S, Walter DO, Karagueuzian HS, Kogan B, Evans SJ, Karpoukhin M, Hwang C, Uchida T, Gotoh M, Nwasokwa O, Sager P, Weiss JN. Quasiperiodicity and chaos in cardiac fibrillation. J Clin Invest 99:305–314, 1997.
39. Qu Z, Weiss JN, Garfinkel A. Cardiac electrical restitution properties and stability of reentrant spiral waves: a simulation study. Am J Physiol 276:H269–H283, 1999.
40. Koller ML, Riccio ML, Gilmour RF Jr. Dynamic restitution of action potential duration during electrical alternans and ventricular fibrillation. Am J Physiol 275:H1635–H1642, 1998.
41. Kim Y-H, Yashima M, Wu T-J, Doshi RN, Chen P-S, Karagueuzian HS. Mechanism of procainamide-induced prevention of spontaneous wave break during ventricular fibrillation. Insight into the maintenance of fibrillation wave fronts. Circulation 100:666–674, 1999.
42. Pastore JM. Girouard SD, Laurita KR, Akar FG, Rosenbaum DS. Mechanism linking T-Wave alternans to the genesis of cardiac fibrillation. Circulation 99:1385–1394, 1999.

43. Kim Y-H, Garfinkel A, Ikeda T, Wu T-J, Athill CA, Weiss JN, Karagueuzian HS, Chen P-S. Spatiotemporal complexity of ventricular fibrillation revealed by tissue mass reduction in isolated swine right ventricle. Further evidence for the quasiperiodic route to chaos hypothesis. J Clin Invest 100:2486–2500, 1997.

44. Rensma PL, Allessie MA, Lammers WJEP, Bonke FIM, Schalji MJ. Length of excitation wave and susceptibility to reentrant atrial arrhythmias in normal conscious dogs. Circ Res 62:395–410, 1988.

45. Garrey WE. The nature of fibrillatory contraction of the heart—its relation to tissue mass and form. Am J Physical 33:397–414, 1914.

46. Wu T-J, Yashima M, Doshi RN, Kim Y-H, Athill CA, Ong JJC, Czer L, Trento A, Blanche C, Kass RM, Garfinkel A, Weiss JN, Fishbein MC, Karagueuzian HS, Chen P-S. Relation between the cellular repolarization characteristics and the critical mass for human ventricular fibrillation. J Cardiovasc Electrophysiol 10:1077–1086, 1999.

47. Vaidya D, Morley GE, Samie FH, Jalife J. Reentry and fibrillation in the mouse heart. A challenge to the critical mass hypothesis. Circ Res 85:174–181, 1999.

48. Voroshilovsky O, Lee M-H, Ohara T, Huang H-LA, Swerdlow CD, Karagueuzian HS, Chen P-S. Mechanisms at ventricular fibrillation in induction by 60 Hz alternating current in isolated swine right ventricle: importance of nonuniform recovery of excitability and the cardiac restitution properties. Circulation 102:1569–1574, 2000.

49. Yashima M, Ohara T, Cao J-M, Kim Y-H, Fishbein MC, Mandel WJ, Chen P-S, Karagueuzian HS. Nicotine increases ventricular vulnerability to fibrillation in hearts with healed myocardial infarction. Am J Phyiol 278:H2124–H2133, 2000.

50. Chen J, Mandapati R, Berenfeld O, Skanes AC, Jalife J. High-frequency periodic sources underlie ventricular fibrillation in isolated rabbit heart. Circ Res 86:86–93, 2000.

51. Kenknight BH, Bayly PV, Gerstle RJ, Rollins DL, Wolf PD, Smith WM, Ideker RE. Regional capture of fibrillating ventricular myocardium: evidence of an excitable gap. Circ Res 77:849–855, 1995.

52. Davy JM, Fain ES, Dorian P, Winkle RA. The relationship between successful defibrillation and delivered energy in open-chest dogs: reappraisal of the "defibrillation threshold" concept. Am Heart J 113:77–84, 1987.

53. Chen P-S, Shibata N, Wolf P, Dixon EG, Danieley ND, Sweeney MB, Smith WM, Ideker RE. Activation during ventricular defibrillation in open-chest dogs. Evidence of complete cessation and regeneration of ventricular fibrillation after unsuccessful shocks. J Clin Invest 77:810–823, 1986.

54. Chen P-S, Wolf PD, Melnick SD, Danieley ND, Smith WM, Ideker RE. Comparison of activation during ventricular fibrillation and following unsuccessful defibrillation shocks in open chest dogs. Circ Res 66:1544–1560, 1990.

55. Chen P-S, Swerdlow CD, Hwang C, Karagueuzian HS. Current concepts of ventricular defibrillation. J Cardiovasc Electrophysiol 9:553–562, 1998.

56. Chen P-S, Shibata N, Dixon EG, Martin RO, Ideker RE. Comparison of the defibrillation threshold and the upper limit of ventricular vulnerability. Circulation 73:1022–1028, 1986.
57. Hwang C, Swerdlow CD, Kass RM, Gang ES, Mandel WJ, Peter CT, Chen P-S. Upper limit of vulnerability reliably predicts the defibrillation threshold in humans. Circulation 90:2308–2314, 1994.
58. Kwaku KF, Dillon SM. Shock-induced depolarization of refractory myocardium prevents wave- front propagation in defibrillation. Circ Res 79:957–973, 1996.
59. Wikswo JP Jr, Wisialowski TA, Altemeier WA, Balser JR, Kopelman HA, Roden DM. Virtual cathode effects during stimulation of cardiac muscle. Two-dimensional in vivo experiments. Circ Res 68:513–530, 1991.
60. Wikswo JP Jr, Lin S-F, Abbas RA. Virtual electrodes in cardiac tissue: a common mechanism for anodal and cathodal stimulation. Biophys J 69:2195–2210, 1995.
61. Knisley SB, Hill BC, Ideker RE. Virtual electrode effects in myocardial fibers. Biophys J 66:719–728, 1994.
62. Ranjan R, Tomaselli GF, Marban E. A novel mechanism of anode-break stimulation predicted by bidomain modeling. Circ Res 84:153–156, 1999.
63. Efimov IR, ChengY, Van Wagoner DR, Mazgalev T, Tchou PJ. Virtual electrode-induced phase singularity: a basic mechanism of defibrillation failure. Circ Res 82:918–925, 1998.
64. Cheng Y, Mowrey KA, Van Wagoner DR, Tchou PJ, Efimov IR. Virtual electrode-induced reexcitation: a mechanism of defibrillation. Circ Res 85:1056–1066, 1999.
65. Weiss JN, Garfinkel A, Kargueuzian HS, Qu Z, Chen PS. Chaos and the transition to ventricular fibrillation: a new approach to antiarrhythmic drug evaluation. Circulation 99:2819–2826, 1999.

10
Noncontact Cardiac Mapping

Anthony W. C. Chow, Richard J. Schilling, David W. Davies, and Nicholas S. Peters
Imperial College School of Medicine and St. Mary's Hospital, London, United Kingdom

I. INTRODUCTION

The electrophysiological mechanisms of a large number of cardiac arrhythmias are now well understood and amenable to ablation therapy. The prerequisite of this therapeutic approach requires accurate localization of critical regions of the arrhythmia circuit by mapping and the delivery of ablative lesions to prevent recurrence. Surgical ablation of arrhythmias is now seldom used, unless there are other reasons for surgical intervention, such as concomitant coronary artery bypass or valve surgery, due to the morbidity and mortality associated with this procedure [1,2]. For the majority of common cardiac arrhythmias, in which the pathological and electrophysiological abnormalities are well characterized, conventional mapping techniques are effective in localizing sites for ablation. However, it may not be possible to map complex cardiac arrhythmias using conventional techniques, because there are a number of limitations of conventional mapping. Contact catheters can only record changes of potential at a single point on the endocardium and there is a limit to the number of catheters that can be used. It is possible to obtain spatial information of activation by sequential point-to-point mapping from different sites and examine the temporal relationship to a reference electrode, but this is time consuming and requires the presence of continuous tachycardia for long periods. This approach may not be feasible for patients with poorly tolerated tachycardias such as fast ventricular tachycardias (VT) or if the arrhythmia is nonsustained. The complex three-dimensional structure of the cardiac chamber, often in the presence of structural heart disease and geometric changes that occur during contraction, makes accurate endocardial mapping difficult. Multielectrode contact catheters have been developed in the form of expandable baskets. These have been limited to 64 electrodes, a proportion of which are functionally redundant due to inadequate endocardial contact, and resolution is limited by the interspline spaces [3,4].

An alternative to direct-contact mapping is the development of noncontact mapping; this is a rapid, high-resolution mapping system capable of simultaneous, global acquisition of endocardial data in the intact heart.

II. HISTORY AND EVOLUTION OF CONCEPTS OF NONCONTACT MAPPING

The concept of noncontact mapping was first described by Taccardi [5] in 1987 in open-chest dog experiments. Olive- and cylindrical-shaped endocavitary probes fitted with 40 silver electrodes placed within the cardiac chamber but not in contact with the ventricular wall were able to record low-amplitude endocardial potentials from ectopic beats. The original study was based on the classical solid-angle theory [6]. When multi-intracavitary electrodes are used, changes in cardiac potential during endocardial activation are detected earliest by the electrode in closest proximity to activation, which has the greatest negative potential change and decreases with increasing distance [7]. If the position and spatial orientation of each recording electrode is known, it is possible to determine the site of origin or progress of sequential activation within the cardiac chamber. Paced beats from over 60 different ventricular points were examined and identified with a theoretical resolution of approximately $1.5\,\mathrm{cm}^2$.

The next key stage in the development of noncontact mapping came from work by Khoury and Rudy [8] using a torso-heart mathematical model to investigate the effect of geometry and conductive parameters on noncontact cavitary potentials. Simulations demonstrated that probe potentials were smoothed out and of low amplitude, with poor spatial resolution to discriminate separate areas of activation. Probe potentials were also out of phase when compared to those at the source. Experimental results showed that the geometry of the probe and its location within the cavity significantly influenced the probe potentials recorded.

Further refinement of raw low-amplitude probe potentials was necessary if noncontact mapping was to have any clinical utility. A method of mathematical reconstruction of the recorded probe potentials was required, such that after computation, the original probe electrode potential would resemble the endocardial potential as it would appear at the source. The process can be divided into two parts. First, the mathematical reconstruction process, and second, the probe–endocardial geometry must be determined in order for the reconstruction to be accurate. The mathematical formulation has been described to resolve electrographic and epicardial potentials [9,10], and this was later applied to noncontact mapping by Khoury et al. [11]. In order to reconstruct endocardial potentials from probe potentials, it is necessary to solve the inverse solution to Laplace's equation in the cavity volume, for known probe and endocardial surfaces. For a known intracavitary volume Ω, probe Sp, and endocardial surface Se, potential V within a known cavity is governed by the following Laplace's equation:

$$\nabla^2 V = 0 \quad \text{in } \Omega$$

Provided boundary conditions are followed and probe potential V_p is known,

$$V = V_p \quad \text{on } S_p$$

and

$$\frac{\partial V}{\partial n} = 0 \quad \text{on } S_p$$

where n is a unit vector normal to the surface, assuming the probe behaves as a nonconductor.

To make the distinction between probe and endocardial surfaces, a standard boundary-element method as previously described for body surface potential [12,13] is applied. This relates probe to endocardial potentials by the matrix equation

$$V_p = A \cdot V_e$$

where the matrix of influence, coefficient A, is determined by the geometric relationship between the probe and endocardial surfaces. V_p and V_e are vectors of potentials at the probe and endocardium, respectively. This technique of reconstruction is, however, inherently ill posed. Errors incurred from creation of geometry and the presence of electrical noise from different sources is systemically amplified, resulting in magnification of inaccuracies. To overcome this problem, a technique of Tikhonov regularization is used:

$$\text{Min}_{V_e}(\|V_p - A \cdot V_e\|2 + t\|V_e\|2)$$

where t is a regularization parameter. This approach stabilizes the solution to endocardial potentials to obtain optimal enhancement of potential reconstruction but limiting errors by imposing a physiological constraint on the solution. Further modifications of this process can be made to enable the computation process to be more rapid and efficient.

III. EARLY STUDIES IN THE BEATING HEART

In order to test these mathematical models, two studies using an isolated canine heart preparation have been conducted [11,14]. In the first, by Khoury et al., two cylindrical noncontact probes containing 65 and 89 electrodes were compared. A total of 52 of 94 plunge electrodes were inserted into the ventricle to record intramural and endocardial potentials, and then compare with reconstructed probe potentials. Chamber geometry was digitally computed by linear extrapolation with the use of metal rods and needles of known lengths as a series of triangles between recording electrodes. The position, lie, and three-dimensional orientation of the noncontact probe relative to the endocardial geometry was configured mathematically by calculating the minimized root-square-mean error of actual endocardial and recorded probe potential over 5 cycles; this method has been previously described by Macchi et al. [15,16] to locate the source of ectopic foci. It is then possible to create isopotential maps by projecting the endocardial potentials onto the three-dimensional model of the cardiac chamber. Close correlation of regions with maximal positive and negative reconstructed potentials was found when compared to those recorded at the endocardial contact electrode. Although reconstructed potentials remained smaller than those recorded at the endocardium, the site of origin and progressive activation patterns recorded by the noncontact probe throughout different time intervals showed good correlation and followed the same spatial and dynamic changes as endocardial mapping data. The location accuracy was found to be within 10 mm. Using this method it was possible to differentiate the presence of two distinct pacing origins that were 10–20 mm apart. Mapping data were compared for 65 and 81 probe electrodes; both were equally good in reconstruction of endocardial potentials and quantitatively no additional improvement or advantage was conferred by the use of greater numbers of electrodes. In the second study, by Lui et al. [14], the same conclusions were reproduced. Correlation coefficients between endocardial and reconstructed potentials >0.9 were found for 56% of all endocardial points recorded, 78% of all electrode sites had >0.75 correlation, and only a few sites had correlation coefficients <0.5.

The next stage of development was to test the utility of the noncontact system in mapping activation in the intact, normally beating canine heart [17]. A probe with 128 electrodes was used and endocardial contact data were recorded over 194 endocardial sites. Probe and chamber geometry were determined by epicardial ultrasound. Localization of activation during 8 pacing protocols, sinus rhythm, and ischaemia-induced ventricular ectopics were assessed. For all activation modalities, correlation of reconstruction was $r = 0.88$, and accuracy of spatial location was similar to those of previous experimental data of 9.2 mm.

Further modifications of the technique for inverse solution have been applied to improve the accuracy of the reconstruction process, where the inverse solution was computed using a higher-order algebraic expression based on Green's second formula [18]:

$$\iint_{\partial D} \left(v\, \frac{\partial w}{\partial n} - w\, \frac{\partial v}{\partial n} \right) dA = \iiint_{D} \left(v\nabla^2 w - w\nabla^2 v \right) dD$$

where D represents the domain, ∂D is the boundary of the domain, ∂/∂ is the outward normal on D, ∇^2 is the Laplacian, dA is the surface area differential, and dD is the volume differential. V is a solution of the Laplacian equation and w is the potential field in free space created by a unit of charge.

A bicubic spline model was used in place of linear splines where sharp triangular points caused significant errors in geometry reconstruction. This model is able to formulate and construct curved endocardial lines from sampled points, in keeping with the true contours of the cardiac chamber. Using this arrangement it is possible to reconstruct 3360 unipolar electrograms of the endocardium simultaneously.

IV. TECHNOLOGICAL DEVELOPMENT OF A MAPPING AND LOCATION SYSTEM FOR CLINICAL USE

In an adaptation for clinical use, the noncontact mapping probe has been modified into a collapsible multielectrode array (MEA), incorporated as a braid of 64 wires woven around an 8-mL balloon on the end of a 9F catheter (Fig. 1). Each 0.0003-in. wire has a laser-etched 0.025-in. break in insulation that allows it to function as a unipolar electrode. A 0.035-in. guide wire passed through the central lumen is used to position the catheter. During transit, the array is maneuvered in the collapsed state and expanded in a stable position

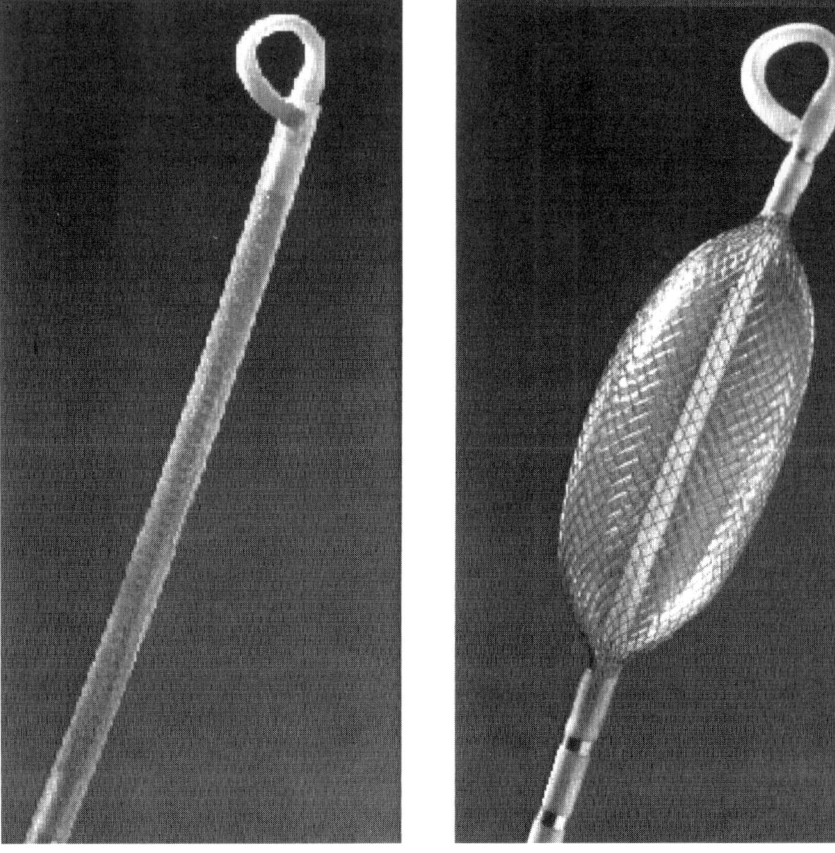

Figure 1   The noncontact catheter is shown with the multielectrode array collapsed within an introducer sheath (left) and fully expanded (right).

inside the cardiac chamber. In order to image the position of the expanded MEA, the balloon is inflated with half-strength radio-opaque contrast (Fig. 2).

For the system to have realistic clinical application, a second mathematical problem of accurate determination of cardiac chamber geometry was required. This had to be a relatively noninvasive, percutaneous, accurate, and clinically acceptable technique. A catheter location system was developed that fulfilled these criteria. A 5.68-kHz low-current locator signal is passed down and emitted from an electrode on any standard electrophysiological catheter, which is detected by the multielectrode array. Given that the positions of the array electrodes are known, the spatial orientation

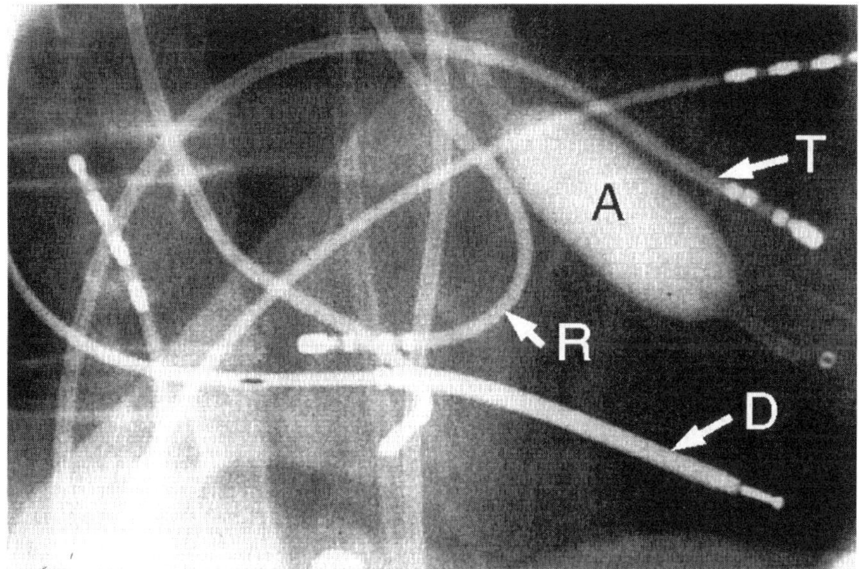

Figure 2 Radiograph in a right anterior oblique projection of a patient undergoing noncontact mapping. The multielectrode array balloon (A) has been filled with half strength contrast and deployed in the left ventricle. Two mapping/ablation catheters are also seen, one by a retrograde transaortic route (R) and the other via a transeptal approach (T). Catheters in the high right atrium, coronary sinus, and right ventricular outflow tract are also present.

of the electrode on the roving catheter can be defined in three-dimensional space, relative to the center of the multielectrode array. To configure the endocardial geometry, the roving catheter is moved around the endocardial chamber to create a sequence of recorded three-dimensional points (Fig. 3a). The maximum probe to endocardial distances are recorded gated at 6 msec before the onset of the surface ECG R-wave. A smoothing process is used to produce contoured cardiac chamber geometry (Fig. 3b); the process typically takes 10 min to create a virtual model of the cardiac chamber. Isopotential activation maps are created by superimposing the reconstructed electrograms onto the virtual model of the cardiac chamber (Fig. 4); the location of any catheter can then be located on this constructed cardiac model and anatomical landmarks and areas of interest can be defined on this virtual map thereafter. The locator system has a further important function of navigating mapping/ablation catheters to sites critical for sustaining the arrhythmia, after noncontact activation maps have identified the target area,

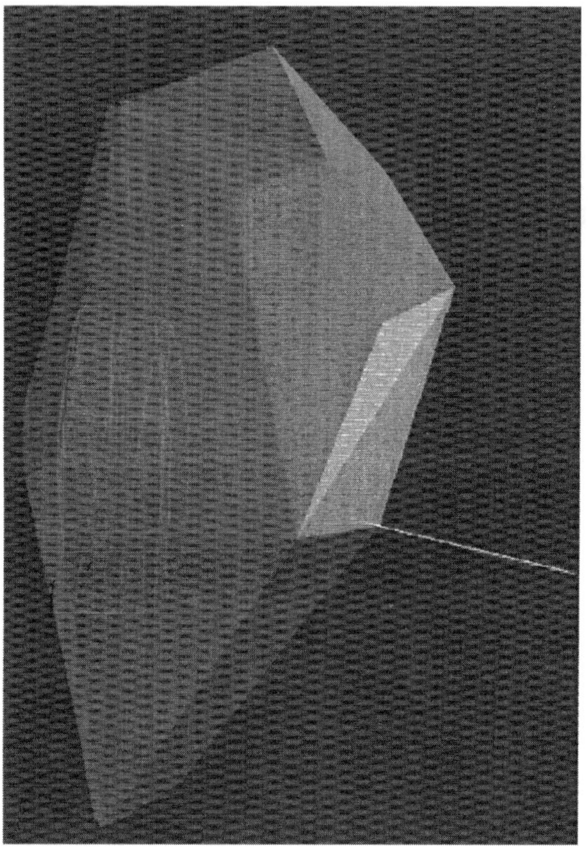

3a

Figure 3 A reconstructed model of a human left ventricle using the noncontact mapping system is shown (a). With the aid of a locator signal emitted from a roving catheter (seen in green), a large number of 3-dimensional endocardial points are recorded as the catheter is dragged along the endocardium to create a "virtual" image. (b) The appearance of the same virtual endocardium after a smoothing process, rotated to a different projection.

so that therapy can be delivered. This nonfluoroscopic guidance system has been shown in dogs to identify catheter positions with an accuracy of within 2 mm [19].

The validation of this catheter location system and the modified inverse solution algorithm was tested in an in-vitro tank model and in

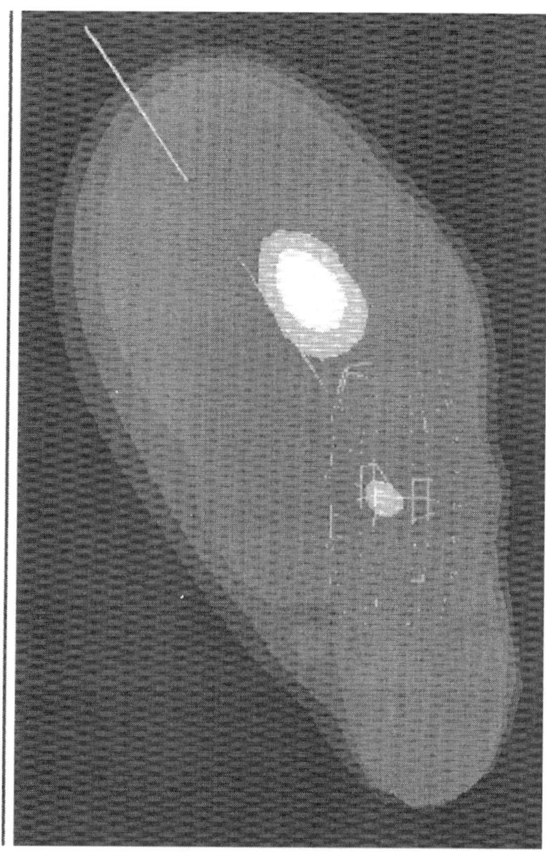

3b

Figure 3 (continued)

the canine LV. Reconstructed potentials were compared to contact data
and sites of activation marked by radiofrequency lesions [20]. Good
accuracy in electrogram reconstruction was found for distances of
< 50 mm from the MEA, but decreased beyond this threshold. The lo-
cation system was able to define and guide the roving catheter to within
2.33 ± 0.44 mm of the focal activation site, and no difference was found
for the polar nonequatorial regions of the MEA. However, with distances > 50 mm, the accuracy of location was reduced to 7.5 ± 1.13 mm.
A total of 17 radiofrequency lesions (mean 5.8-mm diameter) were made

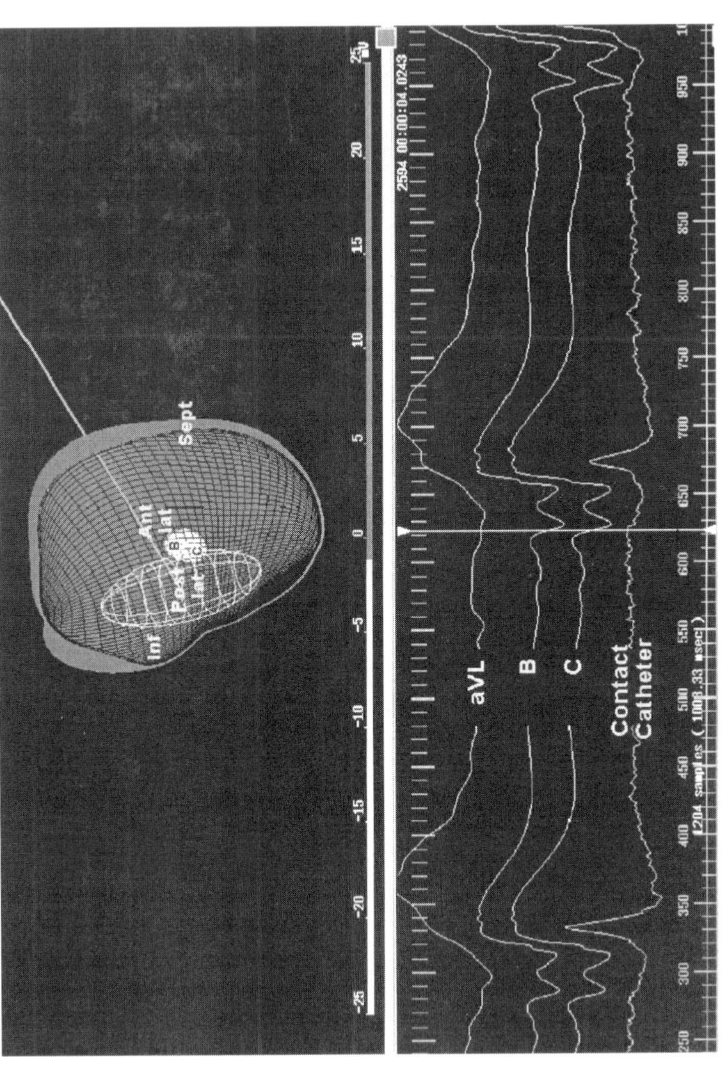

Figure 4 An isopotential map representing the left ventricle recorded during VT. The chamber has been cut along the posterior septum and laid open. Differences in potential are color coded. In this figure activation is seen as a change from resting gray to white on the isopotential map. A surface ECG (aVL), contact catheter electrogram in the anterior basal position and reconstructed electrograms taken from points B and C on the isopotential map are displayed at the bottom of the screen. The vertical white line on the electrogram display represents the point in time that corresponds to the isopotential map shown above. Presystolic activation is seen at positions B and C on the isopotential map and confirmed on the reconstructed electrograms.

in the canine ventricle to assess accuracy of location; the center and edge distances of the RF lesion to target plunge electrode location were found to be 4 ± 3.2 and 1.2 ± 3.2 mm, respectively.

Further validation data on the accuracy of reconstruction and location of the noncontact system has come from canine atrial studies [21]. A contact multielectrode catheter was used to evaluate computed and actual interelectrode distances, and the mean absolute value between was found to be 0.97 ± 0.77 mm of a total of 210 distances measured. The mean error in location was found to be 0.98 ± 0.71 mm during sinus rhythm and 0.93 ± 0.46 m in atrial fibrillation. Correlation coefficients between contact and reconstructed electrograms were 0.8 ± 0.12 in sinus rhythm, 0.85 ± 0.17 in atrial flutter, and 0.81 ± 0.18 in atrial fibrillation. This demonstrates that for different atrial rhythms, including fibrillation there is no attenuation in accuracy of electrogram reconstruction.

V. EARLY CLINICAL EXPERIENCE OF NONCONTACT MAPPING

A. Validation of Mapping Data in the Human Heart

The first clinical experience of noncontact mapping of human arrhythmia was performed in 1995 and reported in patients with ventricular tachycardia (VT) in 1996 [22]. Clinical validation of the system was published on 13 patients with VT [18] in whom mapping of the left ventricle (LV) was performed. All patients had structural heart disease and dilated poorly functioning ventricles (mean LVedd 6.17 cm). In this study, contact endocardial electrograms were compared with reconstructed electrograms from 76 equatorial and 32 nonequatorial LV points during sinus rhythm. Electrogram morphology, polarity, amplitude, and frequency, as well as timing of maximum $-dV/dt$, were examined. An overall difference of -6.44 ± 14.17 msec was found between $-dV/dt$ reconstructed and contact electrograms. For equatorial points, perfect matches could be obtained as far as 52 mm from the center of the multielectrode array but it was noted that beyond 34 mm the reconstructed electrogram timing measuring maximum $-dV/dt$ gradually increased (-1.94 ± 7.12 versus -14.16 ± 19.29 msec, respectively, $p < 0.001$), suggesting that reconstruction made electrograms earlier compared to those recorded by contact catheters. Morphology cross-correlation also deteriorated with increased distance from the MEA, but no clear-cut threshold distance could be identified (0.87 and 0.76, respectively). Data from nonequatorial sites 32.33 ± 10.81 mm from the MEA equator showed good timing reconstruction at distances of up to 44 mm perpendicular from the MEA equator. No statistical difference was found in morphology score with increased perpendicular distance from the equator.

This validation data of the noncontact mapping demonstrates that it is possible to reconstruct electrograms at distances over 50 mm from the center of the MEA, but accuracy decreases significantly after 34 mm in both timing and morphology accuracy. The cause for the fall-off in accuracy beyond 34 mm may be due to changes in the anatomy or function of different parts of the heart, differences in signal-to-noise ratio, or small but significant changes in the endocardial potentials themselves, which cause rejection by the regularization process. This is a complex issue that can potentially be resolved; it may then be possible to further refine the reconstruction algorithms to eliminate errors incurred with increased distances from the MEA.

B. Unique Insight into Arrhythmia Mechanism

In the past there has been little understanding of the electrophysiological behavior behind most complex arrhythmias, because we lacked the tools to map these arrhythmias with sufficient resolution and rapidity. Attempts at catheter ablation of complex cardiac arrhythmias have often failed or incurred high recurrence rates on follow-up. This largely reflects the lack of precision in mapping to guide ablative therapy of appropriate target sites and also our level of understanding of the arrhythmias substrate. With the development of fast global mapping systems this has now become feasible and realistic. Noncontact mapping has been used to map a number of different human cardiac arrhythmias and has provided new insight into the mechanisms of arrhythmogenesis and electrophysiological properties responsible for perpetuation of these arrhythmias.

VI. HUMAN ARRHYTHMIA MAPPING

A. Ventricular Tachycardia

Patients resuscitated from sudden cardiac death have a 10–30% risk of recurrence in the first year [23–25]. Patients with inducible sustained monomorphic VT in the context of structural heart disease not suppressed by antiarrhythmic drug therapy have been found to have high risk of VT recurrence and further fatal cardiac arrhythmias [26–29]. The predominant mechanism that causes VT is reentry [30–32]. Ablation of this arrhythmia is dependent on the ability to identify and ablate the diastolic components critical for maintaining the reentry circuits. Only 10–20% of patients with VT have been considered suitable for ablation, primarily because of poor hemodynamic tolerance of VT or the length of time in tachycardia required for mapping [33]. Although good acute success rates are achieved with conventional catheter mapping techniques, the long-term recurrence rates are still disappointing, ranging from 50% to 85% [34–36]. The clinical efficacy of the noncontact system in mapping and guiding ablation of sustained monomorphic VT has been evaluated in 24 patients [37]. Twenty-one of these patients had structural heart disease and most had poor LV function (mean EF 39%). A total of 81 different VTs were mapped, of which 24 were identified as clinical morphologies. Diastolic activity was identified in 54 (67%) VTs, partial diastolic pathways were seen in 37 VTs constituting $36 \pm 30\%$ of the diastolic pathway, and complete VT circuits mapped in 17 (22%) morphologies (Fig. 5). Based on noncontact data, a total of 38 VTs were successfully ablated, 15 of which were clinical VTs. Four VTs that shared two contrarotating circuits were ablated with two radiofrequency energy applications. The importance of mapping the diastolic activity is reflected in the highest success of ablation achieved at target sites where at least part of the diastolic activity was identified (80%); this compares to the poor results that were achieved at exit sites (21%) and regions remote from the diastolic pathway (9%). During the long-term follow-up of 1.5 years of 20 patients ablated, 14 (70%) were arrhythmia free. Only 2 of 37 (5.4%) targeted VTs recurred; the remaining VT were new morphologies not previously encountered.

Further evidence that noncontact mapping can effectively guide ablation of VT has come from data of patients with implantable defibrillators (ICD) and VT [38]. ICDs have been shown to be effective in reducing the incidence of arrhythmic deaths in patients with ventricular tachycardia. However, these devices are only palliative and do not reduce the frequency of VT occurrence. In a study of 12 patients with ICD and unacceptably frequent device therapy or slow VT, noncontact mapping was used to guide ablation of VT. All patients had structural heart disease and poor left ventricular function. A total of 55 VT were mapped, 13 of which were clinical VTs. Noncontact mapping idendified 23 partial and 12 complete diastolic pathways. Six complete VTs were found to share circuits in contrarotation, a further 8 VTs shared 35% of the diastolic pathway, and 6 further VTs shared common exit sites. A total of 11 clinical and 34 nonclinical VTs were ablated, three of which were nonsustained clinical morphologies. Over a follow-up period of 12.8 ± 15.5 months, no VT has recurred in 7 of 11 patients ablated. Four patients have had further VTs, of which 3 of the acutely ablated VTs morphologies have recurred, giving an overall recurrence rate of ablated VTs of 6.7%; the remainder were due to

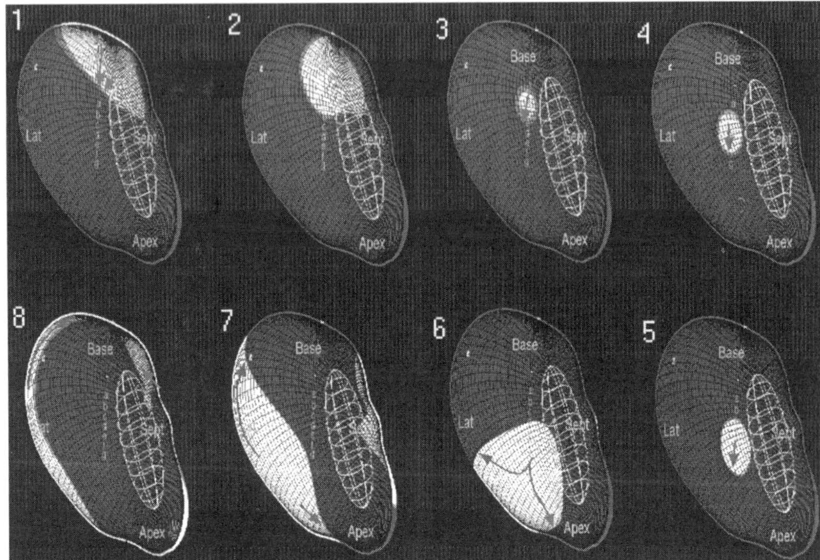

Figure 5 Noncontact data showing a sequence of isopotential maps of a typical figure-of-8 ventricular tachycardia. The left ventricle has been cut along the posterior septum and laid open with anatomical labels as shown. Sept = septum, lat = lateral. The resting endocardial potential is gray and changes to white with activation. Frame 1 shows the end of systole, where activation enters a diastolic pathway beginning at the base of the anterior septum in frame 3. The protected diastolic activity then travels in an apical direction (frames 3–5), and exits at the mid-lower anterior septum to cause systolic activation. The two systolic wavefronts split to activate the rest of the ventricle (frames 6–8) and finally collide in the anterior basal region to complete the figure-of-8 circuit.

new VTs not previously documented. ICD therapies were significantly reduced in all patients. Antitachycardia pacing decreased from 36.8 ± 86.8 to 2 ± 5.7 episodes per month ($p < 0.01$) and defibrillation shocks were reduced from 9 ± 6.3 to 0.2 ± 0.5 episodes per month ($p < 0.005$) after ablation. This provides further clinical evidence of the efficacy of the noncontact system in accurately mapping and guiding catheter ablation of patients with multiple VT morphologies with high success and a low recurrence rate.

B. Ventricular Fibrillation

Ventricular fibrillation is the most common rhythm in sudden cardiac death. There is clinical data to suggest that most episodes of VF arise from acceleration and deterioration of VT [39–41]. Despite its importance, little is known about the properties or mechanisms of VF in the intact human heart, due to the hemodynamic and ethical limitations of studying this arrhythmia. There has been a large volume of experimental data to suggest that a re-entrant mechanism [33,42,43] is involved in the initiation and maintenance of VF [44–46], but limited human data have been available. In a study of 30 consecutive patients with ischemic heart disease undergoing noncontact mapping for VT [47], 7 episodes of VF was initiated by programmed stimulation from sinus rhythm or entrainment pacing from VT. Regional inhomogeneity in recovery of tissue excitability due to progressively premature extrastimuli was found to cause formation of expanding arcs of functional block. This in turn led to the development of areas of late activation, which resulted in reentry and the formation of multiple fibrillation wavelets leading to VF. Two of the patients with multiple episodes of VF induced developed the changes in the same region. When these abnormal areas of functional lock and late activation were analyzed and related to VT circuits, each region was found to contained VT exit sites and a third of all diastolic pathways were located within or immediately adjacent to these regions, thus suggesting that the substrate involved in VF may be shared with those of VT. Further indirect evidence of this hypothesis comes from ablation of VT in ICD patients [47], where a significant proportion of the VF episodes occurred spontaneously and not because of VT acceleration, and where significant reduction of defibrillation shocks for VF was a result of successful VT ablation.

For the first time, using a global noncontact mapping system we have gained direct insight into the mechanism by which VF is initiated and the close association with VT.

C. Atrial Flutter

It is now generally agreed that typical atrial flutter is a macro-reentrant circuit localized within the right atrium and that conduction through the isthmus is necessary for maintenance of the flutter circuit [48–51]. Current catheter ablation strategies target this region to prevent atrial flutter recurring [52–55]. Although much is known about the arrhythmia, there remain a number of unresolved controversies, such as whether an area of slow conduction is essential for flutter to occur and the role of the crista terminalis during atrial flutter.

The noncontact system has been used to map and characterize typical right atrial flutter in 13 patients [56]. Eleven were typical counterclockwise rotation flutters and 2 were atypical clockwise flutters, of whom 5 patients had previous unsuccessful conventional flutter ablations. Noncontact mapping was able to define detailed activation maps of the entire right atrium for the first time. During atrial flutter wavefront slowing was seen in the posterior triangle of Koch in the majority of patients. Variations in activation patterns were seen as the wavefront emerged from the isthmus. Splitting of two wavefronts was seen around the coronary sinus os in 3 of 13 patients. Atrial flutter activation progressed from the right atrium toward the nonisthmus region of the tricuspid annulus in 10 of 12 cases, but activated in a direction away from it in 2 of 12 patients. Interestingly, flutter was found to be slower in the isthmus, but not significantly so, suggesting that this reentry circuit was not critically dependent on a region of slow conduction. For the first time, endocardial conduction velocities were also measured of trabeculated and smooth right atrium in vivo (1.16 ± 0.48 and 1.22 ± 0.65 m/sec, respectively). The crista terminalis formed a barrier to conduction in 12 of 13 patients, and this line of block did not extend along the entire length from SCV to IVC in 2 of 13 patients.

Noncontact mapping has also been used to assess breaks in linear isthmus ablation lines in a study of 12 patients with right atrial flutter [57]. Breaks that allowed continued conduction were rapidly identified and localized by the noncontact system, and the maps were used to guide ablation and successfully achieve bidirectional block across the isthmus of all patients. The system was able to effectively distinguish slow, persistent conduction from complete linear lines of block.

D. Atrial Fibrillation

Atrial fibrillation (AF) is the most common human cardiac arrhythmia. Previously considered to be benign, this arrhythmia has been shown to be associated with increased morbidity and mortality [58–60]. Studies have suggested that the arrhythmias arise from reentry of multiple wavelets [61, 62] and a number of ablation techniques have been used to prevent AF recurrence. Surgical maze operations have been highly successful [63–65]. However, there is a procedure-related morbidity and mortality and a 7% recurrence rate of AF/flutter [66]. Catheter versions of the maze procedure have also been attempted, but with disappointing results and long procedure times [67,68], and the high failure rates reflect our lack of understanding of the electrophysiology of AF. Mapping has previously been limited to the exposed free right atria wall using epicardial multielectrode plaques; little is known of the activation patterns of the entire right atrium. Global noncontact mapping of AF in the right atrium has been described in the canine mode [69]. Recently, AF has been mapped in 11 patients using the noncontact system [70]; 8 had acutely induced and 3 had chronic AF. Activation patterns varied considerably among patients but conformed to the classification proposed by Konings et al.

[71], which categorized AF based on the numbers of wavelets in the atria. No relationship between duration and type of AF was found. A high degree of activation emerged from the septum, suggesting that interatrial conduction may be important in reactivation of unstable reentry to sustain AF. The effects of flecainide in terminating AF were recorded in three patients. Progressive reduction in wavelet number and increasing organization of reentry resulted in restoration of sinus rhythm.

There is increasing interest in ablation of focally triggered AF [72,73], which is thought to arise from focal or microreentry circuits predominantly from the plumonary veins. As a first step in high-resolution mapping, the noncontact system has been deployed in the left atrium of anesthetized dogs to localize simulated ectopic foci produced by pacing [74]. Pulmonary vein origins were correctly identified by the noncontact mapping system and confirmed by intracardiac ultrasound. Eccentric atrial activation patterns were identified in 6 of 22 cases and sites close to but outside the pulmonary veins were correctly located. Noncontact mapping of human left atria in patients with focally triggered atrial fibrillation has already begun, and the results of preliminary clinical studies will shortly be available.

VI. ARRHYTHMIAS IN CONGENITAL HEART DISEASE

With advances in surgical techniques and medical care, increasing numbers of patients with congenital heart disease are reaching adulthood. A number of these patients are prone to develop complex cardiac arrhythmias. The presence of altered cardiac anatomy and geometry, often complicated by the presence of previous surgical correction or palliative procedures, make the prospect of mapping and ablation therapy in these patients more difficult. Because of these factors, experience in this field is limited, and the ability of conventional techniques to map these arrhythmias is particularly limited.

Recently, noncontact mapping has been used to identify atrial arrhythmias in patients with congenital heart disease. Our group has successfully mapped and ablated atrial tachycardias in patients after Sennings and Mustard surgical procedures, and others have used the system to map and ablate atrial tachycardias in patients after Fontans procedures [75]. Although case numbers are small, it is now feasible to map these complex arrhythmias using the noncontact system.

VII. POTENTIAL FOR FURTHER DEVELOPMENT

Validation data have shown that the noncontact system is able to consistently and accurately reconstruct endocardial potentials. Impressive results have been obtained of ablation procedures guided by the noncontact system. These studies have also provided a considerable amount of data that remain to be extracted from extensive off-line analysis of the electrophysiology and pathophysiology of cardiac arrhythmias. This technology has the potential to improve our understanding of fundamental arrhythmic mechanisms and will increase further our ability to treat more complex arrhythmias previously considered unmappable. Continued development and refinement of the system in electrogram reconstruction, catheter location, and development of more rapid and easily interpretable data will see an expanding role of this technology in mapping future arrhythmias.

REFERENCES

1. Mickleborough LL, Usui A, Downar E, Harris L, Parson I, Gray G. Transatrial ballon technique for activation mapping during operations for recurrent ventricular tachycardia. J Thorac Cardiovasc Surg 99(2):227–223, 1990.

2. Mickleborough LL, Harris L, Downar E, Parson I, Gray G. A new intraoperative approach for endocardial mapping of ventricular tachycardia. J Thorac Cardiovasc Surg 95(2):271–280, 1988.

3. Greenspon AJ, Hsu SS, Datorre S. Successful radiofrequency catheter ablation of sustained ventricular tachycardia postmyocardial infarction in man guided by a multielectrode basket catheter. J Cardiovasc Electrophysiol 8(5):565–570, 1997.

4. Schmitt C, Zrenner B, Schneider M, Karch M, Ndrepepa G, Deisenhofer I, Weyerbrock S, Schreieck J, Schömig A. Clinical experience with a novel multielectrode basket catheter in right atrial tachycardias. Circulation 99(18):2414–2422, 1999.

5. Taccardi B, Arisi G, Macchi E, Baruffi S, Spaggiari S. A new intracavitary probe for detecting the site of origin of ectopic ventricular beats during one cardiac cycle. Circulation 75(1):272–281, 1987.

6. Scher A, Spach M. Cardiac depolarization and repolarization and the electrogram. In: Berne RM, ed. Handbook of Physiology, 1979, p. 372.

7. Colli-Franzone P, Guerri L, Viganotti C, Macchi E, Baruffi S, Spaggiari S, Taccardi B. Potential fields generated by oblique dipole layers modeling excitation wavefronts in the anisotropic myocardium. Comparison with potential fields elicited by paced dog hearts in a volume conductor. Circ Res 51(3):330–346, 1982.

8. Khoury DS, Rudy Y. A model study of volume conductor effects on endocardial and intracavitary potentials. Circ Res 71(3):511–525, 1992.

9. Rudy Y, Oster HS. The electrocardiographic inverse problem. Crit Rev Biomed Eng 20:25–45, 1992.

10. Rudy Y, Messinger-Rapport BJ. The inverse problem in electrocardiography: solutions in terms of epicardial potentials. Crit Rev Biomed Eng 16(3):215–268, 1988.

11. Khoury DS, Taccardi B, Lux RL, Ershler PR, Rudy Y. Reconstruction of endocardial potentials and activation sequences from intracavitary probe measurements. Localization of pacing sites and effects of myocardial structure. Circulation 91(3):845–863, 1995.

12. Messinger-Rapport BJ, Rudy Y. Computational issues of importance to the inverse recovery of epicardial potentials in a realistic heart-torso geometry [published erratum appears in Math Biosci 1990 Apr; 99(1):141]. Math Biosci 97(1):85–120, 1989.

13. Messinger-Rapport BJ, Rudy Y. Effects of the torso boundary and internal conductivity interfaces in electrocardiography: an evaluation of the "infinite medium" approximation. Bull Math Biol 47(5):685–694, 1985.

14. Liu ZW, Jia P, Ershler PR, Taccardi B, Lux RL, Khoury DS, Rudy Y. Noncontact endocardial mapping: reconstruction of electrograms and isochrones from intracavitary probe potentials. J Cardiovasc Electrophysiol 8(4):415–413, 1997.

15. Macchi E, Arisi G, Taccardi B. Intracavitory mapping: an improved method for locating the site of origin of ectopic ventricular beats by means of a mathematical model. Proc 10th IEEE Engineering in Medicine and Biology Society, 1988, pp. 187–188.

16. Macchi E, Arisi G, Colli-Franzone P, Guerri L, Olivetti G, Taccardi B. Localization of ventricular ectopic beats from intracavitary potential distributions: an inverse model in terms of source. Proc 11th IEEE/EMBS, 1989, pp. 191–192.

17. Khoury DS, Berrier KL, Badruddin SM, Zoghbi WA. Three-dimensional electrophysiological imaging of the intact canine left ventricle using a noncontact multielectrode cavitary probe: study of sinus, paced, and spontaneous premature beats. Circulation 97(4):399–409, 1998.

18. Schilling RJ, Peters NS, Davies DW. Simultaneous endocardial mapping in the human left ventricle using a noncontact catheter: comparison of contact and reconstructed electrograms during sinus rhythm. Circulation 98(9):887–898, 1998.

19. Jackman WM, Beatty G, Scherlag BJ, Nakagawa H, Arruda M, Widman L, Lazzara R. New noncontact multielectrode array accurately reconstructs left ventricular endocardial potentials(abstr). Pacing Clin Electrophysiol 18(4):898, 1995.

20. Gornick CC, Adler SW, Pederson B, Hauck J, Budd J, Schweitzer J. Validation of a new noncontact catheter system for electroanatomic mapping of left ventricular endocardium. Circulation 99(6):829–835, 1999.

21. Kadish A, Hauck J, Pederson B, Beatty G, Gornick C. Mapping of atrial activation with a noncontact, multielectrode catheter in dogs. Circulation 99(14):1906–1913, 1999.

22. Peters N, Jackman WM, Schilling R, Beatty G, Davies D. Initial experience with mapping human endocardial activation using a novel noncontact catheter mapping system(abstr). Pacing Clin Electrophysiol 19(4), 1996.

23. Baum RS, Alvarez H, Cobb LA. Survival after resuscitation from out-of-hospital ventricular fibrillation. Circulation 50(6):1231–1235, 1974.

24. Goldstein S, Landis JR, Leighton R, Ritter G, Vasu CM, Lantis A, Scrokman R. Characteristics of the resuscitated out-of-hospital cardiac arrest victim with coronary heart disease. Circulation 64(5):977–984, 1981.
25. Myerburg RJ, Kessler KM, Estes D, Conde CA, Luceri RM, Zaman L, Kozlovskis PL, Castellanos A. Long-term survival after prehospital cardiac arrest: analysis of outcome during an 8 year study. Circulation 70(4):538–546, 1984.
26. Roy D, Marchlinski FE, Doherty JU, Buxton AE, Waxman HL, Josephson ME. Electrophysiologic testing of survivors of cardiac arrest. Cardiovasc Clin 15(3):171–177, 1985.
27. Freedman RA, Swerdlow CD, Soderholm-Difatte V, Mason JW. Prognostic significance of arrhythmia inducibility or noninducibility at initial electrophysiologic study in survivors of cardiac arrest. Am J Cardiol 61(8):578–582, 1988.
28. Morady F, Scheinman MM, Hess DS, Sung RJ, Shen E, Shapiro W. Electrophysiologic testing in the management of survivors of out-of-hospital cardiac arrest. Am J Cardiol 51(1):85–89, 1983.
29. Fogoros RN, Elson JJ, Bonnet CA, Fiedler SB, Chenarides JG. Long-term outcome of survivors of cardiac arrest whose therapy is guided by electrophysiologic testing. J Am Cardiol 19(4):780–788, 1992.
30. Ellison KE, Stevenson WG, Couper GS, Friedman PL. Ablation of ventricular tachycardia due to a postinfarct ventricular septal defect: identification and transection of a broad reentry loop. J Cardiovasc Electrophysiol 8(10):1163–1166, 1997.
31. El Sherif N, Gough WB, Restivo M. Reentrant ventricular arrhythmias in the late myocardial infarction period: 14. Mechanisms of resetting, entrainment, acceleration, or termination of reentrant tachycardia by programmed electrical stimulation. Pacing Clin Electrophysiol 10(2):341–371, 1987.
32. Kanaan N, Robinson N, Roth SI, Ye D, Goldberger J, Kadish A. Ventricular tachycardia in healing canine myocardial infarction: evidence for multiple reentrant mechanisms. Pacing Clin Electrophysiol 20:245–260, 1997.
33. Morady F, Harvey M, Kalbfleisch SJ, el Atassi R, Calkins H, Langberg JJ. Radiofrequency catheter ablation of ventricular tachycardia in patients with coronary artery disease. Circulation 87(2):363–372, 1993.
34. Gonska BD, Cao K, Schaumann A, Dorszewski A, zur Mühlen F, Kreuzer H. Catheter ablation of ventricular tachycardia in 136 patients with coronary artery disease: results and long-term follow-up. J Am Coll Cardiol 24(6):1506–1514, 1994.
35. Rothman SA, Hsia HH, Cosss SF, Chmielewski IL, Buxton AE, Miller JM. Radiofrequency catheter ablation of postinfarction ventricular tachycardia: long-term success and the significance of inducible non-clinical arrhythmias. Circulation 96(10):3499–3508, 1997.
36. Stevenson WG, Friedman PL, Kocovic D, Sager PT, Saxon LA, Pavri B. Radiofrequency catheter ablation of ventricular tachycardia after myocardial infarction. Circulation 98(4):308–314, 1998.

37. Schilling RJ, Peters NS, Davies DW. Feasibility of a noncontact catheter for endocardial mapping of human ventricular tachycardia. Circulation 99(19):2543–2552, 1999.
38. Chow A, Schilling RJ, Peters N, Davies DW. Non-contact mapping guided radiofrequency ablation of ventricular tachycardia in patients with frequent defibrillator therapy(abstr). Eur Heart J 20(suppl), 1999.
39. Bluzhas J, Lukshiene D, Shlapikiene B, Ragaishis J. Relation between ventricular arrhythmia and sudden cardiac death in patients with acute myocardial infarction: the predictors of ventricular fibrillation. J Am Coll Cardio 8(1 suppl A):69A–69A, 1986.
40. Pratt CM, Francis MJ, Luck JC, Wyndham CR, Miller RR, Quinones MA. Analysis of ambulatory electrocardiograms in 15 patients during spontaneous ventricular fibrillation with special reference to preceding arrhythmic events. J Am Coll Cardiol 2(5):789–797, 1983.
41. Kowey PR, Friehling T, Meister SG, Engel TR. Late induction of tachycardia in patients with ventricular fibrillation associated with acute myocardial infarction. J Am Coll Cardiol 3(3):690–695, 1984.
42. Pogwizd SM, Corr PB. Mechanisms underlying the development of ventricular fibrillation during early myocardial ischemia. Circ Res 66(3):672–695, 1990.
43. Laurita KR, Girouard SD, Akar FG, Rosenbaum DS. Modulated dispersion explains changes in arrhythmia vulnerability during premature stimulation of the heart. Circulation 98(24):2774–2780, 1998.
44. Kwan YY, Fan W, Hough D, Lee JJ, Fishbein MC, Karagueuzian HS, Chen PS. Effects of procainamide on wave-front dynamics during ventricular fibrillation in open-chest dogs. Circulation 97(18):1828–1836, 1998.
45. Lee JJ, Kamjoo K, Hough D, Hwang C, Fan W, Fishbein MC, Bonometti C, Ikeda T, Karagueuzian HS, Chen PS. Reentrant wave fronts in Wiggers' stage II ventricular fibrillation. Characteristics and mechanisms of termination and spontaneous regeneration. Circ Res 78(4):660–675, 1996.
46. Rogers JM, Huang J, Smith WM, Ideker RE. Incidence, evolution, and spatial distribution of functional reentry during ventricular fibrillation in pigs. Circ Res 84(8):945–954, 1999.
47. Chow A, Schilling R, Davies D, Peters N. Paced initiation of ventricular fibrillation in the infarcted human heart(abstr). Circulation 100(18):I-795, 1999.
48. Cosio FG, Arribas F, López-Gil M, Palacios J. Atrial flutter mapping and ablation. I. Studying atrial flutter mechanisms by mapping and entrainment. Pacing Clin Electrophysiol 19(5):841–853, 1996.
49. Tai CT, Chen SA, Chiang CE, Lee SH, Ueng KC, Wen ZC, Huang JL, Chen YJ, Yu WC, Feng AN, Chiou CW, Chang MS. Characterization of low right atrial isthmus as the slow conduction zone and pharmacological target in typical atrial flutter. Circulation 96(8):2601–2611, 1997.
50. Poty H, Anselme F, Saoudi N. Inferior vena cava-tricuspid annulus isthmus is a critical site of unidirectional block during the induction of common atrial flutter. J Inter Car Electrophysiol 2(1):57–69, 1998.

51. Olgin JE, Kalman JM, Saxon LA, Lee RJ, Lesh MD. Mechanism of initiation of atrial flutter in humans: site of unidirectional block and direction of rotation. J Am Coll Cardiol 29(2):376–384, 1997.

52. Cosio FG, Arribas F, López-Gil M, Gonzilez HD. Atrial flutter mapping and ablation II. Radiofrequency ablation of atrial flutter circuits. Pacing Clin Electrophysiol 19(6):965–975, 1996.

53. Tai CT, Chen SA, Chiang CE, Lee SH, Ueng KC, Wen ZC, Chen YJ, Yu WC, Huang JL, Chiou CW, Chang MS. Electrophysiologic characteristics and radiofrequency catheter ablation in patients with clockwise atrial flutter. J Cardiovasc Electrophysiol 8(1):24–34, 1997.

54. Poty H, Saoudi N, Nair M, Anselme F, Letac B. Radiofrequency catheter ablation of atrial flutter. Further insights into the various types of isthmus block: application to ablation during sinus rhythm. Circulation 94(12):3204–3213, 1996.

55. Saxon LA, Kalman JM, Olgin JE, Scheinman MM, Lee RJ, Lesh MD. Results of radiofrequency catheter ablation for atrial flutter. Am J Cardiol 77(11):1014–1016, 1996.

56. Schilling R, Peters N, Kadish A, Davies D. The conduction properties of the human right atrium during atrial flutter determined by non-contact mapping(abstr). Pacing Clin Electrophysiol 22(4):889, 1999.

57. Schumacher B, Jung W, Lewalter T, Wolpert C, Lüderitz B. Verification of linear lesions using a noncontact multielectrode array catheter versus conventional contact mapping techniques. J cardiovasc Electrophysiol 10(6):791–798, 1999.

58. Kannel WB, Wolf PA, Benjamin EJ, Levy D. Prevalence, incidence, prognosis, and predisposing conditions for atrial fibrillation: population-based estimates. Am J Cardiol 82(8A):2N–2N, 1998.

59. Benjamin EJ, Wolf PA, D'Agostino RB, Silbershatz H, Kannel WB, Levy D. Impact of atrial fibrillation on the risk of death: the Framingham Heart Study [see comments]. Circulation 98(10):946–952, 1998.

60. Krahn AD, Manfreda J, Tate RB, Mathewson FA, Cuddy TE. The natural history of atrial fibrillation: incidence, risk factors, and prognosis in the Manitoba Follow-Up Study. Am J Med 98(5):476–484, 1995.

61. Gray RA, Pertsov AM, Jalife J. Incomplete reentry and epicardial breakthrough patterns during atrial fibrillation in the sheep heart. Circulation 94(10):2649–2661, 1996.

62. Cox JL, Canavan TE, Schuessler RB, Cain ME, Lindsay BD, Stone C, Smith PK, Corr PB, Boineau JP. The surgical treatment of atrial fibrillation. II. Intraoperative electrophysiologic mapping and description of the electrophysiologic basis of atrial flutter and atrial fibrillation. J Thorac Cardiovasc Surg 101(3):406–426, 1991.

63. Isobe F, Kawashima Y. The outcome and indications of the Cox maze III procedure for chronic atrial fibrillation with mitral valve disease. J Thorac Cardiovasc Surg 116(2):220–227, 1998.

64. Cox JL, Boineau JP, Schuessler RB, Kater KM, Ferguson TBJ, Cain ME, Lindsay BD, Smith JM, Corr PB, Hogue CB. Electrophysiologic basis, surgical

development, and clinical results of the maze procedure for atrial flutter and atrial fibrillation. Adv Card Surg 6:1–67, 1995.

65. McCarthy PM, Castle LW, Trohman RG, Simmons TW, Maloney JD, Klein AL, White RD, Cox JL. The Maze procedure: surgical therapy for refractory atrial fibrillation. Cleve Clin J Med 60(2):161–165, 1993.

66. Cox JL, Schuessler RB, Lappas DG, Boineau JP. An 8 1/2-year clinical experience with surgery for atrial fibrillation. Ann Surg 224(3):267–265, 1996.

67. Elvan A, Pride HP, Zipes DP. Replication of the "maze" procedure by radiofrequency cather ablation reduces the ability to induce atrial fibrillation(abstr). Pacing Clin Electrophysiol 17:275, 1994.

68. Avitall B, Hare J, Mughal K, Silverstein E, Krum D, Natale A, Deshpande S, Dhala A, Ahktar M. Ablation of atrial fibrillation in a dog model (abstr). J Am Coll Cardiol 23, 1994.

69. Kadish A, Hauck J, Pederson B, Beatty G, Gornick C. Mapping of atrial activation with a noncontact, multielectrode catheter in dogs. Circulation 99(14):1906–1913, 1999.

70. Schilling RJ, Kadish A, Peters NS, Goldberger J, Davies DW. Endocardial mapping of human atrial fibrillation in the human right atrium using a non-contact catheter. Eur Heart J(in press).

71. Konings KT, Kirchhof CJ, Smeets JR, Wellens HJ, Penn OC, Allessie MA, High-density mapping of electrically induced atrial fibrillation in humans. Circulation 89(4):1665–1680, 1994.

72. Jals P, Halssaguerre M, Shah DC, Chouairi S, Gencel L, Hocini M, Clémenty J. A focal source of atrial fibrillation treated by discrete radiofrequency ablation. Circulation 95(3):572–576, 1997.

73. Halssaguerre M, Jals P, Shah DC, Takahashi A, Hocini M, Quiniou G, Garrigue S, Le Mouroux A, Le Métayer P, Clémenty J. Spontaneous initiation of atrial fibrillation by ectopic beats originating in the pulmonary veins. N Engl J Med 339(10):659–666, 1998.

74. Packer D, Johnson S. Localization of pulmonary vein ectopic activity using a non-contact mapping system(abstr). Circulation 100(8):I-373, 1999.

75. Betts T, Allen S, Salmon A, Edwards T, Morgan JM. Characterization of atrial tachycardia in patients after Fontan surgery using a non-contact mapping system(abstr). Pacing Clin Electrophysiol 22:721, 1998.

11
Electroanatomical Cardiac Mapping

Lior Gepstein
Technion–Israel Institute of Technology and Rambam Medical Center, Haifa, Israel

I. INTRODUCTION

Cardiac mapping was reported as early as 1915, and implies the registration of the electrical activation sequence of the heart by recording of extracellular electrograms [1]. While initially used to study the normal electrical excitation of the heart, the major interest in cardiac mapping soon shifted to analysis of the mechanisms underlying various cardiac arrhythmias, and more recently, to the guidance of curative surgical and catheter ablation procedures [2–4].

The output of the mapping procedure is usually displayed as a spatial representation of activation times, derived from electrograms recorded at a multiplicity of sites. A typical activation map is therefore comprised of several data points, each having two values: (1) the local activation time, and (2) the spatial coordinates of the acquired site within the heart. The latter information can be derived by the use of fixed-shape electrode arrays (epicardial socks and endocardial balloon), usually during open chest surgery [5], or more recently by the percutaneous use of catheters, which are navigated and localized with the use of fluoroscopy [2]. Neverthless, due to the limitation of the two-dimensional nature of fluoroscopy and the fact that the endocardial surface is invisible to X-ray, it is essentially impossible to record the exact coordinates of the recording electrode within the heart.

The recently described nonfluoroscopic, electroanatomical mapping technique (Carto) [5–8] may solve some of the shortcomings of conventional mapping techniques by allowing one to accurately associate endocardial spatial and electrophysiological information in the intact heart. This chapter describes the basic concepts of electroanatomical mapping, details the mapping procedure and the in vivo and in vitro validation studies, and also provides some of the possible research and clinical applications of the technology.

II.　BASIC CONCEPTS

A.　System Components

At its most basic level, the nonfluoroscopic mapping system is comprised of a miniature passive magnetic sensor, an external magnetic field emitter (location pad), and a processing unit (Carto, Biosense-Webster, Israel).

The location sensor is integrated into a standard 7F deflectable-tip electrophysiological catheter and lies just proximal to the tip electrode. The location pad is placed just beneath the operating table and includes three coils, which generate ultralow magnetic fields (0.05–0.5 gauss) the decay as a function of the distance from the sources. The spatial and temporal characteristics of the sensed magnetic fields contain the information necessary to solve a set of overdetermined algebraic equations yielding the location (x, y, and z) and orientation (roll, pitch, and yaw) of the catheter's tip. This allows continuous real-time tracking of the catheter while it is deployed within the heart without the aid of fluoroscopy.

B.　Mapping Procedure

The mapping catheter is introduced, through the appropriate vascular access, under fluoroscopic guidance into the mapped chamber. The mapping procedure involves sequentially dragging the catheter along the endocardium, acquiring the location of the catheter together with the local electrogram recorded from its tip at multiple sites. The location of the mapping catheter is gated to a fiducial point in the cardiac cycle and recorded relative to the location of a second locatable catheter (the reference catheter) fixed at a stable position. This allows compensation for both cardiac and patient motions. By sampling the location of the catheter together with the local electrogram at a plurality of endocardial sites, the 3-D geometry of the chamber is reconstructed in real time (Fig. 1).

The local activation time (LAT) is determined at each sampled site as the time interval between a fiducial point on a fixed reference electrogram (intracardiac of body surface) and the steepest negative intrinsic deflection (dV/dt_{min}) from the unipolar recording. Alternatively, the local bipolar signal (using the onset or maximal or minimal values) can be used.

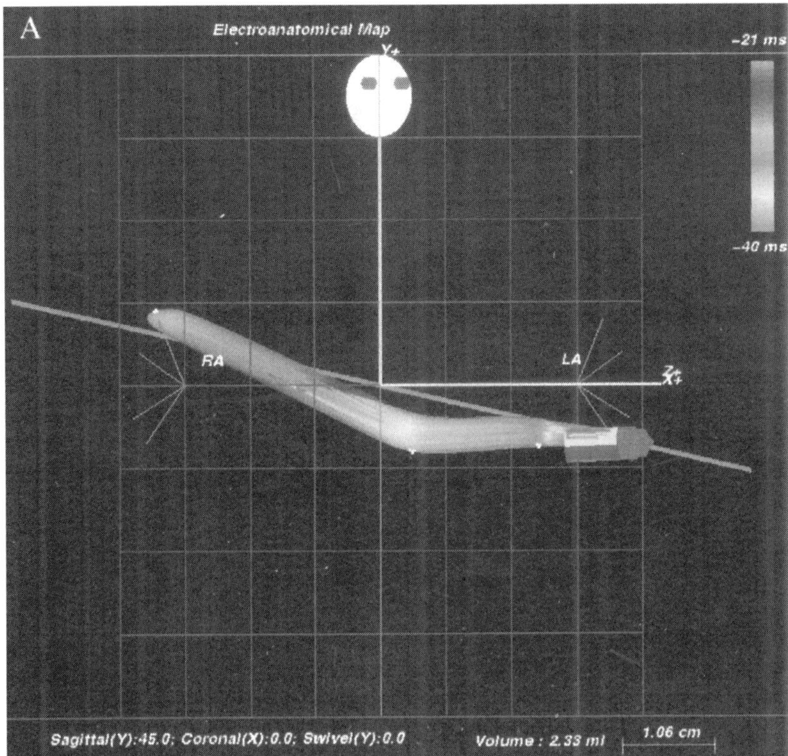

Figure 1 Three progressive stages of the reconstruction of the human left ventricle during sinus rhythm. The map is shown from a right anterior oblique (RAO) view. (A) Initial reconstruction, using four sampled points. (B) Reconstruction from 15 sampled points. Note the catheter icon pointing toward the anterior wall. (C) The complete electroanatomical map, demonstrating the earliest activation (red areas in original) along the superior and inferior parts of the septum.

The electrophysiological information (LAT distribution or other parameters derived from the local electrograms) is color coded and superimposed on the map (electroanatomical map, Fig. 1C).

C. Inherent Limitations of the System

The inherent limitations of the new technique stem from its sequential nature. Like other cardiac mapping techniques which utilize a roving catheter, the mapping procedure is sequential and uses a beat-by-beat approach. This imposes two requirements: (1) the need for a stable rhythm (implying that both the activation sequence and the geometry of the mapped

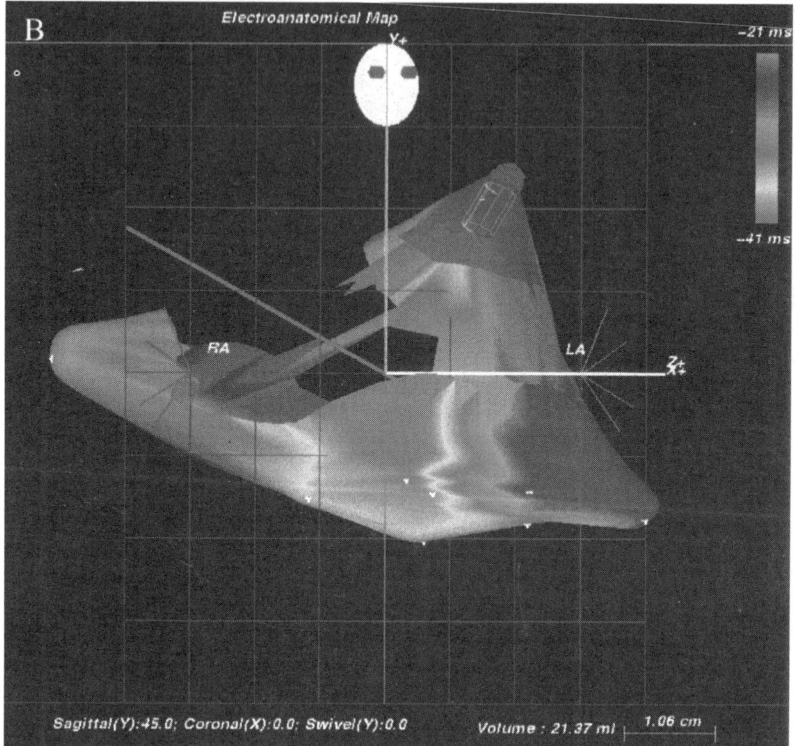

Figure 1 (Continued)

chamber are reproducible on a beat-by-beat basis), and (2) the need for a fixed reference electrogram. This limitation is partially addressed by specific stability criteria that are used before acquisition of new data points.

III. VALIDATION STUDIES

A. In Vitro Studies

The location capabilities of the system were assessed during bench testing and were found to be highly accurate and reproducible. The reproducibility of the system was quantified by measuring the standard deviation (SD) of repeated location determination of the tip of the catheter, using different orientations at various sites. The SD of these repeated measurements was found to be 0.16 ± 0.02 mm (mean $\pm$ SEM), with the maximal range being 0.55 ± 0.07 mm. Similarly, relative distances measured by the system were also found to be highly accurate (mean error, 0.42 ± 0.05 mm).

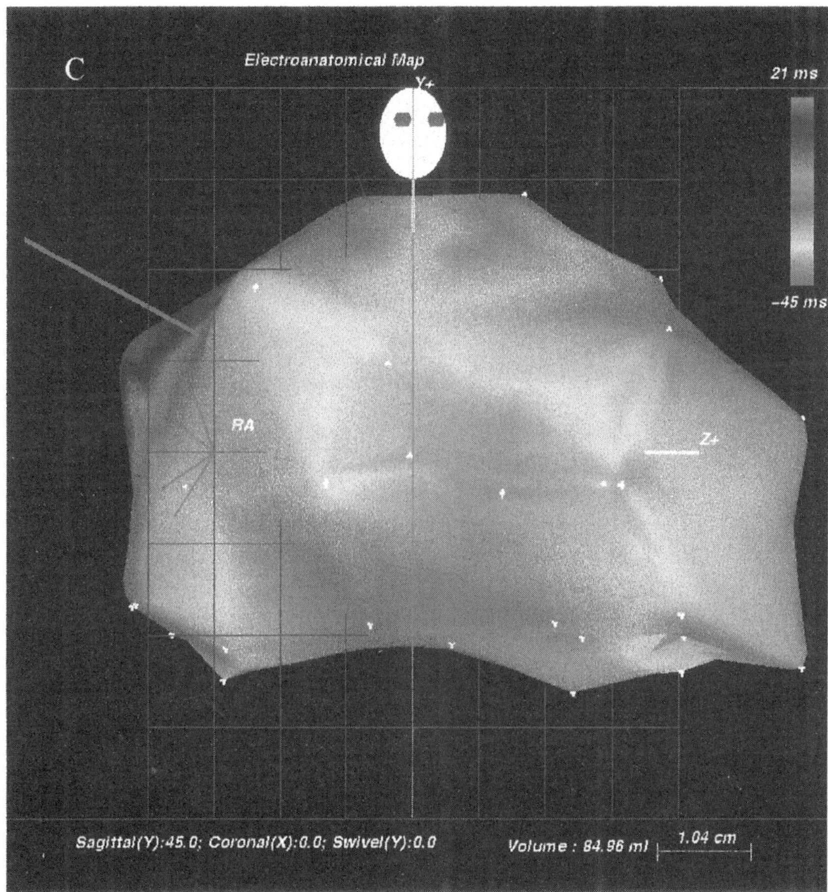

Figure 1 (Continued)

B. In Vivo Studies

The in vivo accuracy of the system was validated both in animals and in clinics in a number of studies.

1. The reproducibility of the system was assessed by measuring the SD of repeated location determination while in contact with the swine's left ventricle (LV) endocardium and was found to be 0.74 ± 0.05 mm with a maximal range of 1.26 ± 0.08 mm [7]. Also determined were the relative distances measured by the system while sequentially withdrawing the mapping catheter inside a long

sheath at 10-mm intervals [7]. The average location error was 0.73 ± 0.03 mm. Similar results were found in humans, with the average location error found to be 0.95 ± 0.8 [9].

2. Repeated electroanatomical maps of the different cardiac chambers demonstrated a reproducible geometry and activation sequence during both sinus rhythm and pacing and also enabled accurate identification of the pacing site in all the animals that were mapped [7].

3. The volumetric measurements of the system (which study both the precision of the location of the system and the reconstruction algorithms) were also found to be highly accurate and reproducible [10]. The high precision and reproducibility was confirmed in simple phantoms, left ventricular casts, in a dynamic test jig, and in the swine's LV. Intraobserver and interobserver variabilities were found to be minimal.

4. The accuracy of the system was also tested by repeatedly applying radiofrequency current to a specific endocardial site, which was tagged on the map, and also by the ability to combine a number of ablation points into a linear lesion [11]. These studies demonstrated that the localization of the catheter is accurate enough to guide delivery of focal RF lesions in an accurate and reproducible manner. Moreover, the combination of accurate navigation and the ability to tag the previously ablated sites on the map enabled us to create long and continuous lesions. A high correlation was found between the computer record of the location, length, and shape of the lesion and the pathological findings.

IV. TYPES OF MAPS

The electroanatomical maps can be presented in several forms. The activation maps (Fig. 1c, Fig. 2a) present the spatial distribution of the LATs, which are determined at each site, color coded, and overlaid on the reconstructed three-dimensional geometry. This enables us to characterize physiological and pathological activation pathways. Note, for example, in Fig. 1C, the normal activation of the human LV with activation (red area in original) originating in the superior and inferior septum, corresponding to the left anterior and posterior bundles, respectively.

The propagation maps display a dynamic color display of the activation wavefront propagation across the reconstructed chamber. For example, Fig. 3 demonstrates six sequential stages in the propagation map of the right

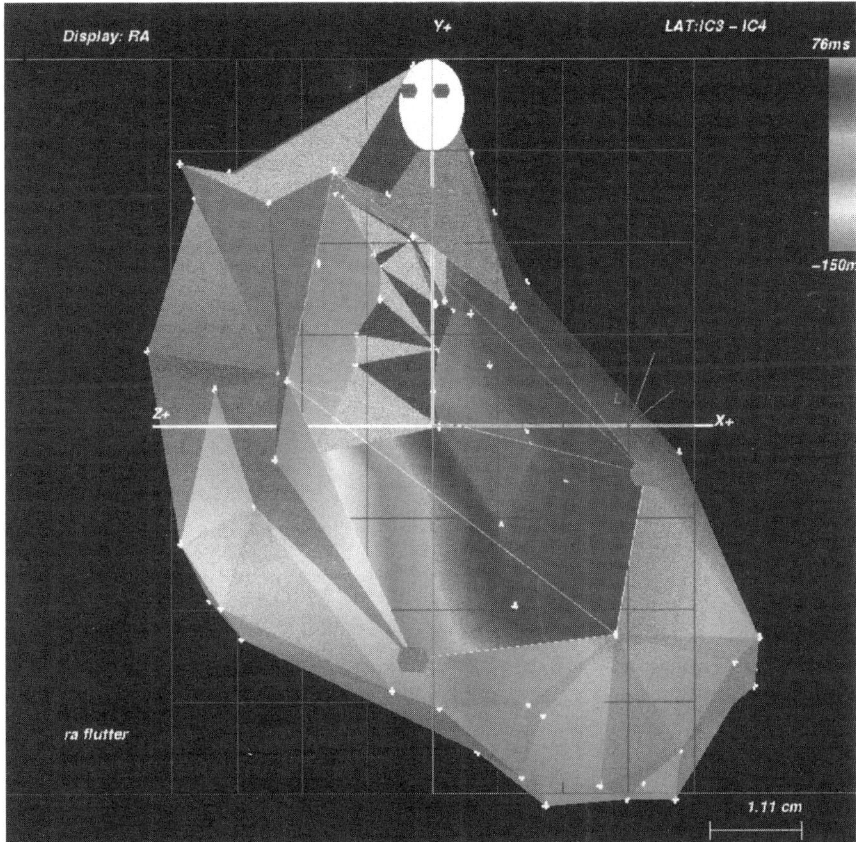

Figure 2 Right atrial activation map (shown from a LAO projection) of a patient with typical atrial flutter. Note the activation propagating around the tricuspid annulus and the close spatial association between "early" (red in original) and "late" sites (purple), indicated by the gray area.

atrium (RA) of a patient with atrial flutter. Note the typical counter-clockwise rotation of the activation wave around the tricuspid annulus.

The voltage maps displays the peak-to-peak amplitude of the local electrogram (unipolar or bipolar) sampled at each site. These values are color coded, with red and purple indicating the lowest and highest electrogram amplitude, respectively. The abnormally low voltage area usually represents scar tissue [12], and may aid in the understanding and treatment of the mechanism of several arrhythmias. Figure 4 demonstrates the voltage map of the LV of a dog, 4 weeks post LAD ligation. Note the presence of

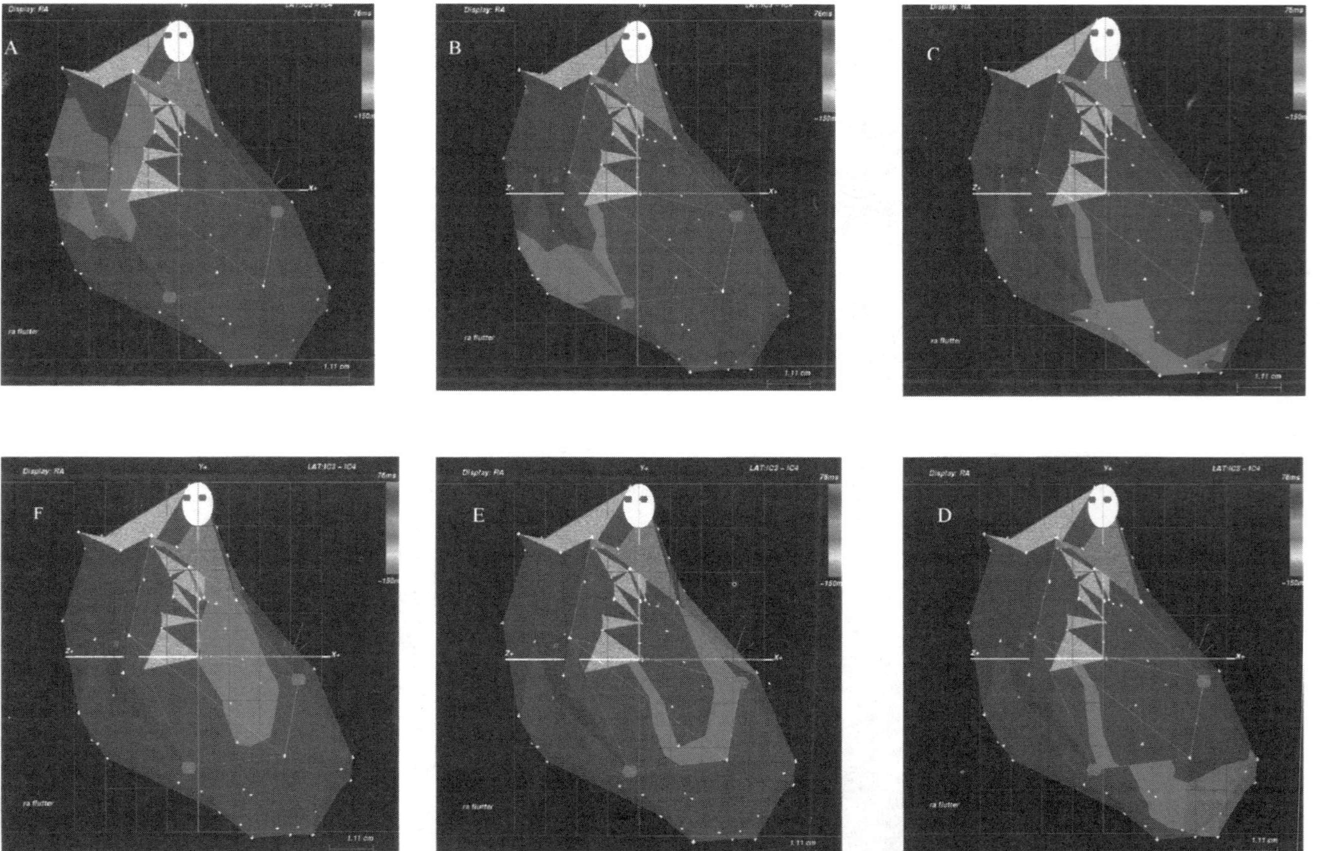

Figure 3 Right atrial propagation map of atrial flutter. Note the counterclockwise propagation (a–f) of the activation wavefront around the tricuspid annulus.

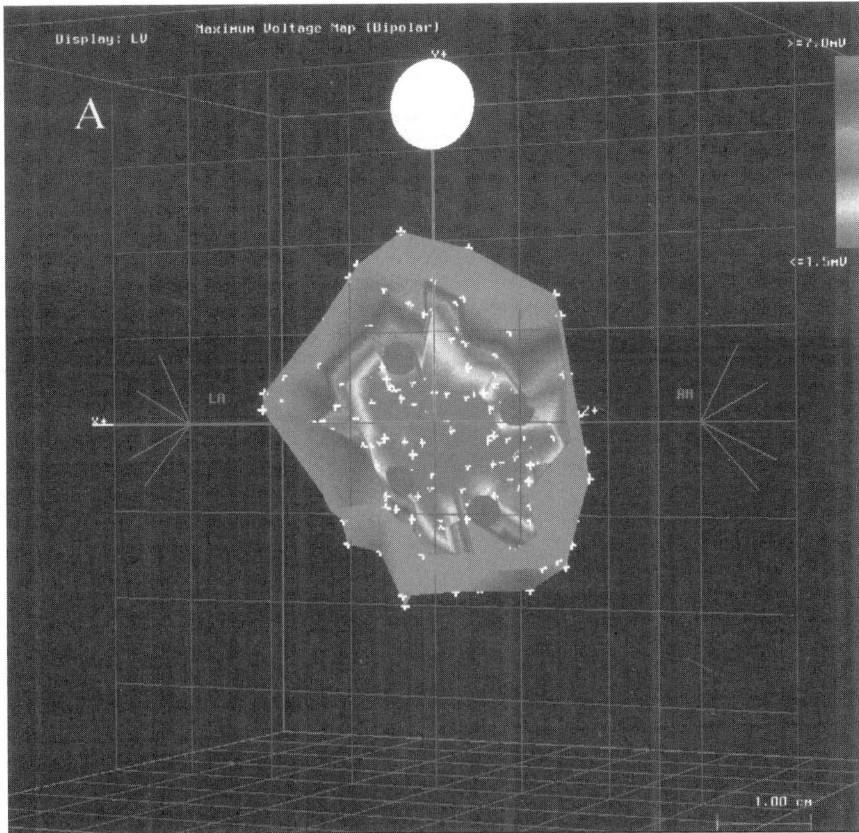

Figure 4 (A) A bipolar voltage map of the canine LV 4 weeks following coronary ligation. Note the presence of an area of low-amplitude electrograms (red area in original) delineating the infarct. The catheter was navigated to the border of the infarct, defined by the steepest voltage gradient where 4 ablation lesions were delivered. (B) The corresponding pathological finding stained with tetrazolium. Note the presence of the infarct and the location of the ablation on the margin.

low-amplitude electrograms (red in original) delineating the infarct (Fig. 4A) and the corresponding TTC-stained pathological specimen.

Besides assessment of the depolarization sequence the spatial dispersion of other properties of the action potential can be analyzed. Thus, maps portraying the repolarization (determined from the timing of local T wave) and activation-recovery interval (ARI, defined as the time interval between the local activation and repolarization timing) patterns can be generated in addition to the conventional activation maps. Using these

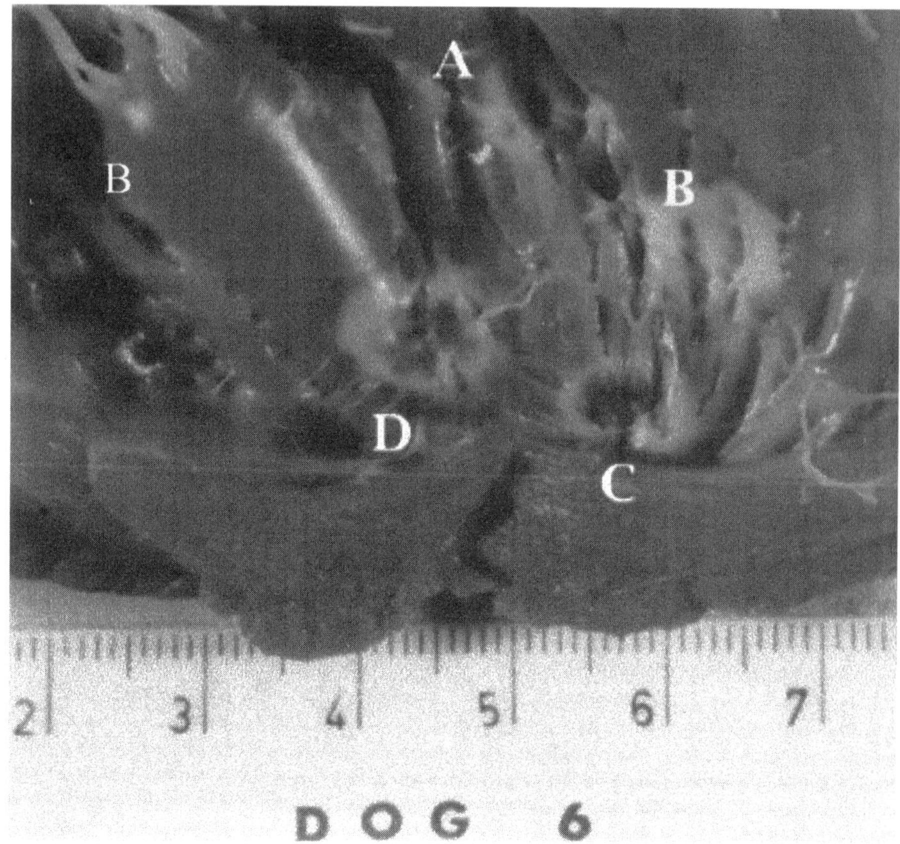

Figure 4 (Continued)

maps we recently found tight spatial correlation between endocardial activation and repolarization in the healthy swine LV [13]. An inverse correlation was noted between activation and ARI patterns, with early and late activation sites associated with the longest and shortest ARI values, respectively. The shortening of ARI (which correlates with action potential duration) as activation proceeds may be the results of electrotonic interactions, and may serve as a novel antiarrhythmic mechanism in healthy tissue since it tends to synchronize repolarization. The ability to characterize the activation and repolarization properties of the heart in a spatially oriented way may also be used in the clinic to evaluate the substrate underlying various arrhythmias and possibly also for identification of patients at risk.

As discussed earlier, a major prerequisite in a sequential mapping technique is the stability of both the geometry and activation sequence on a beat-by-beat basis. Alternatively, statistical methods can be used to identify the electrophysiological properties of each site. For example, in a recent study [14], we mapped the atrial spatial distribution of different stastical organization parameters of atrial fibrillation (AF) in the goat model of AF. Our results demonstrated that significant spatial dispersion exists in atrial fibrillation organization. An example of this spatial heterogeneity can be viewed in Fig. 5. This figure presents the right atrial spatial dispersion of the

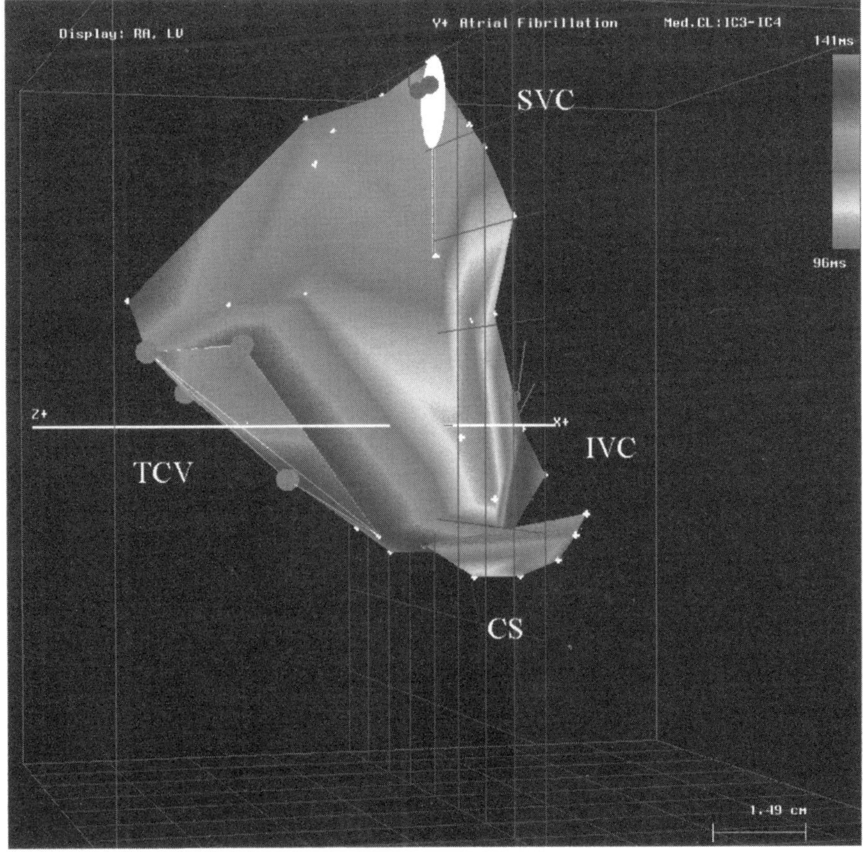

Figure 5 A left lateral electroanatomical map of the goat's right atrium during atrial fibrillation. Colors in original represent the median cycle length (CL) of atrial activations at each site. Note the spatial heterogeneity with the septum characterized by the shortest median CLs (red area).

AF median cycle length (CL) measured at each site. A typical distribution was noted in all animals, with the septum and posterior wall characterized by significantly higher activation frequency (shorter median CLs, red area) when compared to the rest of the atria.

V. CLINICAL APPLICATIONS AND MAPPING OF CARDIAC ARRHYTHMIAS

Since the introduction of the new mapping technique to the clinical arena it has been shown to be effective for the mapping and ablation of a variety of supraventricular and ventricular arrhythmias. These rhythm disturbances include AV nodal reentrant tachycardia [15], AV reentrant tachycardias [16], atrial flutter [17], focal [18].and reentrant atrial tachycardias [19], atrial fibrillation [20], and focal [21] and reentrant ventricular tachycardias [22]. The ability to associate electrophysiological and spatial information, the "bookkeeping" capabilities of the modality, and the ability to navigate the catheter precisely to predetermined sites offer unique advantages for each stage of the ablation procedure. The generated map offers a unique insight to the mechanism underlying the arrhythmia studied, it enables design of an ablation strategy, and finally, one may use the generated electroanatomical map as a road map for delivering the therapeutic energy.

A. Defining the Substrate and the Mechanism of the Arrhythmia

The ability to combine spatial, anatomical, and electrophysiological information may enhance the ability to define the mechanism of the arrhythmia studied, and identify the anatomical and electrophysiological substrate underlying the arrhythmia. Some examples include the following.

1. Differentiation of Focal from Macroreentrant Atrial Arrhythmias

If the entire circuit is mapped, the macroreentrant arrhythmia will usually be characterized by the following features: (1) a range of activation times that will equal or will be slightly shorter than the cycle length (CL) of the tachycardia, and (2) close spatial association between the arbitrary "early" and "late" sites (between the head and tail of the reentrant wavefront). For example, Figs. 2 and 3 depict the corresponding activation and propagation maps of patient with typical atrial flutter. Note the counterclockwise rotation of the activation wavefront with activation traveling around the tricuspid annulus. Total activation time of the atrium was 226 msec, which was slightly sorter than the tachycardia CL (msec). Note the absence of a focal red area and the close spatial association between early (red) and late (purple) activated sites. This phenomenon is demonstrated in the map by the presence of the gray area indicating presence of neighboring early and late activation sites. In contrast, a focal arrhythmia will be characterized by a well-defined early activation site surrounded in all directions by later activation sites without close association of early and late sites. In addition, total activation time of the mapped chamber will usually be significantly shorter than the CL of the arrhythmia.

2. Identifying the Substrate of the Arrhythmia

Abnormalities of the structural or electrophysiological properties of heart usually form the basis for the generation and perpetuation of the arrhythmia. An important example is the presence of nonviable elements (scarred tissue) within the myocardium. Possible arrhythmias in which the presence of such tissue plays an important role include ventricular tachycardia secondary to ischemic cardiomyopathy, postcongenital heart surgery arrhythmias, etc. In these patients, the abnormal substrate could be identified by the presence of extremely low-amplitude electrograms in the voltage maps. These areas may represent scarred tissue (in the ventricle) [12], replacement of myocardial tissue (such as occurs in right ventricular dysplasia) [23], or the presence of past atriotomies, and may aid in understanding the mechanism of the arrhythmia and designing the ablation strategy. A novel approach using this technique was recently proposed by Marchlinski et al. [24], who used the voltage map to identify the scar, and generated linear ablation extending through the scar to control unmappable ventricular tachycardia in patients with ischemic and nonischemic cardiomyopathy.

B. Defining the Ablation Strategy

Following identification of the mechanism involved in the genesis of the arrhythmia and the pathological structural and electrophysiological substrate, the electroanatomical map can be used to design the appropriate ablation strategy. Since ablation procedures are usually based on a combination of anatomical and electrophysiological factors, combining these features may be of major clinical advantage.

Anatomically based linear ablation procedures are usually generated between fixed anatomical structures and may be used for the treatment of variable arrhythmias including atrial flutter, scar-related atrial tachycardias, and atrial fibrillation. The electroanatomical and substrate maps may be used to tag specific anatomical sites and then be used as a "road map" to design these complex ablation procedures.

For focal arrhythmias (such as atrial tachycardia), the earliest activation site is usually defined as the target site for ablation. The investigators addressed these cases by using either global activation mapping or by the regional, stepwise hot–cold approach. Both approaches identified the earliest site of activation only when surrounded in all directions by later-activated sites. The catheter can then be renavigated to the target site for energy delivery.

The ability to associate variable electrophysiological parameters with a specific endocardial site may play a significant role in guiding ablation procedures in reentrant arrhythmias such as ventricular tachycardia, where several criteria guide ablation. Hence, a wide spectrum of electrophysiological information (LAT, the results of entrainment and pace mapping, the morphology of the unipolar and bipolar electrogram, etc.) can be recorded, stored, and associated with a specific "address" in the endocardium. This information is then used to select the possible target sites for ablation.

C. Guiding Ablations

The possible advantages of the mapping technique in guiding ablation result from three unique qualities:

1. The ability to determine the 3-D location and orientation of the ablation catheter with relevance to the generated electro-anatomical map and, thus, the ability to relocate the catheter with great precision to a specific endocardial site
2. The ability to tag and display sites with electrophysiological or anatomical significance and sites where RF energy was already applied
3. The ability to assess the effect of ablation process on the electro-anatomical substrate

The aforementioned capabilities of the mapping technique may play a special role in guiding the creation of complex longitudinal and continuous lesions.

Generation of such lesions, aiming at creating continuous lines of conduction blocks, may be mandatory for the treatment of a variety of arrhythmias such as atrial flutter, reentrant arrhythmias with broad isthmuses, and possibly also for atrial fibrillation. Recent results have demonstrated that the nonfluoroscopic technique may bring a unique value to these procedures. Thus, following establishment of an anatomical shell, the catheter is navigated to the desired area and repetitive applications of RF energy to adjoining sites are used to create the lesion. Each RF application is tagged and added to the map. Recent animal studies have demonstrated that this procedure results in generation of acute and long-term continuous lesions [11,25].

The ability to renavigate the catheter back to the targeted area may allow assessment of the deployed lesion [25]. Lesion continuity may be confirmed by: (1) the presence of conduction block (significant disparities in activation times on opposite sides of the lesion) in the activation and propagation maps as well as by opposite orientation of the wavefront on opposing sides of the lesion; (2) evidence of double potentials along the lesion; and (3) low-amplitude electrograms along the lesion. Using these properties, possible gaps can be identified within the lesion and the catheter renavigated to these sites for delivery of additional energy for the completion of the lesion.

V. SUMMARY AND FUTURE DIRECTIONS

We have presented here the basic concepts of the nonfluoroscopic electroanatomical mapping technique. The ability to associate spatial and electrophysiological information represents a paradigm shift from conventional mapping techniques, and as described above may have significant implications for both basic and clinical electrophysiology. Specifically, these qualities may prove important for tackling these arrhythmias which still possess a mechanistic and therapeutic challenge, such as ventricular tachycardia and atrial fibrillation.

The ability to associate functional and structural information, coupled with a therapeutic modality, may extend to other fields of cardiovascular medicine. For example, we have recently demonstrated that by sampling the location of the catheter throughout the cardiac cycle, the LV regional and global mechanical function can be evaluated [10]. Although beyond the scope of this chapter, the ability to combine spatial, electrophysiological, and mechanical information may provide a useful tool in both research and clinical cardiology. Recent work suggested that the different spectra of ischemic pathologies might be identified, located, and quantified by simultaneous assessment of LV electromechanical properties [12,26]. Hence, healthy myocardial tissue is characterized by normal mechanical and electrical function, irreversibly necrotic tissue can be identified by coupling of low-amplitude electrograms and abnormal mechanics, and chronically ischemic myocardium (hibernation) can be identified by abnormal mechanics with relatively preserved electrical function.

In summary, the new technology described in this chapter provides a unique and innovative approach to cardiovascular research by linking functional and structural information in the in vivo heart. These qualities coupled with accurate targeting of different therapeutic modalities (such as RF ablation) provide a unique tool for both scientists investigating the heart and practicing cardiologists.

REFERENCES

1. Lewis T, Rothschild MA. The excitatory process in the dog's heart, II: the ventricles. Phil Trans R Soc Lond B Biol Sci 206:181–226, 1915.
2. Josephson ME, Horowitz LN, Spielman SR, Waxman HL, Greenspan AM. Role of catheter mapping in the preoperative evaluation of ventricular tachycardia. Am J Cardiol 49:207–220, 1982.
3. Jackman WM, Wang XZ, Friday KJ, et al. Catheter ablation of accessory atrioventricular pathways (Wolf-Parkinson-White syndrome) by radiofrequency current. N Engl J Med 324:1605–1611, 1991.
4. Schienman M, Morady F, Hess D, et al. Catheter-induced ablation of the atrioventricular junction to control refractory supraventricular arrhythmias. JAMA 248:851–855, 1982.
5. Gallegher JJ, Kasell JH, Cox JL, Smith WM, Ideker RE, Smith WM. Techniques of intraoperative electrophysiologic mapping. Am J Cardiol 49:221–240, 1982.
6. Ben-Haim SA, Osadchy D, Schuster I, Gepstein L, Hayam G, Josephson ME. Nonfluoroscopic, in vivo navigation and mapping technology. *Nature Med* 2:1393–1395, 1996.
7. Gepstein L, Hayam G, Ben-Haim SA. A novel method for nonfluoroscopic catheter-based electroanatomical mapping of the heart: in vitro and in vivo accuracy results. *Circulation* 95:1611–1622, 1997.
8. Gepstein L, Evans SJ. Electroanatomical mapping of the heart: basic concepts and implications for the treatment of cardiac arrhythmias. PACE 21:1268–1278, 1998.
9. Smeets JL, Ben-Haim SA, Rodriguez LM, Timmermans C, Wellens HJ. New method for nonfluoroscopic endocardial mapping in humans: accuracy assessment and first clinical results. Circulation 97:2426–2432, 1988.
10. Gepstein L, Hayam G, Shpun S, Ben-Haim SA. Hemodynamic evaluation of the heart with a nonfluoroscopic electromechanical mapping technique. Circulation 96:3672–3680, 1997.
11. Shupun S, Gepstein L, Hayam G, Ben-Haim SA. Guidance of radiofrequency endocardial ablation with real-time three-dimensional magnetic navigation system. Circulation 96:2016–2021, 1997.
12. Gepstein L, Goldin A, Lessick J, Hayam G, Shpun S, Schwartz Y, Shofty R, Turgeman A, Kirshenbaum D, Ben-Haim SA. Electromechanical characterization of chronic myocardial infarction in the canine coronary occlusion model. Circulation 98:2055–2064, 1988.
13. Gepstein L, Hayam G, Ben-Haim SA. Activation-repolarization coupling in the normal swine endocardium. Circulation 96:4036–4043, 1997.

14. Gepstein L, Hayam G, Wolf T, Shofty R, Zaretzki A, Ben-Haim SA. 3D endocardial spatial dispersion of atrial fibrillation organization in the chronic goat model (abstr 1793). Circulation 100(suppl 1):I-342.
15. Cooke PA, Wilber DJ. Radiofrequency catheter ablation of atrioventricular nodal reentry tachycardia utilizing nonfluoroscopic electroanatomical mapping. Pacing Clin Electrophysiol 21:1802–1809, 1998.
16. Worley SJ. Use of a real-time three-dimensional magnetic navigation system for radiofrequency ablation of accessory pathways. Pacing Clin Electrophysiol 21:1636–1645, 1998.
17. Nakagawa H, Jackman WM. Use of a three-dimensional, nonfluoroscopic mapping system for catheter ablation of typical atrial flutter. Pacing Clin Electrophysiol 21:1279–1286, 1998.
18. Kottkamp H, Hindricks G, Breithardt G, Borggrefe M. Three-dimensional electromagnetic catheter technology: electroanatomical mapping of the right atrium and ablation of ectopic atrial tachycardia. J Cardiovasc Electrophysiol 8:1332–7133, 1977.
19. Dorostkar PC, Cheng J, Scheinman MM. Electroanatomical mapping and ablation of the substrate supporting intraatrial reentrant tachycardia after palliation for complex congenital heart disease. Pacing Clin Electrophysiol 21:1810–1819, 1998.
20. Pappone C, Oreto G, Lamberti F, Vicedomini G, Loricchio ML, Shpun S, Rillo M, Calabro MP, Conversano A, Ben-Haim SA, Cappato R, Chierchia S. Catheter ablation of paroxysmal atrial fibrillation using a 3D mapping system. Circulation 100:1203–1208, 1999.
21. Nademanee K, Kosar EM. A nonfluoroscopic catheter-based mapping technique to ablate focal ventricular tachycardia. Pacing Clin Electrophysiol 21:1442–1447, 1988.
22. Stevenson WG, Delacretaz E, Friedman PL, Ellison KE. Identification and ablation of macroreentrant ventricular tachycardia with the CARTO electroanatomical mapping system. Pacing Clin Electrophysiol 21:1448–1456, 1998.
23. Gepstein L, Lashevski I, Reisner S, Boulos M. Electroanatomical mapping of right ventricular dysplasia. J Am Coll Cardiol 38:2020–2027, 2001.
24. Marchlinski FE, Callans DJ, Gottlieb CD, Zado E. Linear ablation lesions for control of unmappable ventricular tachycardia in patients with ischemic and nonischemic cardiomyopathy. Circulation 101:1288–1296, 2000.
25. Gepstein L, Hayam G, Shpun S, Cohen D, Ben-Haim SA. Atrial linear ablations in pigs: chronic effects on atrial electrophysiology and pathology. Circulation 100:419–426, 1999.
26. Kornowski R, Hong MK, Gepstein L, Goldstein S, Ellahham S, Ben-Haim SA, Leon MB. Preliminary animal and clinical experience using an electromechanical endocardial mapping procedure to distinguish infarcted from healthy myocardium. Circulation 98:1116–1124, 1998.

12
Quantitative Analysis of Complex Rhythms

Jack M. Rogers
University of Alabama at Birmingham, Birmingham, Alabama, U.S.A.

Philip V. Bayly
Washington University, St. Louis, Missouri, U.S.A.

I. INTRODUCTION

During cardiac tachyarrhythmias such as fibrillation and polymorphic tachycardia, the normally well-ordered cardiac activation sequence is replaced by an abnormal activation pattern in which wavefronts follow complex, nonrepeating pathways. To understand the mechanisms of these arrhythmias, many investigators have used cardiac mapping techniques. In mapping experiments, spatial and temporal information on cardiac arrhythmias is obtained by simultaneously *recording local electrical activity* from many sites. In electrical mapping, extracellular potentials are recorded. These signals may be unipolar (relative to a distant reference electrode), or bipolar (both poles closely adjacent). Large electrical mapping systems currently have on the order of 500 channels. In optical mapping, myocardium is stained with a dye that fluoresces in proportion to the transmembrane potential when excited by a strong light source. The optical signals can be recorded using a variety of technologies, including laser scanning systems, photodiode arrays, and fast video cameras. Optical mapping systems can record from hundreds to thousands of sites. Temporal sampling rates for cardiac mapping typically range from a few hundred to a few thousand samples per second.

Whichever technology is used, mapping studies collect a large volume of data. This chapter will review techniques that have been developed for quantitative analysis and interpretation of these data in order to reveal the dynamics of the mapped rhythms.

II. ISOCHRONAL MAPPING

The traditional way to analyze cardiac mapping data is to construct isochronal contour maps. In such a map, each contour line connects points that are depolarized at the same time. Successive contour lines track the progression of activation wavefronts across the mapped region (similar maps can also be constructed for repolarization or other events). The first step in constructing an isochronal activation map is to determine the discrete times at which activation occurs at each recording site. How this is done depends on the mapping modality. Unipolar electrograms exhibit a positive deflection as the wavefront approaches the electrode and a negative deflection as it moves away. Thus, for these signals, local activation is usually taken as the time of the maximum negative first derivative [1,2]. Because unipolar electrodes sense activity from distant tissue as well as from tissue in contact with the electrode [3,4], some investigators use bipolar recordings, which record the potential between two closely spaced electrodes, and therefore reject distant activity. In this case, the maximum deflection (of either polarity) is taken as the time of activation. In optical mapping, the time of activation is commonly taken as the peak of the signal's first derivative.

If the mapped region has been activated more than once, the next step in constructing an isochronal map is to group together activations (detected in different channels) that arose from the same beat. This ensures that the contour lines will trace a single wavefront. Finally, the contour lines are generated by interpolating activation times between electrodes. There are a number of ways to do this, including triangulation coupled with linear interpolation [5–7], gridding, and krigging. The latter two methods map the data to a regular grid before drawing the contours and can provide estimates of the error in the map [8,9].

Isochronal maps are ideal for analyzing simple, repetitive rhythms because they compress a great deal of information into a compact, easily interpreted format. However, they have numerous disadvantages when used for complex rhythms such as fibrillation. Because only one activation can be registered on each map, rhythms that change beat-to-beat require multiple maps. A more serious problem involves picking activation times. During complex rhythms, recordings of all modalities often exhibit low amplitudes and multiphasic deflections, making activation picking difficult and ambiguous [10,11]. Although many investigators have studied ways to improve activation picking [12–14], the problem is still unsolved. In addition, during complex rhythms, grouping of activations into beats is not always obvious, and this difficulty is compounded by any errors in activation selection. Grouping errors are likely to cause inappropriate discontinuities or wild excursions of the contours [10]. A final potential problem is that most contour-generation algorithms implicitly assume that it is always permissible to interpolate between recording sites to estimate activation times in between. This is not true in the presence of nonviable tissue or propagation block, in which case activations at neighboring electrodes may be due to different wavefronts. Methods to map around such discontinuities have been developed [15], but require some a-priori knowledge of the activation pattern.

III. COMPUTER ANIMATION

The root of the problem with isochronal mapping is the need to reduce data to one static image. Many of the above difficulties can be avoided if computer animation is used to view the data. The basic idea is to map each recording site to a location on the monitor, color code the variable of interest, and then animate [16]. By playing the mapping data forward and backward at variable speed, the investigator can quickly gain an understanding of the overall pattern and identify important events. For unipolar electrical mapping, the animated variable is typically the extracellular potential or its first temporal derivative. The later is usually preferred because it eliminates baseline drift and sharpens localization of the wavefront [4,17]. In optical mapping, the fluorescence signal is usually animated. Often, to account for spatial heterogeneity in dye distribution, the signals at each recording site are normalized so that a control beat has a common amplitude over the entire mapped region [18]. Another variant is based on picked activation times. When an activation is registered at a recording site, the site cycles through a sequence of colors lasting a fixed total of 50 msec [19]. Thus, the head of a wavefront is represented by one color, while the remaining colors give an indication of where the wavefront has been and how fast it is moving.

Computer animation allows a large volume of data to be analyzed quickly with minimal preprocessing. It is relatively insensitive to noise: spurious activation complexes appear as isolated flashes, while undetected complexes produce a small dropout in the wavefront. Either event is filtered out by the observer's eye and the overall activation pattern is still apparent. However, further analysis is generally required to parameterize the mapping data so that different episodes can be compared quantitatively.

IV. ACTIVATION PATTERN DECOMPOSITIONS

One approach to describing a complex activation pattern quantitatively is to decompose the pattern into some set of building blocks. The number and properties of these blocks then form a basis for parameterization of the rhythm. Several such approaches have been used.

A. The Karhunen-Loeve Decomposition

The Karhunen-Loeve (K-L) decomposition describes an observed data set in terms of a set of basis functions that best approximates the data set for a fixed number of terms [20]. Consider data from a cardiac mapping experiment. At each of N sensors, there is a time series with M samples. After removing the mean from each time series, the snapshot at time m is denoted by the N-dimensionsal vector $\mathbf{s}_m$. This vector can be exactly represented as the weighted sum of N orthonormal basis vectors (patterns or "modes")

$$\mathbf{s}_m = \sum_{n=1}^{N} a_{mn} \mathbf{v}_n \tag{1}$$

The K-L basis vectors, $\mathbf{v}_n$, are chosen to minimize the average mean squared error (or, equivalently, to maximize the variance) in the approximations,

$$\mathbf{r}_m = \sum_{n=1}^{P} a_{mn} \mathbf{v}_n \tag{2}$$

where the number of nodes in the approximation, P, is less than the number of sensors, N. The first K-L mode thus forms the best single-mode approximation to the data set, a weighted sum of the first two K-L modes forms the best two-mode approximation, and so on. How are these optimal basis vectors found? First, the spatial autocovariance matrix, R, is computed:

$$R = [r_{ij}] = \sum_{m=1}^{M} f_{im} f_{jm} \qquad i, j = 1, 2, \ldots, N \tag{3}$$

where f_{mn} is the mth mean-corrected sample from the nth electrode. It can be shown [21] that the K-L modes are simply the eigenvectors of R. The K-L modes are also known as "principal components."

The K-L decomposition has a relatively long history in body-surface potential mapping. In 1971 Barr and co-workers used the K-L basis and the relative variance contributions of each mode to choose measurement locations for body-surface potential maps [22]. Lux et al. represented body-surface potential maps using a relatively small number of K-L modes to reduce the redundancy in their data sets [23]. Claydon et al. used changes in the K-L decomposition to quantify the effects of myocardial infarctions on the body surface map [24]. In the context of epicardial mapping, the K-L procedure has been used to track changes in spatial organization as will be described below. It has also been used to make short-term predictions of epicardial activity during ventricular fibrillation (VF) by extrapolating the observed temporal behavior of the first five modal coefficients [25].

B. Wavefront Isolation and Wavefront Graphs

Another decomposition describes an activation pattern in terms of the individual wavefronts composing the rhythm and how they interact with each other. In this scheme, a wavefront occupies both space and time, and when viewed in a three-dimensional coordinate system with two spatial and one temporal coordinates, fills a volume (Fig. 1). The first implementation of this idea was due to Bollacker et al. [26]. The method was designed for unipolar electrograms recorded from a uniform rectangular array. First, the electrograms are differentiated, and samples (dV/dt measurements indexed by spatial and temporal coordinates) at which $dV/dt < -0.5\,\text{V/sec}$, are marked as active. This particular threshold value was chosen by cryoablating part of the mapped tissue. Any activations detected in the frozen area were attributed to distant activity and not to activation of the tissue under the electrode. With the activation threshold set at $-0.5\,\text{V/sec}$, no false activations were detected in the nonviable region, and so the threshold of $-0.5\,\text{V/sec}$ was considered appropriate for detecting local activation. After identifying active samples, the algorithm scans the data set in a recursive fashion, grouping active samples that are adjacent in time and space to form wavefronts. Small discontinuities in the wavefront volumes are allowed, to account for noise and locally poor recordings. After the isolation, wavefronts below a certain spatiotemporal size (i.e., volume) are regarded as noise and erased. From this decomposition, activation patterns are quantified by the number of wavefronts present, their spatiotemporal size, the area activated by the wavefronts, and the incidence of collision and reentry (reentrant wavefronts are those that activate recording sites more than

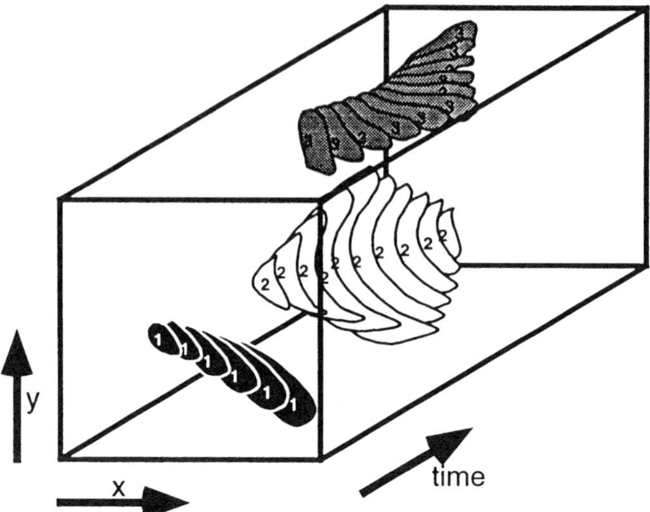

Figure 1   Three isolated wavefronts in a spatiotemporal coordinate system. Each filled shape represents one temporal snapshot of an isolated wavefront. (From Ref. 27.)

once). Bollacker et al. validated their method by comparing the number of wavefronts found by the algorithm with the number counted manually by four observers. The difference between the algorithm's count and that of the humans was similar to the interobserver variability.

Rogers et al. further developed the wavefront isolation idea [27]. In addition to isolating wavefronts, this method identifies collisions, events in which two or more wavefronts coalesce to form a new wavefront, and fractionations, events in which a single wavefront breaks into two or more pieces. Wavefronts are defined to begin when they spontaneously appear in the mapped region (e.g., by propagating in from the edges or by breaking through from below) or when they originate from interwavefront interactions (i.e., one of the child waves of a fractionation, or the single child wave of a collission). Wavefronts end when they propagate out of the mapped region, block, fractionate into a set of new wavefronts, or collide with another wavefront. This definition of a wavefront is distinct from the Bollacker et al. model, in which wavefronts persist through fractionation and collisions events. In the Rogers et al. method, isolated wavefronts and their interrelationships are represented by a directed graph in which the beginning and ending times of wavefronts are the graph's vertices and the wavefronts are edges connecting the vertices [27]. An example wavefront graph derived from 0.5 sec of VF is shown in Fig. 2. In the presence of noise,

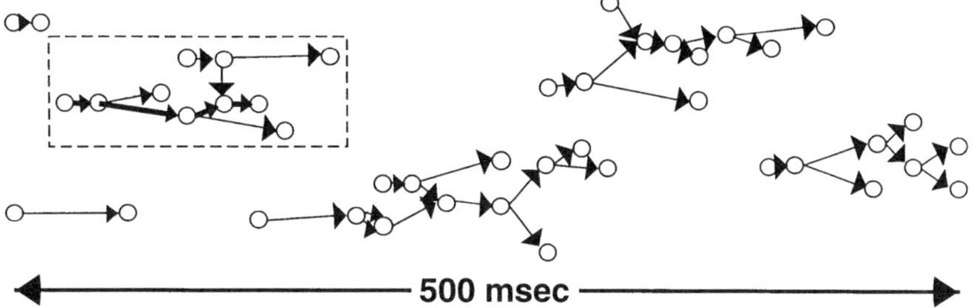

Figure 2 A wavefront graph derived from 0.5 sec of VF. Each arrow represents a wavefront. The horizontal positions of the endpoints locate the wavefront in time. There are six components, one of which is indicated by the dashed box. A single route through this component is in bold. (From Ref. 55.)

the initial wavefront graph is often highly fractionated. To reassemble the wavefront fragments, allowing small discontinuities, but minimizing inappropriate merging of large wavefronts, a filter algorithm guided by the topology of the graph was devised.

Several quantitative descriptors of VF can be derived from this model: the total number of wavefronts, fractionations, and collisions; and the mean wavefront size, area swept out, and duration. Rogers et al. used these parameters to show that the organization of VF increases between 5 and 20 sec after induction [27]. As will be discussed below, a number of additional rhythm descriptors based on this model of wavefront isolation have also been developed.

C. Dominant Frequency Maps

Another approach to rhythm decomposition is based on spectral analysis. In this scheme, the power spectrum of the signal from each recording site is estimated and the largest peak chosen as the local "dominant frequency." The dominant frequency is a measure of the activation rate at each site [28]. During complex rhythms such as VF, dominant frequencies vary spatially across the heart. However, Zaitsev et al. showed that, in the isolated sheep ventricle, they tend to cluster in domains of similar frequency so that the number of domains and the ratios of frequencies in neighboring domains could be used as VF descriptors [29]. Domain patterns were relatively staple temporally, persisting from seconds to minutes. Conduction block frequently occured at domain boundaries. Zaitsev et al. attributed these results to the presence of sustained intramural reentrant sources that drive VF, yet cannot be directly observed by epicardial mapping.

V.   ESTIMATES OF PROPAGATION SPEED AND DIRECTION

The speed and direction of wavefronts in a complex rhythm can be an informative quantitative descriptor. One way to measure propagation velocity is to measure the distance between contours on an isochronal map. However, this requires prior construction of an accurate map. To avoid the uncertainty and ambiguity of this process, several alternative methods have been developed to estimate these quantities in mapping data.

A.   Vector Loop Mapping

An early method for finding the direction of propagation at a site is based on recording from a pair of orthogonal bipolar electrodes [30]. In this method, the outputs of the bipoles are plotted against each other, forming a loop (Fig. 3). Suppose that the two bipoles are aligned with the x and y axes. In this case, a wavefront propagating in the x direction will generate a large signal in the x bipole, but little or no signal in the y bipole. The resulting vector loop will therefore be aligned with the x axis, indicating the direction of propagation. The reverse is true for wavefronts propagating along the y axis. The relative sizes of the x and y signals for arbitrarily oriented wavefronts will have intermediate behavior, but the orientation of the vector loop will in general indicate the propagation direction. Kadish et al. showed that propagation directions computed in this way agreed well with those determined from isochronal maps [30].

Damle et al. used vector loop mapping to quantify the organization of VF in dogs [31]. The animals were instrumented with an 8×14 unipolar electrode plaque (2.5-mm spacing). Vector loops were constructed (by treating the diagonally opposing electrodes from each quartet of electrodes as orthogonal bipoles) and the orientation of each loop was determined. For each vector loop, a multivariate linear regression model was constructed to determine how well the orientation of the loop could be predicted by the orientations of its neighbors in space and time. Significant linking of activation direction was found in both space and time, suggesting that activation during VF was indeed organized. In a subsequent paper, Damle et al. performed a similar linking analysis in dogs with subacute (1-week) healing myocardial infarction (MI), chronic (8-week) healing MI, and no MI. In this study [32], spatial linking with vector loops 2.5 mm away was assessed (as in the previous study [31]) as well as with vector loops 5.0 and 7.5 mm away. It was found that although linking was significant at 2.5 and 5.0 mm in all three groups, linking at 7.5 mm was not significant in the dogs without MI, thus suggesting smaller wavefronts and less organization in these animals.

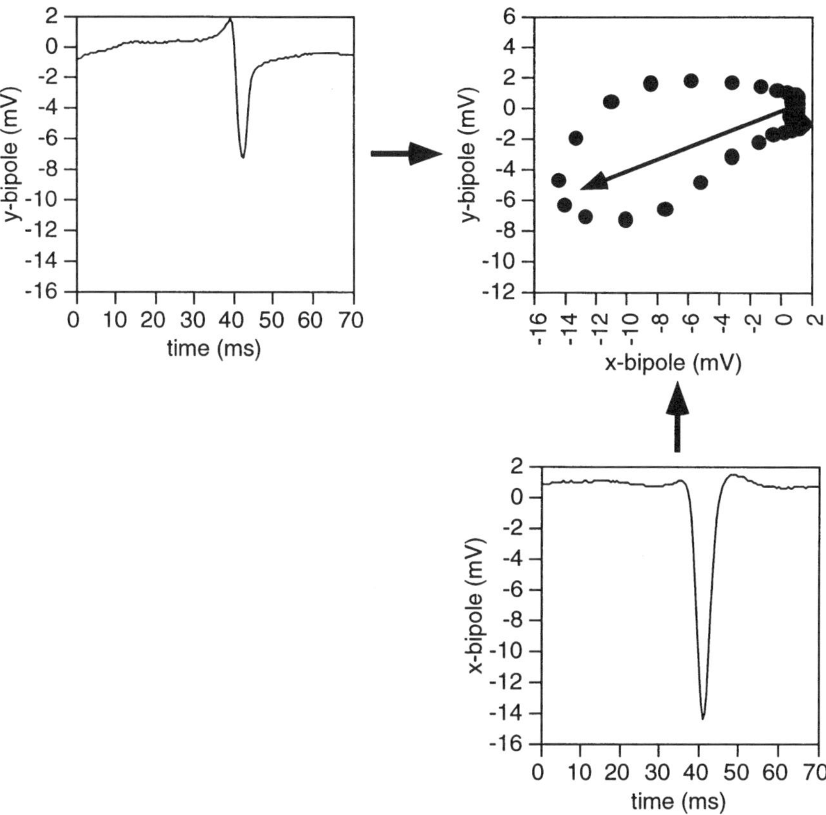

Figure 3 Construction of a vector loop (top right) from two orthogonal bipolar signals. The wavefront in this case propogates approximately west by southwest (long arrow). (From Ref. 60.)

B. The Zero-Delay Wavenumber Spectrum

The variation of potentials on the surface of the epicardium can be modeled as a sum of propagating plane waves of different wavelengths and frequencies [33]. Just as the contribution of each temporal frequency to a time series is represented by the power spectrum, the contribution of each plane wave to the spatiotemporal signal $f(x, y, t)$ is captured by the frequency–wavenumber spectrum (FWS), $S_{ff}(\mathbf{k}, \omega)$. Here ω represents temporal frequency, as usual, and $\mathbf{k} = [k_x, k_y]^T$ is the spatial frequency (or wavenumber). Although it is not difficult to compute the FWS using FFTs in space and time, it is hard to display and interpret the typically large 3-D data set.

An alternative to computing and displaying the full FWS is to focus on the spatial frequencies present in the signal. One approach is to compute the full 3-D FWS and average over all temporal frequencies to obtain average power as a function of 2-D spatial frequency. This spatial power spectrum, $P(\mathbf{k})$, is known as the zero-delay wavenumber spectrum (ZDWS). Nikias et al. [33] describe two methods (using the Bartlett approach and the maximum likelihood method) for computing the ZDWS from an array of sensors. They validated their procedures on simulations, then applied the method to cardiac mapping data acquired from a 3×4 array of electrodes. Nikias et al. found that the maximum likelihood version of the ZDWS (modified to preserve directional information) was useful in determining the number and direction of wavefronts propagating under their array. Bayly and co-workers [34] used the ZDWS to examine data acquired during VF in pigs using a high-resolution array (11×11, 0.28-mm spacing) of epicardial electrodes. Insignificant power was found at wavelengths shorter than 2 mm, which suggested that an electrode spacing of at least 1 mm could be used in further studies. In a variation of the method, Bayly et al. estimated mean propagation speeds during sinus rhythm and VF in pigs from the propagation speed of the dominant plane-wave components in the ZDWS [35].

C. Wavefront Centroid Tracking

Another velocity estimator is based on the wavefront isolation algorithm described above. Once the wavefronts in a rhythm have been isolated from one another, their propagation speed and direction can be estimated in a fairly straightforward way by computing the spatial centroid of each wavefront at each timestep. The velocity of the centroid is the velocity of the wavefront as long as the velocity at all points along the wavefront is uniform and the wavefront has a constant shape (e.g., a planar wavefront rather than a target pattern radiating from a central stimulus). These criteria can be approximately met for fibrillation data by adjusting the wavefront isolation algorithm to prevent recombination of wavefronts fractionated by noise [36]. Using this method, Huang et al. found that the propagation velocity of wavefronts during VF slowed progressively during the first 40 sec after induction [36].

D. Velocity Fields

Bayly and colleagues recently developed a method for constructing velocity vector fields from cardiac mapping data [37]. The algorithm first scans the data set to find *active points*. For the unipolar electrical mapping data that were used in this study, these were defined as points in space and time at

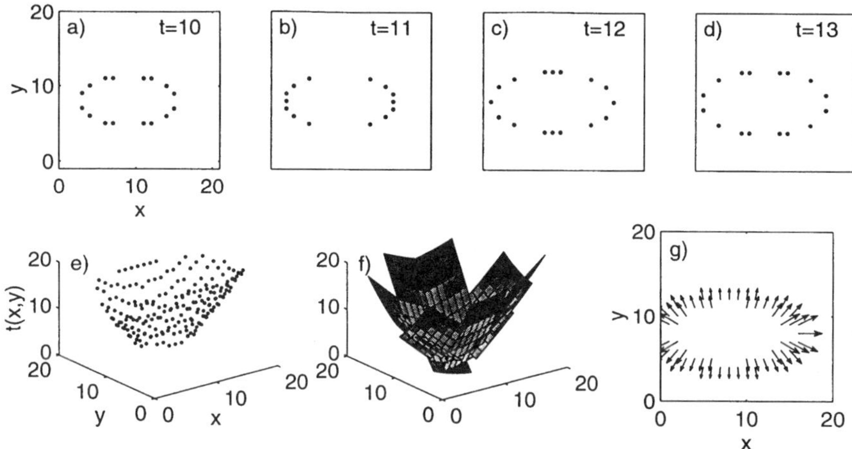

Figure 4 Velocity field estimation of a simulated elliptical wavefront. (a–d) Snapshots of the wavefronts showing active sites. (e) The active sites in x, y, t-space. (f) Polynomial surfaces fitted to the active sites. (g) The computed velocity vector field. (Adapted from Ref. 37.)

which the first derivative of the potential became more negative than -0.5 V/sec after a minimum refractory period of 40 msec. The active points are considered to lie on activation wavefronts. Each active point and its active neighbors (other active points within 5 mm in either direction, or 20 msec in time) are fitted with a smooth quadratic surface, $t(x, y)$, which describes local activation time as a function of position. By taking the gradient of this function and assuming that the direction of propagation is normal to the wavefront, the local velocity, $[\partial x/\partial t, \partial y/\partial t]^T$, is computed at each active point. To improve robustness to noise in the data, velocity vectors are not computed for an active point if there are insufficient points in the neighborhood (<20) or if the residual error of the fit is too large ($r > 0.5$). The method was tested using model data with known velocity fields (Fig. 4) as well as with paced rhythm data from swine epicardium. The method has recently been extended to three-dimensional mapping data collected from arrays of plunge needle electrodes [38].

VI. THE ORGANIZATION OF COMPLEX RHYTHMS

Understanding the underlying organization of fibrillation may provide a means for controlling this arrhythmia and has therefore been a long-standing topic of research. Many of the rhythm descriptors discussed above can be used to quantify organization. For example, the number of K-L modes required to account for a 90% of the variance in an activation pattern was used to track temporal changes in the organization of VF [39]. This study found that VF was more organized 1 min after induction than it was in the first few seconds. In another study, the affect of rapid pacing stimuli on VF was assessed by measuring the cumulative variance contribution of the first two K-L modes. This quantity increased significantly when VF wavefronts were displaced by organized beats emanating from the pacing stimuli [40]. The wavefront isolation algorithm [27] also provides a means to quantify organization. For example, organized rhythms have fewer wavefronts, fractionation, and collision events than disorganized ones [36]. Below we discuss a number of additional algorithms that have specifically developed to quantify the organization of fibrillatory rhythms.

A. Correlation Length

The spatial correlation function describes how the correlation between electrograms decays as a function of their spatial seperation. It is found by estimating cross-correlation functions between signals from locations $\mathbf{x}_i$ and $\mathbf{x}_j$:

$$R(\tau;\ \mathbf{x}_i, \mathbf{x}_j) = 1/M \sum_{m=1} f(\mathbf{x}_i, t_m) f(\mathbf{x}_j, t_m + \tau) \tag{4}$$

The cross-correlation function is then normalized with respect to the variance of the original signals.

$$\hat{R}(\tau; \mathbf{x}_i, \mathbf{x}_j) = \frac{R(\tau; \mathbf{x}_i, \mathbf{x}_j)}{\sqrt{R(0; \mathbf{x}_i, \mathbf{x}_i) R(0; \mathbf{x}_j, \mathbf{x}_j)}} \tag{5}$$

Usually one correlation value $\hat{R}(\tau; \mathbf{x}_i, \mathbf{x}_j)$ is extracted; the correlation at $\tau_0 = 0$ (no relative time shift) [41] or $\tau_0 = \tau_{\max}$ (the time shift with maximum correlation) [42] have been used. The spatial correlation function $R(\delta)$ is obtained by averaging over all correlations $\hat{R}(\tau_0; \mathbf{x}_i, \mathbf{x}_j)$ such that $|\mathbf{x}_i - \mathbf{x}_j| = \delta$.

In many cases the spatial correlation function decays roughly exponentially with sensor separation: $R(\delta) \approx \exp(-\delta/\lambda)$, in which case, spatial correlation can be summarized by the correlation length λ. This characteristic length is a measure of the rhythm's organization. Bayly and co-workers [41] estimated the correlation length during VF in open-chest pigs, using data from a 22×23 epicardial array of unipolar electrodes. They found that the correlation length ranged from 4 to 10 mm, and tended to increase during the first minute of VF. Botteron and Smith [43] used the spatial correlation function to describe spatial organization during AF in humans. Bipolar electrograms were obtained from decapolar catheters placed in the right atrium and coronary sinus. Correlation lengths for AF in humans were found to be between 17 and 52 mm. A physical interpretation of the correlation length was also proposed: the "tissue wavelength" of reentrant circuits was hypothesized to be on the order of $2\pi\lambda$ [43].

B.　Magnitude-Squared Coherence Maps

The magnitude-squared coherence (MSC) is a frequency-domain technique for measuring the similarity between two signals [44,45]. It is defined by the equation [45]

$$\mathrm{MSC}_{fg}(\omega) = \frac{|S_{fg}(\omega)|^2}{S_{ff}(\omega)S_{gg}(\omega)} \tag{6}$$

The function $S_{fg}(\omega)$ is the cross-power spectrum, and $S_{ff}(\omega)$ and $S_{gg}(\omega)$ are the autopower spectra of the signals $f(t)$ and $g(t)$, respectively. The cross- and autopower spectra are the Fourier transforms of the cross-correlation and autocorrelation functions, respectively.

MSC can be estimated from the fast Fourier transforms (FFTs) of N overlapping segments from each signal. The coherence is estimated using

$$\mathrm{MSC}_{fg}(\omega) = \frac{\left|\sum_{n=1}^{N} F_i(\omega)G_i^*(\omega)\right|^2}{\sum_{n=1}^{N} F_i(\omega)F_i^*(\omega) \sum_{n=1}^{N} G_i(\omega)G_i^*(\omega)} \tag{7}$$

where $F_i(\omega)$ and $G_i(\omega)$ are the FFTs of the ith segment of each time series and the asterisk (*) denotes a complex conjugate.

The MSC describes similarity between signals as a function of frequency. Overall similarity between signals can be summarized by the mean coherence over all frequencies or over a frequency range. To describe multichannel data, this mean coherence value may be computed between all pairs and a relationship may be sought between electrode separation, δ, and mean coherence. This is analogous to the spatial correlation function described above. Sih and co-workers [45] performed this calculation using data obtained from a 240-electrode epicardial array during VF in open-chest pigs. They found that mean coherence from 0 to 50 Hz decayed approximately exponentially with a characteristic length $\lambda \approx 9$ mm. Sih et al. also present their results as coherence maps, which show the coherence of each electrogram relative to one or several reference electrodes. This presentation has the advantage that it preserves information on regional variations of complexity.

Coherence and correlation are intimately related and show generally consistent results. MSC has the advantage that if noise or signal are dominant in certain frequency bands, then MSC can be retained or ignored in these frequency bands. The MSC mapping approach can be described as a "hybrid" method: a frequency-domain method is used to analyze time series (MSC), but results are shown as a function of space.

C. Multiplicity and Repeatability

A measure of complexity called *multiplicity* was recently introduced by Rogers et al. Multiplicity is based on the wavefront isolation decomposition described above [27] and quantifies the number of repeating wavefront morphologies that exist in an activation pattern [46]. After isolating the wavefronts in an activation pattern, a cross-correlation technique is used to quantify the similarity between all possible pairs of wavefronts. Using these data, the wavefronts are sorted into a set of clusters, each containing wavefronts that are mutually similar. The number of clusters needed to account for 90% of the total activity in the data set is defined as the multiplicity of the rhythm. Simple rhythms (e.g., epicardial pacing) in which the same wavefront morphology repeatedly activates the mapped region have a multiplicity of 1, while for more complex rhythms such as VF, multiplicity ranges upward. A closely related parameter is *repeatability*. This is the weighted average of the number of wavefronts in each cluster (the weight is the fraction of total activity accounted for by the cluster). Thus, multiplicity and repeatability are complementary measures of organization: multiplicity counts distinct patterns, while repeatability counts how many times the patterns repeat temporally. This analysis, in conjunction with several other quantitative descriptors based on wavefront isolation, was used to show that the organization of VF in unsupported pig hearts evolves in a biphasic way [36]: VF is less organized at 10 sec than 0 sec postinduction, with more, smaller, wavefronts traversing a larger variety of pathways (higher multiplicity) for fewer repetitions. VF patterns then recover organization over the next 30 sec, but by a different mechanism: the spatial size of subpatterns grows, but the dynamics of the rhythm otherwise appears unchanged.

D. Peak Correlation Coefficient

Another method based on cross-correlation was used by Witkowski et al. to characterize the evolution of VF in isolated, perfused canine hearts [47]. VF patterns were optically mapped using a cooled CCD camera. The signals were filtered with a spatial Gaussian filter and a temporal median filter to improve the signal-to-noise ratio. To enhance the locations of wavefronts, the resulting signals were differentiated and all negative derivative values set to 0. The authors computed the cross-correlation coefficient of each frame (the set of derivative values at all sites at one instant) with all other frames in the data set. Reasoning that repeating spatial patterns is a hallmark of an organized rhythm, the peak coefficient was taken as an instantaneous measure of organization. Using this method, the authors found that the presence of transient reentrant circuits on the epicardium during early VF was accompanied by relatively high peak correlation values. After 10 min of VF, epicardial reentry was no longer present and peak correlation had decayed to about half of its previous values.

VII. PROPAGATION BLOCK AND FUNCTIONAL REENTRY

During normal cardiac propagation, wavefronts extend from one tissue boundary to another. Functional reentry occurs when a wavefront breaks, leaving free ends within the bulk of the tissue. For example, if a section of a wavefront propagates over a patch of tissue that has not yet recovered from a previous activation, that section of the wavefront will be blocked, leaving two new wavebreaks, one at the end of each surviving section of the original wavefront. When the refractory patch recovers, the two surviving wavefronts have the potential to pivot around their broken ends forming functionally reentrant circuits. Such circuits are thought to underlie tachyarrhythmias, particularly VF, and a number of methods have been developed to detect their presence in mapping data.

A. Time-Space Plots

Time–space plots (TSPs) [48–50] are constructed by projecting all of the data from a single temporal snapshot of the mapping data set onto one line. For example, for a rectangular recording array, each column of data can be summed to produce a single horizontal row of values. Since all the data in the vertical direction are lumped, the row represents how activity varies in the horizontal direction only. A new row is created for each snapshot in the data set, and the rows are stacked sequentially to form a 2-D image. Similar plots can be constructed by summing across rows of the array instead of down columns. Consider a planar wavefront propagating in the y direction. A TSP constructed by summing rows will contain a diagonal band representing the changing location of the wavefront, while a TSP constructed by summing columns will be more uniformly colored (Fig. 5A).

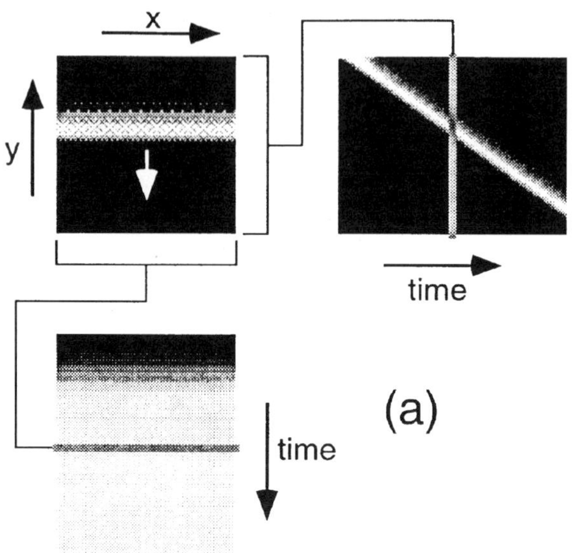

Figure 5 Construction of time-space plots (TSPs) for simulated planar (a) and functionally reentrant (b) wavefronts. The top-left images show a single snapshot of activity. For each such snapshot in the data set, the rows and columns of pixels are summed to produce vertical and horizontal lines, respectively. These lines are stacked to respectively produce the top-right (y-*time*) and bottom-left (x-*time*) images (the TSPs). The white arrows in the top-left images indicate the direction of propagation. The bold gray lines show the lines in the TSPs corresponding to the snapshot shown in the top-left images. The dark region at the top of the x-*time* plot in (a) is an artifact of wavefront initiation. The dashed white lines in (b) show the position of the center of the reentrant wave, which in this example was stationary. (From Ref. 60.)

TSPs have been used to study reentrant activation both in computational models and in data sets collected by optical mapping [48–50]. When reentry is present, the wavefront's orientation changes with respect to the sensitivity of the TSP. This produces a characteristic branching pattern with alternating diagonal bands separated by regions of low amplitude (Fig. 5B). By analyzing the spatial extent and location of the low-amplitude region in two orthogonal TSPs, the size and location of the reentrant wavefront's core (the region circumscribed by the tip of the broken wave) can be tracked. In addition, the cycle length of the spiral wave can be determined from the spacing of the vertical bands along the time dimension.

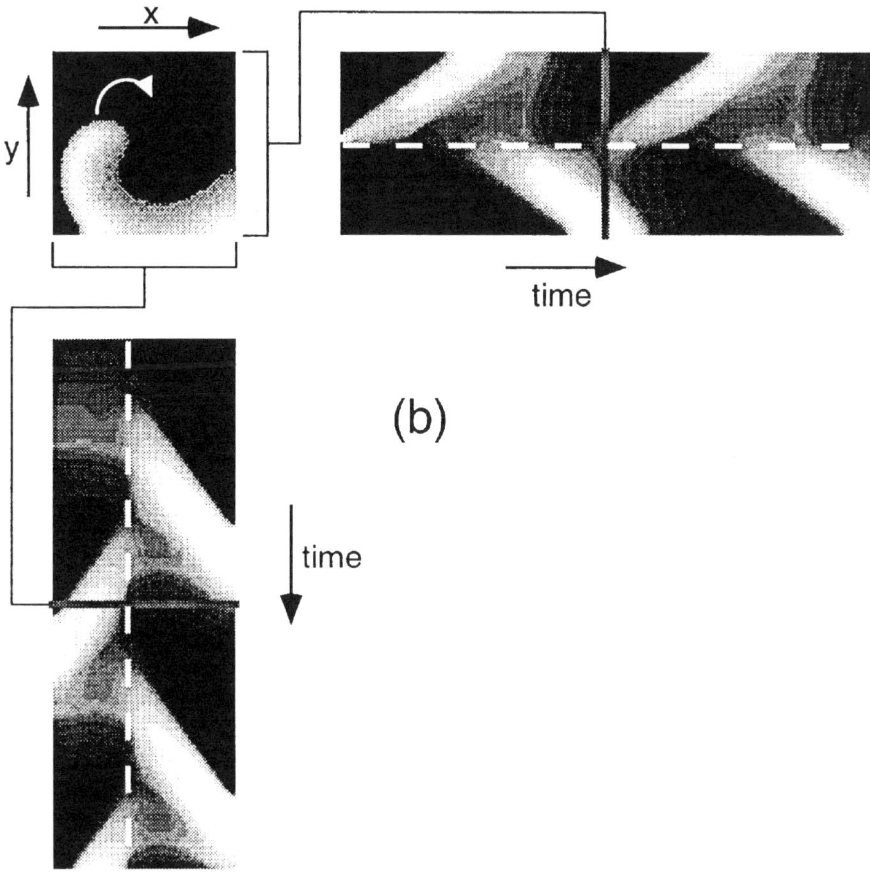

Figure 5 (continued)

B. Detection of Double Potentials

Unipolar electrodes sense electrical activity not only from tissue in contact with the electrode, but from distant tissue as well. Because of this, if an electrode is located near the tip of a reentrant wave, the electrode will register a double deflection, with each deflection corresponding to activation on opposite sides of the "line of block" about which the wave tip circulates [51,52]. This phenomenon has also been reported for optical recordings [53]. Evans and colleagues developed a spectral method to automatically detect double potentials associated with functional block and reentry [54].

The method is based on the short-time Fourier transform (STFT), in which Fourier transforms are computed for sequential, overlapping segments of a time series (a spectrogram). Normally, such a plot shows power residing in a band around the fundamental frequency of the rhythm. However, when double potentials are present, the main peak splits, and a hole appears in the spectrogram (Fig. 6). The algorithm of Evans et al. computes the STFT for each channel in a mapping data set and identifies times and sites where double potentials are present. Instances of block detected in this way in a VF mapping data set were compared to instances found using a propagation delay criterion (>10 msec between activations at neighboring electrodes spaced by 1 mm). Compared to the delay criterion, the STFT algorithm detected conduction block with a sensitivity of 0.74 and a specificity of 0.99.

C. Wavetip Tracking

Rogers et al. recently developed algorithms to identify and quantify re-entrant circuits in VF mapping data using the wavefront isolation method and concepts from graph theory [55]. The first step in the method is to use wavefront isolation to compute a wavefront graph for the VF episode (Fig. 2). Next, families of wavefronts related by fractionation and collision events are identified. These families form subgraphs disconnected from the other wavefronts in the wavefront graph (Fig. 2, boxed region). In the terminology of graph theory, these families are called *components* [56]. Each component contains one or more sequence of wavefronts, or *routes*, that connect the appearance of propagating activity with its disappearance (the succession of bold arrows in the boxed region in Fig. 2 is an example). All routes through a component are examined to determine if they activate the same tissue more than once, i.e., are reentrant. If a reentrant route is found, the entire component is deemed reentrant. In this setting, the incidence of reentry for a VF data set is defined as the fraction of components in the wavefront graph that are reentrant.

Reentry can be further quantified by identifying and tracking the tips of reentrant wavefronts. The *wavetip path* is defined as the shortest possible path connecting active samples in each timestep of a reentrant route (Fig. 7A; recall that an active sample is a dV/dt measurement at a particular time and location that is more negative than $-0.5\,\mathrm{V/sec}$). The wavetip path is found by creating a directed graph [56] in which the nodes are active samples in the route and the edges connect each active sample with all active samples in the succeeding timestep. Each edge is weighted with the distance between the electrodes associated with the nodes connected by the edge. Thus, this graph defines all possible paths through the mapped region that contain one of the route's active samples in each time step. The length of

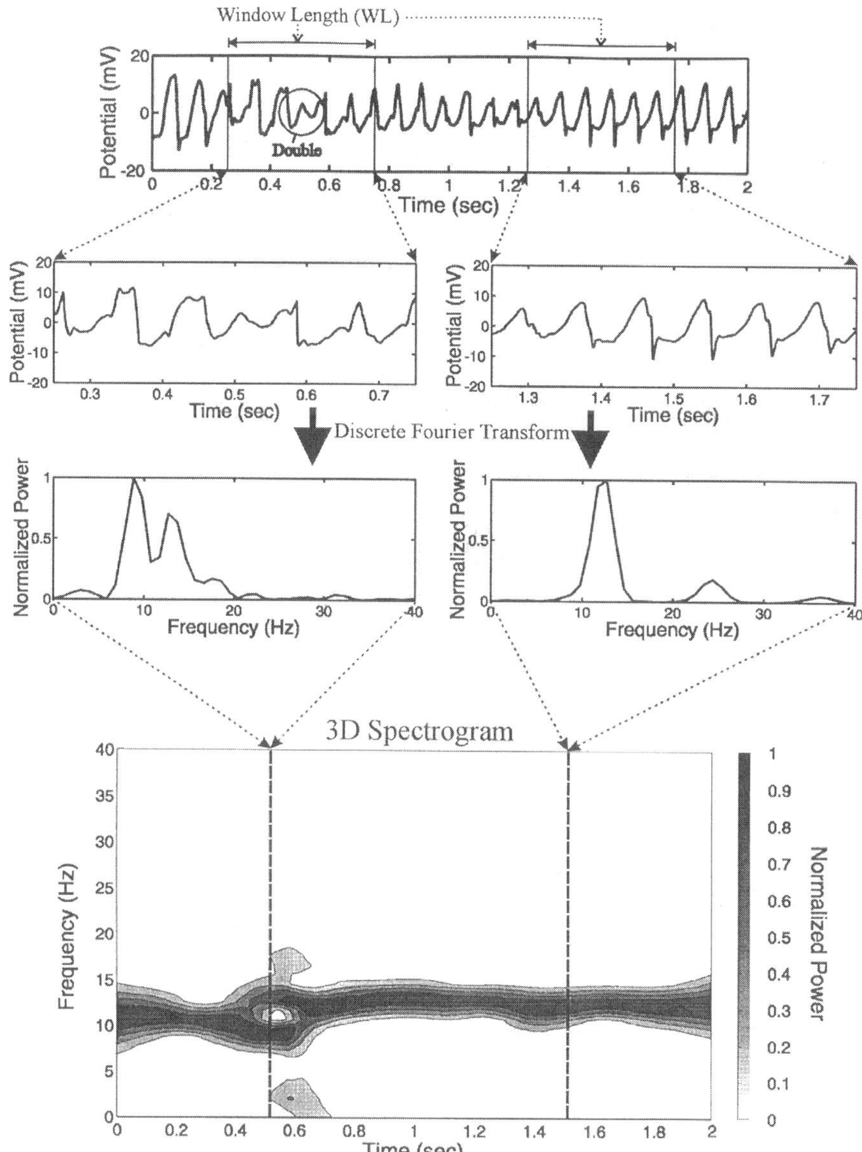

Figure 6 Unipolar electrogram recorded during VF and corresponding STFT spectogram. Two 0.5-sec segments of the elctrogram are magnified, with the power spectrum of each located below. These two spectra are columns in the STFT spectrogram (bottom) located at the time point at the center of the corresponding electrogram segment (0.5 and 1.5 sec). The power at each frequency and time in the spectrogram is rendered in gray scale. The left segment contains a double potential, whereas the right segment does not. The right spectrum has a sharp peak at the fundamental frequency of about 12 Hz, while the left segment has a split peak. (From Ref. 61.)

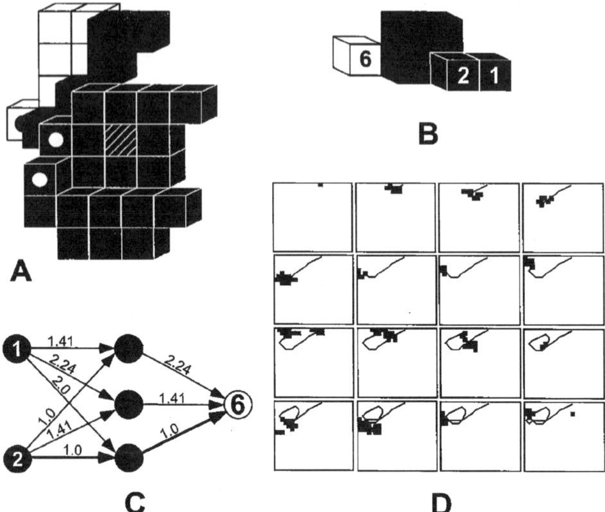

Figure 7 Method for finding the wavetip path of a reentrant circuit. (A) Four snapshots of a reentrant wavefront. Each box is an active sample. The wavetip path (shown by the round dots) is defined as the shortest path that connects recording sites in each snapshot. (B) A wavefront with six active samples in three time steps. (C) Graph and associated edge weights constructed from the wavefront in (B). (D) The wavetip path of a reentrant component during VF. The small squares are active samples and the black line traces the wavetip path. Frames are spaced by 10 msec. (From Ref. 55.)

each path is the sum of the weights of the edges in the path. A hypothetical route containing six active samples in three time steps is shown in Fig. 7B. The graph and associated edge weights constructed from this route are shown in Fig. 7C. The shortest path through such a graph is our desired wavetip path and can be found using a well-known algorithm from graph theory, Dijkstra's shortest-path algorithm [57]. An example of a wavetip path is shown in Fig. 7D.

A wavetip path can be further characterized by breaking the path into closed loops, each of which corresponds to one cycle of reentry, and measuring parameters such as the number of cycles, the area of the loops, and their aspect ratio and orientation with respect to epicardial fibers. Using these methods, Rogers et al. found that epicardial reentry is uncommon and short-lived during VF, but that as VF progresses, reentrant circuits become more common and longer-lived. Neither the orientation of the loops nor the direction of cycle drift for multicycle reentry was well predicted by the fiber orientation. Furthermore, the occurence of reentry was spatially nonuniform within the mapped region [55].

D. Phase and Singularities

Gray et al. recently developed a method that transforms optically mapped transmembrane potential data into a new variable called phase [58]. In this context, phase tracks the progress of the patch of tissue though an action potential (depolarization, plateau, recovery, rest) and is measured in radians. Phase is computed by evaluating the trajectory of each recording site through a reconstructed two-dimensional state-space. The time-delay embedding method [59] is used to reconstruct this space. This simply involves plotting the signal at time t against the signal at time $t + \tau$, where τ is the embedding delay set to be roughly equal to one-quarter of the rhythm's cycle length. In this state-space, trajectories generally circulate clockwise around a central region (Fig. 8A). By transforming each point in the trajectory to a polar coordinate system whose origin is within this central region, the phase of each recording site at each point in time is readily computed. The orientation of the polar coordinate system is indicated by the inset in Fig. 8A.

Much of the phase transformation's utility stems from its ability to reveal wavebreaks in mapping data [58]. During normal propagation, contours of phase do not cross, i.e., all points in the tissue have unique phase. In contrast, the creation of a wavebreak gives rise to a singular point that is

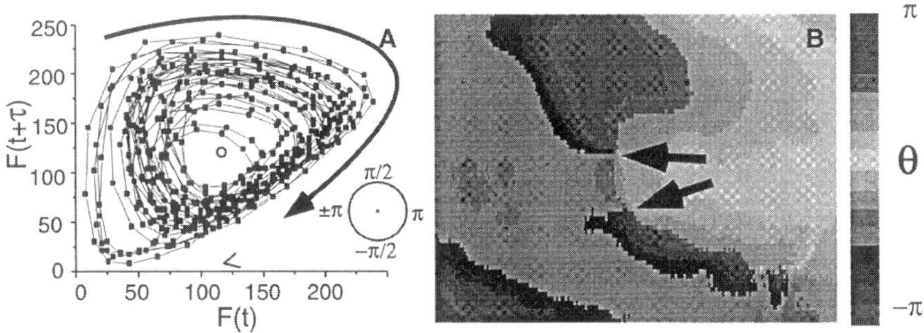

Figure 8 Phase mapping. (A) State-space trajectory of a single recording site. The inset shows the orientation of the polar coordinate system in which phase is computed. (B) A map showing the phase at each point in the mapped region at one instant. Data were optically recording from fibrillating sheep hearts. Two phase singularities are present (arrows).

surrounded by tissue of all phases. The broken wavefronts have the potential to pivot around these singular points and form functionally reentrant circuits. Integrating phase around a phase singularity returns a value of $\pm\pi$, where the sign indicates the direction of rotation. Figure 8B shows a snapshot of a phase map containing two phase singularities. Using phase mapping, Gray et al. showed that phase singularities are created and terminated in oppositely rotating pairs and that phase singularities are connected by isophase lines [58].

VIII. CONCLUSIONS

Quantitative analysis of cardiac mapping data has provided valuable information into the mechanisms of complex cardiac arrhythmias and offers the potential to gain much needed new insight. In such analyses, voluminous electrical and optical measurements of activity are translated into physiologically meaningful parameters. It is likely that rhythms such as ventricular and atrial fibrillation will ultimately be best understood using these objective, statistical approaches. Future work is needed to describe activity in three-dimensional myocardium and in the complex geometry of the atrium. Current methods should be refined to identify which characteristics are essential to the initiation and maintenance of arrhythmia, and which are epiphenomena.

ACKNOWLEDGMENTS

This work was supported in part by biomedical engineering research grants from the Whitaker Foundation, National Science Foundation grant CMS-9625161, and American Heart Association grant 9820030SE.

REFERENCES

1. Lewis T, Rothschild MA. The excitatory process in a dogs heart. Part II. The ventricles. Philo Trans R Soc Lond 206:181–226, 1915.
2. Spach MS, Miller WT, Miller-Jones E, Warren RB, Barr RC. Extracellular potential related to intracellular action potentials during impulse conduction in anisotropic canine cardiac muscle. Circ Res 45:188–204, 1979.
3. Steinhaus BM. Estimating cardiac transmembrane activation and recovery times from unipolar and bipolar extracellular electrograms: a simulation study. Circ Res 64:449–462, 1989.

4. Blanchard SM, Damiano RJ, Asano T, Smith WM, Ideker RE, Lowe JE. The effects of distant cardiac electrical events on local activation in unipolar epicardial electrograms. IEEE Trans Biomed Eng 34:539–546, 1987.

5. Barr RC, Gallie TM, Spach MS. Automated production of contour maps for electrophysiology. I. Problem definition, solution strategy, and specification of geometric model. Comput Biomed Res 13:142–153, 1980.

6. Barr RC, Gallie TM, Spach MS. Automated production of contour maps for electrophysiology. II. Triangulation, verification, and organization of the geometric model. Comput Biomed Res 13:154–170, 1980.

7. Barr RC, Gallie TM, Spach MS. Automated production of contour maps for electrophysiology. III. Construction of contour maps. Comput Biomed Res 13, 1980.

8. Davis JL. Statistics and Data Analysis in Geology. 2nd ed. New York: Wiley, 1986.

9. Robinson JE. Computer Applications in Petroleum Geology. Stroudsburg, PA: Hutchinson Ross, 1982.

10. Ideker RE, Smith WM, Blanchard SM, Reiser SL, Simpson EV, Wolf PD, Danieley ND. The assumptions of isochronal cardiac mapping. PACE 12: 456–478, 1989.

11. Berbari EJ, Lander P, Scherlag BJ, Lazara R, Gesesowitz DB. Ambiguities of epicardial mapping. J Electrocardiol 24(suppl):16–20, 1992.

12. Witkowski FX, Plonsey R, Penkoske PA, Kavanagh KM. Significance of inwardly directed transmembrane current in determination of local myocardial electrical activation during ventricular fibrillation. Circ Res 74:507–524, 1994.

13. Blanchard S, Smith W, Damiano RJ, Molter D, Ideker R, Lowe J. Four digital algorithms for activation detection from unipolar epicardial electrograms. IEEE Trans Biomed Eng 36:256–261, 1989.

14. Anderson K, Walker R, Ershler P, Fuller M, Dustman T, Menlove R, Karwandee S, Lux R. Determination of local myocardial electrical activation for activation sequence mapping. A statistical approach. Circ Res 69:898–917, 1991.

15. Simpson EV, Ideker RE, Kavanagh KM, Alferness CA, Melnick SB, Smith WM. Discrete smooth interpolation as an aid to visualizing electrical variables in the heart wall. Computers in Cardiology, Los Alamitos, CA, 1991.

16. Laxer C, Alferness C, Smith WM, Ideker RE. The use of computer animation of mapped cardiac potentials in studying electrical conduction properties of arrhythmias. Computers in Cardiology, Chicago, IL, 1990.

17. Ershler PR, Lux RL. Derivative mapping in the study of activation sequence during ventricular arrhythmias. Computers in Cardiology, Washington, DC, 1986.

18. Gray RA, Pertsov AM, Jalife J. Incomplete reentry and epicardial breakthrough patterns during atrial fibrillation in the sheep heart. Circulation 94:2649–2661, 1996.

19. Lee JJ, Kamjoo K, Hough D, Hwang C, Fan W, Fishbein MC, Bonometti C, Ikeda T, Karagueuzian HS, Chen PS. Reentrant wave fronts in Wiggers'

stage II ventricular fibrillation. Characteristics and mechanisms of termination and spontaneous regeneration. Circ Res 78:660–675, 1996.

20. Sirovich L. Chaotic dynamics of coherent structures. Physica D 37:126–145, 1989.

21. Therrien CW. Discrete Random Signals and Statistical Signal Processing. Englewood Cliffs, NJ: Prentice-Hall, 1992.

22. Barr R, Spach M, Herman-Giddens G. Selection of the number and positions of measuring locations for electrocardiography. IEEE Trans Biomed Eng 18:125–138, 1971.

23. Lux RL, Evans K, Burgess MJ, Wyatt RF, Abildskov JA. Redundancy reduction for improved display and analysis of body surface potential maps. Circ Res 49:186–196, 1981.

24. Claydon F, Ingram L, Mirvis D. Effects of myocardial infarction on cardic electrical field properties using a numerical expansion technique. J Electrocardiol 24:371–377, 1991.

25. Bayly P, Johnson E, Wolf P, Smith E, Ideker R. Predicting patterns of epicardial potentials during ventricular fibrillation. IEEE Trans Biomed Eng 42:898–907, 1995.

26. Bollacker KD, Simpson EV, Hillsley RE, Blanchard SM, Gerstle RJ, Walcott GP, Callihan RL, King MC, Smith WM, Ideker RE. An automated technique for identification and analysis of activation fronts in a two-dimensional electrogram array. Comput Biomed Res 27:229–244, 1994.

27. Rogers JM, Usui M, KenKnight BG, Ideker RE, Smith WM. A quantitative framework for analyzing epicardial activation patterns during ventricular fibrillation. Ann Biomed Eng 25:749–760, 1997.

28. KenKnight BH, Bayly PV, Chattipakorn N, Windecker W, Usui M, Rogers JM, Johnson CR, Ideker RE, Smith WM. Efficient frequency domain characterization of myocardial activation dynamics during ventricular fibrillation. IEEE Engineering in Medicine and Biology Society, Montreal, Canada, 1995.

29. Zaitsev AV, Berenfeld O, Mironov SF, Jalife J, Pertsov AM. Spatial distribution of frequencies on the endocardium and epicardium of the fibrillating ventricle of the sheep heart. Circulation 100:I-872, 1999.

30. Kadish A, Spear J, Levine J, Hanich R, Prood C, Moore E. Vector mapping of myocardial activation. Circulation 74:603–615, 1986.

31. Damle RS, Kanaan NM, Robinson NS, Ge Y-Z, Goldberger JJ, Kadish AH. Spatial and temporal linking of epicardial activation directions during ventricular fibrillation in dogs: evidence for underlying organisation. Circulation 86:1547–1558, 1992.

32. Damle RS, Robinson NS, Ye DZ, Roth SI, Greene R, Goldberger JJ, Kadish AH. Electrical activation during ventricular fibrillation in the subacute and chronic phases of healing canine myocardial infarction. Circulation 92:535–545, 1995.

33. Nikias C, Raghuveer M, Siegal J, Fabian M. The zero-delay wavenumber spectrum estimation for the analysis of array ECG signals—an alternative to isopotential mapping. IEEE Trans Biomed Eng 33:435–452, 1986.

34. Bayly PV, Johnson EE, Idriss SF, Ideker RE, Smith WM. Efficient electrode spacing for examining spatial organization during ventricular fibrillation. IEEE Trans Biomed Eng 40:1060–1066, 1993.

35. Bayly PV, Hillsley RE, Gerstle RJ, Wolf PD, Smith WM, Ideker RE. Estimation of propagation speed during ventricular fibrillation from frequency-wavenumber power spectra. Computers in Cardiology, Bethesda, MD, 1994.

36. Huang J, Rogers JM, KenKnight BH, Rollins DL, Smith WM, Ideker RE. Evolution of the organization of epicardial activation patterns during ventricular fibrillation. J Cardiovasc Electrophysiol 9:1291–1304, 1998.

37. Bayly PV, KenKnight BH, Rogers JM, Hillsley RE, Ideker RE, Smith WM. Estimation of conduction velocity vector fields from epicardial mapping data. IEEE Trans Biomed Eng 45:563–571, 1998.

38. Barnette AR, Bayly PV, Zhang S, Walcott GP, Ideker RE, Smith WM. Estimation of 3-D conduction velocity vector fields from cardiac mapping data. Submitted.

39. Bayly PV, Johnson EE, Wolf PD, Smith WM, Ideker RE. Measuring changing spatial complexity in VF using the Karhunen-Loeve decomposition of 506-channel epicardial data. Computers in Cardiology, London, U.K., 1993.

40. KenKnight BH, Bayly PV, Gerstle RJ, Rollins DL, Wolf PD, Smith WM, Ideker RE. Regional capture of fibrillating ventricular myocardium. Evidence of an excitable gap. Circ Res 77:849–855, 1995.

41. Bayly PV, Johnson EE, Wolf PD, Greenside HS, Smith WM, Ideker RE. A quantitative measurement of spatial order in ventricular fibrillation. J Cardiovas Electrophysiol 4:533–546, 1993.

42. Smith JM, Botteron GW. Estimation of correlation length of activation processes during atrial fibrillation. Computers in Cardiology, 1993.

43. Botteron GW, Smith JM. A technique for measurement of the extent of spatial organization of atrial activation during atrial fibrillation in the intact human heart. IEEE Trans Biomed Eng 42:579–586, 1995.

44. Ropella K, Sahakian A, Baerman J, Swiryn S. The coherence spectrum. A quantitative discriminator of fibrillatory and nonfibrillatory cardiac rhythms. Circulation 80:112–119, 1989.

45. Sih HJ, Sahakian AV, Arentzen CE, Swiryn S. A frequency domain analysis of spatial organization of epicardial maps. IEEE Trans Biomed Eng 42:718–727, 1995.

46. Rogers JM, Usui M, KenKnight BH, Ideker RE, Smith WM. The number of recurrent wavefront morphologies: a method for quantifying the complexity of epicardial activation patterns. Ann Biomed Eng 25:761–768, 1997.

47. Witkowski FX, Leon LJ, Penkoske PA, Giles WR, Spano ML, Ditto WL, Winfree AT. Spatiotemporal evolution of ventricular fibrillation. Nature 392:78–82, 1998.

48. Davidenko JM, Pertsov AV, Salomonsz R, Baxter WT, Jalife J. Stationary and drifting spiral waves of excitation in isolated cardiac muscle. Nature 355:349–351, 1992.

49. Pertsov AM, Davidenko JM, Salomonsz R, Baxter WT, Jalife J. Spiral waves of excitation underlie reentrant activity in isolated cardiac muscle. Circ Res 72:631–650, 1993.
50. Gray RA, Jalife J, Panfilov A, Baxter WT, Cabo C, Davidenko JM, Pertsov AM. Nonstationary vortexlike reentrant activity as a mechanism of polymorphic ventricular tachycardia in the isolated rabbit heart. Circulation 91:2454–2469, 1995.
51. Konings KT, Smeets JL, Penn OC, Wellens HJ, Allessie MA. Configuration of unipolar atrial electrograms during electrically induced atrial fibrillation in humans. Circulation 95:1231–1241, 1997.
52. Olshansky B, Moriera D, Waldo AL. Characterization of double potentials during ventricular tachycardia. Studies during transient entrainment. Circulation 87:373–381, 1993.
53. Efimov IR, Sidorov V, Cheng Y, Wollenzier B. Evidence of three-dimensional scroll waves with ribbon-shaped filament as a mechanism of ventricular tachycardia in the isolated rabbit heart. J Cardiovas Electrophysiol 10: 1452–1462, 1999.
54. Evans FG, Rogers JM, Smith WM, Ideker RE. Automatic detection of conduction block based on time-frequency analysis of unipolar electrograms. IEEE Trans Biomed Eng 46:1090–1097, 1999.
55. Rogers JM, Huang J, Smith WM, Ideker RE. Incidence, evolution, and spatial distribution of functional reentry during ventricular fibrillation in pigs. Circ Res 84:945–954, 1999.
56. Chachra V, Ghare PM, Moore JM. Applications of Graph Theory Algorithms. New York: North Holland, 1979.
57. Dijkstra EW. A note on two problems in connection with graphs. Numer Math vol. 1, pp. 268–271, 1959.
58. Gray RA, Pertsov AM, Jalife J. Spatial and temporal organization during cardiac fibrillation. Nature 392:75–78, 1998.
59. Takens F. Detecting strange attractors in turbulence. In Rand DA, Young LS, eds. Dynamical Systems and Turbulence, Lecture Notes in Mathematics. Berlin: Springer-Verlag, 1981.
60. Rogers JM, Bayly PV, Ideker RE, Smith WM. Quantitative techniques for analyzing high-resolution cardiac mapping data. IEEE Eng Med Biol Mag 17:62–72, 1998.
61. Evans FG, Rogers JM, Ideker RE. Detection of propagation block in cardiac mapping data. PACE 20(part II):1233, 1997.

13
Quantitative Descriptions of Cardiac Arrhythmias

Kristina M. Ropella
Marquette University, Milwaukee, Wisconsin, U.S.A.

Ziad S. Saad
National Institute of Mental Health, National Institutes of Health, Bethesda, Maryland, U.S.A.

I. INTRODUCTION

Currently, the treatment and prevention of cardiac arrhythmias include antiarrhythmic drug therapy, surgery, ablation, and electronic intervention. Antiarrhythmic drugs, which traditionally have been the only method of therapy for many life-threatening arrhythmias, are often unreliable and unsafe in the prevention of tachyarrhythmias. Surgery, such as excision or cryoblation, is considered appropriate only for a small number of patients and is highly invasive. Ablation has become an excellent means for eliminating a number of arrhythmias with well-defined electrophysiological mechanism. However, there remain hundreds of thousands of individuals for whom the mechanism of life-threatening arrhythmia is ill-defined and thus difficult to treat using traditional measures. For these individuals, implantable electronic antiarrhythmic devices, such as antitachycardia pacemakers, cardioverters, and automatic defibrillators, have become the therapy of choice. Other implantable devices, such as drug-infusion pumps, have also been developed [1] and show promise for future antiarrhythmic therapy. Such devices offer a variety of therapies, each tailored to a specific arrhythmia. Implantable devices are expected not only to administer appropriate therapy in the presence of an arrhythmia, but also to monitor the heart for extensive periods and recognize the need to administer preventive therapy before severe dysfunction occurs [2,3]. Detecting the presence of an arrhythmia requires these devices to have automated detection schemes with high sensitivity and specificity that are immune to changing clinical conditions, such as changes in drug therapy, lead configuration, and physical activity. Automated arrhythmia recognition requires a quantitative description of arrhythmias in terms of standard signal analysis and processing techniques.

To date, numerous signal processing schemes, implemented primarily in the research environment, have proven successful at diagnosing arrhythmias. These arrhythmia detection schemes operate on electrograms, defined as signals collected from electrodes placed inside or directly over the chambers of the heart. However, hardware and power supply limitations in real-world devices prohibit many of these sophisticated schemes from reaching the clinical environment. In practice, primitive algorithms, such as electrogram *rate* and *amplitude distribution*, have been implemented in commercial antitachycardia pacemakers [4–8] and automatic implantable defibrillators [2,9] for some time. *Rate* is typically defined as the frequency with which an electrogram exceeds some preselected amplitude threshold. *Amplitude distribution* is a representation of the *probability density function* of the electrogram. Typically, ventricular fibrillation has higher measured rates and a more diffuse density function than sinus rhythm. Such algorithms are simple to implement but exhibit poor specificity in discrimination of tachycardias [10]. Furthermore, such simple schemes are highly susceptible to noise, large-amplitude, far-field activation (e.g., atrial activity in the ventricular electrogram), and wide activation complexes (e.g., monomorphic ventricular tachycardia). In addition, devices may not be able to differentiate slower ventricular tachycardias with bizarre QRS complexes from ventricular fibrillation on the basis of rate and probability density function alone [2]. The immediate consequence of poor specificity is the unnecessary administration of therapy such as a high-voltage shock. This unwarranted intervention creates excessive pain for the patient and may at times quite the tachycardias that it was designed to prevent [11–13].

If the appropriate modes of therapy (pacing, cardioversion, and high-energy shock) are to be administered for different arrhythmias, more specific arrhythmia classification schemes are required than are present in today's devices. Although current implantable defibrillators are quite effective in terminating potentially lethal ventricular arrhythmias, and more recently atrial fibrillation, the current rate and probability density function estimates result in inappropriate shocks during rapid atrial arrhythmias with accelerated ventricular response and sinus tachycardia [14]. Moreover, electromagnetic noise and myopotential interference [4,7,11,15] may be misinterpreted by the device as ventricular fibrillation. Furthermore, the device may fail to trigger in instances where the rate of the ventricular tachycardia falls below the threshold for administering therapy, despite the need for therapy as indicated by hemodynamic consequences [16]. There is evidence to suggest that within individual patients, following ineffective shocks, changes in signal amplitude and cycle length can lead to postshock oversensing and undersensing, resulting in failure of the device to detect continuing ventricular fibrillation [17].

The following is a review of the signal processing techniques currently used in commercial devices or under investigation for use in the detection and differentiation of cardiac arrhythmias. Note that while this chapter addresses quantitative descriptors of cardiac arrhythmias with respect to intracardiac electrograms, similar descriptors exist for arrhythmias recorded via surface electrocardiograms. Signal processing is not only useful for automated detection schemes, but may, in addition, provide insight into the electrophysiological mechanisms underlying certain arrhythmias, including mechanisms for the initiation and termination of arrhythmias.

II. ELECTROGRAM MEASUREMENT AND CONSIDERATIONS FOR DATA COLLECTION

Before elaborating on the details of the signal processing methods used to quantify cardiac arrhythmias, we must first consider the source of the signals and the manner in which the signals are acquired. Pacemakers and implantable antiarrhythmic devices typically monitor the electrical activity of the heart via fixed or floating electrodes placed in lead wires that are threaded through the venous vasculature extending from the device to either the atrial or ventricular chambers of the heart. In the research environment, electrical signals are typically collected from temporary pacing catheters placed in either humans or animals during electrophysiological investigation. These temporary pacing catheters may have multiple electrodes in a variety of configurations, allowing investigators to influence the characteristics of the signal (frequency content, morphology, and amplitude) being recorded. The electrodes may be designed to optimize specific features of the intracardiac signal, typically referred to as an *electrogram*, thereby contributing to the manner in which arrhythmias are quantified. The initial development and testing of proposed arrhythmia detection schemes is performed via computer simulation using the electrograms collected during electrophysiological investigation or device implant.

A. Typical Electrode Configurations

In both the clinical and research environments, electrograms are typically collected using a unipolar or bipolar lead configuration. As illustrated in

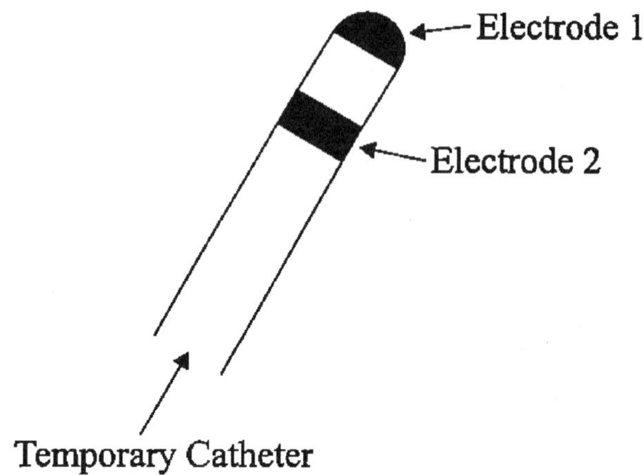

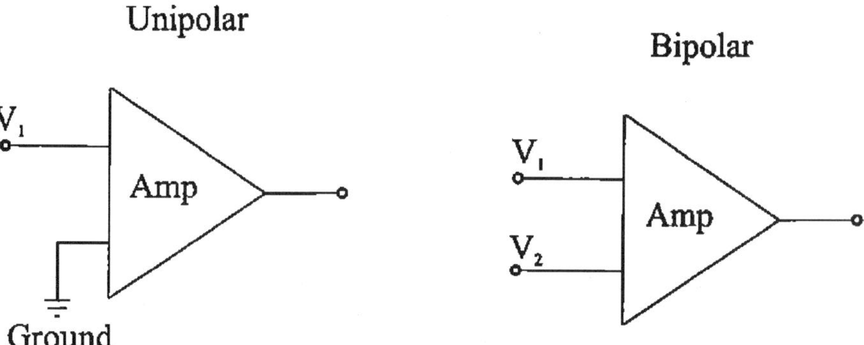

Figure 1 Two electrode elements mounted on a catheter or lead wire may be used to measure electrical activation in surrounding cardiac tissue. When the electrode potential (V_1) is measured with reference to ground, the resulting electrogram is referred to as a unipolar recording. When the difference in electrode potentials $(V_1 - V_2)$ is measured, the resulting electrogram is referred to as a bipolar recording.

Fig. 1, a unipolar configuration, measures the electrical activity (V_1) in the vicinity of one electrode with respect to a reference or ground potential (such a Wilson's central terminal). Thus, the unipolar measurement is not direction sensitive and tends to measure both local and far-field activity.

Conversely, a bipolar electrogram, typically collected from a pair of closely spaced electrodes (Fig. 1), measures the differential voltage ($V_1 - V_2$) between the two electrodes. The bipolar measurement is simply the difference in potential between the two electrodes that comprise the bipolar configuration. The bipolar measurement is sensitive to the direction of nearby depolarization/repolarization wavefronts and typically measures activation that is local to the bipole. This localized measurement results from the subtraction of relatively far-field activity that is common to both electrodes.

One may consider the bipolar measurement to be equivalent to a spatially high-pass-filtered version of the underlying activation passing by the two electrodes. Figure 2 illustrates how the bipolar measurement serves as a high-pass filter. Suppose a uniform wavefront of depolarization, denoted by the positive charges, passes by the two electrodes in the direction indicated. The bipolar signal $V_1 - V_2$, is defined as the difference between potentials V_1 and V_2. Note that the bipolar measurement shows deflections from baseline when potential V_1 differs from V_2. However, when both electrodes are equally surrounded by the same potential, the differential or bipolar measurement is zero. The advantages of bipolar electrograms with respect to unipolar electrograms are the rejection of far-field activity and the sensitivity to the direction of activation propagation. However, this sensitivity to direction may be disadvantageous when wavefronts passing perpendicular to the line of electrodes fail to register in a bipolar configuration. As a consequence, mapping studies aimed at investigating electrophysiological mechanisms often use unipolar electrograms in order to sense all activity and determine time of activation (that is, the actual time at which a depolarizing wavefront arrives at a specific site in the myocardium).

B. Cardiac Mapping

The majority of signal processing research applied to cardiac arrhythmias has been motivated by the need for accurate arrhythmia detection schemes. However, more recent time-series analyses have focused on quantifying the organization (regularity and spatial pattern of depolarization/repolarization) of cardiac arrhythmias and the changes in that organization with onset and termination of an arrhythmia. For such analysis, the electrogram signals are typically acquired from a mesh of epicardial and endocardial electrodes placed at multiple locations on the heart (several hundred to several thousand simultaneous sites) (Fig. 3). These recordings result in

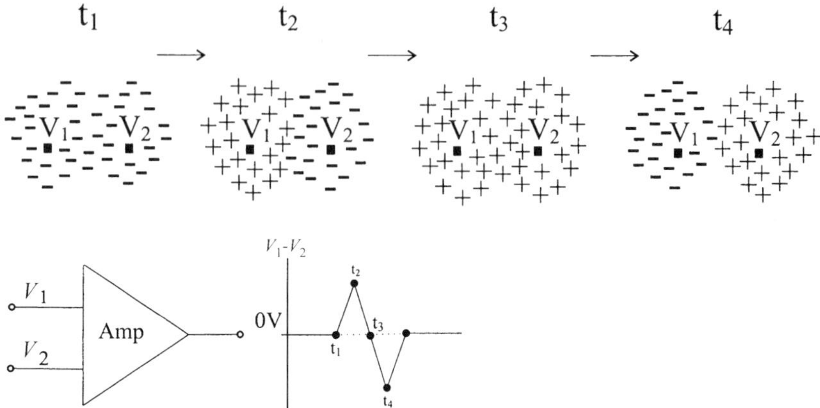

Figure 2   Illustrated is the bipolar electrogram, $V_1 - V_2$, that results as a wavefront of depolarization passes from left to right across two electrodes from which a differential potential is being measured. Note that the electrogram registers a nonzero potential only when there is a difference in potential between V_1 and V_2.

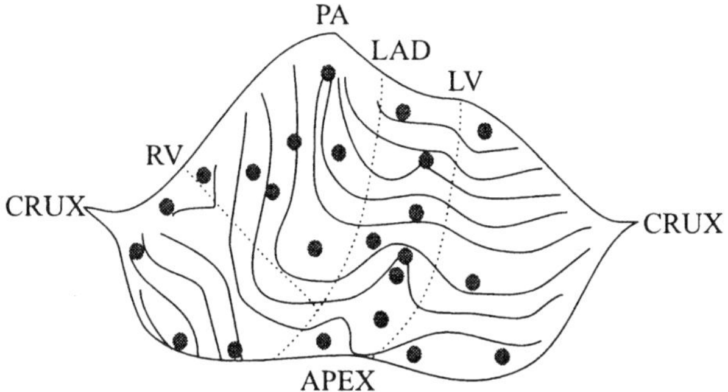

Figure 3   Multiple electrode arrays (filled circles) may be placed on the epicardium to record activation simultaneously from several hundred sites. Such activation maps are used to study patterns of activation (isochronal lines) with high spatial resolution, enabling investigators to study evolution and termination of arrhythmias.

cardiac electrogram maps that are typically unipolar in nature and capture the patterns of activation as waves of depolarization and repolarization traverse the myocardium. The mapping studies are used to investigate the mechanisms of fibrillation and defibrillation [18–20]. An important issue with regard to understanding the mechanisms of fibrillation and defibrillation is the minimum electrode spacing necessary to faithfully capture the underlying electrophysiology and avoid spatial aliasing [21].

Some of the signal processing schemes used to summarize cardiac maps include temporal activation maps [19], correlation and coherence maps [22,23], and more recently, state-phase map based on optical mapping data [24]. Quantitative descriptors derived from spatial mapping data to study the mechanisms of cardiac rhythms are covered in other chapters of this volume.

C. Sampling Electrograms: Digital Data

Until recently, pacemakers and implantable antiarrhythmic devices used analog signals to perform arrhythmia detection, due to the simplicity and low power requirements of analog hardware. However, analog circuitry is rather limited with regard to sophistication of signal processing methods that may be used to differentiate arrhythmias. As digital circuitry is incorporated into implantable devices, more sophisticated signal processing techniques may be used to quantify electrogram data. At this point, we review a few concepts with regard to digital (discrete-time and discrete-amplitude) signals. The majority of research in the differentiation of arrhythmias has been performed on digital data, due to the computational efficiency and ease with which signal processing may be implemented on a computer. Moreover, digital data is easily stored, faithfully reproduced, and transferred to a computer for further manipulation.

In analog-to-digital (A/D) conversion, a continuous-time and continuous-amplitude signal is converted into a discrete-time and discrete-amplitude signal. An A/D conversion scheme of a continuous-time and continuous-amplitude electrogram is illustrated in Fig. 4. Let us sample the amplitude or value of a continuous signal, $x(t)$, at discrete instances of time, NT, where N is an integer and T is the time between samples. Such a sampled signal may be denoted by $x(n)$ and is said to have a sampling frequency of $f_s = 1/T$ samples/second or hertz. In order to faithfully represent $x(t)$ with a sampled version, $x(n)$, the sampling frequency, f_s, must satisfy the Nyquist criterion [25]. Simply stated, the Nyquist criterion requires that $x(t)$ be sampled at a frequency that exceeds twice the highest frequency component of the continuous signal, $x(t)$. In Section VIII we will discuss frequency content of electrograms is greater detail, but suffice it to say that we must sample fast enough to capture information about the most rapidly changing parts of the signal.

Figure 4 shows the loss of information that occurs when we undersample the data. As illustrated in the bottom panel, if we sample the electrogram recording only once every 1/12 sec, we miss a great deal of information about the changes in the analog signal in between the sampled

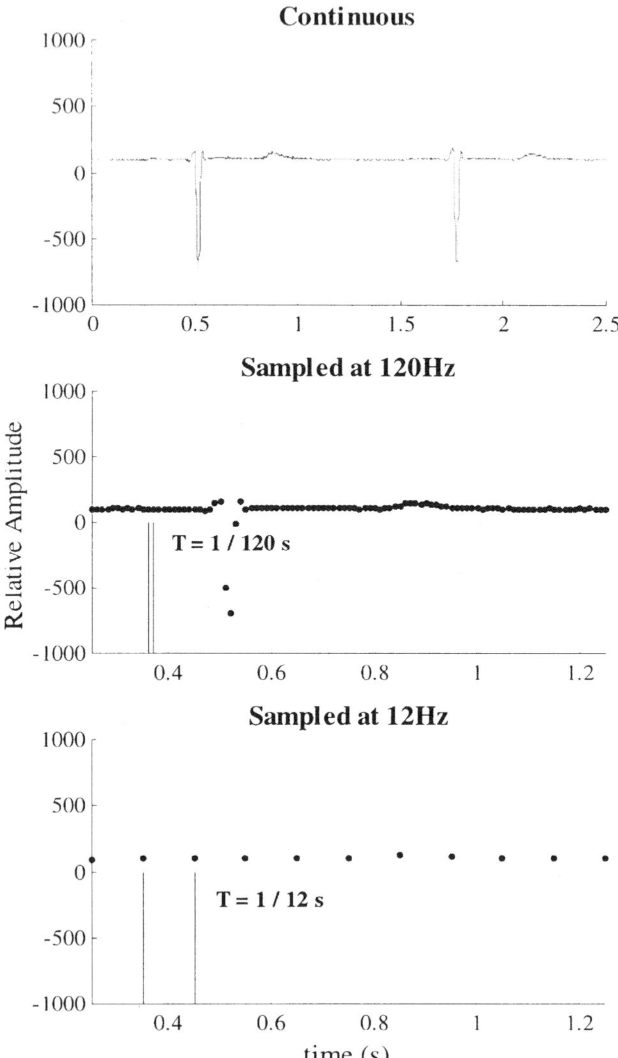

Figure 4 A continuous-time, continuous-amplitude electrogram (top tracing) is sampled at regular, discrete time intervals (T) to produce discrete-time, continuous-amplitude electrograms (middle and bottom tracings). Lengthy time intervals between consecutive samples (bottom tracing) fail to capture information about relatively short-duration changes in electrogram activity.

data points. Thus, when examining the digital signal in the bottom panel, we are falsely led to believe that the electrical activity did not change much in the 1-sec recording.

Insufficient sampling rates result in high-frequency (rapid, $> f_s/2$) signal changes aliasing or mimicking lower-frequency components. All frequency components present in the analog signal and higher than $f_s/2$ will appear somewhere between 0 and $f_s/2$ [25] in the digitized signal. To illustrate the aliasing phenomenon, we look at the sampled signals in Fig. 5, where we have sampled an analog sine wave, $x(t)$, with a single frequency component of 20 Hz, at sampling frequencies, f_s, of 30 and 60 Hz. In the left panel, we see the time-domain representation of both sampled signals, and in the right panel we find the frequency spectra (see Section II.D) corresponding to each of the sampled signals. Note that for a sampling frequency of 30 Hz, the sampled 20-Hz signal looks like a 10-Hz sine wave. In contrast, for a sampling frequency of 60 Hz, the sampled signal looks like the 20-Hz signal from which it was derived. Note also that the frequency spectra of the sampled signals have some additional spectral lines neighboring the fundamental frequency. These lines result from spectral leakage, which is addressed in Section II.D.

For both clinical and experimental data, the frequency content of intracardiac activity is typically in the 0–500-Hz range [26]. To avoid aliasing of frequencies larger than $f_s/2$, we must anti-alias filter the analog data (with an analog filter) prior to sampling. Therefore, once the bandwidth of the desired signal is determined, the sampling frequency f_s is chosen to be somewhat larger than twice the highest frequency in the desired bandwith. Prior to sampling, the investigator must anti-alias filter the signal prior to digitization to eliminate frequency content beyond $f_s/2$.

Another issue for concern in performing analog-to-digital (A/D) conversion is the quantization of a continuous-amplitude signal into a discrete set of amplitudes limited by the number of bits in the A/D converter. In other words, the computer must register amplitude information using a finite range of numbers. More specifically, the A/D board represents signal amplitude using a finite number of bits (sequence of 0's and 1's). The number of bits used to represent signal amplitude limits the number of discrete amplitudes that can be represented. For example, if there are 3 bits, then 2^3 or eight discrete numbers (amplitudes) can be represented:

$$000 \quad 001 \quad 010 \quad 011 \quad 100 \quad 101 \quad 110 \quad 111$$

When the data are converted from analog to digital form, the analog amplitudes that take on a continuum of values, let us say from 0 to 5 V, must now be mapped to 8 discrete levels. Thus the amplitude resolution of the

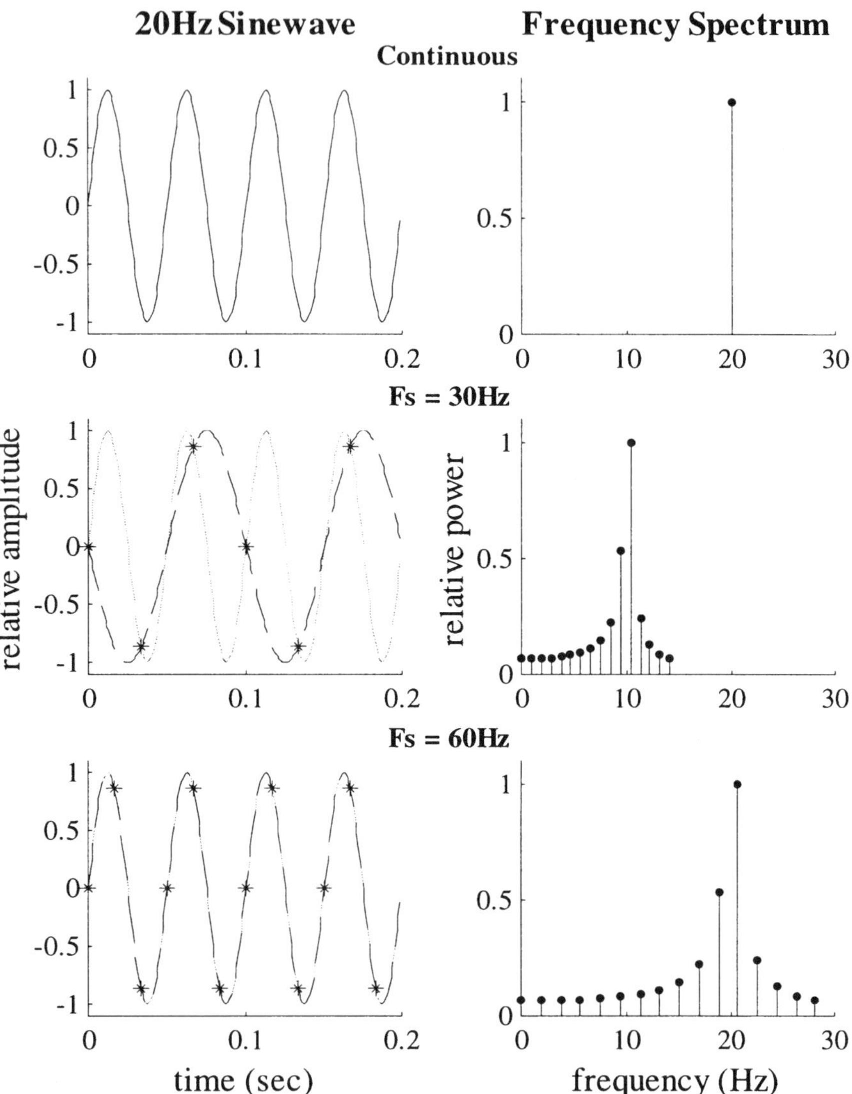

Figure 5 A 20-Hz sine wave (top left panel) is sampled at two different sampling frequencies, 30 and 60 Hz (middle and bottom panels). The 30-Hz sampling frequency leads to aliasing in which the 20-Hz sine wave mimics a 10-Hz sine wave. The alias may be seen in both the time domain (middle left panel) and the frequency domain (middle right panel). Note that sampling also leads to leakage of spectral energy into frequencies adjacent to 20 Hz.

A/D converter in this case in 5/7 or 0.71 V per discrete step. In other terms, the amplitude of the analog signal must change by an amount of 0.62 V from one sample to the next in order for the A/D converter to register the change in voltage as a change in bit. The fact that there are a finite number of bit sequences to represent an infinite number of voltage values results in an artifact termed "quantization error." Figure 6 shows the "steplike" discontinuities that result from representing an analog signal with a digital signal. In general, for a fixed amplitude range, the greater the number of bits in the A/D converter, the less the quanitization error. The effective amplitude resolution is equal to the amplitude range of the A/D converter divided by number of discrete levels. To maximize amplitude resolution, care should be taken to amplify or reduce the amplitude of the analog signal to closely match the range of the analog signal to the voltage input range of the A/D converter. Finally, care should be taken to avoid saturating the A/D converter, which occurs when the amplitude of the analog signal exceeds the input voltage range of the A/D converter, causing clipping or railing at the extreme amplitudes (Fig. 6).

Both sampling frequency and quantization may affect the signal processing methods used to differentiate cardiac arrhythmias. For example, clipping can introduce high-frequency components to an electrogram signal that are not contained in the original signal. Figure 6 demonstrates the high-frequency components introduced to the eletrogram signal by clipping.

D. Time-Domain Versus Frequency-Domain Representation of Electrograms

We may examine electrogram signals in the time domain, frequency domain, or both. In the time domain, we represent the signal amplitude, $X(t)$, as a function of time (Fig. 7). In the time-domain representation, we are interested in the signal amplitude, how rapidly the signal changes with time (slope), the sequence of signal changes (morphology), and the relative timing between multiple signals (time delay). In addition, we look for repeatability or predictability in the signal (autocorrelation).

Alternatively, we may examine an electrogram signal in the frequency domain, whereby the energy of the signal is now represented as a function of frequency (Fig. 7). To represent a signal in the frequency domain, we use the Fourier transform. For a continuous-time, transient signal, $x(t)$, the Fourier Transform, $X(f)$, is defined as

$$X(f) = \int_{-\infty}^{\infty} x(t)e^{-j2\pi ft}\, dt \tag{1}$$

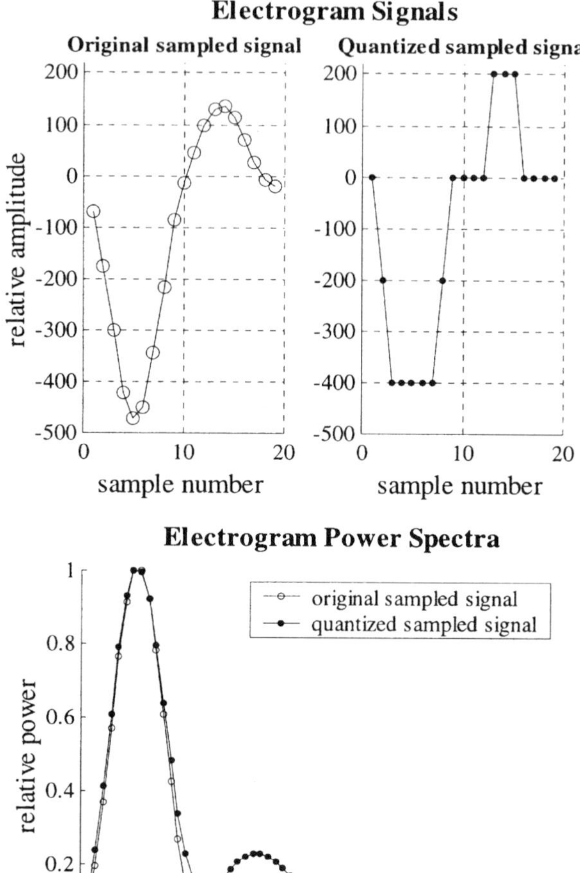

Figure 6 Discrete-time (sampled) electrograms (left) are subject to quantization (right) during analog-to-digital conversion. Quantization results when the continuous range of electrogram amplitude must be represented by a finite number of bits in the computer. As a result, electrogram amplitudes that fall between quantization levels (in this case, -400, -200, 0, and 200) must be rounded to the nearest quantization level. Furthermore, electrogram amplitudes that exceed the input voltage range of the A/D converter are clipped to the largest amplitudes (-400 and 200) that may be represented by the A/D converter. The bottom panel shows the frequency-domain representation of the original sampled signal (open circles) and the quantized sampled signal that has been clipped (filled circles). Note that clipping introduces high-frequency harmonics to the frequency spectrum.

$X(f)$ is a complex function of frequency, f, and describes the relative complex voltages (amplitudes and phases) as a function of f that are present in $x(t)$. Like the time domain, the frequency domain may be used to convey the morphological and periodic information contained in a signal. For digital, finite-duration signals, we use the discrete Fourier transform (DFT) to examine the frequency content of the signal [Eq. (2)]. In simple terms, we use the DFT to represent any digital signal, $x(n)$, of length N, as a sum of k distinct, harmonically related complex exponentials (sinusoidal waveforms),

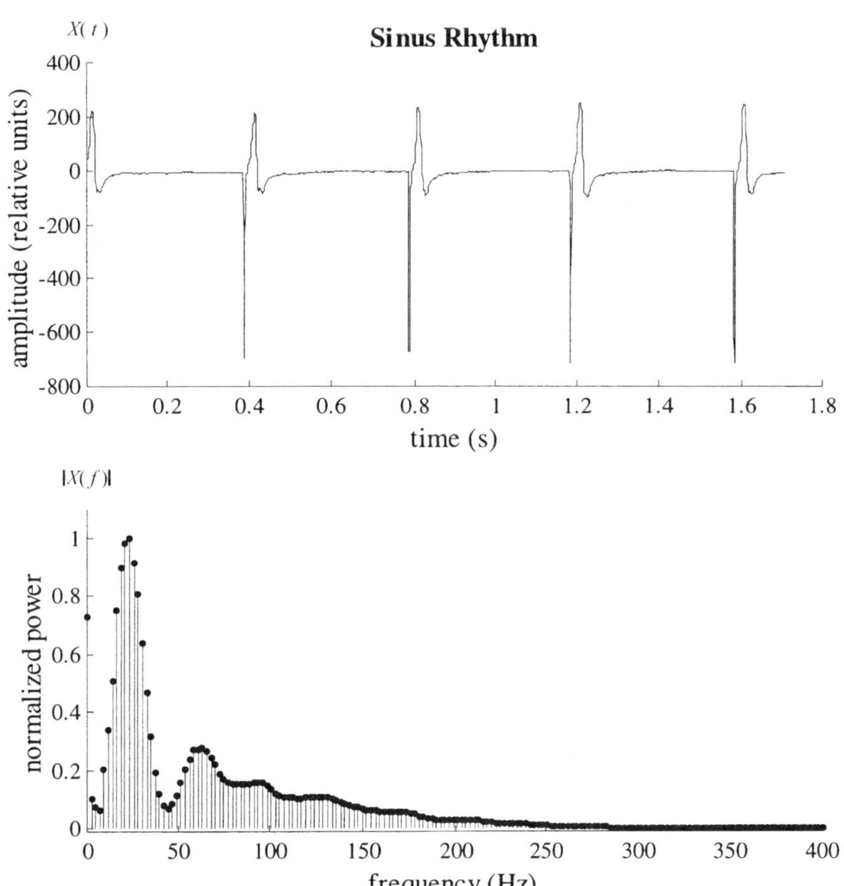

Figure 7 Electrogram, $X(t)$, shown as a function of discrete time, t. $X(f)$, is a frequency-domain representation of electrogram, $X(t)$, shown as a function of discrete frequency, f.

where each exponential, $X(k)$, will have a single frequency, amplitude, and phase.

$$X(k) = \sum_{n=0}^{N-1} x(n)e^{-j(2\pi nk)/N} \qquad 0 \leq K \leq N - 1 \qquad (2)$$

Note that a simple sinusoid is represented, in the frequency domain, by a line in the magnitude spectrum and a line in the phase spectrum. In both the magnitude and phase spectra, the lines occur at the frequency of the sinusoid.

Electrogram signals are much more complex than a simple sinusoid, but the DFT may be used to model the electrogram as a sum of simple sinusoidal signals, thereby transforming the electrogram into a sequence of spectral lines. The magnitudes of these spectral lines quantify the relative contribution of different frequencies to the electrogram signal. In Section VIII the use of the frequency spectrum and the power spectrum, a related quantity, in differentiating cardiac arrhythmias will be addressed.

E. Arrhythmias as Random Processes

Recall that our electrogram signals are projections of the cardiac electrical activity onto some electrode configuration. The electrogram signals represent the outcomes of some random process controlling the electrical activation of the heart. The random process(es) controlling the activation of the heart may be described by a *probability density function* that, in theory, allows one to predict the likelihood of a future state of the heart given the present state. In differentiating cardiac arrhythmias, we are often looking for differences in the probability density function(s) underlying the various arrhythmias. We can consider cardiac arrhythmias to be random processes, where a future value cannot be exactly predicted by a mathematical expression. However, a random process may be characterized by a number of statistical descriptors that describe the probability density function underlying the random process (Fig. 8, right panel) [25]. A probability density function may be described by its mean, variance, and higher-order moments, such as skew and kurtosis. If the underlying random process is Gaussian (or normal), the probability density function is completely described by its mean and variance. The electrograms acquired during an arrhythmia are assumed to be sample functions of a random process. In trying to differentiate various arrhythmias, we are seeking to quantify those aspects of the underlying random process that uniquely characterize an arrhythmia.

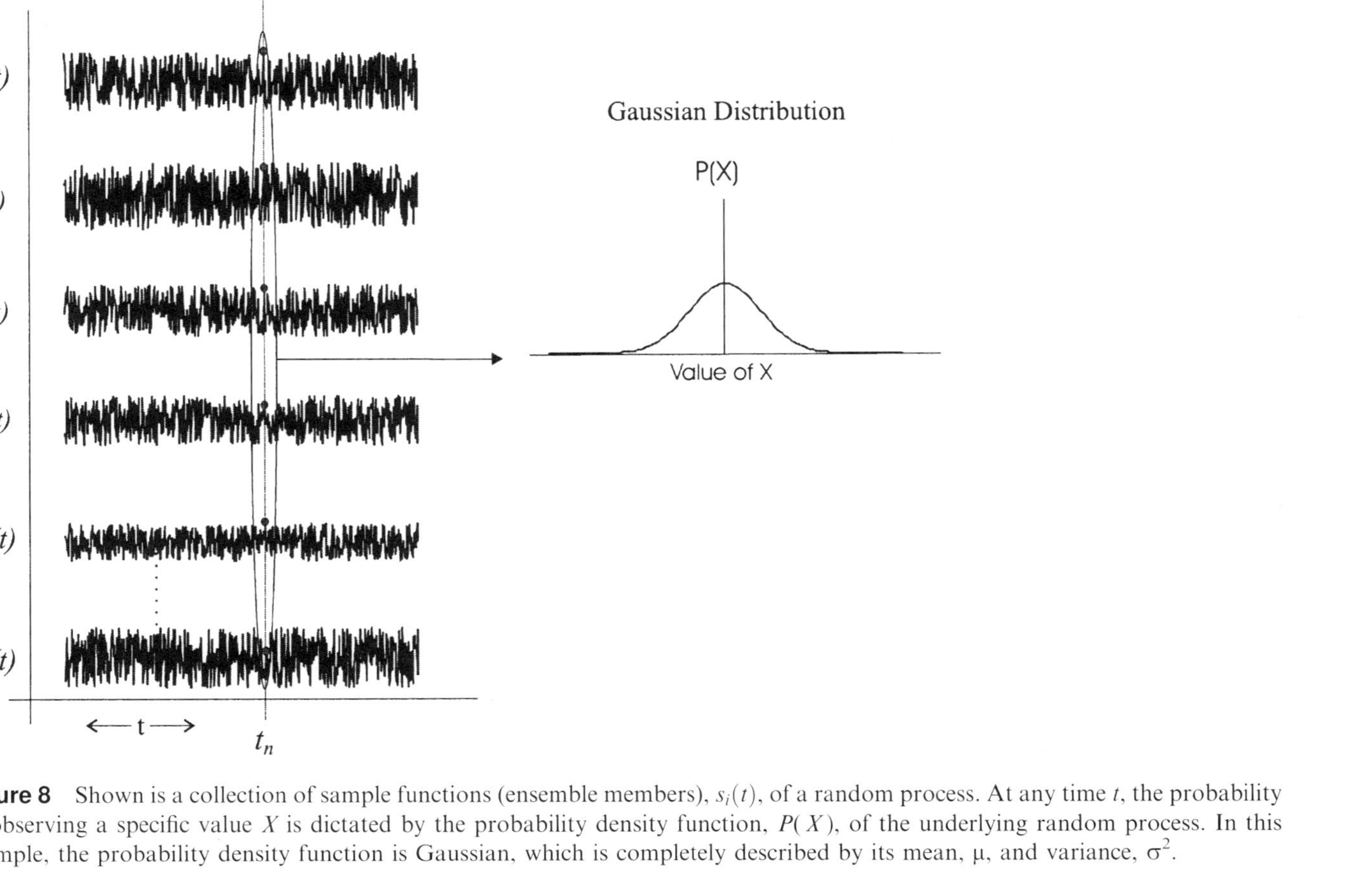

Figure 8   Shown is a collection of sample functions (ensemble members), $s_i(t)$, of a random process. At any time t, the probability of observing a specific value X is dictated by the probability density function, $P(X)$, of the underlying random process. In this example, the probability density function is Gaussian, which is completely described by its mean, μ, and variance, σ^2.

The probability density function describes the likelihood that a random variable, such as the amplitude of the electrogram, will take on a specific value, A, at some time, t. When estimating the statistical properties of an arrhythmia, such as its mean, variance, autocorrelation (section V.A), and power spectrum (Section VIII), we often assume that the underlying random process controlling the arrhythmia is stationary. This means that the probability density function, and consequently the statistics, of the random process do not change during the sampling period of interest. Thus, the quantities that we estimate from the electrogram over some time interval will be equal to the same quantities estimated over a different interval of time. In other words, if the process controlling the arrhythmia is stationary, the time of day, week, or month during which we record the electrogram should not influence the statistical descriptors that we are trying to estimate.

In addition to stationarity, we often assume that the random process controlling the arrhythmia is ergodic. To understand stationarity and ergodicity, we must first introduce the concept of sample functions. The random process controlling an arrhythmia may be represented by a collection of sample function $S_i(t)$ (Fig. 8). Each sample function represents a time sequence of possible outcomes of the random process. In other words, at any instant in time, t_n, the magnitude of the measured electrogram could take on a multitude of values, where the likelihood of observing a specific value is governed by some probability density function (Fig. 8, right). If the probability density function governing the amplitude of the electrogram is constant over time, we say that the random process is stationary. In theory, the probability density function for the random process, at any instant in time, may be determined from the collection of all possible sample functions. From the collection of sample functions, we may define the statistical mean, μ, the statistical variance, σ^2, and other statistical properties of the random process (Fig. 8). In practice, we cannot observe all possible sample functions for a random process. In fact, we are typically restricted to observing a single sample function (i.e., a finite-duration electrogram signal). We may use this single function of length N to estimate time averages, such as the sample mean [Eq. (3)] and sample variance [Eq. (4)]. Ideally, we would like the time averages, (such as $\bar{X}$ and S^2) estimated from a single sample function to be representative of the statistical averages (such as μ and σ^2) of the underlying random process giving rise to the electrogram signal. If the time averages estimated from any sample function equal the true statistical averages of the random process, we say that the random process is *ergodic*. The assumption of ergodicity is important to our quantification of cardiac arrhythmias, because often we have only brief electrogram recordings available from which to study the random processes driving an arrhythmia. Note, however, that some arrhythmia discrimination schemes actually make use of the nonstationarity of an arrhythmic process to differentiate one arrhythmia from another [27].

$$\bar{X} = \frac{1}{N} \sum_{n=0}^{N-1} x(n) \tag{3}$$

$$S^2 = \frac{1}{N-1} \sum_{n=0}^{N-1} [x(n) - \bar{X}]^2 \tag{4}$$

Another useful descriptor for describing random processes is the *joint probability density function*. The joint probability density function may be used to quantify the statistical dependence of one random process on another random process. As described in Sections V.B and VIII.C, statistics such as the cross-correlation function and coherence function may be used to examine the correlation between two or more electrogram signals, which may, in turn, be used to differentiate between various arrhythmias.

F. Filtering Electrogram Signals

Many modes of signal analysis involve some sort of digital filtering or transformation of the electrogram signal. Linear digital filters (filters which satisfy superposition) involve a combination of three basic operations:

Addition of sequences
Multiplication by a constant
Time delay

These operations are illustrated in combination in the following low-pass Hanning filter, which operates on the original signal, $x(n)$, to produce a filtered signal, $y(n)$, that has an overall bandwidth that is less than the original signal.

$$y(n) = \frac{1}{4} [x(n) + 2x(n-1) + x(n-2)] \tag{5}$$

This type of filter is also known as a moving-average filter because the output signal is a weighted average of several input signal elements, each weighted by a unique factor and each delayed in time with respect to other samples by an integer number of sampling intervals (nT). Like any linear, time-invariant system, a digital filter may be fully described by its impulse response. The impulse response describes the output of the filter given a unit impulse input (Fig. 9). To estimate the impulse response for practical filters,

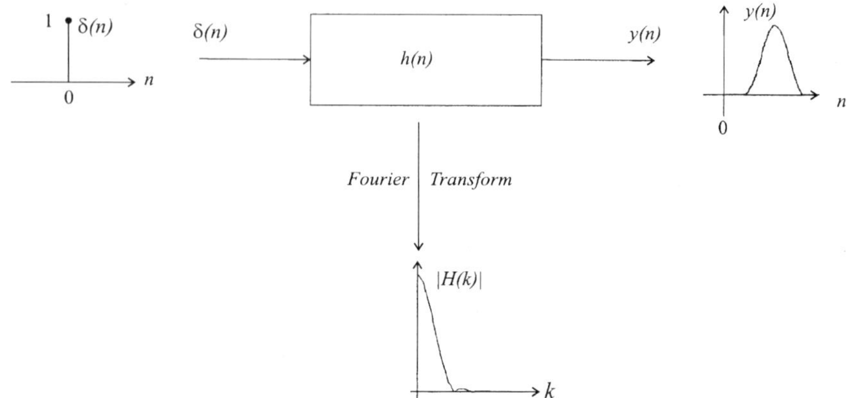

Figure 9 The filter output, $y(n)$, produced by a unit impulse input, $\delta(n)$, is known as the impulse response, $h(n)$, of the digital filter. The transfer function, $H(k)$, is the discrete Fourier transform of $h(n)$, and represents the frequency response for the digital filter. Note that the filter is a low-pass filter.

we may let the input to the filter, $x(n)$, be a unit impulse sequence, $\delta(n)$, given by

$$\delta(n) = 1 \quad \text{if } n = 0$$
$$ = 0 \quad \text{if } n \neq 0 \tag{6}$$

The resulting output, $y(n)$, is also known as the impulse response of the filter and is often denoted as $h(n)$. For a linear, time-invariant system, the impulse response completely characterizes the filter and any output, $y(n)$, may be determined for any input, $x(n)$, using a property known as convolution:

$$y(n) = \sum_{k=-\infty}^{\infty} h(k)x(n-k) \tag{7}$$

It is often more intuitive to consider the filtering operation in the frequency domain. For a linear, time-invariant digital filter, the filtering operation in the frequency domain is written

$$Y(Z) = X(Z)H(Z) \quad \text{where } Z = e^{-jw} \tag{8}$$

In words, the frequency content of the output signal, $Y(Z)$, is equal to the frequency content of the original signal, $X(Z)$, weighted (multiplied) by the transfer function, $H(Z)$, of the filter. In practical applications, the transfer function, $H(Z)$, is simply the discrete Fourier transform $[H(k)]$ of the filter's impulse response, $h(n)$. Given the frequency spectrum for a specific input signal, a direct examination of the filter's transfer function can inform us of the bandwidth of the output signal. Figure 9 shows the transfer function, $H(k)$, for a low-pass filter.

Filters are typically categorized as finite impulse response (FIR), meaning that $h(n)$ has a finite number of terms. Thus, the effects of transient inputs on the output fade after a finite time period. Alternatively, filters may have an infinite impulse response (IIR), meaning that $h(n)$ has an infinite number of terms. Thus, transient inputs may affect the output of the filter at all future times.

Typically, FIR filters are simpler to design, are always stable, and have a linear phase response. IIR are more complex to design, may be unstable, and may have nonlinear phase response. However, IIR typically have steeper roll-off, are more efficient, and allow for more flexibility in design. Nonlinear, nonstationary filters may also be used to process the data. Their design and implementation tend to be more complex and not as well understood in terms of performance or response to various inputs. Adaptive filters, which will be addressed in section VII, are a type of nonstationary filter in which the coefficients vary as a function of time. More complex coverage of digital filters and their application to biomedical signals may be found in Tompkins et al. [28].

III. MORPHOLOGICAL DESCRIPTION OF ARRHYTHMIAS

Differences in the underlying probability density functions governing a random process (such as an arrhythmia) are manifested in the morphology (size, shape, and sequence of slopes) and periodicity of the sampled signal (i.e., electrogram). Morphological differences may be easily quantified in the time domain through measurements of amplitude, slope (derivatives), zero crossings, histograms, autocorrelation functions, and even wavelet analysis [29]. Periodicity, or predictability, in morphology may be quantified with measures such as rate and the autocorrelation function.

The majority of arrhythmia detection schemes make use of signal morphology to differentiate various arrhythmias. Changes in morphology stem from abnormal alterations in the path of electrical activation in the myocardium. Typically, ventricular activation complexes during normally conducted sinus rhythm differ in morphology from those complexes observed during ventricular tachycardia. If we examine recordings of sinus rhythm and monomorphic ventricular tachycardia recorded from bipolar electrograms in the right ventricular apex (Fig. 10), we note that a number of morphological features differ between the two electrogram recordings. Differences are manifested in the signal amplitude (arrow A), beat-to-beat intervals (arrow I), width (w), slope, and amplitude distribution during some time period (histogram on right panel). The challenge remains in finding a robust, automatic way of quantifying those morphological changes.

A. Probability Density Functions

The simplest arrhythmia detection schemes used in commercial antiarrhythmic devices use the *amplitude probability density function* (PDF). PDF is a morphology-based method for discriminating fibrillatory from nonfibrillatory rhythms. Basically, the PDF is a measure of the amplitude

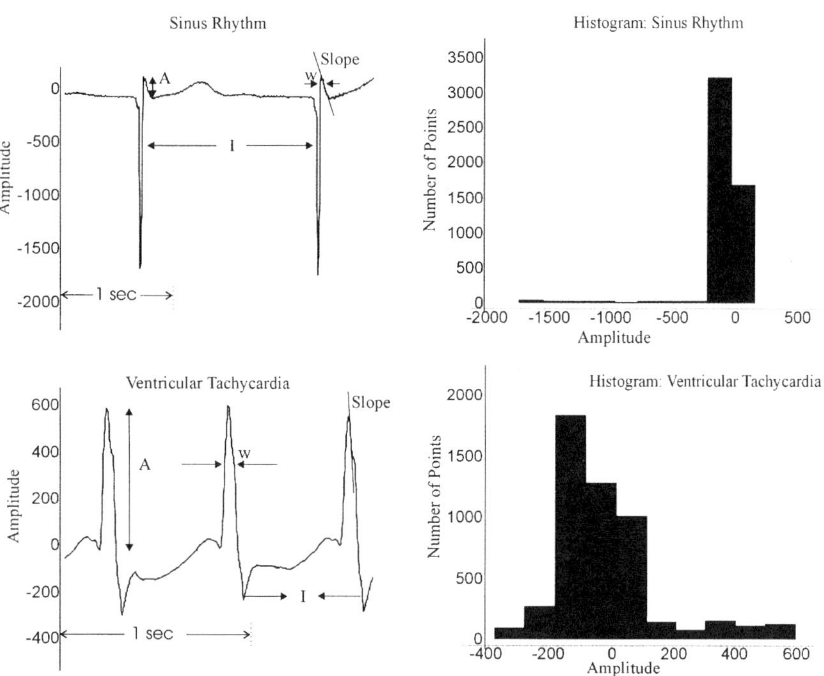

Figure 10 Illustrated are bipolar electrogram recordings (left) of sinus rhythm and ventricular tachycardia, both recorded in the ventricle. The electrograms are characterized by amplitude (A), duration (W), beat-to-beat intervals (I), slopes, and morphology. To the right are amplitude histograms for each of the electrogram recordings.

distribution or variability of the electrogram signal over a finite time interval. In essence, the PDF quantifies some attribute of the amplitude histogram of the electrogram. Electrogram signals acquired during fibrillation typically spend little time in the isoelctric region (that portion of the electrogram during which a chamber of the heart is quiescent or refractory) (Fig. 11). In contrast, electrogram signals acquired during sinus rhythm and regular tachycardias typically have well-defined isoelectric regions between consecutive activation complexes. Slocum et al. [30] and Jenkins et al. [26] have demonstrated the ability to differentiate atrial fibrillation from sinus rhythm and regular atrial tachycardias using a PDF measure. Both groups calculate an amplitude histogram from a predefined time interval of the digitized data and subsequently count the number of data points in that time interval that have an amplitude within a specific distance from the isoelectric region.

If we were to compute the PDF for the digitized data in Fig. 11, we would first normalize the data to a predetermined range (for example, ± 1000 units). Normalization allows the PDF criterion to be independent of actual signal amplitude, which may vary significantly with electrode location (which is not determined until time of implant), electrode configuration, drug regimen, posture, and patient activity. Following normalization, one selects a bin width and the number of bins in which to sort the signal samples by amplitude. Note that to make the algorithm robust in the face of patient-to-patient differences in electrogram amplitude, the bin width is allowed to be a function of the standard deviation of the individual patient electrogram. Histograms for sinus rhythm, ventricular tachycardia, and ventricular fibrillation show significant differences in widths.

Fibrillation is typically characterized by rapid, frequent, and unpredictable changes in signal slope compared to sinus rhythm. Thus, rather than use the amplitude of the raw electrogram signal for PDF analysis, commercial devices use the slope of the raw electrogram signal. The slope is obtained by taking the derivative of the waveform (see Section III.C). In commercial devices, the PDF of slope is estimated by first filtering the electrogram to obtain its first derivative and then measuring the statistical distribution of the filtered signal. The PDF algorithm then measures the percent of time that the derivative signal spends at high slopes. The PDF criterion reportedly differentiates ventricular fibrillation and ventricular tachycardia from sinus rhythm in about 50% of episodes tested [31–35].

During sinus tachycardia or rapid supraventricular tachycardias with wide ventricular complexes, the potential for overlap in PDF with that of ventricular fibrillation is increased, thereby increasing the tendency for false alarm. Atrial fibrillation with a rapid, irregular ventricular response may be particularly difficult for PDF to differentiate from ventricular fibrillation

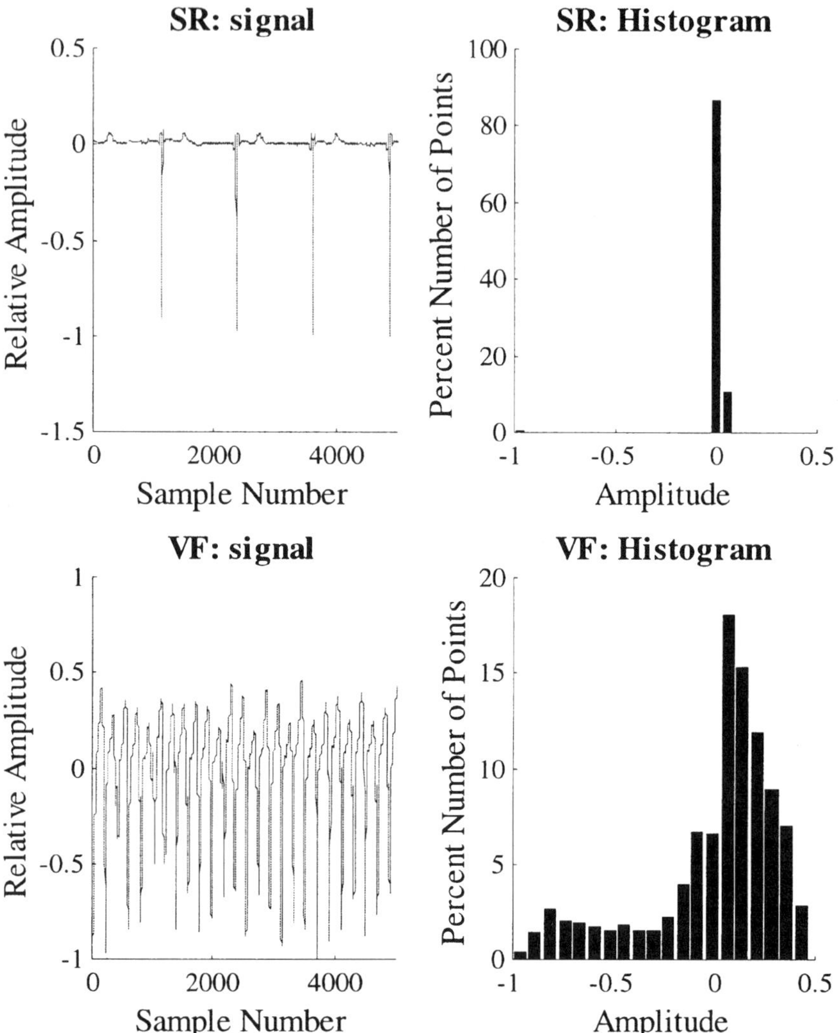

Figure 11 During ventricular fibrillation (lower left), the electrogram signal spends little time in the isoelectric region. In contrast, electrograms recorded during sinus rhythm (upper left) spend a much greater portion of time in the isoelectric region. Amplitude histograms corresponding to the electrograms as shown to the right. Note that ventricular fibrillation is characterized by a highly variable probability density function compared to sinus rhythm.

[3,26]. Furthermore, monomorphic ventricular tachycardias with accelerated rates and wide complex morphology have PDF overlapping with that of ventricular fibrillation [31]. Conversely, slower-rate ventricular tachycardia with narrow ventricular complexes may be interpreted as physiological sinus tachycardia.

The ability of PDF to differentiate fibrillation from nonfibrillatory rhythms is highly dependent on amplitude threshold [36]. As a result, PDF measures typically require automatic gain control in commercial devices to compensate for the dramatic decrease in electrogram amplitude often accompanying onset of ventricular fibrillation.

B. Detection and Description of Atrial and Ventricular Activations

Most morphology-based arrhythmia discrimination schemes require that we accurately detect the presence and temporal placement of atrial and/or ventricular activations. In the case of unipolar electrograms located in the high right atrium or right ventricular apex, such detection may be fairly simple because the electrogram typically registers activation primarily from one chamber, with little far-field activity from the adjacent chamber. However, unipolar electrograms typically contain unwanted far-field artifact from the neighboring chamber. Thus, to facilitate event detection for the activation in a single atrial or ventricular chamber, arrhythmia detection schemes often use bipolar electrogram recordings. Typically, bipolar electrogram recordings during nonfibrillatory rhythms are characterized by regularly repeating, discrete, high-frequency complexes alternating with fairly quiet isoelectric regions. For such electrogram recordings, event detection may be accomplished through simple amplitude threshold and blanking period algorithms.

1. Amplitude Threshold and Blanking Period Algorithms

A simple method for detecting atrial and ventricular activation that may be performed in hardware or software establishes an amplitude threshold (as denoted by the horizontal lines in Fig. 12) and searches for those regions of the electrogram signal that exceed threshold (denoted by the open circles). Of course, the actual amplitude of the signal will vary from patient to patient, due to a number of clinical factors. To avoid complications due to these variations in amplitude, we often establish an amplitude threshold that is tailored to the individual patient and allows for automatic gain control. (Note that even within a single patient, the amplitude of the electrogram signal may change dramatically with arrhythmia onset.) For example, during a specific time period, we may search for the maximum amplitude in

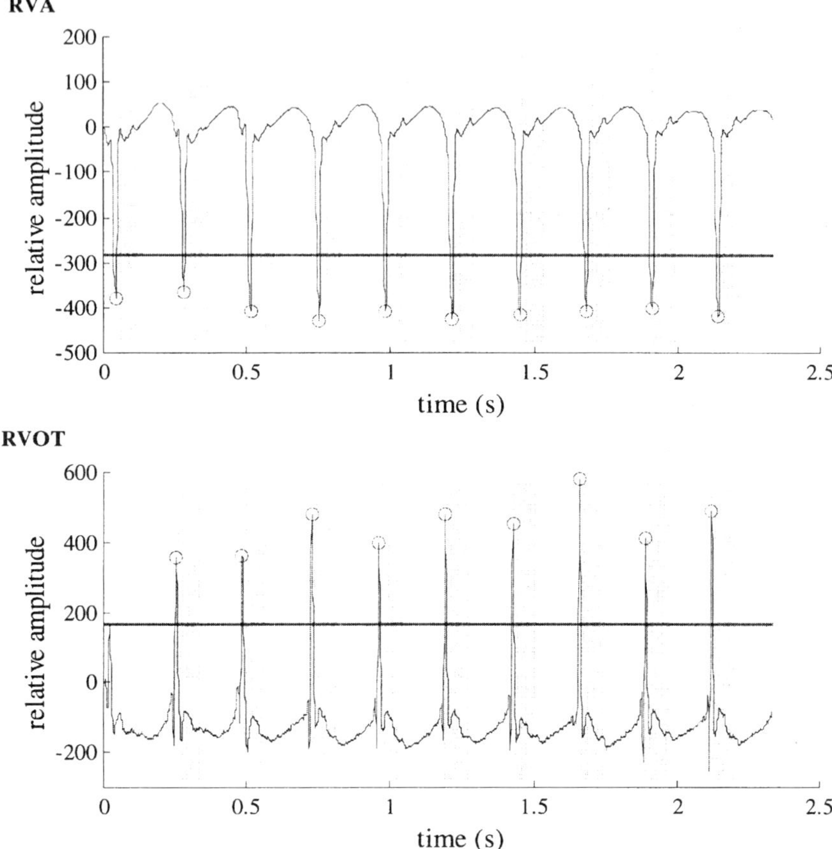

Figure 12 Bipolar electrogram recordings of monomorphic ventricular tachycardia recorded from the rate leads of an ICD in the right ventricular apex (RVA) and right ventricular outflow tract (RVOT). Ventricular activation may be automatically detected in electrograms by using an amplitude threshold (thick horizontal line) in combination with a blanking period (shade region). Any portion of the electrogram exceeding the amplitude threshold and lying outside the blanking region will be classified as a ventricular event (open circles).

the electrogram signal, and consequently, set the amplitude threshold for subsequent event detection to be a percentage of that maximum amplitude. If the device continually monitors the changing amplitude of the signal sand automatically updates the amplitude threshold to compensate for changes in overall signal gain, the chances of the device missing an event will be greatly reduced.

In addition to using an amplitude threshold criterion to detect an event, we typically incorporate a "blanking period" into the detection scheme. The blanking period is a fixed period of time following the crossing of an amplitude threshold or the detection of an event during which event detection is temporarily halted. In Fig. 12, the blanking period is indicated by the shaded vertical gray bar following each detected event. This blanking period prevents detection of multiple amplitude threshold crossings during a single event or activation. The blanking period may be set to coincide with physiologically accepted values for atrial refractory periods (AA intervals) or ventricular refractory periods (VV intervals), or may be set to be some fraction of the heart rate or previous AA or VV interval.

Blanking periods also prevent detection of far-field activation by halting detection during a time period when far-field activity is expected to appear in the electrogram. Of course, sudden changes in rate and frequency of activation can alter the effectiveness of these blanking periods.

As seen in Fig. 12, the polarity (positive and negative signal excursions) of activation may vary from patient to patient or with actual lead placement. To prevent sensitivity of the event detection scheme to actual polarity of the detected activation, the algorithm may take the absolute value of the signal prior to detecting threshold crossings.

2. Estimation of Rate and Onset Criteria

Simple event detection may be used to differentiate arrhythmias by simply quantifying the frequency of occurrence of the detected events. The simplest methods, in terms of hardware and computation, for discriminating atrial and ventricular tachycardias from sinus rhythm and bradycardias use measures of atrial or ventricular *rate*. The first few generations of devices recorded electrical activation from a single lead in the ventricle. Newer devices are designed to record activity from both the ventricle and the atrium. A rate estimate is simply a measure of the frequency with which an electrogram signal exceeds some predetermined amplitude threshold. Ventricular fibrillation and ventricular tachycardia are typically characterized by rates that are much faster than that of sinus rhythm or sinus tachycardia. However, for many patients, there is overlap between these rhythm classes which results in false rhythm classification. Furthermore, the estimated rate might or might not have meaning in terms of a physiological rate. For example, the sinus rhythm electrogram shown in Fig. 13 has regularly occurring activations that are relatively constant in amplitude, timing, and morphology, corresponding to the depolarization and subsequent repolarization of the ventricle. The physiological rate is defined by the frequency with which those activations occur. In the case of sinus rhythm, typical

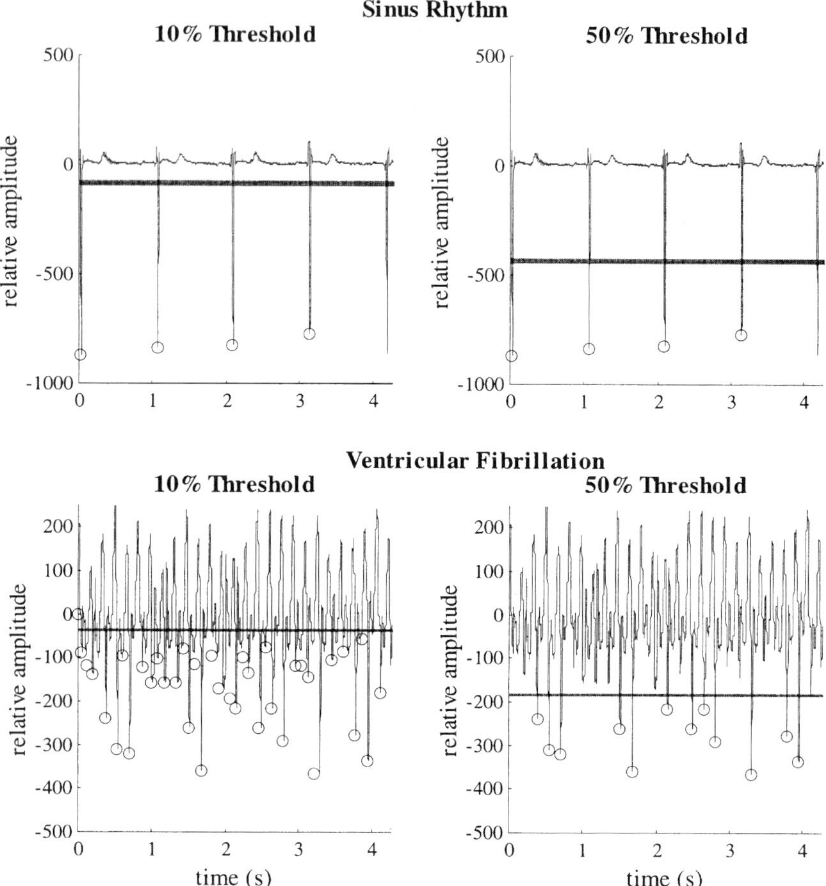

Figure 13 An example of sinus rhythm (top) and ventricular fibrillation (bottom) analyzed for rate using 10%(left) and 50%(right) amplitude thresholds. Note that for sinus rhythm, the number of detected events (open circles) is the same across several amplitude thresholds. Conversely, for ventricular fibrillation, the absence of discrete regular activations renders an arbitrary rate that is highly dependent on the amplitude threshold.

normal rates are in the range of 60–100 depolarizations per minute. These rates are determined by the refractory period of the cardiac tissue, which for the ventricles ranges from 150 to 300 msec under normal conditions.

The electrogram during ventricular fibrillation, shown in the lower panel of Fig. 13, fails to show regular, discrete activations of constant amplitude, timing, and morphology. For such an arrhythmia, the meaning of rate is unclear. The physiological mechanism underlying fibrillation is thought to consist of multiple circulating wavelets or rotors [18,21,37]. The speed, size, and duration of these rotors are determined by the refractory period and conduction velocity of the underlying tissue. The electrogram morphology observed during fibrillation typically reflects a summation of these multiple wavelets that are continually changing in magnitude, speed, and direction. The resultant electrogram is highly irregular, and thus the estimated rate is somewhat arbitrary and is highly variable depending on the algorithm used to estimate rate. Furthermore, the estimated rate is highly dependent on the amplitude threshold chosen for event detection.

A simple rate criterion has been used to separate ventricular tachycardia and ventricular fibrillation from sinus rhythm [2,26,30,31] and to separate atrial fibrillation and atrial flutter from sinus rhythm. A rate estimate is simple to implement with analog circuitry, quick to calculate (1–5 sec), and effective in differentiating pathological tachycardias form nonpathological arrhythmias. Rate criterion fails, however, for ventricular electrograms when there is double sensing of atrial activation or ventricular repolarization [30,33,38], and undersensing due to insufficient signal amplitude [26,30]. Whenever amplitude threshold methods are used to detect complexes, the rate estimate is greatly dependent on the chosen amplitude threshold, and it may be shown that adjusting the threshold may give rise to significant overlap between regular tachycardias and fibrillatory rhythms [30]. Even with correct rate estimation, false classification is possible since nonpathological tachycardias often have rates similar to the electrically terminable arrhythmias [39]. Such overlap of physiological rates prevents discrimination of supraventricular tachycardia, ventricular tachycardia, and accelerated ventricular rates due to atrial fibrillation and atrial flutter [4,15,26]. Furthermore, pathological tachycardias may vary in rate due to autonomic changes [3], rendering them slower than physiological sinus tachycardia. Moreover, both myopotential interference and electromagnetic interference may trigger rate criteria.

Simple rate criteria have been enhanced through the addition of *rapidity of onset* and *rate stability* criteria [40,41]. For example, sudden-onset criteria (the rapidity with which the rate changes) have been shown to separate sinus tachycardia from ventricular tachycardia when the estimated rates overlap [41]. Theoretically, onsets of ventricular tachycardia are quite sudden, developing in a matter of two to three beats. In contrast, sinus tachycardia typically develops more gradually. Thus, detection schemes look at the rapidity with which consecutive VV intervals shorten. If the shortening occurs faster than a predetermined threshold, the arrhythmia is classified as ventricular tachycardia.

Another factor to consider in arrhythmia classification is the stability of rate over time. To quantify the stability of rate, one may simply estimate the variance of VV intervals in a specified time interval and seek those instances in which the intervals become highly variable. Bardy and Olson [40] use onset and stability criterion to reject isolated premature ventricular contractions from ventricular tachycardia. A stability criterion rejects accelerated rates due to atrial fibrillation and polymorphic ventricular tachycardia.

The latter two enhancements of the rate criterion are still susceptible to error. Fisher et al. [42] show that sinus rhythm can onset as rapidly as ventricular tachycardia and ventricular fibrillation. In addition, some tachycardias commence with premature systoles followed by compensatory pauses, which may be interpreted as a gradual onset. Moreover, Geibel et al. [43] suggest that the onset of monomorphic ventricular tachycardia may be associated with irregular cycle lengths (variations greater than 5%), violating the stability criterion.

C. Slope-Based Arrhythmia Discrimination and Event Detection

Another means for arrhythmia discrimination and the detection of atrial and ventricular activations is through detection of *slope* changes in the electrogram signal. Derivatives of the electrogram signal may be estimated at specific points in time, and the magnitude and polarity of those slopes may be used to detect an event.

1. Arrhythmia Discrimination

We may differentiate arrhythmias using explicit slope information. *Slope* represents the change in signal amplitude during a specified time period. The slope of the electrogram, defined at a specific point in time, is simply the derivative of the elctrogram at that specific point in time. Arrhythmia differentiation may be based on a sequence or pattern of slope changes in a predefined time period.

There are a number of numerical methods for estimating the slope (derivative) of a digital signal. For example, if $x(n)$ is our electrogram signal, we may calculate the slope or derivative of $x(n)$ at an instant of time, n, using one of the following formulas. The computationally simplest slope estimate, $y(n)$, at a point $x(n)$ may be given by

$$y(n) = \frac{x(n+1) - x(n)}{h} \tag{9}$$

where h is the step size in time between two consecutive data points (may be assumed to be unity for practical implementations). This estimate is sensitive to outlying data points and noise because it uses only two consecutive data points. The inclusion of more information (data points) in the derivative (slope) estimate should yield an estimator that is less sensitive to noise. A more popular slope estimate is the central differences estimator, given by

$$y(n) = \frac{x(n+1) - x(n-1)}{2h} \tag{10}$$

Figure 14 shows an example of monomorphic VT and its derivative waveform (lower panel). The derivative waveform quantifies the sequences of slopes in the original electrogram recordings.

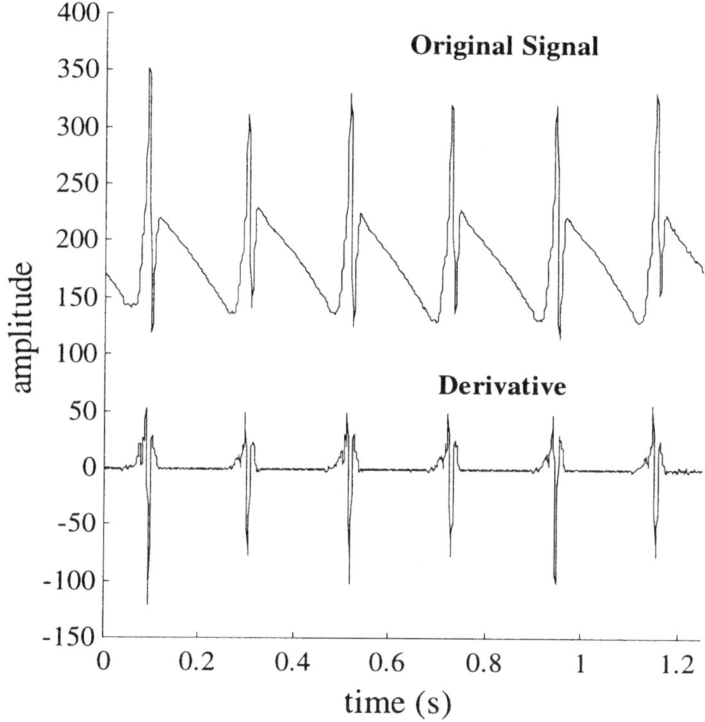

Figure 14 Derivative waveform of an example of monomorphic ventricular tachycardia is shown beneath the original waveform. Gradient patterns (−and + slopes) reflect differences in patterns of activation and may be used to differentiate arrhythmias.

The estimators given is Eqs. (9) and (10) assume a straight-line fit between the two data points used in the estimate. The assumption of a straight-line fit between two consecutive data points is satisfactory if the data are sampled frequently. However, there are portions of the electrogram (for example, slow ventricular depolarization) where a straight-line fit may be inappropriate, due to the curvature of the signal. Such curvature in the signal requires a slope estimate based on the fit of a second-order or higher polynomial to the data points [Eq. (11)]. For example, cubic splines may be used to model the data between four or more consecutive samples that appear to behave like a cubic function. Once the data points are fit to a polynomial of appropriate order, the derivative of the polynomial is determined, and the value of that derivative in the vicinity of the desired data point may be estimated. More detail regarding the numerical estimation of derivatives (slopes) may be found in [44].

$$y(n) = \frac{x(n-2) - 8x(n-1) + 8x(n+1) - x(n+2)}{12h} \tag{11}$$

In electogram analysis, one is often interested in quantifying the maximum and minimum amplitudes of atrial and ventricular activations. A robust method for automatically detecting minimal and maximal amplitudes of a signal is to perform a second derivative analysis directly from the data using the following formula:

$$y(n) = \frac{x(n+1) - 2x(n) + x(n-1)}{h^2} \tag{12}$$

Some algorithms differentiate arrhythmias using the sequence of slope changes in a signal. Theoretically, arrhythmias such as ventricular tachycardia should have ventricular activation patterns that differ from sinus rhythm or supraventricular tachycarida. Changes in the path of activation typically alter the direction and velocity with which a wavefront passes an electrode. These changes in direction and velocity alter the sequence of slopes, particularly for a bipolar electrogram. For fibrillation, the absence of orderly conduction and the continually changing patterns of activation should be reflected as a continually changing, unpredictable sequence of positive and negative turning points in the electrogram signal. In addition to defining a sequence of slopes for arrhythmia classification, differentiation schemes may impose a magnitude threshold on the slope criteria, requiring that the negative or positive slopes exceed some preset magnitude for a specified time period. These magnitude and duration threshold criteria reduce susceptibility to high-frequency, low-amplitude noise (e.g., electromagnetic interference or electromyographic artifact) and sudden artifact surges in the electrogram signal due to instrumentation error.

Davies, Wainwright, and colleagues use the sequence of positive and negative turning points in an electrogram signal to differentiate sinus rhythm from both atrial arrhythmias and ventricular arrhythmias [13,45–49]. These algorithms are sometimes referred to as gradient pattern detection (GDP) algorithms. Gradient pattern detection is designed for atrial electrograms to discriminate sinus rhythm from retrograde atrial depolarization, atrioventricular reentrant tachycardia, AV nodal reentrant tachycardia, and ventricular tachycardia [13,39]. For ventricular signals, the GPD has been used to discriminate ventricular tachycardia and AV reentrant tachycardias from sinus rhythm. These GPD algorithms operate on the estimated derivative of the measured electrogram signal. For bipolar recordings (0.5 to 1-cm bipoles; DC,250 Hz), a first derivative is estimated from the original digital electrogram signal. A gradient pattern representing the sequence of turning points in the derivative signal is then determined and compared to the reference pattern obtained during sinus rhythm. A turning point is defined as a change of direction in the signal, or a change in sign of the derivative. The classification criterion consists of the initial deflection (absolute value) of the derived signal being greater than some amplitude thershold and maintenance of this derivative amplitude for a specified window of time. This time window ensures that noise will not be misinterpreted as signal. An arrhythmia is detected when the polarity or the amplitudes of the derivatives differ from the sinus rhythm reference.

GPD is reported resistant to change in respiration, posture, rate, electrogram amplitude, antiarrhythmic drugs, and ST segment alterations. The GPD allows differentiation of multiple arrhythmias with similar rates, it is simple to implement with hardware, and it may be executed in real time. Moreover, clinical variables to do not appear to affect GPD performance adversely. However, like other morphology-based algorithms, the stability of electrogram morphology is crucial to GPD performance.

2. Detection of Atrial and Ventricular Activations

First and second derivatives are used in surface ECG applications for automatically detecting QRS complexes [28]. For example, to detect R wave complexes in the surface ECG automatically, Pan and Tompkins (50) use first and second derivatives of the ECG to obtain a obtain the sequence of rectangular pulses from which to detect QRS complexes. In this event detection scheme, the rising edge of the rectangular pulses coincides with the peak of the R wave in the original ECG. To detect the rising edge, one simply applies a simple amplitude threshold and blanking period algorithm to the rectangular pulses. Because of patient-to-patient variability in QRS morphology and the frequent occurrence of large-amplitude T waves,

R-wave detection is often more consistent using amplitude threshold schemes on the rectangular pulses than on the original ECG. There is less sensitivity to low-frequency, high-amplitude T waves and to variations in peak R-wave amplitudes that result from sampling the ECG signal.

D. Area Measures

Another simple method for comparing electrogram morphology between various arrhythmias involves measuring the total *area* in atrial or ventricular complexes. Often, the detection scheme will quantify the area of an atrial or ventricular complex for a rectified portion (absolute value) of the electrogram signal. Such area schemes require reliable detection of atrial and ventricular activations. Santel et al. [51] demonstrated that the ventricular activation complex during ventricular tachcardia had a significantly greater area than the ventricular complexes seen during sinus rhythm (ratio of areas ranged from 1.8 to 5.2) in human subjects. However, complication arose when T waves increased in area during sinus rhythm (a problem sometimes resolved by band-pass filtering) and when low-amplitude or narrow ventricular complexes occurred during ventricular tachycardia.

E. Templates and Signal Averaging

Morphology schemes may be made more robust in the face of subtle nonstationarities in signal amplitude, slope, morphology, and noise by using signal-averaging techniques. A number if arrhythmia differentiation schemes involve *signal averaging* and, subsequently, template analysis. A template may be used to represent a frequently occurring event, such as an atrial or ventricular complex. *Templates* are typically created by averaging repeated measurements of a specific event. Alternatively, templates may be artificially designed to resemble a desired event. Once created, templates may be used to detect future events in the electrogram signal and to differentiate between different activation patterns or sequences. Templates may also be used to remove unwanted information from a signal in order to unveil the desired signal or enhance the signal-to-noise ratio.

As stated previously, templates for a specific event in the electrogram are typically determined using signal-averaging methods. Signal-averaging takes advantage of the fact that white noise and events uncorrelated with the signal of interest, averaged over a sufficiently large number of signal measurements, average to zero. Thus, the premise behind obtaining a representation (template) of a desired signal or event by averaging repeated measurements of the desired event is that each measured event is comprised of an unchanging desired signal and additive random (unpredictably changing) white noise. In addition, we assume that the noise is uncorrelated in time with the desired event. Consequently, by averaging multiple measured events, the noise cancels out and only the desired event signal remains. To illustrate this notion, we synthesized a series of samples functions, $S_i(t)$ shown in Fig. 15. Each synthesized signal consists of the desired signal (shown in the upper right) and additive white noise that is uncorrelated with the desired signal. Moreover, the additive noise in $S_i(t)$ is independent of the additive noise in $S_j(t)$. If we now average all 100 synthesized signals on a point-by-point basis and then divide the sum by 100, we end up with the resultant signal-averaged waveform in the lower right corner. Note how the

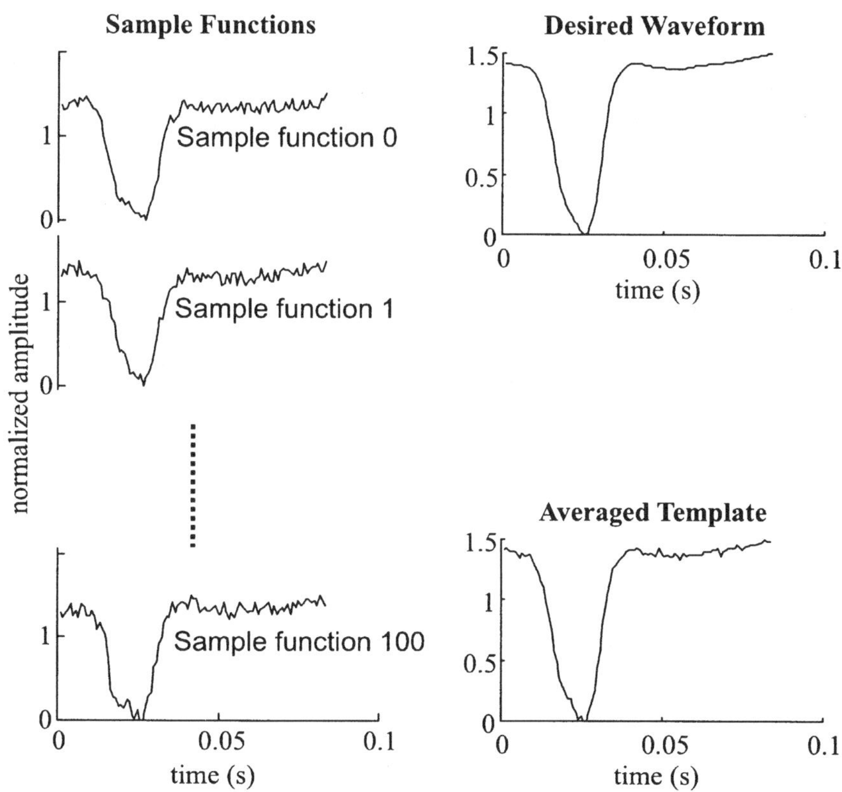

Figure 15 A collection of signals (left) each containing a desired waveform, (upper right), and additive white noise. (Note that the noise in sample function i is independent of the noise in sample function j). When the sequences are added together and averaged on a point-by-point basis, the resultant average template signal reveals the desired waveform while suppressing the additive noise.

desired signal remains in the averaged waveform, while the noise averages to zero. Such a filtering method is often useful when the frequency content of both the desired signal and the noise overlap.

There is always some physiological variation (noise) from desired event to desired event (or from beat to beat in the case of the heart), making precise slope, amplitude, and width measurements difficult to use for arrhythmia differentiation. Furthermore, changes in clinical variables such as pharmacology, patient activity, and metabolic state may all influence these morphological parameters. Thus, finding schemes that are fairly robust in the face of these clinical changes is critical. Morphology schemes based on templates may allow for more robust comparisons because the template is allowed to adapt over time to compensate for slowly changing signal characteristics as well as patient-to-patient variability. Once a template has been established, we may use the template to detect future events through cross-correlation [52, 53] or area-of-difference (AOD) methods [52,53] (see below) or to subtract out undesirable signal elements in order to enhance smaller amplitude features. Templates have been used in surface ECG applications to detect AV dissociation [54] and atrial fibrillation [55], where atrial activation is often masked by ventricular activity.

Template formation begins by first automatically detecting the events for which one desires a template and then combining these detected events in some fashion to produce a template. To automatically detect events, we may use something as simple as an amplitude threshold and blanking period algorithm or a more sophisticated first-and second-derivative [50,55] approach presented earlier. Once events have been detected, fiducial points (points of reference for each event) must be defined. The events may then be aligned with respect to the fiducial points, added together on a point-by-point basis, and then divided by the total number of events used in the summation. The resultant signal represents the average template of the detected events.

In clinical applications of template-based arrhythmia detection schemes, there is often a learning period when the patient is first connected to the device which allows the device a finite period of time to detect normal or resting-state events and, consequently, form a template of normal activation. Once the template for normal activation is established, the template may be used to automatically detect and classify future activation complexes.

Once a template is created, we may use the template to detect future events in an electrogram signal. An event is detected when a portion of the measured signal and the template are matched in morphology. To establish the goodness-of-match between the measured signal and the template, a number of pattern-matching schemes may be used, such as *correlation waveform analysis* (using a correlation coefficient) or *bin area methods* (using area of difference) [52,53,56]. The simplest template-based event detection schemes utilize a synthesized waveform to represent the desired event. For example, a triangle waveform may be used as a rough representation, or template, of the ventricular complexes. We then perform pattern matching (such as a moving cross-correlation function, see following paragraphs) between the synthesized template waveform and the electrogram to detect the presence of ventricular complexes. Note that while such a template is simple to define and implement, it is not highly specific and may match equally well with a number of ventricular complexes of variable morphology. Furthermore, if we were to use this template to subtract ventricular activation from the original electrogram, we would find a significant amount of ventricular artifact remaining after the subtraction of the template. Thus, the triangular waveform template may serve well for simple event detection, however, its lack of specificity for more complex morphology leads to false detection, especially in the presence of multiple morphologies.

Whether the template is created from a synthesized waveform or an average of detected events in a learning period, future events may be detected by using a pattern-matching criterion, such as a *correlation coefficient estimate*, between the template and a window of signal equal in length to the template. The correlation coefficient may take on a value between -1 and 1. A crosscorrelation coefficient of 1(or -1) indicates perfect positive (or negative) correlation between the two signals being compared. Conversely, a value of 0 indicates the absence of correlation between the two signals. Note that the cross-correlation coefficient is insensitive to changes in gain in either of two signals.

The correlation coefficient, ρ, between two finite-duration signals, $x(n)$ and $y(n)$, over the range of samples, $n = [a,\ b]$, is defined by

$$\rho = \frac{\sum_{i=a}^{b}(x_i - \bar{x})(y_i - \bar{y})}{\sqrt{\sum_{i=a}^{b}(x_i - \bar{x})^2 \sum_{i=a}^{b}(y_i - \bar{y})^2}} \tag{13}$$

where $\bar{x}$ and $\bar{y}$ are the mean values of sequences, $x(n)$ and $y(n)$, respectively, on the interval $[a, b]$.

Figure 16 illustrates how a moving correlation coefficient may be used to quantify the similarity between template and a portion of the electrogram signal in order to detect ventricular activations. When a ventricular activation is encountered, the correlation coefficient between the template and the electrogram is closest to 1, indicating a good match between the template and the electrogram event. These template schemes are relatively insensitive to actual electrogram amplitude and baseline fluctuations.

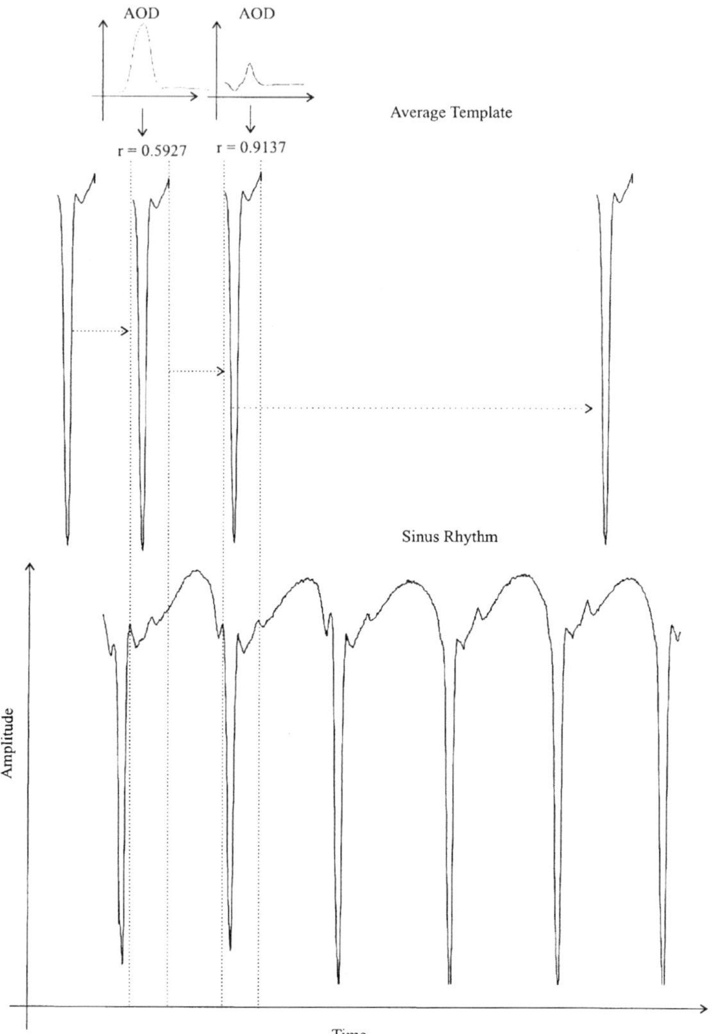

Figure 16 To detect events in an electrogram recording, a template may be moved in a sliding fashion across the electrogram signal. As the template moves to a new window of data, a correlation coefficient, r, is computed between the template and the corresponding window of data. Events are detected when the correlation coefficient exceeds some threshold value. Events may also be detected by finding significant differences in area (AOD) between the template and the corresponding window of data being searched for an event.

Rather than use a correlation coefficient, an *area-of-difference* (AOD) criterion [57] such as the bin area method [52] between the template and the electrogram signal may be use to detect desired events. As for the correlation coefficient method, the template is moved, with respect to time, across the electrogram signal. At each time step, the absolute area between the template and a window of electrogram signal is determined (Fig. 16) and compared to a threshold criterion. The template is then shifted in time along the electrogram signal in search of new complexes. Theoretically, when an event is encountered, the AOD should be close to zero. However, unlike the correlation coefficient, the AOD is sensitive to changes in electrogram signal gain and baseline offset.

For both the correlation coefficient and AOD methods, the detection of an event depends on the setting of a threshold criterion. Thresholds for event detection are typically established from a population of training electrogram data, and those thresholds are then used on new electrogram data to find desired events.

Cross-correlation or AOD require a larger number of computations than simpler rate and slope methods. To reduce power consumption and speed up detection in implantable devices, one may reduce the number of computations by using a blanking scheme. Rather than use a continually sliding cross-correlation or AOD search for desired events, one may reduce the computations by first detecting events using a simple amplitude threshold and blanking period scheme. Events detected using this simple scheme are then aligned with the template of the desired events and, subsequently, a correlation coefficient or AOD measure is used to indicate the similarity between the template and detected event. For patients who frequently exhibit a multitude of atrial and ventricular morphologies, a commercial device often stores multiple templates. One template may represent a normally conducted beat and other templates may represent frequently occurring irregular morphologies such as premature ventricular complexes. With multiple templates available, a finer classification of rhythms is possible by finding the best match between the observed electrogram and the various templates stored in the device. Subsequently, arrhythmia classification is performed by examining sequences of normal and abnormal complexes. For example, some arrhythmias, such as ventricular tachycardia, are defined in terms of the frequency of occurrence and the timing of premature ventricular complexes. Note that as events are detected in an electrogram signal, the template(s) may be updated with each subsequent detected event if the event is determined to be of the same class as the template. The new event is simply added to the sum of other events, thereby allowing the template to be updated to compensate for subtle changes in electrophysiology, lead placement, patient activity, and drug therapy.

While the template algorithms are easily performed in real time, they require a learning period to create templates of atrial and ventricular activation during sinus rhythm.

Multiple template matching is used to separate sinus rhythm and sinus tachycardia from ventricular tachycardia [35,58–62]. Langberg and Griffin [59], using the AOD between a sinus rhythm template and a detected event, found that the smallest area of difference for ventricular tachycardia was 5 times greater than the area-of-difference found for sinus rhythm. Tomaselli et al. [60] reported areas of difference for ventricular tachycardia and bundle branch block complexes to be 100–170% greater than that of sinus beats. In some instances, however, variations in the AOD due to drug administration or autonomic changes lead to overlap in AOD between sinus rhythm and ventricular tachycardia.

Template methods are also used to unveil small-amplitude events concealed by larger signals. For example, atrial activation may be occasionally hidden within ventricular activation [54,55]. Slocum et al. [54] showed that small-amplitude P waves concealed by QRS complexes during AV dissociated rhythms could be unmasked using a signal-averaged template scheme, where an average QRS complex was subtracted from the dominant ventricular activity in the surface ECG. If the atrial and ventricular activations are not correlated in time, subtraction of the template signal from each occurrence of ventricular activation will unveil underlying atrial activity. As another example, templates may be used to isolate far-field ventricular events in intra-atrial electrograms by removing large-amplitude, dominant atrial activity in the unipolar electrograms (Fig. 17). Once the atrial activity is removed, we may use a simple amplitude threshold scheme or derivative scheme to detect the remaining ventricular activity.

Template schemes may be used to detect events, unveil hidden events, improve signal-to-noise ratio, and identify arrhythmias. Template schemes rely on stationarity of the signal, regularity of desired events, and constancy in electrogram morphology. Disadvantages of template-based schemes include the need for a learning period in order to create a template, and the computation and storage requirements for point-to-point subtraction. Furthermore, estimating an appropriate time window for the template and the fiducial points with which to align each detected event with the template may be cumbersome and continually changing. Another disadvantage of template-based methods arises from subtle nonstationarities in the original electrogram signal. These nonstationarities produce a template that is a low-pass-filtered (smoothed) version of the events being detected. If the smooth template is used to remove activation complexes from the electrogram, pieces of activation will be left in the remainder electrogram.

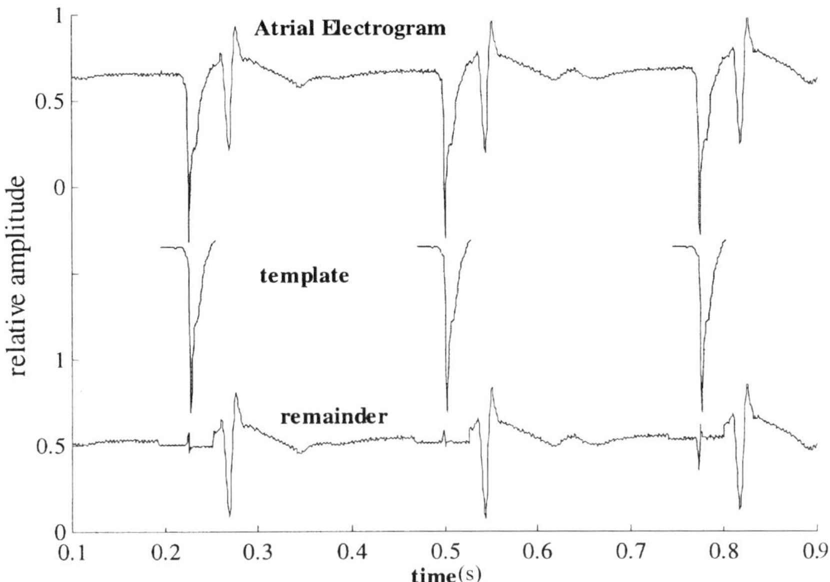

Figure 17   An example of sinus rhythm recorded with a unipolar electrode configuration (top tracing) in the high right atrium. Note the large atrial activity (shaded region) that dominates the ventricular activity. A template (insert) created by signal averaging may be used to subtract atrial activity from an electrogram (top trace) to isolate ventricular activity (bottom trace).

Template-based arrhythmia classification, like all morphology-based discrimination schemes, are challenged by the presence of ventricular tachycardias of multiple morphology in the same patient [63]. Moreover, morphology schemes are highly dependent on lead configuration and lead position in the atrium or ventricle [45,64,65]. Most morphology measures require explicit detection of specific events in the electrogram (atrial and/or ventricular activation), which can be challenging, especially in the presence of fibrillation, where discrete complexes are absent.

IV. CONTEXTUAL METHODS

To facilitate arrhythmia classification, most morphology-based arrhythmia classification schemes are used in conjunction with contextual analysis. *Contextual analysis* differentiates arrhythmias by detecting individual atrial and ventricular events, measuring the amplitude, width, and polarities of these events, and than evaluating the position or timing of each event with respect to neighboring ventricular and atrial events. For example, in the case of a normally conducted sinus beat, the sequence of activations should be A-V-A-V-A-V-, etc. The typical duration between A and V is on the order of 10–20 msec. In the case of atrial flutter, we often see the following series of activations: AA-V-AA-V-AA-V, etc. In other words, there are two atrial activations (flutter waves) for each ventricular activation. To implement contextual analysis, specific amplitude, duration, and polarity criteria must be established for each type of event being examined (atrial depolarization and repolarization, ventricular depolarization and repolarization, etc.) Assuming that events are accurately identified, the time intervals and order of activations becomes important to the classification of arrhythmias.

Most arrhythmia discrimination schemes suffer in their ability to differentiate supraventricular tachycardias (SVT), AV reentrant tachycardias (AV), and ventricular tachycardias (VT) because the electrogram timing and morphologies measured in a single chamber do not differ sufficiently between these different arrhythmias. Where these arrhythmias differ is in the occurrence of ventricular events with respect to atrial events. To reliably discriminate between SVT, AV reentrant tachycardias, and VT, simultaneous recordings from multiple chambers are required [66,67]. Traditionally, implantable antiarrhythmic devices offer only a single lead from a single chamber for arrhythmia discrimination. Recently, however, antiarrhythmic devices have introduced multiple leads in both atrial and ventricular chambers for contextual arrhythmia discrimination schemes. These arrhythmia classification schemes rely on accurate event detection simultaneously from both atrial and ventricular electrograms. Once the atrial and ventricular events are detected, the arrhythmia discrimination algorithms measure AA, VV, AV, and VA intervals. The sequence of these intervals within specified time intervals, as well as the length of these intervals, are used in combination to classify arrhythmias [68]. For example, if the tachycardic events are primarily atrial, then the rate of the tachycardia is computed and the diagnosis becomes atrial fibrillation (rate>330/min), atrial flutter (240<rate< 330/min), or atrial tachycardia with second-degree atrioventricular block [15]. To differentiate sinus tachycardia from other tachycardias with 1:1 AV relationships, speed of tachycardia onset is used. Furthermore, the device may deliver a premature atrial stimulus and note the timing of the subsequent ventricular activations to distinguish sinus rhythm from pathological tachycardias with 1:1 AV relationships [69]. Accuracy rates of 99% have been reported when differentiating physiological tachycardias from pathological tachycardias using interval measurement-based [15,70] classification algorithms. Theoretically, such dual-chamber contextual analysis could be performed in a single chamber with a single unipolar electrogram having sufficient far-field activity to provide sufficient registration of both atrial and ventricular activation. However, during accelerated rates, the activation in one chamber (e.g., ventricular activation) will likely conceal the events (e.g., atrial activation) in the neighboring chamber.

In designing contextual schemes, the primary element of the classification scheme is determining the length and sequence of AA, AV, VA, and VV intervals specific to a particular arrhythmia. These intervals are determined by the underlying electrophysiology (refractory periods and conduction velocities). There is variability in these intervals from patient to patient and with metabolic state and pharmacology. Thus, when the devices are implanted, various arrhythmias are induced in a controlled clinical setting to allow the parameters to be tailored to the individual patient. Furthermore, the patient has regular follow-ups for adjusting the parameters as patient activity, health, and pharmacological therapy are altered.

V.　CORRELATION METHODS

A.　Autocorrelation

Fibrillation is characterized by electrograms that continually vary in morphology, timing, and polarity. There is a lack of regular, repetitive, organized activity that is typically seen during nonfibrillatory rhythms. The random, chaotic nature of the electrograms stems from the disorganized patterns of activation that occur during fibrillation, where multiple wavelets or rotors are thought to constitute the underlying electrophysiological mechanism [37]. The lack of predictability or periodicity in the electrograms can be used to detect fibrillation. A time-domain method for quantifying the predictability or periodicity in a signal, which is relatively independent of specific signal morphology and amplitude, is the autocorrelation function. The *autocorrelation function*, R_{xx}, for a continuous, wide-sense stationary signal, $x(t)$, is defined by [71]

$$R_{xx}(\tau) = E[x(t)x(t + \tau)] \tag{14}$$

where $E[\ ]$ is the expectation operation (statistical average) and τ is the time lag. In words, the autocorrelation function is a measure of the match between the signal and a delayed (or time-shifted, τ) version of itself. If a signal has a periodicity of T, there is a perfect match at $\tau = nT$ (n being an integer).

The autocorrelation function has the following properties:

The autocorrelation function has a maximum at zero lag.

The autocorrelation function at zero lag is equal to the mean square value of $x(t)$.

If $x(t)$ has a periodic component, R_{xx} will also have a periodic component of the same frequency.

If $x(t)$ has a mean value (A) not equal to zero, R_{xx} will have a constant component equal to A^2.

Thus, the autocorrelation function may be used to unveil periodicities or predictability in the original signal. For digital data, $x(n)$, the autocorrelation function may be estimated using a biased estimator,

$$\hat{R}_{xx}(m) = \frac{1}{N} \sum_{n=1}^{N-m} x(n)x(n+m) \tag{15}$$

or an unbiased estimator,

$$\hat{R}_{xx}(m) = \frac{1}{N-m} \sum_{n=1}^{N-m} x(n)x(n+m) \tag{16}$$

where N is the length of the data and m is the lag.

For short-duration time series, such as those encountered during fibrillation, the biased estimator provides a more stable estimate of the autocorrelation function. Because we have finite-duration data, as a rule of thumb, we do not estimate R_{xx} for lags that exceed one-third the total number of data points, N, in the electrogram being evaluated. For lags that extend beyond $N/3$ edge effects due to the finite duration of the original electrogram signal significantly reduce the autocorrelation estimate. Thus, even for a periodic-like signal, the magnitude of the autocorrelation will decline with increased lag, due to the finite duration of the sampled data.

For example, the autocorrelation function of a finite-duration, discrete sinusoid is also a discrete sinusoid of the same period (Fig. 18) that tapers off with increasing lag. In contrast, for a white-noise signal, which has no periodicity, the autocorrelation function is a unit impulse with amplitude of 1 at τ equal to 0 (Fig. 18) and 0 at all other lags.

The autocorrelation has been used to differentiate ventricular tachycardia from ventricular fibrillation [72] and sinus rhythm from ventricular fibrillation [73] using intracardiac electrogram recordings. Figure 19 illustrates examples of autocorrelation functions for examples of monomorphic ventricular tachycardia (MVT) and ventricular fibrillation (VF). Note that R_{xx} for monomorphic ventricular tachycardia exhibits regular, periodic behavior. Conversely, for ventricular fibrillation, R_{xx} is much less regular,

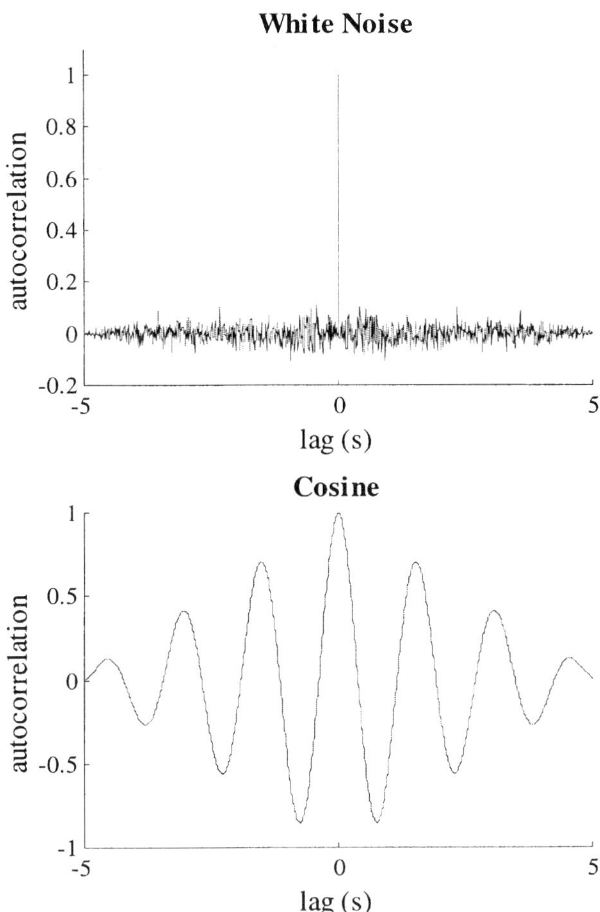

Figure 18 The autocorrelation function for finite-duration sampled signals of white noise (top) and a single cosine (bottom). For white noise, the autocorrelation is an impulse at zero lag. For the cosine function, the autocorrelation functioning is a decaying sinusoid. For finite-duration sampled electrograms, the amplitude of the autocorrelation function declines as τ increases.

with a large peak at zero lag (equal to the signal variance) and a more rapid decrease in autocorrelation value with increasing lag. Moreover, there is a mix of periodic-like components. Chen et al. [72] fitted a line to the roll-off in R_{xx} as a function of τ and used the slope of this line to differentiate ventricular fibrillation from monomorphic ventricular tachycardia.

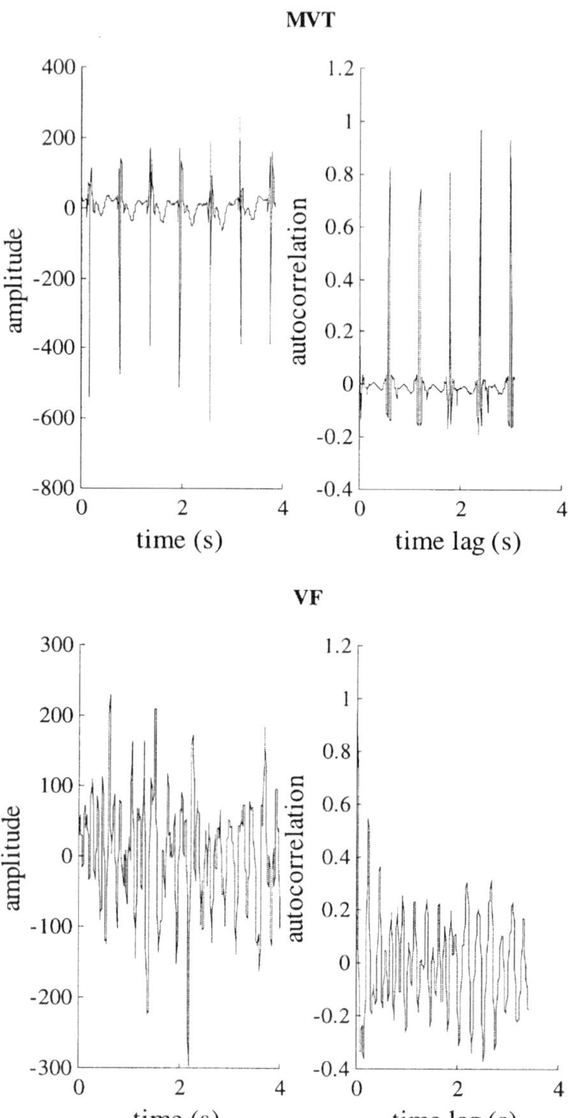

Figure 19 Autocorrelation function for an example of monomorphic ventricular tachycardia (MVT) (top) and ventricular fibrillation (VF) (bottom). For monomorphic ventricular tachycardia, the periodic-like behavior of the autocorrelation function reflects periodicities in the original electrogram. In contrast, the autocorrelation function for ventricular fibrillation shows less regularity across lags, yet there is some periodicity for the autocorrelation function during fibrillation.

Drawbacks of autocorrelation schemes include large numbers of computations and the need for lengthy data segments to obtain statistically sound estimates. Moreover, the autocorrelation estimate is basically another method for estimating regularity of cycle length without the need for explicit discrete event detection. Thus, it does not add information beyond that obtained using simpler rate and rate stability measures and thus is equally susceptible to confounding physiological tachycardias and pathological tachycardias.

B. Cross-Correlation

While autocorrelation may be used to quantify the predictability or periodicity at a single site in the heart, cross-correlation may be used to quantify the linear relationship (or correlation) between two or more sites in the heart. The *cross-correlation function* may be used to quantify the correlation or linear relation between two signals as a function of lag or time delay between the two signals. The cross-correlation function for continuous, jointly wide-sense stationary [71] signals, $x(t)$ and $y(t)$, is defined by

$$R_{xy}(\tau) = E[x(t)y(t + \tau)] \tag{17}$$

Note that the cross-correlation is similar to the autocorrelation except that the signal is no longer correlated with itself, but with another signal. The cross-correlation function has the following properties.

$R_{xy}(-\tau)$ equals $R_{yx}(\tau)$.
The magnitude of the cross-correlation at any lag is equal to or less than one-half the sum of the variances of $x(t)$ and $y(t)$.

For finite-duration, digital data, $x(n)$ and $y(n)$, the cross-correlation function may be estimated from the following equation:

$$\hat{R}_{xy}(m) = \frac{1}{N - m} \sum_{n=1}^{N-m} x(n)y(n + m) \tag{18}$$

The cross-correlation function is a measure of the match between two signals at different time delays (m). If the two signals are similar except for a time delay Δt, the cross-correlation will be maximal at $\tau = \Delta t$. Thus, by finding the maximum of the cross-correlation function, one may estimate the time delay between two signals. For example, let discrete signals, $x(n)$ and $y(n)$, be defined by

$$x(n) = \sin(\omega_1 n) + 2\sin(w_2 n) + 2\sin(w_3 n) + \mathrm{noise}_1(n)$$
$$y(n) = \sin[\omega_1(n+m)] + 2\sin[w_2(n+m)] + 2\sin[w_3(n+m)] + \mathrm{noise}_2(n)$$

If we estimate R_{xy} between the original signal, $x(n)$, and the delayed signal, $y(n)$, we obtain the function illustrated in Fig. 20. Note that the peak of the cross-correlation function occurs at a time lag of 2.5 sec, which corresponds to the time delay between $x(n)$ and $y(n)$. The more uncorrelated noise is added to the signals, the larger is the variability in the estimates of the time delay between the two signals.

The cross-correlation function, performed between pairs of electrogram signals, may be useful for differentiating fibrillatory rhythms from nonfibrillatory rhythms. In the case of fibrillation, multiple circulating wavelets make the activation at neighboring sites relatively uncorrelated. In contrast, the orderly conduction of a single activation front during nonfibrillatory rhythms makes the activity at two distant sites highly coordinated or correlated. Figure 21 illustrates the cross-correlation functions for examples of atrial flutter (AFLUT) and atrial fibrillation (AF), each recorded in the atrium. In each example, one electrogram was obtained in the high right atrium and the second electrogram was obtained in the midright atrium. Both electrogram recordings were bipolar, with an interelectrode spacing of 1 cm. Compared to autocorrelation, cross-correlation may better differentiate fibrillation from nonfibrillatory rhythms because it captures spatial organization. There is some evidence [27] that during fibrillation, an electrogram from one site may show very regular, organized activity while a neighboring electrogram shows irregular, disorganized activity. In such a case, the autocorrelation function for the organized electrogram would appear more like that of a nonfibrillatory rhythm, while the cross-correlation would resemble that typically seen during fibrillation.

VI. MULTIPLE ELECTROGRAM METHODS

Arrhythmia discrimination algorithms using a single electrogram may be problematic. At times, during fibrillation, a portion of the atrium or ventrical where the electrode is placed shows organized, regular depolarization, while neighboring areas show fibrillatory-type activity [27]. However, due to the hardware and surgical implant constraints of commercial devices, the majority of arrhythmia detection schemes use information from a single intracardiac lead or electrogram. Nonetheless, a number of discrimination algorithms make use of the timing and phase information between two or more simultaneously recorded electrograms. Both the contextual methods

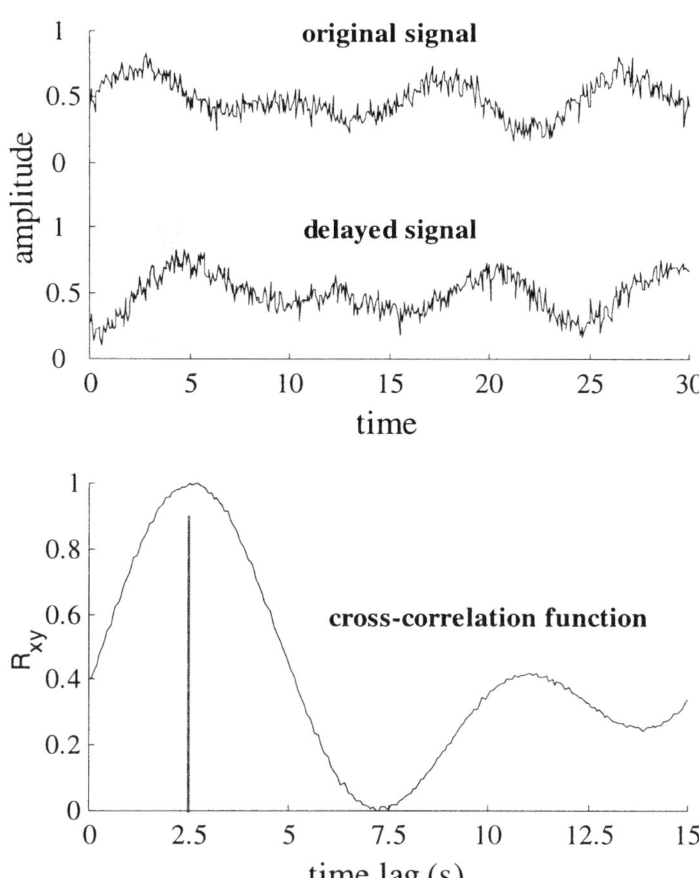

Figure 20 For two synthesized signals (original and delayed) that differ only in their time delay, the cross-correlation coefficient function shows a peak value at the time delay and high correlation at integer multiples of that time delay. In this instance, each signal consists of three sinusoids, 1/20, 1/8, and 1/13 Hz, with additive white noise. The delay difference was 2.5 sec.

and template methods described previously may take advantage of multiple electrogram recordings [56,66,67]. Likewise, the cross-correlation functions defined in the previous section quantify the timing and phase information between multiple electrograms, but require considerable computation, even with the use of the fast Fourier transform. One computationally simple method for evaluating the degree of correlation between activation at multiple sites in the heart is simply to use event detection and relative timing

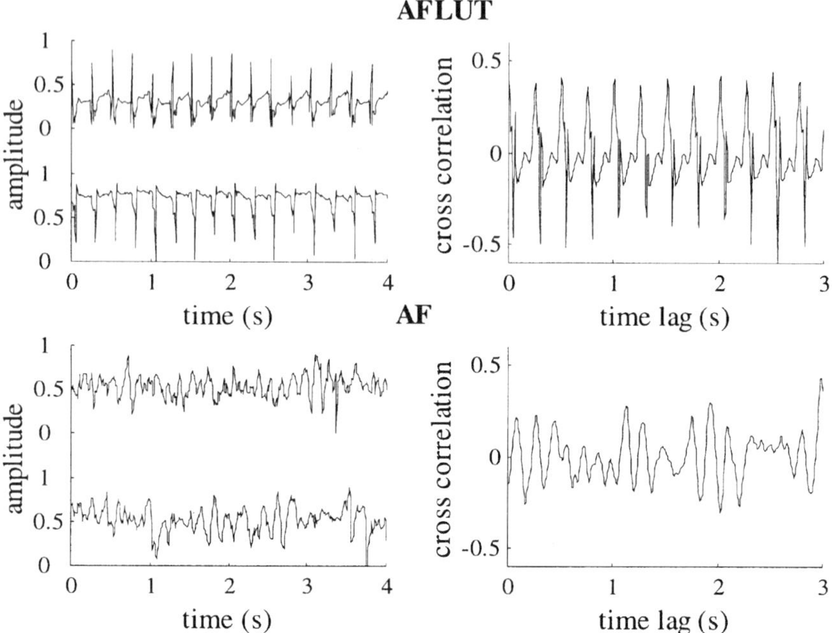

Figure 21 Cross-correlation function (right trace) estimated between the two electrogram recordings (left traces). The rhythms are atrial flutter (AFLUT) (top panel) and atrial fibrillation (AF) (bottom panel), and the electrograms were recorded from two bipoles located in the high atrium separated by a distance of 2 cm.

information from multiple electrograms. These multiple electrogram algorithms explicitly detect atrial and/or ventricular activation from three or four sites simultaneously and use a sequence of decision rules for determining whether a sequence of activations from all sites belongs to one rhythm class or another. Mercando and Furman [39,74–76] proposed an algorithm which relies on electrograms recorded simultaneously from two sites in the ventricle to differentiate sinus rhythm from ventricular and supraventricular ectopic activity. During sinus rhythm, there is typically synchrony and a repeatable pattern of activation from one site to the next. Conversely, during ventricular fibrillation, the activity varies from site to site, with electrograms at each site characterized by rapid, irregular, multimorphic activity. The algorithm measures the sequence and timing of intrinsic deflection in each lead [39,74]. In most cases, for a particular individual, the timing between two electrograms differs for premature ventricular complexes or ventricular tachycardia compared to sinus rhythm, even when the morphology of the premature complexes is similar to that of sinus rhythm. The mean difference in interelectrogram timing between sinus complexes and ventricular ectopic complexes is on the order of 2–5 msec. However, such differences are highly dependent on lead placement and location. This multiple-electrogram timing algorithm requires explicit event detection and sequential rule-based testing of interelectrogram intervals. The explicit sequence of activation may vary with age, vagal tone, posture, and other clinical factors.

VII. ADAPTIVE FILTERS

Whether using timing information from multiple electrograms or event-to-event correlation in electrogram morphology, adaptive filters may be used to differentiate arrhythmias without the need for explicit event detection and detailed morphology measures. Adaptive filters have been proposed for a number of arrhythmia detection schemes [77,78] to quantify the synchrony between multiple electrograms or the cycle-to-cycle similarity in atrial and ventricular complexes. The advantage of adaptive filters is their ability to adapt to gradual changes in electrogram morphology and provide a time-varying template of sinus rhythm. Using simple-to-compute, least-mean-square error adaptation, these adaptive signal processing schemes provide a simple statistic [the mean-square error, Eq. (19)] with which to detect an arrhythmia and adjust to patient specific electrogram morphology. More specifically, given the linear, adaptive filters illustrated in Fig. 22, the error, $e(n)$, between the actual filter output, $y(n)$, and the desired filtered output, $d(n)$, is minimized in the mean-squared sense:

$$\text{MSE} = [E[e^2(n)]] \qquad \text{where } e(n) = d(n) - y(n) \tag{19}$$

The adaptive filter algorithms make use of discrete-time, finite impulse-response filters with time-varying filter coefficients to predict future values of a signal based on present and past samples (Fig. 22). The filter coefficients, $w(n)$, are allowed to vary with time. These filter parameters are iteratively adjusted to minimize the mean-square error between the predicted value and true value of the electrogram at any point in time. In practical applications, the MSE is estimated using Eq. (20):

$$E(k) = \sum_{k=-M}^{M} |d(k) - y(k)|^2 \tag{20}$$

In Finelli [77], a series of LMS filters is used to predict the data points of the kth electrogram cycle using a linear combination of data points from

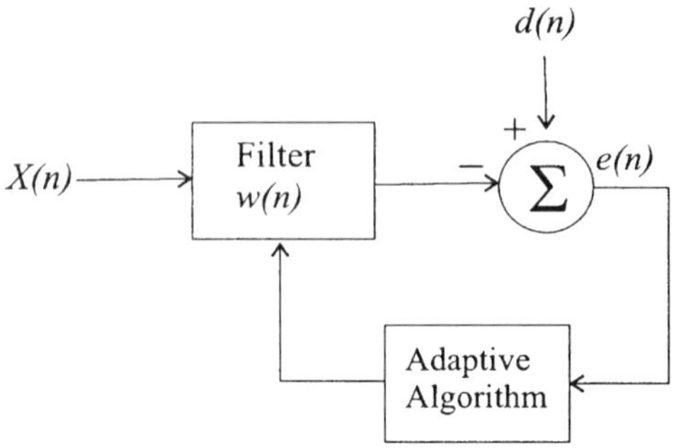

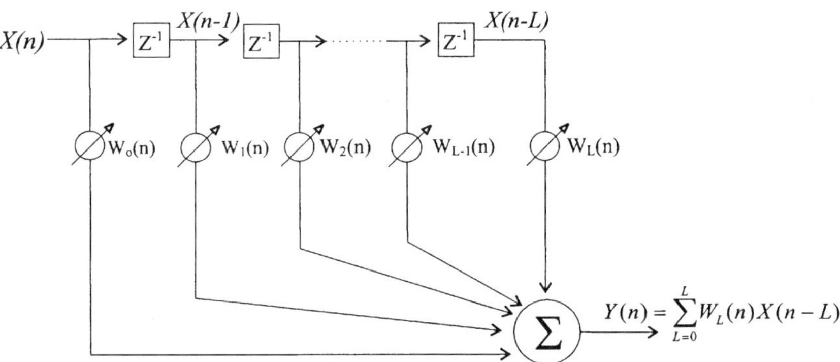

Figure 22 Illustrated is an adaptive least-mean-square filter. The weights, $W_i(n)$, of the filter are allowed to vary as a function of the error, $e(n)$, between the predicted filter output, $Y(n)$, and the desired filter output, $d(n)$.

the $(k-1)$st cycle. The error between the predicted electrogram and actual electrogram at the kth event tend to increase very abruptly with the onset of a cardiac arrhythmia. The LMS adaption uses a gradient descent method to update the filter coefficients, $w(n)$, such that the MSE will be minimum [79]. The LMS equations may be summarized by

$$w(k + 1) = w(k) + \mu \nabla_k \tag{21}$$

where

$$\nabla_k = -2e(k)x(k)$$

at each time step, k.

The user must specify two parameters in implementing an adaptive system, the logarithmic update, μ, and the length of the adaptive filter, L (number of filter coefficients). Arryhythmias may be differentiated by placing a threshold on the MSE.

DuFault and Wilcox [80] use a two-lead algorithm in combination with a pair of adaptive filters to differentiate ventricular fibrillation from sinus rhythm and ventricular tachycardia. In this method, pairs of adaptive filters are used to approximate the transfer function between the electrogram signals recorded at two distant sites. Two sets of filters are used, one for each of sinus rhythm and ventricular tachycardia. During a learning period, the filters adapt to ("learn") the electrogram properties for sinus rhythm and for ventricular tachycardia. Once the filters have adapted, the filter coefficients remain fixed. Future signals from one site are then fed through each of the two adapted filters, and the output signal from each of the filters is compared to the actual recorded signal at the second site. The difference between the filtered signal and the actual measured signal is considered the error signal. It is this error signal that is used to classify the arrhythmia.

If the rhythm is sinus rhythm or ventricular tachycardia, the output of the appropriate filter should closely resemble the actual measured signal at the second site, resulting in an error signal close to zero. However, during ventricular fibrillation, when there is little predictability between two signals at different sites, the error between the filtered (predicted) signal and true measured signal should be much greater than for sinus rhythm or ventricular tachycardia. The drawback for this adaptive filter method is the need for a different filter for each morphology of ventricular tachycardia and every regular, nonfibrillatory rhythm. Also, an adaptation time is required for each nonfibrillatory rhythm the patient may potentially experience that requires differentiation from ventricular fibrillation.

VIII. SPECTRAL METHODS

Another means for conveying the morphological and periodic information contained in a signal is to transform the data to the frequency domain. As described in Section II.D, the transformation is achieved through the use of Fourier transform, which allows us to look at signal energy as a function of frequency.

Data acquired in the clinical or research environment is random in nature. It may be shown [71] that the Fourier transform may not exist for some random data. However, it may also be shown [71] that the *power density spectrum*, S_{xx}, does exist for all random signals, and is defined by

$$S_{xx}(f) = \lim_{T \to \infty} \frac{E[|X_T(f)|^2]}{2T} \tag{22}$$

where f is frequency and T is the finite time interval for which a sample function, $x_T(t)$, of the random process is defined. $X_T(f)$ is the Fourier transform of $x_T(t)$.

A. Practical Estimation of Power Spectrum

For real-world digital signals that are finite in duration, we may estimate the power spectrum using a variety of estimation schemes. The method of spectral estimation used most often for differentiating cardiac arrhythmias is the periodogram estimate. The *periodogram* estimate is based on the discrete Fourier transform (DFT) described in Section II.D.

$$X(k) = \sum_{n=0}^{N-1} x(n)e^{(-2\pi nk/N)} \qquad 0 \leq K \leq N - 1 \tag{23}$$

In other words, the discrete-time, finite-duration signal, $x(n)$, may be modeled as a sum of discrete, harmonically related complex exponential (sinusoids). If there are N data points in the signal, $x(n)$, the DFT will result in N discrete, harmonically related sinusoids. Furthermore, only the first $N/2$ discrete frequencies will be unique, with the $N/2$ to N frequencies being mirror images of the first $N/2$ spectral lines. The spectral lines will occur at the fundamental frequency $(1/N)$ and integer multiples (harmonics) of this fundamental frequency. The periodogram estimate is derived from Parseval's theorem [25] and may be written as

$$\hat{S}_{xx}(k) = \frac{1}{N}|X(k)|^2 \tag{24}$$

Simply stated, the power spectrum is estimated from the squared-magnitude of the DFT.

Figure 23 shows a power spectrum estimate using Eq. (24) for an example of atrial flutter. Note the striking presence of harmonics throughout the spectrum. Because the DFT imposes a periodic extension on the data, discontinuities (truncation artifacts) occur at the ends of the finite-duration sampled data (Fig. 24) and result in spectral leakage in the power

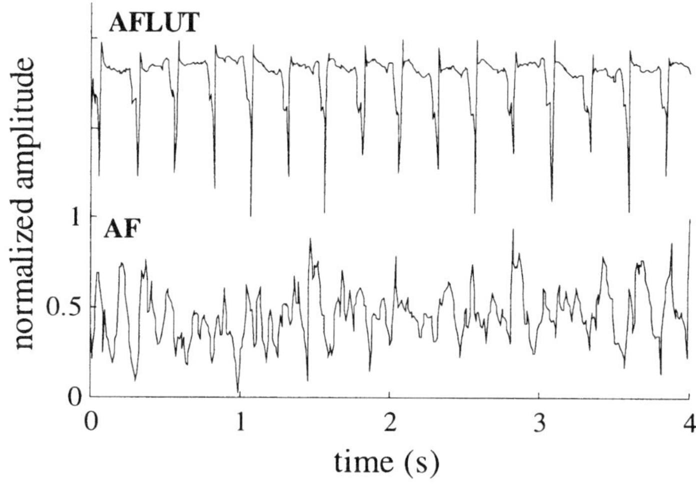

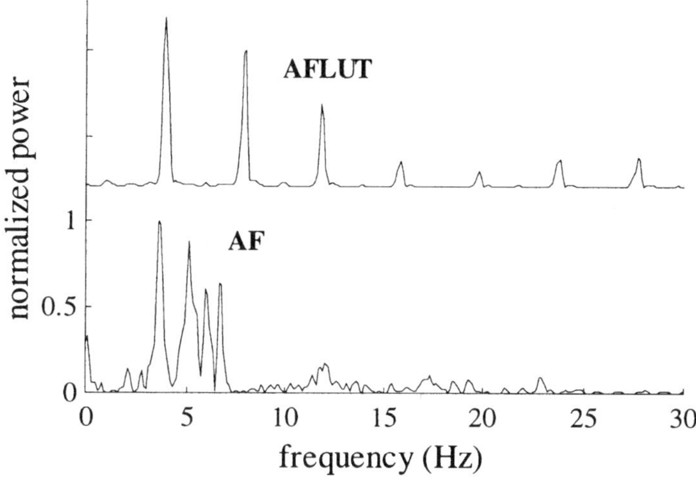

Figure 23 Illustrated are electrogram recordings for atrial flutter (AFLUT) and atrial fibrillation (AF) (top tracings) and their corresponding autopower spectra (bottom tracings). Note that atrial flutter is characterized by broad-band, harmonic spectra, whereas atrial fibrillation is characterized by more narrow-band, non-harmonic spectra.

Truncation Artifacts

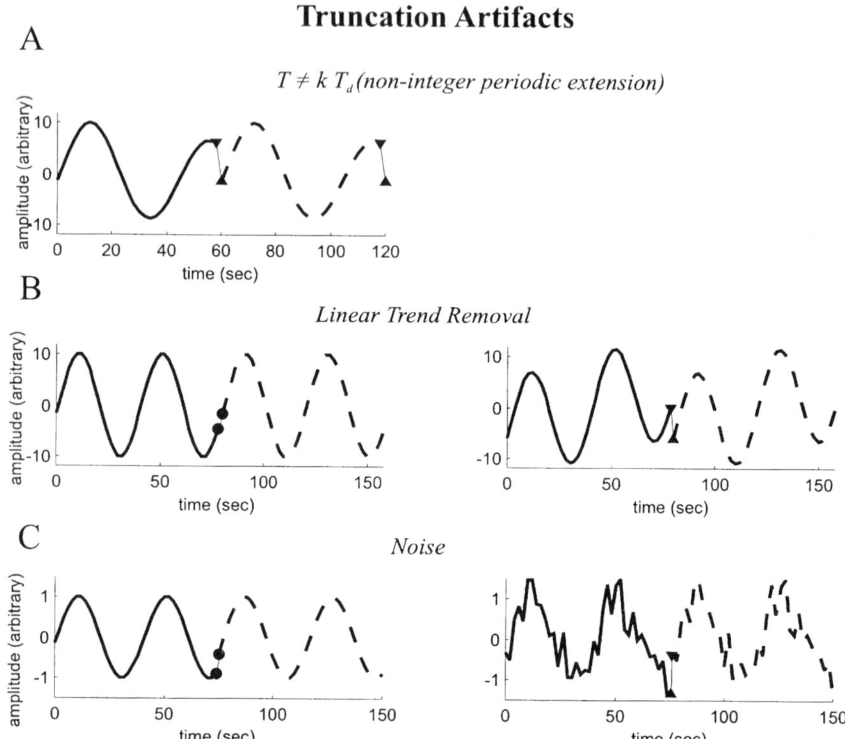

Figure 24 The discrete Fourier transform imposes a periodic extension of the data being analyzed for spectral content. Truncation artifacts caused by: (A) sampling for a duration T which is not an integer multiple of the signal's period, Td; (B) removing the linear trend; and (C) noise. Black circles and triangles indicate the absence and presence of truncation artifacts, respectively.

spectrum [25]. In effect, the periodic extension gives rise to high-frequency spectral components that occur at integer multiples of the fundamental frequency (which is determined by the length of the DFT). To reduce spectral leakage, we typically apply windows, such as a Hanning window (Fig. 25), to the discrete signal prior to DFT computation. The purpose of the window is to reduce side-lobe magnitudes (hence, reduced leakage from the main lobe to neighboring frequencies). The trade-off is a loss of spectral resolution due to the widened main lobe width of these alternative windows.

Leakage errors can be reduced by using different windows, and spectral resolution can be improved by using longer data segments. However, neither of these manipulations reduces the variance (random error) of the

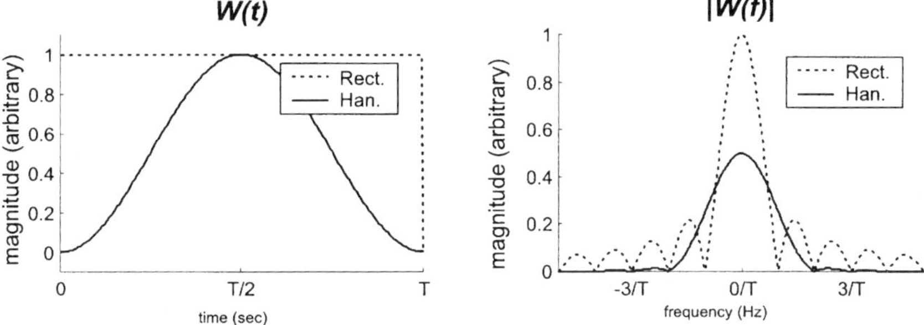

Figure 25 Rectangular (Rect.) and Hanning (Han.) windows, $W(t)$, and their corresponding Fourier transform magnitude spectra, $W(f)$.

periodogram estimate. We may decrease the variance of the periodogram and consequently improve the statistical accuracy of the power spectrum estimate by segmenting the sampled data into subsegments, estimating the spectrum for each of the subsegments and then averaging the power spectrum over all the subsegments. Equation (25) defines the modified, windowed periodogram estimate for power spectrum estimation. For a signal $x(n)$ of total duration N, we may segment the data into M subsegments, $x_i(n)$, each of length L, with intersegment overlap of length D. Each subsegment is multiplied by a window, $w(n)$, prior to DFT estimation. The final power spectrum estimate, $\hat{S}_{xx_\text{welch}}$, is estimated from an average of the power spectra, $\hat{S}_{xx_i}$, estimated from each subsegment.

$$x_i(n) = x(n + (i+1)D) \qquad \text{from} = 0, 1, \ldots, L-1$$

$$\hat{S}_{xx_i}(k) = \frac{1}{LU} \left| \sum_{n=0}^{L-1} x_i(n) w(n) e^{-j(2\pi kn/L)} \right|^2 \tag{25}$$

$$\hat{S}_{xx_\text{welch}}(k) = \frac{1}{M} \sum_{i=1}^{M} S_{xx_i}(k)$$

where

$N =$ total data points
$L =$ number of data in each subsegments
$D =$ overlap between subsegments
$M =$ number of subsegments

$$U = \frac{1}{L} \sum_{n=0}^{L-1} w^2(n)$$

Alternative methods of spectral estimation that have been used to investigate cardiac arrhythmias require parametric modeling of the electrogram data. Unlike the periodogram estimate, which is nonparametric (assumes no underlying statistical model of the data), parametric estimates assume some underlying model for the origin of the time series. A popular parametric spectral estimator is the *autoregressive* (AR) spectral estimator. This estimator assumes that the data is the output of a linear system with a white-noise input. It may be shown [81] that with enough model coefficients, any signal may be modeled as an AR process. However, an excessive number of model parameters indicates that the AR model is not a good description of the random process being evaluated. The equation for an AR process, $x[n]$, may be expressed as

$$x[n] = -\sum_{k=1}^{p} a[k]x[n-k] + u[n] \tag{26}$$

where $a[k]$ are the model coefficients and $u[n]$ is a white-noise excitation.

In other words, if the electrogram data behaves as an autoregressive process, the electrogram data should be well described by a few parameters, $a[n]$. These parameters, $a[n]$, may be found using least-squares minimization methods and fast-recursive algorithms, such as Levinson's recursion [81]. Once the model parameters are determined, the power spectrum for the original electrogram signal may be determined indirectly from the model parameters using

$$S_{xx_{AR}}(f) = \frac{\sigma^2}{\left| 1 - \sum_{k=1}^{p} a[k]e^{-j2\pi fk} \right|^2} \tag{27}$$

where f is frequency and σ^2 is the variance (power) of the white-noise excitation.

Theoretically, if the model accurately describes the electrogram data, the spectral resolution should be infinite using Eq. (27). A number of investigators have used autoregressive models to characterize ventricular fibrillation [82,83] and atrial fibrillation [83,84].

B. Differentiation of Arrhythmias with Spectral Parameters

A number of investigators have used power spectrum analysis to differentiate cardiac rhythms. Theoretically, those arrhythmias which produce electrograms with discrete events that are relatively constant in morphology, polarity, and timing should produce power spectra with discrete peaks of power at harmonic frequencies that may cover a broad band of frequencies (Fig. 23, AFLUT). In contrast, those arrhythmias that produce electrograms of constantly changing morphology and timing should exhibit power spectra that lack harmonic character and discrete peaks of power (Fig. 23, AF).

Spectral analysis has been used to examine both atrial and ventricular electrograms recorded from bipolar catheters [36,85,86]. These early studies indicated that no significant spectral energy existed beyond 250 Hz. Thus with adequate antialias filtering, a sampling frequency slightly larger than 500 Hz is adequate for capturing the spectral characteristics of the electrograms.

A number of parameters are used to summarize the spectral characteristics of an arrhythmia. These parameters include *peak power*, *percent power*, *median frequency*, and *bandwidth*. Peak power is simply the magnitude of the largest peak in the spectrum (Fig. 26). This parameter is extremely sensitive to signal gain, duration of the underlying electrogram (determines locations of spectral lines, which may or may not align exactly with the true frequency content), and electrode configuration and location. Another related parameter is the peak frequency, or the frequency at which the maximum spectral peak occurs. Again, such a parameter is sensitive to electrode configuration and location, heart size, etc. To eliminate the sensitivity to some of these clinical parameters, parameters such as percent power and median frequency have been proposed. Median frequency, defined as the frequency that divides in half the power within a certain frequency band, has been used to track changes in atrial fibrillation during drug administration [36]. As illustrated in Fig. 26, percent power is simply the ratio of power in a narrow band (dark gray shading) with respect to the power in a much broader band. For arrhythmias, such as atrial fibrillation and ventricular fibrillation, the power spectrum is characterized by a large peak of narrow power in the 4–12-Hz frequency band, with little power in the neighboring frequency bands. Thus, the ratio of power in the 4–9-Hz band with respect to power in the 1–60 Hz band is large. Conversely, a rhythm such as atrial flutter exhibits peaks of power at harmonic frequencies throughout the 0–60-Hz band. For this example of atrial flutter, the percent power in the same 4–9-Hz band is much lower than for atrial fibrillation. Thus, a percent power parameter has been used to differentiate atrial fibrillation from atrial flutter and other regular atrial arrhythmias [30].

Slocum et al. [30] have used power spectrum analysis to distinguish atrial fibrillation from sinus rhythm and regular atrial tachycardias. Jenkins et al. [26] also report significant differences in spectral content in isolated frequency bands for atrial fibrillation compared to sinus rhythm. However, no spectral statistics allowed complete separation of atrial fibrillation from

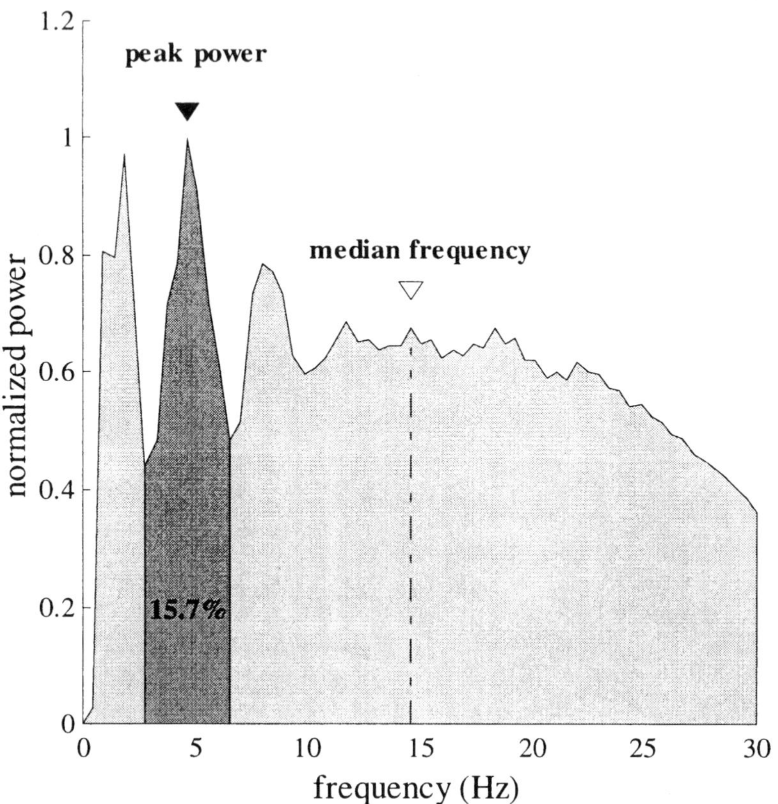

Figure 26 Parameters such as peak power, percent power in a specified frequency band (%p), and median frequency as illustrated on this spectrum may be used to differentiate arrhythmias.

sinus rhythm, especially sinus rhythm with highly irregular AA intervals. Furthermore, ST elevation adversely affected the discriminatory abilities of power spectral discriminants.

Power spectra have also been used to differentiate sinus rhythm from ventricular fibrillation [32,87]. Results for such discrimination schemes are mixed; the locations of the spectral peaks vary with species and lead configuration. In some cases, ventricular fibrillation exhibits harmonics in the power spectrum. Furthermore, there is a fair amount of overlap in the frequency band of maximum energy between sinus rhythm and ventricular fibrillation [88]. In addition, spectral content is sensitive to changes in electrode configuration, drugs, and metabolic state.

Frequency-domain algorithms have been used to a limited extent in differentiating ventricular tachycardia from sinus rhythm [34,74,86,89–91]. Considerable overlap exists between spectra for these two rhythm classes. Within individuals, there are reported differences in peak frequency between sinus rhythm and ventricular tachycardia, but there is no separation for the population as a whole [86,89]. Too much overlap has rendered this method of discrimination relatively useless for these two rhythm classes [74,90,91].

Brachman et al. [92] and Stroobandt et al. [93] report some difference in frequency spectra between ventricular fibrillation and ventricular tachycardia using a monophasic action potential (MAP) catheter.

As for morphology-based discrimination algorithms, simple spectral analysis is sensitive to changes in lead configuration and position [94], changes in vagal tone and body temperature [85], and pharmacological agents [36]. While these sensitivities may be undesirable for automated detection schemes, they may be useful for probing electrophysiologic mechanism during the onset, maintenance, and termination of an arrhythmia.

C. Coherence Spectrum

As described previously, fibrillation is characterized by disorganized, continually changing patterns of activation and the absence of a constant temporal relationship between multiple sites on the heart. This continually changing temporal or phase relationship may be quantified in the frequency domain by magnitude-squared coherence [25]. *Magnitude-squared coherence* (MSC) is defined as

$$\text{MSC}(f) = \frac{|S_{xy}(f)|^2}{S_{xx}(f)S_{yy}(f)} \tag{28}$$

where $x(t)$ and $y(t)$ are two simultaneous electrogram recordings, S_{xy} is the cross-power spectrum between signals x and y, and S_{xx} and S_{yy} are the individual power spectra for signals x and y, respectively. MSC is a measure of the linear relation between signals as a function of frequency, f and is a real quantity with value between 0 and 1. In other terms, MSC measures the constancy of the time delay (phase) at a specific frequency between signals x and y. Two linearly related signals (in the absence of noise) will have an MSC function equal to 1 at all frequencies present in both signals, while two random, uncorrelated signals will have an MSC equal to 0 at all frequencies. MSC is similar in concept to the cross-correlation coefficient except that it is insensitive to actual phase difference. It is only sensitive to the constancy of phase between the two signals. Any linear operation (multiplication by a constant or addition of a constant) on one or both of the signals will not alter the MSC between x and y. However, additive, uncorrelated noise and system nonlinearities will reduce MSC for two similar signals.

MSC may be estimated for sampled electrogram data using a method of overlapped and averaged FFT spectral estimates [95]. Basically, estimates of S_{xx}, S_{yy}, and S_{xy} are determined using a periodogram technique (see Section VIII. A), and their estimates are then used in the definition of MSC [Eq. (28)]. As defined in Eqs. (29), (30), and (31), the autopower spectra and cross-power spectra of two sequences, $x(n)$ and $y(n)$, are estimated and subsequently averaged over several segments of sequences x and y. MSC is then calculated from these averaged spectra.

$$\hat{S}_{xx}(f) = m \sum_{i=1}^{k} |X_i(f)|^2 \tag{29}$$

$$\hat{S}_{yy}(f) = m \sum_{i=1}^{k} |Y_i(f)|^2 \tag{30}$$

$$\hat{S}_{xy}(f) = m \sum_{i=1}^{k} X_i(f) Y_i^*(f) \tag{31}$$

where $m = 1/kpf_s$ and f_s is the sampling frequency. More specifically, $x(n)$ and $y(n)$ are each divided into k segments, each p points long. Thus, $kp = N$ is the total number of data points in each of sequences $x(n)$ and $y(n)$. Each p-point sequence is weighted by an appropriate window (i.e., Hanning), and a p-point DFT [$X_i(f)$ and $Y_i(f)$] is then performed. This estimate for MSC is biased. If the number of segments, k, equals 1, the MSC will be unity at all frequencies regardless of the true MSC between x and y [95]. To reduce this bias in estimating the true MSC value, one needs to increase the number of segments. This results in a trade-off between spectral resolution and accuracy of the MSC estimate. For finite-length (N) data sequences, a large p (small k) provides good spectral resolution but poor statistical accuracy of the estimate, and vice versa. Carter et al. [95] provide a statistical framework for determining the appropriate number of segments to be used in an MSC estimate.

Ropella et al. [27,96] have used MSC to discriminate fibrillatory from nonfibrillatory rhythms. Nonfibrillatory rhythms typically exhibit moderate to high levels of MSC throughout the 0–60-Hz band (Fig. 27, SR). Unlike rate and PDF discrimination schemes, MSC can discriminate rapid, but organized ventricular tachycardia from ventricular fibrillation. Sinus rhythm and regular tachycardias typically exhibit peaks of coherence at the rhythm's fundamental frequency and its harmonics. Conversely, fibrillatory rhythms typically exhibit low MSC throughout the same 0–60-Hz band (Fig. 27, AF). Furthermore, for the fibrillatory rhythms, there is an absence of harmonic behavior for MSC.

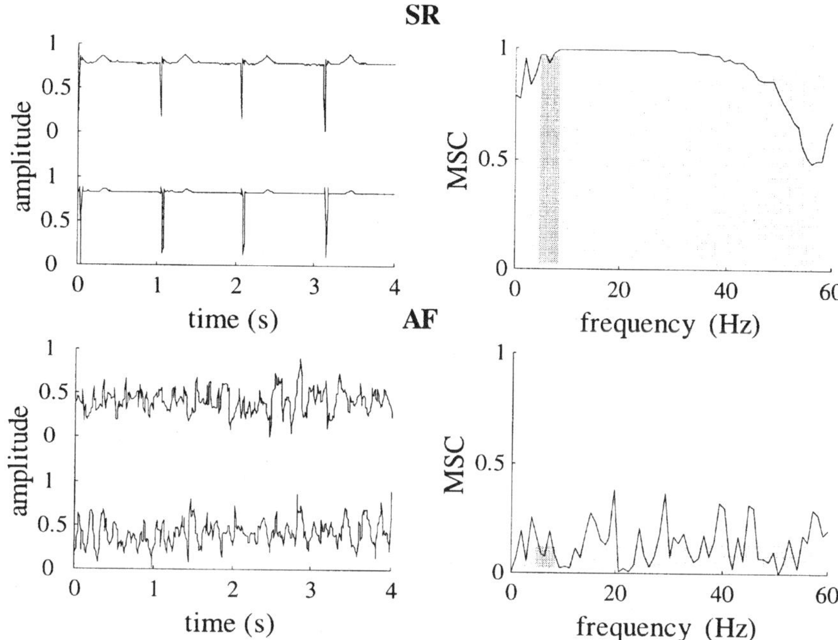

Figure 27 MSC in the 0–60-Hz band is illustrated for example of sinus rhythm (SR) (upper tracings) and atrial fibrillation (AF) (lower tracings), each recorded in the high right atrium. Note that sinus rhythm has moderate to high MSC throughout the 0–60-Hz band. Conversely, atrial fibrillation has low MSC throughout the same frequency band. Mean MSC is illustrated for both the 1–60-Hz band (light gray shade) and the 4–9-Hz band (dark gray shade).

Unlike power spectrum analysis and other morphology-based arrhythmia discrimination schemes, the ability of MSC to differentiate fibrillatory from nonfibrillatory rhythms is relatively immune to changing lead configurations [96], changes in signal gain, and the specific morphology of the ventricular tachycardia [27]. However, MSC requires considerable more computation than the simpler morphology-based discrimination schemes.

In addition to arrhythmia discrimination, MSC has been shown to be a quantitative descriptor of rhythm organization. In recent years, there has been a growing interest in quantitatively investigating the "organization" (spatially and temporally) of fibrillatory rhythms. A number of studies have tried to capture the organization of a rhythm by quantifying the temporal synchrony and harmony between multiple sites in the heart [22,23,27,96,97].

Such measures of organization may be useful for probing the underlying mechanisms for the initiation, sustenance, and termination of fibrillatory rhythms [98].

IX. TIME-FREQUENCY ANALYSIS

Traditional methods of power spectrum analysis assume that the signal being analyzed is stationary over the period of analysis. However, one hallmark of some arrhythmias, such as fibrillation, is the nonstationary behavior of the underlying electrophysiology and, hence, electrogram recordings. When the spectral estimates are obtained from electrogram recordings spanning a long period of time, the presence of nonstationary behavior is unnoticeable and can significantly distort the spectral estimates. Figure 28 shows an electrogram recording spanning 7 sec during which the signal was nonstationary. To observe changes in spectral properties over time, *time-frequency methods* may be applied. In their simplest form, these methods consist of estimating the spectral properties using short-duration electrogram segments and the *short-time Fourier transform* (STFT) [99]. By monitoring the changes in spectral content from consecutive segments, one may detect nonstationary behavior. Figure 28 (bottom) illustrates how the spectral content of the signal varies over time. Phase spectra and coherence spectra may be estimated in a similar manner, allowing investigators to track changes in interelectrogram phase delay as a function of time [98]. In fact, it is these temporal changes in interelectrogram phase delay that are quintessential to fibrillation. A major drawback for the STFT method of time-frequency analysis is the poor spectral resolution that results from estimating the power spectra over short-duration signals. In other words, as one attempts to increase the time resolution of the spectral analysis by using very-short-duration segments of signal, the spectral resolution is greatly compromised. Other time-frequency methods offer improved time resolution and spectral resolution compared to the STFT time-frequency representations [99]. The Wigner distribution, the Choi-Williams distribution, multitaper magnitude-squared coherence [98] and the wavelet decomposition all transform electrogram signals into descriptions that have high resolution in both time and frequency. Drawbacks for these time-frequency estimators include cross-terms leading to false peaks of power and complexity of estimation compared to the STFT. The complexity of time-frequency analysis may prohibit its use in implantable devices. However, such transformations may be particularly useful in the research environment, where investigators studying underlying physiological mechanisms controlling an arrhythmia wish to observe nonstationary behavior.

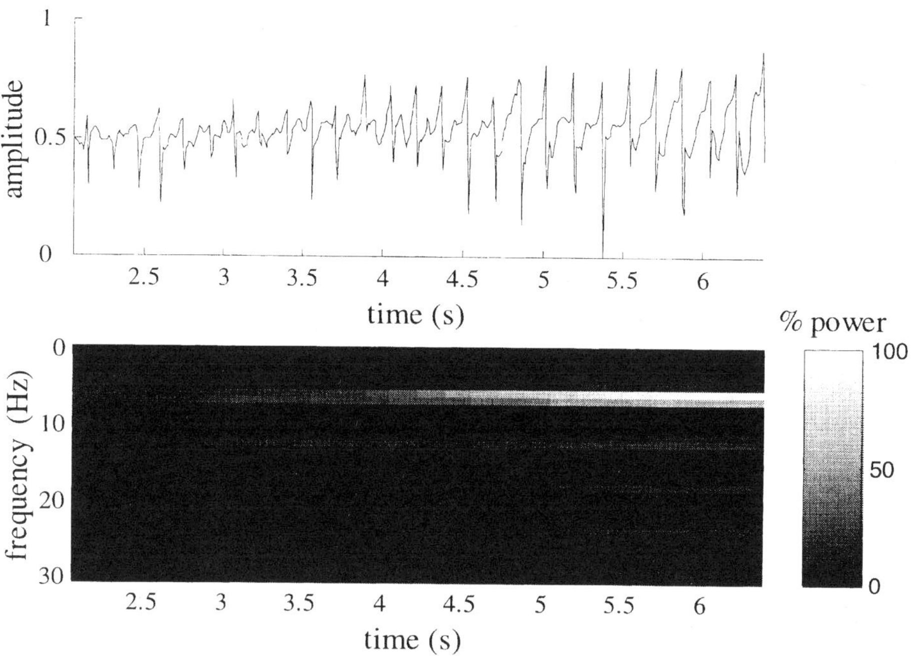

Figure 28 Shown is a time-frequency representation of an electrogram recorded during polymorphic ventricular tachycardia. Power spectra are estimated using the short-time Fourier transform. Note the evolution of power around 5 Hz over the 6-sec time interval.

X. WAVELET TRANSFORMS

Like Fourier analysis, wavelet decompositions [100] may be used to extract features from an electrogram. While the Fourier transform models an electrogram signal as a linear combination of sines and cosines (basis function), the *wavelet transform* models the electrogram using a set of basis functions that bear a greater resemblance (in both time and frequency content) to atrial and ventricular activation. Moreover, Fourier transforms model electrograms as narrow-band signals all of equal bandwidth, while wavelets model the same electrograms as a mix of narrow-band and wide-band signals. The wavelet model seems more appropriate for signals that change in a rapid fashion (over very brief time periods) and, consequently, have a relatively wide bandwidth. In essence, while the Fourier transform offers a frequency resolution that is constant (individual sinusoids) across the frequency spectrum, the wavelet transform offers variable spectral resolution across the spectrum. In wavelet analysis, the electrogram is decomposed into a set of orthonormal (nonredundant) basis functions that are limited in both time and frequency. The wavelet transform decomposes a signal into a family of functions obtained by dilating and translating a function known as a mother wavelet, $\Omega(t)$. In most applications, the mother wavelet is simply chosen to be the impulse response of a bandpass filter. Both dilations and translation of mother wavelet provide variable time-frequency resolution. As is illustrated in Fig. 29, the basis function may be designed to capture those portions of the data that are transient or brief with respect to time ($i = -1$) or capture those portions of the signal that have very narrow band of frequency content ($j = -6$). Such a decomposition is well suited to electrograms, where activation complexes tend to be localized in both time and frequency. The wavelet transformation is equivalent to passing the electrogram though a bank of bandpass filters of varying bandwidth and center frequency, where the frequency response of the filters is ultimately determined by the mother wavelet. Moreover, the filter banks may be designed such that the electrogram may be perfectly reconstructed from the wavelet set. Wavelets have been used to detect abnormal QRS complexes and premature ventricular contractions in surface ECG applications [101]. Similarly, wavelets have been used to discriminate atrial arrhythmias as well as detect life-threatening arrhythmias [29,102]. The reader may find more detailed information on wavelet transforms in Ref. 100.

XI. HEMODYNAMIC CONSEQUENCES OF ARRHYTHMIAS

To date, the majority of arrhythmia detection algorithms use only the electrical activation recorded from the atrium and ventricle. Recently, however, detection schemes have attempted to monitor the hemodynamic consequences of a tachyarrhythmia [31,33,103–107]. These discrimination schemes rely on measurements of intracardiac impedance (shown to be correlated with mean arterial pressure [103]) and/or intramyocardial tissue pressure. These tissue pressure measurements are highly sensitive to the myocardial contractile status associated with life-threatening arrhythmias [106]. The addition of a hemodynamic measure to other signal criteria improves the classification of ventricular tachycardia and ventricular fibrillation [26,107]. Despite improvement in rhythm classification, the hemodynamic measures can still lead to unwarranted treatment. For example, poor left ventricular function can result in pressure drops during supraventricular tachycardias as well as ventricular tachycardias. Currently, the difficulty in monitoring the

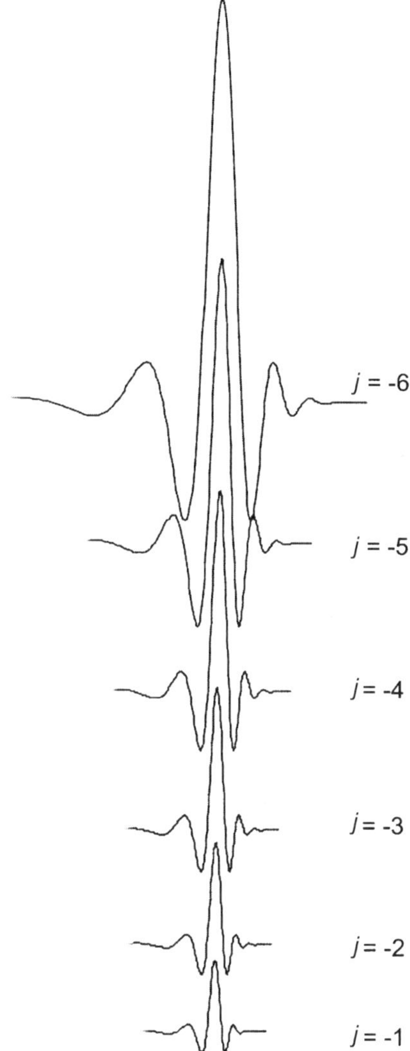

Figure 29 Illustrated are basis functions for a wavelet decomposition. Each wavelet is a dilated and time-shifted version of some mother wavelet. j represents the resolution level, where increases in time resolution result in decreases in spectral resolution.

hemodynamics stems from the lack of sensors with long-term stability, biocompatibility, and low energy consumption.

XII. NEURAL NETWORKS

While early generations of implantable devices and computerized ECG interpretation systems used single parameters to identify an arrhythmia, advances in hardware, power supplies, and computer technology have lead to much more sophisticated algorithms that use a combination of parameters. In the future, implantable devices, like existing computerized ECG interpretation schemes, will combine several quantitative measures to improve both specificity and sensitivity of arrhythmia discrimination. The automatic use of various quantitative measures for rhythm classification can be done using neural networks [108]. Neural networks are designed to combine information from several parameters and produce a decision or classfication based on those inputs. In short, neural networks are mapping functions that map one or more inputs (such as slopes, amplitudes, power, rates, etc.) to one or more outputs (tachychardia, fibrillation, etc.). The mapping functions (or decision rules) between the input and output are not explicitly defined in the network; rather, they are adaptively formed using a training set. During the training period, a set of inputs is applied to the neural network and the mapping functions automatically modified such that the output of the network matches the desired output. In this manner, the network is trained to recognize a set of patterns or specific features of a class of arrhythmia. Once trained, the networks are able to classify new data based on previous training. Because of their adaptive nature, neural networks may allow the mapping functions to adjust to modifications in the input (such as signal amplitude and morphology) without changes in the output. A more thorough treatment of neural networks and biomedical applications may be found in [108] and is beyond the scope of this chapter.

XIII. SUMMARY

The ability to describe cardiac arrhythmias quantitatively is critical to the clinical setting, where accurate diagnosis and treatment depend on the reliable identification and discrimination of arrhythmias. Arrhythmia identification is important to the functioning of automatic external defibrillators, critical care monitoring in the hospital environment, Holter monitoring, and implantable antiarrhythmic devices. In life-threatening situations such as ventricular fibrillation, the rapidity with which the arrhythmia is detected and treated is of utmost priority. In such cases, sensitivity is favored over specificity, and fast detection schemes are used. Conversely, in cases that are not immediately life-threatening, rapidity of diagnosis is not as critical as accuracy, and slower, more sophisticated schemes may be implemented to improve the specificity as well as the sensitivity of the detection scheme. We presented a number of time-domain and frequency-domain methods that have been proposed or implemented for the discrimination of cardiac arrhythmias. For each method, there is trade-off in the complexity of parameter estimation, sensitivity and specificity for specific arrhythmias, dependence on electrode configuration, dependence on signal amplitude and morphology, dependence on stationarity of the signal, and the correlation with underlying electrophysiology.

Rate parameters have the advantage of being simple to estimate in terms of hardware and battery power requirements. They tend to have excellent sensitivity for rapid, pathological tachycardias that require immediate intervention. However, the trade-off for the simplicity and high sensitivity is poor specificity, which results in inappropriate therapy being administered during nonpathological tachycardias or myopotential interference.

Morphological methods are designed to evaluate electrogram activity on an event-by-event basis. Such methods of analysis allow for detailed examination of individual atrial and ventricular events and, consequently, have good time resolution. Such methods ultimately use a sequence of rules based on the morphology and frequency of occurrence of individual events. The strength of these methods is in the ability to detect very short runs or occurrences of premature ventricular or atrial contractions or the absence of atrial or ventricular activations. In the case of multiple sensing leads, these morphological methods can also differentiate supraventricular arrhythmias from ventricular arrhythmias. Morphological methods are also relatively simple to implement in hardware. However, the strength of these simple methods is also their weakness; they are sensitive to specific morphology, which is highly dependent on the individual heart and the electrode configuration. While morphological measures can be tuned to the individual patient during time of device implant, drug therapy, patient activity, electrode drift, and changes in the electrophysiological substrate will continue to change following implant. Furthermore, most patients have tachycardias of multiple morphology. Morphological methods typically require a learning period to allow the algorithm to define normal and abnormal complexes for each patient. The proper functioning of these algorithms depends on repeated initiations of each of the arrhythmias in the clinical environment, and tuning the device to recognize the arrhythmia.

Frequency-domain methods, such as power spectrum analysis and coherence analysis, examine the energy distribution of the electrogram as a function of frequency. Such methods examine longer time series or runs of activation rather than individual events, which makes spectral analysis less sensitive to beat-to beat variations in the electrogram. Given that many arrhythmias are defined as runs of abnormal atrial or ventricular events, such forms of analysis are more appropriate than beat-to-beat classification. Frequency-domain methods are particularly useful for quantifying periodicities in an electrogram. They tend to be less sensitive to specific morphology of an arrhythmia and do not require explicit event detection, which often requires a sequence of decision statements. Furthermore, while some events may overlap in time, they may be well separated in frequency. Thus, differentiation may be easier in the frequency domain when it is not possible in the time domain. Drawbacks of spectral analysis include greater requirements for computation and power than rate and morphology methods. Moreover, the power spectrum is sensitive to changes in lead configuration. Nonetheless, spectral measures seem to outperform rate parameters in differentiating monomorphic ventricular tachycardias from ventricular fibrillation [27]. While both arrhythmias have a high rate, the former exhibits spatially and temporally organized patterns that require a less dramatic therapy than fibrillation.

Arrhythmia discrimination algorithms that use simultaneous recordings from multiple leads quantify the spatial uniformity and temporal evolution of the electrogram between multiple sites on the heart. These schemes, such as cross-correlation and MSC, are designed to detect changes in wavefront orientation and velocity. Such changes reflect alterations in foci, conduction velocity, refractoriness, and pathway of conduction. Thus, the spatial information provided by multilead measures is particularly useful in studying mechanisms and complexity of arrhythmias. Multiple sites allow us to examine complexity of an arrhythmia over space. This is particularly important for fibrillation, in which one site may show fairly organized, periodic activity while other sites show comparatively less organized and or periodic activity [27]. It is this breakdown in spatial organization that is quintessential to fibrillation. MSC completely separates fibrillatory and nonfibrillatory rhythms when rate and morphology measures fail. Moreover, MSC is less sensitive to lead configuration and specific morphology than the power spectrum. However, the computation required for MSC analysis is considerable compared to simple rate and morphology measures. Continued advances in computer hardware and battery power will eventually alleviate these computational burdens for implantable devices.

While most quantitative analysis of intracardiac electrograms has been motivated by the need for reliable detection schemes for use in implantable devices, more recent methods of quantitative analysis have been steeped in quantifying the spatial and temporal organization (and thus underlying mechanisms) of cardiac arrhythmias. The study of the mechanisms leading to arrhythmias would make early preventive intervention possible. Electrophysiological mapping and computer models suggest that fibrillation is rather complex. However, over relatively short time intervals and small regions in space, there is evidence that regular, organized activation exists during fibrillation. The greatest obstacle to overcome in studying the electrophysiological mechanisms of fibrillation is the ability to record simultaneously from many sites over sufficiently long time periods in order to obtain sufficient spatial and temporal resolution.

Better arrhythmia detection schemes will ultimately surface with a better understanding of the heart's electrophysiology. Most methods of arrhythmia analysis assume an underlying stationary, linear, deterministic system, which is a simplified view of the true electrophysiology. In reality, many arrhythmias, such as fibrillation, are nonstationary and may result from nonlinear, deterministic processes leading to a chaotic rhythm. Nonstationary behavior can be explored through a number of time-frequency representations. Such representations have enabled researchers to study the evolution of heart signals during various arrhythmias [83,99,101]. These time-frequency distributions allow more detailed study of the onset and termination of arrhythmias than traditional periodogram analysis. Time-frequency methods of MSC have also been introduced to study time-varying phase relations between multiple electrograms [98] during drug administration. Chaotic signals may be analyzed using dimensional analysis and Lyapunov exponents [109,110]. The goal of current research is to determine whether complex arrhythmias, such as fibrillation, may be described by a low-dimension dynamical system. Lower-dimensional systems lend themselves to identification and, consequently, control. Kaplan et al. [109,110] found that using methods of dimensional analysis on surface ECG failed to describe ventricular fibrillation as a low-dimension system, yet microelectrode recordings and optical mapping experiments reveal low-order dynamics during ventricular fibrillation [24,111]. Thus, at the cellular level, the dynamics may be fairly deterministic. Other chapters of this volume review methods for dynamic systems analysis of cardiac arrhythmias.

A number of clinical factors, such as lead configuration and patient activity, alter the signal characteristics measured during arrhythmias. Some of these factors are accounted for at the time the device is implanted in the patient, or during postimplant follow-up. However, one of the most important clinical factors to consider is antiarrhythmic drugs. The majority of patients who receive an implanted device also receive pharmacological therapy. Antiarrhythmic drugs have been shown to alter signal characteristics to the extent that arrhythmia interpretation schemes fail [36]. Thus the electrogram may change over time, and failure to account for the pharmacological influence may severely hinder the performance of an antiarrhythmic device. On a more positive note, changes in signal characteristics may be used to elucidate underlying mechanism of antiarrhythmic drugs [98].

In using quantitative measures to identify cardiac arrhythmias, it is important to recognize that there is not always a clear distinction between rhythms such as atrial flutter and atrial fibrillation or ventricular tachycardia, polymorphic ventricular tachycardia, and ventricular fibrillation. Rather, there appears to be a continuous spectrum of complexity from monomorphic tachycardia to fibrillation. Furthermore, rapid monomorphic ventricular tachycardia may require the same therapy as ventricular fibrillation, despite seemingly different mechanisms. Consequently, the appropriate treatment for a specific arrhythmia may not be solely dependent on the electrophysiological substrate of the arrhythmia but also on the hemodynamic consequences and other physiological parameters.

ACKNOWLEDGMENTS

The authors would like to express their sincere gratitude to Olga Yakubovich for assistance with illustrations and to Steven Swiryn and James A. Roth for making available the electrogram data used in figures throughout the chapter.

REFERENCES

1. Arzbaecher R, Bump T, Munkenbeck F, Brown J, Yurkons C. An algorithm for automatic infusion of procainamide in acute management of paroxysmal atrial fibrillation. Computers in Cardiology, 1984, pp 57–82.
2. Mirowski M, Mower MM, Reid PR, Watkins L, Langer A. The automatic implantable defibrillator. New modality for treatment of life-threatening ventricular arrhythmias. Pacing Clin Electrophysiol 5(3):384–401,1982.
3. Camm AJ, Davies DW, Ward DE. Tachycardia recognition by implantable electronic devices. Pacing Clin Electrophysiol 10(5):1175–1190,1987.
4. Griffin JC, Sweeney M. The management of paroxysmal tachycardias using the Cybertach-60. Pacing Clin Electrophysiol 7(6 pt 2):1291–1295,1984.
5. Spurrell RA, Nathan AW, Bexton RS, Hellestrand KJ, Nappholz T, Camm AJ. Implantable automatic scanning pacemaker for termination of supraventricular tachycardia. Am J Cardiol 49(4):753–760,1982.
6. Rothman MT, Keefe JM. Clinical results with Omni-Orthocor, an implantable antitachycardia pacing system. Pacing Clin Electrophysiol 7(6 pt 2):1306–1312,1984.

7. Sowton E. Clinical results with the Tachylog antitachycardia pacemaker. Pacing Clin Electrophysiol 7(6 pt 2):1313–1317,1984.

8. Zipes DP, Prystowsky EN, Miles WM, Heger JJ. Initial experience with Symbios Model 7008 pacemaker. Pacing Clin Electrophysiol 7(6 pt 2):1301–1305,1984.

9. Langer A, Heilman MS, Mower MM, Mirowski M. Considerations in the development of the automatic implantable defibrillator. Med Instrum 10(3):163–167,1976.

10. Pannizzo F, Mercando AD, Fisher JD, Furman S. Automatic methods for detection of tachyarrhythmias by antitachycardia devices. J Am Coll Cardiol 11(2):308–316,1988.

11. Den Dulk K, Bertholet M, Brugada P, Bar FW, Demoulin JC, Waleffe A, Bakels N, Lindemans F, Bourgeois I, Kulbertus HE. Clinical experience with implantable devices for control of tachyarrhythmias. Pacing Clin Electrophysiol 7(3 pt 2):548–556,1984.

12. Lerman BB, Waxman HL, Buxton AE, Sweeney M, Josephson ME. Tachyarrhythmias associated with programmable automatic atrial antitachycardia pacemakers. Am Heart J 106(5 pt 1):1029–1035,1983.

13. Davies W, Wainwright R, Tooley M, Lloyd D, Nathan A, Spurrell R, Camm A. Arrhythmia recognition from electrogram morphology. PACE 8:48, 1985.

14. Poole JE, Troutman CL, Anderson J, Bardy GH, Greene HL. Inappropriate and appropriate discharges of the automatic table cardioverter defibrillator. J Am Col Cardiol 11:210A, 1988.

15. Arzbaecher R, Bump T, Jenkins J, Glick K, Munkenbeck F, Brown I, Nandhakumar N. Automatic tachycardia recognition. Pacing Clin Electrophysiol 7(3 pt 2):541–547,1984.

16. Kadri N, Niazi I, Elkhatib I, Jazayeri M, Decker S, Mahmud R, Tchou P, Akhtar M. Automatic implantable cardioverter defibrillator: problems and complications. J Am Coll Cardiol 9:142A, 1987.

17. Bardy GH, Olson WH, Ivey TD, Johnson G, Greene L. Does unsuccessful defibrillation adversely effect subsequent AICD sensing of ventricular fibrillation? PACE 11:485, 1988.

18. Allessie M, Lammers W, Bonke F, Hollen J. Experimental evaluation of Moe's multiple wavelet hypothesis of atrial fibrillation. In: Zipes D, Jalife J, eds. Cardiac Electrophysiology and Arrhythmias. Orland, FL Grune & Straton, 1985, pp. 265–275.

19. Ideker R, Klein G, Smith W, Harrison L, Kasell J, Wallace A, Gallagher J. Epicardial activation sequences during onset of ventricular tachycardia and ventricular fibrillation. In: Kulbertus H, Wellens H, eds. Sudden Death. The Hague: Martinus Nijhoff, 1980, pp. 165–185.

20. Josephson M, Spiciman S. Greenspan A, Horowitz L. Mechanisms of ventricular fibrillation in man. Am J Cardiol 44:623–631,1979.

21. Bayly P, Johnson EE, Idriss SF, Ideker RE. Smith WM. Efficient electrode spacing for examining spatial organization during ventricular fibrillation. J IEEE Trans Biomed Eng 40(10):1060–1066,1993.

22. Bayly P, Johnson E, Wolf P, Geenside H, Smith W, Ideker R. A quantitative measurement of spatial order in ventricular fibrillation. J Cardiovasc Electrophysiol 4:533–546,1993.

23. Sih H, Sahakian A, Arentzen C, Swiryn S. A frequency domain analysis of spatial organization of epicardial maps. IEEE Trans Biomed Eng 42:718–727,1995.

24. Gray R, Pertsov A, Jalife J. Spatial and temporal organization during cardiac fibrillation. Nature 392:75–78,1998.

25. Bendat JS, Piersol GA. Random Data Analysis and Measurement Procedures. 2nd ed. New York: Wiley, 1986.

26. Jenkins J, Noh KH, Guezennec A, Bump T, Arzbaecher R. Diagnosis of atrial fibrillation using electrograms from chronic leads: evaluation of computer algorithms. Pacing Clin Electrophysiol 11(5):622–631,1988.

27. Ropella KM. Baerman JM, Sahakian AV, Swiryn S. Differentiation of ventricular tachyarrhythmias. Circulation 82(6):2035–2043,1990.

28. Tompkins W. Biomedical Digital Signal Processing. Englewood Cliffs, NJ: Prentice-Hall, 1993.

29. Khadra L, Al-Fahoum A, AL-Nashash H. Detection of life-threatening cardiac arrhythmias using the wavelet transformation. Med Biol Eng Comput 35:626–632,1997.

30. Slocum J, Sahakian A, Swiryn S. Computer discrimination of atrial fibrillation and regular atrial rhythms from intra-atrial electrograms. Pacing Clin Electrophysiol 11(5):610–621,1988.

31. Aubert AE, Goldreyer BN, Wyman MG, Ector H, DeGeest H. Detection of ventricular fibrillation during AICD implantation using electrogram analysis. PACE 11:524, 1988.

32. Herbschleb JN, Heethar RM, Van Der Tweel I, Zimmerman ANE, Meijler FL. Signal analysis of ventricular fibrillation. Computers in Cardiology, 1979, pp. 49–54.

33. Winkle RA, Bach SM Jr, Echt DS, Swerdlow CD, Imran M, Mason JW, Oyer PE, Stinson EB. The automatic implantable defibrillator: local ventricular bipolar sensing to detect ventricular tachycardia and fibrillation. Am J Cardiol 52(3):265–270,1983.

34. Lin DP, DiCarlo LA, Jenkins JM. Identification of ventricular tachycardia using intracavitary ventricular electrograms: analysis of time and frequency domain patterns. Pacing Clin Electrophysiol 11(11 pt 1):1592–1606,1988.

35. DiCarlo L, Jenkins JM, Throne R, Mays C, Lin D. Classification of arrhythmias using atrial and ventricular endocardial electrograms. J Electrocardiol 22(suppl):230,1989.

36. Ropella KM. Sahakian AV, Baerman JM, Swiryn S. Effects of procainamide on intra-atrial electrograms during atrial fibrillation: implications for detec-

tion algorithms [published erratum appears in Circulation 1988 Jun; 77(6):1344]. Circulation 77(5):1047–1054,1988.

37. Moe G, Abildskov J. Atrial fibrillation as a self-sustaining arrhythmia independent of focal discharge. Am Heart J 58:59–70,1959.

38. Barold SS, Falkoff MD, Ong LS, Heinle RA. Double sensing by atrial automatic tachycardia-terminating pulse generator. Pacing Clin Electrophysiol 10(1 pt 1):58–64,1987.

39. Mercando AD, Furman S. Measurement of differences in timing and sequence between two ventricular electrodes as a means of tachycardia differentiation. Pacing Clin Electrophysiol 9(6 pt 2):1069–1078,1986.

40. Bardy GH, Olson WH. Tachyarrhythmia detection algorithm for implantable cardioverter and defibrillator. PACE 10:641, 1987.

41. Olson WH, Bardy GH, Mehra R, Keimel JG, Huberty KP, Almquist C, Biallas RM. Onset and stability for ventricular tachyarrhythmia detection in an implantable pacer-cardioverter-defibrillator. Computers in Cardiology, 1986, pp 167–170.

42. Fisher JD, Goldstein M, Ostrow E, Matos JA, Kim SG. Maximal rate of tachycardia development: sinus tachycardia with sudden exercise vs. spontaneous ventricular tachycardia. Pacing Clin Electrophysiol 6(2 pt 1):221–228,1983.

43. Geibel A, Zehender M, Brugada P. Changes in cycle length at the onset of sustained tachycardias—importance for antitachycardiac pacing. Am Heart J 115(3):588–592,1988.

44. Gerald CF, Wheatley PO. Applied numerical analysis. 3rd ed. Reading, MA: Addison-Wesley, 1984.

45. Wainwright R, Davies W, Tooley M. Ideal atrial lead positioning to detect retrograde atrial depolarization by digitization and slope analysis of the atrial electrogram. Pacing Clin Electrophysiol 7(6 pt 2):1152–1158,1984.

46. Davies DW, Wainwright RJ, Tooley MA, Lloyd D, Nathan AW, Spurrell RA, Camm AJ. Detection of pathological tachycardia by analysis of electrogram morphology. Pacing Clin Electrophysiol 9(2):200–208,1986.

47. Davies DW, Tooley MA, Cochrane T, Wainwright RJ, Camm AJ. Analysis of electrogram morphology in real-time for diagnosis and management of tachycardia. PACE 10:664, 1987.

48. Davies W, Tooley M, Cochrane T, Wainwright R, Camm AJ. Real time automatic diagnosis and treatment of tachycardia by recognition of electrogram morphology. PACE 9:305, 1986.

49. Davies DW, Nathan AW, Wainwright RJ, Large SR, Edmondson SJ, Rees GM, Camm AJ. Recognition of ventricular tachycardia and fibrillation from epicardial electrogram timings. Circulation 72:III-475, 1985.

50. Pan J, Tompkins W. A real-time QRS detection algorithm. IEEE Trans Biomed Eng 32:230–236,1985.

51. Santel D, Mehra R, Olson W, Bardy G. Integrative algorithm of ventricular tachyarrhythmias from the intra-cardiac electrogram. Computers in Cardiology, 1986, pp 175–177.

52. DiCarlo LA, Throne RD, Jenkins JM. A time-domain analysis of intracardiac electrograms for arrhythmia detection. Pacing Clin Electrophysiol 14(2 pt 2):329–336,1991.

53. Morris MM, Jenkins JM, diCarlo LA. Band-limited morphometric analysis of the intracardiac signal: implications for antitachycardia devices. Pacing Clin Electrophysiol 20(1 pt 1):34–42,1997.

54. Slocum J, Byrom E, McCarthy L, Sahakian A, Swiryn S. Computer detection of atrioventricular dissociation from surface electrocardiograms during wide QRS complex tachycardias. Circulation 72(5):1028–1036,1985.

55. Sadek L, Ropella K. Detection of atrial fibrillation from surface ECG using magnitude-squared coherence. Int Conf for IEEE Engineering in Medicine and Biology Society, Montreal, 1995, pp 1–4.

56. Caswell SA, Klugs KS, Chiang CM, Jenkins JM, DiCarlo L. Pattern recognition of cardiac arrhythmias using two intracardiac channels. Computers in Cardiology, 1993, pp 181–184.

57. Throne RD, Jenkins JM, DiCarlo LA. The bin area method: a computationally efficient technique for analysis of ventricular and atrial intracardiac electrograms. Pacing Clin Electrophysiol 13(10):1286–1297,1990.

58. DiCarlo L, Lin D, Jenkins J. Analysis of time and frequency domain patterns to distinguish ventricular tachycardia from sinus rhythm using endocardial electrograms. Circulation 76:IV-280, 1987.

59. Langberg JJ, Griffin JC. Arrhythmia indentification using the morphology of the endocardial electrogram. Circulation 72:III-474, 1985.

60. Tomaselli GF, Gibb WJ, Langberg JJ, Chin MC, Griffin JC. In vivo testing of a morphology based approach to cardiac rhythm identification using the endocardial electrograms. Circulation 76:IV-280, 1987.

61. Pannizzo F, Wanliss M, Furman S. Discrimination of ventricular tachycardia electrograms by syntactic methods. PACE 10:A-725, 1987.

62. Pannizzo F, Furman S. Pattern analysis in tachycardia detection: a comparision of algorithms. Circulation 76:IV-281, 1987.

63. Lux RL, Nussbaum JA, Mannis D, Freedman RA, Mason JW. Dissimilarity in morphology of premature ventricular complexes and ventricular tachycardia. Circulation 74 (suppl II):187, 1986.

64. Amikam S, Furman S. A comparison of antegrade and retrograde atrial depolarization in the electrogram. PACE 6:111, 1983.

65. Davies DW, Wainwright RJ, Tooley M, Lloyd D, Camm AJ. Electrogram recognition by digital analysis: relevance to pacemaker arrhythmia control? J Am Coll Cardiol 5:570-A, 1985.

66. Schuger C, Jackson K, Steinman R, Lehman M. Atrial sensing to augment ventricular tachycardia detection by the automatic implantable cardioverter defibrillator: a utility study, PACE 11:1456–1464,1988.

67. Walsh C, Singer L, Mercando A, Furman S. Differentation of arrhythmias in the dog by measurement of activation sequence using an atrial and two ventricular electrodes. PACE 11:1732–1738,1988.

68. DiCarlo LA, Lin D, Jenkins JM. Automated interpretation of cardiac arrhythmias. Design and evaluation of a computerized model. J Electrocardiol 26(1):53–67,1993.

69. Munkenbeck F, Bump T, Arzbaecher R. Differentiation of sinus tachycardia from paroxysmal 1:1 tachycardias using single late diastolic atrial extrastimuli. PACE 9:53,1986.

70. Arzbaecher R, Bump T, Jenkins J, Munkenbeck F, Brown J. Automatic tachycardia detection and distinction in anti-tachycardia pacing. PACE 8:A-48, 1985.

71. Peebles PZ. Probability, random variables, and random signal principles. 3rd ed. New York: McGraw-Hill, 1993, pp xxiii, 401.

72. Chen S, Thakor NV, Mower MM. Analysis of ventricular arrhythmias: a reliable discrimination technique. Computers in Cardiology, 1986, pp 179–182.

73. Aubert AE, Denys BG, Ector H, DeGeest H. Fibrillation recognition using autocorrelation analysis. Computers in Cardiology, 1982, pp 477–480.

74. Davies DW, Wainwright RJ, Tooley MA, Nathan AW, Camm AJ. Endocardial electrogram analysis for the automatic recognition of ventricular tachycardia. Circulation 72:III-474, 1985.

75. Mercando A, Gabry M, Klemetowicz P, Furman S. Detection of ectopy by measurement of ventricular activation sequence using two electrodes. J Am Coll Cardiol 7:184A, 1986.

76. Mercando AD, Furman S, Fisher JD, Kim SG. Stability of activation sequence measured by two ventricular electrodes during supraventricular tachycardia. J Am Coll Cardiol 9:199A, 1987.

77. Finelli CJ. The time-sequenced adaptive filter for analysis of cardiac arrhythmias in intraventricular electrograms. IEEE Trans Biomed Eng 43(8):811–819,1996.

78. Ropella K, Lovett E. Parametric approaches to coherence estimation for intracardiac electrograms. Proceedings of the Int Conf for the Engineering in Medicine and Biology Society, 1993, pp 707–708.

79. Haykin SS. Adaptive filter theory. 3rd ed. Upper Saddle River, NJ: Prentice-Hall, 1996, pp xvii, 989.

80. Dufault RA, Wilcox AC. Dual lead fibrillation detection for implantable defibrillators via LMS algorithm. Computers in Cardiology, 1987, pp 163–166.

81. Marple SJ. Digital Spectral Analysis with Applications. Englewood Cliffs, NJ: Prentice-Hall, 1987.

82. Throne R, Wilbur B, Blakeman B, Arzbaecher RC. Autoregressive modeling of epicardial electrograms during ventricular fibrillation. Computers in Cardiology, 1991, pp 197–200.

83. Lovett E, Ropella K. Autoregressive spectral analysis of intra-cardiac electrograms: comparison to fourier analysis. Computers in Cardiology, 1992, pp 503–506.

84. Bloem D, Arzbaecher R. Discrimination of atrial arrhythmias using autoregressive modelling. Computers in Cardiology, 1992, pp 235–238.

85. Oberg A, Samuelsson RG. Fourier analysis of cardiac action potentials. J Electrocardiol 14(2):139–142,1981.

86. Pannizzo F, Furman S. Frequency spectra of ventricular tachycardia and sinus rhythm in human intracardiac electrograms—application to tachycardia detection for cardiac pacemakers. IEEE Trans Biomed Eng 35(6):421–425,1988.

87. Aubert AE, Goldreyer BN, Wyman MG, Ector H, DeGeest H. Frequency analysis of VF episodes during AICD implantation. PACE 11:891, 1988.

88. Morkrid L, Ohm OJ, Engedal H. Time domain and spectral analysis of electrograms in man during regular ventricular activity and ventricular fibrillation. IEEE Trans Biomed Eng 31(4):350–355,1984.

89. Craelius W, Hussain SM, Pantapoulos D, Saksena S, Parsonnet, V. Intraoperative spectral analysis of ventricular potentials during sinus rhythm and ventricular tachycardia. PACE 6:321, 1983.

90. Pannizzo F, Furman S. Optimal tachycardia sensing for cardiac pacemakers. PACE 8:785, 1985.

91. Lin D, Jenkins JM, Wiesmeyer MD, Jadvar H, DiCarlo LA. Analysis of time and frequency domain patterns of endocardial electrograms to distinguish ventricular tachycardia from sinus rhythm. Computers in Cardiology, 1986, pp 171–174.

92. Brachman J, Stroobandt R, Aidonidis I, Senges J, Kubler W. Analysis of monophasic action potentials facilitates differentiation of ventricular tachyarrhythmias. PACE 9:308, 1986.

93. Stroobandt R, Smits K, Wielders P, Bourgeois I, Brachmann J, Kubler W, Senges J. Algorithm for the detection of sustained ventricular tachycardia and ventricular fibrillation. PACE 8:48, 1985.

94. Goldreyer BN, Almquist CK, Beck RG, Olson WH. Waveform and frequency analysis of unipolar, bipolar, and orthogonal atrial electrograms. PACE 9:283, 1986.

95. Carter C, Knapp CH, Nuttall B. Estimation of the magnitude-squared coherence function via overlapped fast Fourier transform processing. IEEE Trans Audio Electroacoustics 21:337–344,1973.

96. Ropella KM, Sahamian AV, Baerman JM, Swiryn S. The coherence spectrum. A quantitative discriminator of fibrillatory and nonfibrillatory cardiac rhythms. Circulation 80(1):112–119,1989.

97. Botteron G, Smith J. Quantitative assessment of the spatial organization of atrial fibrillation in the intact human heart. Circulation 93:513–518,1996.

98. Lovett EG, Ropella KM. Time-frequency coherence analysis of atrial fibrillation termination during procainamide administration. Ann Biomed Eng 25(6):975–984,1997.

99. Alfonso V, Tompkins W. Detecting ventricular fibrillation. IEEE Eng Med Biol Mag 14:153–159,1995.

100. Thakor N, Sherman D. Wavelet (time-scale) analysis in biomedical signal processing. In: Bronzino J, ed. The Biomedical Engineering Handbook. Boca Raton, FL: CRC Press, 1995, pp 886–906.
101. Senhadji L, Carrault G, Bellanger J, Passariello G. Comparing transforms for recognizing cardiac patterns. IEEE Eng Med Biol Mag 14:167–173,1995.
102. Jung J, Strauss D, Sinnwell T, Hohenberg R, Fries R, Wern H, Schieffer H, Heisel A. Assessment of intersignal variability for discrimination of atrial fibrillation from atrial flutter. PACE 21:2426–2430,1998.
103. Bardy GH, Olson WH, Fishbein DP, Fellow CL, Coltorti F, Weaver WD, Greene HL. Transvenous right ventricular impedance spontaneous ventricular arrhythmias in man. Circulation 72:III-474, 1985.
104. Olson WH, Miles WM, Zipes DP, Prystowsky EN. Intracardiac electrical impedance during ventricular tachycardia and ventricular fibrillation in man. J Am Coll Cardiol 5:506A, 1985.
105. Olson WH, Bennett TD, Huberty KP, Anderson KM. Automatic detection of ventricular fibrillation with chronically implanted pressure sensors. J Am Coll Cardiol 7:182A, 1986.
106. Kresh JY, Brockman SK. Arrhythmia detection and discrimination in man by monitoring intramyocardial pressure. Computers in Cardiology, 1986, pp 159–162.
107. Verrydt W, Van den Bossche J, Van den Bossche A, Van de Voorde P, Witters E, Aubert AE, Sansen W, Ector H, Degeest H. Automatic defibrillator, antitachy pacemaker and cardioverter. Computers in Cardiology, 1986, pp 45–48.
108. Nazeran H, Behbehani K. Neural networks in processing and analysis of biomedical signals. In: Akay M, ed. Nonlinear Biomedical Signal Processing. Piscataway, NJ: IEEE Press, 2000, pp 69–97.
109. Kaplan D, Smith J, Saxberg B, Cohen R. Nonlinear dynamics in cardiac conduction. Math Biosci 90:19–48,1988.
110. Kaplan D, Cohen R. Is fibrillation chaos? Circ Res 67:886–892,1990.
111. Gray R, Jalife J, Panfilov A, Baxter WT, Cabo C, Davidenko JM, Pertsov AM. Non-stationary vortex-like reentry as a mechanism of polymorphic ventricular tachycardia in isolated rabbit heart. Circulation 91:2454–2469,1995.

14
Optical Mapping of Microscopic Impulse Propagation

Stephan Rohr
University of Bern, Bern, Switzerland

I. CARDIAC FUNCTION AND MICROSCOPIC ACTIVATION PATTERNS

It is imperative for the efficient pump function of the heart that the electrical impulse generated by the sinus node propagates in a regular manner to the ventricular tissue. Along its journey from the sinus node, the impulse invades the atria, the AV node, and the fast intraventricular conduction system before it ultimately reaches the ventricles. There, it is crucial for the proper function of the heart that activation of the ventricular wall occurs almost simultaneously. This synchronicity of excitation is guaranteed by a highly effective transmission of the impulse among the cardiomyocytes, which represent a three-dimensional network of electrically coupled excitable elements.

In the past, it has been recognized that both the macroscopic and microscopic structures of this network of coupled excitable cells play an important role in shaping the characteristics of impulse conduction, as illustrated by the following selected experimental findings. (1) It was found several decades ago that impulse propagation is locally delayed or even fails at the site of an abrupt tissue expansion as represented, e.g., by the Purkinje fiber–ventricular junction [1,2]. This local impairment of conduction is due to the presence of a so-called current-to-load mismatch at the site of the tissue expansion, i.e., the depolarizing current generated by the tissue in front of the expansion (the Purkinje fiber) is barely large enough to drive the tissue of the expansion (bulk of the working myocardium) to threshold. (2) It has been known for many years that anisotropic conduction in the myocardium is the result of the specific cellular architecture of cardiac tissue, consisting of parallel aligned rod-shaped individual cells. This type of cytoarchitecture combined with an extensive longitudinal electrical coupling by gap junctions causes conduction to be significantly faster in the direction of the general fiber axis than it is in the transverse direction [3,4]. (3) Slow conduction velocities were observed in both chronically infarcted and aged ventricular myocardium. This tissue showed, histologically, an increase in collagenous septa, resulting in a reduction in the spatial frequency of lateral electrical coupling of muscle bundles. When these preparations were activated in a direction perpendicular to the bundles, the infrequent lateral coupling resulted in a "zigzag" course of activation and, thus, in macroscopically slow conduction [5–7]. (4) It was shown that surviving tissue in an infarct scar can exhibit a highly complex microarchitecture, with small tissue strands connecting islands of intact tissue ("mottled myocardium"), and it was suggested that such structures, due to their capacity for inducing slow conduction and unidirectional conduction blocks, contribute to the arrhythmogenicity of the border zone of healed infarcts [8]. While all of these findings underline the importance of the cellular microarchitecture of cardiac tissue for impulse propagation, they are based on experiments which were conducted with spatial resolutions considerably larger than the dimensions of single cells. Thus, it remained elusive to what extent individual cardiomyocytes were involved in the respective activation patterns. Owing to the absence of suitable experimental systems, questions regarding the involvement of individual cells in impulse conduction were addressed in the past nearly exclusively by computer simulations. These theoretical studies showed, e.g., that impulse propagation in a chain of single cells is discontinuous at the cellular level, due to the recurrent increases in axial resistance at the cell-to-cell borders [9,10], that the calcium inward current can be crucial for supporting conduction at sites of an impedance mismatch [11,12], and that, paradoxically, the safety of conduction in both linear cell strands [10] and at the site of a tissue expansion [13,14] can be increased by partial gap junctional uncoupling.

For all of these examples, where the microscopic architecture of cardiac tissue is likely to influence the characteristics of impulse propagation, it would obviously be desirable to investigate the underlying structure–function relationships directly in "real" tissue. Ideally, such an experimental system would permit the correlation of cellular microarchitecture and microscopic impulse propagation in a preparation with a defined cellular structure. It is the goal of this chapter to describe such an experimental system, with emphasis placed on methodological aspects of measuring impulse propagation at the cellular/subcellular scale. After defining the general requirements for measuring microscopic impulse propagation, there will be a short paragraph on the fundamentals of optical recording of transmembrane voltage. This will be followed by a discussion of issues relevant for the measurement of optical signals from subcellular regions. Finally, an implementation of a suitable recording system is presented, and its capabilities are illustrated by two examples of measurements of impulse propagation in cardiac tissue with cellular/subcellular resolution.

II. MEASURING MICROSCOPIC IMPULSE PROPAGATION

In order to investigate the question of how the cellular architecture of cardiac tissue influences propagation, it is required that (1) the cellular architecture of the tissue under investigation is known precisely and (2) propagation in a given preparation can be followed with cellular or subcellular resolution. The latter requirement implies that the spatial resolution of the recording system is smaller than the dimensions of single cells and that the temporal resolution is sufficiently high to permit the accurate tracking of activation delays between and within individual cells. Considering that, e.g., cultured cardiomyocytes measure $\sim 15\,\mu m \times 60\,\mu m$ [15], the spatial resolution as defined by the detector size referred to the object plane should be $\leq 10\,\mu m$ (Fig. 1A). At this resolution, activation delays between adjacent detectors at normal conduction velocities of $0.5\,m/sec$ are $20\,\mu sec$ for the case of continuous ("axonlike") conduction (Fig. 1B). If, however, macroscopic conduction velocities of $0.5\,m/sec$ are based on microscopically discontinuous conduction, interdetector activation times within a given cell will be even shorter. According to previous computer simulation studies of conduction along a chain of single cardiomyocytes [9,10,16], it can be expected that propagation times along individual cells ($\sim 60\,\mu sec$) are similar to propagation times across cell-to-cell borders ($\sim 60\,\mu sec$) under conditions of normal gap junctional coupling. Thus, for conduction velocities of $0.5\,m/sec$ and for a spatial resolution of $10\,\mu m$, intracellular activation delays between contiguous detectors will be $\sim 10\,\mu sec$, whereas intercellular activation delays will be of the order of $60\,\mu sec$ (Fig. 1C). Intracellular activation delays will decrease further if gap junctional resistance and therefore the degree of discontinuity increases [10]. In order to resolve such small temporal differences, it must be known with high precision ($\sim 1\,\mu sec$) when a given membrane patch is activated.

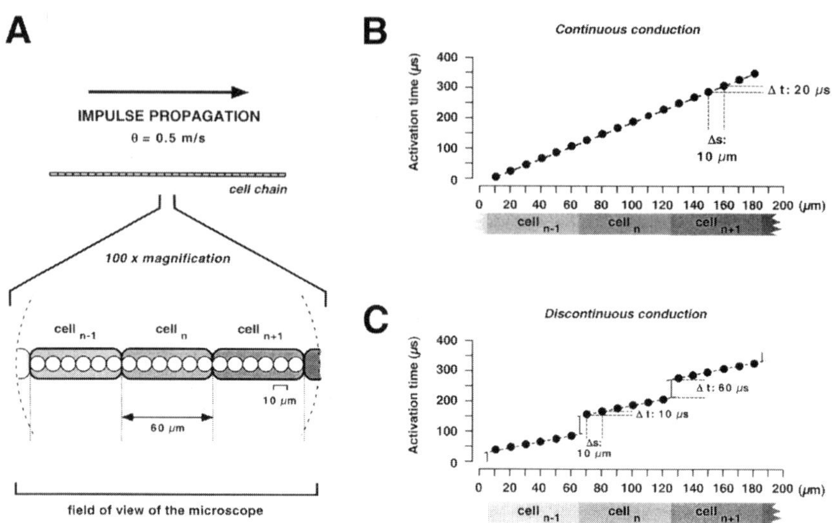

Figure 1 Spatiotemporal requirements for the assessment of microscopic impulse propagation. (A) Spatial requirements: because intermediate image planes of microscopes measure typically 20 mm in diameter, a magnification of 100× permits the imaging of 3 full cell lengths (shaded in gray; cell length = 60 μm). The circles indicate the positioning and size of detectors used to measure microscopic propagation. (B) Temporal requirements: at conduction velocities of 0.5 m/sec, activation delays between adjacent detectors amount to 20 μsec during continuous conduction. (C) During discontinuous conduction exhibiting the same overall velocity (0.5 m/sec), intracellular activation delays decrease to 10 μsec while cell-to-cell activation delays amount to 60 μsec.

A. Suitability of Different Types of Preparations

Even though intact cardiac tissue would be preferable in many ways as an experimental preparation for establishing a microscopic structure–function relationship, it is not yet amenable to this type of study, because presently available recording techniques are not yet sufficiently advanced to permit the characterization of activation within a complex three-dimensional cellular network with microscopic resolution. Even thin epicardial preparations obtained by cryoablation are still several cell layers thick, thus rendering the identification of single cells very difficult if not impossible [17]. An alternative approach consists of reducing the deminsionality of the preparation, i.e., of using two-dimensional monolayer cultures of cardiomyocytes instead. In this instance, each and every cell involved in the activation process can be readily identified, and impulse propagation can be followed at the level of single cells by using appropriate recording techniques. Moreover, cell cultures have the advantage that photolithographic patterning techniques permit the construction of virtually any two-dimensional tissue geometry [15] and, thus, the systematic investigation of the dependence of impulse propagation on the two-dimensional geometry of the tissue.

B. Suitability of Different Recording Techniques

In the past, three types of recording techniques have been commonly used to determine patterns of activation in cardiac tissue: (1) measurement of conduction between two or more intracellular electrodes, (2) multiple-site extracellular recordings, and (3) multiple-site optical recording of transmembrane voltage (MSORTV). While obviously, multiple intracellular recordings cannot be used for assessing conduction at the cellular/subcellular level in monolayer cultures because of practical reasons, the use of multiple-site extracellular recording appears to be an appropriate approach. There, photolithographic techniques permit the construction of closely spaced extracellular electrodes which can report activation patterns from the monolayer cultures growing on top of the electrodes [18,19]. The major disadvantages of this method in the context of measuring impulse propagation at the subcellular scale are twofold: first, even the smallest interelectrode distance of $20\,\mu m$ achieved so far is not small enough for obtaining a detailed picture of impulse propagation within single cells and across cell-to-cell borders. Second, computer simulations have suggested that, even at a hypothetical interelectrode spacing of $5\,\mu m$, extracellular electrodes might fail to detect propagation delays across cell-to-cell borders [9]. These disadvantages can be overcome by optical recording methods. These techniques are based, as outlined in more detail below, on the use of so-called voltage-sensitive dyes, which, after insertion into the cell membrane, report local changes in transmembrane voltage by changing their optical properties. Thus, these dyes act like localized voltmeters which can be "interrogated," using appropriate recording techniques, as to how and when during a propagated impulse a given patch of cell membrane is activated. It is the goal of the remainder of this chapter to give an introduction to this technique and to discuss its advantages and limitations in the context of measuring impulse propagation at the cellular/subcellular level in monolayer cultures of cardiomyocytes.

III. FUNDAMENTALS OF OPTICAL RECORDING OF TRANSMEMBRANE VOLTAGE

Optical recording of transmembrane voltage is based on the use of voltage-sensitive dyes. These molecules are capable of sensing local changes in electric fields, and react with a change in their optical properties which can be recorded using appropriate detectors. Since the first descriptions of the use of voltage-sensitive dyes in neuronal preparations in the early 1970s by Larry Cohen and Brian Salzberg, and their colleagues (for review cf. [20]), roughly 2000 dyes have been tested for their ability to report transmembrane voltage changes in excitable cells. These dyes can be categorized based on how they react to a change in electric field (change in absorption, birefringence, or fluorescence) and how rapidly they react to such a change (fast versus slow).

A. How Do Voltage-Sensitive Dyes Work?

Among the different types of potentiometric probes, fast fluorescent dyes have found widespread application in cardiac optoelectrophysiology, and the structure of a representative of this class of indicators (di-8-ANEPPS) is illustrated in Fig. 2A. While it is beyond the scope of this chapter to go into details of the molecular mechanisms of voltage sensitivity (cf. [21]), there is a series of interesting aspects to this dye which have implications for its use. As shown in Fig. 2A, the dye molecule has an amphiphilic structure which favors its insertion into the phospholipid bilayer of the cell membrane, where the chromophore is aligned such that the naphthalene is pointing toward the cell interior (Fig. 2B). It was suggested that changes in the electric field across the cell membrane interact with the chromophore such that they induce intramolecular shifts of electrical charge (electrochromism [22,23]). In addition, it was proposed that the dye is slightly dislocated during a change in membrane potential [24]. These effects of a change in transmembrane voltage manifest themselves, as shown in Fig. 2C, as a blue-shift of both the excitation and emission spectra during depolarization. In addition, there is a slight decrease of the peaks of both spectra (in the range of a few percent) and a slight broadening of the emission spectrum (not indicated in the figure; for details cf. [24]). From these spectral shifts it follows that the choice of excitation/emission filters determines whether a depolarization $(+V_m)$ is followed by a decrease or an increase in the intensity of the emitted light. As an example, given a fixed broad-band excitation, placing the emission filter at the right ("red") wing of the emission spectrum will result in a decrease in emitted fluorescence upon depolarization, whereas placement of the filter at the left ("green") side of the spectrum will have the opposite effect, i.e., depolarization will be accompanied by an increase in emitted fluorescence. Normally, di-8-ANEPPS and related dyes are used in the "inverse" mode (reduction in emitted fluorescence during depolarization), because the red region of the spectrum is sufficiently remote from the excitation range to permit a clear separation of the two wavelengths by appropriately chosen excitation and emission filters.

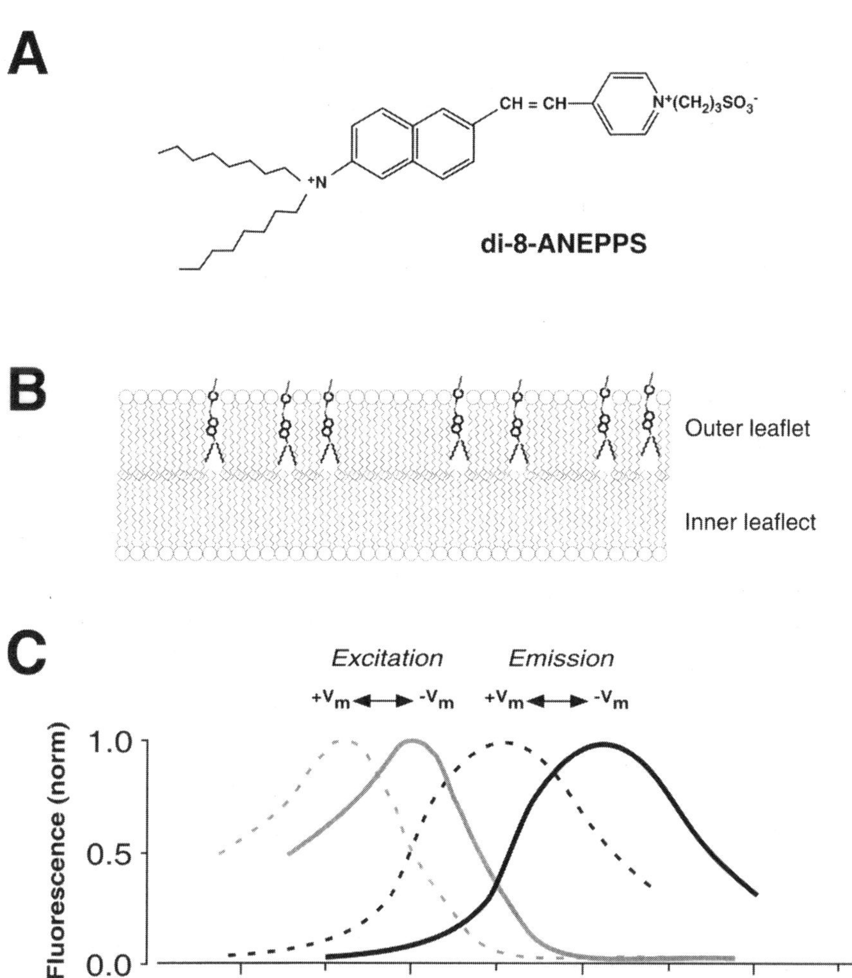

Figure 2 Properties of the voltage-sensitive dye di-8-ANEPPS. (A) Molecular structure of di-8-ANEPPS (di-8-butyl-amino-napthyl-ethylene-pyridinium-propyl-sulfonate). (B) Schematic drawing of the insertion of dye molecules into the outer leaflet of the phospholipid bilayer constituting the cell membrane. (C) Shifts in excitation and emission spectra upon polarization (V_{m-}) and depolarization (V_{m+}) of the cell membrane. (Redrawn with permission from Ref. 25. Copyright 1999. Biophysical Society.)

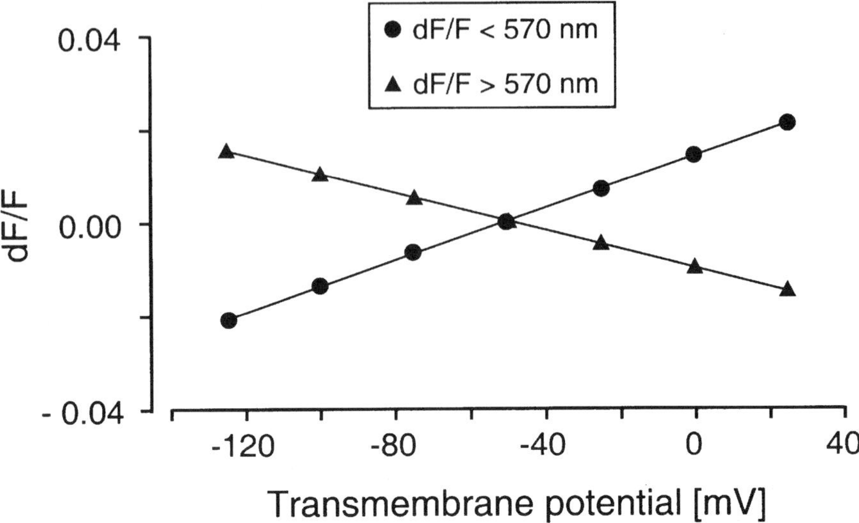

Figure 3 Correlation between transmembrane potential and fractional fluorescence changes ($\Delta F/F$) for the potentiometric dye di-8-ANEPPS. At wavelengths of emission <570 nm, depolarization results in an increase of $\Delta F/F$, whereas at wavelengths >570 nm, depolarization is followed by a decrease of $\Delta F/F$. In both cases, the fractional changes in fluorescence are linearly related to membrane potential. (Redrawn with permission from Ref. 25. Copyright 1999. Biophysical Society.)

The dependence of the change in emitted fluorescence on transmembrane voltage is shown in Fig. 3 for both the emission set to >570 nm and that set to <570 nm ([25]; cultured neurons stained with di-8-ANEPPS). In either case, there exists a linear dependence between transmembrane voltage and change in fluorescence within the range of voltages occurring during a normal action potential. Thus, like many other dyes, di-8-ANEPPS behaves like a linear voltage sensor and therefore produces signals with a shape identical to a conventional microelectrode recording. An example for this congruence between electrically and optically recorded action potentials is illustrated in Fig. 4, which shows complete superposition of the two signals during the action potential upstroke, whereas the repolarization phase in the optically recorded signal is distorted by the contraction of the cell ("motion artifact").

While optical recordings faithfully reproduce the temporal course of transmembrane voltage changes, they generally do not allow the determination of absolute values of transmembrane voltages [26]. While this does not pose a problem for following impulse propagation in excitable tissue,

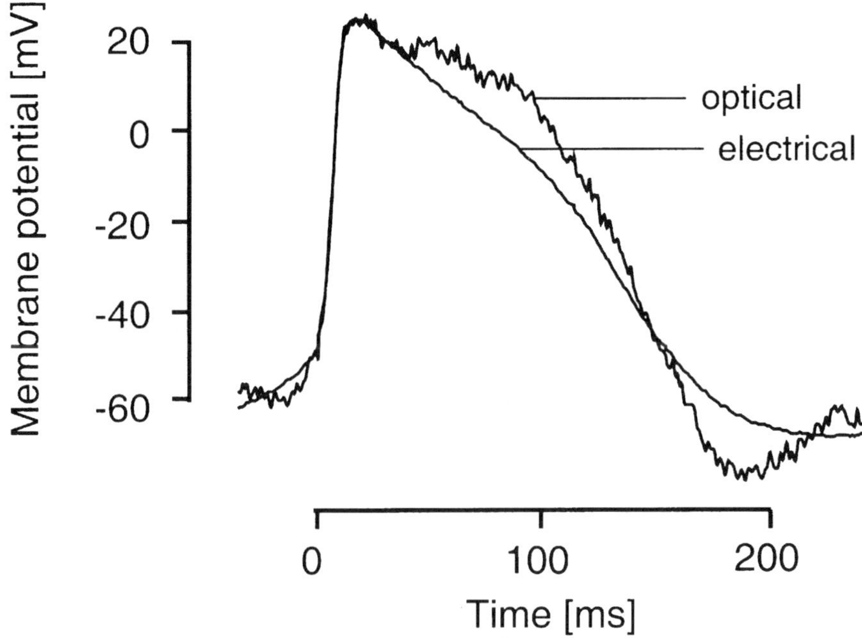

Figure 4 Comparison between optical and electrical measurements of an action potential in cultured cardiomyocytes. While the time course of the action potential upstroke is identical for the two types of measurements, the repolarization phase is distorted in the optical recording due to the motion artifact. (Redrawn with permission from Ref. 34. Copyright 1994. Biophysical Society.)

there exist instances where it would be desirable to know the spatial pattern of absolute voltages present in a given preparation. Because both di-8-ANEPPS and the closely related dye di-4-ANEPPS display a shift in excitation and emission spectra upon a change in transmembrane voltage, it has been suggested that estimates of absolute values of transmembrane voltage can be obtained by using a ratiometric approach. Accordingly, dual-wavelength excitation ratiometry was shown to produce estimates of transmembrane voltage with a precision of 10 mV in lipid vesicles stained with di-4-ANEPPS [27]. On the other hand, it was recently reported using a dual-wavelength emission approach that absolute values for membrane potentials in cultured neurons can be assessed with a precision of 5 mV [25]. These measurements, however, each require individual calibrations for different tissues.

B.   Measuring Small Optical Signals

Signals from voltage-sensitive dyes tend to be small, thus resulting generally in modest signal-to-noise ratios (SNRs). This disadvantage is further aggravated when assessing microscopic impulse propagation, because only minute cell membrane areas contribute to the signal. From this and from the fact that the precise determination of activation times at the subcellular scale requires high bandwidth, and is also critically dependent on adequate SNRs, it is evident that successful measurements of microscopic impulse propagation necessitate a careful optimization of signal sizes and a minimization of noise. In the following, common sources of noise and factors determining the signal size in optical recording systems are briefly reviewed, whereas details concerning optimization of SNRs in the context of measuring impulse propagation in cultured cardiomyocytes are treated in Sec. IV.C.

1.   Sources of Noise

a. Photon Noise.   The accuracy of any given measurement of light is limited by the stochastic nature of photon emission by the light sources. This results in fluctuation of the number of photons emitted over time and sets a physical limit to maximally achievable signal-to-noise ratios (SNRs). For an ideal light source (tungsten filament), the magnitude (RMS) of the fluctuations corresponds to the square root of the average number of photons emitted over time. As an example, if a light source emits 10,000 photons/nsec, the fluctuation (RMS) amounts to 100 photons/nsec and maximal SNR is 100. This square-root relationship between signal-to-noise ratios and light intensities at the detector is illustrated by the line indicating photon noise (shot noise)-limited measurements in Fig. 5. The relationship implies that, even in the hypothetical case of a completely noiseless detector with a quantum efficiency of 100% (i.e., each photon impinging on the detector is converted into an electron), SNRs can be very low when the photon flux at the level of the detector is low, because of the increasing relative size of the photon noise. Conversely, the relationship also implies that, during photon noise-limited measurements, any measure increasing the intensity of the light at the level of the detector will result in an increase in SNR following a square-root relationship (i.e., doubling of the intensity will result in an improvement of SNR by ~40%).

b. Dark Noise.   As the name implies, dark noise denotes the level of noise present in an optical recording system under conditions of complete darkness. This "electronic" noise is generated both at the level of the photo-detector and by the amplifiers themselves and sets a floor below which no signals can be detected. Because, unlike photon noise, dark noise has a fixed

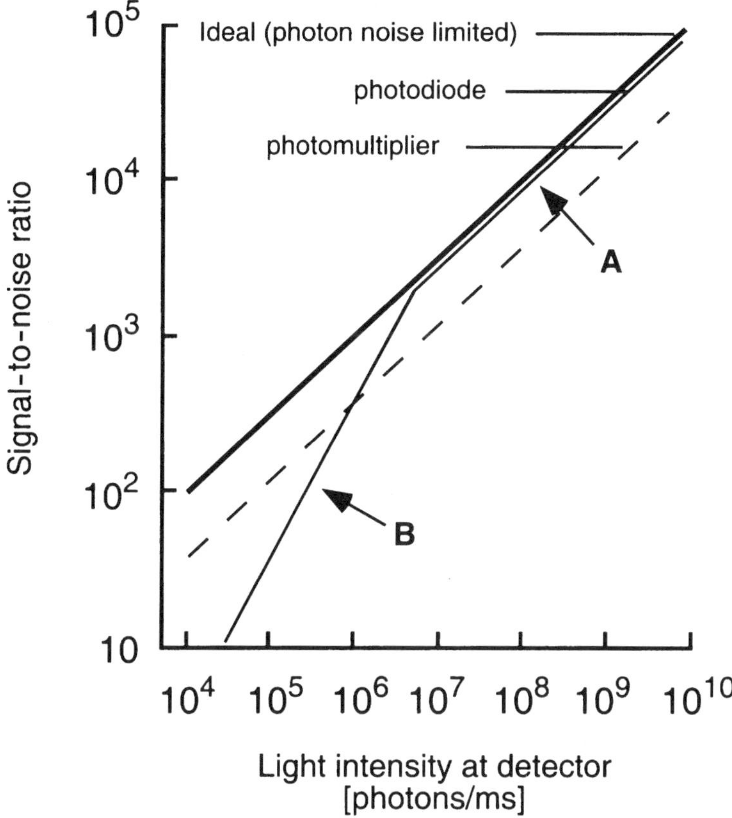

Figure 5 Dependence of signal-to-noise ratios (SNRs) on light intensity at the detector. Bold line: SNR as a function of light intensity during photon noise-limited measurements using an ideal detector (quantum efficiency Q.E. = 100%, no dark noise). Fine line: maximal SNRs obtained with a silicon photodiode. Segment A indicates the region of photon noise-limited measurements. Compared to the ideal case, SNRs at comparable light intensities are slightly lower because the Q.E. is <100%. Segment B indicates the region of dark noise-limited measurements where SNR decreases linearly with decreasing light intensity at the detector. Dashed line: maximal SNRs obtained with a photomultiplier tube (Q.E. 15%). Because dark noise is virtually absent, the curve runs in parallel to the curve, indicating ideal photon noise-limited measurements over the entire range of light intensities.

size, SNRs in the case of dark noise-limited measurements will decrease linearly with decreasing signal size. This is indicated schematically in Fig. 5 by segment B of the line characterizing photodiode performance.

c. Movement. Movements of extrinsic (vibrational noise) or intrinsic (contraction of cells) origin have the capacity to significantly distort signals recorded with voltage-sensitive dyes [28]. Noise originating from vibrations can be counteracted by adequate measures such as placing the recording setup on a vibration isolation table or, if necessary, by moving the setup to a location with a stable floor (basement). In regard to contraction-related signal artifacts, it depends on the goals of a given experiment whether they have to be suppressed. If, in a given experiment, only the spatial pattern of activation is of interest, contraction artifacts are generally of no concern because they start with a latency of several milliseconds with respect to the action potential upstroke. If, on the other hand, the repolarization phase needs to be assessed, either pharmacological or optical means have to be used to suppress the artifact, as outlined in more detail below.

d. Instabilities of Light Sources. Light sources themselves can contribute to the noise level in optical signals. Ripple produced by power supplies produces ripple in the light level which is easy to spot due to its regularity. This type of noise can be suppressed by using either low-ripple power supplies or by driving the lamp with batteries. In arc lamps, the so-called arc wander can induce sporadic fluctuations in light intensity. This type of noise can be reduced by using short arc lamps, in which arc wander is generally less of a problem. Moreover, xenon arc lamps seem to be somewhat quieter than mercury arc lamps.

2. Factors Determining Signal Size

a. Light Sources. Light sources generally used for exciting voltage-sensitive dyes include tungsten halogen lamps and mercury or xenon arc lamps. While tungsten lamps in conjunction with low-ripple power supplies produce a very stable light output, arc lamps produce higher light flux densities per watt, because the size of their source is considerably smaller (e.g., $0.4\,mm^2$ for a 100-W xenon arc lamp) than the filament size of a tungsten halogen lamp (e.g., $9.5\,mm^2$ for a 100-W lamp). It has been reported that a 150-W xenon arc lamp yields two to three times more light than a comparable tungsten-filament lamp [29]. Moreover, the smaller size of the arc lamp source permits more efficient light collimation for high-magnification objectives. Practically, short arc lamps in the 100 to 250-W range in conjunction with low-ripple power supplies produce adequate energy in the range of wavelengths used to excite commonly used voltage-sensitive dyes, to obtain useful signal-to-noise ratios during microscopic measurements of transmembrane voltage changes.

b. Microscope Optics. The suitability of a given microscope for measuring microscopic impulse propagation is determined foremost by the efficiency of the light throughput. While light throughput is increased in system designs with as few optical elements in the optical path as possible (e.g., microscopes with a bottom exit which have no light-deflecting elements in the optical path), the most decisive element is the objective. In epifluorescence measurements, both the intensity of excitation light reaching the preparation and the amount of emitted fluorescence from the preparation is a square function of the numerical aperture (N.A.) of the objective used. This implies that the total intensity of the emitted light of a given objective is related to the fourth power of the N.A. In practical terms this means that the emission intensity obtained with a $100\times$, 1.4-N.A. objective is roughly four times as bright as that obtained with a $100\times$, 1.0-N.A. objective and, consequently, SNRs will be increased by a factor of 2. In addition, the number of optical elements in highly corrected objectives must also be considered, as two objectives having the same N.A. may have very different transmission ratios.

c. Photodetectors. Various types of photodetectors are currently used in fluorescence imaging. Among these, CCD imagers and photodiode arrays are the imagers of choice for optical recordings of spatial patterns of transmembrane voltage changes using voltage-sensitive dyes, whereas photomultiplier tubes serve to record changes in fluorescence intensity in confocal or random-access scanning systems. In order to understand which detector system is best suited for recording microscopic impulse propagation in cardiac tissue, it is necessary to review briefly the characteristics of these detector types with respect to their spatial resolution, their quantum efficiency, their dark noise, and their speed (for a more exhaustive discussion of the characteristics of photodetectors, cf. [30]). As outlined above, it is essential for measuring microscopic propagation that signals are obtained with high enough spatial resolution (10 µm or better), with sufficient temporal resolution ($\sim$1 µsec), and with adequate SNRs. In the context of SNRs, the efficiency with which photons are converted into electrons (quantum efficiency, Q.E.) by any type of photodetector deserves special attention, because SNRs increase with the square root of Q.E. up to the point (Q.E. = 100%) where optical recordings become photon noise-limited (cf. Fig. 5). Thus, a detector with a Q.E. four times better than another detector will improve SNRs by a factor of 2.

CCD Cameras. The pixel size of commonly available CCD imagers is in the few micrometers range. Thus, these detectors offer a very high spatial resolution when coupled to a microscope. Moreover, the quantum efficiency (Q.E.) in the 600- to 700-nm range, which corresponds to the range of emission spectra of commonly used voltage-sensitive dyes, is 40% or better. This number is even higher when using recently available thin back-illuminated CCDs, which exhibit Q.E.s of up to 90% at 600 nm. While both of these characteristics would make CCD imagers the ideal choice for measuring impulse propagation at the cellular scale, the small pixel size implies that the number of collected photons per individual detector during microscopic measurements is small and, hence, SNRs are modest. The main drawback of commonly available CCD imagers, however, is their slow readout speed, which prevents their use in measuring impulse propagation at the cellular scale. The frame rate of "fast" CCD systems is in the 100-Hz range which implies, under the assumption of a conduction velocity of 0.5 m/sec and a cell length of 100 µm, that the impulse would encompass 50 cells between frames. Even faster systems developed specifically for the measurement of impulse propagation in excitable tissues using voltage-sensitive dyes, which run at up to 2 kHz ([31]; commercial systems from RedShirtImaging LLC), are too slow to capture details of propagation at the subcellular scale (roughly two cells are activated between consecutive frames). However, and as described in detail in other chapters of this book, this type of detector is well suited for recording macroscopic activation in intact cardiac tissue because (1) signal sizes are relatively large, as many stacked cell layers contribute to the signal, and (2) activation times are relatively long, thereby permitting the determination of activation patterns at the "slow" frame rates typical for these devices.

Photodiodes and Photodiode Arrays. There are several reasons why photodiodes are the detectors of choice for measuring signals produced by voltage-sensitive dyes with microscopic resolution. Besides the fact that photodiodes are relatively inexpensive and rugged, they have a number of specific advantages whenever available light intensities at the level of the detector fall into the range where measurements are photon noise-limited (segment A of the photodiode curve in Fig. 5). (1) Quantum efficiency: photodiodes have high Q.E.s in the range where commonly used voltage-sensitive dyes emit light ($\sim$80% at 600–700 nm). This gives them a considerable advantage over, e.g., photomultipliers, which have Q.E.s at the same wavelengths of only up to $\sim$15%. This difference results, at a given emission intensity, in an SNR of photodiodes which is more than twice as large as that of photomultipliers. (2) Dynamic range: Photodiodes have a large dynamic range, i.e., they can be used to measure light intensities ranging from below $10^{-13}\,\mathrm{W/cm^2}$ to intensities above $10^{-1}\,\mathrm{W/cm^2}$. This makes them ideally suited for measuring signals from voltage-sensitive probes because saturation, even in the case of absorption measurements, is generally not a problem. In this respect, photodiodes again differ from photomultiplier tubes, which can easily be overexposed and (and damaged). (3) Speed of response: When operated in biased photoconductive mode (fast but "noisy"), photodiodes can react within picoseconds to changes in light intensity, which explains their everyday use in fiber optic communication networks. But even if they are used in the "silent" nonbiased photo-conductive mode, they can, depending on the amplifier layout, still be orders of magnitude faster than any change in transmembrane voltage occurring in biological systems. (4) In contrast to the serial readout of CCD chips, signals from photodiodes can be read out in parallel, thus permitting "frame rates" adequate for the precise assessment of microscopic impulse propagation. (5) Packaging: Commercially available arrays of silicon photodiodes are fabricated on single chips with pixel counts ranging up to 34×34 [32]. Alternatively, arrays are manufactured by coupling discrete photodiodes rigidly to optical fibers which form the input window of the detector [33]. Pixel sizes of these arrays measure $\sim 1\,\text{mm} \times 1\,\text{mm}$, thus resulting in a spatial resolution of $\sim 10\,\mu\text{m}$ when used in conjunction with a $100 \times$ objective.

Avalanche Photodiodes.   Avalanche photodiode detectors can essentially be regarded as photodiodes operated at very high bias voltages (up to 2.5 kV). This voltage accelerates electrons generated during the absorption of photons up to the point where they start to induce an electron multiplication process similar to that in a photomultiplier tube. This results in a modest internal gain of a few hundred, which, together with their high Q.E. and their robustness, gives these devices some advantages over photomultiplier tubes in certain applications ("solid state PMTs"). However, even though such detectors with an inherent gain of a few hundred in combination with a high Q.E. would seem, at first, an ideal combination for measuring signals from voltage-sensitive dyes, avalanche photodiodes exhibit a substantial dark current which increases with increasing bias voltage. Because of this dark noise, the advantage of internal gain is partially offset and SNRs in the range of signal levels normally encountered during measurements with voltage-sensitive dyes are only slightly improved (unpublished observation). Furthermore, because these devices have to be driven with high voltages, the costs of complete systems are rather high.

Photomultiplier Tubes (PMT).   For low-light-level measurements, PMTs are still the preferred choice because of their very high sensitivity. In these detectors, electrons liberated from the photocathode by impinging photons are accelerated by an electric field and strike a series of dynodes from which they liberate additional electrons which are finally collected by the anode. This amplification process by electron multiplication is virtually noiseless and permits the identification and the counting of individual photons. While this noiseless type of signal amplification seems to render a PMT attractive for measuring signals from voltage-sensitive dyes, because measurements are photon noise-limited over a large range of light intensities, there is a significant drawback compared to photodiodes. Q.E.s at wavelengths of interest (600–700 nm) are lower by a factor of ≥ 4, thus resulting, at a given intensity of emission, in a reduction of SNRs by 50%. This is indicated by the dashed line in Fig. 5, which runs in parallel to but at a lower level to the line indicating photon noise-limited measurements. The crossover between the curves for photodiodes and PMTs indicates the light intensity at the level of the detector, below which PMTs will operate with a higher SNR than photodiodes. Thus, unless available light levels are very small and, consequently, SNRs in photodiode measurements are governed by the detector noise rather than the photon noise, PMTs will produce lower SNRs than the much less expensive photodiodes.

In summary, light intensity at the level of the photodetector and speed requirements ultimately determine which type of detector is best suited for measuring microscopic impulse propagation. At the low end of light intensities, PMTs are the detectors of choice, while at the high end, photodiodes are superior in noise performance. In between, there is a narrow window of intensities where avalanche photodiodes produce better SNRs than either of the two other types of detectors.

IV. SPECIFIC ASPECTS OF MEASURING MICROSCOPIC IMPULSE PROPAGATION IN CULTURED CARDIOMYOCYTES

While the previous section dealt with fundamental aspects of optical measurements of transmembrane voltage changes applicable to different types of preparations, the following sections deal with issues specifically related to the measurement of microscopic impulse propagation in cultured cardiomyocytes. In particular, it will give indications as to which dyes and which imaging systems are appropriate for this purpose and how SNRs, which represent the most critical point in measurements of microscopic impulse propagation, can be optimized.

A. Selection of Voltage-Sensitive Dyes

Among the many dyes tested for their ability to report transmembrane voltage changes in cultured cardiomyocytes (5 absorption dyes,

15 potentiometric dyes of the styryl class; for list, cf. [34]), only a few fluorescent dyes produced signals with acceptable signal-to-noise ratios, whereas absorption dyes produced no measurable signals at all. All of the fluorescent indicators were "fast" dyes, i.e., they react within microseconds to changes in transmembrane voltage [35]. They are therefore ideally suited to reproduce the action potential upstroke with high fidelity and, thus, permit a precise determination of the activation time of a given membrane patch. Given that several of the fluorescent dyes tested produced a signal in response to a transmembrane voltage step, what were the criteria for the selection of the most suitable dye?

1.　Fractional Fluorescence Change

Because signals recorded from microscopic areas on single cells are inherently small, it is necessary to evaluate dyes with the highest possible fractional fluorescence change ($\Delta F/F$: change in fluorescence intensity during a given voltage step divided by background fluorescence) in order to optimize the SNR. Among the dyes tested, the highest values of average $\Delta F/F$ for an action potential (corresponding to a transmembrane voltage change of $\sim$100 mV [15]) were 10% for di-8-ANEPPS, 8.5% for RH423, and 6.8% for di-4-ANEPPS [34]. These values were obtained with broad-band excitation (full width at half-maximum, FWHM, 90 nm) and broad-band emission. When both the ranges of excitation and emission were narrowed, $\Delta F/F$ increased up to 22.3% for di-8-ANEPPS.

2.　Internalization of the Dyes

Any process removing dye molecules from the sarcolemma will reduce the signal size. While some of the dye molecules are probably continuously lost to the superfusion medium, others are internalized into the cells. As illustrated by the fluorescence micrographs in Figs. 6A and 6B, the internalization of dyes is characterized by a decrease in membrane staining and a concomitant increase in staining of intracellular membranes. Even though the mechanism of dye internalization has never been investigated in detail (endocytotic membrane cycling?), the rate of this process is clearly dependent on the type of dye. As an example, di-4-ANEPPS, which differs from di-8-ANEPPS only with respect to the length of the di-alkyl chains (four carbon atoms instead of eight), is internalized at a substantially faster rate. This is illustrated in Fig. 6C, which shows the development of the ratio of perinuclear to nuclear fluorescence over time for cardiomyocytes stained with di-4-ANEPPS and di-8-ANEPPS. Clearly, the initial ratio is considerably lower in the case of di-8-ANEPPS and reaches values comparable to di-4-ANEPPS only after 40 min [34]. This finding suggests that the

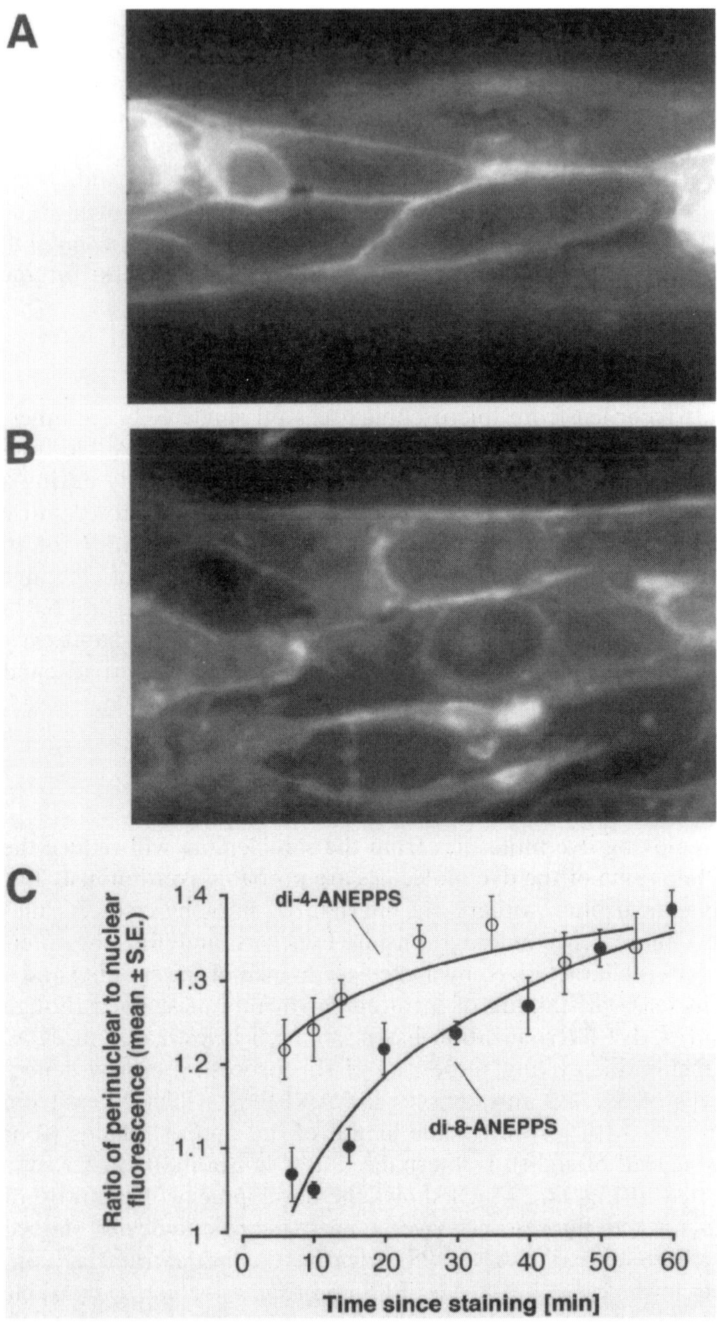
A
B
C
di-4-ANEPPS
di-8-ANEPPS
Ratio of perinuclear to nuclear fluorescence (mean ± S.E.)
1.4
1.3
1.2
1.1
1
0
10
20
30
40
50
60
Time since staining [min]

anchoring of dye molecules in the sarcolemma is improved with increasing length of the alkyl chains and it gives di-8-ANEPPS a clear advantage over di-4-ANEPPS in cultured cardiomyocytes.

3. Phototoxicity

A further decisive factor for dye selection is determined by the degree of photodynamic damage introduced by a given indicator. Unfortunately, and in contrast to absorption dyes (some of which permit measuring periods of tens of minutes without detectable damage to the preparations [36]), fluorescent potentiometric dyes are well known to exhibit substantial phototoxicity. Because the severity of phototoxic effects is related directly to the light levels incident on the preparation, they increase with increasing magnification due to the increase in light intensities necessary to achieve acceptable signal-to-noise ratios. This largely explains why phototoxic effects are virtually absent during macroscopic measurements in intact preparations, whereas during microscopic measurements in cultured cell monolayers, they are substantial and can limit optical recording to a single short exposure (at 100× magnification). While it is known that free-radical formation during illumination of the dye molecules is central to phototoxicity, the exact mechanisms and the variability in phototoxicity as a function of the type of dye have never been investigated systematically. Our own findings suggested that internalized dye molecules might aggravate phototoxicity. This conclusion is based on the finding that di-4-ANEPPS led to a substantially faster decay of action potential amplitudes in cultured cardiomyocytes than did di-8-ANEPPS (Fig. 7). This result might be explained by the possibility that radical formation from internalized fluorophores damages the cells more effectively than radicals released at the level of the surface membrane.

Based on all of these criterions, di-8-ANEPPS proved to be the most suitable dye for recording transmembrane voltage changes in cultured

Figure 6 Internalization of voltage-sensitive dyes. (A) Fluorescence micrograph recorded 8 min after staining cultured cardiomyocytes with di-8-ANEPPS. (B) Fluorescence micrograph of the same culture (different location) 38 min after staining. Clearly, there is an increase in staining of intracellular structures which is especially prominent in the perinuclear region. (C) Time course of dye internalization for di-4-ANEPPS (open circles, $n = 8$) and di-8-ANEPPS (filled circles, $n = 8$). Preparations were stained with 20 µg/mL of each dye, and the ratio of perinuclear to nuclear fluorescence was monitored for a period of 60 min thereafter. (Redrawn with permission from Ref. 34. Copyright 1994. Biophysical Society.)

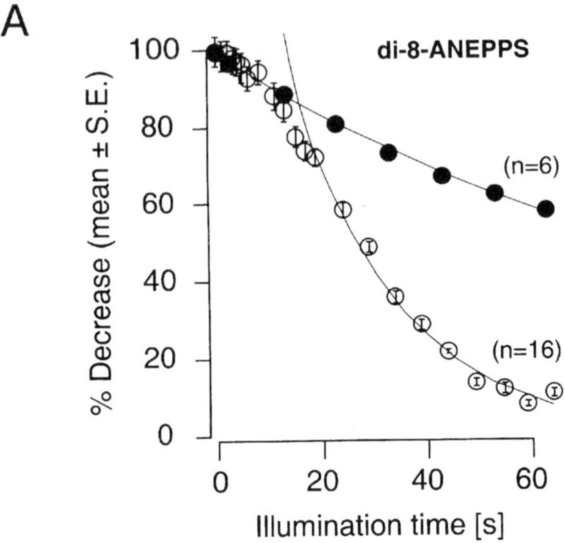

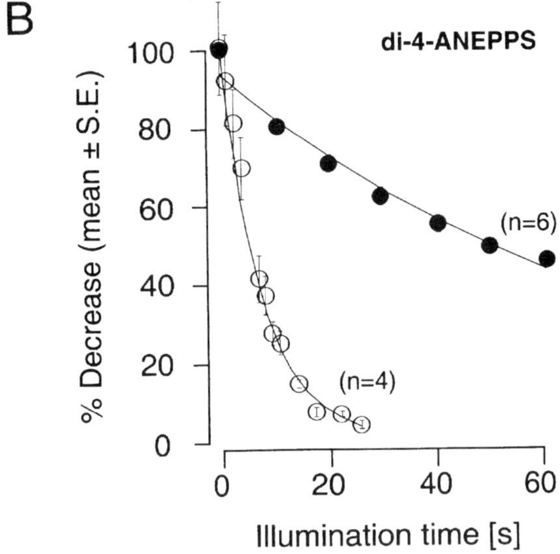

Figure 7 Comparison between signal decay and decrease in resting fluorescence during continuous illumination of cultured cardiomyocytes stained with di-8-ANEPPS (A) and di-4-ANEPPS (B). (A) During the initial 15 sec of illumination, signal amplitude decay (open circles) and decrease of resting fluorescence (filled circles) share a common time course, suggesting that the decrease in signal size is due

cardiomyocytes, and most of the example recordings shown below were obtained with this particular dye.

B. Temporal Resolution of the Recording System

Given that fast voltage-sensitive dyes react in a virtually instantaneous manner to changes in transmembrane voltage, the temporal precision of tracking propagated activity is dependent on SNRs, on the mode of determination of local activation times, and on the temporal fidelity (bandwidth) of the optical recording system.

1. Signal-to-Noise Ratios

Obviously, the higher the noise content of a given signal, the lower will be the precision with which activation delays between adjacent detectors can be determined. Moreover, at a given SNR, the mode of calculation of local activation times influences the precision of the measurement. Local activation times can be determined either (1) as the point in time when the action potential upstroke of a given membrane patch reaches half of the entire signal amplitude (at$_{50}$) [34] or (2) as the point in time of occurrence of the maximal upstroke velocity of the action potential (at$_{dV/dt_{max}}$) [16]. While the precision of both types of determinations is compromised by noise present in the signals, determinations based on at$_{50}$ are more robust. This is illustrated in Fig. 8, which shows the dependence of the variability in the determination of activation times on SNRs for both modes of calculation. The values were obtained by superposition of an optically recorded noise trace on a "noiseless" simulated smooth action potential upstroke, and defined SNRs were achieved by appropriate scaling of the noise amplitude. Independent of how activation times were determined, variations thereof increased drastically with decreasing SNR and with decreasing upstroke velocity. Moreover, the determination based on at$_{dV/dt_{max}}$ was less accurate than the determination based on at$_{50}$ at SNRs typically encountered in

---

solely to bleaching of the dye. Thereafter, signal amplitudes decay more rapidly than overall fluorescence, indicating onset of photodynamic damage. (B) In preparations stained with di-4-ANEPPS, the decay of signal amplitudes (open circles) was substantially faster than that of resting fluorescence (filled circles), indicating a rapid onset of photodynamic damage after the beginning of illumination which led to irreversible cell damage after 25 sec. (Redrawn with permission from Ref. 34. Copyright 1994. Biophysical Society.)

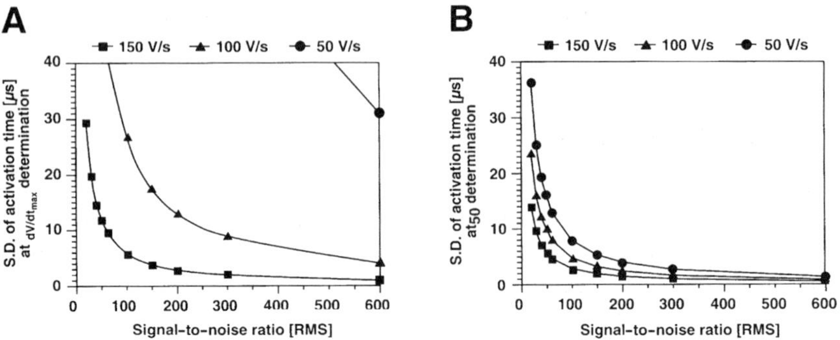

Figure 8 Accuracy of the determination of activation times as a function of signal-to-noise ratios. (A) Dependence of the variability in determinations of activation times based on the point in time of occurrence of the maximal upstroke velocity (at$_{dV/dt_{\max}}$) on SNR. With decreasing SNR and $dV/dt_{\max}$, the accuracy in the determination of activation times is increasingly reduced. (B) When activation times are determined using the point in time of 50% depolarization (at$_{50}$), similar relationships apply, but the accuracy at any given SNR and upstroke velocity is considerably increased when compared to determinations based on at$_{dV/dt_{\max}}$. (Reproduced with permission from Ref. 43. Copyright 1998. Biophysical Society.)

cultured cardiomyocytes (100–250). This suggests that an accurate measurement of microscopic activation is critically dependent on adequate SNRs and that preference should be given to the determination of local activation times based on at$_{50}$ in the case of monotonically rising action potential upstrokes.

2. Variances in Bandwidth Among the Detectors

Because the bandwidth of the photodiode–amplifier combination is, due to noise considerations (cf. Sec. IV.C.2), close to the bandwidth of fast-rising action potential upstrokes, it is essential that the frequency response of all amplifiers be identical. If this prerequisite is not met, the time course of the measured action potential upstrokes will vary according to the bandwidth differences among the amplifiers, thus introducing errors in the determination of local activation times.

3. Sampling Scheme

The sampling frequency of the analog-to-digital conversion system should be high enough to permit an accurate tracking of the action potential upstroke. Based on the Nyquist criterion, this implies that data are sampled at a frequency which is at least double the bandwidth of the optical amplifiers. Moreover, because the time of acquisition of each measurement point has to be known with high precision, all channels should be equipped with a sample-and-hold amplifier, thus ensuring synchronous data acquisition. Alternatively, if the channel scanning sequence of the multiplexer feeding the signals to the analog-to-digital converter (ADC) is known, signals can be corrected offline for the time skew introduced by the sequential channel scanning of the ADC.

If all of these prerequisites are fulfilled, the rising phase of each signal is exactly defined in time and delays between signals of interest can be calculated with high precision. In other words, even though optical amplifiers might exhibit rise times as slow as 100 μsec, their identical bandwidths in conjunction with a simultaneous sampling scheme permits the calculation of activation delays between signals of interest with a temporal precision in the few microsecond range.

C. Strategies for Maximizing Signal-to-Noise Ratios

As outlined above, the assessment of microscopic impulse propagation using voltage-sensitive dyes is critically dependent on obtaining optical signals exhibiting adequate signal-to-noise ratios in order to track accurately cellular/subcellular activation patterns during single-shot experiments. Often enough, SNRs are not satisfactory and means must be devised to improve the quality of the signals. In this situation, it is necessary to identify the primary source(s) of noise which can interfere with the signals generated by fluorescent voltage-sensitive dyes (for a more exhaustive discussion of noise sources, cf. [26,33,37]).

When, in a given situation, signal-to-noise ratios have to be improved, it first has to be known whether the recordings are photon noise or dark noise limited. One way to decide this question is to measure the noise level during illumination and during complete darkness. Alternatively, the photon flux can be systematically varied with neutral-density filters in order to decide whether SNRs change with the square root of the photon flux. If the noise level is independent of the photon flux, the measurements are most likely dark noise limited. On the other hand, knowing the fractional fluorescence change of a given dye and the overall gain, the photon flux at the level of the detector can be calculated, and from this the SNR for a photon noise-limited measurement can be estimated. If the measured SNR deviates substantially from this estimate, the measurement is again likely to be dark noise limited. Depending on whether the measurements are dark noise or photon noise limited, the following strategies may help to improve the SNR.

1. Photon Noise-Limited Measurements

In the case of photon noise-limited measurements, the two main strategies to improve SNRs consist of either increasing the signal size or of decreasing the bandwidth of the optical recording system. Either measure will cause an increase in SNR according to a square-root relationship.

a. Increase of the Intensity of Emitted Light

Dye-Related Issues. Maximizing SNRs involves a systematic dye screening with the aim of selecting the compound exhibiting the highest fractional fluorescence change ($\Delta F/F$). If such a dye is found, the staining protocol has to be adjusted such that the greatest possible amount of dye molecules is inserted into the surface membrane without compromising cell function (or producing inner filter effects; B. M. Salzberg, personal communication). Attention to these two details ensures that emitted light intensities and, hence, signal sizes are optimized.

Light Source. Obviously, the higher the intensity of the excitation light, the larger the emitted fluorescence and, thus, the larger the signal. Appropriately light levels can be obtained by switching from tungsten halogen lamps to arc lamps. In the latter case, it may be worthwhile to compare the excitation spectrum of the dyes used with the emission spectrum of the lamps in order to decide which type of arc lamp (xenon or mercury) delivers more energy in the excitation wavelength range of interest.

Numerical Aperture of the Objective. As discussed above, the N.A. of the objectives used in epifluorescence measurements is the main determinant for light throughput in the optical system. As an example, a change from a 100×, 1.0-N.A. objective to a 100×, 1.4-N.A. objective will increase emission intensity by a factor of 4 and, consequently, SNR by a factor of 2 in the case of photon noise-limited measurements, assuming the two objectives are otherwise similar in their transmission characteristics.

b. Bandwidth of the Recording System.

If it is consistent with the goals of a given experiment, decreasing the bandwidth of the recording system by analog or digital filtering techniques will increase the SNR according to a square-root relationship. However, as illustrated by the simulations shown in Fig. 8, decreasing the bandwidth to an extent which compromises maximal upstroke velocities will not necessarily increase the precision of determination of local activation times. Thus, one must carefully evaluate, in the context of a given experimental question, whether a reduction in bandwidth will actually benefit the analysis of the data.

2. Dark Noise-Limited Measurements

In the case of dark noise-limited measurements, SNRs can be improved by increasing the amplitude of the signal, by decreasing the bandwidth of the recording system, or by reducing amplifier noise.

a. Increase of the Signal Amplitude. Basically, all of the issues discussed above (Secs. IV.C.1–IV.C.3) photon noise-limited measurements apply for dark noise-limited measurements as well. The only difference concerns the efficiency of these measures, which will be higher in dark noise-limited measurements, because SNRs will increase proportionally with signal size (cf. Fig. 5). As an example, using a $100\times$, 1.4-N.A. objective instead of a $100\times$, 1.0-N.A. will increase SNR by a factor of 4 as opposed to a factor of 2 in photon noise-limited measurements. Another issue related to increasing SNR in dark noise-limited measurements concerns the choice of excitation and emission filters: while maximal $\Delta F/F$ values for a given dye are obtained by exciting the dye at its excitation maximum and measuring the emitted light at the emission maximum with narrow-band filters, this approach may not necessarily result in the best possible SNRs in dark noise-limited measurements. Instead, using broad-band excitation/broad-band emission will increase the signal size and, thus, increase SNR proportionally [34].

b. Reduction of the Bandwidth of the Recording System. The same issues as discussed in the previous section apply.

c. Noise Reduction in the Recording System. During optical measurements of voltage signals from microscopic membrane patches, the minute photocurrents generated by photodiodes have to be amplified in excess of 10^{10} in order to match the input ranges of commonly available analog-to-digital converters. For reasons of noise reduction, most of this large gain is normally implemented in the current-to-voltage converters (IVCs), which are connected directly to the photodiodes and in which a feedback resistor (R_f) determines the gain. Using sufficiently quiet operational amplifiers, a fundamental limit for the noise performance of an IVC is the Johnson noise arising from R_f. The current noise amplitude (RMS) produced by this resistor is $i_j = (4kTB/R_f)^{1/2}$, where k is the Boltzmann constant, T is the absolute temperature, and B is the bandwidth [38]. From this formula, it can easily be seen that the noise is minimized by a reduction of the bandwidth and by an increase in the value of R_f. From this, it would obviously be advantageous for the noise performance of the system to choose a feedback resistor which is sufficiently large to obtain the entire desired gain in the first amplification stage, i.e., in the IVC ($\sim$2–50 GΩ for microfluorescence measurements). However, an increase in the value of the feedback resistor is accompanied by a substantial decrease in the bandwidth of the amplifier, due to the increasing susceptibility of large-feedback resistors to stray capacity effects (corner frequency, f_o, typically around 400 Hz for $R_f = 5\,G\Omega$ [33]. Thus feedback resistors have to be chosen that represent a reasonable compromise between noise and bandwidth performance of the recording system. By calculating the current noise amplitude for a given recording system using the formula above and by comparing the results to actual measured values, it can be decided whether the selection of different amplifier–photodiode combinations might improve SNRs (theoretical value $\ll$ measured value) or whether improvements in SNR will only be possible by switching to more sensitive light detectors (theoretical value $\approx$ measured value).

Another technique used to increase SNR in both photon noise- and dark noise-limited experiments consists of signal-averaging multiple sweeps. In general, however, this technique cannot be applied when investigating impulse propagation at the cellular/subcellular scale because of differences in the timing and spatial pattern of activation from sweep to sweep. Thus, in order to preserve the details of microscopic activation in a given preparation, it is essential that SNRs are optimized to the point where characteristics of activation can be extracted from single-sweep recordings.

3. Extraneous Noise

In cultured cardiomyocytes, noise related to contraction generally increases with increasing spatial resolution. This interdependence is most likely explained by the fact that contraction artifacts do not originate from the cardiomyocytes themselves, but are produced by brightly stained debris sticking to their surface (remnants of dead cells with variable sizes in the few micrometers range). At high spatial resolution, these small pieces of debris, which contribute substantially to the resting fluorescence, are likely to move in and out of the area imaged by a given detector and, consequently, give raise to large motion artifacts [34]. On the other hand, at low spatial resolution, this effect is much less pronounced and motion artifacts are virtually absent (own observations with a tandem-lens macroscope, where each detector images a region in the preparation with a diameter of 600 μm). Methods to suppress this type of noise are based on pharmacological suppression of contraction with either butanedionemonoxime (BDM) [39,40] or verapamil [41,42]. Alternatively, motion artifacts can be eliminated directly from optical recordings by using a dual-wavelength emission approach (cf. Sec. V.F).

V. IMPLEMENTATION OF AN OPTICAL RECORDING SYSTEM SUITABLE FOR THE ASSESSMENT OF MICROSCOPIC IMPULSE PROPAGATION

In the following, an example of an optical recording system is presented which fulfills the requirements discussed in the previous sections, i.e., which permits the assessment of impulse propagation in cultured cardiomyocytes with a spatial resolution in the few micrometers range and a temporal resolution in the few microseconds range [43].

A. System Overview

As illustrated in Fig. 9, the recording system is built around a commercially available inverted microscope equipped for epifluorescence (Axiovert 135 M, Zeiss, Switzerland). The excitation light is provided by a 150-W short-arc xenon lamp, which is connected to a low-ripple power supply (Optiquip, New York, NY). A shutter (D122, Vincent Assoc., Rochester, NY) mounted between the lamp housing and the microscope permits illumination of the preparations to be kept as short as possible in order to minimize phototoxic damage. Because the opening of the shutter induces vibrations which tend to distort the initial phase of the optical recordings, a custom-built vibration isolator is inserted between the lamp housing and the microscope. This isolator consists of an aluminum cage holding the spring-suspended shutter. After passing a cutoff filter, the excitation light is deflected toward the objective by the dichroic mirror of the microscope. Generally, objectives with high numerical apertures are used during the experiments in order to increase the signal-to-noise ratio by maximizing light throughput (Fluar $5\times$, N.A. 0.25; Fluar $10\times$, N.A. 0.5; Fluar $20\times$, N.A. 0.75; Fluar $40\times$, N.A. 1.3; Plan-Apochromat $100\times$, N.A. 1.4, all from Zeiss, Switzerland). After passing the emission filter, the image of the preparation can be enlarged beyond the specifications of a given objective by a built-in magnifying lens (additional magnification by a factor of $1.6\times$ or $2.5\times$).

The microscope used is equipped with an optical port at the bottom which, in essence, converts the instrument into a straight optical bench, therefore offering an efficient light throughput because of the absence of additional deflecting optical elements in the lightpath. A custom-built attachment to this port permits the emitted light to be filtered according to the goals of a given experiment and to relay the image of the preparation to separate detector arrays or to a CCD camera.

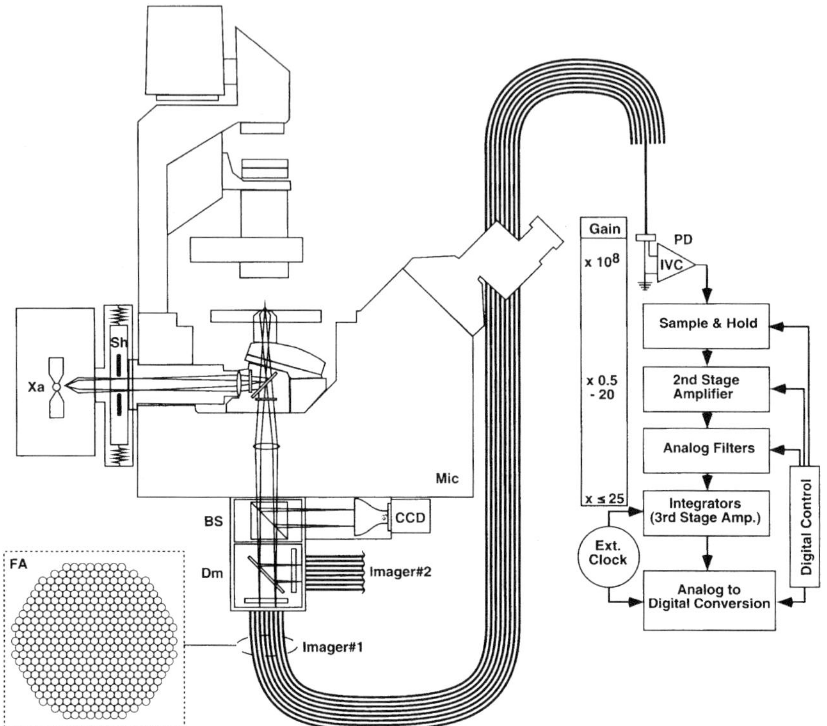

Figure 9 Schematic drawing of the optical recording system. Abbreviations: BS, beamsplitter; CCD, video camera; Dm, dichroic mirror; FA, arrangement of fibers in the face plate of the fiber optic imager; IVC, current-to-voltage converter; Mic, inverted microscope equipped for epifluorescence; PD, photodiode; Sh, shutter; Xa, xenon short-arc lamp. For a detailed description of the setup, cf. text. (Reproduced with permission from Ref. 43. Copyright 1998. Biophysical Society.)

B. Fiber Optic Image Conduit

In the case of commercially available multisite detectors with parallel readout (single-chip photodiode arrays or discrete photodiodes coupled rigidly to fiber optics), the layout of the detectors is fixed, i.e., each element records changes in light from a defined region within the field of view of the microscope. In the setup presented, a variation of the fiber optic approach was implemented. Instead of a fixed attachment between fiber optic cables and detectors, each detector is equipped with a connector which permits fiber optic cables and photodiodes to be combined freely. This approach has the following advantages. (1) The useful spatial resolution is not limited by the physical dimensions of the photodetectors used, but is defined solely by the diameter of the individual optical fibers, by the optical magnification used, and by SNR considerations. (2) The possibility of rearranging the spatial pattern of detectors has the unique advantage that the location of recording sites can be adjusted to regions of interest in a given preparation, thereby circumventing the problem of "wasted" photodetectors. (3) Because the photodetectors are noncommitted, it is possible to assign individual detectors to different ports of the microscope, thus enabling dual-emission-wavelength measurements. Moreover, it is possible to monitor simultaneously other parameters relevant for a given experiment such as, e.g., light intensity fluctuations of the lamp, which might serve to correct the optical signals for light ripple or arc wander. (4) Finally, using discrete photodetectors permits the selection of the most appropriate types of photodiodes during construction of the recording setup. In the present system, the fiber optic signal conduit consists of a custom-built hexagonal array of 379 plastic fiber optic cables with an active diameter of 1 mm each (for layout, cf. inset FA in Fig. 9). Depending on the magnification used, the fiber dimensions translate into spatial resolutions ranging from $4\,\mu m$ ($250\times$) to $200\,\mu m$ ($5\times$), thus permitting the assessment of impulse propagation at the subcellular to the multicellular scale.

C. Signal Conditioning

Three electronic stages serve to convert and amplify the minute photocurrents produced by the photodiodes into signals of suitable size for the digital data acquisition system.

1. *Current-to-voltage conversion stage.* According to Sec. IV.C.2, most of the overall gain of the system is implemented in the first amplification stage, i.e., in the current-to-voltage converters connected to the photodiodes (overall gain: 10^9 to 5×10^{10}; gain of IVC $= 10^8$). With this gain, the bandwidth of the IVC ($f_o \approx 1.6$ kHz) is still sufficiently high to resolve maximal frequencies reached during the upstroke of the propagated action potential (maximal upstroke velocities of 100–200 V/sec [15]). As shown experimentally below, the corner frequency of 1.6 kHz actually permits recording dV/dt_{max} values of up to 500 V/sec. All of the components of the IVC are assembled on individual printed circuit boards which are mounted into individual brass casings in order to minimize noise pickup. One end of the brass casing is designed as a fiber optic receptacle, which permits the reversible coupling of the optical fibers to the photodiodes.

2. *Analog signal conditioning stage.* The second amplification stage serves to further condition the raw signals produced by the IVCs. Sample-and-hold amplifiers at the input of this stage permit the subtraction of background fluorescence before the signal is further amplified (additional gain of 0.5, 1, 2, 5, or 20×) using either a DC or an AC coupling mode (time constants for AC coupling of 60 msec, 750 msec, or 9 sec). Finally, the signals are passed through RC low-pass filters (f_o, of 0.5, 1, 2, or 3 kHz).
3. *Integrator stage.* The final amplifier stage consists of integrators (ACF2101BP, Burr-Brown) whose gain is inversely proportional to their clock frequency: usually, experiments are performed at 20 kHz, resulting in an additional gain of 25×. The outputs of the integrators are fed to sample-and-hold stages which store the signals during the scanning cycle of the analog-to-digital converters.

The detectors and their circuitries are mounted in groups of 12 on printed circuit boards, which are connected to a digital control bus. This modular design permits upgrading the total number of channels by simply adding additional boards to the bus and by expanding the digitization capabilities of the system. Presently, signals of 80 detectors are acquired by two 12-bit ADCs (PC20501C, Burr-Brown, installed in a personal computer), which scan 40 channels each with a frame rate of 20 kHz, resulting in 1.6 Msamples/sec (3.2 MB/sec of data).

D. Temporal Accuracy

As outlined in detail above, the measurement of microscopic impulse propagation implies that differences in local activation times can be determined with a precision in the range of a few microseconds. In order to achieve this degree of temporal resolution, the optical amplifiers have to exhibit highly similar bandwidths, the signals have to be sampled simultaneously, and the action potential upstrokes have to be digitized at adequate rates. These three requirements are addressed in the recording system as follows.

Similar bandwidths among optical amplifiers. During construction of the current-to-voltage converters, time constants of the amplifiers ranged from ∼70 to ∼90 μsec. This variability was compensated by introducing an adjustable stray capacitance into the IVC by soldering a fine Teflon-insulated wire to one terminal of the feedback resistor and by varying the distance between the wire and the resistor such that the time constant of each amplifier was exactly 100 μsec ($f_o = 1.6$ kHz [34]).

Simultaneous sampling. Because the A/D conversion is initiated with the same clock driving the sample-and-hold output stages of the integrators, all signals recorded originate from exactly the same point in time. Therefore, it is known with submicrosecond precision when each data point is acquired.

Analog-to-digital conversion rates. The sampling frequency of the analog-to-digital conversion is 20 kHz. This value is well above the Nyquist criterion (twice f_o) and corresponds to ~10 measurement points along the action potential upstroke under the assumption of a duration of the upstroke of 500 μsec.

Based on these measures, it was expected that temporal delays in the microsecond range could be detected between fast-rising optical signals. This assumption was tested by applying a square pulse of a light-emitting diode (LED) to the array and by measuring the temporal dispersion of "activation times" of the light transient at each recording site based on at_{50}. These measurements showed that there remained a temporal dispersion of "activation times" with a standard deviation of ± 2.1 μsec (range -6 to 6 μsec; $n = 75$) [43]. In order to compensate for this dispersion, which was most likely both due to slight inaccuracies in the "Teflon-wire" procedure and to variations in the temporal responses of the second amplification stage, a short LED pulse was routinely recorded with each experiment. This permitted the determination of the temporal deviation of each individual channel and, subsequently, the correction of activation times obtained in biological preparations. Using this approach with an LED pulse simulating an action potential upstroke, the standard deviation of activation times was reduced to ± 0.4 μsec ($n = 75$; range -0.8 to 1.2 μsec). Based on these specifications, it could be expected that propagating events in the range of 1 m/sec could be resolved at a spatial resolution of ≤ 10 μm. This was tested by simulating a propagating light-intensity change with a rotating steel blade (2000 RPM) positioned in the object plane of a $20\times$ objective and by recording, in transillumination mode, the shuttering of the field of view. The result of such an experiment is illustrated in Fig. 10. Signals were recorded with the spatial arrangement of detectors shown in Fig. 10A. Figure 10B depicts the isochrones of the light-intensity change as the blade swept over the objective. Individual signals recorded along the center row of detectors are shown in Fig. 10C. A linear fit of the at_{50} values of these signals yielded a velocity of the blade of 4.25 m/sec which closely matched the theoretically predicted value of 4.19 m/sec. From the finding that a light-intensity change propagating with a velocity of 4.25 m/sec can be measured with a spatial resolution of 50 μm ($20\times$ objective), it can be inferred that the system is capable of

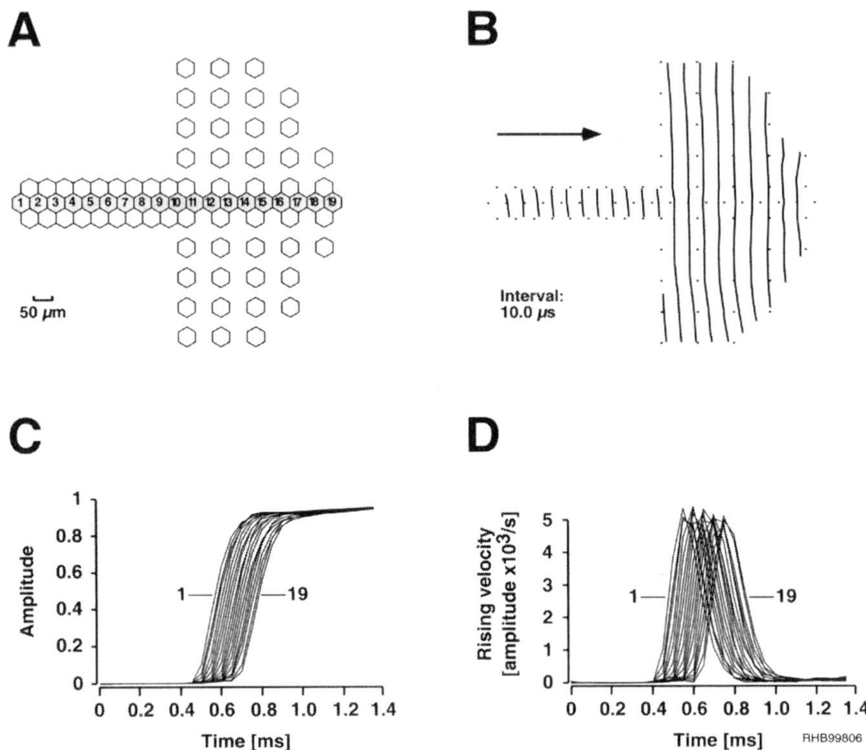

Figure 10   Test of the temporal resolution of the recording system by tracking a fast-moving shutter in the object plane of the microscope (steel blade rotating at 2000 RPM). (A) Spatial arrangement of the detectors. (B) The parallel and evenly spaced isochrones correspond to the shuttering of the field of view and indicate that a moving light signal with a velocity of ~4 m/sec can be accurately tracked with a spatial resolution of 50 µm. (C) Plot of individual signals along the center row of detectors. The numbers correspond to the numbering in A. (D) First derivative of the signals shown in C. (Reproduced with permission, from Ref. 43. Copyright 1998. Biophysical Society.)

tracking events propagating at up to 0.8 m/sec with a spatial resolution of 10 µm or less, i.e., with cellular/subcellular resolution.

E.   Spatial Resolution

The useful optical resolution of the recording system was tested by measuring propagated action potentials in linear strands of cultured neonatal rat ventricular myocytes (width: 50/100 µm) at magnifications ranging from 5× to 250× [43]. The preparations were mounted in an experimental chamber and stained for 3–4 min with 135 µmol/L di-8-ANEPPS at room temperature. Thereafter, they were continuously superfused with Hanks' balanced salt solution (HBSS) at 36°C. After identification of a region of interest, an extracellular stimulation electrode was positioned at a distance ≥ 1 mm from the recording site and the preparations were paced at 2 Hz. The distance of 1 mm was chosen in order (1) to prevent electrotonically mediated stimulation artifacts from distorting the signals of interest and (2) to permit activation to reach steady-state conditions at the recording site. Results of such experiments are shown in Fig. 11. While action potentials could be recorded at all magnifications corresponding to spatial resolutions ranging from 4 to 200 µm, signals at either extreme yielded poor SNRs of approximately 30, because light

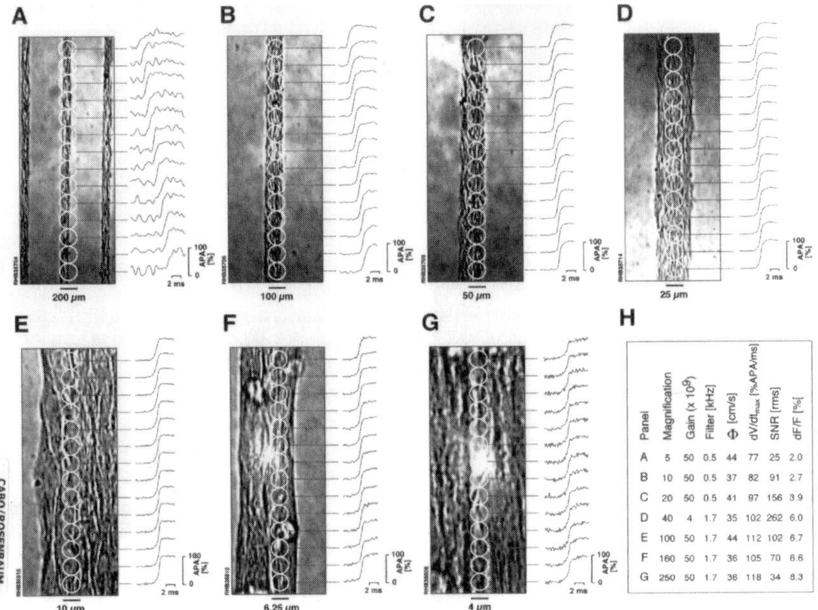

Panel	Magnification	Gain ($\times 10^9$)	Filter [kHz]	Θ [cm/s]	dV/dt_{max} [%APA/ms]	SNR [rms]	dF/F [%]
A	5	50	0.5	44	77	25	2.0
B	10	50	0.5	37	82	91	2.7
C	20	50	0.5	41	97	156	3.9
D	40	4	1.7	35	102	262	6.0
E	100	50	1.7	44	112	102	6.7
F	160	50	1.7	36	105	70	6.6
G	250	50	1.7	36	118	34	8.3

Figure 11 Quality of optical signals as a function of spatial resolution. (A–G) Illustration of action potential upstrokes recorded along linear cell strands at increasing optical magnifications. The corresponding spatial resolutions are indicated by the bars below the photomicrographs. While signal-to-noise ratios were poor at both very low and very high spatial resolution (SNRs ≈ 30), intermediate resolutions resulted in SNRs ranging from ~70 to 260. Parameters relevant to the recordings are summarized in H. Abbreviations: θ, conduction velocity; dV/dt_{max}, maximal upstroke velocity; SNR, signal-to-noise ratio; $\Delta F/F$, fractional fluorescence change. (Reproduced with permission from Ref. 43. Copyright 1998. Biophysical Society.)

levels and therefore signal amplitudes were small. At low magnification (200 µm; Fig. 11A), this was due primarily to the small numerical aperture of the objective, while at high magnification (4 µm; Fig. 11G), light intensities were low due to the small size of the area imaged. Between these two extremes, objectives offering highly efficient light throughputs improved SNRs substantially. The highest values (SNRs between 200 and 300) could be recorded with a 40×, 1.4-N.A. objective, as illustrated by the example shown in Fig. 11D. These experiments illustrate that the recording system is capable of resolving transmembrane voltage changes with spatial resolutions ranging from the subcellular to the multicellular level.

F. Optical Motion Artifact Removal

Optical recordings of transmembrane voltage changes in contractile tissues have the disadvantage that contraction-induced light scattering distorts the repolarization phase of the action potentials. For cultured cells, this effect is especially pronounced when measuring propagation at high spatial resolution (cf. Sec.IV,C.3). This artifact can be removed by profiting from the fact that the emission spectrum of di-8-ANEPPS contains two distinct regions responding either with a decrease (longer wavelengths) or an increase (shorter wavelengths) in fluorescence intensity to depolarizations of the membrane (cf. Fig. 3). Because, obviously, the motion artifact is not dependent on wavelength, a dual-emission-wavelength approach will therefore permit the removal of this nuisance in optical recordings from contractile tissue. The result of such an experiment is illustrated in Fig. 12. The preparation was broadly excited (excitation <500 nm, dichroic mirror; 505 nm, emission >515 nm) and the emitted light was split by a dichroic mirror (590 nm) and directed to two fiber optic arrays which were exactly matched in space. Accordingly, one array recorded light with wavelengths >590 nm (Fig. 12B; positive-going action potential upstrokes; signals scaled to resting fluorescence), while the other received light with wavelengths between 515 and 590 nm (Fig. 12C; negative-going action potential upstrokes; signals scaled to resting fluorescence). Signals recorded at either wavelength showed a considerable distortion of the action potential due to the motion artifact. As shown in Fig. 12D, this distortion was completely eliminated after subtracting the signals in Fig. 12C from those in Fig. 12B. In addition to the elimination of motion artifacts, dual-emission-wavelength measurements also tended to increase the SNR because the signal sizes and the level of common-mode noise rejection were increased [25,43]. These results illustrate that it is feasible to record optically, with cellular resolution, spatial patterns of action potential repolarization without using drugs which might interfere with the normal electrophysiological properties of the tissue.

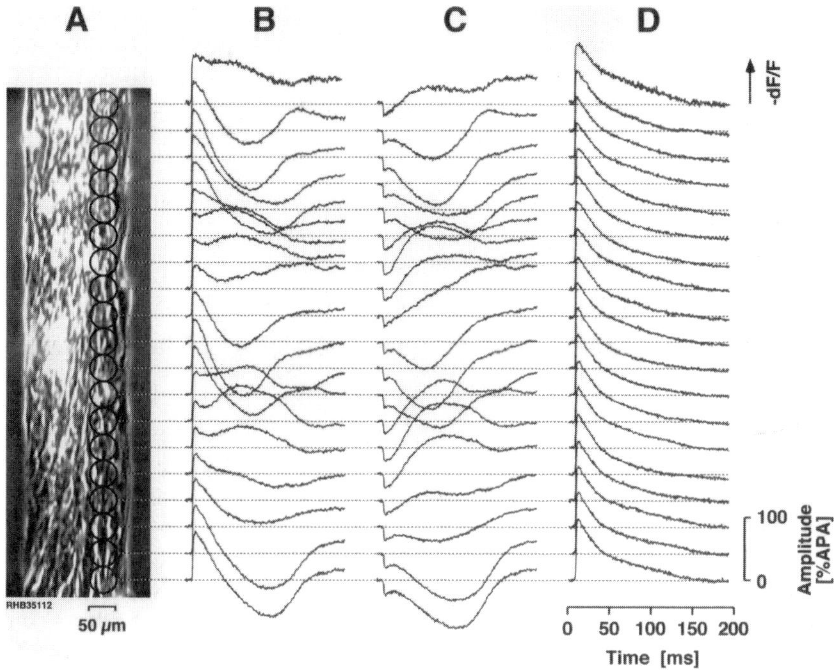

Figure 12 Optical motion artifact subtraction. (A) Phase-contrast image of the preparation with overlaid black circles indicating the positions of the photodetectors. (B) Signals recorded at wavelengths at which the potentiometric dye di-8-ANEPPS responds to membrane depolarization with a decrease of emitted fluorescence. Following the general convention, signals are drawn with reversed sign, resulting in positive-going action potential upstrokes. (C) Signals recorded simultaneously at identical sites, but at wavelengths at which di-8-ANEPPS responds to membrane depolarization with an increase of emitted fluorescence (negative-going action potentials). (D) Subtraction signals: the subtraction of signals shown in B and C (both normalized to resting fluorescence) resulted in action potential shapes (amplitude normalized to 100%) which were virtually free of motion artifacts. (Reproduced with permission from Ref. 43. Copyright 1998. Biophysical Society.)

VI. EXAMPLES OF MEASUREMENTS OF MICROSCOPIC IMPULSE PROPAGATION

A. Microscopic Impulse Propagation in Narrow Strands of Cardiomyocytes

One of the first problems ever investigated with the combination of patterned-growth myocyte cultures and optical high-resolution mapping of impulse propagation was the question of whether impulse propagation in cardiac tissue is discontinuous at the level of single cells due the recurrent increases in longitudinal resistance at sites of cell-to-cell abutments [44]. Such discontinuities occurring under conditions of normal cell-to-cell coupling had been described earlier on the basis of computer simulations [9,45]. These studies suggested that conduction times along a given cell were roughly equal to conduction times across the cell-to-cell border.

The question of microscopically discontinuous conduction was addressed with patterned growth preparations which consisted of single cell chains of cardiomyocytes. In these one-cell-wide strands, impulse propagation was followed at high spatiotemporal resolution, permitting the determination of activation delays both within individual cardiomyocytes and across cell-to-cell borders. An example of such a measurement is shown in Fig. 13. As shown schematically in Fig. 13A, the region of the preparation selected for the optical recording consisted of cardiomyocytes that were abutted in the center of the field of view. Typically, cells were not joined in the bricklike manner typical for adult tissue, but rather showed some form of partial overlap as indicated by the medium-gray area. The extent and position of these overlaps was determined by tracking the intensely stained sarcolemma while stepping the focal plane through the preparation. During action potential propagation from left to right, the simultaneously recorded action potential upstrokes shown in Fig. 13B revealed an activation gap between detectors #5 and #7, thus demonstrating that conduction was discontinuous at the cellular level. The action potential upstroke recorded by detector #6 showed an intermediate timing, which is explained by the circumstance that this detector received input simultaneously from the left and the right cell. From the distribution of activation delays measured between neighboring detectors along the preparation (Fig. 13C), it is clearly evident that conduction was highly "saltatory." In contrast to this spatially contiguous measurement of microscopic impulse propagation, another study performed with the same type of preparation used an elegant noncontiguous approach [16]. In that study, conduction times were assessed between three linearly arranged photodiodes whose "receptive fields" in the preparation were spaced 30 µm apart. The detectors were placed such that two of them recorded activation within a given cell while the third recorded activation in a neighboring cell. Assuming constant conduction velocities within individual cells, intercellular conduction delays were calculated by subtracting intracellular from intercellular conduction times. Based on this procedure, conduction delays across cell-to-cell borders were estimated to be ~80 µsec. Given the width of the gap junctional complex of ~15 nm [46], this local activation delay translates into a virtual

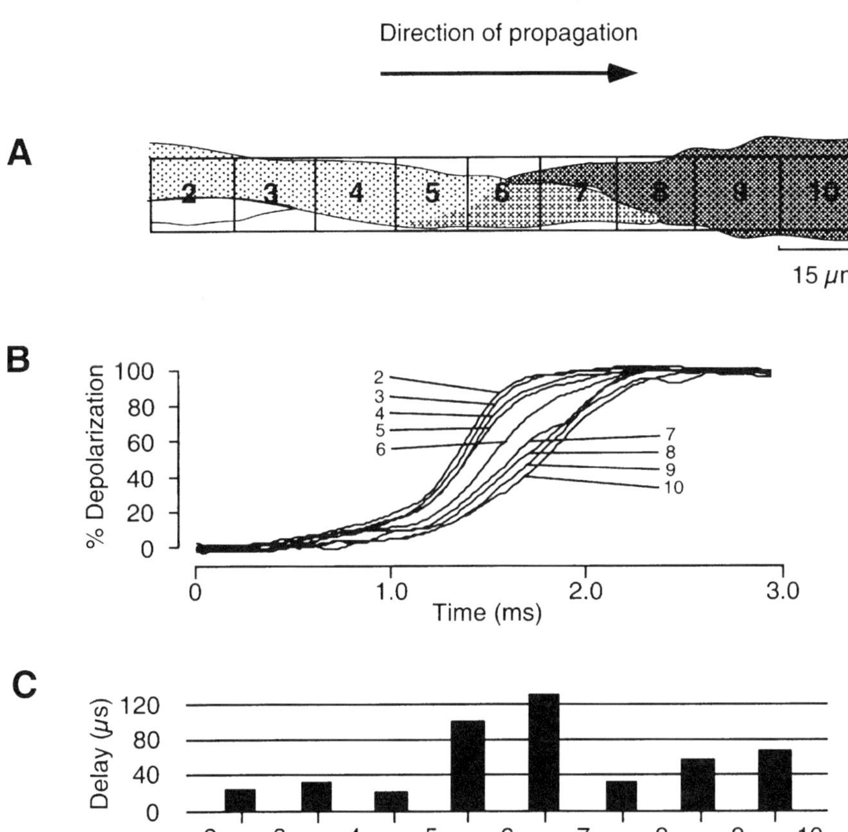

Figure 13 Discontinuous impulse propagation in single-cell chains of cardio-myocytes. (A) Schematic drawing of the imaged region of the preparation consisting of two slightly overlapping cardiomyocytes (light gray and dark gray). The squares indicate the positions of individual photodetectors. (B) Action potential upstrokes recorded during propagation from left to right. Numbers correspond to the numbering of photodetectors in A. (C) Local activation delays along the preparation. (Redrawn with permission from Ref. 44. Copyright 1992. Marine Biological Laboratories.)

local conduction velocity of ∼0.2 mm/sec, which is roughly three orders of magnitude lower than intracellular conduction velocities.

The finding of discontinuous conduction at the cellular level was pertinent to single-cell chains of cardiomyocytes. When the same type of experiment was conducted with several-cell-wide strands, discontinuities at

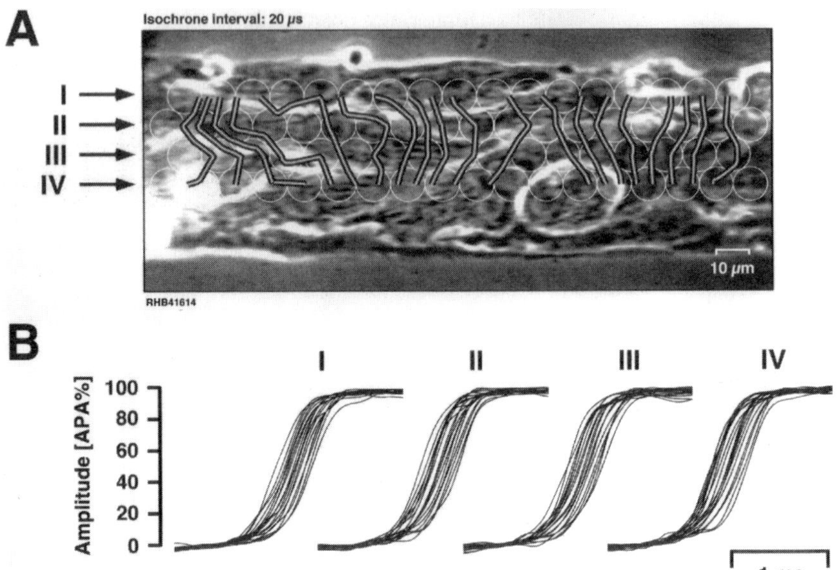

Figure 14 Microscopic impulse propagation in a several-cell-wide linear strand. (A) Phase-contrast picture of the preparation with overlaid circles indicating the positions of the photodetectors. The spatial pattern of activation during impulse propagation from left to right is indicated by isochrones spaced 20 μsec apart. (B) Plot of action potential upstrokes recorded simultaneously from the four rows of photodetectors numbered I through IV in A. (Redrawn with permission from Ref. 47. Copyright 1998. American Heart Association.)

the sites of cell-to-cell appositions were smaller [16] or could no longer be observed [44,47]. An example of such a recording is shown in Fig. 14. The preparation consisted of a linear strand (3–4 cells wide; Fig. 14A) in which the characteristics of impulse propagation were assessed at high spatial resolution (10 μm). As indicated by the even spacing between the activation isochrones, activation in this wide strand occurred in a mostly continuous manner. The absence of major local discontinuities in conduction is furthermore illustrated by the rather uniform spacing between action potential upstrokes recorded along the preparation (Fig. 14B; signals grouped according to detector rows). The difference in activation patterns between single-cell chains of cardiomyocytes and wider cell strands can be explained by the rather intense lateral gap junctional coupling observed in several-cell-wide strands [48], which tends to smooth differences in local activation times ("lateral averaging" [16,44]). This averaging occurs because, as shown in computer simulation studies, the staggered arrangement of laterally

connected myocytes offers the excitatory current a collateral pathway around
a given end-to-end connection. This delays conduction along the cytoplasm
while speeding up conduction across cell junctions situated at end-to-end-
abutted cells [16]. Since, in three-dimensional ventricular tissue an individual
myocyte is coupled, on average, to 9 surrounding cells [49], versus 6 cells in
the simulation study or 4–6 cells in monolayer cultures, it can be expected
that the averaging effect of lateral connections is further increased, thus
leading to a highly uniform activation wavefront in intact tissue.

B. Microscopic Impulse Propagation During Partial Gap Junctional Uncoupling

While, as shown above, activation in several-cell-wide strands under conditions of normal intercellular coupling is continuous, this situation changes
drastically during gap junctional uncoupling: as shown in computer simulations, a reduction of intercellular conductance will result (1) in an increase
of activation delays across the cell-to-cell borders and (2) in the confinement
of depolarizing current to individual cells, leading to an increasingly discontinuous type of conduction [10,45]. In order to investigate this type of
conduction experimentally, several-cell-wide strands were partially uncoupled with palmitoleic acid while impulse propagation was monitored optically at the cellular level [47]. The results of such as experiment are shown in
Fig. 15. The 4–5-cell-wide preparation was uncoupled to a degree nearly
inducing conduction block, and action potential upstrokes recorded during
propagation from left to right are shown, in superimposed form, in Fig.
15B. When compared to control recordings, conduction was not only slowed substantially during critical uncoupling (decrease of 97.5%, from 43 to
1.1 cm/sec), but activation became highly discontinuous. This is indicated
by the clustering of optically recorded action potential upstrokes which
pointed to a stepwise advancement of excitation. Activation delays
among the clustered action potential upstrokes ranged from 0.5 to 4.5 msec,
while the activation of the clusters themselves took only 80 to 450 μsec
to complete. The origin of the clustered signals within the preparation is
illustrated in Fig. 15C, which shows the projection of all recording sites
onto a schematic drawing of the preparation with highlighted borders of
individual cells. The correlation of sites being activated in a near-simulta-
neous manner with the cellular structure of the preparation reveals that
clustered activity originated from small patches of the preparation consisting of one to three cells. Thus, during critical uncoupling, conduction
invaded the preparation in a saltatory fashion where the patches were
activated sequentially with variable delays. As indicated qualitatively by the
dashed arrows, the activation path was tortuous due to the presence of a

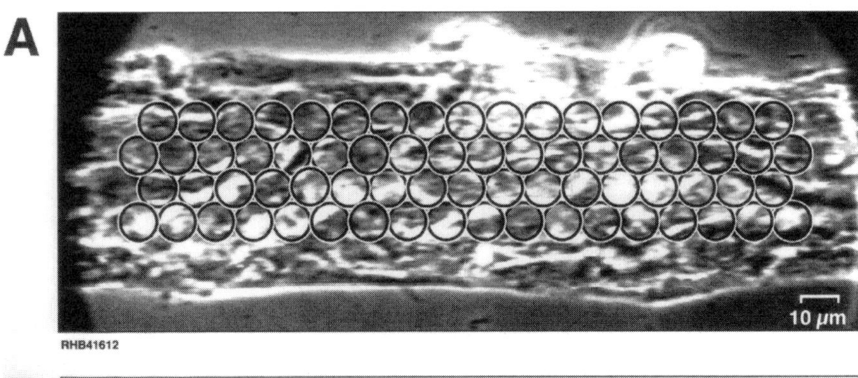

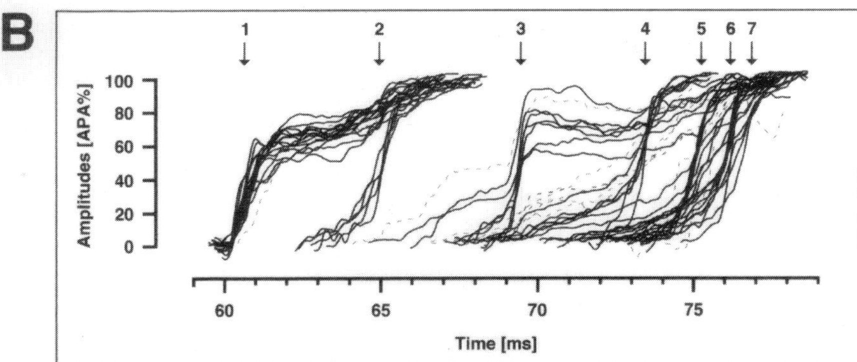

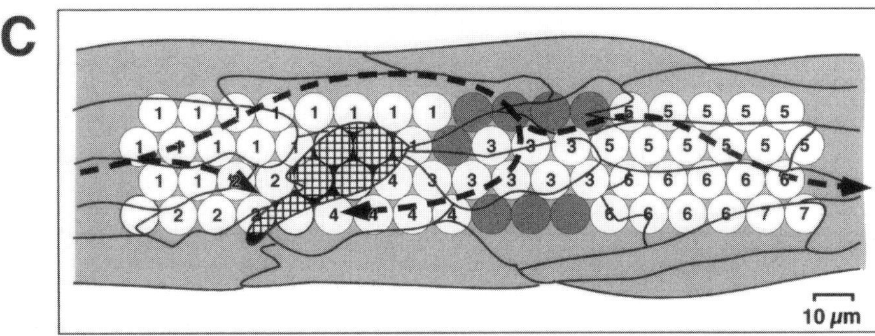

Figure 15 Highly discontinuous microscopic conduction during critical gap junctional uncoupling. (A) Phase-contrast picture of the preparation with overlaid white circles indicating the positions of individual photodetectors. (B) Plot of all action potential upstrokes recorded simultaneously from all photodetectors during impulse propagation from left to right under conditions of critical gap junctional uncoupling. Under these conditions, action potential upstrokes are not evenly spaced in time anymore, but occur in clusters indicated by numbered arrows. (C) Schematic

central obstacle consisting of a single myocyte (cross-hatched outline). This cell, which was completely uncoupled at the time of the measurement, forced an activation detour resulting in a region of the preparation exhibiting backward propagation.

These findings show that critical uncoupling induces a highly discontinuous type of propagation and results in overall conduction velocities as low as 1 cm/sec or less [47]. The potential relevance of this finding relates to the question of the minimal area of cardiac tissue which may host an anatomically fixed reentrant pathway. Under the assumption of a refractory period of 100–200 msec, a conduction velocity of 1 cm/sec would result in a minimal path length of 1–2 mm, which would occupy an area as small as 1 mm². Thus, it seems feasible that reentrant excitation can occur in areas of cardiac tissue which are so small that they escape presently available recording techniques used in intact tissue to assess impulse propagation.

VII. LIMITATIONS

Limitations of optical recordings of microscopic impulse propagation include (1) constraints regarding the temporal precision of the measurements due to insufficient SNRs, (2) the inability to perform long-term experiments because of photodynamic damage, and (3) limitations regarding the extrapolation of results obtained with cultured cardiomyocytes to intact tissue.

A. Temporal Precision

During measurements of fluorescence changes from microscopic areas on single cells, SNRs tend to be small due to the low intensity of emitted fluorescence. As outlined in more detail above, this compromises the

---

drawing of the cellular architecture of the imaged region of the preparation. The overlaid disks are numbered according to the numbers assigned in B to clustered upstrokes. Identical numbers therefore indicate regions being activated nearly simultaneously. The disks without numbers correspond to regions displaying non-monotonically rising action potentials which could not be attributed unequivocally to given clusters of action potential upstrokes (dashed signals in B). The dashed line describes qualitatively the path of activation of the preparation. (Redrawn with permission from Ref. 47. Copyright 1998. American Heart Association.)

temporal precision with which activation can be tracked at the subcellular level during single-shot measurements. If appropriate in the context of the goals of a given experiment, a statistical approach might help to increase the precision of the determination of intracellular or intercellular activation delays [16], but only the development of new voltage-sensitive dyes with improved fractional fluorescence changes will lead ultimately to an increase in SNRs and thus increase the precision of the determination of microscopic activation patterns during single-trial experiments.

B. Phototoxicity

An ever-present problem, which is especially pronounced for fluorescent voltage-sensitive dyes, is the photodynamic damage exerted on the cells during repeated or prolonged illumination. This effect is especially pronounced during measurements of microscopic impulse propagation because light intensities have to be high in order to achieve acceptable SNRs. Photodynamic damage is characterized by a decrease in signal amplitude, maximal upstroke velocity, and conduction velocity as well as an increase in action potential duration until, finally, inexcitability ensues [34,50]. While photodynamic damage cannot be entirely eliminated, it can be minimized to a certain extent by reducing light intensities to levels still compatible with desired SNRs, by keeping illumination times as short as possible, by selecting appropriate dyes (e.g., di-8-ANEPPS instead of di-4-ANEPPS), and, potentially, by using antioxidant drugs (e.g., catalase, [50]). However, the possibility of performing long-term experiments with no or little phototoxicity will have to await the development of new classes of dyes.

C. Cultured Cells Versus Intact Tissue

The most obvious difference from intact tissue is the absence of a third dimension in monolayer cultures of cardiomyocytes. While individual cells in monolayer cultures are electrically coupled to ~4–6 neighboring cells, cardiomyocytes in intact tissue are coupled to approximately twice as many cells [49]. Obviously, this difference affects the passive electrical properties of a given cellular network, which need to be taken into account by using appropriate scaling procedures [14] whenever extrapolations are to be made from findings in two-dimensional monolayers to intact three-dimensional tissue. A second difference from intact tissue reflects issues related to the extracellular space. While the extracellular space beneath the cultured cardiomyocytes is very small and is formed by a diffusion barrier (= glass substrate), the upper sides of the cells face an "infinite" extracellular space.

This results not only in differences regarding the extracellular longitudinal resistance, but the restricted diffusion of ions below the cells might additionally influence the electrophysiology of individual cells. It is presently not clear to which extent (if any) this polar environment might influence passive and/or active membrane properties. Finally, because cultured cells grow in an artificial environment devoid of neurohumoral influences, the cellular development might be altered by medium composition and extracellular matrix as compared to the situation in vivo [51]. Nevertheless, because the general electrophysiological properties of cultured cardiomyocytes are very similar to those of intact tissue with respect to upstroke velocity, conduction velocity, and action potential duration [15], it seems to justified to extrapolate to intact tissue at least the qualitative conclusions obtained with this experimental preparation.

VIII. PERSPECTIVES

Future improvements in the optical measurement of microscopic impulse propagation in cardiac tissue might come from some or all of the following.

1. *Determination of absolute transmembrane voltages.* As mentioned above, dual-wavelength excitation or dual-wavelength emission measurements are capable of reporting absolute changes in transmembrane voltage when used in conjunction with appropriate voltage-sensitive dyes and calibration procedures [25,27]. While the relatively low degree of precision in indicating absolute voltages ($\sim$10 mV) of the earlier study might have prevented a broader use of this technique, the increased precision of the dual emission approach ($\sim$5 mV) might render this approach more attractive to get a detailed picture of spatial heterogeneity in transmembrane voltage distributions in cardiac tissue.

2. *Optical measurements with dual indicators.* The combination of voltage-sensitive dyes with optical indicators of ion concentrations would represent a powerful tool for understanding the correlation between specific spatial activation patterns and concomitant changes in ion transients. As an example, the simultaneous monitoring of calcium transients and membrane voltages during early afterdepolarizations (EADs) might help to elucidate the spatiotemporal patterning of this event at the level of a few cells.

3. *Correlation of microscopic impulse propagation with the microtopography of membrane channels.* It has recently been shown that sodium channels co-segregate with gap junctions not only in intact tissue but also in cultured cardiomyocytes [52,53]. This finding raises the question of whether the specific distribution of ion channels might contribute to the shaping of microscopic impulse propagation. While there are as yet no experimental

indications as to whether such a distribution has any functional implications, the question might be approached by correlating the characteristics of microscopic impulse propagation with the underlying spatial distribution of channels obtained by immunocytochemistry.

4. *Measurement of microscopic impulse propagation in intact tissue.* While the perspective of measuring microscopic impulse propagation in intact cardiac tissue seemed quite remote in the past, the recent development of a fast random-access laser scanning system might provide a foundation for conducting such studies [54]. In principle, this system could be modified so as to include confocal detection schemes, thus permitting the assessment of activation in intact three-dimensional networks of cardiomyocytes.

5. *Development of new dyes.* As described in detail above, the successful measurement of microscopic impulse propagation is critically dependent on obtaining signals with adequate SNRs. Among all of the factors influencing SNR, the voltage-sensitive dyes themselves have remained the weakest link, as their fractional fluorescence change for a given change in transmembrane voltage is generally quite small. After a decade in which virtually no new dyes were developed, it might be hoped that the interest in searching for new indicators with improved fractional fluorescence changes (and decreased phototoxic side effects) might be sparked de-novo, as the use of these dyes has spread beyond the neurosciences and has become very popular in cardiac sciences.

6. *Local superfusion and transfection.* In the past, the combination of optical recording and spatially defined superfusion of the preparations with substances affecting passive and/or active properties of the cells have permitted the assessment of the relationship between specific cell architectures and local alterations in the functional state of the tissue [55,56]. Using the same approach, it might be possible to change locally the function of a given tissue by altering the composition of ion channels using transfection techniques. Such investigations could yield important information as to the extent to which defined alteration of the cellular ion channel repertoire influences microscopic impulse propagation in a geometrically defined cellular structure.

ACKNOWLEDGMENTS

I would like to thank Brian M. Salzberg for having introduced me years ago to MSORTV and for his invaluable comments on the manuscript. This work was supported by the Swiss National Science Foundation.

REFERENCES

1. Mendez C, Mueller WJ, Merideth J, Moe GK. Interaction of transmembrane potentials in canine Purkinje fibers and at Purkinje fiber-muscle junctions. Circ Res 24:361–372, 1969.
2. Overholt ED, Joyner RW, Veenstra RD, Rawling D, Wiedmann R. Unidirectional block between Purkinje and ventricular layers of papillary muscle. Am J Physiol 247:H584–H595, 1984.
3. Sano T, Takayama N, Shimamoto T. Directional difference of conduction velocity in the cardiac ventricular syncytium studied by microelectrodes. Circ Res 7:262–267, 1959.
4. Clerc L. Directional differences of impulse spread in trabecular muscle from mammalian heat. J Physiol (London) 255:335–346, 1976.
5. Gardner PI, Ursell PC, Fenoglio JJ, Wit AL. Electrophysiologic and anatomic basis for fractionated electrograms recorded from healed myocardial infarcts. Circulation 72:596–611, 1965.
6. Spach MS, Dolber PC, Heidlage JF. Influence of the passive anisotropic properties on directional differences in propagation following modification of the sodium conductance in human atrial muscle. A model of re-entry based on anisotropic discontinuous propagation. Circ Res 62:811–832, 1988.
7. de Bakker JMT, van Capelle FJL, Janse MJ, Tasseron S, Vermeuleu JT, de Jonge N, Lahpor JR. Slow conduction in the infracted human heart: "Zigzag" course of activation. Circulation 88:915–926, 1993.
8. Ursell PC, Gardner PI, Albala A, Fenoglio JJ, Wit AL. Structural end electrophysiological changes in the epicardial border zone of canine myocardial infarcts during infarct healing. Circ Res 56:436–451, 1985.
9. Rudy Y, Quan W. Propagation delays across cardiac gap junctions and their reflection in extracellular potentials: a simulation study. Cardiovasc Electrophysiol 2:299–315, 1991.
10. Shaw RM, Rudy Y. Ionic mechanisms of propagation in cardiac tissue: roles of the sodium and L-type calcium currents during reduced excitability and decreased gap-junction coupling. Circ Res 81:727–741, 1997.
11. Sugiura H, Joyner RW. Action potential conduction between guinea pig ventricular cells can be modulated by calcium current. Am J Physiol 263:H1591–H1604, 1992.
12. oyner RW, Kumar R, Wilders R, Jongsma HJ, Verheijek EE, Golod DA, van Ginneken ACG, Wagner MB, Godsby WN. Modulating L-type calcium current affects discontinuous cardiac action potential conduction. Biophys J 71:237–245, 1996.
13. Leon LJ, Roberge FA. Directional characteristics of action potential propagation in cardiac muscle. A model study. Circ Res 69:378–395, 1991
14. Fast VG, Kléber AG. Block of impulse propagation at an abrupt tissue expansion: evaluation of the critical stand diameter in 2- and 3-dimensional computer models. Cardiovasc Res 30:449–459, 1995.

15. Rohr S, Schölly DM, Kleber AG. Patterned growth of neonatal rat heart cells in culture. Morphological and electrophysiological characterization. Circ Res 68:114–130, 1991.

16. Fast VG, Kléber AG. Microscopic conduction in cultured strands of neonatal rat heart cells measured with voltage-sensitive dyes. Circ Res 73:914–925, 1993.

17. Girouard SD, Pastore JM, Laurita KR, Gregory KW, Rosenbaum DS. Optical mapping in a new guinea pig model of ventricular tachycardia reveals mechanisms for multiple wavelengths in a single reentrant circuit. Circulation 93:603–613, 1996.

18. Israel DA, Edell DJ, Mark RG. Time delays in propagation of cardiac action potentials. Am J Physiol 258:H1906–H1917, 1990.

19. de Bakker JMT, van Capelle FJI, Tasseron SJA, Janse MJ. Load mismatch as a cause of longitudinal conduction block in infracted myocardium (abstr). Circulation 96(8):I–497, 1997.

20. Cohen LB, Salzberg BM. Optical measurement of membrane potential. Rev Physiol Biochem Pharmacol 83:35–88, 1978.

21. Rosenbaum DS, Jalife J, eds. Optical Mapping of Cardiac Excitation and Arrhythmias. Armonk, NY: Futura, 2001.

22. Platt JR. Electrochromism, a possible change of color producible in dyes by an electric field. J Chem Phys 34:862–871, 1962.

23. Loew LM, Bonneville GW, Surow J. Charge shift optical probes of membrane potential. Biochemistry 17:4065–4071, 1978.

24. Fromherz P, Lambacher, A. Spectra of voltage-sensitive fluorescence of styryl-dye in neuron membrane. Biochem Biophys Acta 1068:149–156, 1991.

25. Bullen A, Saggau P. High-speed, random-access fluorescence microscopy: II. Fast quantitative measurements with voltage-sensitive dyes. Biophys J 76: 2272–2287, 1999.

26. Salzberg BM. Optical Recording of electrical activity in neurons using molecular probes. In: Barker JL, McKelvy JF, eds. Current Methods in Cellular Neurobiology. New York: Wiley, 1983, pp. 139–187.

27. Montana V, Farkas DL, Loew LM. Dual-wavelength ratiometric fluorescence measurements of membrane potential. Biochemistry 28:4536–4539, 1989.

28. Salzberg BM, Grinvald AL, Cohen LB, Davila HV, Ross WN. Optical recording of neuronal activity in an invertebrate central nervous system: Simultaneous monitoring of several neurons. J Neurophysiol 40:1281–1291, 1977.

29. Wu JY, Cohen LB, Falk CX. Fast multisite optical measurement of membrane potential, with two examples. In: Mason WT, ed. Fluorescent and luminescent probes for biological activity. London: Academic, 1999, pp. 222–237.

30. Inoué S, Spring KR. Videomicroscopy. The Fundamentals. New York and London: Plenum, 1997.

31. Iijima T, Witter MP, Ichikawa M, Tominaga T, Kajiwara R, Matsumoto G. Entorhinal-hippocampal interactions revealed by real-time imaging. Science 1996; 272(5265):1176–1179.

32. Hirota A, Sato K, Momose-Sato Y, Sakai T, Kamino K. A new simultaneous 1020-site optical recording system for monitoring neural activity using voltage-sensitive dyes. J Neurosci Meth 56(2):187–194, 1995.

33. Chien CG, Pine J. An apparatus for recording synaptic potentials from neuronal cultures using voltage-sensitive fluorescent dyes. J Neurosci Meth 38(2–3):93–105, 1991.

34. Rohr S, Salzberg BM. Multiple site optical recording of transmembrane voltage in patterned growth heart cell cultures: assessing electrical behavior, with microsecond resolution, on a cellular and subcellular scale. Biophys J 67: 1301–1315, 1994.

35. Salzberg BM, Obaid AL, Bezanilla F. Microsecond response of a voltage-sensitive merocyanine dye: fast voltage-clamp measurements on squid giant axon. Jpn J Physiol 43(suppl 1):37–41, 1993.

36. London JA, Zecevic D, Cohen LB. Simultaneous optical recording of activity from many neurons during feeding in Navanax. J Neurosci 7:649–661, 1987.

37. Cohen LB, Lesher S. Optical monitoring of membrane potential: methods of multisite optical measurement. In: DeWeer P, Salzberg BM, eds. Optical Methods in Cell Physiology. New York: Wiley, 1986, pp. 71–99.

38. Horowitz P, Hill W. The Art of Electronics. Cambridge, UK: Cambridge University Press, 1989.

39. Pertsov AM, Davidenko JM, Salomonsz R, Baxter WT, Jalife J. Spiral waves of excitation underlie reentrant activity in isolated cardiac muscle. Circ Res 72:631–650, 1993.

40. Gray RA, Jalife J, Panfilov A, Baxter WT, Cabo C, Davidenko JM, Pertsov AM. Nonstationary vortexlike reentrant activity as a mechanism of polymorphic ventricular tachycardia in the isolated rabbit heart. Circulation 91:2454–2469, 1995.

41. Dillon SM. Optical recordings in the rabbit heart show that defibrillation strength shocks prolong the duration of depolarization and the refractory period. Circ Res 69:842–869, 1991.

42. Kwaku KF, Dillon SM. Shock-induced depolarization of refractory myocardium prevents wave-front propagation in defibrillation. Circ Res 79:957–973, 1996.

43. Rohr S, Kucera JP. Optical recording system based on a fiber optic image conduit: assessment of microscopic activation patterns in cardiac tissue. Biophys J 75:1062–1075, 1998.

44. Rohr S, Salzberg BM. Discontinuities in action potential propagation along chains of single ventricular myocytes in culture: multiple site optical recording of transmembrane voltage (MSORTV) suggests propagation delays at the junctional sites between cells. Biol Bull Mar Biol Lab 183:342–343, 1992.

45. Murphy CR, Clark JW, Giles WR, Rasmusson RL, Halter JA, Hicks K, Hoyt B. Conduction in bullfrog atrial strands: simulations of the role of disc and extracellular resistance. Math Biosci 106:39–84, 1991.

46. Sosinsky GE, Jesior JC, Caspar DLD, Goodenough DA. Gap junction structures. VIII. Membrane cross-sections. Biophys J 53:709–722, 1988.

47. Rohr S, Kucera JP, Kleber AG. Slow conduction in cardiac tissue: I. Effects of a reduction of excitability vs. a reduction of electrical coupling on microconduction. Circ Res 83:781–794, 1998.

48. Darrow BJ, Fast VG, Kléber AG, Beyer EC, Saffitz JE. Functional and structural assessment of intercellular communication: increased conduction velocity and enhanced connexin expression in dibutyryl cAMP-treated cultured cardiac myocytes. Circ Res 79:174–183, 1996.

49. Hoyt RH, Cohen ML, Saffitz JE. Distribution and three-dimensional structure of the intercellular junctions in canine myocardium. Circ Res 64: 563–574, 1989.

50. Schaffer P. Ahammer H, Muller W, Koidl B, Windisch H. Di-4-ANEPPS causes photodynamic damage to isolated cardiomyocytes. Pflügers Arch 426: 548–541, 1994.

51. Simpson DG, Terracio U, Terracio M, Price RL, Turner DC, Borg TK. Modulation of cardiac myocyte phenotype in vitro by the composition and orientation of the extracellular matrix. J Cell Physiol 161:89–105, 1994.

52. Cohen SA. Immunocytochemical localization of rH1 sodium channel in adult rat heart atria and ventricle. Circulation 94:3083–3086, 1996.

53. Rohr S, Flückiger R, Cohen SA. Immunocytochemical localization of sodium and calcium channels in cultured neonatal rat ventricular cardiomyocytes. Biophys J 76:A366, Tu-Pos515, 1999.

54. Bullen S, Patel S, Saggau P. High-speed, random-access fluorescence microscopy: I. High-resolution optical recording with voltage-sensitive dyes and ion indicators. Biophys J 73:477–491, 1997.

55. Rohr S, Kucera JP. Involvement of the calcium inward current in cardiac impulse propagation: induction of unidirectional conduction block by nifedipine and reversal by Bay K 8644. Biophys J 72:754–766, 1997.

56. Rohr S, Kucera JP, Fast VG, Kléber AG. Paradoxical improvement of impulse conduction in cardiac tissue by partial cellular uncoupling. Science 275:841–844, 1997.

15

The Electrophysiological Substrate for Reentry: Unique Insights from High-Resolution Optical Mapping with Voltage-Sensitive Dyes

David S. Rosenbaum and Fadi G. Akar
Case Western Reserve University, Cleveland, Ohio, U.S.A

I. INTRODUCTION

Sudden cardiac death due to ventricular arrhythmias is a major public health problem, claiming over 350,000 lives annually in the United States alone [1]. Arrhythmias can be either focal in nature or reentrant. Focal arrhythmias are those arising from a single cell or a population of cells and exhibiting abnormally rapid firing, which can override the natural rhythm of the heart. On the other hand, reentrant excitation, which underlies the vast majority of lethal arrhythmias, is based on a process by which abnormal electrical circuits develop in the heart, driving it at fast and highly irregular rates. This process can then lead to hemodynamic deterioration and death if normal cardiac rhythm is not restored promptly by electrical defibrillation. A fundamental requirement for the initiation and maintenance of reentry is the presence of electrical heterogeneities in the heart, which can result in variable degrees of myocardial excitability. This can cause a cardiac impulse to block selectively in certain regions, but to propagate (usually slowly) in others, thereby allowing the formation of a reentrant circuit. Unlike arrhythmia mechanisms that are dependent on focal sources, reentrant arrhythmias are by definition multicellular processes, involving conduction disturbances between cells. Therefore, a complete understanding of mechanisms underlying the development and early maintenance of reentrant arrhythmias requires the measurement of electrical properties from many cells across intact heart (i.e., cardiac mapping).

Traditional multisite cardiac mapping techniques are powerful tools for investigating arrhythmia mechanisms in experimental models of heart disease, and in guiding therapy for patients. Although mapping systems vary considerably in design and implementation, they are rather similar in principle. The basic premise of any cardiac electrophysiological mapping system involves recording electrical activity either simultaneously or sequentially from multiple sites on the heart. Typically, an extracellular electrogram is recorded from each electrode, and a local activation time is estimated. Subsequently, a series of activation times (i.e., time domain) are mapped onto the location of each electrode in space (i.e., space domain) in order to construct maps of activation, which are typically displayed as isochrone or contour plots. Such plots are useful for depicting propagating waves during normal and abnormal cardiac rhythms.

Using optical mapping, one can overcome several limitations inherent to extracellular mapping. Because the time course of membrane potential is registered at every recording site, optical mapping provides an additional dimension (i.e., the voltage domain) that is not attainable by conventional mapping techniques. Therefore, using optical mapping it is possible to relate complex propagation patterns to voltage changes occurring at the level of individual cells. This has many important implications. For instance, the timing of local propagation is determined *directly* from the upstroke of the optically recorded action potential, eliminating ambiguity associated with estimating local activation time from extracellular signals. Also, because optical mapping is an imaging modality, it eliminates the need for physical electrodes, thereby permitting essentially unlimited spatial resolution. In this chapter, we discuss recent applications of optical mapping, specifically with regard the measurement of repolarization and cell-to-cell coupling in the intact heart. We focus on how such measurements have advanced our understanding of basic arrhythmia mechanisms.

II. HIGH-RESOLUTION OPTICAL ACTION POTENTIAL MAPPING

To investigate the electrophysiological substrate for reentry, a system capable of recording transmembrane potentials with high spatial, voltage, and temporal resolutions from hundreds of sites across the epicardial or transmural surfaces was needed [2–7]. A schematic diagram of this system is shown in Fig. 1. Guinea pig hearts [5,8,9] (for the purpose of epicardial

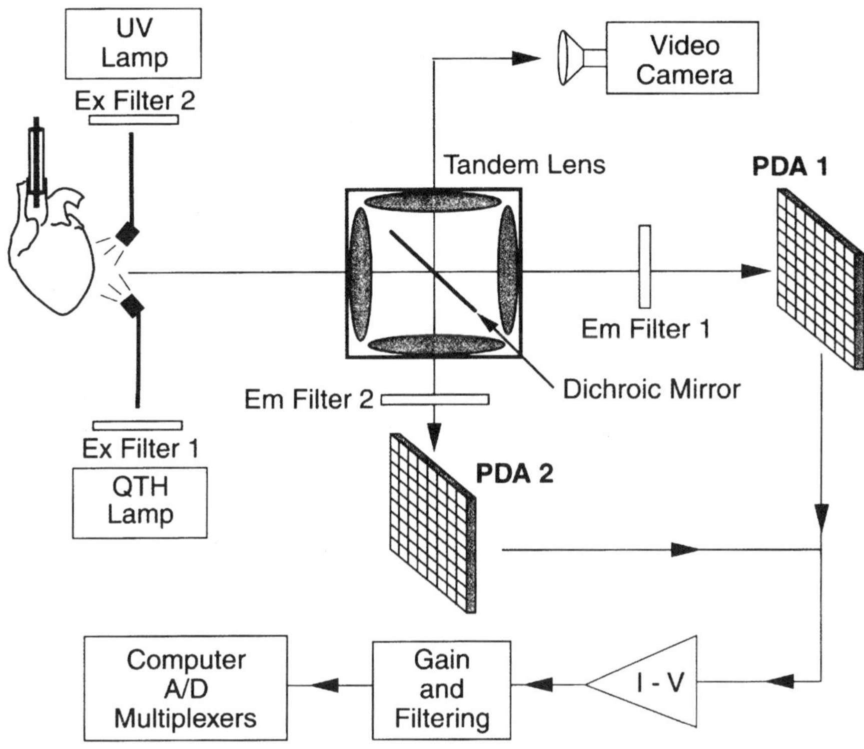

Figure 1 Optical action potential mapping system using tandem lens configuration.

mapping) or canine wedge preparations [10] (for the purpose of transmural mapping) were isolated and perfused via a cannula inserted into the aorta (guinea pigs) or a secondary branch of the left anterior descending coronary artery (canine wedges). Preparations were then perfused with normal Tyrode's solution at a constant flow rate using a digital flow pump. Perfusion pressure was continuously monitored and kept within a normal range for each type of preparation (40–60 mmHg) by adjusting perfusion flow rate (15–25 mL/min).

Preparations were placed in a custom-built imaging chamber and stabilized against an imaging window by applying gentle constant pressure via a movable piston. Previously, we showed that this procedure effectively eliminates motion-related artifacts, allowing accurate measurement of cellular repolarization without altering electrophysiological characteristics of the heart [5]. By fully submerging these preparations in temperature-controlled Tyrode's solution, intramyocardial temperature gradients were eliminated. Preparations were then stained with 100 mL of the voltage-sensitive dye, di-4-ANEPPS (15 µM) by direct arterial perfusion. Fluorescence was excited using a 270-W tungsten light source filtered at $\sim$500 nm, because in preliminary experiments we found that excitation spectra of di-4-ANEPPS bound to myocytes peaked at this wavelength. This strategy enhanced fluorescence signals considerably (over 1.5-fold) compared to standard filters used previously [11]. Moreover, we found that the bandwidth of the excitation light contributes importantly to the level of contraction artifacts present in optical action potentials. Since such artifacts often contaminate or even mask the measurement of repolarization, their suppression was necessary in studies aimed at investigating the dynamics and gradients of repolarization across the heart. This was achieved by using a relatively narrow-band ($\pm$5-nm) excitation filter, which prevented detection of changes in light reflection caused by contraction. In other studies where the accurate detection of the time course of repolarization was not required, a wide-band ($\pm$25-nm) filter was used instead. This further enhanced (by over 4$\times$) fluorescence intensity, which was useful in studies where a relatively high ($<$1-mV) voltage resolution was required, such as during the measurement of the cardiac space constant [3].

Using optical emission filters, fluoresced light was separated from excitation light, and then collected by a custom-built tandem lens imaging system (Fig. 1). The tandem lens configuration consisted of a pair of single-lens-reflex photographic lenses, focused at infinity and placed with the bayonet mounts facing outward. The preparation was placed in the focal plane of the front ("objective") lens, while the detector (photodiode array) was placed at the focal plane of the back ("detector") lens. The optical magnification in a tandem lens system is determined by the ratio of focal lengths of detector to objective lenses. Hence, by simply changing the lens combination, one can easily obtain a wide range of magnifications (0.8$\times$ to 4.4$\times$) using this system. Fluoresced light exiting the detector lens was then filtered ($>$610 nm) and focused onto the photodiode array. Importantly, we found that tandem lens imaging substantially enhanced signal intensity (by over 3$\times$) at most magnifications compared with standard single-lens optics [2]. Another advantage of the tandem lens configuration is that it provides collimated light, which can then be directed onto multiple detectors. This is especially useful for the simultaneous measurement of calcium transients and action potentials in the same heart [12] and for obtaining high-quality images of the preparation using a CCD camera (Fig. 1).

The photodetector is a 256-element photodiode array. The elements, each consisting of a PIN silicon photodiode, are arranged in a 16 $\times$ 16 square grid. Each element features a fast response time and high sensitivity in the visible-to-near-infrared range and has an active area of 1.1 mm $\times$ 1.1 mm.

Therefore, using the aforementioned range of magnifications, each pixel has an effective imaging area ranging from 0.23 mm × 0.23 mm to 1.4 mm × 1.4 mm. The photodiode array has a strong spectral response from 400 to 1000 nm, with the peak response occurring at 800 nm. When light strikes a photodiode, it is converted into photocurrent whose amplitude is dependent on the fluorescence intensity and wavelength of the incident light, and the spectral response of the photodiode. Each photodiode is coupled to a current-to-voltage amplifier with feedback resistance of 100 MΩ. This corresponds to an amplifier gain of 10^8 V/A, and a feedback capacitance of 1 pF, which, in turn, results in a low-pass filtering effect with a frequency response that is dependent on the amplifier gain. The frequency response is relatively constant from DC to the cutoff frequency (1.0 kHz). The output of the current-to-voltage amplifier is fed to a second stage of amplifiers. The voltage signal is AC coupled with a variable time constant (1.8, 2.2, 10 sec, or ∞) and amplified with variable gain (1×, 50×, 200×, or 1000×). The signal is then low-pass, anti-alias filtered with variable cutoff frequency.

A computer-based digitization system for acquiring optical signals with high throughput was also designed. It consisted of 8 analog input expansion boards, each capable of multiplexing 64 analog inputs, for a total of 512 channels (Microstar Labs, Bellevue, WA). The analog input expansion boards are connected to a data acquisition board, which is designed for fast analog sampling at 50-ns time resolution, resulting in a total throughput of approximately 2.0×10^8 samples/sec. This translates into 3400, 6800, and 54,400 samples/sec for 256, 128, and 8 channels, respectively. The data are sampled with 12-bit precision. The data acquisition board is connected directly to the data acquisition computer motherboard through a PCI slot.

Data acquisition is controlled by a software package custom developed using Labview (National Instruments, Austin, TX). Through a menu and toolbar system, the user can specify the sampling rate and the duration of each acquisition, and the number of channels being sampled. For example, by selecting a subset of the 256 available channels, the user can achieve higher sampling rates, and therefore greater temporal resolution. The software also allows the continuous monitoring of pressure, temperature, flow, and the ECG in real time, which is crucial in determining the stability of the preparation. The software also controls a stimulator and the light shutter, thereby allowing the investigator to focus on the preparation at all times during an experiment. Furthermore, the software aids the investigator in improving signal fidelity by providing the dynamic range for each channel, and thereby guiding the adjustment of excitation light intensity and position. A running circular buffer prevents the "loss" of important electrophysiological events by providing the investigator with the option of retrieving a predefined amount of data preceding each recording.

Using this system, optical action potentials could be recorded with high spatial (0.36-mm), voltage (0.5-mV), and temporal (0.3-msec) resolutions; thereby allowing the accurate measurement of key electrophysiological parameters including depolarization and repolarization times, action potential duration, conduction velocity, repolarization gradients, as well as the cardiac wavelength and space constant.

III. OPTICAL MEASUREMENT OF CELLULAR REPOLARIZATION IN INTACT HEARTS

Numerous studies have established a close association between spatial heterogeneity of repolarization and arrhythmogenesis [13–15]. Recently, this association has been underscored by a growing interest in investigating the rich heterogeneity of cell types present across the ventricular wall. For example, both epicardial and mid-myocardial (M) but not endocardial cells exhibit a distinct spike-and-dome morphology, due to a strong expression of the transient outward potassium current (I_{to}) in these cells [16]. Also, isolated M cells exhibit a longer action potential duration (APD) and a more enhanced sensitivity to class III agents than other cell types, due to reduced expression of the slowly activating delayed rectifier potassium current (I_{Ks}) in these cells [17].

In addition to heterogeneities of repolarizing currents present across different layers of myocardium, we have recently found that such heterogeneities are also present within even a single myocardial layer (epicardium), despite strong-cell-to-cell coupling within that layer [18]. As will be discussed in this chapter, the functional presence of even minor heterogeneities of repolarization can have profound implications to arrhythmogenesis.

Traditionally, critical components of the electrophysiological substrate for reentry such as spatial dispersion of repolarization [19] were classically thought to be static [14]. However, because ion channels governing repolarization are time dependent [20], one would predict otherwise [21]. In fact, regional diversity of ion channels may be expected to influence the pattern and spatial synchronization of ventricular repolarization dynamically on a beat-by-beat basis.

Using conventional recording techniques, it is difficult to track the dynamic beat-by-beat changes in spatial synchronization of repolarization, and their effect on the electrophysiological substrate for reentry. Although microelectrode and monophasic action potential recordings faithfully reproduce beat-by-beat changes in cellular repolarization, they are limited to a few selected sites on the heart, and therefore are not suitable for a comprehensive understanding of the dynamic spatiotemporal electro-

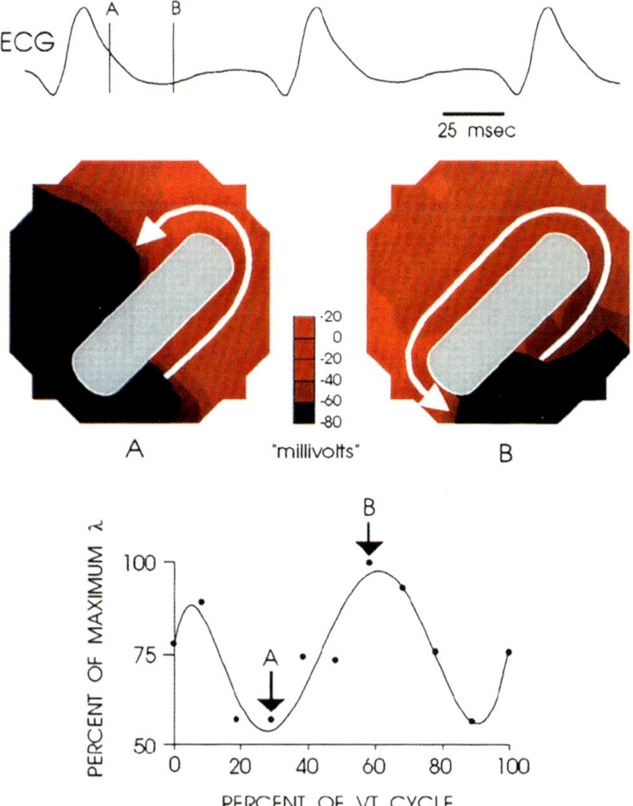

Figure 15.3 Graphical representation of wavelength from transmembrane potential maps generated by optical mapping. Top panel shows an ECG recording for three beats of sustained monomorphic VT in the guinea pig ventricle. Below are two snapshots taken at two time points (A and B) during a single reentrant cycle. The action potential amplitude is mapped to a pseudo-voltage color scale. As the leading edge of the wavefront propagates around the reentrant circuit, the magnitude of wavelength changes. Bottom panel plots the cyclic variation of wavelength during one beat of VT. (From Ref. 23.)

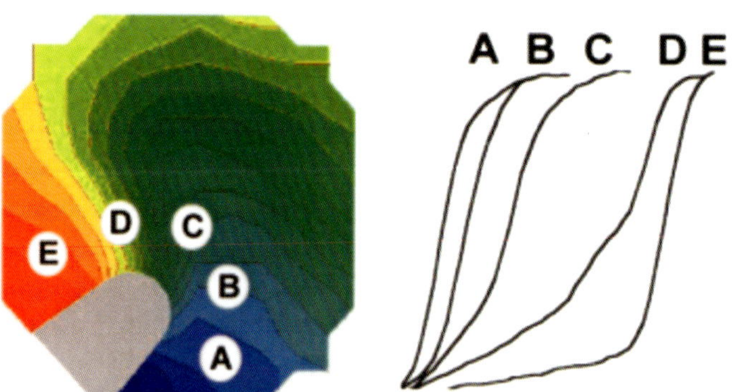

Figure 15.4 High-spatial-resolution map (400-μm resolution) of depolarization (1-msec isochrones) around the basal tip of the epicardial obstacle (hatched area) during established reentrant VT. Pivoting around the obstacle requires that the reentrant wavefront first propagate transverse to fibers (blue region), then turn parallel to fibers (blue > green > yellow), and finally turn transverse to fibers again (red region). As the wavefront pivots, conduction slows (i.e., crowding of isochrones) paradoxically as propagation turns parallel to fibers. Action potential upstrokes recorded from five evenly spaced sites around the pivot point are shown on the right. While the wavefront enters the pivot point (A and B), conduction is relatively fast and upstrokes are sharp. However, as the wavefront pivots (C and D), action potential upstrokes become increasingly slowed. After pivoting is complete (E), conduction and action potential upstroke velocity are again normal. (From Ref. 23.)

physiological changes that underlie arrhythmias. On the other hand, extra-cellular mapping is capable of simultaneously recording electrograms from hundreds of sites, but direct information regarding repolarization is lost. Alternatively, optical action potential mapping with voltage-sensitive dyes makes it possible to measure action potentials simultaneously from hundreds of sites with high voltage, temporal, and spatial resolution. This has led to the exciting possibility of measuring the dynamic gradients of repolarization across the heart, and their role in the formation of arrhythmia substrates.

A. Measurement of Repolarization Time

To quantify spatial heterogeneities of repolarization, objective algorithms that accurately and reproducibly determine repolarization times from hundreds of optical action potentials were developed [5]. Shown in Fig. 2 is a schematic of an action potential along with its first and second derivatives. As illustrated, depolarization time is defined as the time when the first derivative is maximum, and repolarization time as the time during final repolarization when the second derivative is maximum. Unlike repolarization algorithms based on an absolute threshold or first derivative, this is a robust technique that is insensitive to artifacts of contraction or baseline drift, which often contaminate optical signals.

IV. OPTICAL MEASUREMENT OF THE CARDIAC WAVELENGTH

One of the earliest observations made on reentry was that the spatial extent of the circulating wave (wavelength) must somehow fit into the available reentrant path for the reentry to persist. It was further hypothesized that interventions resulting in the advancement of the head or the extension of the tail may extinguish the reentrant circuit. However, despite being conceptually straightforward, quantitative measurements of cardiac wavelength were difficult. The accurate depiction of the spatial and temporal behavior of wavelength requires that both the depolarizing head and repolarizing tail of the propagating wave be recorded simultaneously. This can only be accomplished by simultaneously recording cellular action potentials from multiple sites on the heart. Herein lies the rationale for using optical mapping to measure the cardiac wavelength.

When the entire wave is contained within the mapping field, wavelength can be measured at any point in time as the distance between the head of the wavefront and its recovering tail. In this manner, wavelength can be measured directly as a function of time and space [22]. In general, whenever

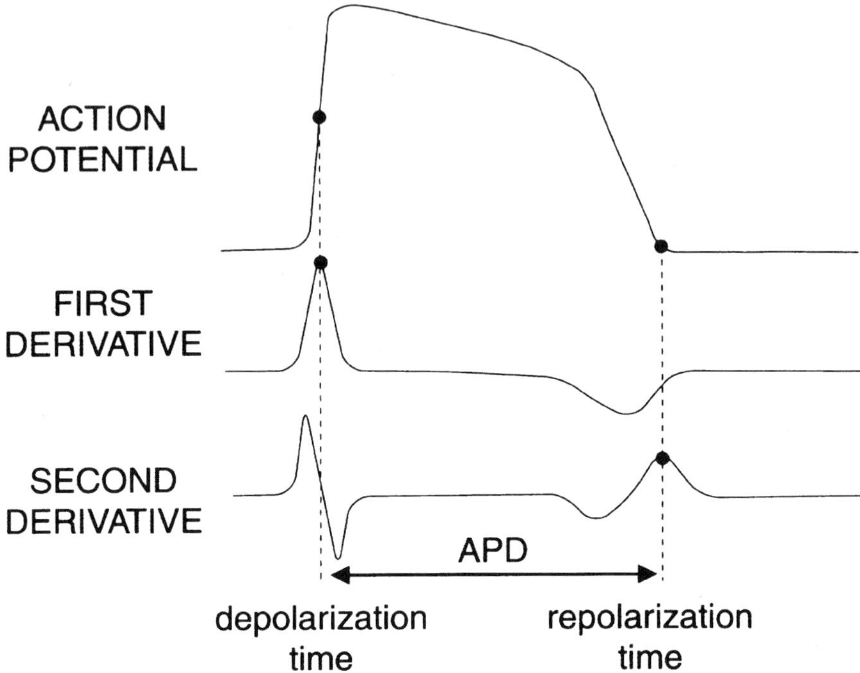

Figure 2 Schematic representation of an action potential and its first and second derivatives. Depolarization time is defined as the time when the first derivative of the action potential is maximum (first dashed line). Repolarization time is defined as the time during final repolarization when the second derivative of the action potential is maximum (second dashed line). Repolarization time defined using this technique is less sensitive to the presence of motion artifact and baseline drift, unlike algorithms based on the absolute threshold or first derivative. (From Ref. 34.)

the head of the wave accelerates relative to its tail, wavelength increases. Conversely, if the head of the wave decelerates relative to its tail, wavelength shortens.

A.   Dependence Wavelength on Fiber Structure

Previous measurements of wavelength in canine atria were made from a limited number of extracellular recordings of conduction velocity and the effective refractory period, and hence did not take into account the anisotropic properties of myocardium. We found that the cardiac wavelength measured during plane-wave propagation was highly dependent on fiber structure. Although tissue anisotropy markedly affects conduction, APD was not dependent on the direction of propagation, fiber orientation, or anisotropy. As a result, wavelength was consistently longer during propagation in the longitudinal direction compared to the transverse direction due to anisotropic conduction properties (not shown).

During steady-state reentry, wavelength is readily measured from isopotential maps constructed from optically recorded action potentials, as depicted in Fig. 3 (Fig. 3; see color plate). By computing the wavelength at discrete times throughout a reentrant cycle, we demonstrated that the wavelength of a reentrant impulse varies considerably (by ~50%) as it traverses a single, fixed reentrant circuit [23]. Since wavelength varies continuously in time, the excitable gap also varies within the circuit. This finding has important implications, as it may explain why responses of clinical VT to programmed electrical stimulation are often dependent on the region of the reentrant circuit where stimuli are delivered.

B. Optical Recordings Reveal Mechanism of Conduction Slowing at Pivot Points

In addition to the influence of fiber structure on wavelength, we found that conduction velocity slowing near the pivot points of a reentrant circuit was an important factor for shortening wavelength during VT. At the cellular level, slow conduction at pivot points was associated with a reduction in action potential upstroke velocity. This is illustrated in Fig. 4 (Fig. 4; see color plate), where action potential upstrokes recorded from five uniformly spaced sites around a pivot point are shown. As the wavefront initially approached the pivot point (lower right corner), action potential upstrokes were rapid (Fig. 4, potentials A and B) and were comparable to those in areas where the wavefront geometry was planar. However, as the wave rotated, action potential upstrokes became progressively slower. Typically, multiple notches were observed (Fig. 4, potentials C and D). In contrast, during plane-wave stimulation, action potentials recorded from the same sites (C and D) exhibited normal upstrokes, indicating that conduction slowing during reentry was not due to intrinsically depressed excitability. These data are consistent with findings that link the curvature of a wavefront to its propagation velocity [24]. As a reentrant wavefront takes on increased curvature, the advancing wave of depolarization encounters a greater mass of downstream, unexcited tissue, resulting in source–sink mismatch. Since all types of reentry require wavefront rotation, conduction slowing and subsequent shortening of wavelength at pivot points may be present in most forms of reentry [25].

Data derived from optical maps during pacing and spontaneous reentrant activation suggest that the wavelength of a propagating impulse adapts to the local electrophysiological environment. Thus, wavelength can shorten or lengthen dynamically as local conditions change spatially or temporally. Wavelength dynamics are readily observed from isopotential plots obtained from high-resolution optical mapping. Finally, it is important to emphasize how optical recordings of membrane voltage specifically provided insights into the biophysical basis for pivot point conduction slowing.

C. Wavelength Adaptation as a Mechanism of Reentry: Dynamic Arrhythmia Substrates

Because of difficulties in measuring wavelength using traditional extracellular mapping techniques, beat-by-beat changes in wavelength have not been quantified during the dynamic events leading to reentry. Following unidirectional block the wavelength must be shorter than the reentrant pathlength in order for reentry to ensue. The critical membrane events leading to the development of reentry are illustrated in Fig. 5. Isochrone lines depicting the activation patterns during the initiation of reentry are shown (Fig. 5A). Unidirectional block develops following the last paced beat (S4) and stable VT ensues. The membrane potential responses during the steady-state drive train, premature stimuli, and initial reentrant beats are shown in Fig. 5B. During steady-state pacing, action potentials demonstrate normal characteristics (i.e., sharp upstrokes, plateau and rapid repolarization phase). In contrast, following each premature stimulus, APD and conduction velocity decreased and wavelength shortened. The last premature stimulus (S4) stimulus captured the tissue closest to the pacing site (potentials A and AA) and the impulse propagated in the orthodromic direction (potentials AA–FF). Propagation along this direction was characterized by a progressive increase in action potential amplitude and upstroke velocity at more distal sites (i.e., incremental conduction). In contrast, the antidromic wave (potentials A–F) propagated decrementally until conduction failed (site D). However, the returning orthodromic wave propagated unimpeded (V1) at sites where conduction previously failed, since sufficient time passed for excitability to be restored (i.e., wavelength less than pathlength). Indeed, because of the shortened wavelength following S4, on the first reentrant beat, the sites that previously blocked were the most excitable as evidenced by the long diastolic intervals at sites A–F prior to V1.

Since baseline wavelength is much longer than any reentrant pathlength, adaptation of wavelength to dimensions shorter than the pathlength is a requirement for the initiation of VT. Therefore, using optical mapping, we demonstrated how the measurement of wavelength can advance our understanding of basic mechanisms of reentry initiation and maintenance.

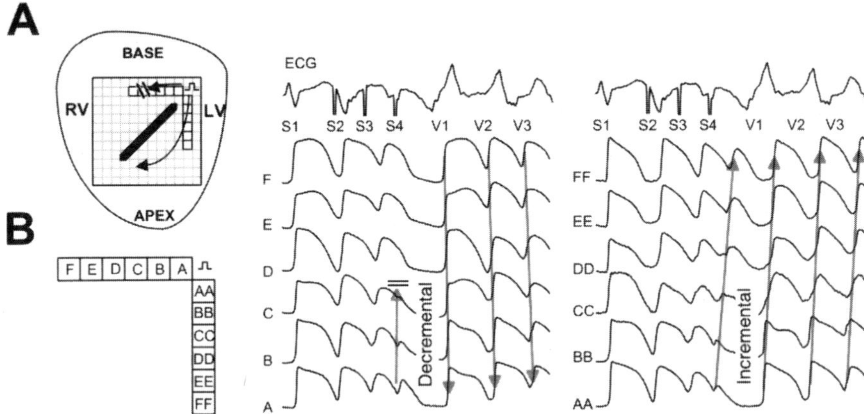

Figure 5 Initiation of reentry by three premature stimuli. Top: Stimuli S1–S3 propagate with progressive slowing (note relative crowding of 10-msec isochrones) away from the pacing site. S4, however, encounters unidirectional block at the LV base while the remaining wave propagates with sufficiently short wavelength that the head of depolarization reenters the circuit and stable reentry ensues (V1). Lower left panel: Membrane responses leading to unidirectional block. Action potentials are shown along two orthogonal directions from the pacing site (shown schematically on the lower right) for propagation on the left hand side of the lesion (A–F), and propagation on the right hand side of the lesion (AA–FF). During baseline pacing, action potentials are relatively long and conduction is rapid in both directions. With increasing prematurity, APD rapidly shortens at all sites indicating wavelength adaptation. Following the S4 stimulus, unidirectional block develops. The S4 stimulus barely captures the tissue at the pacing site (A, AA). Propagation along the left side of the lesion proceeded decrementally (A–C), with decreased action potential amplitude and dV/dt, resulting in unidirectional block. Simultaneously, propagation along the right side of the lesion proceeded incrementally, and the wavefront reentered the previous site of block (gray arrows). (From Girouard SD, Rosenbaum DS, J. Cardiovasc Electrophysiol 12:697–707, 2001.)

V. OPTICAL MEASUREMENT OF EPICARDIAL REPOLARIZATION KINETICS

The kinetics of repolarization can be characterized, in part, by the response of APD to a premature stimulus, referred to as APD restitution [26,27]. Typically, APD shortens exponentially as the diastolic interval shortens with shortening of the premature coupling interval [21,28]. Restitution reflects the time-dependent kinetics of membrane and/or intracellular ionic currents that govern repolarization [29]. By focusing on restitution properties across the epicardial surface of the guinea pig ventricle, the direct influence of *cellular* restitution heterogeneities on the substrate for reentry could be investigated at the level of the *whole heart*.

APD restitution was measured by delivering a premature stimulus (S2) over a broad range of diastolic intervals as shown in Fig. 6A. For the majority of recording sites, restitution followed a single exponential. However, nonexponential behavior was also observed [30]. In fact, the characteristics of restitution curves varied significantly across the ventricular surface. Therefore, we did not assume a predefined mathematical relationship between APD and diastolic interval. Instead, an empirical restitution rate constant, R_K, was calculated at each site [18]. Greater values of R_K indicated a faster time course of restitution and a greater degree of APD adaptation for a given change in diastolic interval. Shown in Fig. 6B are restitution curves measured from two ventricular sites, one where R_K was slow and the other where R_K was fast, illustrating the differential response of those two sites to premature stimulation.

The spatial variation of restitution across the epicardial surface is shown in Fig. 6C. R_K was markedly heterogeneous, varying by as much as 500% (range 0.04–0.24) within 1 cm^2 of epicardium. Moreover, spatial heterogeneity of R_K was not random; rather, there was an organized pattern of R_K across the epicardial surface. In particular, the gradient of R_K was oriented parallel to cardiac fibers [23], despite strong cell-to-cell coupling in that direction [31], thereby suggesting the presence of considerable heterogeneity of cellular ionic processes.

Figure 6 (A) Schematic representation of an action potential during the last beat of a 50-beat baseline drive train (S1) and a single premature beat (S2). Superimposed are: APD_b, APD of the baseline beat; DI, diastolic interval; and APD_p, APD of the premature beat. (B) Two restitution curves calculated from action potentials recorded in guinea pig, one at a site where APD_b was longest (filled circles) and the other where APD_b was shortest (open circles). Shown are the parameters used to estimate the rate constant of restitution (R_K) at the site where R_K was smallest, where ΔAPD is the extent of APD_p shortening over the range of diastolic intervals tested (ΔDI). (C) Diagram of the mapping field (1-cm^2 grid) and its position relative to the intact heart preparation (top, left). Spatial dispersion of restitution kinetics (R_K). Shown to the right of the contour map is a gray scale with corresponding numerical values in normalized units (R_K). RA, right atrium; LA, left atrium; RV, right ventricle; LV, left ventricle; LAD, left anterior descending coronary artery. (From Ref. 18.)

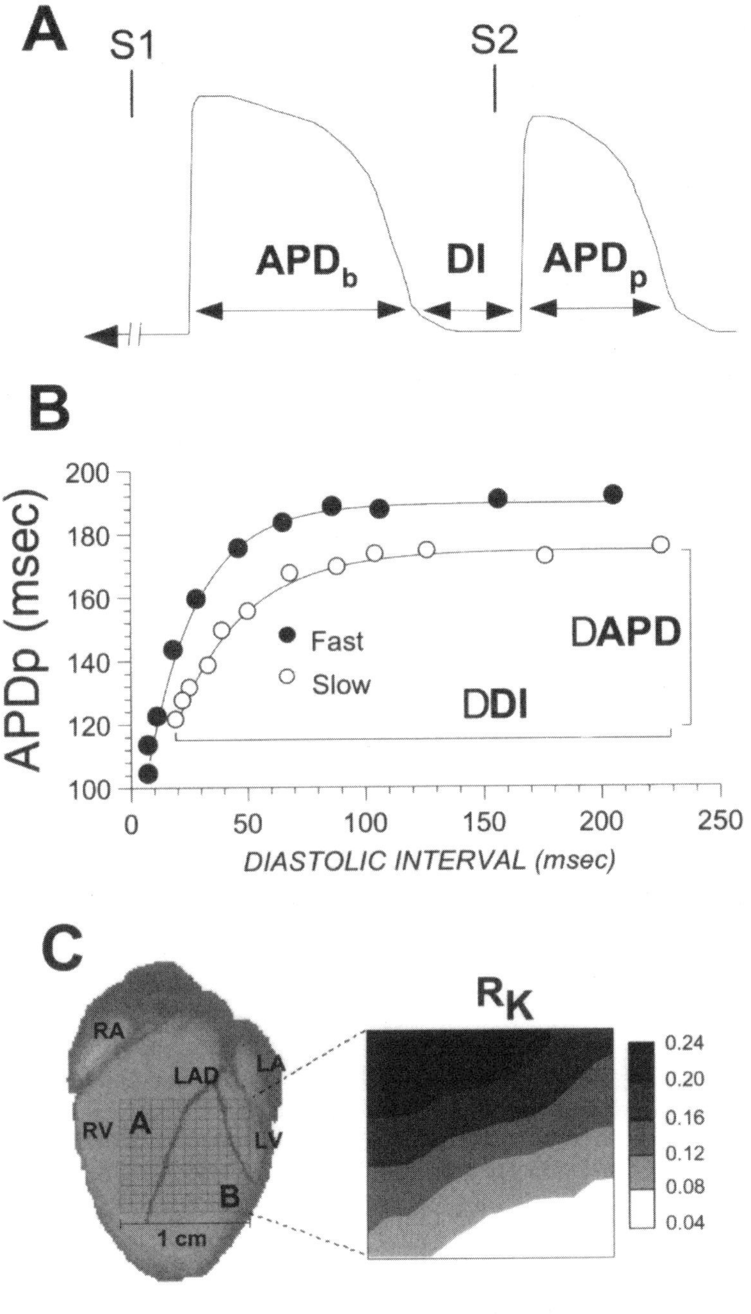

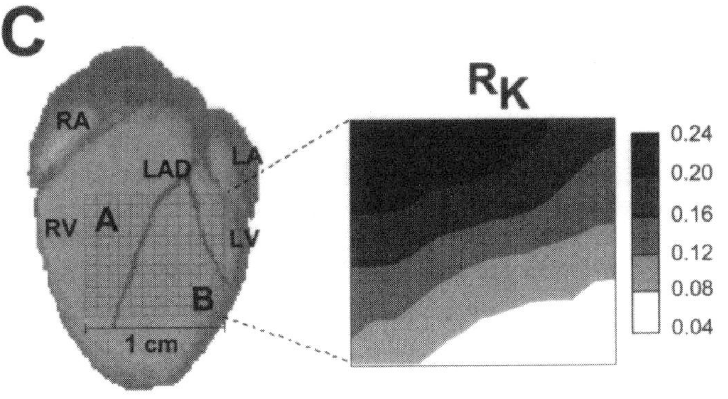

A
S1
S2
APD_b
DI
APD_p

B
200
180
160
140
120
100
APDp (msec)
Fast
Slow
DAPD
DDI
0
50
100
150
200
250
DIASTOLIC INTERVAL (msec)

C
RA
LAD
LA
RV
A
LV
B
1 cm
R_K
0.24
0.20
0.16
0.12
0.08
0.04

A. Heterogeneity of Epicardial Restitution Causes Modulated Dispersion of Repolarization

Heterogeneity of restitution kinetics across the epicardial surface is expected to alter significantly the sequence and pattern of repolarization during premature stimulation of the heart. Shown in Fig. 7 is the pattern of repolarization during stimulation at a coupling interval equal to the baseline pacing cycle length, an intermediate coupling interval, and slightly above the effective refractory period of the baseline beat. During baseline pacing, a significant gradient of repolarization was present, with latest repolarization occurring near the base of the heart and earliest repolarization near the apex. In general, the repolarization gradient during baseline pacing was oriented in an apex-to-base fashion, parallel to the cardiac muscle fibers (dashed line) [23]. A premature stimulus introduced at an intermediate coupling interval eradicated the gradient of repolarization present during baseline pacing. When a premature stimulus was introduced at a very short coupling interval, repolarization was once again altered. The new gradient had a comparable

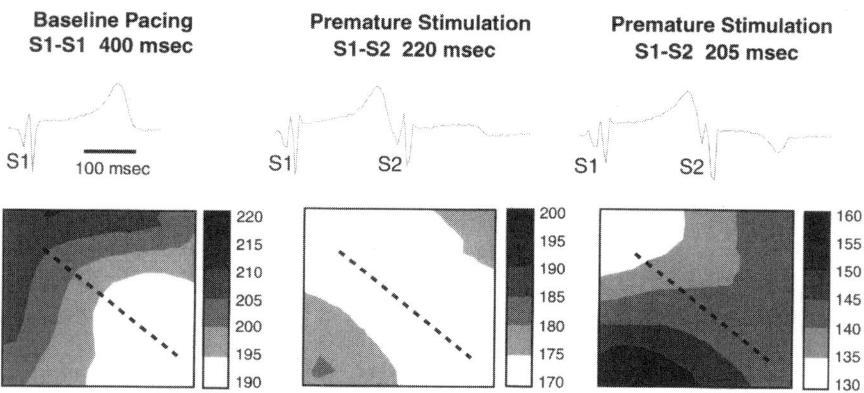

Figure 7 Contour maps of repolarization at a coupling interval equal to the baseline pacing cycle length (left), a premature stimulus at an intermediate coupling interval (middle), and a premature stimulus at a coupling interval near the refractory period (right), measured from the epicardial surface of a guinea pig. The ECG recorded during the last two baseline beats (S1) and the premature stimulus (S2) is shown across the top. Depolarization and repolarization times are in milliseconds. The site of pacing was identical for all recordings. The dashed lines indicate epicardial fiber direction. The gradient of repolarization (solid arrow) was markedly influenced by a premature stimulus. Reduced heterogeneity (B) and inversion of repolarization gradients (C) are reflected in the ECG by T-wave flattening and a change in T-wave polarity, respectively. (Modified from From Ref. 9.)

magnitude but opposite orientation to the one present during baseline pacing. The eradication and subsequent reversal of the repolarization gradient by intermediate and short premature coupling intervals were closely paralleled by the flattening and subsequent inversion of the ECG T wave of the premature beat, indicating that similar cellular processes were occurring throughout the heart and not just within the mapping field.

The initial decrease and subsequent increase (i.e., modulated dispersion) of repolarization gradients with shortening of premature coupling interval can be explained by heterogeneity of cellular restitution kinetics across the epicardial surface. Where APD during baseline pacing (APD_b) was longest, R_K was fastest and vice versa [18]. Since the restitution rate constant was faster at sites having longer APD_b, APD of the premature beat (APD_p) shortened more rapidly at these sites compared to sites with shorter APD_b, effectively eliminating repolarization heterogeneity across the epicardium. With further shortening of S1S2 coupling interval, cells initially having the longest APD exhibited the shortest APD due to their relatively fast restitution kinetics.

B. Modulated Dispersion Forms Substrate for Reentry

A premature impulse is traditionally viewed as a "trigger" for arrhythmias, which, in the presence of a suitable "substrate," can provoke reentry [32]. Our data demonstrate that the trigger and the substrate are not mutually exclusive, since a premature stimulus actively modulates the electrophysiological properties of the heart. Traditionally, premature stimuli delivered at progressively shorter coupling intervals shorten refractoriness at the stimulus site, allowing capture of subsequent stimuli at increasing degrees of prematurity and thereby increasing the likelihood of inducing reentry [33]. An alternative hypothesis, referred to as the modulated dispersion hypothesis, is presented here. In addition to shortening refractoriness, a premature beat also changes the underlying arrhythmogenic substrate by modulating spatial dispersion of repolarization in a coupling interval dependent manner.

Vulnerability to VF following a premature beat was measured. As coupling interval of the premature stimulus was shortened, dispersion of repolarization was modulated in a biphasic fashion (Fig. 8). For S1S2 coupling intervals near the baseline pacing rate, dispersion of repolarization was relatively high; however, as S1S2 coupling interval was shortened, dispersion of repolarization decreased until a critical coupling interval was reached (255 msec; Fig. 8, dashed arrow). With further shortening of S1S2 coupling interval, dispersion of repolarization rose abruptly to higher than baseline levels.

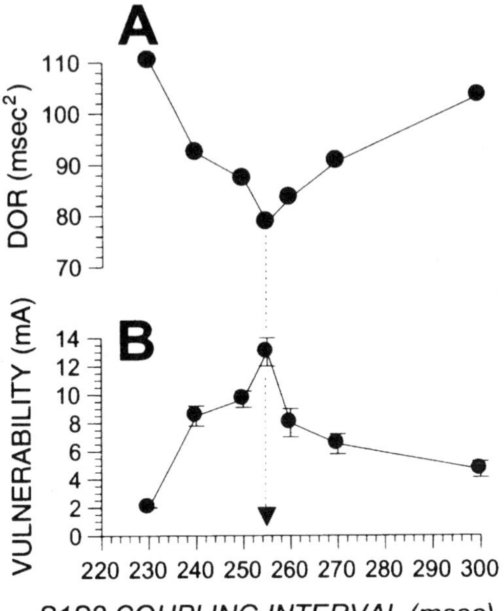

Figure 8 (A) Dispersion of repolarization of an S2 premature beat as a function of S1S2 coupling interval. These values were calculated from 128 optical action potentials recorded from the epicardial surface of guinea pig ventricle. Dispersion of repolarization was calculated by the variance of repolarization times measured over the entire mapping field. (B) Changes in vulnerability to VF induced by an S3 pulse train in the wake of repolarization patterns induced by various S1S2 coupling intervals. Dispersion of repolarization (A) and vulnerability to fibrillation (B) were modulated in a similar biphasic fashion, with minimum vulnerability and minimum dispersion occurring at the same S1S2 coupling interval (255 msec, dashed arrow). (From Ref. 9.)

The effect of cycle length-dependent modulation of repolarization on susceptibility to VF is illustrated in Fig. 8. It is evident that vulnerability to VF was also modulated in a biphasic fashion, in parallel with dispersion of repolarization. As S1S2 coupling interval was shortened to a critical value (Fig. 8, dashed arrow), vulnerability decreased. With further shortening of S1S2, vulnerability increased to levels below those present at baseline pacing. These data indicate that the electrophysiological substrate for VF is dynamic and can form, disappear, and re-form in a predictable fashion [34].

Figure 8 illustrates a paradoxical decrease in arrhythmia vulnerability as premature stimulus coupling interval was initially shortened over a broad range of coupling intervals. The attenuation of repolarization gradients by a premature stimulus may serve as a protective mechanism in electrophysiologically normal myocardium. On the other hand, the rapid increase in vulnerability at very short coupling intervals may explain why multiple, closely coupled, premature stimuli are typically required to initiate VF in relatively normal hearts. These findings highlight the importance of investigating changes in dispersion of repolarization throughout the heart, and not just refractoriness at one site to obtain a more comprehensive understanding of arrhythmia mechanisms.

VI. TRANSMURAL HETEROGENEITIES OF REPOLARIZATION

As mentioned above, the ventricular myocardium, which until recently was thought to be electrically homogeneous, has now been shown to comprise a rich variety of cell types. The ionic basis for the distinct electrophysiological properties of cells isolated from epicardial, mid-myocardial, and endo-cardial layers has been studied in detail [35]. Of particular importance was the discovery of M cells, which are characterized by a longer APD, a steeper rate dependence of APD, and a stronger sensitivity to class III antiarrhythmic agents compared to other myocardial cell types. These characteristics have implicated M cells in the development of transmural dispersion of repolarization and, in turn, to the pathogenesis of intramural reentry and Torsade de Pointes [10].

Despite inherent differences in the ionic characteristics of cell types spanning the transmural wall, electrotonic flow of current between cells is expected to reduce the functional expression of electrical heterogeneities across the normal heart [31]. Hence, the extent to which properties of M cells may functionally influence dispersion of repolarization and arrhythmogenesis remains controversial [36].

A. Transmural Optical Action Potential Mapping

In order to investigate the functional presence and significance of transmural dispersion of repolarization and its relationship to arrhythmogenesis, transmural optical mapping was developed by applying high-resolution optical action potential mapping to the transmural surface of a multi-cellular, three-dimensional preparation (canine wedge) where the influence of cell-to-cell coupling is present. This approach provided a quantitative, beat-by-beat assessment of transmural dispersion of repolarization under various electrophysiological conditions including a surrogate model of LQTS [10]. Briefly, transmural wedges (approximately $3 \times 2 \times 1.5$ cm) of myocardium surrounding secondary branches of the circumflex or the left anterior descending coronary arteries were dissected from the anterior, anterolateral, or posterior free walls of the canine left ventricle. Wedges were perfused via a cannula inserted under microscopic guidance into the small arterial branch. Wedges were stained with the voltage-sensitive dye, di-4-ANEPPS (15 μM) by direct coronary perfusion for 5 min, which allowed the transduction of cellular transmembrane potentials into optical fluorescence [37].

B. Repolarization Heterogeneities Are Functionally Expressed Across the Ventricular Wall

Using the technique of transmural optical mapping, fundamental differences in the characteristics of action potentials recorded from different layers of myocardium were evident. For example, action potentials recorded from epicardial and subepicardial cells (sites A and B, Fig. 9) exhibited a distinct spike and dome morphology, whereas endocardial cells (sites D and E) lacked the dome and exhibited a negatively sloping phase 2. M cells were characterized by a relatively longer APD at baseline, and a disproportionate prolongation of APD in response to conditions that prolong the QT interval, and generally resided in deep subepicardial to subendocardial muscle layers.

In control, a difference in APD of ~ 30–40 msec was measured between populations of cells displaying the longest and shortest action potentials. On the other hand, in LQTS, the transmural APD gradient was much larger (100 msec). Since cellular repolarization time is defined as the sum of activation time and APD [8], heterogeneity in APD of cells spanning the ventricular wall had a profound influence on the ensuing transmural repolarization gradients. Under control conditions, a relatively minor transmural gradient of repolarization (<4 msec/mm) existed, whereas in LQTS, transmural dispersion of repolarization, as measured by the maximum spatial gradient of repolarization times, was markedly enhanced (>12 msec/mm).

C. Transmural Dispersion of Repolarization Affects Susceptibility to Reentrant VT

The M-cell zone in LQTS markedly enhanced transmural dispersion of repolarization by producing regions of highly refractory tissue bordered by

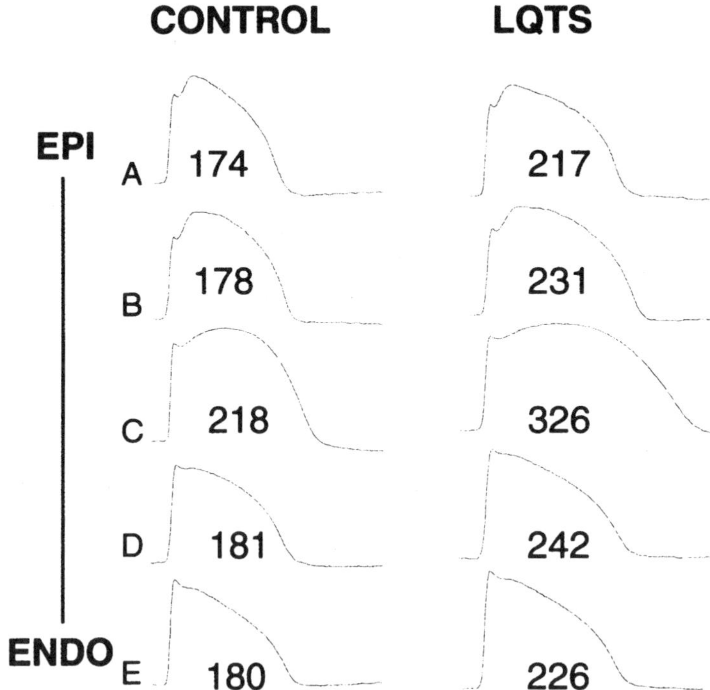

Figure 9 Optical action potentials recorded from the transmural wall of the canine wedge preparation from epicardium (top) to endocardium (bottom) during normal perfusion of the wedge (control) and during QT interval prolonging conditions (LQTS). (From Ref. 10.)

areas of excitable myocardium. The large transmural dispersion of repolarization created by M cells in LQTS formed arcs of block extending from the mid-wall toward the epicardial or endocardial borders when encountered by appropriately timed premature stimuli. In LQTS, M-cell zones resulted in conduction block and the formation of self-sustained intramural reentrant circuits having electrocardiographic characteristics typical of torsade de pointes [10].

VII. OPTICAL MEASUREMENT OF CELL-TO-CELL COUPLING IN INTACT HEART

Dispersion of repolarization depends on two factors: (1) the extent to which repolarization properties of neighboring cells vary, and (2) the extent to which cells are electrically coupled to one another. Using optical mapping, we have demonstrated thus far how spatial heterogeneities in repolarization properties of cells result in the formation of dispersion of repolarization across the apico-basal and transmural axes. In what follows, we will illustrate how optical mapping can be used to obtain direct functional measurements of cell-to-cell coupling in the intact heart. Since cell-to-cell coupling plays a critical role in propagation, repolarization, and arrhythmias [38], measurement of cell-to-cell coupling in the intact ventricle is important for understanding arrhythmia mechanisms. Previously, functional measurements of cell-to-cell coupling in intact myocardium have not been feasible, and thus the exact role of cell-to-cell coupling in arrhythmogenesis remained unclear.

The advent of voltage-sensitive-dye techniques led to the exciting possibility of recording transmembrane voltages (V_m) free of stimulus artifacts from hundreds of sites across the intact heart. This feature was widely applied to the investigation of V_m during relatively large defibrillatory shocks [39–41]. However, the distribution of V_m during subthreshold (ST) electrical stimuli is more difficult to investigate because it requires a high density of recording sites within a relatively small distance (<2 mm) and a high degree of sensitivity to relatively small changes in V_m. Therefore, we developed a high-resolution optical action potential mapping system capable of measuring V_m with sufficient fidelity to calculate λ from the decay of ST V_m in space, yielding a functional index of cell-to-cell coupling in the intact guinea pig heart [12].

A. Measurement of Space Constant

In one-dimensional cable theory, V_m caused by unipolar stimulation from a point source decays exponentially with distance from the site of stimulation. The space constant of this decay, λ, reflects, the combined influences of membrane (R_m), intracellular (R_i), and extracellular (R_o) resistances [43]. In heart, R_i, reflects the sum of gap junctional and cytoplasmic resistances. Since membrane and cytoplasmic resistances are relatively constant in space (i.e., between cells) and over time throughout diastole (i.e., in the absence of an action potential), a change in λ indicates a change in cell-to-cell electrotonic interactions of which gap junctional and extracellular resistances are major determinants. In this study, the decay of ST V_m along each of multiple linear paths directed away from the site of ST stimulation was fitted to a monoexponential for each path. λ along any given path was defined as the normalized rate of decay of ST V_m in that direction.

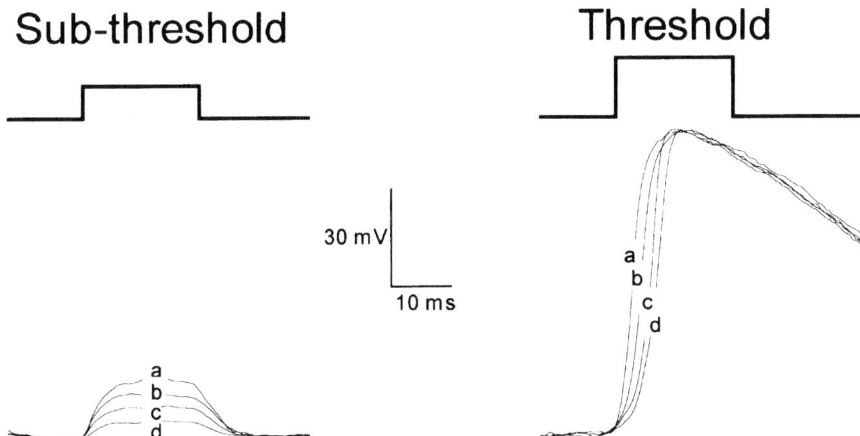

Figure 10   Optical recording of transmembrane voltage (V_m) induced by a stimulus above (right) and below (left) diastolic threshold from four sites 0 to 1.5 mm away from the electrode (sites a–d, respectively). Shown on top is a schematic representation of the stimulus pulse in both cases illustrating time course and relative amplitude. (From Ref. 42.)

B.   Optically Recorded Subthreshold Membrane Responses

V_m resulting from stimuli delivered above and below diastolic threshold are compared in Fig 10. In this example, recordings are shown from equally spaced sites (a–d) at increasing distances from the stimulus electrode. ST V_m (left panel) was characterized by depolarizing and repolarizing phases that exactly followed the timing of the stimulus waveform. Both phases followed exponentials having similar time constants that were not affected by the cell's distance from the site of stimulation. In contrast to the time course of ST V_m responses, the amplitude of ST V_m varied considerably in space, decaying with increasing distances from the site of stimulation (sites a–d). Several clear distinctions between action potentials (right panel) and ST V_m (left panel) are illustrated in Fig. 10. (1) Because they arise from regenerative active ionic processes, action potentials did not decay in amplitude at sites distal to the electrode. (2) Action potential repolarization far outlasted the stimulus pulse, whereas, the onset of ST V_m repolarization coincided exactly with the stimulus pulse due to its passive membrane nature. (3) Action potential depolarization, plateau, and repolarization were generated by active ionic currents, giving the action potential its distinctive shape, whereas ST V_m had a symmetric morphology, typical of the charging and discharging of a resistive-capacitive network which characterizes passive properties of myocytes. (4) Finally, while action potential depolarization was associated with step delays as the impulse propagated from one site to the next (sites a–d) ST V_m depolarization and repolarization were essentially simultaneous at all sites.

C. Space Constant Follows Tissue Anisotropy

Figure 11 illustrates the effect of tissue anisotropy on the decay of ST V_m across the epicardial surface from a representative experiment. ST V_m recorded from sites along the longitudinal and transverse fiber axes are shown in Fig. 11A. While ST V_m decayed away from the site of stimulation in both directions, the decay was faster transverse compared to longitudinal to cardiac fibers. The isopotential map (Fig. 11B) shows that the distribution of ST V_m was anisotropic, and closely followed fiber orientation (dotted lines). Consequently, λ was longer longitudinal compared to transverse to cardiac fibers (Fig. 11C).

D. Pharmacological Alteration of Space Constant

Conduction velocity is dependent on the extent of cell-to-cell coupling and the availability of inward sodium current. While a given change in the former is expected to influence λ, a change in the latter is not. Therefore, we investigated the effect of reducing cell-to-cell coupling by heptanol and blocking sodium current by flecainide on conduction velocity and λ. As shown in Fig. 12, while conduction velocity slowing (by 60%) with heptanol was associated with a 50% reduction in λ, conduction velocity slowing with flecianide occurred without a change in λ, reaffirming that λ as measured by optical mapping is an index of passive membrane properties and is independent of conduction velocity slowing.

Figure 11 Space constant (λ) follows tissue anisotropy. (A) ST V_m shown for sites 0 to 2.5 mm from the stimulating electrode along the longitudinal (Long) and transverse (Trans) axes of propagation. (B) Distribution of ST V_m surrounding the site of ST stimulation shown as isopotential plot. (C) Decay of ST V_m in space plotted along and transverse to fiber orientation. λ was calculated from the exponential decay constant in each direction. λ_{Trans} is approximately 50% of λ_{Long}, reflecting directional differences in intercellular coupling. ST V_m: subthreshold membrane voltage. (From Ref. 42.)

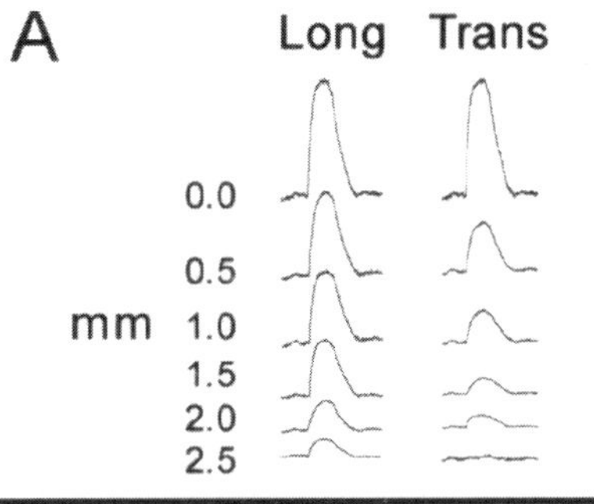
A
Long   Trans
mm
0.0
0.5
1.0
1.5
2.0
2.5

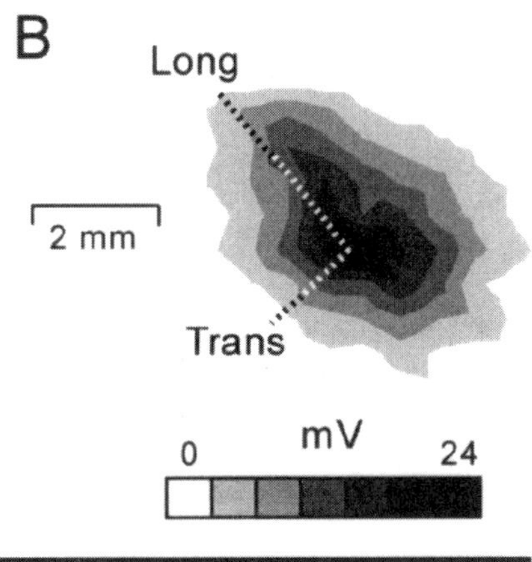
B
Long
2 mm
Trans
0   mV   24

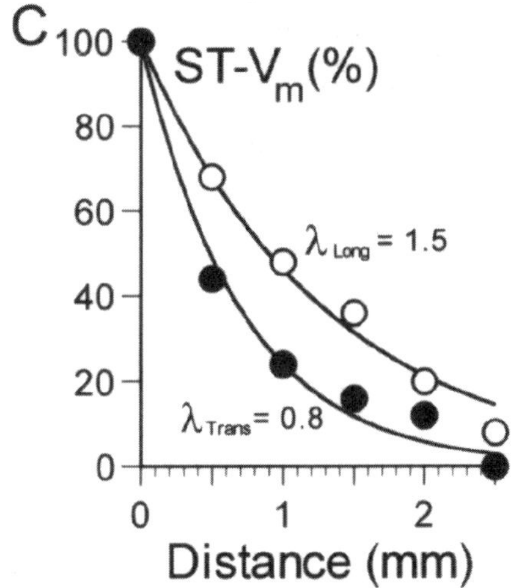
C
100
80
60
40
20
0
ST-V$_m$(%)
λ_{Long} = 1.5
λ_{Trans} = 0.8
0   1   2
Distance (mm)

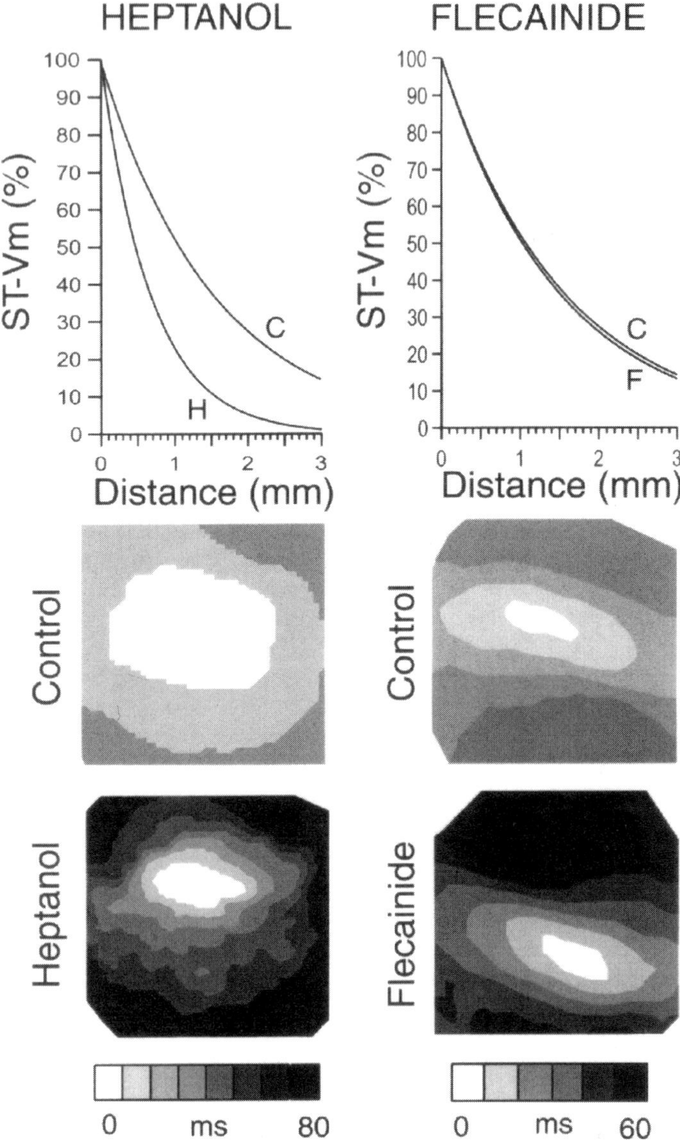

Figure 12   Effect of heptanol on CV and λ in a representative experiment. Heptanol (H) reduced both CV (as indicated by crowding of isochrones) and λ by $\sim 50\%$. Effect of flecainide on λ (top) and CV (bottom) is shown. Although CV was significantly slowed with flecainide, no significant effect on λ was detected. CV, conduction velocity; λ, cardiac space constant; ST V_m, subthreshold membrane voltage. (From Ref. 42.)

VIII. SUMMARY

Optical mapping allows the detection of cellular membrane potential with high spatial and temporal resolutions, thereby allowing investigators for the first time to investigate cell physiology in the intact heart. Optical mapping has been instrumental in advancing our knowledge of the kinetics of cellular repolarization and their gradients across the heart. Direct mapping of those gradients during the induction of arrhythmias has illustrated many basic principles of reentry initiation, maintenance, and termination. More recently, optical mapping was also used to assess cell-to-cell coupling directly, and thus promises to further our understanding of electrophysiological substrates that promote arrhythmias.

REFERENCES

1. Myerburg RJ, Kessler KM, Bassett AL, Castellanos A. A biological approach to sudden cardiac death: structure, function, cause. Am J Cardiol 63:1512–1516, 1989.
2. Laurita K, Libbus I. Optics and detectors used in optical mapping. In: Rosenbaum DS, Jalife J, eds. Optical Mapping of Cardiac Excitation and Arrhythmias. New York: Futura, 2001, pp. 61–78.
3. Akar FG, Roth BJ, Rosenbaum DS. Optical measurement of cell-to-cell coupling in intact heart using subthreshold electrical stimulation. Am J Physiol Heart Circ Physiol 281:H533–H542, 2001.
4. Eloff BC, Lerner DL, Yamada KA, Schuessler RB, Saffitz JE, Rosenbaum DS. High resolution optical mapping reveals conduction slowing in connexin43 deficient mice. Cardiovasc Res 51:681–690, 2001.
5. Girouard SD, Laurita KR, Rosenbaum DS. Unique properties of cardiac action potentials recorded with voltage-sensitive dyes. J Cardiovasc Electrophysiol 7:1024–1038, 1996.
6. Laurita KR, Akar FG, Girouard SD, Rosenbaum DS. Modulated dispersion explains changes in arrhythmia vulnerability during premature stimulation of the heart (abstr). PACE 19:643, 1994.
7. Pastore JM, Girouard SD, Laurita KR, Akar FG, Rosenbaum DS. Mechanism linking T-wave alternans to the genesis of cardiac fibrillation. Circulation 99:1385–1394, 1999.
8. Pastore JM, Rosenbaum DS. Role of structural barriers in the mechanism of alternans-induced reentry. Circ Res 87:1157–1163, 2000.
9. Laurita KR, Girouard SD, Akar FG, Rosenbaum DS. Modulated dispersion explains changes in arrhythmia vulnerability during premature stimulation of the heart. Circulation 98:2774–2780, 1998.

10. Akar FG, Yan G, Antzelevitch C, Rosenbaum DS. Unique topographical distribution of M cells underlies reentrant mechanisms of torsade de pointes in the Long QT syndrome. Circulation 105:1247–1253, 2002.

11. Efimov IR, Huang DT, Rendt JM, Salama G. Optical mapping of repolarization and refractoriness from intact hearts. Circulation 90:1469–1480, 1994.

12. Laurita KR, Singal A. Mapping intracellular calcium and transmembrane potential in the same heart (abstr). Ann Biomed Eng 26:S-18, 1998.

13. Mines GR. On dynamic equilibrium in the heart. J Physiol (Lond) 46:349–383, 1913.

14. Allessie A, Bonke FI, Schopman FJG. Circus movement in rabbit atrial muscle as a mechanism of tachycardia: the role of nonuniform recovery of excitability in the occurrence of unidirectional block as studied with multiple microelectrodes. Circ Res 39:169–177, 1976.

15. Han J, Moe G. Nonuniform recovery of excitability in venticular muscle. Circ Res 14:44–60, 1964.

16. Litovsky SH, Antzelevitch C. Transient outward current prominent in canine ventricular epicardium but not endocardium. Circ Res 62:116–126, 1988.

17. Liu D-W, Antzelevitch C. Characteristics of the delayed rectifier current (I_{Kr} and I_{Ks}) in canine ventricular epicardial, midmyocardial, and endocardial myocytes: a weaker I_{Ks} contributes to the longer action potential of the M cell. Circ Res 76:351–365, 1995.

18. Laurita KR, Girouard SD, Rosenbaum DS. Modulation of ventricular repolarization by a premature stimulus: role of epicardial dispersion of repolarization kinetics demonstrated by optical mapping of the intact guinea pig heart. Circ Res 79:493–503, 1996.

19. Kuo C, Munakata K, Reddy CP, Surawicz B. Characteristics and possible mechanisms of ventricular arrhythmia dependent on the dispersion of action potential durations. Circulation 67:1356–1357, 1983.

20. Carmeliet E. K^+ channels and control of ventricular repolarization in the heart. Fund Clin Pharmacol 7:19–28, 1993.

21. Boyett MR, Jewell BR. A study of the factors responsible for rate-dependent shortening of the action potential in mammalian ventricular muscle. J Physiol 285:359–380, 1978.

22. Girouard S, Rosenbaum D. Mapping arrhythmia substrates related to repolarization: 2. Cardiac wavelength. In: Rosenbaum DS, Jalife J, eds. Optical Mapping of Cardiac Excitation and Arrhythmias. New York: Futura, 2001, pp. 61–78.

23. Girouard SD, Pastore JM, Laurita KR, Gregory KW, Rosenbaum DS. Optical mapping in a new guinea pig model of ventricular tachycardia reveals mechanisms for multiple wavelengths in a single reentrant circuit. Circulation 93:603–613, 1996.

24. Cabo C, Pertsov AM, Baxter WT, Davidenko JM, Gray RA, Jalife J. Wavefront curvature as a cause of slow conduction and block in isolated cardiac muscle. Circ Res 75:1014–1028, 1994.

25. Reiter MJ, Zetelaki Z, Kirchhof CJH, Boersma L, Allessie MA. Interaction of acute ventricular dilatation and d-sotalol during sustained reentrant ventricular tachycardia around a fixed obstacle. Circulation 89:423–431, 1994.

26. Bass BG. Restitution of the action potential in cat papillary muscle. Am J Physiol 228:1717–1724, 1975.

27. Carmeliet E. Repolarization and frequency in cardiac cells. J Physiol (Paris) 73:903–923, 1977.

28. Saitoh H, Bailey J, Surawicz B. Action potential duration alternans in dog Purkinje and ventricular muscle fibers. Circulation 80:1421–1431, 1989.

29. Sanguinetti MC, Jurkiewicz NK. Two components of cardiac delayed rectifier K^+ current: differential sensitivity to block by class III antiarrhythmic agents. J Gen Physiol 96:195–215, 1990.

30. Laurita KR, Girouard SD, Rudy Y, Rosenbaum DS. Role of passive electrical properties during action potential restitution in the intact heart. Am J Physiol 273:H1205–H1214, 1997.

31. Lesh MD, Pring M, Spear JF. Cellular uncoupling can unmask dispersion of action potential duration in ventricular myocardium: a computer modeling study. Circ Res 65:1426–1440, 1989.

32. Myerburg RJ, Kessler KM, Castellanos A. Sudden cardiac death: structure, function, and time-dependent risk. Circulation 85(suppl I):1-2–1-10, 1992.

33. Moe GK, Childers RW, Merideth J. An appraisal of supernormal A-V conduction. Circulation 38:5–28, 1968.

34. Laurita K, Pastore J, Rosenbaum D. Mapping arrhythmia substrates related to repolarization: 1. Dispersion of repolarization. In: Optical Mapping of Cardiac Excitation and Arrhythmias. New York: Futura, 2001.

35. Sicouri S, Antzelevitch C. Electrophysiologic characteristics of M cells in the canine left ventricular free wall. J Cardiovasc Electrophysiol 6:591–603, 1995.

36. Anyukhovsky EP, Sosunov EA, Gainullin RZ, Rosen MR. The controversial M cell. J Cardiovasc Electrophysiol 10:244–260, 1999.

37. Akar FG, Laurita KR, Rosenbaum DS. Cellular basis for dispersion of repolarization underlying reentrant arrhythmias. J Electrocardiol 33(suppl):23–31, 2000.

38. Jongsma HJ, Wilders R. Gap junctions in cardiovascular disease. Circ Res 86:1193–1197, 2000.

39. Kwaku KF, Dillon SM. Shock-induced depolarization of refractory myocardium prevents wave-front propagation in defibrillation. Circ Res 79:957–973, 1996.

40. Knisley S, Smith W, Ideker R. Effect of field stimulation on cellular repolarization in rabbit myocardium: implications for reentry induction. Circ Res 70:707–715, 1992.

41. Efimov IR, Cheng Y, Van Wagoner DR, Mazgalev T, Tchou PJ. Virtual electrode-induced phase singularity—a basic mechanism of defibrillation failure. Circ Res 82:918–925, 1998.

42. Akar F, Roth B, Rosenbaum D. Optical measurement of cell-to-cell coupling in intact heart using subthreshold electrical stimulation. Am J Physiol 281:H533–H542, 2001.

43. Plonsey R, Barr R. The four-electrode resistivity technique as applied to cardiac muscle. IEEE Trans Biomed Eng 29:541–546, 1982.

16

Optical Mapping of Cardiac Stimulation: Fluorescent Imaging with a Photodiode Array

Igor R. Efimov
Case Western Reserve University, Cleveland, Ohio, U.S.A.

Yuanna Cheng
The Cleveland Clinic Foundation, Cleveland, Ohio, U.S.A.

Optical mapping of electrical activity in the heart employing imaging techniques based on voltage-sensitive dyes has become an increasingly common research tool in basic cardiac electrophysiology. This was prompted by the failure of conventionally used intra- or extracellular recordings to provide high-resolution spatiotemporal maps of electrical activity, especially during application of electric stimuli. Despite a century of evolution, conventional techniques have failed to work in at least two key areas of research: in the study of the role of spatiotemporal organization of repolarization in arrhythmogenesis and in the study of the effects of external electric shocks on cellular electrical activity.

Optical recordings of transmembrane potentials can be performed in a wide range of spatial resolutions, from the subcellular level to the whole heart. The response time of fast voltage-sensitive dyes lies in the microsecond range, and the temporal resolution of the technique can potentially exceed that of conventional microelectrode recordings. Progress in modern computer technology permits simultaneous optical recordings from multiple sites with individual signal conditioning, providing high-resolution spatiotemporal maps of electrical activity.

In this chapter we review current technological approaches developed in the area of fluorescence imaging and describe the new cardiac fluorescence imaging systems based on photodiode arrays.

I. INTRODUCTION

Investigation of the effects of externally applied electric stimuli on the heart has a long history, starting from the nineteenth century. In 1850, Ludvig and Hoffa [1] demonstrated that an electrical discharge applied directly to the heart induced ventricular fibrillation. In 1899, Prevost and Battelli [2] discovered an opposite effect of electrical discharge, which restored a normal sinus rhythm in a fibrillating canine heart. These two effects have been puzzling researchers for generations. Unfortunately, the inability of conventional electrophysiological techniques to record electrical activity of the heart during electrical stimuli impeded attempts to unravel this problem.

The breakthrough in the field occurred during last decade of the twentieth century, after the introduction of potentiometric probes, which were envisioned and developed by neuroscientist Lawrence Cohen and his colleagues [3,4], and applied to the heart by Salama and Morad [5]. Using potentiometric probes, Stephen Dillon [6,7] demonstrated the possibility of observing electrical activity in the heart during electric shocks, free of the overwhelming artifacts always present in conventional electrograms. Later, this experimental approach was adopted with some modifications by others involved in investigation of cardiac stimulation and defibrillation [8–13]. Optical mapping turned out to be nearly a perfect tool for the task.

Once again, the possibility of recording dynamic changes in the transmembrane potential of excitable cells by optical means was first suggested in 1968 by a group of neuroscientists, Lawrence B. Cohen and coworkers [3], who discovered potential-dependent changes in the intrinsic optical properties of squid giant axons. It took nearly a decade until the first optical action potentials were recorded from giant axons [4] and mammalian hearts by means of voltage-sensitive dyes [5]. The first cardiac application of this method was the localization of pacemaker activity in embryonic heart preparations in 1981 [14]. Although in the 1980s optical mapping of the heart was mostly restricted to a few non-cardiac electrophysiology laboratories, widespread application of these techniques to problems unsolvable by other means began in the 1990s [6,9,10,15–22].

Is optical mapping a technique for a problem, or do we have to invent problems for the technique? What is its niche in cardiac electrophysiology? One of the pioneers of fluorescent methods in neurophysiology, B. M. Salzberg, predicted that voltage-sensitive dyes "could, we believe, provide a powerful new technique for measuring membrane potential in systems where, for reasons of scale, topology, or complexity, the use of electrodes is inconvenient or impossible" [23]. Based on our current experience in cardiac electrophysiology, Salzberg's list needs to be extended to recordings of action potentials in the presence of external electrical fields during stimulation and defibrillation, which were impossible with extracellular and intracellular electrodes. This application will be the focus of our chapter.

After decades of innovations and development, conventional electrode techniques have been brought to the limits of perfection. The intracellular microelectrode technique still represents the "gold standard" for recording transmembrane action potentials, and only recently has the optical method approached comparable signal-to-noise ratios. The chief disadvantage of the microelectrode technique lies in the impossibility of maintaining stable recordings over longer time periods from more than two or three sites, especially if the preparation is moving. Similarly, monophasic action potential recordings can be maintained only at a few sites for short periods of time [24]. Thus spatiotemporal mapping by either method requires the use of a roving probe, limiting the application of either method to the study of periodic activation patterns. While extracellular contact and noncontact multielectrode arrays are excellent for recording activation maps from the heart in vitro, they also remain the only techniques for in vivo mapping. Problems do exist, however, concerning the precise interpretation of the electrogram data [25,26], the determination of repolarization times being particularly unreliable [27]. Thus, optical mapping is the only technique capable of recording high-resolution maps of cardiac repolarization [28]. Finally, optical mapping is the only method that allows uninterrupted and artifact-free recordings of the transmembrane potentials during pacing stimuli [9,10,20] and defibrillation shocks [6,12].

Despite impressive success of optical mapping, its technical details remain poorly described in the literature. In this chapter we therefore not only review current technological approaches developed in the area of fluorescence imaging, but also describe the design and implementation of our cardiac fluorescence imaging systems, based on arrays of photosensitive diodes. These, in our opinion, represent the best current solutions for macroscopic mapping of cardiac electrophysiological activity in terms of image quality, cost and labor of setting up, and ease of operation. A more comprehensive review of optical mapping techniques and areas of application was presented in a recently published book [29].

II. PHYSICAL PRINCIPLES OF FLUORESCENCE RECORDINGS

The general mechanism underlying fluorescence is the absorption of photons of certain energy by a fluorescent compound, which is then excited from the

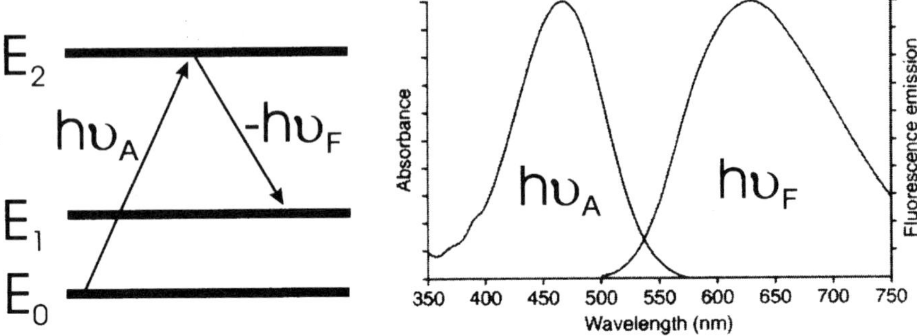

Figure 1 Quantum mechanism of the difference between the absorbance and fluorescent emission spectra (see text for detail).

ground state to an unstable energy-rich state (see Fig. 1). When the compound falls back to an intermediate, lower energy level, the compound fluoresces by emitting a photon of a lower energy than the exciting photon. The wavelength of a photon being a function of its energy, the emitted light always has a longer wavelength that the exciting light, the so-called Stoke's shift.

Potential-sensitive fluorescence can be the consequence of one of several mechanisms, which are related to voltage gradient-dependent intra- and extramolecular rearrangements. Cohen and Salzberg [30], introduced a simple classification of voltage-sensitive dyes into two groups, fast and slow dyes, based on their response times and presumed molecular mechanism of voltage sensitivity. Only the fast probes are used in cardiac electro-physiology, due to their ability to follow electrical responses on a time scale of microseconds [19,31]. The precise mechanisms underlying the voltage-dependent spectroscopic properties of fast voltage-sensitive dyes are still not fully understood. According to the *electrochromic mechanism*, one of the most commonly accepted theories [32], the spectral shift in a chromophore's properties is voltage dependent if two conditions are met: (1) the light photon-produced excitation of the chromophore molecule from the ground to an excited state is accompanied by large shift in electronic charge; (2) the vector of the intramolecular charge movement is oriented parallel to the electric field gradient. If the charge movement in the dye molecule occurs perpendicular to the cellular membrane of a cardiomyocyte, a dye's fluor-escence will be sensitive to the transmembrane potential. An alternative theory is the *solvatochromic mechanism*, which is related to electric field-induced reorientation of the dye molecule [33]. Dye molecules experience a change in the polarity of the lipid environment during reorientation pro-duced by the voltage gradient. Therefore, energy needed for excitation from the ground state to the first excited state is released during transition in the opposite direction and will be voltage dependent. This dependency causes the spectral voltage dependence of the chromophore.

Designers of voltage-sensitive dyes had to solve several problems:

1. To find a chromophore that is capable of producing the largest movements of charges during quantum transition from the ground to the first excited state, and therefore the largest measurable spectral changes with a change in the external electric field
2. To assure the possibility of delivering dye molecules to the cellular membrane
3. To assure the proper orientation of the dye molecule perpendicular to the membrane
4. To maximize duration of stay of the dye molecules in the desired position
5. To minimize photobleaching of the dye
6. To minimize side effects of the dye on the preparation in the presence and absence of light

In tests of over 1500 different compounds, several useful classes of chromophores have emerged, including merocyanine, oxonol, and styryl dyes. Styryl dyes represent the most popular family of dyes, RH-421, di-4-ANEPPS, and di-8-ANEPPS being the most important members of this family. The spectroscopic properties of these dyes have been shown to change linearly with membrane potential changes in the normal physiological range of transmembrane voltages in axons [34] and heart [35]. The orientation of the molecules of these dyes in the cell membrane is assured by the presence of lipophilic and hydrophilic groups at opposite ends of the molecule. While the hydrophilic, negatively charged sulphonyl group anchors the dye molecule in the aqueous extracellular space, the highly lipophilic hydrocarbon chains at the other end of the molecule hold it within the bilayer lipid membrane.

The stability of the position of dye molecule in the membrane can be improved by increasing of the length of the hydrocarbon tails as was done in the ANEPPS family (di-4-ANEPPS, di-8-ANEPPS, di-12-ANEPPS, nomenclature described by Loew) [36]. In single-layer cell culture preparations, Rohr et al. [19] demonstrated a significantly retarded decay of the amplitude of optical action potentials in preparations stained with di-8-ANEPPS as compared with di-4-ANEPPS on continuous illumination for 60 sec, presumably on the basis of the slower translocation of di-8-ANEPPS into the cell. The price for the improved lipophilic properties of di-8-ANEPPS, however, is decreased water solubility; this necessitates the use of surfactants, such as pluronic (F127, BASF Corp.), which may not be free of toxic side effects [37]. In a series of Langendorff-perfused and superfused heart preparations, we were able to achieve superior signal-to-noise ratios with di-8-ANEPPS and pluronic as compared to di-4-ANEPPS in only 2 of 8 experiments, while in the other 6 experiments the signal-to-noise ratio was significantly worse (unpublished personal data). In our opinion, di-4-ANEPPS remains the dye of choice for whole-heart and tissue preparations.

The signal-to-noise ratio in stained cardiac preparations is dependent not only on the dye itself, but also on its mode of delivery. Salama [35] noted that the signal-to-noise ratio in optical action potentials in frog hearts stained by injection of merocyanine-540 into the aortic root during Langendorff perfusion was 10-fold higher than in similar preparations stained by superfusion with a bath solution containing the voltage-sensitive dye. In the rat papillary muscle, Müller et al. [38] were able to demonstrate by fluorescence microscopy that superfusion with $10–25\,\mu M$ di-4-ANEPPS stained no more than one or two surface cell layers, whereas preparations stained by arterial perfusion are expected to show a more or less homogeneous distribution of dye throughout the wall. The precise depth of optical recordings in preparations stained by perfusion has been the subject of debate. Model calculations by Salama based on the depth of field of the optical system predicted a depth of $144\,\mu m$ [35]. Direct measurement by Knisley et al. [39] in a tapering wedge of tissue showed that the intensity of optical action potentials ceased to increase if the thickness of the tissue was larger than $300\,\mu m$. Based on measurements of the absorption coefficient of myocardium for the excitation and emission spectrum, Girouard et al. [40] predicted that 95% of the signal energy originates from a tissue depth of $500\,\mu m$ or less. Our recent estimation of the depth of tissue contributing to the fluorescent signals in the rabbit atrioventricular nodal preparation [41] even exceed these estimates. We believe the signals may originate as deep as 1–2 mm from the surface of the preparation. That appears to agree with similar measurements of Choi and Salama [42].

III. PHARMACOLOGICAL EFFECTS OF VOLTAGE-SENSITIVE DYES

The application of voltage-sensitive dyes has been demonstrated to cause side effects on preparations, which calls for careful control experiments in in-vitro studies. So called photodynamic damage or phototoxic effects have been documented under intense illumination, with alterations of electric activity both in neurons [43] and in isolated cardiac myocytes [44]. In the isolated cardiac myocytes, 1 min of illumination with $1\,W/cm^2$ will first cause a gradual depolarization of the membrane resting potential and a decay of the action potential amplitude. This is followed by after-depolarizations, occasionally triggered activity, and finally cell death. Recently, the voltage-sensitive dye RH421 has been shown to increase the contractility of isolated rat cardiac cells and Langendorff-perfused hearts [45]. Similar effects were observed during staining with di-4-ANEPPS and di-8-ANEPPS. The low level of light used in this study ($< 1 \, \text{mW}/\text{cm}^2$) suggests that a mechanism different than the formation of free radicals may be involved.

Another pharmacological side effect of potential-sensitive styryl dyes is vasoconstriction. This effect was first described for the dye RH-414 by Grinvald et al. in 1986 [46].

The exact mechanism of these effects of voltage-sensitive dyes remains unknown. Formation of free radicals, sensitized by the dye molecules in the presence of photons, has been implicated in causing the phototoxic effect [47], and the use of radical scavengers (antioxidants) has been shown to reduce the phototoxic effect in isolated cardiac cells [44]. The voltage-sensitive dye may interact directly with voltage-gated channels, perhaps L-type Ca^{2+} or K^+ channels, and may alter the conductivity and time-dependent gating of these channels.

This hypothesis is supported by data obtained in bilayer preparations, which suggested an interaction of voltage-sensitive dyes with a number of channels. Rokitskaya and co-workers [48] have recently shown that RH421 increased the dissociation constant of gramicidin in bilayer preparations, proposing that this was due to modification of the dipole potential of the bilayer membrane by RH421. Data from several laboratories have suggested an interation of RH421 with Na^+, K^+-ATPase [49–51], though there is yet no consensus regarding the mechanism(s) underlying this effect. Frank et al. [51] implicated an RH421-induced change in membrane fluidity in the inhibition of the hydrolytic activity of the Na^+, K^+-ATPase. Fedosova and co-workers [50], on the other hand, proposed an electrostatic mechanism of interaction between potential-sensitive styryl-based dyes (including RH421) and Na^+, K^+-ATPase. Finally, RH421 has been shown to interact with the water-soluble protein ribulose-1,5-bisphosphate carboxylase/oxygenase as well as with polyamino acids (tyrosine, lysine, and arginine residues) [51].

The voltage-sensitive dye di-4-ANEPPS has not been studied as deeply as RH421, and therefore no effects of di-4-ANEPPS on ionic channels and pumps have been documented so far. However, two lines of evidence suggest possible side effects. Schaffer et al. [44] demonstrated that di-4-ANEPPS causes photodynamic damage to isolated cardiomyocytes. We observed increase in contractility caused by both RH421 and di-4-ANEPPS [45].

IV. MECHANICAL MOTION ARTIFACT

Faithful optical recordings of cardiac action potentials require immobilization of the preparation, unless the study is limited to the analysis of action potential upstrokes only [18,19]. Movement artifacts in optical action potentials tend to be most pronounced during the action potential plateau, when contraction reaches its peak amplitude, and during the action potential downstroke, when relaxation sets in while action potential upstrokes are generally well preserved. Movement artifacts can occur due to one of several mechanisms. First, physical movement of the tissue from the field of view of one detector element into the field of view of a neighboring element may lead to artifactual recordings of transmembrane potentials from different but neighboring cells in different phases of the action potential. This kind of motion artifact tends to be most pronounced at the edges of the preparation. Second, action potentials can be distorted due to the modulation of light scattering by mechanical contraction, a phenomenon that can be observed in monolayer cell cultures even in the absence of gross movement of cells across the photodetector array [19].

There are three basic approaches for coping with motion artifacts: (1) mechanical immobilization, (2) use of motion-artifact insensitive signal analysis algorithms, and (3) pharmacological immobilization.

Preparations can be mechanically immobilized by restricting the movement of the muscle between the walls of the tissue chamber and/or pressure pads [35,52] or by stretching the tissue [35]. While mechanical immobilization is free of pharmacological side effects on the electric activity of the preparation, there are several limitations to the technique. This method can only be used in preparations with moderate amplitude of contractions, otherwise the pressure needed to eliminate movements can cause ischemia. This technique has only been applied successfully to guinea pig heart preparations [21,28]. In addition to external compression of the heart by pressure pads, Girouard et al. [21] used endocardial cryoablation [53], which leaves a thin epicardial rim of viable muscle attached to a noncontracting core of dead myocardium [54]. Despite efforts, mechanical immobilization does not fully eliminate motion artifacts in all channels even in a guinea pig heart. Therefore, special signal processing techniques based on the maximum of the second derivative of the optical action potential [16,28] are usually employed to measure action potential duration instead of standard criteria such as APD_{90}. The disadvantage of this technique, however, is that it requires very good signal-to-noise ratios in the optical recordings. A new approach has been recently proposed by the group of Knisley [55]. It is based on ratiometric measurements of optical action potentials.

Pharmacological methods of producing mechanically quiescent preparations may significantly affect electrical activity. Several methods are being employed: perfusion with low calcium [5,10], Ca-channel blockers [6], 2,3-butanedione monoxime (BDM) [56], and, most recently, cytochalasin D [57–59], Low-calcium solutions and calcium channel blockers are rarely used in fluorescent recordings of transmembrane voltage because of their effect on crucial calcium-dependent cellular processes [5,40,60].

BDM, also known as diacetyl-monoxime (DAM), has effects on a variety of channels [61–63] and gap junctions [64]. The effect of BDM on action potential duration is species dependent. A shortening of action potential duration in rabbits [65], sheep [62], guinea pig [62], and dog hearts is observed, while action potential duration increases it in the rat heart [63]. This species dependence is a consequence of the reduction of both the calcium current and net potassium outward current, two antagonistic currents with opposing influences on action potential plateau development. Thus, even though the above pharmacological methods effectively suppress contractions of the cardiac muscle, they significantly alter the electrical activity of cardiac cells, limiting their usefulness in studies related to repolarization and requiring carefully designed control experiments.

Cytochalasin D appears to be a more promising agent in some species [58,59], but it requires further investigation.

V. APPROACHES TO EXPERIMENTAL DESIGN

Every engineering approach to the design of an optical system has to address one major problem: improving the signal-to-noise ratio in the optical recordings at the required spatial and temporal resolution. This is achieved by decreasing system noise and/or improving the amplitude of the signal.

An example of a raw optical signal is shown in Fig. 2. Cellular depolarization during an action potential causes a reduction in fluorescence of 1–10% of the total fluorescence signal. Background fluorescence, which thus accounts for up to 99% of the fluorescence signal, is caused by accumulation of dye in nonexcitable cells, in the inner layer of the lipid membrane, or the lipid membranes of intracellular organelles across which there is no potential change during excitation [19]. To reconstruct the intracellular action potential it is therefore necessary (1) to subtract the background fluorescence, (2) to invert the signal, and (3) to normalize the signals to uniform amplitudes assuming a homogeneity of action potential amplitudes in the imaged area of the preparation.

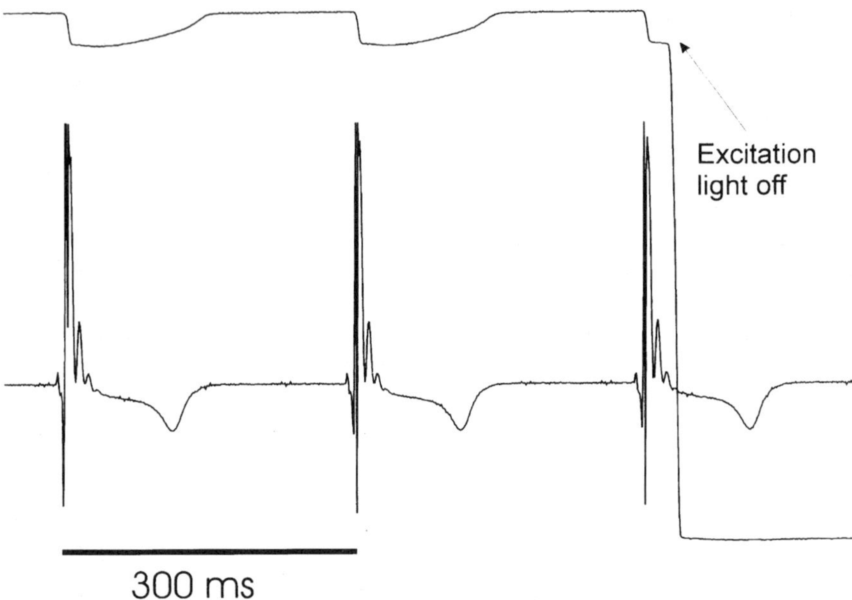

Figure 2 Optical recordings of action potentials. An epicardial fluorescence re-cording from a rabbit Langendorff-perfused he stained with di-4-ANEPPS (top trace) is juxtaposed with a simultaneous bipolar electrogram from the apex of the same heart. Fluorescence was excited at 520 ± 45 nm, collected above 610 nm, am-plified, filtered at 500 Hz, and sampled at 1000 Hz. The light was turned off before the end of the recording to demonstrate the amount of background fluorescence in fluorescence signals before DC offset subtraction.

A. Noise

There are three major sources of noise in optical recordings: shot noise, dark noise, and extraneous noise.

Shot noise is the natural fluctuation in the number of photons detected by a photodetector, caused by the quantum statistical nature of photon emission and detection. Thus shot noise occurs even in the presence of an ideal noise-free light source and ideal noise-free detector and cannot be eliminated. Shot noise is estimated by the root-mean-square deviation of the number of photons hitting a photodetector per unit time and is equal to the square root of their number. A typical tungsten lamp filament ($1800°C$) emits 10^{14} photons/sec. Only a small fraction of these, at best about 10^{10} photons/sec [66], will reach the photodetector, due to significant losses in the illumination optics with their narrow-band excitation filter and dichroic mirror, and due to losses in the imaging optics with their long-pass emission filter. The number of photons detected by the photodetector will finally be reduced by the quantum efficiency of the photodetector, which is defined as the number of photoelectrons per photon. Of the two types of photo-detectors used for optical recordings, photodiodes and photocathodes, the former have quantum efficiencies of 0.8–1.0 and the latter of only 0.15. Thus, photodiodes have a significant advantage over photocathodes, potentially yielding a nearly 6 times higher signal-to-noise ratio.

Dark noise is the noise signal emitted by a photodetector in the absence of light. Photodiodes tend to have much higher dark noise than photo-cathodes, which despite their much higher quantum efficiency limits their utility at low light intensities such as in single-cell measurements, in which case the dark noise can be comparable to or even higher than the shot noise. Thus photocathodes may yield higher signal-to-noise ratios than photo-diodes at very low light levels, and a direct experimental comparison between the two detectors is required in each case.

Extraneous noise is caused by noise sources in the laboratory environment. The following measures serve to cut extraneous noise down to acceptable levels: using a dark or DC-light illuminated room to eliminate stray light from noisy sources, an antivibration table to isolate the setup from mechanical vibration generated in the building and the rest of the experimental setup, a Faraday cage to reduce radiofrequency noise, grounding equipment to a common isolated ground to eliminate 50- or 60-Hz noise picked up by ground loops, using a low-noise light source, and isolating power supplies and amplifiers and from computing equipment.

B. Light Source and Filters

Three types of excitation light sources are used in optical recordings: *lasers, arc lamps, and tungsten lamps.* The choice of a light source will depend on the required spatial resolution of the optical recordings. Lasers and arc lamps are typically used in micrometer-scale measurements, while tungsten lamps are most commonly used in macroscopic preparations, where a resolution of hundreds of micrometers is sufficient.

Lasers can provide intense illumination, which can be easily and rapidly delivered to a small spot. However, lasers have a 1–5% variation in the beam intensity, which is comparable to the average signal intensity recorded from most voltage-sensitive dyes [67]. Ratio-calculation feedback signal processing techniques have been applied during recordings to eliminate signal noise related to laser light intensity variability [67,68].

While the intensities of arc lamps have been reported to exceed 50–100 times that of tungsten lamps, this large difference in intensity cannot be translated directly into better signal-to-noise ratios, as arc lamps have a significant intensity only at distinct narrow lines of the spectrum. Unfortunately, the excitation frequency of the most efficient dyes which produce the largest fractional change in fluorescence, such as di-4-ANEPPS and RH421, do not overlap with spectral lines of arc lamps. Since the absorption peaks of most of the voltage-sensitive dyes are relatively wide, the lower light output of tungsten lamps at a given spectral line can be compensated by the use of wider-bandpass excitation filters.

Tungsten lamps remain the most popular light source for optical recordings. They provide a very stable output of light over a very wide range of the spectrum without the sharp peaks observed in arc lamp spectra. The choice of excitation filters with half-width of 90 nm can provide significant improvement in signal quality over narrow-width filters. We recorded action potentials from a Langendorff-perfused rabbit heart ($n = 3$) stained with di-4-ANEPPS using several narrow- and wide-band excitation filters. There was no significant difference in signal quality observed between the two narrow-band filters (520 ± 10 and 540 ± 10 nm), while recordings with wide-bandwidth filters (520 ± 45 and 545 ± 30 nm) yielded 3.1 and 2.4 times better signal-to-noise ratio, respectively. The increase in excitation bandwidth did not result in a measurable increase in phototoxicity or photobleaching [12].

C. Photodetector Design

Three engineering solutions have been used over the last two decades in cardiac electrophysiology for optical multiple-channel recording systems: photodiode arrays [69], laser scanner systems [70], and video [71], or CCD [56] cameras. They differ not only by their way of collecting the fluorescent output of the specimen but also by the way the excitation light is delivered.

Photodiode arrays (PDAs) have been used for optical mapping studies in neurophysiology and cardiology since 1981 [72,73]. With a quantum efficiency above 0.8 [66], photodiodes are the most sensitive sensors for medium to large light intensities, their main drawback being the size of their dark current, which may limit their usefulness at very low light intensities as in neurophysiological applications. Photodiodes are packaged in arrays of 100, 144 or 256, 464, or more, and each photodiode will record from a large enough surface area of the preparation to receive enough photons per unit time for the accurate representation of an optical action potential. Each recording channel requires independent signal conditioning before analog-to-digital (A/D) conversion by the computer. Subtraction of background fluorescence in optical action potentials is possible on a per-channel basis either before or after A/D conversion, which can be with a resolution of 12–16 bits or more at sample rates in excess of 2–3 kHz.

Video/CCD cameras were introduced into optical mapping to achieve higher spatial resolutions and avoid the complications of setting up PDA systems. However, the accuracy of mapping data with video/CCD cameras is limited by a number of factors. (1) The signal-to-noise ratio in signals from video/CCD cameras is usually very poor, each small pixel being hit by very few photons per sample interval. Only a few CCD cameras allow the so-called binning of pixels, i.e., aggregating several pixels into one, in order to improve signal-to-noise ratios. (2) The time resolution is generally 16.7 msec, which is the NTSC standard, 4 msec in the faster cameras and 1 msec only in dedicated high-speed cameras. (3) Many cameras do not allow subtraction of the background fluorescence, and if they do, it is a uniform offset potential subtracted off all channels. (4) Amplitude resolution is usually only 8 bits, so that assuming a fractional fluorescence of the dye of 10%/100 mV, rarely more than 10–20 of the 256 gray levels will be available for encoding an action potential. Wikswo and colleagues [74] presented details of a new mapping system, in which a cooled CCD camera system was able to achieve a single-pixel signal-to-noise rate of 5–10 at a spatial re-solution of 128×127 pixels and 1.2 msec with 12-bit A/D conversion.

Laser scanning systems represent an entirely different approach, which has been described in detail elsewhere [75]. The output of a single laser is acoustico-optically deflected to scan some 100 sites of the whole prepara-tion. The fluorescence emitted by each site at the time of illumination is collected by a single photodiode, which thus sequentially records the optical signals from all sites scanned by the laser beam. The advantage of laser scanning systems over PDAs is that they can cover a wider area, and flatness of the preparation is not an optical requirement. However, the time re-solution is limited to some 1 msec per 100 scanning sites. The major dis-advantage of the laser scanning techniques is the considerable photobleaching at the light intensity levels required for reconstruction of optical action potentials, which will result in a significant decrease of the level of fluorescence on a beat-to-beat time scale and necessitate a recali-bration of the signals [76,77]. There was no statistically significant decrease in fluorescence on the same time scale through photobleaching in experi-ments using a tungsten lamp in conjunction with a photodiode array de-tection system [12].

The main drawback of photodiode arrays was that until recently the difficulties in building PDA-based mapping systems were formidable, as there were no ready-made products on the market. This approach could only be chosen by a few groups with access to advanced engineering resources. However, the situation has changed with the advent of the

Hamamatsu C4675 photodetector (Hamamatsu, Japan), a 256-element photodiode array manufactured in a compact enclosure complete with 256 current-to-voltage converters, with the availability of affordable bioamplifiers ($\sim$\$30 per channel) developed in the laboratory of Lawrence B. Cohen of Yale University [66] or, more recently, by our group in collaboration with Innovative Technologies, Inc., and with powerful multichannel A/D conversion boards for personal computers. While previous optical mapping systems had to be custom-designed from the ground up, most hardware components can now be bought. In the following text we describe the setup of our current mapping system, which was developed in the Department of Cardiovascular Medicine at the Cleveland Clinic Foundation.

VI. PHOTODIODE ARRAY SYSTEM

The core of the optical mapping system (Fig. 3) is the Hamamatsu C4675 photodiode array detector, which combines a 16×16 element photodiode array (chip size $17.45 \times 17.45\,\mathrm{mm}^2$ with 256 square $0.95 \times 0.95\,\mathrm{mm}^2$ photodiodes spaced $0.15\,\mathrm{mm}$ apart) with 256 current-to-voltage converters in a single compact ($136 \times 136 \times 154\,\mathrm{mm}$) enclosure. The optical system is built around a central beamsplitter cube (Oriel Corp.) bearing the illumination system, a bellows apparatus with the imaging lens, ground-glass screen and reticule for focusing the imaging optics, and the above photodetector. The bellows, beamsplitter cube, and photodetector are all mounted on an optical rail (Nikon) borne by a ball-bearing boom stand (Diagnostic Instruments, Sterling Heights, MI), which permits easy readjustments of the detector in all three dimensions, including a change in orientation between vertical and horizontal preparations [12,22]. The excitation light is produced by a 250-W quartz tungsten halogen lamp (Oriel Corp.) powered by a low-noise direct current (DC) power supply (Oriel Corp. or Power One Corp.). After cooling by means of a cold mirror, which lets the infrared spectrum pass into a finned heat sink (Oriel Corp.), the light path is controlled by an electronic shutter (Oriel Corp.), which opens for only a few seconds during each scan. The light beam is made quasi-monochromatic by passing it through an infrared filter (KG1, Schott) and $520 \pm 45\,\mathrm{nm}$ interference filter (Omega Optical). A 585-nm dichroic mirror (Omega Optical) held in the beamsplitter cube deflects the excitation light into the imaging lens, which then focuses the excitation light onto the preparation. Our original optical design kept the illumination optics separate from the imaging optics, using a liquid light guide (Oriel Corp.) to direct the illumination light onto the preparation. A comparison of the illumination systems showed that epi-illumination through the imaging optics produced $3\times$ better signal-to-noise ratios and signal amplitudes as well as improved homogeneity of illumination. Since then the use of the liquid light guide has been discontinued.

The imaging optics consists of a Nikon photographic lens attached to the central beam splitter cube (Oriel Corp.) by the means of an f-mount adapter ring (Newport) and a pair of Nikon bellows. The magnification of the system is set by expanding or shrinking the bellows, while focusing is performed by adjusting the distance between the imaging lens and the preparation. In imaging mode, the fluorescent light emitted from the preparation passes the dichroic mirror without reflection and is then filtered by a long-pass filter (>610 nm, Schott) before hitting the sensing area of the 16×16 photodiode array.

In focusing mode, the dichroic mirror inside the beam splitter cube is replaced by a front-surfaced plain mirror which deflects the image of the preparation onto a ground-glass reticule which bears a 1:1 representation of the outline of the photodiode array at precisely the same distance from the center of the beamsplitter cube as the photodetector.

We compared several Nikon lenses: 85-mm $f/1.4$ AF Nikkor; 50-mm $f/1.2$ Nikkor, 50-mm $f/1.4$ Nikkor, and 28-mm $f/2.8$ AF Nikkor. Table 1 summarizes the range of field of view seen by the detector and by a single photodiode (pixel resolution), as well as the the corresponding working distances for different expansion states of the bellows.

The photocurrent produced by each photodiode is first-stage amplified by its own low-noise operational amplifier inside the compact housing of the Hamamatsu C4675 detector (feedback resistors 10–100 MΩ, resulting in a gain of 10^7–10^8 V/A). Increasing the first-stage amplification has been reported to improve signal-to-noise ratios in optical signals [78]. However, increasing the feedback resistors in the C4675 camera reduced its frequency response from 15 to 1.5 kHz, which may be undesirable in some applications. The outputs of the first-stage amplifiers were connected to 256 second-stage amplifiers, four 64-channel cards developed at, and available from, Yale University, which offer DC coupling and AC coupling with several time constants (short time constant for DC offset subtraction, time constant of 30 sec during data acquisition). A computer-driven TTL pulse is used to reset the second-stage amplifiers immediately before data acquisition in order to remove the DC offset of the optical signals caused by background fluorescence. Signals were filtered by Bessel filters with a cutoff frequency of 500–2000 Hz, depending on the sampling rate of the data acquisition system.

The signals were fed to a multiplexer and A/D converter boards. We used two types of boards: (1) 12-bit A/D boards with on-board memory DAP 3200/415e from Microstar Laboratories; (2) 16-bit A/D boards PCI-6033E or 6031E from National Instruments, which allow real-time uninterrupted logging to hard disk. Sampling was performed at a rate of

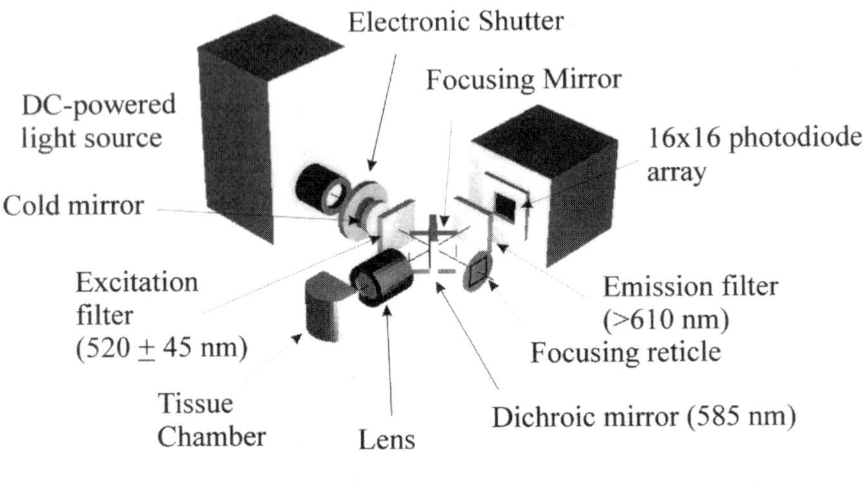

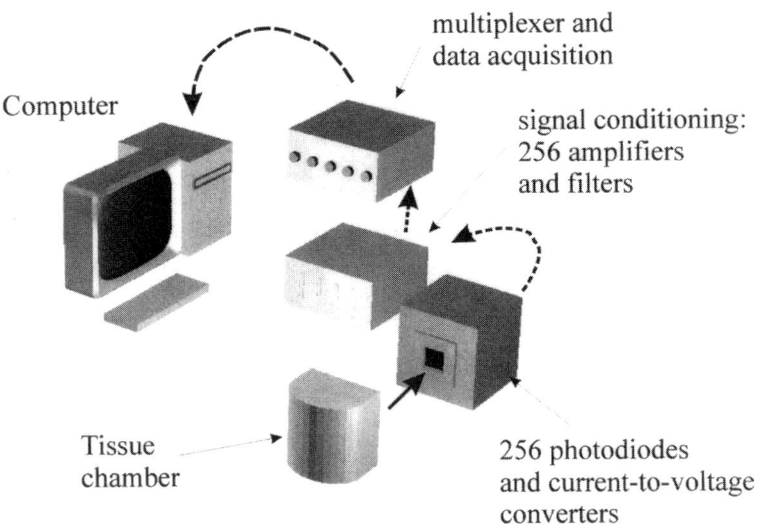

Figure 3 Experimental setup. Fluorescence in cardiac preparation stained with the potential-sensitive dye di-4-ANEPPS is excited by the light of a 250-W DC-powered tungsten lamp. After passing a cold mirror and an electric shutter controlling the light beam, the beam is made quasi-monochromatic by means of an infrared filter and a 520 ± 45 nm interference filter. It is then deflected into the light path of the illumination optics by means of a dichroic mirror and finally focused on the preparation by the imaging lens. Light emitted from the preparation is focused on the light-sensitive area of a 16×16 element photodiode array (PDA) by the imaging lens after passing the dichroic mirror and filtering with a long-pass (>610-nm) colored

Table 1 Field of View and Working Distances for Different Nikon Photographic Lenses at 0–200 mm Expansion of the Bellows

| | Field of view of | | Working |
Lens	PDA array (mm)	Single photodiode (µm)	distance (mm)
$f/1.2$, 50 mm	4.0–18	250–1120	5–46
$f/1.4$, 50 mm	3.8–19	230–1190	13–49
$f/1.4$, 85 mm	6.0–31	380–1940	70–200
$f/2.8$, 28 mm	5.5–11	340–690	2–10

1000–3000 frames/sec. The sampling rate S and the cutoff frequency f_c of the filters were always kept to satisfy Nyquist criterion ($S = 2f_c$) in order to avoid aliasing errors introduced by digitization of the signals. Each frame included 256 optical channels and 8 instrumentation channels including surface electrogram, stimulation and defibrillation triggers, and aortic pressure, which were stored on a hard disk for off-line analysis.

A. Computer-Controlled Instrumentation Interface, Data Acquisition, and Analysis

One of the biggest problems in mapping system design is the development of user-friendly and efficient software for data acquisition and analysis.

The chief problem in electrical mapping is optimizing system performance to allow continuous data acquisition to a high-capacity storage medium such as a hard disk or a digital tape over long time intervals. In optical mapping, it is also essential to provide integrated and easy-to-use control over the complex instrumentation. To reduce problems of phototoxicity and/or photobleaching, the exposure of preparations to intense light must be kept to a minimum. Data are typically acquired only in short

---

glass filter. The photocurrents of all 256 photodiodes are converted into voltages by 256 first stage amplifiers integrated in the housing of the PDA. The signals are then amplified and filtered in parallel by a 256-channel second-stage amplifier and finally multiplexed and analog-to-digital converted together with 8 additional instrumentation channels by a PC-based data acquisition and analysis system. To focus the optics on the preparation, a front-surfaced mirror replaces the dichroic mirror, deflecting the light from the preparation on a ground-glass reticule bearing a scale 1:1 outline of the photodiode array. (Reproduced with permission from Ref. 85.)

bursts of few seconds, and care needs to be taken that data acquisitions are not repeated unnecessarily. An optical mapping system control unit must (1) precisely synchronize the initiation of a mapping sequence with the stimulation of the preparation, (2) provide pulses for the control of the electric shutter of the light source, for the DC offset subtraction of the signal amplifiers and the start of the data acquisition by the A/D boards, (3) allow the fast logging of data to disk, and (4) give the operator near-instant feedback on the quality of the logged data so that the mapping sequence can be repeated if necessary.

The consequence of the first requirement is to control both stimulation and all other components of the mapping system by the same timing device. We therefore chose to integrate the programmable stimulator into the mapping system using a PC-TIO-10 timer board (National Instruments), which provides 10 programmable 16-bit counters. The front end of our data acquisition system and data analysis package was programmed under LabView 5.0 for Windows NT, and integrates a full-featured programmable stimulator, control of the mapping system, CyberAmp amplifier (Axon Instruments), and extensive data processing, visualization, and analysis package. Figure 4 shows the main panel of the data acquisition program.

Supported stimulation protocols include burst and continuous pacing S_1 with up to three premature pulses (S_2, S_3, S_4) out of the first output channel and a single pulse triggered on any previous pulse out of a second output channel for cross-field stimulation or control of another device, such as an external stimulator. Analysis algorithms include various filtering, calibration, reconstruction of activation, repolarization and action potential duration maps, subtraction of two corresponding responses in order to visualize the differences between them caused by electric shocks, etc [12,79,80].

B. Experimental Preparations

Experiments can be performed in vitro on Langendorff-perfused whole-heart preparations and isolated superfused or coronary-perfused atrial and ventricular preparations in rabbits, dogs, guinea pigs, rats, and humans. The precise details of animal preparation protocols are published elsewhere [12,22].

For staining, a stock solution of 5 mg di-4-ANEPPS (Molecular Probes) is prepared in 4 ml dimethyl sulfoxide (DMSO, Fisher Scientific), and is stored frozen at $-20°C$. After gentle rewarming immediately before the experiment, a syringe is filled with 300–500 µL of the stock solution, which is then gradually injected into an injection port (Radnoti Glass) above the bubble trap of the perfusion system mannually or by means of an infusion pump. The method of gradual hand injection of dye into the injection port

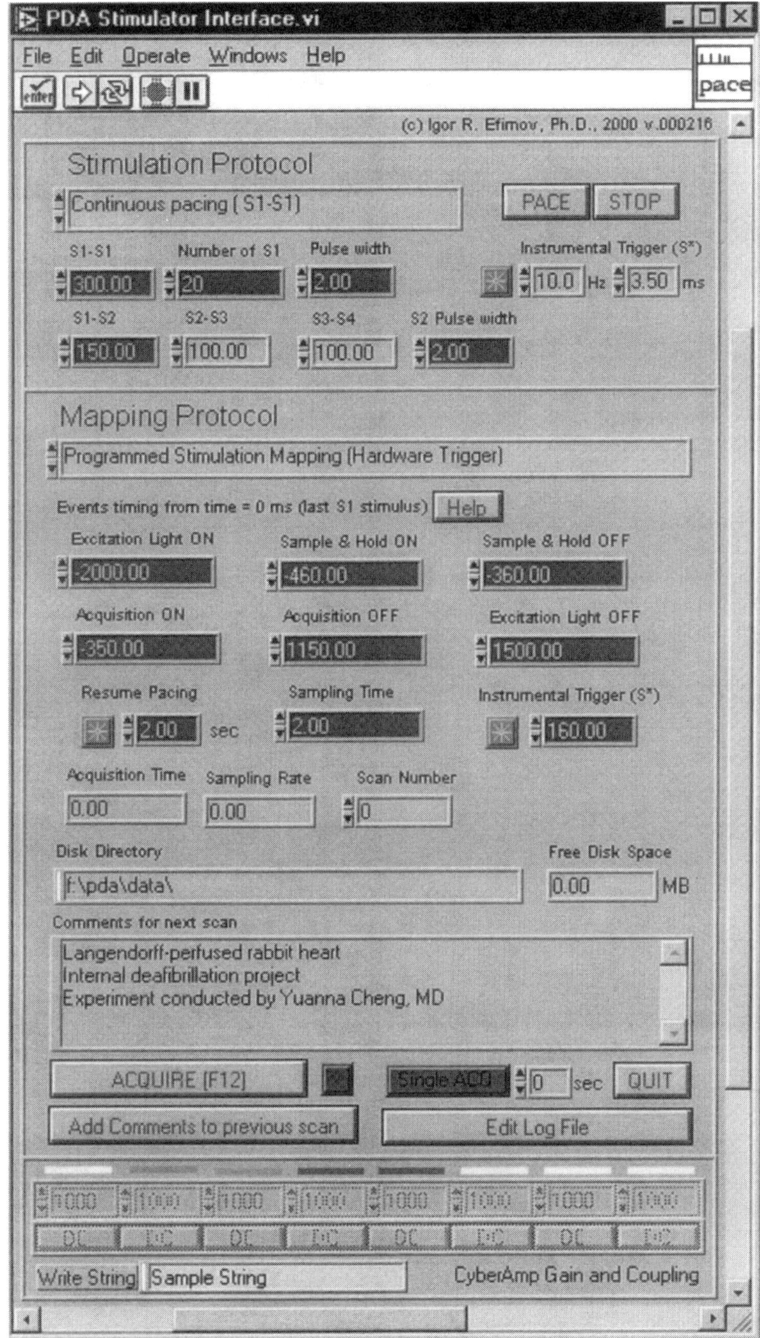

Figure 4 The main panel of data acquisition program, which controls pacing protocol, various instruments, and data acquisition.

in the bubble-trap [28] is less cumbersome but less reproducible. Staining procedure takes 10–40 min, depending on the preparation.

Levels of optical signals and signal-to-noises ratios (SNRs) decrease over time, presumably due to the translocation of the voltage-sensitive dye molecules into the cell interior, as illustrated in Fig. 5. We monitored signal-to-noise ratio in all 256 channels during a 3-hr mapping experiment in a Langendorff-perfused rabbit heart. Signal-to-noise ratios were defined as the ratio between optical action potential amplitude and peak-to-peak noise amplitude as measured during diastolic intervals, while the mean value and standard deviation were as calculated from all 256 recordings. Using a single exponential function approximation $(SNR = SNR_0\, e^{-t/\tau})$, we estimated half-life of the signal $\tau = 105.2 \pm 31.2$ min. Restaining was done of some preparations if needed.

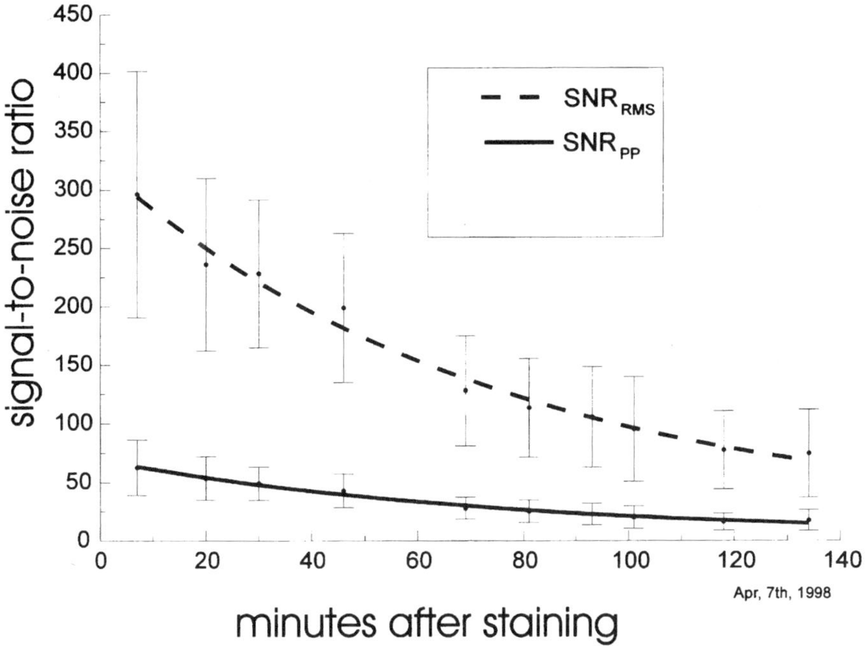

Figure 5   Decline of signal-to-noise ratios over time. The figure plots mean $\pm$ SD of the signal-to-noise ratios in all 256 fluorescence recordings from the epicardial surface of a Langendorff-perfused rabbit heart stained with di-4-ANEPPS versus time. The preparation was kept in a darkened room and was illuminated only during data acquisition for 1–2 sec at a time.

VII. MAPPING THE STIMULATION OF THE HEART

Optical recordings of electrical activity are the only methods which resolve questions related to the interaction of strong electrical fields with excitable cells, conventional electrode recordings being distorted by large-amplitude artifacts. Using fluorescent mapping techniques we have been able to optically

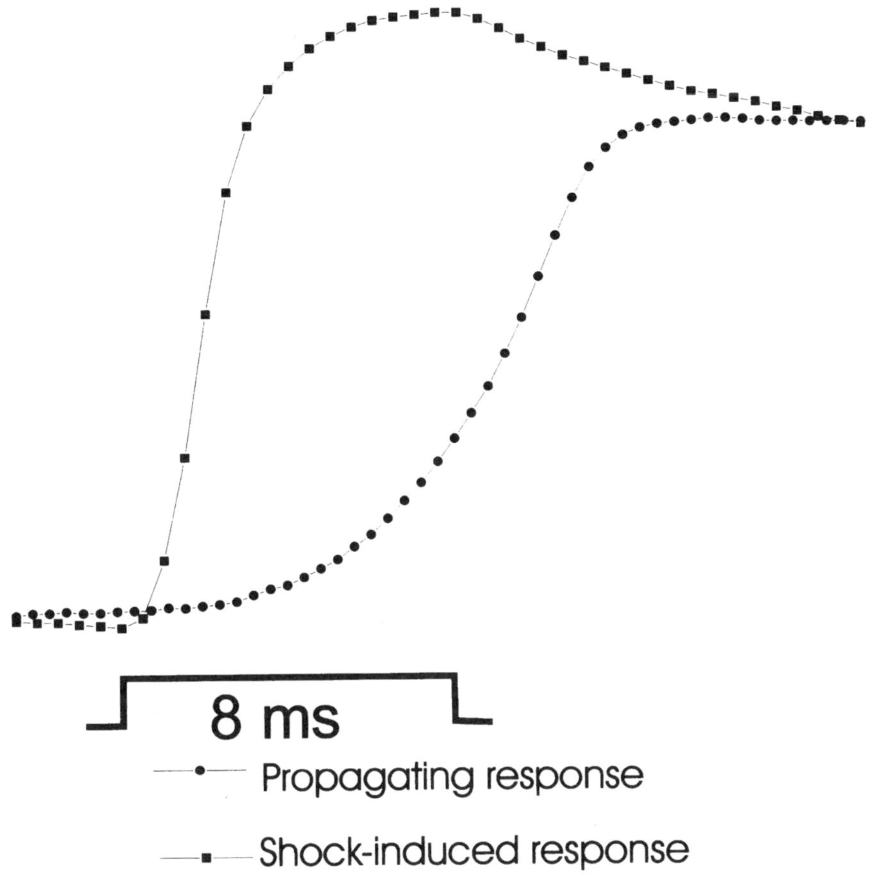

Figure 6 Optical action potential upstrokes of a propagated response during epicardial pacing (circles) and a response induced by an 8-msec 100-V monophasic electric shock (squares) indicated by the time bar. The data was amplified after DC-offset subtraction, filtered at 1 kHz, sampled at 1.9 kHz (sample interval 528 μsec and represents the summed response of the $650 \times 650\,\mu m$ region of epicardial myocardium within the field of view of one photodiode.

 Efimov and Cheng

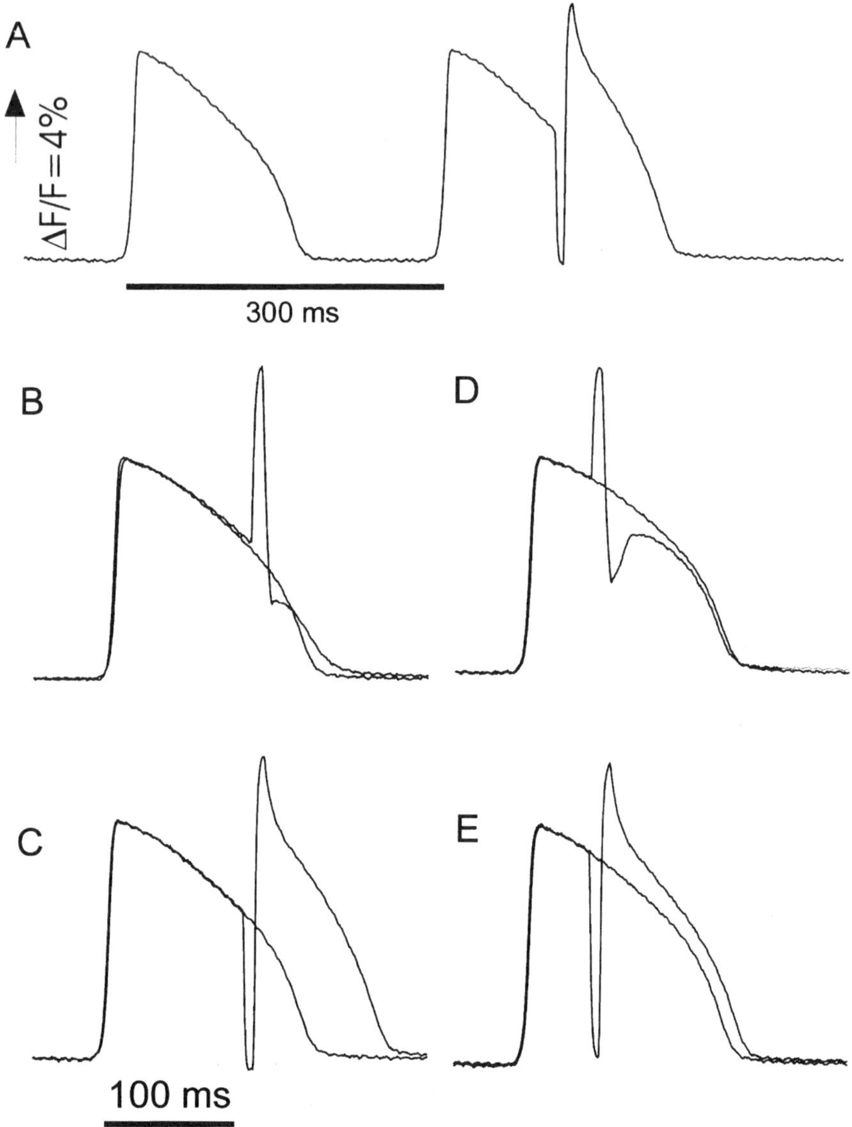

Figure 7 Optical recordings during defibrillation-strength shocks. Traces A–E were recorded from the same $650 \times 650\,\mu m$ area of the left ventricular epicardium 8 mm from the defibrillation electrode, which was positioned close to the septum of the right ventricular cavity in a Langendorff-perfused rabbit heart. This area corresponds to the "virtual electrode" area [12,79]. Optical signals were amplified, filtered at 1 kHz, and sampled at 2 kHz. No additional software filtering was applied.

record changes in transmembrane voltage with a high signal-to-noise ratio from all 256 channels. This has permitted us to acquire two-dimensional patterns of shock-induced polarization with submillisecond resolution.

Figure 6 shows upstrokes from two optical action potentials recorded from the same 650×650-µm area of right ventricular epicardium. As can be seen, recordings during defibrillation shocks, however, have much faster components [12] than normal propagated responses, the frequency content of which has been reported to be limited to 150 Hz [40]. Therefore, faster sampling is required. In our experiments, signals were conditioned by a low-pass 1-kHz Bessel filter and sampled at 2 kHz without any additional filtering by software in order to preserve the frequency content. The reason for the slower rise time of propagated responses is that optical action potentials represent spatially integrated responses of cells confined within the three-dimensional field of view of each photodiode, which extends some 100–2000 µm in depth [28,39–41]. Thus optical action potential upstroke rise times are a function of both conduction velocity and the size of the field of view [35,40].

A. Investigation of the Effects of Extracellularly Applied Stimuli

Figure 7 presents several examples of typical cellular responses during biphasic shocks applied in the plateau phase of a propagated action potential. A data scan included the last normal propagated action potential before the shock (left waveform on trace A) and the action potential altered by the shock (right waveform on trace A). Figure 7B–7E show two such action potentials from different regions of the mapping array which are superimposed and aligned by their upstrokes. As can be seen, postshock action potential prolongation is strongly dependent on the polarity of the shock and its timing in respect to the phase of the action potential. Figure 8 shows

During stimulation at 300 msec from the RV apex, shocks were applied during the AP plateau. Trace A shows a sequential record of a normal action potential (AP) followed by the AP during and after application of the shock, while in traces B–E the two APs are superimposed, being aligned on their upstrokes. Traces B and C were recorded during shocks applied 100 msec after the upstroke, while traces D and E were recorded during shocks applied 50 msec after the upstroke. Traces B and D were recorded during $+150/-100$ biphasic shocks applied 100 and 50 msec after the upstroke, while traces C and E were recorded during biphasic shocks of the opposite polarity 100 and 50 msec after the upstroke. As can be seen from the recordings, the AP prolongation through biphasic defibrillation shocks is strongly dependent on both timing and polarity of the shock.

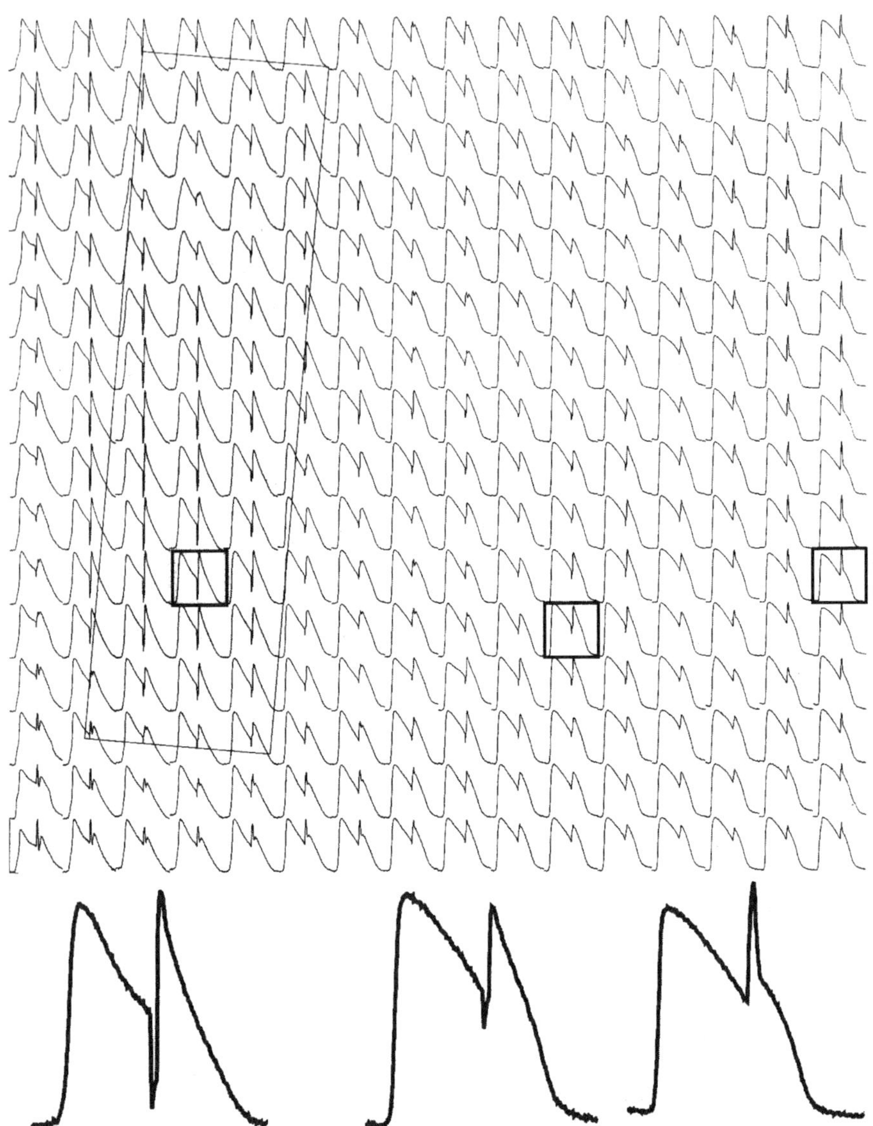

Figure 8 Map of 256 optical APs recorded from anterior epicardium during application of a biphasic $+150/-100$-V defibrillation shock applied during the plateau phase of cardiac action potentials in a Langendorff-perfused rabbit heart. The gray rectangular area indicates the position of the distal defibrillation electrode inside the RV cavity. The data show dramatic differences between recordings performed near and far from electrode. The bottom traces represent enlarged views of the highlighted recordings.

the entire mapping array with all 256 potential waveforms recorded from an 11×11 mm^2 area of the left ventricular anterior wall of a rabbit heart during application of a $+150/-100$-V 8/8-msec biphasic shock 90 msec after the onset of the action potential in the middle of the array. Recordings such as these cannot be achieved by any other known methods. As can be seen, cellular responses recorded at different location with respect to the electrode can differ dramatically by amplitude and polarity of cellular polarization [12]. Most important, these experimental techniques allowed us to link together several important parameters which are responsible for arrhythmogenesis during failed defibrillation shocks: virtual electrode dispersion of transmembrane polarization, followed by dispersion of repolarization, and new wavefront formation, followed by formation of phase singularities and reentry [79–81].

B. Virtual Electrode Effects During Stimulation/ Defibrillation

The dispersion of repolarization shown previously was due to so-called virtual electrode effect [82]. Figure 9 presents virtual electrode polarization patterns produced by 2-msec unipolar and bipolar stimuli applied at the epicardium of a rabbit heart. This figure exemplifies a common observation of virtual electrode polarization during externally applied stimuli, e.g., pacing [9] or defibrillation [12]. Virtual electrode polarization is characterized by simultaneous development of positive and negative polarizations in adjacent areas, presumably due to asynchronous redistribution of charges in the intracellular versus extracellular spaces.

As during pacing, the virtual electrode patterns develop during large-scale internal [12] or external [83], defibrillation shocks. The exact polarization pattern depends on electrode configuration and the shock waveform. Regardless of the pattern, however, there are common effects associated with virtual electrode polarization. Figure 10 shows three types of effects [80]: de-excitation, prolongation, and re-excitation. De-excitation develops in an area of negative polarization and results in shortening of the preshock action potential. If such polarization is strong enough, it can completely de-excite cells, fully restoring excitability in the myocardium, which was refractory before shock. Positive polarization results in an extension of action potential duration and the refractory period [6]. And finally, re-excitation occurs in de-excited cells, which sustain postshock wave of break excitation.

Indeed, as shown in Fig. 11, the negatively polarized region is invaded by a wavefront of activation upon shock withdrawal. This wavefront originates at the boundary between areas of negative and positive polarizations

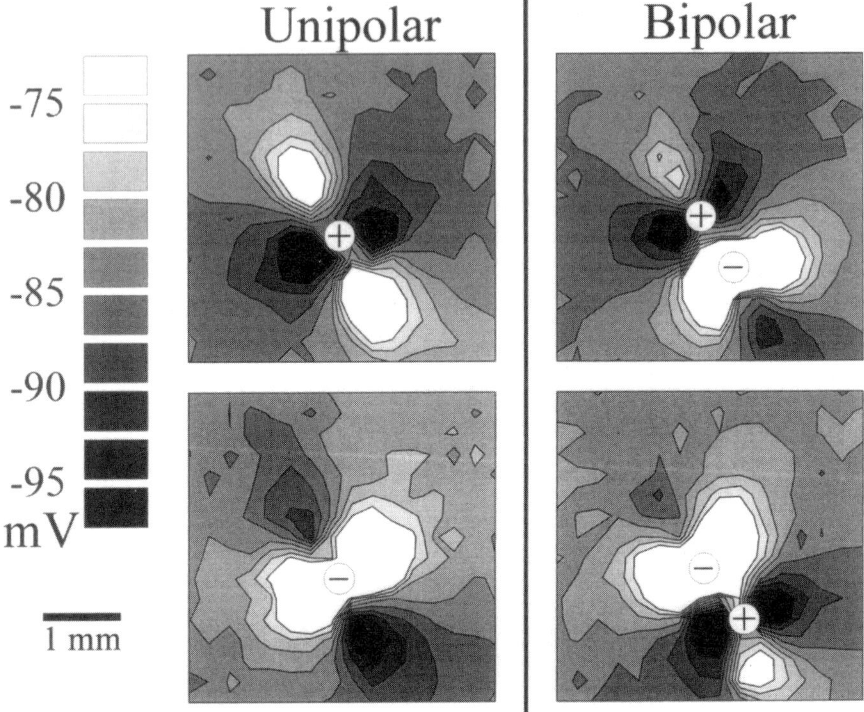

Figure 9 Virtual electrode polarization patterns produced by unipolar and bipolar epicardial stimulation. The stimuli were applied during diastolic interval at the anterior epicardium of the rabbit heart. The field of view was 4 × 4 mm. Shown maps of transmembrane voltage were recorded 1 msec after the onset of 2-msec pulses. Data represent a single 528-μsec frame. No averaging or filtering was applied to optical recordings. (Reproduced with permission from Ref. 80.)

and then spreads across the negatively polarized region. This conduction depends on the degree of de-excitation, presumably due to different degrees of recovery from inactivation of sodium and calcium channels by the negative polarization. Figure 12 shows an example of different rate of conduction in the same heart in response to three shocks of different intensities. Progressive increase in shock intensity resulted in progressive increase of de-excitation, and as a result in progressive acceleration of wave of re-excitation. And, oppositely, progressive decrease in shock intensity resulted in progressive decrease in de-excitation and slowing of the conduction, until a wavefront could not be generated.

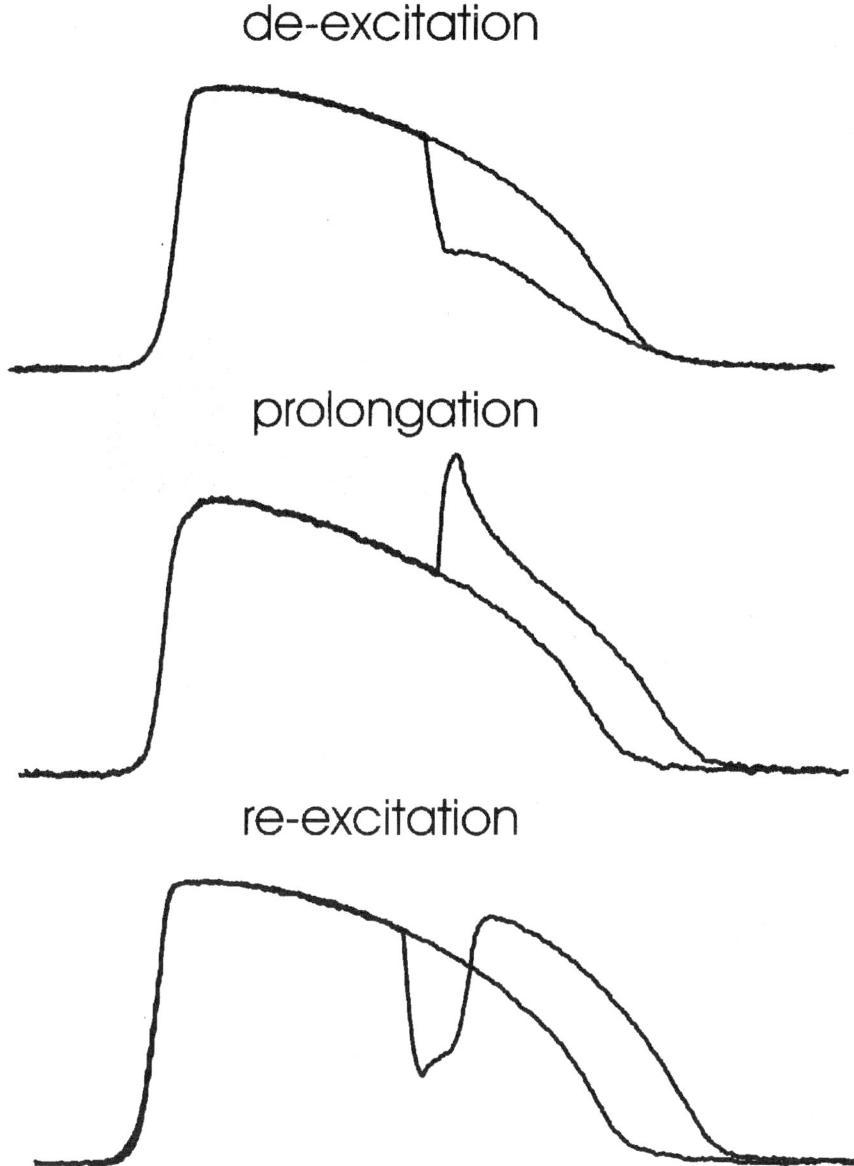

Figure 10 Three effects of virtual electrode polarization: de-excitation, prolongation, and re-excitation (see text for details).

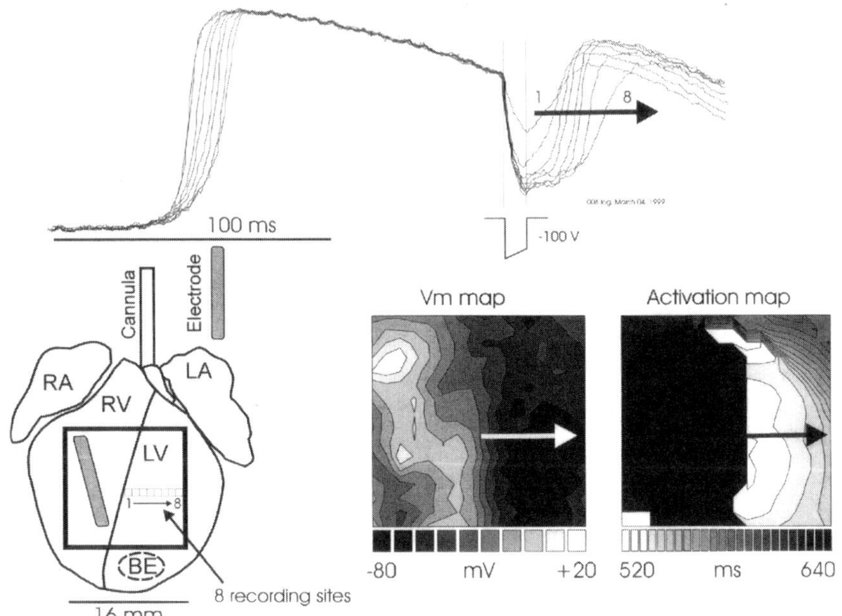

Figure 11 Genesis and conduction of virtual electrode induced wavefront of excitation. Upper panel shows 8 representative traces recorded at the anterior epicardium during the application of a monophasic shock (-100 V, 8 msec). The shock produced negative polarization in all 8 sites. The location of the recording sites are shown in the lower left panel. The lower middle panel shows the map of transmembrane polarization at the end of shock. The lower right panel shows the map of activation (5-msec isochrone lines) after the shock withdrawal at 520 msec. This shock resulted in arrhythmia. The direction and location of arrows in all panels correspond to the direction of conduction and location of recording sites. (Reproduced with permission from Ref. 80.)

C.   Virtual Electrode-Induced Phase Singularity

Genesis of wavefront of re-excitation is a complex three-dimensional phenomenon. It occurs along a steep gradient between the positive and negative polarizations of virtual electrode polarization. Heterogeneity of the latter may lead to wavebreaks in a generated wavefront. Such wavebreaks are known to be associated with phase singularities leading to sustained reentry [84]. Figure 13 illustrates the onset of reentry after a shock applied at the T wave from an internal electrode placed in the right ventricular cavity [80]. This mechanism, first described by our group [79], is known as *virtual*

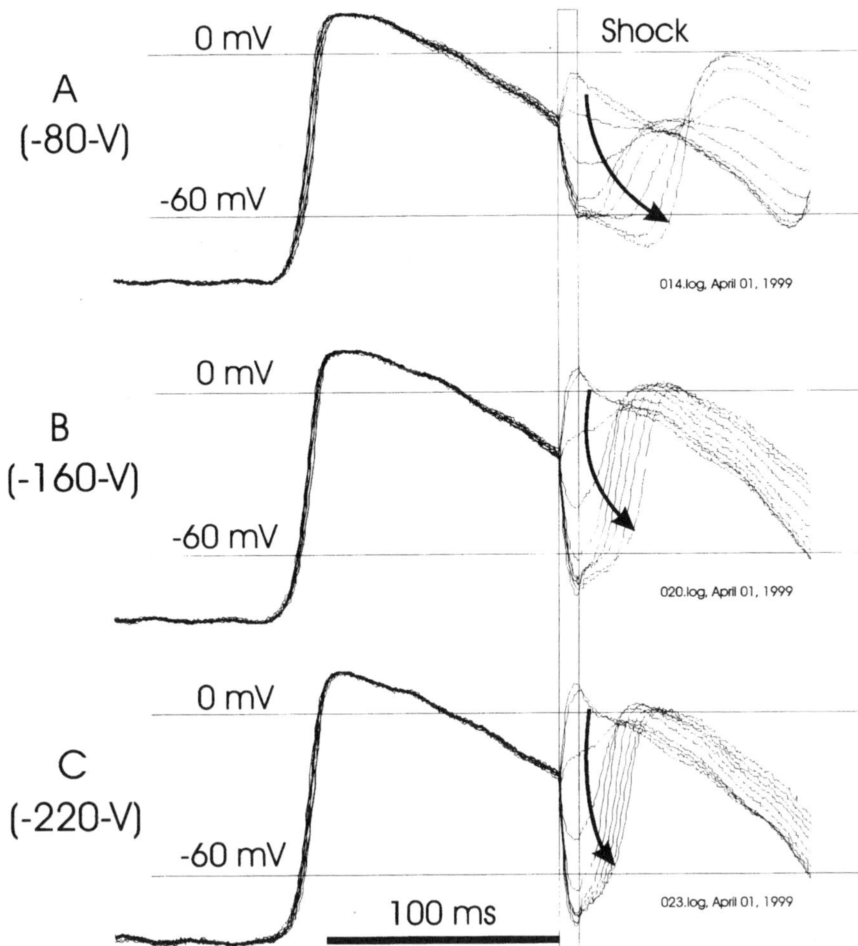

Figure 12 Virtual electrode polarization and postshock excitation: modulation of conduction velocity and VEP amplitude by shock intensity. Optical recording during virtual electrode induced wavefront generation and propagation. (A–C) ten representative traces recorded from sites along a gradient of virtual electrode polarization, similar to Fig. 11 but in a different heart. Shock intensities were −80, −160, and −220 V, respectively. Shock duration was 8 msec. (Reproduced with permission from Ref. 80.)

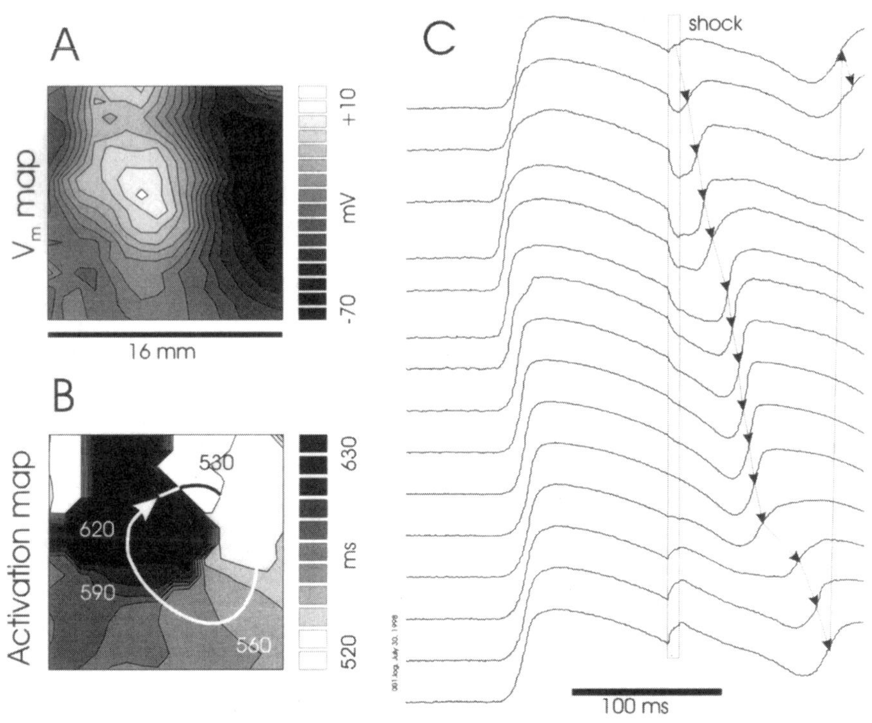

A
Vm map
+10
mV
-70
16 mm
B
Activation map
530
620
590
560
630
ms
520
C
shock
100 ms
D
475 ms

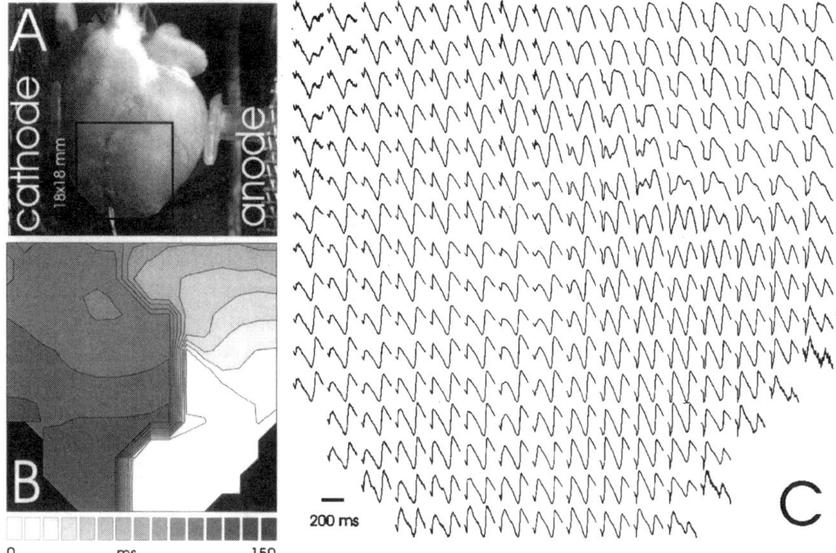

Figure 14 Virtual electrode induced phase singularity during external shock. (A) Digital images of the preparation, the pacing electrode, and the shock electrodes. Data were collected from an 18×18 mm area of anterior epicardium including the apex. (B) Isochronal map of activation. Lines are drawn 10 msec apart. The time scale starts at the end of the shock. (C) Optical recordings collected before, during, and after the monophasic shock (100 V/8 msec). The records shown start 10 msec before the shock. These data were used to plot the map of activation shown in B. (Reproduced with permission from Ref. 83.)

electrode induced phase singularity. A similar effect was documented during shocks applied externally [83]. Figure 14 shows onset of arrhythmia resulting from the virtual electrode induced phase singularity, during externally applied shock. Arrhythmias which result from the shock are sustained by three-dimensional scroll waves with a ribbon-shaped filament [83,85]. Figure 15 shows an example of the signature of the filament of the scroll wave, which is evident in optical recordings.

◀━━

Figure 13 *Virtual electrode* induced arrhythmia. (A) VEP developed at the end of monophasic cathodal shock (-100 V, 8 msec). (B) Postshock activation resulting from VEP. Shock lasted from 512 to 520 msec. (C) Raw optical recordings collected around the virtual electrode induced phase singularity area. Recording sites were sequentially selected along the circular arrow in B. (D) 256 optical recordings during two initial periods of arrhythmia resulted from virtual electrode induced phase singularity. Bold traces are shown in C. (Reproduced with permission from Ref. 80.)

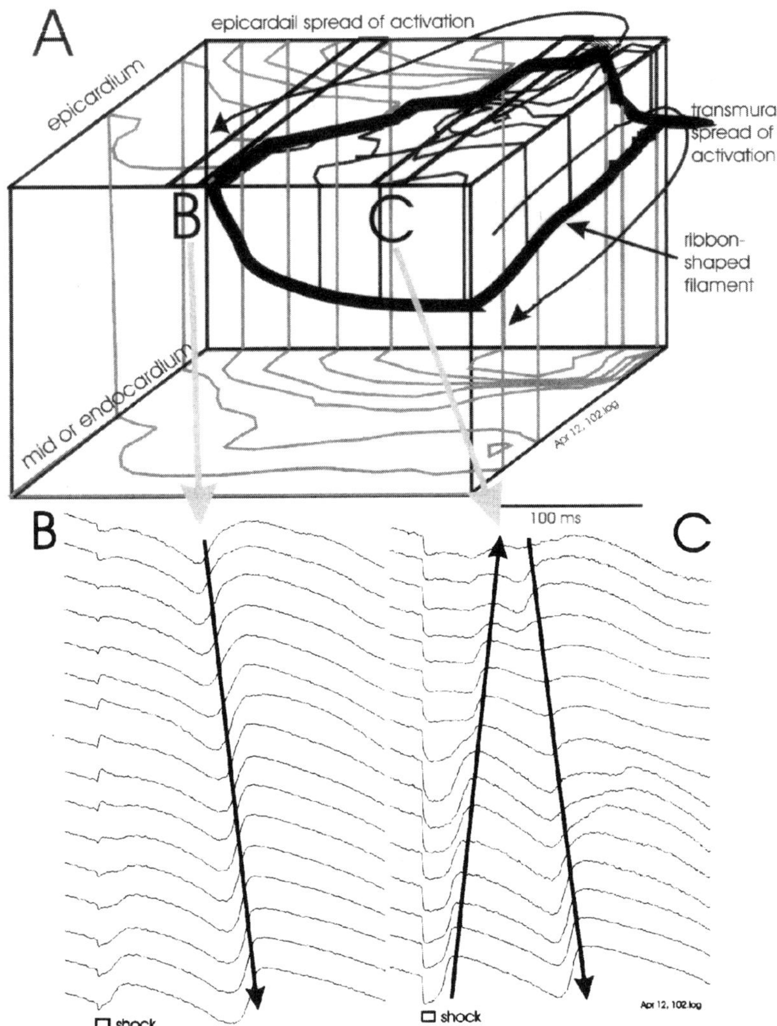

Figure 15 Qualitative reconstruction of a scroll wave with ribbon-shaped filament resulting from a monophasic shock (120 V/8 msec) applied externally. (A) Isochrone map of activation (10 msec) constructed on the basis of the largest postshock $(dV/dt)_{\mathrm{max}}$ peaks only (epicardial spread of activation) or $(dV/dt)_{\mathrm{max}}$ peaks detected after 580 msec (mid or endocardial map). The thick black line traces the boundaries of the ribbon-shaped filament. The arrows indicate spread of activation in the epicardial and transmural directions. The shock ended at 520 msec. B and C select the two columns of signals illustrated in panels B and C. (B) Only one wave of excitation propagated from base to apex was observed after the shock withdrawal after a delay. (C) Right column of recording sites. Two waves of excitation propagated though the area: one above the filament (from apex to base) and one below the filament (from base to apex). (Reproduced with permission from Ref. 83.)

Virtual electrode induced phase singularity has been shown to underlie mechanisms of both vulnerability and defibrillation failure [79,80,83,86]. This area of research is now developing at a very rapid pace, due entirely to fluorescent imaging techniques employing voltage-sensitive dyes.

VIII.　CONCLUSIONS

Fluorescent imaging methods developed and refined in a number of laboratories [6,9,10,15–22] have proven to be valuable tools extending the frontiers in cardiac electrophysiology. We believe that recent technological advances will make these experimental techniques available for a larger number of research institutions. However, future development of fast fluorescent mapping will have to solve several limitations and challenges.

A major limitation of current techniques of potential imaging is the lack of absolute calibration. Unlike many ratiometric fluorescent probes for calcium imaging, voltage-sensitive dyes can only provide relative information about changes in transmembrane voltage. Although the changes in the absolute amount of fluorescence excited at one wavelength depend linearly on the transmembrane voltage of the viewed cells, accurate calibration has so far been impossible because the number of cells contributing to the signal is unknown. Monatana et al. [87] demonstrated that measurements of fluorescence ratios excited at two wavelengths can provide such information. They have shown that the ratio between di-4-ANEPPS fluorescence levels excited at 440 and 505 nm depends linearly on transmembrane potential in a lipid vesicle in the range from $-125\,\text{mV}$ to $+125\,\text{mV}$. Similarly, the ratio of di-8-ANEPPS fluorescence from N1E-115 neuroblastoma cells excited at 450 and 530 nm depends linearly on transmembrane voltage. Similar efforts are now being made to measure absolute change in membrane potentials in the heart [55].

REFERENCES

1. Hoffa M, Ludwig C. Einige neue Versuche uber Herzbewegung. Z Rationelle Med 9:107–144, 1850.
2. Prevost JL, Battelli F. Sur quel ques effets des dechanges electriques sur le coer mammifres. C R Seances Acad Sci 129:1267, 1899.
3. Cohen LB, Keynes RD, Hille B. Light scattering and birefringence changes during nerve activity. Nature 218:438–441, 1968.
4. Davila HV, Salzberg BM, Cohen LB, Waggoner AS. A large change in axon fluorescence that provides a promising method for measuring membrane potential. Nat New Biol 241:1594–1560, 1973.

5. Salama G, Morad M. Merocyanine 540 as an optical probe of transmembrane electrical activity in the heart. Science 191:485–487, 1976.

6. Dillon SM. Optical recordings in the rabbit heart show that defibrillation strength shocks prolong the duration of depolarization and the refractory period. Circ Res 69:842–856, 1991.

7. Dillon SM. Synchronized repolarization after defibrillation shocks. A possible component of the defibrillation process demonstrated by optical recordings in rabbit heart. Circulation 85:1865–1878, 1992.

8. Knisley SB, Afework Y, Li J, Smith WM, Ideker RE. Dispersion of repolarization induced by a nonuniform shock field. Pacing, Clin Electrophysiol 14:1148–1157, 1991.

9. Wikswo JP, Lin S-F, Abbas RA. Virtual electrodes in cardiac tissue: a common mechanism for anodal and cathodal stimulation. Biophys J 69:2195–2210, 1995.

10. Neunlist M, Tung L. Spatial distribution of cardiac transmembrane potentials around an extracellular electrode: dependence on fiber orientation. Biophys J 68:2310–2322, 1995.

11. Gillis AM, Fast VG, Rohr S, Kleber AG. Spatial changes in transmembrane potential during extracellular electrical shocks in cultured monolayers of neonatal rat ventricular myocytes. Circ Res 79:676–690, 1996.

12. Efimov IR, Cheng YN, Biermann M, Van Wagoner DR, Mazgalev T, Tchou PJ. Transmembrane voltage changes produced by real and virtual electrodes during monophasic defibrillation shock delivered by an implantable electrode. J Cardiovasc Electrophysiol 8:1031–1045, 1997.

13. Gray RA, Ayers G, Jalife, J. Video imaging of atrial defibrillation in the sheep heart. Circulation 95:1038–1047, 1997.

14. Kamino K, Hirota A, Fujii S. Localization of pacemaking activity in early embryonic heart monitored using voltage-sensitive dye. Nature 290:595–597, 1981.

15. Davidenko JM, Kent PF, Chialvo DR, Michaels DC, Jalife J. Sustained vortex-like waves in normal isolated ventricular muscle. Proc Natl Acad Sci USA 87:8785–8789, 1990.

16. Rosenbaum DS, Kaplan DT, Kanai A, Jackson L, Garan H, Cohen RJ, Salama G. Repolarization inhomogeneities in ventricular myocardium change dynamically with abrupt cycle length shortening. Circulation 84:1333–1345, 1991.

17. Knisley SB, Hill BC. Optical recordings of the effect of electrical stimulation on action potential repolarization and the induction of reentry in two-dimensional perfused rabbit epicardium. Circulation 88:2402–2414, 1993.

18. Fast VG, Kleber AG. Microscopic conduction in cultured strands of neonatal rat heart cells measured with voltage-sensitive dyes. Circ Res 73:914–925, 1993.

19. Rohr S, Salzberg BM. Multiple site optical recording of transmembrane voltage (MSORTV) in patterned growth heart cell cultures: assessing electrical behavior, with microsecond resolution, on a cellular and subcellular scale. Biophys J 67:1301–1315, 1994.

20. Windisch H, Ahammer H, Schaffer P, Muller W, Platzer D. Optical multisite monitoring of cell excitation phenomena in isolated cardiomyocytes. Pflugers Archiv Eur J Physiol 430:508–518, 1995.

21. Girouard SD, Pastore JM, Laurita KR, Gregory KW, Rosenbaum DS. Optical mapping in a new guinea pig model of ventricular tachycardia reveals mechanisms of multiple wavelengths in a single reentrant circuit. Circulation 93:603–613, 1996.

22. Efimov IR, Fahy GJ, Cheng YN, Van Wagoner DR, Tchou PJ, Mazgalev TN. High resolution fluorescent imaging of rabbit heart does not reveal a distinct atrioventricular nodal anterior input channel (fast pathway) during sinus rhythm. J Cardiovas Electrophysiol 8:295–306, 1997.

23. Cohen LB, Lesher S. Optical monitoring of membrane potential: methods of multisite optical measurement. In: De Weer P, Salzberg BM, eds. Optical Methods in Cell Physiology. New York: Wiley-Interscience, 1984.

24. Franz MR. Method and theory of monophasic action potential recording. Prog Cardiovasc Dis 33:347–368, 1991.

25. Berbari EJ, Lander P, Geselowitz DB, Scherlag BJ, Lazzara R. The methodology of cardiac mapping. In: Shenasa M, Borggrefe M, Breithardt G. eds. Cardiac Mapping. Mount Kisko, NY: Futura, 1993.

26. Biermann M, Shenasa M, Borggrefe M, Hindricks G, Haverkamp W, Breithardt G. The interpretation of cardiac electrograms. In: Shenasa M, Borggrefe M, Breithardt G, eds. Cardiac Mapping, Mount Kisko, NY: Futura, 1993.

27. Steinhaus BM. Estimating cardiac transmembrane activation and recovery times from unipolar and bipolar extracellular electrograms: a simulation study. Circ Res 64:449–462, 1989.

28. Efimov IR, Huang DT, Rendt JM, Salama G. Optical mapping of repolarization and refractoriness from intact hearts. Circulation 90:1469–1480, 1994.

29. Rosenbaum DS, Jalife J, eds. Optical Mapping of Cardiac Excitation and Arrhythmias. Armonk, NY: Futura, 2002.

30. Cohen LB, Salzberg BM. Optical measurement of membrane potential. Rev Physiol Biochem Pharmacol 83:35–88, 1978.

31. Lev-Ram V, Grinvald A. Ca^{2+}- and K^+-dependent communication between central nervous system myelinated axons and oligodendrocytes revealed by voltage-sensitive dyes. Proc Natl Acad Sci USA 83:6641–6655, 1986.

32. Loew LM. Spectroscopic Membrane Probes. Boca Raton, FL: CRC Press, 1988.

33. Clarke RJ, Zouni A, Holzwarth JF. Voltage sensitivity of the fluorescent probe RH421 in a model membrane system. Biophys J 68:1406–1415, 1995.

34. Ross WN, Salzberg BM, Cohen LB, Grinvald A, Davila HV, Waggoner AS, Wang CH. Changes in absorption, fluorescene, dichorism, and birefringence in stained giant axons: optical measurement of membrane potential. J Membr Biol 33:141–183, 1977.

35. Salama G. Optical measurements of transmembrane potential in heart. In: Loew LM, ed. Spectroscopic Membrane Probes. Boca Raton, FL: CRC Press, 1988.

36. Loew LM. Voltage-sensitive dyes: measurement of membrane potentials induced by DC and AC electric fields [Review]. Bioelectromagnetic Suppl 1:179–189, 1992.
37. Cohen LB, Salzberg BM, Davila HV, Ross WN, Landowne D, Waggoner AS, Wang CH. Changes in axon fluorescene during activity: molecular probes of membrane potential. J Membr Biol 19:1–36, 1974.
38. Muller W, Windisch H, Tritthart HA. Fast optical monitoring of microscopic excitation patterns in cardiac muscle. Biophys J 56:623–629, 1989.
39. Knisley SB. Transmembrane voltage changes during unipolar stimulation of rabbit ventricle. Circ Res 77:1229–1239, 1995.
40. Girouard SD, Laurita KR, Rosenbaum DS. Unique properties of cardiac action potentials recorded with voltage-sensitive dyes. J Cardiovasc Electrophysiol 7:1024–1038, 1996.
41. Efimov IR, Mazgalev TN. High-resolution three-dimensional fluorescent imaging reveals multilayer conduction pattern in the atrioventricular node. Circulation 98:54–57, 1998.
42. Choi BR, Salama G. Optical mapping of atrioventricular node reveals a conduction barrier between atrial and nodal cells [see comments]. Am J Physiol 274:H829–H845, 1998.
43. Grinvald A, Hildesheim R, Farber IC, Anglister L. Improved fluorescent probes for the measurement of rapid changes in membrane potential. Biophys J 39:301–308, 1982.
44. Schaffer P, Ahammer H, Muller W, Koidl B, Windisch H. Di-4-ANEPPS causes photodynamic damage to isolated cardiomyocytes. Pflugers Arch 426:548–551, 1994.
45. Cheng YN, Mazgalev T, Van Wagoner DR, Tchou PJ, Efimov IR. Voltage-sensitive dye RH421 increases contractility of cardiac muscle. Can J Physiol Pharmacol 76:1146–1150, 1998.
46. Grinvald A, Anglister L, Freeman JA, Hildesheim R, Manker A. Real-time optical imaging of naturally evoked electrical activity in intact frog brain. Nature 308:848–850, 1984.
47. Grinvald A, Segal M, Kuhnt U, Hildesheim R, Manker A, Anglister L, Freeman JA. Real-time optical mapping of neuronal activity in vertebrate CNS in vitro and in vivo. In: De Weer P, Salzberg BM, eds. Optical Methods in Cell Physiology. New York: Wiley, 1986.
48. Rokitskaya TI, Antonenko YN, Kotova EA. Effect of dipole potential of a bilayer lipid membrane on the gramicidin channel dissociation kinetics. Biophys J 73:850–854, 1997.
49. Clarke RJ, Schrimpf P, Schoneich M. Spectroscopic investigations of the potential-sensitive membrane probe RH421. Biochim Biophys Acta 1112:142–152, 1992.
50. Fedosova NU, Cornelius F, Klodos I. Fluorescent styryl dyes as probes for Na,K-ATPase reaction mechanism: significance of the charge of the hydrophilic moiety of RH dyes. Biochemistry 34:16806–16814, 1995.
51. Frank J, Zouni A, van Hoek A, Visser AJ, Clarke RJ. Interaction of the fluorescent probe RH421 with ribulose-1,5-bisphosphate carbox-

ylase/oxygenase and with $Na^+K^{(+)}$-ATPase membrane fragments. Biochem Biophys Acta 1280:51–64, 1996.

52. Efimov IR, Ermentrout B, Huang DT, Salama G. Activation and Repolarization patterns are governed by different characteristics of ventricular myocardium: experimental study with voltage-sensitive dyes and numerical simulations. J Cardiovasc Electrophysiol 7:512–530, 1996.

53. Schalij MJ, Lammers WJ, Rensma PL, Allessie MA. Anisotropic conduction and reentry in perfused epicardium of rabbit left ventricle. Am J Physiol 263(pt 2):H1466–H1478, 1992.

54. Allessie MA, Schalij MJ, Kirchhof CJ, Boersma L, Huybers M, Hollen J. Experimental electrophysiology and arrhythmogenicity. anisotropy and ventricular tachycardia. Eur Heart J 10(suppl E):2–8, 1989.

55. Knisley SB, Justice RK, Kong W, Johnson PL. Ratiometry of transmembrane voltage-sensitive fluorescent dye emission in hearts. Am J Physiol 279:H1421–H1433, 2000.

56. Davidenko JM, Pertsov AV, Salomonsz R, Baxter W, Jalife J. Stationary and drifting spiral waves of excitation in isolated cardiac muscle. Nature 355:349–351, 1992.

57. Undrovinas AI, Maltsev VA. Cytoskeleton disruption results in electromechanical dissociation in rat ventricular cardiomyocytes. J Am Coll Cardiol 29:404A–405A, 1997.

58. Biermann M, Rubart M, Wu J, Moreno A, Josiah-Durant A, Zipes DP. Effects of cytochalasin D and 2,3-butanedione monoxime on isometric twitch force and transmembrane action potentials in isolated canine right ventricular trabecular fibres. J Cardiovasc Electrophysiol 9:1348–1357, 1998.

59. Wu J, Biermann M, Rubart M, Zipes DP. Cytochalasin D as excitation-contraction uncoupler for optically mapping action potentials in wedges of ventricular myocardium. J Cardiovasc Electrophysiol 9:1336–1347, 1998.

60. Niedergerke R, Orkand RK. The dual effect of calcium on the action potential of the frog's heart. J Physiol (Lond) 184:291–311, 1966.

61. Chapman RA. The effect of oximes on the dihydropyridine-sensitive ca current of isolated guinea-pig ventricular myocytes. Pflugers Arch 422:325–331, 1993.

62. Liu Y, Cabo C, Salomonsz R, Delmar M, Davidenko J, Jalife J. Effects of diacetyl monoxime on the electrical properties of sheep and guinea pig ventricular muscle. Cardiovasc Res 27:1991–1997, 1993.

63. Coulombe A, Lefevre IA, Deroubaix E, Thuringer D, Coraboeuf E. Effect of 2,3-butanedione 2-monoxime on slow inward and transient outward currents in rat ventricular myocytes. J Mol Cell Cardiol 22:921–932, 1990.

64. Verrecchia F, Herve JC. Reversible blockade of gap junctional communication by 2,3-butanedione monoxime in rat cardiac myocytes. Am J Physiol 272(pt 1):C875–C885, 1997.

65. Cheng Y, Mowrey KA, Efimov IR, Van Wagoner DR, Tchou PJ, Mazgalev TN. Effects of 2,3-butanedione monoxime on the atrial-atrioventricular nodal conduction in isolated rabbit heart. J Cardiovasc Electrophysiol 8:790–802, 1997.

66. Wu JY, Cohen LB. Fast multisite optical measurement of membrane potential. In: Mason WT, ed. Fluorescent and Luminiscent Probes for Biological Activity: A Practical Guide to Technology for Quantitative Real-Time Analysis. San Diego, CA: Academic, 1993.

67. Bullen A, Patel SS, Saggau P. High-speed, random-access fluorescene microscopy: i. high-resolution optical recording with voltage-sensitive dyes and ion indicators. Biophys J 73:477–491, 1997.

68. Dillon SM. Optical mapping. In: Shenasa M, Borggrefe M, Breithardt G, eds. Cardiac Mapping. Mount Kisko, NY: Futura, 1993.

69. Fujii S, Hirota A, Kamino K. Optical indications of pace-maker potential and rhythm generation in early embryonic chick heart. J Physiol (Lond) 312:253–263, 1981.

70. Dillon S, Morad M. A new laser scanning system for measuring action potential propagation in the heart. Science 214:453–456, 1981.

71. Blasdel GG, Salama G. Voltage-sensitive dyes reveal a modular organization in monkey striate cortex. Nature 321:579–585, 1986.

72. Grinvald A, Cohen LB, Lesher S, Boyle MB. Simultaneous optical monitoring of activity of many neurons in invertebrate ganglia using a 124-element photodiode array. J Neurophysiol 45:829–840, 1981.

73. Fujii S, Hirota A, Kamino K. Optical recording of development of electrical activity in embryonic chick heart during early phases of cardiogenesis. J Physiol (Lond) 311:147–160, 1981.

74. Wikswo JP, Lin SF, Abbas RA. Virtual electrodes in cardiac tissue: a common mechanism for anodal and cathodal stimulation. Biophys J 69:2195–2210, 1995.

75. Morad M, Dillon S, Weiss J. An acousto-optically steered laser scanning system for measurement of action potential spread in intact heart. Soc Gen Physiol Ser 40:211–226, 1986.

76. Dillon SM, Mehra R. Prolongation of ventricular refractoriness by defibrillation shocks may be due to additional depolarization of the action potential. J Cardiovasc Electrophysiol 3:442–456, 1992.

77. Zhou X, Ideker RE, Blitchington TF, Smith WM, Knisley SB. Optical transmembrane potential measurements during defibrillation-strength shocks in perfused rabbit hearts. Circ Res 77:593–602, 1995.

78. Farber IC, Grinvald A. Identification of presynaptic neurons by laser photostimulation. Science 222:1025–1027, 1983.

79. Efimov IR, Cheng Y, Van Wagoner DR, Mazgalev T, Tchou PJ. Virtual electrode-induced phase singularity: a basic mechanism of failure to defibrillate. Circ Res 82:918–925, 1998.

80. Cheng Y, Mowrey KA, Van Wagoner DR, Tchou PJ, Efimov IR. Virtual electrode induced re-excitation: a basic mechanism of defibrillation. Circ Res 85:1056–1066, 1999.

81. Lin FC, Roth BJ, Wikswo JP Jr. Quatrefoil reentry in myocardium: an optical imaging study of the induction mechanism. J Cardiovasc Electrophysiol 10:574–586, 1999.

82. Sepulveda NG, Roth BJ, Wikswo JP. Current injection into a two-dimensional anisotropic bidomain. Biophys 55:987–999, 1989.

83. Efimov IR, Aguel F, Cheng Y, Wollenzier B, Trayanova N. Virtual electrode polarization in the far field: implications for external defibrillation. Am J Physiol 279:H1055–H1070, 2000.

84. Winfree AT. When Time Breaks Down: The Three-Dimensional Dynamics of Electrochemical Waves and Cardiac Arrhythmias. Princeton, NJ: Princeton University Press, 1987.

85. Efimov IR, Sidorov VY, Cheng Y, Wollenzier B. Evidence of 3D scroll waves with ribbon-shaped filament as a mechanism of ventricular tachycardia in the isolated rabbit heart. J Cardiovasc Electrophysiol 10:1452–1462, 1999.

86. Efimov IR, Grey RA, Roth BJ. Virtual electrodes and de-excitation: new insights into fibrillation induction and defibrillation. J Cardiovasc Electrophysiol 11:339–353, 2000.

87. Montana V, Farkas DL, Loew LM. Dual-wavelength ratiometric fluorescence measurements of membrane potential. Biochemistry 28:4536–4539, 1989.

88. Nikolski V, Efimov IR. Virtual electrode polarization of ventricular epicardium during bipolar stimulation. J Cardiovasc Electrophysiol 11:605, 2000.

17
Video Imaging of Fibrillation and Defibrillation

Richard A. Gray and Isabelle Banville
University of Alabama at Birmingham, Birmingham, Alabama, U.S.A.

I. INTRODUCTION

Cardiac fibrillation is characterized by rapid, irregular electrical activity as recorded by an electrocardiogram (ECG) from the body surface [1]. Ventricular fibrillation (VF) is the leading cause of death in the industrialized world, claiming the lives of more than 1000 Americans each day [2]. Atrial fibrillation (AF) is the most common sustained cardiac arrhythmia and often leads to stroke [3]. The application of high-energy electric fields is the most effective method to terminate fibrillation (this process is called defibrillation).

The activity of the heart is monitored via the ECG, and these recordings have led to many diagnostic advances in cardiology. However, since the ECG is recorded from the body surface at a distance from the heart, it reveals very little about the events occurring during complex cardiac rhythms. For example, during VF, ECG deflections continuously change in shape, magnitude, and direction, which has led to the idea that fibrillation is the result of disorganized, highly complex, perhaps even random activation of the heart. The inefficient and asynchronous contractions that occur during fibrillation are the result of spatiotemporal patterns of electrical activity in the heart. Many local recordings (i.e., cardiac mapping) are required to reveal the electrical activity throughout the heart during rapid cardiac rhythms (tachyarrhythmias) [4,5].

"Traditional" cardiac mapping involves recording extracellular potentials and calculating activation times from up to 512 sites [6], allowing a spatiotemporal description of the propagating waves of electrical activity in the heart. Another form of cardiac mapping is to use "voltage-sensitive dyes" to "optically map" the electrical activity of the heart. These "potentiometric" dyes bind to the membrane of cardiac cells. When these dyes are excited with photons at a particular wavelength (excitation wavelength), photons are emitted at another wavelength (emission wavelength). The number of the emitted photons (i.e., the intensity of the emitted light) is linearly related to the transmembrane potential [7]. The emitted light is captured with either photodiodes or a charged-coupled device (CCD). A single photodiode can be used in conjunction with a scanning laser [8,9], or a photodiode array can be used with a static light source [10,11].

A unique advantage of using voltage-sensitive dyes is that transmembrane activity is recorded, allowing the analysis of both the depolarization and repolarization processes. In addition, using voltage-sensitive dyes allows the ability to record transmembrane activity during the application of an electric field. Optical mapping is the only technique to provide the ability to record repolarization events simultaneously from many sites. The spatial distribution of action potential duration (APD) and the dynamics of the repolarization are phenomena that are well suited to study via optical mapping [12–19]. This chapter focuses on the unique features of optical mapping with a CCD camera [20], a technique called video imaging.

II. DATA ACQUISITION

A. Overview

Either one or two complete video imaging systems are used to record from the various cardiac preparations. Two systems are utilized to record simultaneously from two different areas of the heart (e.g., opposite surfaces of the preparation [19,21], or the right and left atrium [22]). Figure 1 shows a schematic diagram of a single video imaging setup. The light from a powerful light source is passed through a heat filter, a collimating lens, and a bandpass excitation filter. The light is then reflected 90° from a dichroic mirror onto the heart surface. If a monochromatic light source is used, such as a laser [8,9,23], or light-emitting diodes [24], neither the bandpass filter nor the dichroic mirror is necessary. The light emitted from the heart representing the transmembrane activity occurs at a different wavelength than the excitation light. The emitted light is transmitted through an emission

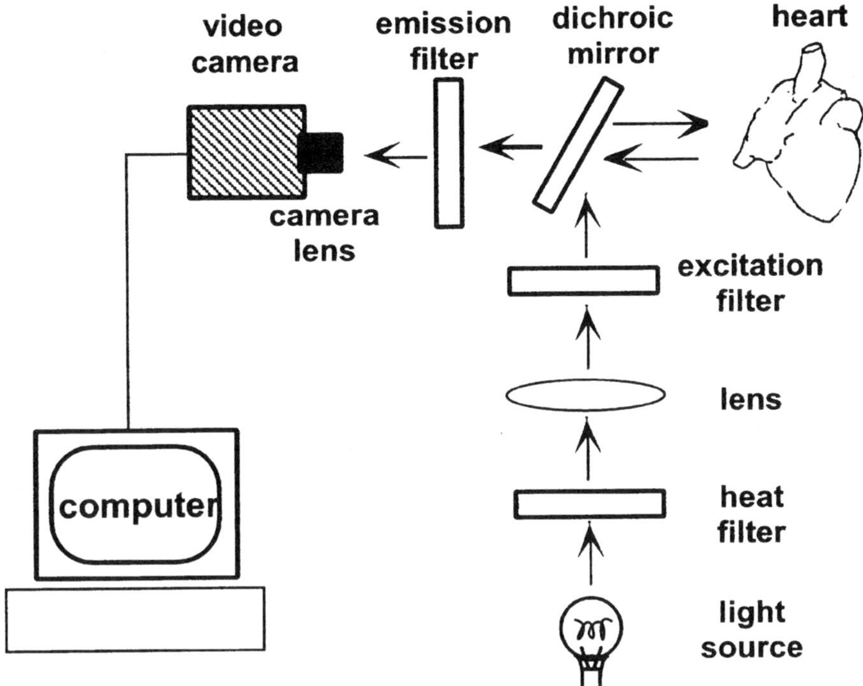

Figure 1 Video imaging system. A CCD camera is used to map transmembrane activity from the heart using a voltage-sensitive dye. The heart tissue is usually connected to a perfusion system and maintained at a constant temperature. The voltage-sensitive dye is injected into the perfusate and binds to the membrane of cardiac cells. A powerful light source is used to provide the light required to excite the dye. The light is typically filtered such that only photons at the appropriate excitation wavelength reach the heart surface. A dichroic mirror reflects the excitation light onto the heart. The excitation light causes photons to emitted from the dye at a wavelength determined by the emission spectrum of the dye. The final optical signal passing through the emission filter is linearly related to the transmembrane potential. After reaching the CCD camera, the signal is digitized and saved in computer memory.

filter and projected onto a CCD video camera. The analog-to-digital conversion is accomplished within the camera for digital cameras or within the computer for analog cameras. A "frame grabber" board in the computer accepts each image and passes sequential images to either onboard memory or random-access memory (RAM) on the computer. Most often, drugs are used to eliminate the contractions of the heart that interfere with the data acquisition and processing [25–27]. These "uncoupling" agents disrupt the normal transduction of the action potential to the mechanical contraction within the cardiac cells.

B. CCDs

CCD chips contain an array of photosensitive elements (pixels) as shown in Fig. 2. When photons hit a pixel, electrons are released, thus increasing the charge contained on that pixel. This process has often been compared to buckets or wells filling with rain drops. From this analogy comes the term "full-well capacity," meaning the maximum charge (number of electrons) a pixel can hold without "spilling" charge onto adjacent pixels. Each pixel in the array accumulates charge simultaneously, based on the number of photons hitting that particular element. Pixels are not necessarily square, and the ratio of pixel height to width is called the "aspect ratio." The entire pixel array is rectangular, and most CCDs are manufactured so that there is no dead space between columns or rows. The charge from each pixel is moved horizontally and then vertically using shift registers (see Fig. 2B). Using our analogy, this is like passing buckets in a "bucket brigade." Although the charge on each pixel is acquired simultaneously, the data is read out sequentially via a horizontal readout register. Since this sequential readout is the rate-limiting step in data acquisition, many CCDs contain a second identical array of pixels shielded from light. A high-speed parallel transfer from the active photosensitive region to this light-shielded region allows the CCD to acquire data while data readout is occurring from the light-shielded region. In cameras without a light-shielded region, data acquisition is halted during readout. Therefore, the exposure time (the time that pixels integrate charge from incoming photons) is usually the inverse of the frame rate minus either the high-speed transfer time or the time for an entire array to be read out. Since readout and transfer times are extremely rapid, it is common to consider the integration time as the inverse of the frame rate. However, in some applications the light is "strobed" or the camera shutter is not open continuously, resulting in shorter exposure times to localize events more precisely in time [28]. The charge contained on each pixel is passed along the readout registers and converted to voltage via a capacitor. Therefore, the image is converted to a time-varying voltage signal. This voltage signal is amplified and then converted to digital numbers (DNs) via an analog-to-digital (A/D) converter and these DNs are stored in memory on the computer (see Fig. 2A). The number of bits used to represent the intensity at each pixel is determined by the A/D converter. The CCD and the computer communicate vital information (triggering information, frame valid, line valid, etc.) via analog signals. A master clock which is either

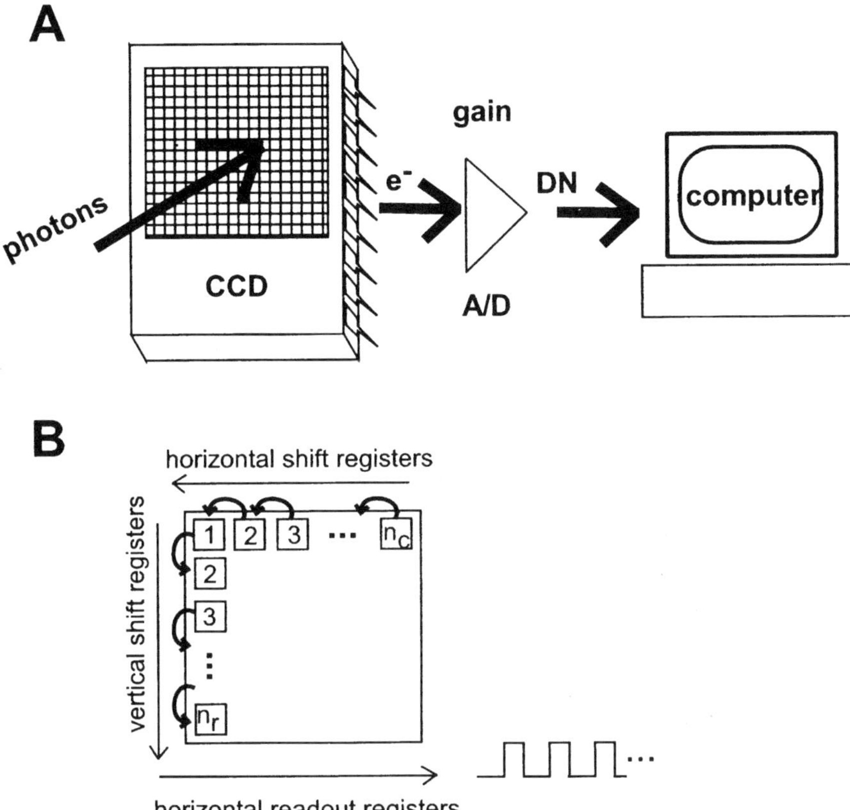

Figure 2 Charge-coupled device. (A) The CCD array collects photons emitted from the heart and outputs an analog signal that is amplified, digitized, and then stored in computer memory. The precision of the digitized signal is determined by the analog-to-digital (A/D) converter. The number of bits used to represent each pixel will result in a certain range of possible values or levels (in digital number, DN). Electrons are represented by e^-. (B) CCDs are made up of a rectangular array of pixels; we denote by n_c and n_r the number of columns and rows, respectively. After exposure of the CCD array to photons, each pixel is read out sequentially by horizontal and vertical shift registers. The charge contained on each pixel is passed along the readout registers and converted to voltage via a capacitor. Therefore, the "image" has been converted into a stepwise, time-varying voltage signal.

on the CCD or on the frame grabber board ensures synchronization and determines the speed at which the camera outputs data.

The speed at which the data can be read out from the camera into memory is the factor that limits the spatial resolution, frame rate, and number of bits of digitization (N_{bits}). The standard video rate is 30 frames per second (fps), which is too slow to record accurately the changes in transmembrane potential occurring in the heart. Higher speeds are achieved, in part, by reducing the spatial resolution. Improvements in computer technology are leading to very high data transfer rates (e.g., 80–100 megabytes per second).

C. Dynamic Range

One important way to analyze a camera's performance is to quantify the number of levels or DNs that can be used to record accurately the intensity of light falling on the CCD. When a black lens such as a lens cover is placed over the camera, the output is not zero. Some noise results from the A/D and gain processes and some noise results from photons hitting the CCD. The overall noise is called the read noise (RN) and is quantified in terms of DN. The mean value of read noise (RN_0) is always greater than zero due to the integrative nature of CCDs. The speed of the A/D conversion affects noise levels, and hence read noise is a function of the frame rate (see Fig. 3). Multiple A/D converters can be connected to a single CCD array to reduce this source of noise (Pixel Vision, Beaverton, OR). RN_0 determines the minimum DN and hence the number of levels available to record light intensity (i.e., $2^{N_{bits}} - RN_0$). However, this does not represent the precision of the camera, since the fluctuations in RN may be greater than one level. Therefore, the number of levels that can be used to record accurately the intensity of light falling on the CCD, i.e., the dynamic range (DR), is

$$DR(DN) = \frac{2^{N_{bits}} - RN_0}{RN_{rms}} \tag{1}$$

where RN_{rms} is the root-mean-square of $RN(t)$. Therefore, the number of bits used in the A/D conversion and the read noise are the two main factors that determine the camera's dynamic range (not to be confused with signal-to-noise ratio, SNR). The dynamic range for three different cameras as calculated from measurements in our laboratory are shown in Fig. 4.

D. Shot Noise

Now, let us consider the physical events associated with recording fluorescence at a single picture element (i.e., pixel) within the CCD. Let $\bar{F}$ be the flux density of photons leaving the heart in photons/mm^2/sec (see Fig. 5). It is important to know that an inherent feature of light is "shot noise." Shot noise results from the quantile nature of light and is unavoidable.

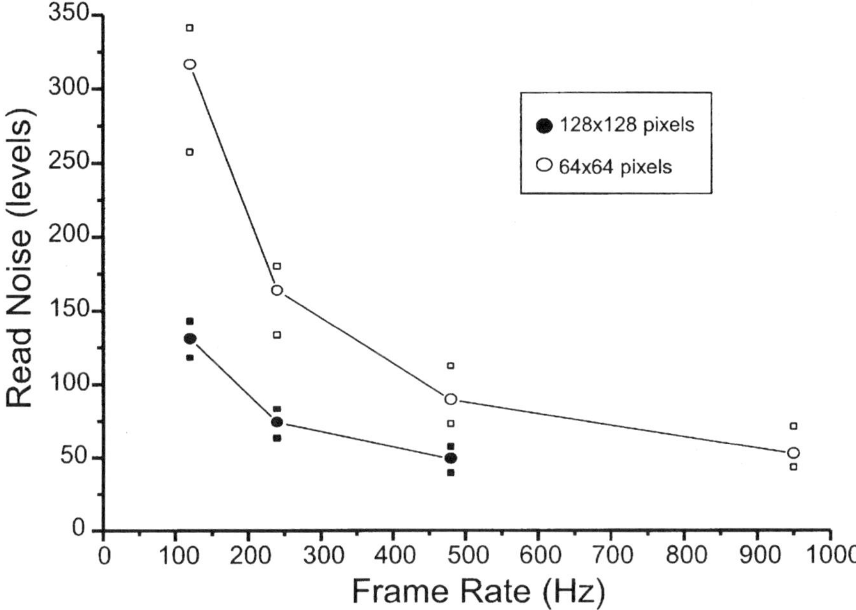

Figure 3   Effect of frame rate on read noise. Read noise is a function of the camera electronics, but is also affected by a small number of photons hitting each pixel even in total darkness. Due to the integrative nature of CCDs, the read noise decreases as the frame rate increases. This curve was generated using a Dalsa (model 128ST, Ontario, Canada) camera with binning capabilities. At full spatial resolution (128×128 pixels), the read noise is less than at low spatial resolution (64×64 pixels). The small square symbols represent the minimum and maximum values in the entire CCD array. At low spatial resolution, the effective pixel size on the CCD array is bigger, therefore increasing the read noise. The read noise is minimized at the fastest frame rate (480 Hz at full resolution and 960 Hz at low resolution).

The magnitude of shot noise is equal to the square root of the intensity of the light. Therefore, increasing the flux density $\bar{F}$ will act to reduce the relative fraction of shot noise (i.e., $\sqrt{\bar{F}}/\bar{F}$). Remember, full-well capacity provides an upper limit on the number of photons that can be integrated on a single pixel. The amount of charge that a CCD can store in each pixel depends largely on the physical size of the pixel. "Scientific" CCDs have relatively large pixels ($> 10\,\mu m$) that allow the user to increase $\bar{F}$ and hence decrease the influence of shot noise. Since the cost of producing CCDs is strongly area dependent, nonscientific CCDs have very small pixels sizes, typically $8\,\mu m$ or less [29]. It is important to remember that individual pixel sizes in CCDs are

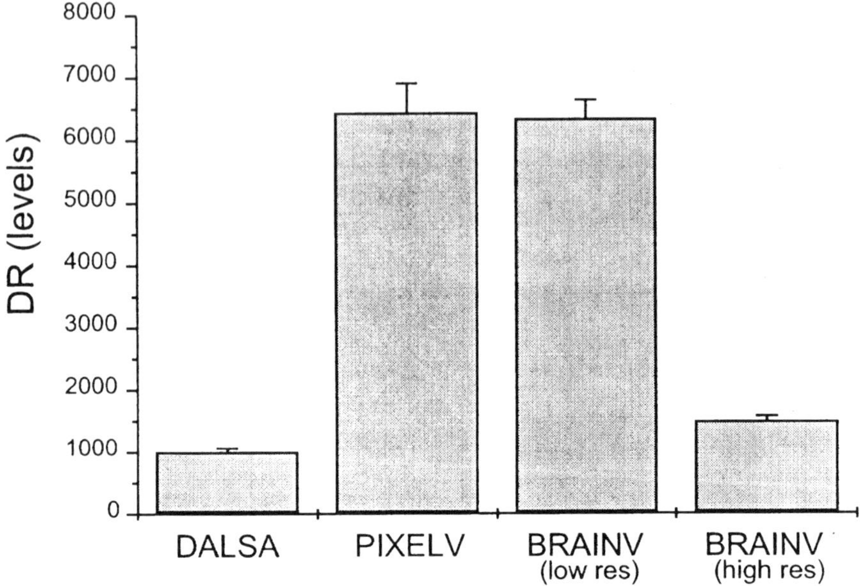

Figure 4 Dynamic range. Dynamic range (DR), as defined in Eq. (1), is the number of levels available to record optical signals. It is a function of the number of bits used to digitize each pixel and the read noise. DR varies dramatically among CCD cameras. DR_S for the following cameras are displayed. DALSA: Dalsa (model 128ST, Ontario, Canada) with 128×128 resolution running at 480 Hz with 12-bit digitization. PIXELV: Pixel Vision (FastOne model, Pixel Vision, Beaverton, OR) with 80×80 resolution running at 282 Hz with 14-bit digitization. BRAINV (MiCam01 ICX082, Sci-Media Ltd, Tokyo, Japan) with 96×64 resolution running at 500 Hz with 14-bit digitization (low res) and with 192×128 resolution running at 286 Hz with 13-bit digitization (high res). Eight-bit cameras would have even lower DR values.

much smaller than the size of an element in a photodiode array (~ 1 mm). A properly designed camera will be "shot noise limited," that is, the maximum signal-to-noise ratio will be limited by the inherent statistical nature of light rather than the read noise floor of the camera electronics.

E. Signal-to-Noise Ratio

In video imaging, the majority of levels are used to represent the "background fluorescence" (see Fig. 6A). The background fluorescence (F_0) represents the photon flux ($\bar{F}$) when there are no changes in transmembrane potential occurring throughout the heart (i.e., diastole). This can be thought

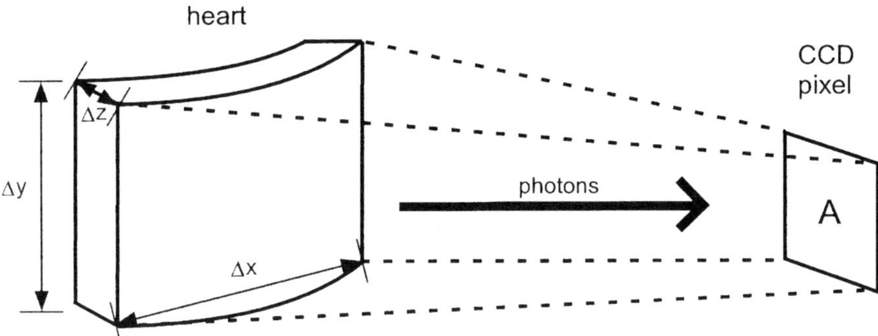

Figure 5 Photon flux. Each pixel of the CCD array collects photons emitted from a portion of the heart. The portion of tissue that contributes to the optical signal is three-dimensional, with horizontal (Δx) and vertical (Δy) spread as well as depth (Δz). The size of the region mapped onto each pixel is a function of the optical magnification (zoom) and the pixel aspect ratio. The surface area of each pixel (A) determines the full-well capacity (described in the text).

of as the "DC offset" portion of the signal and is unavoidable because of the integrative nature of CCDs. In contrast, the amplifiers for photodiode arrays can be "AC coupled" to eliminate this DC offset prior to the A/D conversion [10,11]. The signal of interest is transmembrane potential, and it is important to recognize that optical recordings do not represent true transmembrane potential (see below). In particular, shot noise is usually the dominant source of noise in optical recordings, and it cannot be eliminated by improving the equipment characteristics. As discussed above, increasing the numbers of photons hitting the CCD will act to decrease the relative amount of shot noise.

The transmembrane potential changes by approximately 100 mV during the upstroke of an action potential. The ability to transduce membrane potential to fluorescence depends on the specific voltage-sensitive dye. We use di-4-ANEPPS, which results in a fluorescence change during the upstroke (δF_{max}) equal to about 8% of the background fluorescence [19,30]. For di-4-ANEPPS, an increase in transmembrane potential results in a decrease in emitted fluorescence (see Fig. 6), but this depends on the spectral properties of the specific voltage-sensitive dye. For optical recordings it is customary [19,20,31,32] to define SNR as

$$\mathrm{SNR} = \frac{\delta F_{max}}{F_{0,rms}} \qquad (2)$$

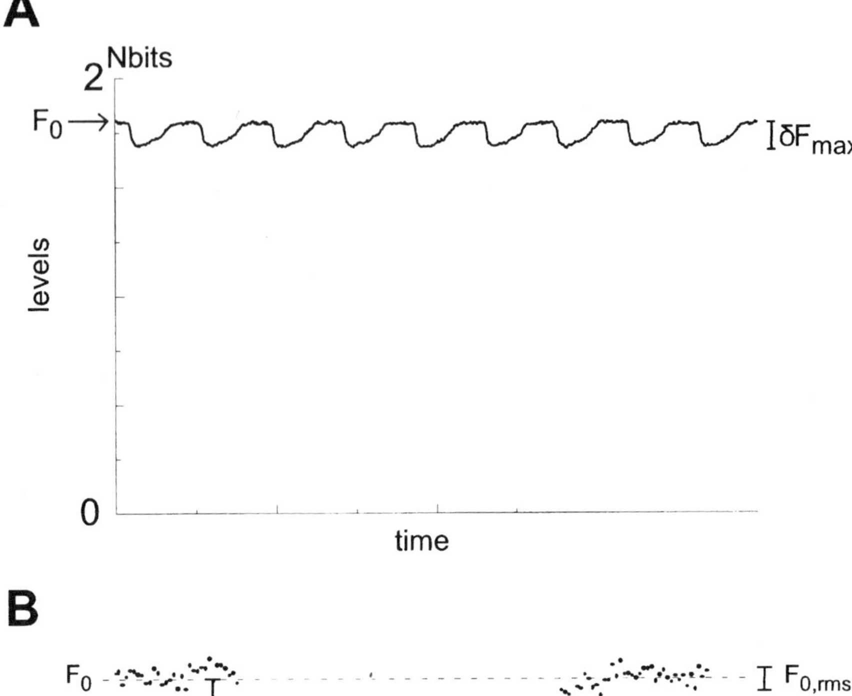

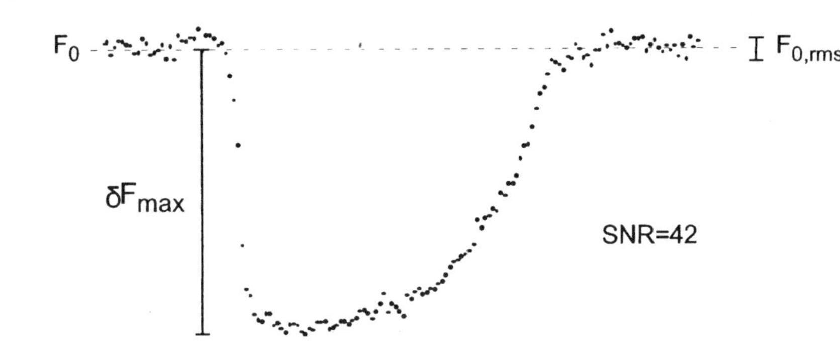

Figure 6 Optical signals. (A) Fluorescence signal from a pixel recording from a heart stained with di-4-ANEPPS. Transmembrane potential depolarization (increase) causes a decrease in the fluorescent optical signal. F_0 is the background fluorescence and δF_{max} is the change in fluorescence associated with the entire depolarization process (i.e., the upstroke of the action potential). (B) Unprocessed signal from one pixel recorded with a Brainvision camera (MiCam01 ICX082, Sci-Media Ltd, Tokyo, Japan) at 500 Hz with 96 × 64 resolution. Signal-to-noise ratio (SNR) is measured as $\delta F_{max}/F_{0,rms}$, where $F_{0,rms}$ is the standard deviation of the noise during diastole. Here SNR is equal to 42.

(see Fig. 6B), although peak-to-peak noise [33] has also been used in the denominator. High-intensity excitation light causes "dye bleaching" to occur, which results in a decrease in SNR over time whose time constant can be very rapid (4–24 sec) for cell monolayers [34] or much slower ($\sim$76 min) for whole hearts [35]. If one assumes an action potential amplitude of 100 mV, SNR can be thought of as the voltage resolution (e.g., SNR = 20 represents a voltage resolution of 5 mV).

F. Nyquist Criterion

A fundamental danger of A/D conversion is that high-frequency components of the recorded signal can be "aliased" and not digitized properly. Typically, a low-pass analog filter is inserted immediately preceding the A/D converter with a cutoff frequency at half the digitization rate (the Nyquist frequency). If frequencies greater than the Nyquist frequency exist and are not filtered out, the resulting digital signal differs from the true signal (i.e., aliasing). For video imaging, the potential exists for both temporal and spatial aliasing in the digital recording of dynamic spatial patterns. In order to avoid spatial aliasing, the intensity of each image must not fluctuate by more than one cycle for every two samples (pixels in the case of CCDs). Accordingly, careful attention to the spatial resolution of a mapping system and the spatial frequency of the signal being recorded is important [36]. Because of the integrative nature of CCD cameras (see above), it is impossible to insert an analog filter to prevent temporal aliasing. Most often, aliasing results in artifacts that are easy to distinguish from electrophysiological signals. Nevertheless, it is prudent to record the same spatiotemporal event with various spatial and temporal resolutions to ensure that the recorded dynamic patterns are the same and that neither spatial or temporal aliasing is occurring. Figure 7 shows raw and averaged (see below) action potentials recorded at the same site with various temporal and spatial resolutions. The fact that the action potential shapes are identical ensures that no spatial or temporal aliasing was occurring.

G. Spatial Summation of Fluorescence

As shown in Fig. 5, Δx and Δy are the horizontal and vertical spatial dimensions of the heart area mapped onto each pixel; Δz represents the depth of tissue contributing fluorescence. The values for Δx and Δy are a function of the optical magnification; these values are usually similar (or identical), and this value is typically called the spatial resolution. Typically, optical recordings represent a weighted average of the transmembrane potential over many cells [32], although some investigators use high magnification

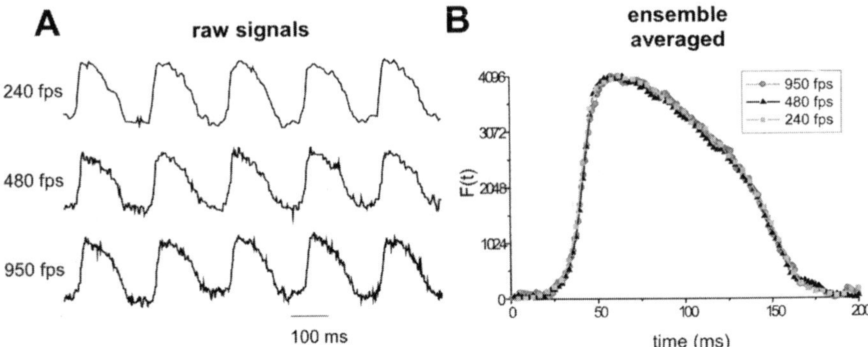

Figure 7   Ensemble averaging. (A) Action potentials recorded from the anterior surface of a rabbit heart paced at a cycle length of 200 msec. Recordings are from a Dalsa camera (model 128ST, Ontario, Canada) at three frame rates: 240 Hz (128 × 128 resolution); 480 Hz (128 × 128 resolution); and 950 Hz (64 × 64 resolution). (B) Ensemble averaging was applied such that 25 sequential action potentials were averaged together. The ensemble averaged beats for all three frame rates show identical depolarization and repolarization sequences, indicating that no blurring was evident.

to study subcellular transmembrane potential dynamics [31,37]. In general, it has been believed that optical recordings from the heart surface represent a measure of transmembrane potential from a thin, essentially two-dimensional, epicardial layer. This thickness has been quantified by various techniques and is estimated to be ∼0.3 mm [32,38]. However, recent data suggest that optical signals may include fluorescence from much deeper layers [39–41]. In fact, these studies suggest that it may be possible to reconstruct electrical impulse propagation in multiple layers of tissue from a single array of photodetectors! If fluorescence contributions come from layers deeper than a space constant (∼1 mm), then the interpretation of the recorded signals is complicated [42].

III.   DATA PROCESSING

A.   Initial Steps

The first digital processing step is to subtract the background fluorescence for each site, $F_0(i,j)$, from each frame as follows:

$$F_{\text{sub}}(i,j,n) = F_0(i,j) - F_{\text{raw}}(i,j,n) \tag{3}$$

and the sign can be adjusted to ensure upright action potentials even if they were inverted in the raw signals (see Fig. 6). At this stage, the data are "scaled" such that further processing is not unduly compromised by quantization effects [20,32]. Often signals at each site are "normalized" such that the action potential amplitude during a paced beat (or sinus rhythm) at each site is the same, to correct for spatial nonuniformities in fluorescence intensity, $F_0(i,j)$. This normalization procedure is required in order to compare the magnitude of changes at various sites and also relate changes to membrane potential. This normalization is justified (given the limitations of spatial summation of fluorescence described above) if the regions from which we are recording have similar electrophysiology (i.e., resting potential and action potential amplitude).

B. Ensemble Averaging

In some situations where the activation patterns are repetitive, multiple beats can be averaged (ensemble averaging) to increase the SNR [13,19]. Ensemble averaging improves the signal quality significantly (see Fig. 7) because it results in a decrease of noise by a factor of $\sqrt{p}$, where p is the number of beats averaged together. If the period of activity is not a multiple of the inverse of the frame rate (Δt), a sequence of k beats can be averaged such that k^* (period of activity) is a multiple of Δt. In our laboratory, we select pacing cycle lengths such that they are multiples of Δt.

C. Filtering

Spatial and temporal filters can be applied, usually by convolving a filter kernel with the data set (linear filtering), resulting in the final "movie." Linear filters can be completely characterized in the frequency domain, which allows one to assess the effect of the filters on the underlying signals [5]. The frequency response of a median cannot be determined analytically, but median temporal filters cause less blurring than averaging filters and have been used to improve signal quality [19,43]. Wavelet filters also might prove useful for the analysis of spatiotemporal data sets [43]. Spatio-temporal filters have been employed by convolving a three-dimensional (two spatial dimensions and time) kernel with the data set. Most often, low-pass spatial and temporal filters are used to eliminate high-frequency components of the signal that are typical of noise. The size of the filter kernels are largely dependent on the SNR and the specific scientific question to be addressed. It is desirable to keep filter sizes small; with smaller kernels, events can be localized better in time and space.

D. Blurring

Due to the integrative nature of CCDs, moving objects, including propagating waves, are blurred in the resulting movies [20]. The spatial extent of this blur is the distance the wave travels during one frame (i.e., the product of Δt and the speed of propagation). It should be noted that spatial filtering will not affect the digitized frequency content of the signal if the kernel size is less than the blur amount. Similarly, the amount of blur should be considered when selecting the spatial resolution. For example, we suggest that the wave front should move approximately 1 pixel per frame. Therefore, for a wave propagation speed of 30 cm/sec, using a camera running at 500 Hz, the spatial resolution should be approximately 0.6 mm. Since the speed of propagation is dependent on the cycle length of activation and is faster along fibers compared to across them (see below), the amount of blurring may vary in space and time.

E. Depolarization

Activation times for each site are determined most commonly by a threshold criterion [13,17–19,28], although maximum derivatives have also been used [31,44]. Generally, each site is analyzed and the first frame where the signal is greater than the threshold value while the previous frame was less than the threshold is called the activation or depolarization time for that site. Most often a cutoff value of 50% calculated from the maximum and minimum in each time series is used as the threshold value, denoted as F_{50}. However, it is very important to remember that the fluorescence signal does not represent true transmembrane potential and the normalization procedure described above is sensitive to the cardiac electrophysiological status of the heart. For example, during fibrillation, the take-off potential (i.e., the potential from which the action potential begins) is elevated from resting levels [45]. Thus, a cutoff level of 50% may correspond to various transmembrane potential values depending on the underlying activity. Alternatively, the maximum change in F between two frames (dF/dt_{max}) can be used to identify activation times. Activation times calculated with these two methods are nearly identical for paced beats, but during arrhythmias the upstroke velocity of action potentials decreases and discrepancies can occur [18]. Fast and Kléber have used computer simulations to show that, in regions of slow conduction, calculating activation times using a threshold value is more appropriate than using the maximum derivative [46]. These methods provide activation times with a resolution of Δt, although interpolation algorithms can be used to refine the precision of the calculation of activation times [47]. This interpolation procedure is similar, in principle, to the spatial interpolation applied to electrical mapping data [48].

F. Repolarization

Repolarization times are calculated in a similar manner to depolarization times, and both threshold [17,18] and derivative [12] methods are used. Various threshold values are used, and it is customary to represent the cutoff values in terms of a percentage of the height of the previous action potential. For example, repolarization times are often presented in terms of 50%, 75%, or 95% recovery to baseline. Once again, during arrhythmias and rapid rates, take-off potentials may vary, making this computation difficult. Since repolarization is a much slower process than depolarization, sometimes additional temporal filtering is applied in an effort to decrease the influence of noise on the calculation of repolarization times [17]. Rosenbaum et al. [12] used the maximum in the second derivative of the fluorescence signal ($d^2F/dt^2_{\max}$) to compute repolarization time, and Efimov et al. showed that this was valid under a variety of situations [49]. With the improvements of SNR achievable with new cameras and ensemble averaging in combination with the filtering steps suggested by Rosenbaum et al. [12], this method may prove useful for video imaging.

G. Action Potential Duration Maps

Action potential duration (APD) is the difference between the repolarization time and the depolarization time. Although APD at many sites can be analyzed similar to other data sets, in terms of a mean and standard deviation, many analyses do not account for the spatial distribution of APD. For video imaging, the spatial distribution of APD is typically presented as a color-coded image, where the APD for each pixel is represented by a color or shade of gray [13,15].

H. Isochrone Maps

Isochrone maps can be generated from the spatial distribution of depolarization or repolarization times. Multiple isochrones (lines denoting events that occur at the same time) are displayed together to illustrate the spatial and temporal sequence of events. Most often, isochrone maps are used to display the sequences of depolarization, but they can be used to display sequences of repolarization as well [19,16,49]. Due to the integrative nature of CCDs, motion-induced blurring can occur, resulting in a region of pixels, perpendicular to the motion, activated in a single frame. Thus, depolarization and repolarization maps can be comprised of bands, not lines. For video imaging, isochrone maps are often displayed as color-coded images similar to APD maps [13,50]. However, lines can be drawn between bands

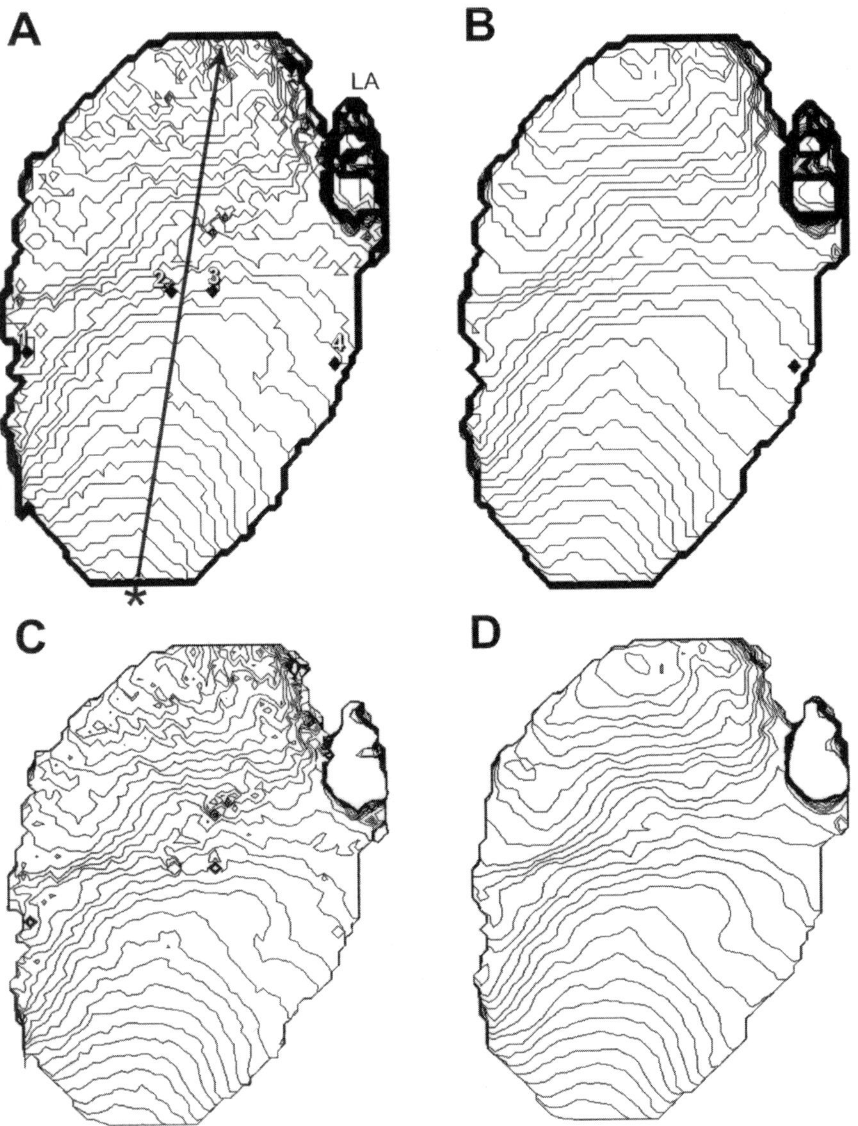

Figure 8 Isochrone maps. The depolarization sequence on the anterior surface of an isolated rabbit heart was recorded with a Brainvision camera (MiCam01 ICX082, Sci-Media Ltd, Tokyo, Japan) running at 500 Hz with 96 × 64 resolution. The heart was paced from the apex (*) at a cycle length of 200 msec. Isochrone maps illustrating the depolarization sequence are shown in A–D. The isochrone maps were computed using a contour algorithm incorporating linear interpolation between pixels. The

manually [20], or contour algorithms can be applied to the color-coded images [19] to generate isochrone maps with labeled lines similar to the traditional presentation of electrical mapping data. Investigators should specify what interpolation algorithm is used to generate contours and what additional processing steps are accomplished. Four isochrone maps generated from the same set of raw data are shown in Fig. 8. In our laboratory, we apply a spatial median filter to the depolarization/repolarization map and then compute contours using linear interpolation [19].

Now a few words of caution. First, polynomial interpolation may result in highly curved contours that may not reflect the events occurring between sampling points. Second, a limitation of isochrone maps is that if a site exhibits an event (e.g., depolarization or repolarization) twice during the time interval the map was computed, only one event will be displayed. Third, displaying data with isochrone maps often implies that there was smooth and continuous spatial progression of events. While this is well accepted for propagating wave fronts and depolarization maps, it is less clear for the repolarization process.

I. Other Ways to Represent Dynamic Spatial Patterns

Due to the limitations of isochrone maps, alternative methods have been used to describe the sequence of depolarization and repolarization. One such method is the time–space (or frame–stack) plot. Time–space plots (TSPs) show, in a single picture, the evolution of fluorescence over time for a given region of the heart [50]. A single line (or a two-dimensional region projected onto a line) from successive frames are stacked to form an intensity image whose axes are comprised of one spatial dimension and one time dimension. Although this process eliminates (or compresses) one spatial dimension, it is

---

gray arrow in A shows the general direction of propagation and the numbers 1–4 represent sites that were saturated, hence activation times could not be computed at these sites. (A) Activation times were discretized in terms of frame number and contours were generated. (B) Activation times were discretized in terms of frame number and contours were generated after a 3×3 median spatial filter was applied to the activation time map. (C) Activation times were computed using linear interpolation between frames and contours were generated. (D) Activation times were computed using linear interpolation between frames and contours were generated after a 3×3 median spatial filter was applied to the activation time map. In all panels, no spatial or temporal filtering was applied to the original data set (F). Linear interpolation improved the temporal resolution and a spatial median filter applied to the map of activation times removed outliers (such as saturated sites).

useful for calculating conduction speeds and identifying certain spatio-temporal patterns such as rotating spiral waves. (More details are provided in Ref. 50.) Another method to represent dynamic spatial patterns is to identify the position of wave fronts and wave tails in each video frame. Since the temporal derivative during depolarization is much greater than during repolarization (and of opposite sign), movies of the temporal derivative, dF/dt, have been used to describe the sequence of wave front propagation [28,51]. Alternatively, the full data set, F, can be binarized, such that all data points greater than some cutoff value are set equal to one value (e.g., 1) and all other points are set to another value (e.g., 0). Hence, the percentage of the heart that is depolarized at any moment can be computed (see Fig. 9); we have denoted this number as $\%F_{50}$ because we use a 50% cutoff value for the activation threshold [52]. Therefore, the spatial excitable gap (EG_x) is equal to $100\%F_{50}$. After binarization, a temporal derivative can be applied so that all data points are either equal to 1, 0, or -1. In a given frame, the sites that are equal to 1 are part of a wave front and the sites that are equal to -1 are part of a wave tail (see Fig. 10) [53]. (More details are provided in Ref. 53.) This procedure simplifies the display and analysis of the complete spatiotemporal data set, F. It should be noted that these spatial measures are strongly dependent on the cutoff threshold, and more emphasis should be placed on relative comparisons rather than absolute values.

J. Conduction Velocity

Impulses propagate throughout the three-dimensional heart and conduction speed is directionally dependent. Therefore, conduction velocity is a vector, not a scalar, quantity. In addition, the sequence of activation depends on a variety of factors such as the pacing site and rate. Accordingly, one must be very careful in the interpretation and presentation of conduction velocities in cardiac tissue. Although most often conduction speeds are presented, a few investigators have provided a measurement of the "conduction velocity vector field" on the surface of the heart [47,54,55]. Even this approach is limited, because propagation beneath the surface is not accounted for; however, these fields have recently been computed in three dimensions [56].

K. Cardiac Phase

The transmembrane potential (V_m) alone does not represent the "state" of the heart accurately at each site. This is not surprising, since V_m does not uniquely describe the cardiac action potential; during one beat V_m is equal to a certain value (e.g., $-20\,\mathrm{mV}$) twice, once during depolarization and once during repolarization. At least one additional variable, such as dV_m/dt,

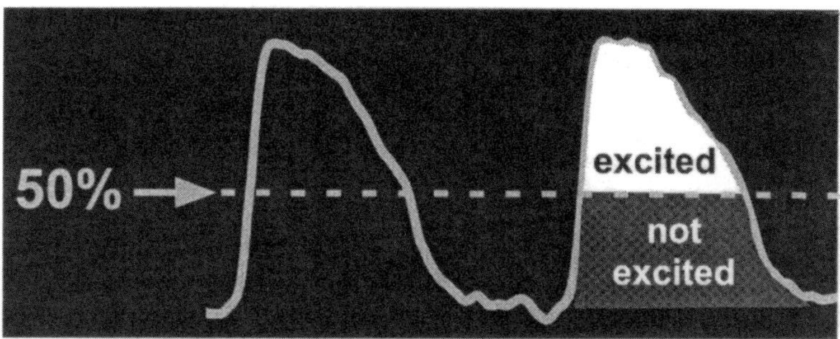

Figure 9 Spatial excitable gap. Using the 50% threshold method, at each instant each pixel is classified as excited ($F > 50\%$) and colored white or not excited ($F < 50\%$) and colored gray (see top panel). A snapshot of the resulting binarized image is shown in the bottom panel. The extent of the spatial excitable gap, EG_x, is the percentage of pixels in the recording array (and on the heart) that are not excited (gray). The portion of the heart that is excited (white) propagates into the nonexcited (gray) regions as shown by the black arrow in the bottom panel. Since EG_x can be computed for each frame, its transients can be studied.

is required to identify the state of each site [57]. Consider the multidimensional monodomain cable equation with N variables: V_m and typically $N - 1$ gating variables. In a computer model we have access to all state variables; however, in experiments we usually record only one state variable. For example, using optical mapping we record fluorescence, which

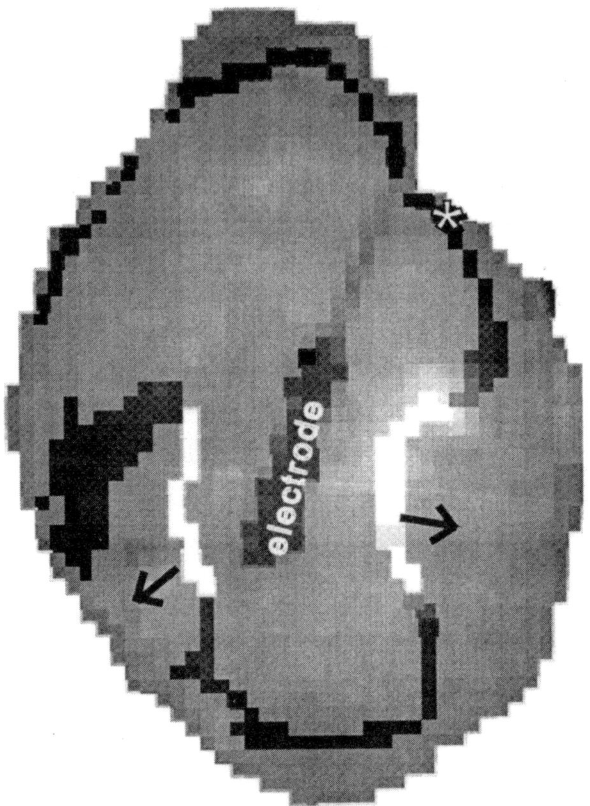

Figure 10   Quatrefoil reentry. Recordings were made from the anterior surface of an isolated rabbit heart and recorded with a Brainvision camera (MiCam01 ICX082, Sci-Media Ltd, Tokyo, Japan) running at 500 Hz with 96 × 64 resolution. A 3 × 3 × 3 spatiotemporal averaging filter was applied to the raw data. The heart was paced from the left ventricle (*) at a cycle length of 200 msec. A 2-A, 10-msec-duration, cathodal stimuli was applied 120 msec after the pacing stimuli via the electrode (1-cm coil) placed on the heart surface. At this time the whole anterior surface was depolarized and the shock resulted in quatrefoil reentry (four reentrant pathways comprised of a pair of figure-of-eight patterns) [77]. The position of the wave fronts (white) and wave tails (black) 2 msec after the end of the shock are shown superimposed on the gray-scale image of the heart. The snapshot illustrates two wave fronts parallel to the electrode propagating away from the electrode. In addition, four sites can be identified where the wave fronts meet the wave tails, indicating quatrefoil reentry.

is a measure of $V_m(t)$. Fortunately, we can create a *reconstructed* state space that is *topologically* equivalent to the *true* state space [58]. The reconstructed state variables can be represented as

$$V_m, \frac{\partial V_m}{\partial t}, \frac{\partial^2 V_m}{\partial t^2}, \ldots, \frac{\partial^{N-1} V_m}{\partial t^{N-1}} \qquad \text{or}$$

$$V_m(t), V_m(t+\tau), V_m(t+2\tau), \ldots, V_m[t+(N-1)\tau] \qquad (4)$$

By translating to a two-dimensional polar coordinate system using the recorded variable and a reconstructed variable (e.g., V_m or dV_m/dt), it is possible to create a phase variable (θ) that uniquely describes each portion of the cardiac action potential as shown in Fig. 11. This is strictly true only if the trajectory of the true high-dimensional system projects to an open loop in the two-dimensional state-space reconstruction. Even though many state variables are used to represent the dynamics of heart cells [59], the results of previous studies show that the essential heart dynamics during fibrillation are well represented in this reconstructed 2-D state space [4].

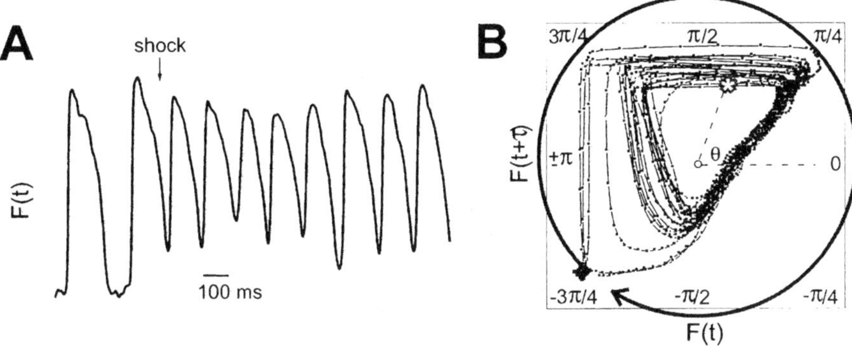

Figure 11 Cardiac phase. (A) Fluorescence recording, $F(t)$, from a site taken from the episode shown in Fig. 10. (B) Data in A plotted in reconstructed state space using $F(t)$ and $F(t+\tau)$. The phase variable (θ) is calculated by translating to a polar coordinate system with the origin denoted by the small open circle. θ is a function of frame number and hence is computed for each datum (small filled symbols) as the angle between the line connecting the datum to the origin and the horizontal line passing through the origin (i.e., the angle between the two dashed lines for the enlarged gray*). In reconstructed state space, the trajectories encircle the origin clockwise and it is easy to differentiate the depolarization ($-3\pi/4$ to $\pi/4$) and the repolarization processes ($\pi/4$ to $-3\pi/4$) of the action potential.

In the past we have formed a reconstructed state space using the method of time-delay embedding [59] to avoid the increased effects of noise when taking derivatives.

$$0(i,j,n) = \arctan 2[F(i,j,n+\tau) - F^*, F(i,j,n) - F^*] \tag{5}$$

where F^* is the average fluorescence value over a 2-sec interval calculated for each site (i,j) and τ is a delay (lag) factor, typically of the order of 5–20 msec [4]. Here, we specifically denote the *arctan* 2 function to clarify that the resulting value must span a range of 2π and uniquely describe all four quadrants of the newly formed polar coordinate system. (The *arctan* function generally returns a value spanning a range of only π.) The noise in new scientific-grade CCDs is considerably reduced, so we can calculate the phase variable using the temporal derivative of the fluorescent signal (F).

$$\theta(i,j,n) = \arctan 2\left[\frac{\partial F(i,j,n)}{\partial t}, F(i,j,n) - F^*\right] \tag{6}$$

L. Phase Mapping

Analyzing the dynamic spatial patterns of the cardiac phase variable has many advantages. First, the analysis of phase (phase singularities in particular) has a firm mathematical basis, and has been used to describe physical phenomena in physics [60], chemistry [61,62], and biology [63]. Second, certain analyses can only be accomplished using a phase variable. For example, the cyclic nature of phase allows the existence of "topological defects" or "topological invariants" which are equivalent to a phase singularity [the mathematical description of a phase singularity is provided below in Eq. (7)] [64,65]. Third, self-sustaining rotating waves cannot exist in excitable or oscillatory media in the absence of a phase singularity [66,67]. Fourth, spatial phase singularities can be identified at each instant, if all of the state variables are known, or with as few as three video frames (6 msec) from our experimental data. This ability to localize phase singularities in time is advantageous compared to other methods, which require a complete rotation (>100 msec) and stationarity to function appropriately [68].

The mathematical description for a spatial phase singularity is

$$\oint dr \nabla \theta = \pm 2\pi \tag{7}$$

where the sign indicates its chirality [65]. In other words, a phase singularity occurs when the line integral of the *change* of phase around a site is equal to $\pm 2\pi$. Phase singularities are not physical entities but are defined based on the value of phase in a nearby region. Therefore, phase singularities are zero-dimensional and, for spatially discrete data, occur at the junctions of pixels. The existence of a spatial phase singularity is a necessary, although not sufficient, condition for sustained rotation of impulse propagation in the heart [4]. An example of a phase map where a pair of figure-of-eight reentrant waves (and hence four phase singularities) are present is shown in Fig. 12. The transmembrane potential near a phase singularity exhibits only

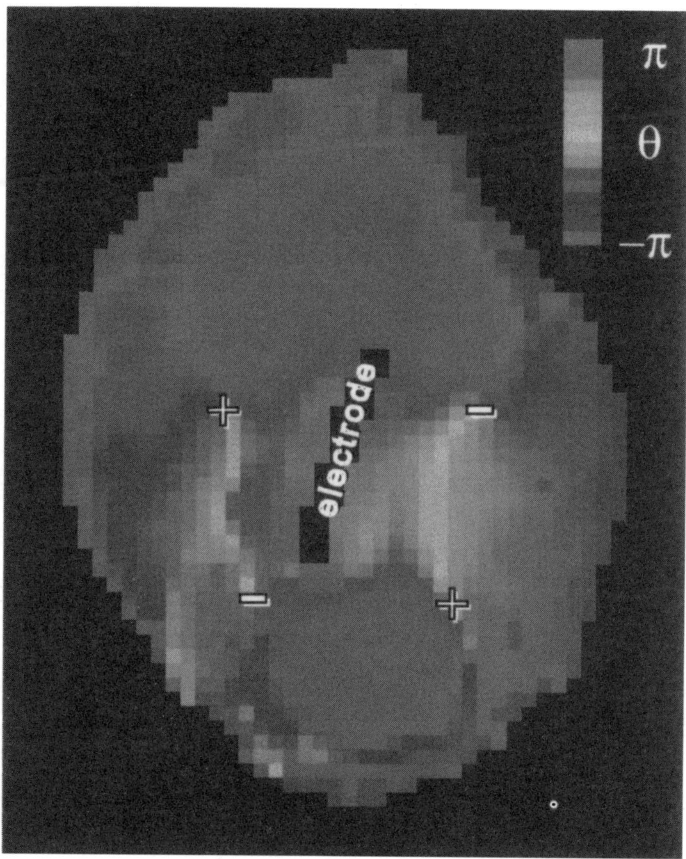

Figure 12   Phase mapping. A phase map at the same instant as the snapshot in Fig. 10. The phase variable was computed using Eq. (6) with $F^* = F_{50}$. Phase singularities are easily identified as sites where all phase values converge. Phase singularities on the heart surface can be classified according to their chirality; clockwise and labeled with $+$ and counterclockwise with $-$ [4]. The phase singularities occur where the wave fronts and wave tails meet (see Fig. 10), as ensured by the identical parameter values (i.e., cutoff threshold and F^* both equal to F_{50}).

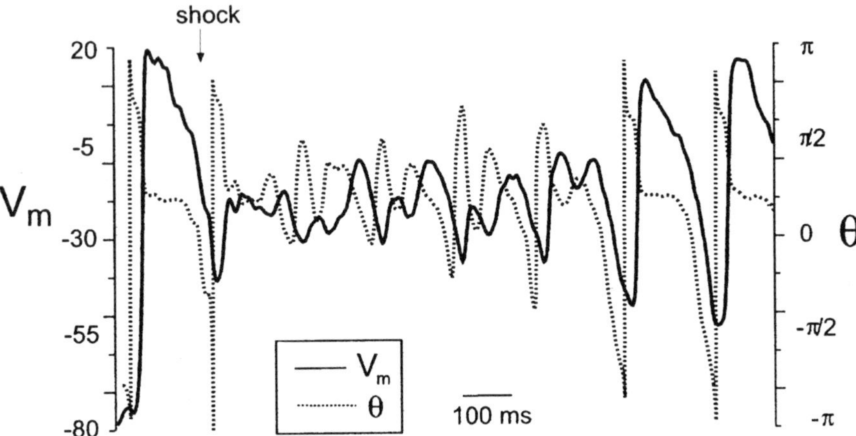

Figure 13 Transmembrane potential and cardiac phase. A fluorescent recording normalized and scaled to represent transmembrane potential [2] (V_m, solid line) and the corresponding phase variable (θ, dashed line) computed using Eq. (6). This recording is taken from a site near the center of reentry in the episode shown in Fig. 10. During full action potentials such as the one before the shock, the phase spans the entire 2π interval. The phase variable is cyclic in nature and $+\pi = -\pi$; therefore the large transition in $\theta(\sim 2\pi)$ that occur during one frame actually represent small changes. Near the center of reentry (i.e., the phase singularity), the phase variable oscillates over a smaller interval, as does the transmembrane potential recording which no longer has an amplitude of 100 mV. At the end of recording, the phase singularity has moved away from this site, causing the return of large action potentials, and θ goes through a range of 2π (indicating a complete rotation in reconstructed state space around the origin) for the last two beats.

small oscillations and "full-blown" action potentials are not evident, as shown in Fig. 13.

M. Shock-Induced Changes in Transmembrane Potential

Many investigators have used optical recordings to study how electric fields alter transmembrane potential. Shock-induced changes in fluorescence should be measured in relation to a control action potential because transmembrane potential may change spontaneously during the time interval of the shock, had no shock been given. This measure may not be possible to compute during fibrillation, but is easily achieved for shocks delivered during pacing or monomorphic tachycardia (i.e., repetitive rhythms). Shock-induced hyperpolarizationl/depolarization is typically defined as a decrease/increase of F at the end of a shock compared to the value of F of the previous action potential at the same time, relative to the onset of the pacing stimulus [69]. The "uniformity index" is a global measure of the effect of the shock on F [69]. For each site (i, j), let F_{diff} be the difference between the shock-induced change in F (F at the beginning of the shock subtracted from F at the end of the shock) and the change in F normally occurring during the same interval when no shock is given (measured during the previous beat). The uniformity index is the sum of the number of sites having an $F_{\text{diff}} > 5\,\text{mV}'$ minus the number of sites having an $F_{\text{diff}} > -5\,\text{mV}'$ divided by the total number of sites with $|F_{\text{diff}}| > \text{mV}'$. A positive uniformity index indicates that the depolarized region was larger than the hyperpolarized region during the shock, while a negative index represents the opposite. A spatial map of F_{diff} in millivolts is shown in Fig. 14.

IV. CONCLUDING REMARKS

A. Remember What You Are Studying!

At this point it is wise to remember the electrophysiological question you wish to answer. If you are interested in the spatiotemporal dynamics of transmembrane potential (V_m) in the heart, optical mapping will provide a measure of these dynamics, but remember that the fluorescence measurements do not represent true V_m. A variety of factors, including dye binding, excitation and emission spectra of the dye, transmission of light through the heart, recording system characteristics, etc., are involved in the transduction of transmembrane potential to the recorded digital fluorescence signal. All optical mapping modalities have unique advantages and limitations associated with them, and it is important to keep the characteristics of the recording system in mind when planning experiments and analyzing data. In the past, the most significant limitations of video imaging of cardiac tissue have been the poor SNR and the slow speeds. For example, in 1991, frame rates were 60 fps and SNR was $\sim$1 [70]. Less than 10 years later, SNR and frame rates have increased dramatically. SNR values greater than 40 have been achieved (see Fig. 6) and frame rates $> 1\,\text{kHz}$ have been reported in the literature [47], and even faster ones have been developed. Remember that, for optical mapping, SNR is related to many factors such as optical magnification, sensor size, frame rate (for CCDs), etc. Imaging with high-speed scientific-grade CCDs offers much flexibility in the design of a recording system. Most notable, binning (adding the value of neighboring pixels) can be accomplished via software or hardware and increases the SNR and can lead to increased frame rates (if binning is a feature of your CCD camera). Similar to ensemble averaging, binning decreases the relative noise by the

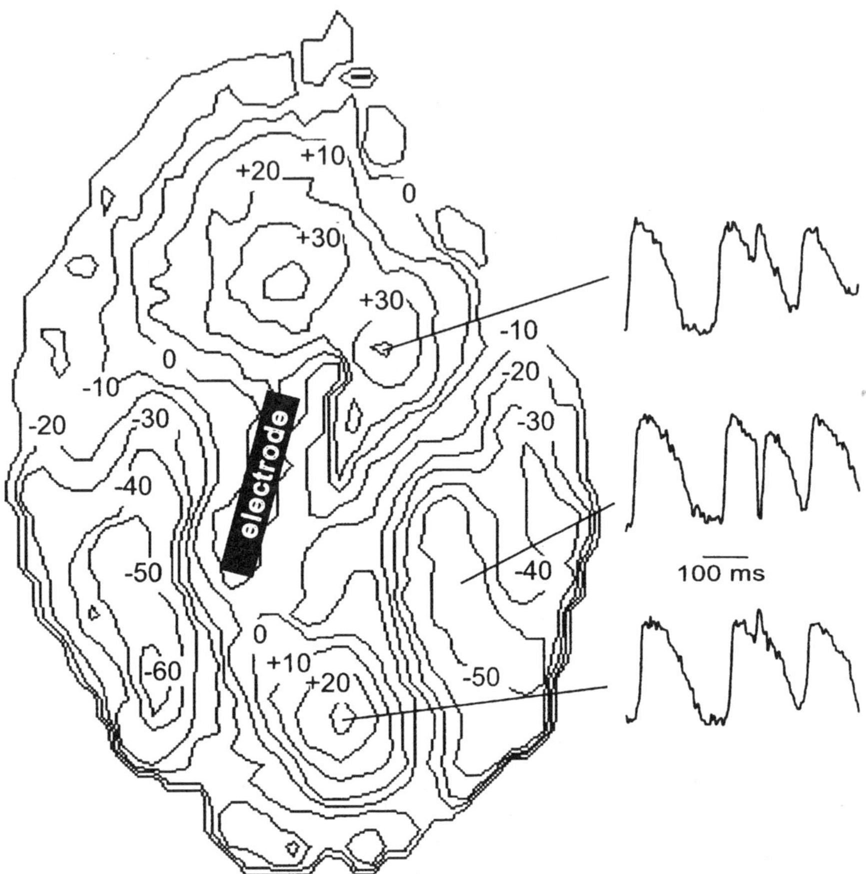

Figure 14 Shock-induced changes in transmembrane potential. A map of the relative change in transmembrane potential caused by a shock applied during pacing (from the episode shown in Fig. 10, although no spatial filter was applied but a three point median temporal filter was applied). Numbers represent a percentage of the amplitude of the last paced action potential as computed for each site. The changes are measured at the end of the shock relative to the previous paced action potential (the pacing stimulus serves as the timing reference). Shock-induced depolarization is evident as positive values (above and below the electrode), while shock-induced hyperpolarization results in negative values (to the right and left of the electrode). Time series from three sites are shown to the right. The uniformity index [69] computed from this image as described in the text is −0.37.

square root of the number of pixels added together (e.g., 2×2 binning increases SNR by a factor of 2). One must identify the resolution of the measured voltage required to address your question of interest (e.g., V_m precision of 1 mV). Low SNRs are acceptable for studying wave front propagation, but larger SNRs are required to analyze repolarization events. As a rule of thumb, we suggest that a voltage resolution of $\sim$5 mV (this corresponds roughly to a SNR of 20) is adequate for most studies. Investigators should optimize the speed and spatial resolution of their recording system, keeping in mind the spatial and temporal frequency content of the signal being studied.

B. Previous Studies of Fibrillation and Defibrillation Using Video Imaging

Video imaging allows the simultaneous recording from up to 100,000 sites [20], compared to a maximum of less than 500 with photodiode arrays. (Recently, a laser scanning technique has been developed to record from up to 10,000 sites, although the sites are recorded sequentially due to the scanning procedure of the laser beam [71]. This unique high spatial resolution has led to some important findings regarding the characteristics of wave propagation in the heart. Video imaging technology has been used to study the speed of wave front propagation and its role in reentrant arrhythmias. Recently, video imaging technology has been combined with molecular biology to study conduction velocity in connexin 43-deficient mice [47]. Other studies have shown that the propagation speed in cardiac tissue is linearly related to the curvature of the wave front and a critical curvature for propagation exists [72]. During reentry, the shape of the wave front is curved, with increasing curvature toward the center of rotation resulting in a spiral shape [68,70].

C. Video Imaging of Fibrillation

Video imaging has revealed that rotating waves of electrical activity propagate in the ventricles during tachyarrhythmias, including fibrillation. [4,50,73], In the ventricles, stationary rotating waves result in monomorphic tachycardias, while polymorphic tachycardias result when they move [50,74]. In fact, if a single reentrant wave moves rapidly through the heart, the ECG resembles fibrillation [73]. Video imaging has been used to show that global ischemia in the isolated rabbit heart leads to a decrease in fibrillation rate resulting from an increase in the organizing center (i.e., core) of reentrant waves [53]. In addition, calcium channel blockade in isolated rabbit hearts results in a conversion of fibrillation to monomorphic tachycardia as a result of a decrease in wave front fragmentation [75]. Recently, Vaidya et al. showed that reentry and fibrillation could occur in the isolated mouse heart, which challenges the general belief that only large hearts can sustain fibrillation [55]. Finally, Zaitsev et al. used two CCDs to map from both endo- and epicardial surfaces simultaneously [21]. In this isolated ventricle preparation they found that reentrant waves were rarely seen on the heart surface during fibrillation and that activation patterns on opposite surfaces were not correlated.

Video imaging technology has been used to study the patterns of wave propagation in the atria as well. In the atria, complete reentrant waves have not been observed in the right atrium during sustained fibrillation in isolated sheep hearts [13], and the complicated atria structure has been shown to play an important role in wave propagation, especially during fast rates such as tachyarrhythmias [76]. Mandapati et al. used video imaging to map from both atria in isolated sheep hearts and found that rapid regular activity in the left atrium appeared to be sustaining atrial fibrillation [22].

D. Video Imaging of Repolarization Dynamics

The major advantage of optical mapping is that provides a means to study the spatiotemporal evolution of transmembrane potential in the heart, thus allowing the analysis of repolarization dynamics as well as the propagation of wave fronts. Video imaging has been used to quantify and study the excitable gap during reentry in cardiac tissue [68,70]. In addition, the spatial distribution of APD and its rate dependence as well as repolarization patterns in isolated sheep and rabbit hearts have been studied using video imaging [13,18,19]. By combining measures of APD and conduction velocity, a measure of wavelength can be computed [13,55]. A cardiac phase variable (see above) can be calculated from optical signals, and video imaging allows the analysis of the spatial distribution of phase with a high degree of accuracy [4,5]. The fact that this phase variable constructed from experimental data obtained during fibrillation is consistent with theoretical predictions provides very strong support for the rationale and importance of the phase analysis [4].

E. Video Imaging of Defibrillation

Another advantage of optical mapping is that transmembrane activity can be recorded during the application of high-energy electric fields (i.e., "shocks"). Video imaging has been utilized to study the transmembrane patterns preceding, during, and following shocks applied to cardiac tissue. Wikswo et al. characterized how transmembrane patterns around the stimulating electrode were influenced by electrode polarity and the state of the tissue near the electrode at the time of the shock during pacing [23]. In addition, the same group showed that a critically timed shock could initiate an arrhythmia via the simultaneous formation of four reentrant waves (i.e., "quatrefoil reentry") [77]. Banville et al. showed that the shock-induced transmembrane changes over the surface of the ventricles in the isolated rabbit heart during pacing were largely affected by shock timing, and that reentrant waves were initiated only at relatively short coupling intervals [19].

Video imaging has also been used to study atrial and ventricular defibrillation. Gray et al. found that atrial defibrillation shocks depolarized the entire atrial surface in isolated sheep hearts [17]. Near the defibrillation threshold they found that asynchronous repolarization occurred following the shocks, followed by a quiescent period and organized activation patterns even when the shock was unsuccessful. In a follow-up study, Skanes et al. used video imaging to investigate the possibility of a "hybrid" therapy for atrial defibrillation by studying the role of pacing the atria after a defibrillation shock [78]. Gray et al. also found that inappropriately timed atrial defibrillation shocks could induce ventricular fibrillation via multiple focal beats that produced waves that propagated but "broke down" into reentrant waves in regions exhibiting large repolarization gradients before the shock [18]. By mapping from the ventricles of arrhythmic isolated rabbit hearts, Gray et al. recently confirmed the long-standing belief that the spatial excitable gap immediately preceding and immediately following defibrillation shocks affects the outcome [52].

F. Future Directions for Video Imaging

A significant limitation of optical mapping is the motion of cardiac tissue. This motion is reflected in the recorded optical signals and presents a significant problem. The most important issue is that a single recording site will be collecting data from various regions of the heart as it moves. This fact is particularly important for video imaging, where high spatial resolution is achieved and very small movements are problematic. One study has utilized a "morphing" algorithm to convert video imaging fluorescence movies of the contracting heart to movies of a motionless heart with propagation patterns still intact [79]. We believe that this approach is very promising, and the current speeds of desktop computers may make this technique more widespread in the future. Often, the entire depolarization sequence precedes the motion of the heart but repolarization events tend to be contaminated with "motion artifact." During rapid activation where no diastolic interval is present, motion is still a problem; however, applying a temporal derivative is a good means to identify wave fronts in the presence of motion. Although repolarization time [49] as well as depolarization times [79] can be computed in the presence of motion, the spatial resolution of the recorded data may be compromised for the reason mentioned above. Most often, drugs are added to the perfusate to eliminate the mechanical contractions. Of course, these drugs will have some effect on the heart and its electrical system and will vary among uncoupling agents [25,26]. It is important to acknowledge this fact and consider this limitation (as well as the effects of the voltage-sensitive dye and variability among species) when considering optical mapping studies in relation to clinical issues and other studies.

The "Holy Grail" of cardiac mapping would be to record transmembrane potential throughout the entire three-dimensional beating heart without toxic probes. While this may not be achieved in our lifetimes, amazing developments are underway toward this end. Video imaging has been used extensively to record transmembrane activity from large regions of the heart surface. One advantage of optical mapping is that the recording device is generally not in direct contact with the heart, and this provides great flexibility. This "advantage" means that the two-recorded images are two-dimensional projections from the three-dimensional heart. It is difficult to accurately record signals near the edges of highly curved surfaces or surfaces "hidden from view." Clever techniques incorporating mirrors [28] and prisms [80] hold much promise for recording from those "hard-to-reach" areas. Lin and Wikswo have used two mirrors and image processing algorithms to visualize propagating waves on the entire ventricular surface of the isolated rabbit heart by projecting planar images back onto the curved heart surface [28]. Akar et al. have used prisms pressed against the cut (transmural) edges of ventricular slab preparations to reflect excitation and emission light 90° allowing simultaneous mapping from the epicardial and transmural surfaces [80]. Recent studies have shown that its is possible to record the sequence of wave propagation from several depths simultaneous using only a single array of photodetectors [39–41]. This is accomplished by separately analyzing various distinct morphological characteristics in the optically recorded "action potentials" that represent fluorescence from various depths of tissue. The first studies were carried out in the atrio-ventricular nodal region, where the global sequence of activation is known and there is some knowledge regarding distinct layers of cardiac tissue [39,40]. A more recent study extended this technique to the ventricular wall of the isolated rabbit heart, showing that "dual-humped" optical "action potentials" were indicative of transmural reentry within the ventricular wall [41]. A new optical mapping technique called "transillumination" involves recording emitted photons from the *opposite* side of the tissue, where the excitation light originates [13,81,82]. Many more details need to be addressed, but it is clear that transillumination provides information regarding transmembrane potential from within the heart wall.

Optical mapping of variables beside transmembrane potential is possible; in fact, many investigators use video cameras to record calcium dynamics. However, only recently have researchers begun to record and analyze the dynamic spatial patterns of voltage and calcium [83–87]. Recently, video imaging with two CCDs has been used to simultaneously map transmembrane activity and intracellular calcium during fibrillation in the isolated rabbit heart [88]. The motivation for this work was to allow the construction of a phase variable based on two state variables of the heart (voltage and calcium) and alleviate the need for computing a reconstructed variable. This new phase variable provides a means for relating the phase variable to electrophysiological quantities for the first time. New optical probes most probably will be developed to allow the investigation of the spatiotemporal behavior of even more variables.

ACKNOWLEDGMENTS

We would like to thank David Gardner for his comments on the Nyquist criterion and CCDs, Kyle Justice for the development of the video imaging data analysis software (VIDAS), and Fred Evans for providing experimental data for Figs. 6 and 8.

REFERENCES

1. Robles de Medina EO, Bernard R, Coumel P, Damato AN, Fisch C, Krikler D, Mazur NA, Meijler FL, Morgensen L, Moret P, Pisa Z, Wellens HJ. Definition of terms related to cardiac rhythm, WHO/ISFC Task Force. *Eur J Cardiol* 8:127–144, 1978.
2. Myerburg RI, Kessler KM, Interian A, Fernandez P, Kimura S, Kozlovskis PL, Furukawa T, Bassett AL, Castellanos A. Clinical and experimental pathophysiology of sudden cardiac death. *In*: Zipes DP, Jalife J, eds. *Cardiac Electrophysiology, From Cell to Bedside*. Philadelphia: Saunders, 1990, pp. 666–678.
3. Wolf PA, Dawber TR, Thomas E Jr, Kannel WB. Epidemiologic assessment of chronic atrial fibrillation and risk of stroke: the Framingham study. *Neurology* 28:973–977, 1978.
4. Gray RA, Pertsov AM, Jalife J. Spatial and temporal organization during cardiac fibrillation. *Nature* 392:675–678, 1998.
5. Gray RA, Jalife J. Video imaging of cardiac fibrillation. *In*: Rosenbaum DS, Jalife J, eds. Optical Mapping of Cardiac Excitation and Arrhythmias. New York: Futura, 2001, pp. 245–264.

6. Smith WM, Wharton JM, Blanchard SM, Wolf PD, Ideker RE. Direct cardiac mapping. *In*: Zipes DP, Jalife J, eds. Cardiac Electrophysiology. From Cell to Bedside. Philadelphia: Saunders, 1993, pp. 849–858.

7. Salama G, Morad M. Merocyanine 540 as an optical probe of transmembrane electrical activity in the heart. *Science* 191:485–487, 1976.

8. Kwaku KF, Dillon SM. Shock-induced depolarization of refractory myocardium prevents wave-front propagation in defibrillation. Circ Res 79:957–973, 1996.

9. Knisley SB, Hill BC. Optical recordings of the effect of electrical stimulation on action potential repolarization and the induction of reentry in two-dimensional perfused rabbit epicardium. *Circulation* 88(1):2402–2414, 1993.

10. Rohr S. Optical mapping of microscopic impulse propagation. Chapter 14 in this volume.

11. Efimov I. Optical mapping of cardiac stimulation. Chapter 16 in this volume.

12. Rosenbaum DS, Kaplan DT, Kanai A, Jackson, L, Garan H, Cohen RJ, Salama G. Repolarization inhomogeneities in ventricular myocardium change dynamically with abrupt cycle length shortening. *Circulation* 84:1333–1345, 1991.

13. Gray RA, Pertsov, AM, Jalife J. Incomplete reentry and epicardial breakthrough patterns during atrial fibrillation in the sheep heart. *Circulation* 94:2649–2661, 1996.

14. Girouard SD, Pastore JM, Laurita KR, Gregory KW, Rosenbaum DS. Optical mapping in a new guinea pig model of ventricular tachycardia reveals mechanism for multiple wavelengths in a single reentrant circuit. *Circulation* 93:603–613, 1996.

15. Laurita KR, Girouard SD, Rosenbaum DS. Modulation of ventricular repolarization by a premature stimulus. *Circulation* 79:493–503, 1996.

16. Kanai A, Salama G. Optical mapping reveals that repolarization spreads anisotropically and is guided by fiber orientation in guinea pig hearts. *Circ Res* 77:784–802, 1995.

17. Gray RA, Ayers G, Jalife J. Video imaging of atrial defibrillation in the sheep heart. *Circulation* 95:1038–1047, 1997.

18. Gray RA, Jalife J. Effects of atrial defibrillation shocks on the ventricles in isolated sheep hearts. *Circulation* 97(16):1613–1622, 1998.

19. Banville I, Gray RA, Ideker RE, Smith WM. Shock-induced figure-of-eight reentry in the isolated rabbit heart. *Circ Res* 85:742–752, 1999.

20. Baxter WT, Davidenko JM, Loew LM, Wuskell JP, Jalife J. Technical features of a CCD video camera system to record cardiac fluorescence data. *Ann Biomed Eng* 25:713–725, 1997.

21. Zaitsev AV, Berenfeld O, Mironov SF, Jalife J, Pertsov AM. Distribution of excitation frequencies on the epicardial and endocardial surfaces of fibrillating ventricular wall of the sheep heart. *Circ Res*, in press.

22. Mandapati R, Skanes A, Chen J, Berenfeld O, Jalife J. Stable microreentrant sources as a mechanism of atrial fibrillation in the isolated sheep heart. *Circulation* 101:194–199, 2000.

23. Wikswo JP, Lin SF, Abbas RA. Virtual electrodes in cardiac tissue: a common mechanism for anodal and cathodal stimulation. *Biophys J* 69(6):2195–2210, 1995.
24. Sakuma I, Dohi T, Mikami M, Ohuchi K, Fukui Y, Shibata N, Honjo H, Kodama I. A new multi-channel optical system to record cardiac action potentials utilizing high-power blue light emitting diodes and fiber optics (abstr). *PACE* 22(4II):702, 1999.
25. Liu Y, Cabo C, Salomonsz R, Delmar M, Davidenko J, Jalife J. Effects of diacetyl monoxime on the electrical properties of sheep and guinea pig ventricular muscle. *Cardiovasc Res* 27:1991–1997, 1993.
26. Wu J, Biermann M, Rubart M, Zipes DP. Cytochalasin D as an excitation-contraction uncoupler for optically mapping action potentials in wedges of ventricular myocardium *J Cardiovasc Electrophysiol* 9:1336–1347, 1998.
27. Biermann M, Rubart M, Moreno A, Wu J, Josiah-Durant A, Zipes DP. Differential effects of cytochalasin D and 2,3 butanedione monoxime on isometric twitch force and transmembrane action potential in isolated ventricular muscle. *J Cardiovasc Electrophysiol* 9:1348–1357, 1998.
28. Lin SF, Abbas A, Wikswo JP. High-resolution high-speed synchronous epi-fluorescence imaging of cardiac activation. *Rev Sci Instrum* 68:213–217, 1997.
29. Gardner D. Demystifying high-performance CCD camera specs and terms *Advanced Imaging* April:64–67, 1999.
30. Fluhler E, Burnham VG, Loew LM. Spectra, membrane binding, and potentiometric responses of new charge shift probes. *Biochemistry* 24:5749–5755, 1985.
31. Fast VG, Kléber AG. Microscopic conduction in cultured strands of neonatal rat heart cells measured with voltage-sensitive dyes. *Circ Res* 73:914–925, 1993.
32. Girouard SD, Laurita KR, Rosenbaum DS. Unique properties of cardiac action potentials recorded with voltage-sensitive dyes. *J Cardiovasc Electrophysiol* 7:1024–1038, 1996.
33. Efimov IR, Cheng YN, Biermann M, Van Wagoner DR, Mazgalev TN, Tchou PJ. Transmembrane voltage changes produced by real and virtual electrodes during monophasic defibrillation shock delivered by an implantable electrode. *J Cardiovasc Electrophysiol* 8:1031–1045, 1997.
34. Rohr S, Salzberg BM. Multiple site optical recording of transmembrane voltage (MSORTV) in patterned growth heart cell cultures: assessing electrical behavior, with microsecond resolution, on a cellular and subcellular scale. *Biophys J* 67:1301–1315, 1994.
35. Efimov IR, Biermann M, Zipes D. Fast fluorescent mapping of electrical activity in the heart: Practical guide to experimental design and applications. *In*: Cardiac Mapping, 2nd ed. Futura, in press.
36. Bayly PV, Johnson EE, Wolf PD, Greenside HS, Smith WM, Ideker RE. Efficient electrode spacing for examining spatial organization organization during ventricular fibrillation. *J Cardiovasc Electrophysiol* 4:533–546, 1993.
37. Windisch H, Ahammer H, Schaffer P, Müller W, Platzer D. Optical multisite monitoring of cell excitation phenomena in isolated cardiomyocytes. *Pflügers Arch Eur J Physiol* 430:508–518, 1995.

38. Knisley SB. Transmembrane voltage changes during unipolar stimulation of rabbit ventricle. *Circ Res* 77:1229–1239, 1995.

39. Choi B, Salama G. Optical mapping of atrioventricular node reveals a conduction barrier between atrial and nodal cells. *Am J Physiol* 274:H829–H845, 1998.

40. Efimov IR, Mazgalev TN. High-resolution, three-dimensional fluorescent imaging reveals multilayer conduction pattern in the atrioventricular node. *Circulation* 98:54–57, 1998.

41. Efimov IR, Sidorov V, Cheng Y, Wollenzier BS. Evidence of three-dimensional scroll waves with ribbon-shaped filament as a mechanism of ventricular tachycardia in the isolated rabbit heart. *J Cardiovasc Electrophysiol* 10(11):1452–1462, 1999.

42. Gray RA. What exactly are optically recorded "action potentials"? *J Cardiovasc Electrophysiol* [Editorial]. 10(11):1463–1466, 1999.

43. Witkowski FX, Leon LJ, Penkoske PA, Clark RB, Spano ML, Ditto WL, Giles WR. A method for visualization of ventricular fibrillation: design of a cooled fiberoptically coupled image intensified CCD data acquisition system incorporating wavelet shrinkage based adaptive filtering. *Chaos* 8(1):94–102, 1998.

44. Efimov IR, Cheng Y, Van Wagoner DR, Mazgalev T, Tchou PJ. Virtual electrode-induced phase singularity: a basic mechanism of defibrillation failure. *Circ Res* 82:918–925, 1998.

45. Xhou X, Guse P, Wolf PD, Rollins DL, Smith WM, Ideker RE. Existence of both fast and slow channel activity during the early stages of ventricular fibrillation. *Circ Res* 70:773–786, 1992.

46. Fast VG, Kléber AG. Cardiac tissue geometry as a determinant of unidirectional conduction block: assesment of microscopic excitation spread by optical mapping in patterned cell cultures and in a computer model. *Cardiovasc Res* 29:697–707, 1995.

47. Morley GE, Vaidya D, Samie FH, LO C, Delmar M, Jalife J. Characterization of conduction in the ventricles of normal and heterozygous Cx43 knockout mice using optical mapping. *J Cardiovasc Electrophysiol* 10:1361–1375, 1999.

48. Ideker RE. The assumptions of isochronal cardiac mapping. *PACE* 12:456–478, 1989.

49. Efimov IR, Huang DT, Rendt JM, Salama G. Optical mapping of repolarization and refractoriness from intact hearts. *Circulation* 90:1469–1480, 1994.

50. Gray RA, Jalife J, Panfilov AV, Baxter WT, Cabo C, Davidenko JM, Pertsov AM. Non-stationary vortex-like reentry as a mechanism of polymorphic ventricular tachycardia in the isolated rabbit heart. *Circulation* 91:2454–2469, 1995.

51. Witkowski FX, Leon LJ, Penkoske PA, Giles WR, Spano ML, Ditto WL, Winfree AT. Spatiotemporal evolution of ventricular fibrillation. *Nature* 392:78–82, 1998.

52. Gray RA, Banville I. Video imaging of cardioversion in the rabbit heart (abstr). *PACE* 22(4II):703, 1999.

53. Mandapati R, Asaon Y, Baxter WT, Gray RA, Davidenko J, Jalife J. Quantification of the effects of global ischemia on the dynamics of ventricular fibrillation in the isolated rabbit heart. *Circulation* 98:1688–1696, 1998.

54. Bayly PV, KenKnight, BH, Rogers JM, Hillsley RE, Ideker RE, Smith WM. Estimation of conduction velocity vector fields from 504-channel epicardial mapping data. *IEEE Trans Biomed Eng* 45(5):563–571, 1998.

55. Vaidya D, Morley GE, Samie FH, Jalife J. Reentry and fibrillation in the mouse heart: a challenge to the critical mass hypothesis. *Circ Res* 85:174–181, 1999.

56. Barnette AR, Bayly PV, Zhang S, Walcott GP, Ideker RE, Smith WM. Estimation of 3-D conduction velocity vector fields from cardiac mapping data. In: Murray A, ed. Computers in Cardiology. IEEE Computer Society Press, 1998.

57. Dorian P, Penkoske PA, Witkowski FX. Order in disorder: effect of barium on ventricular fibrillation. *Can J Cardiol* 12(4):399–406, 1996.

58. Takens F. Detecting strange attractors in turbulence. In: Rand DA, Young LS, eds. Dynamical Systems and Turbulence. Lecture Notes in Mathematics, Vol. 898. Berlin: Springer-Verlag, 1981, pp. 366–381.

59. Luo C, Rudy Y. A dynamic model of the ventricular cardiac action potential: I. Simulation of ionic currents and concentration changes. *Circ Res* 74:1071–1096, 1994.

60. Coullet P, Frisch T, Gilli JM, Rica S. Excitability in liquid crystal. *Chaos* 4(3):485–489, 1994.

61. Winfree AT. Scroll-shaped waves of chemical activity in three dimensions. *Science* 181:937–939, 1973.

62. Muller SC, Plesser T, Hess B. The structure of the core of the spiral wave in the Belousov-Zhabotinsky reagent. *Science* 230:661–663, 1985.

63. Goldbeter A. Mechanism for oscillatory synthesis of cAMP in *Dictyostelium discoideum*. *Nature* 253:540–542, 1975.

64. Cross MC, Hohenberg PC. Pattern formation outside of equilibrium. *Rev Mod Phys* 65(3):851–1112, 1993.

65. Walgraef D. *Spatio-Temporal Pattern Formation*. New York: Springer, 1997.

66. Winfree AT. *When Time Breaks Down*. Princeton, NJ: Princeton University Press, 1987.

67. Zel'dovich YB, Malomed BA. Topological invariants and strings in distributed active dynamical systems. *Sov Phys Dokl* 25(9):721–723, 1981.

68. Pertsov AM, Davidenko JM, Salomonsz R, Baxter WT, Jalife J. Spiral waves of excitation underlie reentrant activity in isolated cardiac muscle. *Circ Res* 72:631–650, 1993.

69. Knisley SB, Baynham TC. Line stimulation parallel to myofibers enhance regional uniformity of transmembrane voltage changes in rabbit hearts. *Circ Res* 81:229–241, 1997.

70. Davidenko JM, Pertsov AM, Salomonsz R, Baxter WT, Jalife J. Stationary and drifting spiral waves of excitation in isolated cardiac muscle. *Nature* 355:349–351, 1991.

71. Bove RT, Dillon SM. Optically imaging cardiac activation with a laser system. *IEEE Eng Med Biol* 17(1):84–94, 1998.

72. Cabo C, Pertsov AM, Baxter WT, Davidenko JM, Gray RA, Jalife J. Wavefront curvature as a cause of slow conduction and block in isolated cardiac muscle. *Circ Res* 75:1014–1028, 1994.

73. Gray RA, Jalife J, Panfilov AV, Baxter WT, Cabo C, Davidenko JM, Pertsov AM. Mechanisms of cardiac fibrillation: drifting rotors as mechanism of cardiac fibrillation. *Science* 270:1222–1223, 1995.

74. Davidenko J. Spiral wave activity: a possible common mechanism for polymorphic and monomorphic ventricular tachycardia. *J Cardiovasc Electrophysiol* 4:730–746, 1993.

75. Samie FH, Madapati R, Gray RA, Watanabe Y, Zuur C, Beaumont J, Jalife J. A mechanism of transition from ventricular fibrillation to tachycardia: effect of calcium channel blockade on the dynamics of rotating waves. *Circ Res*, in press.

76. Gray RA, Takkellapati K, Jalife J. Dynamics and anatomical correlates of atrial flutter and fibrillation. *In*: Zipes DP, Jalife J, eds. Cardiac Electrophysiology: From Cell to Bedside. Philadelphia: Saunders, 1999, pp. 432–439.

77. Lin SF, Roth BJ, Wikswo JP. Quatrefoil reentry in myocardium: an optical imaging study of the induction mechanism. *J Cardiovasc Electrophysiol* 10:574–586, 1999.

78. Skannes AC, Gray RA, Zuur CL, Jalife J. Spatio-temporal pattern of atrial fibrillation: role of the subendocardial structure. *In*: Interventional Treatment of Cardiac Arrhythmias. Ruskin J, Keane, eds. Philadelphia: Saunders, 1997, pp. 185–193.

79. Asano Y, Davidenko JM, Baxter WT, Gray RA, Jalife J. Optical mapping of drug-induced polymorphic arrhythmias and torsade de pointes in the isolated rabbit heart. *J Am Coll Cardiol* 29:831–842, 1997.

80. Akar FG, Rosenbaum DS. Multi-surface optical mapping of cardiac tissue with prisms. *Ann Biomed Eng* 26 (suppl 1):S19, 1998.

81. Baxter WT, Pertsov A, Berenfeld O, Mironov S. Demonstration of three-dimensional reentry in isolated sheep right ventricle (abstr). *PACE* 20(4,II): 1080, 1997.

82. Baxter WT. Intramural optical recordings of cardiac electrical activity via transillumination. Ph.D. thesis, State University of New York at Syracuse, 1999.

83. Johnson PL, Smith WM, Baynham TC, Knisley SB. Errors caused by combination of di-4-ANEPPS and Fluo 3/4 for simultaneous measurements of transmembrane potentials and intracellular calcium. *Ann Biomed Eng* 27(4): 563–571, 1999.

84. Laurita KR, Singal A, Pastore JM, Rosenbaum DS. Spatial heterogeneity of calcium transients may explain action potential dispersion during T-wave alternans (abstr). *Circulation* 98(17):I-187, 1998.

85. Fast VG, Ideker RE. Fast co-local optical recordings of transmembrane potential and intracellular calcium in myocyte cultures (abtr). *PACE* 22:702, 1999.

86. Clusin W, Han J, Quan Y. Simultaneous recordings of calcium transients and action potentials from small regions of the perfused rabbit heart (abstr). *PACE* 22:834, 1999.

87. Choi BR, Salama G. Spatio-temporal relationship between action potentials and Ca^{2+} transient in anterior region of guinea pig hearts (abstr). *PACE* 22:702, 1999.
88. Gray RA, Wikswo JP, Lin SF, Baudenbacher F. Phase mapping using both transmembrane potential and calcium (abstr). *PACE* 23:608, 2000.

Index